Chemistry
in
Context

A Project of the American Chemical Society

Seventh Edition

Chemistry
in
Context

Applying Chemistry to Society

Catherine H. Middlecamp
University of Wisconsin–Madison

Steven W. Keller
University of Missouri

Karen L. Anderson
Madison Area Technical College

Anne K. Bentley
Lewis & Clark College

Michael C. Cann
University of Scranton

Jamie P. Ellis
The Scripps Research Institute

A Project of the American Chemical Society

*Connect
Learn
Succeed™*

The McGraw·Hill Companies

Mc Graw Hill — Connect Learn Succeed™

CHEMISTRY IN CONTEXT: APPLYING CHEMISTRY TO SOCIETY, SEVENTH EDITION

Published by McGraw-Hill, a business unit of The McGraw-Hill Companies, Inc., 1221 Avenue of the Americas, New York, NY 10020. Copyright © 2012 by the American Chemical Society. All rights reserved. Previous editions © 2009, 2006, and 2003. No part of this publication may be reproduced or distributed in any form or by any means, or stored in a database or retrieval system, without the prior written consent of The McGraw-Hill Companies, Inc., including, but not limited to, in any network or other electronic storage or transmission, or broadcast for distance learning.

Some ancillaries, including electronic and print components, may not be available to customers outside the United States.

This book is printed on acid-free paper.

1 2 3 4 5 6 7 8 9 0 DOW/DOW 1 0 9 8 7 6 5 4 3 2 1

ISBN 978–0–07–337566–3

MHID 0–07–337566–7

Vice President, Editor-in-Chief: *Marty Lange*
Vice President, EDP: *Kimberly Meriwether David*
Senior Director of Development: *Kristine Tibbetts*
Publisher: *Ryan Blankenship*
Sponsoring Editor: *Todd L. Turner*
Developmental Editor: *Jodi Rhomberg*
Executive Marketing Manager: *Tamara L. Hodge*
Lead Project Manager: *Sheila M. Frank*
Senior Buyer: *Kara Kudronowicz*
Senior Media Project Manager: *Sandra M. Schnee*
Designer: *Tara McDermott*
Cover Designer: *Christopher Reese*
Cover Image: © *Getty Images/Grant Faint*
Lead Photo Research Coordinator: *Carrie K. Burger*
Photo Research: *Jerry Marshall/pictureresearching.com*
Compositor: *Aptara, Inc.*
Typeface: *10/12 Times Roman*
Printer: *R. R. Donnelley*

All credits appearing on page or at the end of the book are considered to be an extension of the copyright page.

Library of Congress Cataloging-in-Publication Data

Chemistry in context : applying chemistry to society / Catherine H. Middlecamp ... [et al.]. -- 7th ed.
    p. cm.
  Includes index.
  "A Product of the American Chemical Society."
  ISBN 978–0–07–337566–3 — ISBN 0–07–337566–7 (hard copy : alk. paper)
  1. Biochemistry. 2. Environmental chemistry. 3. Geochemistry. I. Middlecamp, Catherine.
QD415.C482 2012
540--dc22

2010024045

SUSTAINABLE FORESTRY INITIATIVE
Certified Fiber Sourcing
www.sfiprogram.org

Printed with inks containing soy and/or vegetable oils

www.mhhe.com

Brief Contents

Contents

Contents

Dear Readers,

When first published in 1993, *Chemistry in Context* was "the book that broke the mold." Unlike the books of its time, it did not teach chemistry in isolation from people and the real-world issues they were facing. Similarly, it did not introduce a fact or concept for the sake of "covering it" as part of the curriculum. Rather, *Chemistry in Context* carefully matched each chemical principle to a real-world issue such as air quality, energy, or water use. Each was introduced on a need-to-know basis; that is, at the point in the book at which there was a demonstrated need for the principle. Most importantly, the book presented chemistry in the *context* of significant social, political, economic, and ethical issues.

Context! The word derives from the Latin word meaning "to weave." The spider web motif on the *Chemistry in Context* cover exemplifies the complex connections that can be woven between chemistry and society. In the absence of the real-world issues, there could be no *Chemistry in Context*. Similarly, without teachers and students who were willing (and brave enough) to engage in these issues, there could be no *Chemistry in Context*. Together we weave chemistry into the issues that we face in our lives.

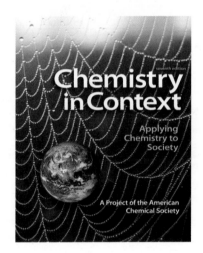

Context! Today we also know that teaching in context is a high-impact practice backed by the research on how people learn. *Chemistry in Context* uses real-world contexts that engage students on multiple levels: their individual health and well-being, the health of their local communities, and the health of wider ecosystems that sustain life on this planet.

As we planned this edition, the writing team members questioned how a tradition of "breaking the mold" might best be continued today. The team raised the question not only for the sake of keeping with tradition (and for the fun of breaking molds), but also for a compelling reason: the needs of our readers. We wanted to continue to find ways of communicating chemistry that served our students, given the challenging issues they face today, the complex needs of the societies in which they live, and the changing landscapes on which they will work in the future.

Teaching (and Learning) in Context

The organization of *Chemistry in Context* has remained the same in every edition. The first six chapters form a core through which basic chemical principles are introduced. These chapters provide a coherent strand of topics that focus on a single theme—the environment. They develop a foundation of chemical concepts that can be expanded in subsequent chapters. Chapters 7 and 8 consider alternative (non-fossil fuel) energy sources—nuclear power, batteries, fuel cells, and the hydrogen economy. The remaining chapters are carbon-based, focusing on polymers, drugs, food production, and genetic engineering. They provide students with the opportunity to explore interests, as time permits, beyond the core topics.

Sustainability—New Content

Global sustainability is not just a challenge; rather, it is the defining challenge of our century. The seventh edition of *Chemistry in Context* is designed to help students better meet this challenge. With its new Chapter 0, **"Chemistry for a Sustainable Future,"** our intent was to establish sustainability as a core, normative part of the chemistry curriculum and part of the foundational learning.

Sustainability adds a new degree of complexity to *Chemistry in Context*. In part, the complexity arises because sustainability can be conceptualized in two ways: as a topic worth studying *and* as a problem worth solving. As a topic, sustainability provides a new

body of content for students to master. For example, the *tragedy of the commons*, the *Triple Bottom Line*, and the concept of *cradle-to-cradle* are part of this new body of content and are introduced in Chapter 0. As a problem worth solving, sustainability generates new questions for students to ask—ones that help them to imagine and achieve a sustainable future. For example, students will find questions about the risks and benefits of both acting and of *not* acting on behalf of future generations. To incorporate sustainability, then, requires more than a casual rethinking of the curriculum.

How do you teach and learn about something as complicated as sustainability? In responding to this question, the author team realized that it was necessary both to update the material and to recast it in a new light. Here are some examples of the changes that the team made:

WATER FOR LIFE
2005-2015

Chapter 3, The Chemistry of Global Climate Change, was updated in the light of new developments in climate change science. It now clearly outlines the consequences of climate change, introducing the sustainability concept of external costs.
Chapter 5, Water for Life, now connects to the "Water for Life" decade themes of the United Nations: the scarcity of fresh water, sustainable management of water resources, and water contamination.
Chapter 8, Energy from Electron Transfer, was recast to better show the match between our energy needs and the available technologies. The sustainability concept of cradle-to-cradle, introduced earlier in Chapter 0, is connected to battery design.
Chapter 11, Nutrition, Food for Thought, points out that what you eat affects both your health *and* the health of the planet.

In addition, here is a listing of the sustainability concepts and the chapters in which they can be found:

Tragedy of the commons:	Chapters 0, 1, 2, 5, and 6
Triple Bottom Line:	Chapters 0, 1, 4, 5, 6, and 7
Cradle-to-cradle:	Chapters 0, 7, and, 8
External costs:	Chapters 0, 3, and, 8
Environmental footprints:	Chapters 0, 1, 4, 5, 11, and 12

Green chemistry, a means to sustainability, continues to be an important theme in *Chemistry in Context*. As in previous editions, examples of green chemistry are highlighted in each chapter. In this new edition, look for even more examples. This expanded coverage offers the reader an even better sense of the need for and the importance of greening our chemical processes. For easier access, the principles of green chemistry are now listed on the inside front cover of the text.

Updates to Existing Content

People sometimes ask us "Why do you release new editions so often?" We also hear the question "Would it work to keep using the older edition with our students?" Indeed, we are on a fast publishing cycle, turning out a new version every three years. We do this because *Chemistry in Context*, with its current real-world focus, risks being out-of-date the very day it is published. Given this, we strongly urge instructors to switch to the new edition immediately, given how sensitive the real-world content is to the passing of time.

With each new addition, the author team reworks practically every chapter, updating its content and focus to reflect new scientific developments, changes in policies, energy trends, and current world events. These updates are nontrivial to implement. They involve writing new content as well as producing new graphs and data tables. The issues that we select to "hook" the reader at the start of the chapter also are recast from edition to edition.

For example, in this new edition, the "Water Chapter," Chapter 5, underwent a significant revision. This chapter has been on our list for a makeover for many years, and each successive author team has puzzled over how best to refocus it. Changes in earlier editions had switched the chapter hook to the issue of bottled water vs. tap water; but even with this change, we knew there was more work to be done. In the seventh edition, we finally reworked Chapter 5 from start to finish. The chapter now follows the United Nations theme of *"Water for Life,"* highlighting that fresh water is indeed limited on our planet and all need access to it. Safe drinking water, water footprints, and water for agriculture and industry are the keys to our health and prosperity. From this point of departure, all of the chemical principles fall nicely into place.

The "Food Chapter," Chapter 11, also underwent a major revision. In past editions, we have employed several different hooks to open the chapter. For example, we talked about different diets, including the high protein one. We also talked about the dizzying array of food recommendations. But for this new edition, we went with a new theme that couples food, personal health, and the health of the planet. This theme better connects to issues of energy consumption, water use, land use, and public health. The questions of sustainability also more naturally arise. In addition, this new theme well sets the stage for Chapter 12 on genetic engineering.

Speaking of Chapter 12, we did a careful technical review of this chapter, reworking its content to better reflect what we know about the genetic code and recent developments in the field. Chapter 8 on batteries and alternative energy sources also received careful attention to its technical content. Chapter 3 was updated to reflect the latest climate change science, including that we changed its title from "The Chemistry of Global Warming" to "The Chemistry of Climate Change." Chapter 6, the "Acid Rain" chapter, now opens with ocean acidification, connecting emissions from combustion to the coral reefs and changes in sea water. And Chapter 1 now has an expanded section on indoor air quality, including a green chemistry section about paints that emit fewer volatile organic compounds while drying.

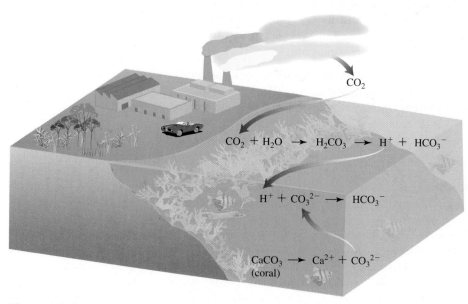

Figure 6.6

Chemistry of CO_2 in the ocean.

Updates to the Laboratory Manual

As we rethought the chapters of *Chemistry in Context* with an eye to sustainability, we recognized that we also needed to rethink the accompanying Laboratory Manual. Both the introductions to the experiments and the post-lab questions were revised in order to emphasize and reinforce the environmental issues and the sustainability concepts presented in the textbook. The experiments were redesigned to include "greener" procedures, such as a microwave synthesis of aspirin and an investigation of the properties of a biodegradable polymer. Additionally, pre-lab questions were added and expanded data sheets encourage students to form and test hypotheses during each experiment.

The New Edition—A Team Effort

It is always a pleasure to bring a new textbook or new edition to fruition. But the work is not done by just one individual; rather, it is the work of many talented individuals. The seventh edition builds on the tradition of prior author teams led by A. Truman Schwartz, Macalester College, Conrad L. Stanitski, University of Central Arkansas, and Lucy Pryde Eubanks, Clemson University. This edition, we were fortunate to have the leadership and encouragement of Mary Kirchhoff, the Director of the ACS Education Division. We also recognize the able assistance of ACS staff members Marta Gmurczyk, Michael Mury, Jerry Bell, and Corrie Kuniyoshi.

This new edition was prepared by a team of writers: Cathy Middlecamp, Steve Keller, Karen Anderson, Anne Bentley, Michael Cann, and Jamie Ellis. The laboratory manual to accompany it was revised and updated by Jennifer Tripp. Each of us brought different expertise to the project. In common, though we brought our goodwill, hopes, dreams, and seemingly boundless enthusiasm to bring real-world chemistry into the classroom.

The McGraw-Hill team was superb in all aspects of this project. Marty Lange (Vice President, Editor-in-Chief), Ryan Blankenship (Publisher), Todd Turner (Sponsoring Editor), and Jodi Rhomberg (Developmental Editor) led this outstanding team. Tami Hodge served as the Executive Marketing Manger. The Lead Project Manager was Sheila Frank, who coordinated the production team of Carrie Burger (Lead Photo Research Coordinator), Jerry Marshall (Photo Research), Tara McDermott (Designer), Kara Kudronowicz (Senior Buyer), and Mary Jane Lampe (Project Coordinator). The Digital Product Manager was Daryl Bruflodt, and Sandra Schnee served as Senior Media Project Managers. The team also benefited from the careful editing of Linda Davoli.

The author team truly benefitted from the expertise of a wider community. We extend our thanks to the following individuals for the technical expertise they provided to us in preparing the manuscript:

Mark E. Anderson, University of Wisconsin–Madison
David Argentar, Sun Edge, LLC
Marion O'Leary, Carnegie Institution for Science
Ross Salawitch, University of Maryland
Kenneth A. Walz, Madison Area Technical College

We also acknowledge those who served as reviewers for the new edition:

Bill Blanken	*Chapman University, Lemoore Campus*
Michael Doescher	*Benedictine College*
Jeannine Eddleton	*Virginia Tech*
Denise V. Greathouse	*University of Arkansas, Fayetteville*
Susan J. Glenn	*University of South Carolina, Aiken*
Narayan S. Hosmane	*Northern Illinois University*
Donna K. Howell	*Park University*
Richard Kashmar	*Wesley College*
John Kirk	*University of Iowa*
Amy Lumley	*Coffeyville Community College*
Mya A. Norman	*University of Arkansas–Fayetteville*
Marion O'Leary	*Carnegie Institution for Science*
Heather U. Price	*Highline Community College*
Victor Ryzhov	*Northern Illinois University*
William L. Schreiber	*Monmouth University*
Lee Alan Shaver	*Washburn University*
Jeff Shen	*Park University*

Carnetta Skipworth	*Western Kentucky University*
Laura K. Stultz	*Birmingham–Southern College*
Sara J. E. Sutcliffe	*University of Texas, Austin*
Burton Tropp	*Queens College, CUNY*

Finally, we would like to thank Andy Jorgensen, University of Toledo, for his expertise developing Connect problems and to Maggie Phillips, Great Lakes Bioenergy Research Center College of Agricultural, and Life Sciences, University of Wisconsin–Madison, for her work developing Figures Alive! Connect problems.

Wishing Our Readers Well

We are very excited by the features of this new edition that continue to "break the mold" in bringing chemistry to you, our reader. We selected engaging and timely topics that we hope will serve you not only today, but also in the years to come. At the same time, we strove to be honest to the science behind these topics.

We wish you well as you read, explore the issues, argue with each other (and with the authors) and, most importantly, as you use what you learn to bring your dreams to reality.

Sincerely, and with all good wishes from the author team,

Cathy Middlecamp
Senior Author and Editor-in-Chief
October 2010

From the Publisher

McGraw-Hill Education, a division of The McGraw-Hill Companies, is proud to bring you this new edition of *Chemistry in Context*. We are a leading innovator in the development of teaching and learning solutions for the 21st century. Through a comprehensive range of traditional and digital education content and tools, McGraw-Hill Education empowers and prepares professionals and students of all ages to connect, learn, and succeed in the global economy.

Chemistry in Context is supported by a complete package for instructors and students. Several print and media supplements have been prepared to accompany the text and make learning as meaningful and up-to-date as possible.

McGraw-Hill Connect Chemistry

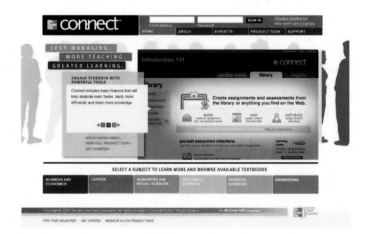

McGraw-Hill Connect™ Chemistry (www.mcgrawhillconnect.com), is a web-based assignment and assessment platform that gives students the means to better connect with their coursework, with their instructors, and with the important concepts that they will need to know for success now and in the future.

With Connect Chemistry, your instructor can deliver assignments, quizzes, and tests online. Questions from the text are presented in an auto-gradable format and tied to the text's learning objectives. Instructors can edit existing questions and author entirely new problems. They also can track individual student performance—by question, assignment, or in relation to the class overall—with detailed grade reports. Integrate grade reports easily with Learning Management Systems (LMS) such as WebCT and Blackboard. And much more.

By choosing Connect Chemistry, instructors are providing their students with a powerful tool for improving academic performance and truly mastering course material. Connect Chemistry allows students to practice important skills at their own pace and on their own schedule. Importantly, students' assessment results and instructors' feedback are all saved online—so students can continually review their progress and plot their course to success.

Like Connect Chemistry, **McGraw-Hill ConnectPlus™ Chemistry** provides students with online assignments and assessments, plus 24/7 online access to an eBook—an online edition of the text—to aid them in successfully completing their work, wherever and whenever they choose.

McGraw-Hill Higher Education and Blackboard®

McGraw-Hill Higher Education and Blackboard have teamed up! What does this mean for you?

1. **Your life, simplified.** Now you and your students can access McGraw-Hill's Connect and Create™ right from within your Blackboard course – all with one single sign-on. Say goodbye to the days of logging in to multiple applications.
2. **Deep integration of content and tools.** Not only do you get single sign-on with Connect and Create, you also get deep integration of McGraw-Hill content and content engines right in Blackboard. Whether you're choosing a book for your course or building Connect assignments, all the tools you need are right where you want them – inside of Blackboard.
3. **Seamless Gradebooks.** Are you tired of keeping multiple gradebooks and manually synchronizing grades into Blackboard? We thought so. When a student completes an integrated Connect assignment, the grade for that assignment automatically (and instantly) feeds your Blackboard grade center.
4. **A solution for everyone.** Whether your institution is already using Blackboard or you just want to try Blackboard on your own, we have a solution for you. McGraw-Hill and Blackboard can now offer you easy access to industry leading technology and content, whether your campus hosts it, or we do. Be sure to ask your local McGraw-Hill representative for details.

Website to Accompany the Text

Throughout *Chemistry in Context*, you will see references to the textbook website, www.mhhe.com/cic. For example, many of the Consider This activities embedded in each chapter reference technical websites with data, such as the stratospheric ozone data from NASA (U.S. National Aeronautics and Space Administration) and the air quality data from the EPA (U.S. Environmental Protection Agency). On the *Chemistry in Context* website, students and instructors can access these websites with a single click.

Figures Alive!

Figures Alive! is a set of interactive, web-based activities, also available on the textbook website. Each one is keyed to a figure in *Chemistry in Context* and leads the student through the discovery of various layers of knowledge inherent in the figure. Look for this icon 📍 in the textbook. The self-testing segments built into Figures Alive! are based on the same categories as the chapter-end problems—*Emphasizing Essentials, Concentrating on Concepts, and Exploring Extensions.*

Digital Lecture Capture: Tegrity

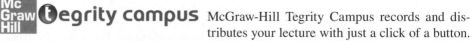 McGraw-Hill Tegrity Campus records and distributes your lecture with just a click of a button. Students can view anytime/anywhere via computer, iPod, or mobile device. Tegrity indexes as it records your slideshow presentations and anything shown on your computer, so students can use keywords to find exactly what they want to study.

Also for the Instructor

Test Bank: The electronic test bank offers a bank of questions that can be used for homework assignments or the preparation of exams. The test bank can be utilized to quickly create customized exams. It allows instructors to sort questions by format or level of difficulty; edit existing questions or add new ones; and scramble questions and answer keys for multiple versions of the same test.

McGraw-Hill Presentation Center: Build instructional material wherever, whenever, and however you want! McGraw-Hill Presentation Center is an online digital library containing assets such as photos, artwork, and other media types that can be used to create customized lectures, visually enhanced tests and quizzes, compelling course websites, or attractive printed support materials. The McGraw-Hill Presentation Center Library includes thousands of assets from many McGraw-Hill titles. This ever-growing resource gives instructors the power to utilize assets specific to an adopted textbook as well as content from all other books in the library. The Presentation Center can be accessed from the instructor side of your textbook's Connect website, and the Presentation Center's dynamic search engine allows you to explore by discipline, course, textbook chapter, asset type, or keyword. Simply browse, select, and download the files you need to build engaging course materials. All assets are copyrighted by McGraw-Hill Higher Education but can be used by instructors for classroom purposes.

Chemistry
in
Context

A Project of the American Chemical Society

0 Chemistry for a Sustainable Future

The "blue marble," our Earth, as seen from outer space.

"The first day or so, we all pointed to our countries. The third or fourth day, we were pointing to our continents. By the fifth day, we were aware of only one Earth."

Prince Sultan bin Salman Al Sa'ud, Astronaut, Saudi Arabia, 1985.

Only one Earth. From the vantage point of outer space, the planet we call home is truly magnificent—a "blue marble" of water, land, and clouds. In 1972, the crew of the Apollo 17 spacecraft photographed the Earth at a distance of about 28,000 miles (45,000 kilometers). In the words of Soviet cosmonaut Aleksei Leonov, "the Earth was small, light blue, and so touchingly alone."

Are we alone in the universe? Possibly. Clearly, though, we are not alone on our planet. We share it with other creatures large and small. Biologists estimate that upwards of 1.5 million species exist in addition to our own. Some help to feed and sustain us. Others contribute to our well-being. Still others (like mosquitoes) annoy and perhaps even sicken us.

We also share the planet with just short of 7 billion other people. Over the past century, the human population on Earth has more than tripled, an unprecedented growth spurt in the history of our planet. By 2050, the population may grow by another 2 to 3 billion.

Large and small, all of the species on our planet somehow connect. Exactly how this happens, however, may not be so obvious to us. For example, unseen microorganisms shuttle nitrogen from one chemical form to another, providing nutrients for green plants to grow. These plants harness the light energy of the Sun during the process of photosynthesis. Using this energy, they convert the compounds carbon dioxide and water to that of glucose, their food source. At the same time, they release the element oxygen into the air that we breathe. And we humans are host to countless microorganisms that have taken up residence in our skin and internal organs.

Chapter 6 describes the nitrogen cycle. Chapter 4 describes photosynthesis.

Large and small, these connections are breaking at an alarming rate. Our grandparents used to be able to tell stories of "how it was" when they fished and hunted. Consider, for example, that salmon once ran the rocky streams of the North Atlantic coast. Sea captains once reported large fish in such abundance that they literally could be taken by hand. Abalone used to be present everywhere off the Pacific coasts of North America, rather than in the few good spots boats can find today. However, nobody now living remembers fish in this abundance.

A change has occurred in our perspective. Today, we are accustomed to reading reports of declining fish populations and of endangered species. Have you run across the term **shifting baselines?** This refers to the idea that what people expect as "normal" on our planet has changed over time, especially with regard to ecosystems. The abundance of fish and wildlife that once was normal is no longer carried in the memories of those living today. Similarly, many of us have no memory of cities unclogged by vehicles. Fewer people remember clear summer days when it really did seem that you could see forever.

Clearly we humans are industrious creatures. We grow crops, dam rivers, burn fuels, build structures, and jet across time zones. When we carry out such activities by the million, we change the quality of the air we breathe, the water we drink, and the land on which we live. Over time, our actions have changed the face of our planet. What was it once like? See what you can find out by doing the next activity.

Consider This 0.1   Shifting Baselines

Seek out one of the elders in your community. This person may be a friend, relative, or possibly even a community historian.

a. Think about the current price of a loaf of bread, a gallon of gas, or a candy bar. Then inquire about what things used to cost. How has what people expect as "normal" shifted?

b. Now a more difficult task. Think about the local rivers, air quality, vegetation, or wildlife. In talking with an elder, see if you can identify at least one case in which the perception of what is "normal" has shifted. It may be that nobody remembers.

Consider This activities appear in all chapters. These activities give you a chance to use what you are learning to make informed decisions. For example, they may require you to consider opposing viewpoints or to make and defend a personal decision. They may require additional research.

The bottom line? The things we perceive today as "normal" were not normal in the past. Although we cannot turn back the clock, we still can make choices that promote our health and the health of our planet today and in the future. A knowledge of chemistry can help. The global problems that we face—and their solutions—are intimately linked with chemical expertise and good old human ingenuity.

0.1 | The Choices We Make Today

Individually, it may seem that our actions have little effect on a system as large as our planet. After all, compared with a hurricane, a drought, or an earthquake, what we do every day can seem pretty inconsequential. What difference could it possibly make if we biked to work instead of driving, used a reusable cloth bag instead of discarding a plastic one, or ate foods grown locally instead of consuming those shipped from hundreds or even thousands of miles away?

Most human activities—including biking, driving, using bags of whatever sort, and eating—have two things in common: *They require the consumption of natural resources, and they result in the creation of waste.* Driving a car requires gasoline (refined from crude oil), and burning gasoline sends waste products out the exhaust pipe. Although riding a bike is a more ecological choice, all bicycles, just like automobiles, still require the manufacture and disposal of metals, plastics, synthetic rubbers, fabrics, and paints. Shopping bags, whether paper or plastic, require the materials to produce them. Later down the line, these bags become a waste product. And growing food requires water and energy to harvest and transport to market. In addition, food production may require fertilizer and involve the use of insecticides and herbicides.

You can see where this is going. Any time we manufacture and transport things, we consume resources and produce waste. Clearly, though, some activities consume fewer resources and produce less waste than others. Biking produces less waste than driving; reusing cloth bags produces less waste than continually throwing plastic ones away. Although what you do may be negligible in the grand scheme of things, what 7 billion people do clearly is not. Our collective actions not only cause local changes to our air, water, and soil but also hurt regional and global ecosystems.

We need to think by the billion. A single cooking fire? No problem; well, unless it accidentally burns down a dwelling. But imagine a few billion people across the planet each tending an individual cooking fire. Add in fires from those who cook using stoves, brick ovens, and outdoor grills. Now you have a lot of fuel being burned! Each fuel releases waste products into the atmosphere as it is burned. Some of these waste products—better known as air pollutants—are *highly* unfriendly to our lungs, our eyes, and of course to our ecosystems.

Today, the waste products we release are unprecedented in their scale and in their potential to lower our quality of life and even shorten it. For example, in a large city such as New York, Atlanta, Mexico City, or Beijing, you will find that hospital admissions and death rates correlate with air pollution levels. Although the health risks are smaller than those caused by obesity or smoking, the issues of public health loom large because people are exposed to air pollutants, both indoors and out, over a lifetime.

Also worrisome is that our actions (by the billion) release waste products that destroy the habitats of other species on the planet. Extinction, of course, is a natural phenomenon. But today the rate is many times faster than would be expected from natural causes. Our destruction of local specialized habitats, particularly those of plants, has led to these extinctions.

Underpinning much of our waste production is *energy*. The need to find energy sources that are both clean and sustainable is arguably the major challenge of our century. Currently we are consuming nonrenewable and renewable resources and adding waste to our air, land, and water at a rate that cannot be sustained. This should

Look for more about crude oil and how it is refined in Chapter 4.

Look for more about what comes out a tailpipe in Chapter 1.

Chapters 1, 3, 4, and 6 all explore the connections between fuels and the waste products they release when burned.

Nonrenewable resources, such as crude oil and metal ores, are finite in supply. Once we use them up, they are gone.

come as no surprise. From what you have learned from your other studies, most likely you are well aware of this.

With any problem comes the opportunity to find creative solutions. We hope you are asking yourself "What can I do?" and "How can I make a difference in my community?" As you ask questions such as these, remember to include chemistry in your deliberations. Indeed, chemistry is well named the "central science." Today, chemists are at the center of the action when it comes to the sustainable use of resources. Chemists are challenged to use what they know, to do so responsibly, and to proceed with reasonable haste. The same, of course, is true for you. In this book, we will support you as you learn and encourage you to use what you learn to act responsibly and with reasonable haste.

"…No matter what, the road to better health, better materials, and better energy sources goes through chemistry."

Bill Carroll, 2006
American Chemical Society
Past President

0.2 | The Sustainable Practices We Need for Tomorrow

What does it mean to use the resources of our planet in a sustainable manner? We hope that you can answer this question—at least in part—from what you already have learned in other classes. People who study many other disciplines, including economics, political science, engineering, history, nursing, and agriculture, have a stake in developing sustainable practices. And as you will learn in this text, those of us in chemistry also have a stake in developing sustainable practices.

Because the term **sustainability** is used by so many groups of people, it has taken on different meanings. We have selected one that is frequently quoted: "Meeting the needs of the present without compromising the ability of future generations to meet their needs." This definition is drawn from a statement written over two decades ago, a 1987 report, *Our Common Future,* written by the World Commission on Environment and Development of the United Nations. Even so, it has stood the test of time. In Table 0.1, we reprint excerpts from the foreword to *Our Common Future* so that you can read its challenging words in their original context.

Brundtland's words carry a message to those who teach and learn. She writes: "In particular, the Commission is addressing the young. The world's teachers will have a crucial role to play in bringing this report to them." We agree. To this end, we hope that your chemistry course will stimulate conversations both inside and outside of the classroom. One such conversation is about practices that are *not* sustainable. For example, you will study fossil fuels and learn why their use is not sustainable (Chapter 4). But don't stop here. You also need to discuss what you can do to *solve* the problems we face today. Use what you learn about air quality to act to improve local air quality *and* to make informed decisions as a citizen to improve it more widely (Chapter 1). Similarly, use what you learn about aqueous solubility and waste water to evaluate public policies that relate to water quality (Chapter 5).

Our Common Future is also called the *Brundtland report*. It was named after Gro Harlem Brundtland, the woman who chaired the commission.

Any discussion of sustainability needs to include conservation. Although people may equate conservation with undue self-sacrifice, the equation is far more complex than this. To conserve, we need the vigorous development of community-based technologies that improve

Table 0.1	Our Common Future (excerpts from the Foreword)

"A global agenda for change"—this was what the World Commission on Environment and Development was asked to formulate. It was an urgent call by the General Assembly of the United Nations.

In the final analysis, I decided to accept the challenge. The challenge of facing the future, and of safeguarding the interests of coming generations.

After a decade and a half of a standstill or even deterioration in global co-operation, I believe the time has come for higher expectations, for common goals pursued together, for an increased political will to address our common future.

The present decade has been marked by a retreat from social concerns. Scientists bring to our attention urgent but complex problems bearing on our very survival: a warming globe, threats to the Earth's ozone layer, deserts consuming agricultural land.

The question of population—of population pressure, of population and human rights—and the links between these related issues and poverty, environment, and development proved to be one of the more difficult concerns with which we had to struggle.

But first and foremost our message is directed towards people, whose well being is the ultimate goal of all environment and development policies. In particular, the Commission is addressing the young. The world's teachers will have a crucial role to play in bringing this report to them.

If we do not succeed in putting our message of urgency through to today's parents and decision makers, we risk undermining our children's fundamental right to a healthy, life-enhancing environment.

In the final analysis, this is what it amounts to: furthering the common understanding and common spirit of responsibility so clearly needed in a divided world.

Gro Harlem Brundtland, Oslo, 1987.

efficiency, promote the use of renewable resources, and minimize or prevent waste. All of these require ingenuity and chemical know-how.

> Renewable resources are expected always to be available. They include sunlight, wind and wave energy, and timber.

Any discussion of sustainability usually is accompanied by a sense of urgency as well. We currently are not using the resources of our Earth in a sustainable manner. We need to change our practices and we cannot delay. This sense of urgency has put sustainability squarely on the radar screens of chemists, other scientists, and the professional societies of which they are members. For example, botanist Peter H. Raven, former president of the American Association for the Advancement of Science, spoke about sustainability in his 2002 address as president titled *Science, Sustainability and the Human Prospect*. He called for nothing short of new ways of thinking.

> *"New ways of thinking—an integrated multidimensional approach to the problems of global sustainability—have long been needed, and it is up to us to decide whether the especially difficult challenges that we are facing today will jolt us into finding and accepting them."*
>
> *Peter H. Raven, 2002*

Not only must we find these ways, but we will also be judged by our success—or lack thereof—in employing them. Bill Carroll, a recent president of the American Chemical Society, clearly points this out: "By 2015, the chemistry enterprise will be judged under a new paradigm of sustainability. Sustainable operations will become both economically and ethically essential." These words came from a report he coauthored in 2005, *The Chemistry Enterprise in 2015*. Carroll also pointed out that "No matter what, the road to better health, better materials, and better energy sources goes through chemistry." In this text, we will be following this road. The next section describes the Triple Bottom Line, an important way to set our bearings as we travel.

> This report, *The Chemistry Enterprise in 2015*, also points out that most issues of sustainability ultimately come down to questions of energy, a topic of several chapters in this book.

0.3 | The Triple Bottom Line

Scientists aren't the only ones. If you are a business or economics major, you may be well aware that people in the business sector have put sustainability on the corporate agenda. In fact, sustainable practices now offer a competitive advantage in the marketplace.

In the world of business, the bottom line always has included turning a profit, preferably a large one. Today, however, the bottom line includes more than this. For example, corporations are judged to be successful when they are fair and beneficial to workers and to the larger society. Another measure of their success is how well they protect the health of the environment, including the quality of the air, water, and land.

Taken together, this three-way measure of the success of a business based on its benefits to the economy, to society, and to the environment has become known as the **Triple Bottom Line.** One way to represent the Triple Bottom Line is with the overlapping circles shown in Figure 0.1. The economy must be healthy, that is, the annual reports need to show a profit. But no economy exists in isolation; rather, it connects to a community whose members also need to be healthy. In turn, communities connect to ecosystems that need to be healthy. Hence the figure includes not one, but three connecting circles. At the intersection of these circles lies the "Green Zone." This represents the conditions under which the Triple Bottom Line is met.

Harm that occurs in any of the circles of Figure 0.1 ultimately translates into harm for the business. Conversely, achieving success can provide a competitive advantage, both immediately and in the years to come. Businesses can turn a profit; at the same time, they can get good publicity (and minimize any harm) by using less energy, consuming fewer resources, and creating less waste. A triple win!

Recent news articles document the changes that are occurring. For example, read this excerpt from a news article about Clorox, a company that produces the bleach that you are likely to find with the cleaning supplies in your local supermarket. The source is *Chemical & Engineering News (C&EN)*, the weekly publication of the American Chemical Society. One sentence was underlined for emphasis.

The Triple Bottom Line sometimes is shortened to the 3Ps: Profits, People, and the Planet.

Greener Cleaners: Consumer demand for environmentally friendly cleaning products has changed the game for chemical suppliers

"This month, Clorox, a company almost synonymous with the environmental lightning rod chlorine, is going national with what might seem like an unlikely product line: a family of natural cleaners sold under the Earth-friendly name Green Works.

That a consumer products giant like Clorox would venture into the market for so-called green cleaning products says a lot about how much the home care industry has changed in the past two years. Once solely the province of fringe players, green or sustainable cleaners are attracting the interest of big corporations in America and elsewhere. In such products, companies see both a growing market and a way to burnish their environmental credentials." (*C&EN*, Jan. 21, 2008)

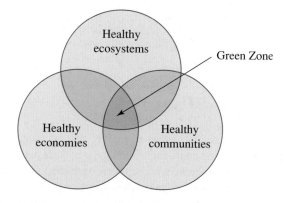

Figure 0.1

A representation of the Triple Bottom Line. The "Green Zone" (where the Triple Bottom Line is achieved) lies at the intersection of these three circles.

At the time the article was published, Clorox posted a video on its corporate website about new green cleaning products. The sound track described the "natural plant-based cleaners… without harsh chemical fumes or residues." The website also included strongly worded environmental statements indicating that the corporation is working toward "using as little packaging material as needed to do the job" and "using recycled material wherever it is environmentally and economically sound to do so." Clearly Clorox is responding to the negative image alluded to in the news article. At issue here is the chlorine-containing bleach, sodium hypochlorite ($NaClO$). We will revisit this compound in the next section.

Before we leave the green cleaning scene, consider one more news article, this time on laundry detergents. The article describes the placards in hotels that give you the option of *not* having the sheets and towels changed daily. What happens, though, when people who market soap products stay at these hotels? Do they choose the option that means selling less detergent? Again, one sentence was underlined for emphasis.

Having It All: Chemical makers supplying the detergents industry seek both sales and sustainability

"Although it's lush to the point of excess, the Boca Raton Resort & Club in Florida does demonstrate a modicum of environmental responsibility with a card placed on nightstands informing guests that bedsheets won't be removed and washed unless requested.

This card no doubt presented a conundrum to chemical industry executives who were at the resort earlier this month for the Soap & Detergent Association's annual conference: Should they or shouldn't they participate in a program intended to cut consumption of the very products they are there to sell?

But then, the challenges of good environmental behavior have been on the minds of all participants in the cleaning products industry lately. Everyone from the government to retailers to consumers seems to be demanding environmentally sustainable products…. Still, the chemical companies that supply ingredients to the cleaning products industry see robust sales and environmental stewardship as mutually obtainable. Rather than cut back on surfactants or other cleaning chemicals, they are advising their customers to formulate products with ingredients that have less of an environmental impact." (*C&EN*, February 18, 2008)

Look for more about sodium hypochlorite in Chapter 5 in connection with water purification.

Consider This 0.2   Green Conundrums

Put yourself in two different roles. First, you are a manufacturer attending the annual Soap & Detergent Association conference. You find a card on your pillow saying that your bedsheets will not be washed unless you request it. Argue for a course of action, either way. Second, you are the hotel manager. Do you remove the card from the pillows, knowing that Soap & Detergent Association people are coming to town? Again make your case.

The point? The debates over how to be "green" are likely to continue over your lifetime. The issues are not new; they are not likely to be resolved anytime soon. *Chemistry in Context* will encourage you to explore these issues, arming you with the knowledge you need and enabling you to more creatively formulate your own response.

"A sustainable society is one that is far-seeing enough, flexible enough, and wise enough not to undermine either its physical or its social systems."

Donella Meadows, 1992

Donella Meadows (1941–2001) was a scientist and writer. Her books include *The Limits to Growth* and *The Global Citizen*.

0.4 | Cradle-to-Where?

You may have heard the expression **cradle-to-grave;** that is, an approach to analyzing the life cycle of an item, starting with the raw materials from which it came and ending with its ultimate disposal someplace, presumably on Earth. This catchy phrase offers a frame of reference from which to ask questions about consumer items. Where did the item come from? And what will happen to it when you are finished with it? More than ever, individuals, communities, and corporations are recognizing the importance of asking these types of questions.

For example, the website for Clorox asserts that the bleach you purchase, sodium hypochlorite, "starts as salt and water and ends as salt and water." We agree. However, we urge you to consider the entire life cycle of a product like Clorox. One chemical used in the manufacture of Clorox is an element named chlorine (Cl). Chlorine will receive mention in almost every chapter of *Chemistry in Context!* How was the chlorine produced and transported? What waste products were created in its manufacture? Cradle-to-grave means thinking about *every* step in the process.

Companies should take responsibility—as should you—for items from the moment the natural resources used to make them were taken out of the ground, air, or water to the point at which they were ultimately "disposed of." Think of items such as batteries, plastic water bottles, T-shirts, cleaning supplies, running shoes, cell phones—anything that you buy and eventually discard.

Look for more about batteries in Chapter 8 and more about plastics in Chapter 9.

Cradle-to-grave thinking clearly has its limitations. As an illustration, let us follow one of the plastic bags that supermarkets provide for your groceries. The raw material for these bags is petroleum. Accordingly, the "cradle" of this plastic bag most likely was crude oil somewhere on our planet, for example, the oil fields of Canada. Let's assume that the oil was pumped from a well in Alberta and then transported to a refinery in the United States. At the refinery, the crude oil was separated into fractions. One of the fractions was then cracked into ethylene, the starting material for a polymer. Ethylene next was polymerized and formed into polyethylene bags. These bags were packaged and then trucked (burning diesel fuel, another refinery product) to your grocery store. Ultimately, you purchased groceries and used one of the bags to carry them home.

Chapter 4 explains how and why crude oil is separated into fractions.

As stated, this is not a cradle-to-grave scenario. Rather, it was cradle-to-your-kitchen, definitely several steps short of any graveyard. So what happened to this plastic bag after you used it? Did it go into the trash? The term *grave* describes wherever an item eventually ends up. One trillion plastic bags, give or take, are used each year in supermarkets. Only about 5% are recycled. The rest end up in our cupboards, our land-fills, or littered across the planet. As litter, these bags begin a 1000-year cycle, again give or take, of slow decomposition into carbon dioxide and water.

Chapter 9 explains how polyethylene is made from ethylene (and why).

Cradle-to-a-grave-somewhere-on-the-planet is a poorly planned scenario for a supermarket bag. If each of the trillion plastic bags instead were to serve as the starting material for a new product, we then would have a more sustainable situation. **Cradle-to-cradle,** a term that emerged in the 1980s, refers to a regenerative approach to the use of things in which the end of the life cycle of one item dovetails with the beginning of the life cycle of another, so that everything is reused rather than disposed of as waste. In Chapter 9, we will examine different recycle-and-reuse scenarios for plastic bottles. But right now, you can do your own cradle-to-cradle thinking in the next activity.

The term *cradle-to-cradle* caught on with the publication of a book of this same title published in 2002.

Your Turn 0.3   The Can That Holds Your Beverage

People tend to think of an aluminum can as starting on a supermarket shelf and ending in a recycling bin. There is more to the story!

a. Where on the planet is aluminum ore (bauxite) found?

b. Once removed from the ground, the ore usually is refined to alumina (aluminum oxide) near the mining site. The alumina is then transported to a production facility. What happens next to produce aluminum metal?

c. The metal is then shaped into a beverage can. See if you can find where the can was filled and how far it was transported to land on the shelves of your neighborhood store.

d. What happens to the can after you recycle it?

Answers

a. Bauxite is mined in several places, including Australia, China, Brazil, and India.

b. The ore must be refined electrolytically to produce aluminum metal. This process is energy-intensive and carried out in many locations worldwide.

Note: the textbook's website provides some helpful links.

Your Turn activities appear in all chapters. They provide an opportunity for you to practice a new skill or calculation that was just introduced in the text. Answers are given either following the Your Turn activity or in Appendix 4.

Some activities are marked with a Web icon. This cues you to use the resources of the Internet as you work.

As you can tell from these examples, it is not just the decisions of manufacturers that matter. Your decisions do as well. What you buy, what you discard, and how you discard it all warrant attention. The choices that we make—individually and collectively—matter.

At the risk of repeating ourselves, we remind you that the current state of affairs in which we consume the nonrenewable resources of our planet and add waste to our air, land, and water is *not* sustainable. With a sense of urgency, we recall the words of botanist Peter Raven quoted earlier: "We must find new ways to provide for a human society that presently has outstripped the limits of global sustainability." In the next section, we explain why, setting the stage for the topics in chemistry that you will explore in this text.

0.5 | Your Ecological Footprint

You already may know how to estimate the gas mileage for a vehicle. Likewise, you can estimate how many calories you consume. How might you estimate how much of the Earth's natural capital it takes to support the way in which you live? Clearly, this is a far more difficult task. Fortunately, other scientists already have grappled with how to do the math. They base the calculations on the way in which a person lives coupled with the available renewable resources needed to sustain this lifestyle.

Consider the metaphor of a footprint. You can see the footprints that you leave in sand or snow. You also can see the muddy tracks that your boots leave on the kitchen floor. Similarly, one might argue that your life leaves a footprint on planet Earth. To understand this footprint, you need to think in units of hectares or acres. A hectare is a bit more than twice the area of an acre. The **ecological footprint** is a means of estimating the amount of biologically productive space (land, water, and sea surface) necessary to support a particular standard of living or lifestyle.

For the average U.S. citizen, the ecological footprint was estimated in 2005 to be about 9.7 hectares (24 acres). In other words, if you live in the United States, on average it requires 9.7 hectares to provide the resources to feed you, clothe you, transport you, and give you a dwelling with the creature comforts to which you are accustomed. The people of the United States have relatively big feet, as you can see in Figure 0.2. The world average in 2003 was estimated to be 2.2 hectares per person; today the value is believed to be closer to 2.7.

How much biologically productive land and water is available on our planet? We can estimate this by including regions such as croplands and fishing zones, and omitting regions such as deserts and ice caps. Currently, the value is estimated at about 11 billion hectares (roughly 27 billion acres) of land, water, and sea surface. This turns out to be about a quarter of the Earth's surface. Is this enough to sustain everybody on the planet with the lifestyle that people in the United States have? The next activity allows you to see for yourself.

> Although you may not know how much air you breathe in a day, Chapter 1 will help you to estimate this.

> Carbon footprint is a subset of ecological footprint. Look for more about carbon footprints in Section 3.9.

> A hectare is 10,000 square meters, or 2.471 acres.

🕸 Your Turn 0.4   Your Personal Share of the Planet

As stated earlier, an estimated 11 billion hectares (~27 billion acres) of biologically productive land, water, and sea is available on our planet.

a. Find the current estimate for the world population. Cite your source.
b. Use this estimate together with the one for biologically productive land to calculate the amount of land theoretically available for each person in the world.

Answers
a. In 2010, the population of the Earth was between 6.8 and 6.9 billion.
b. About 1.6 hectares or ~4 acres per person.

Bottom line? A nation whose people have an average footprint greater than about 1.6 hectares is exceeding the "carrying capacity" of the Earth. Using the United States as an example, let's do one more calculation to see by how much.

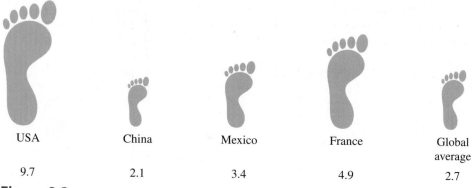

USA	China	Mexico	France	Global average
9.7	2.1	3.4	4.9	2.7

Figure 0.2

A comparison of ecological footprints, in global hectares per person.

Source: Global Footprint Network, 2008 data tables.

Your Turn 0.5   How Many Earths?

In 2009, the United States had an ecological footprint of about 9.7 hectares (~24 acres) per person.

a. Find an estimate of the current population of the U.S. Cite your source.
b. Calculate the amount of biologically productive space (land, water, and sea surface) that the U.S. currently requires for this population.
c. What percent is this amount of the biologically productive space (about 11 billion hectares or ~27 billion acres) available on our planet?

Answer

b. Estimating the U.S. population at 310 million and using the estimate of 9.7 hectares per person, the United States required about 3 billion hectares (7.3 billion acres).

Let's now say that *everyone* on the planet lived like the average citizen in the United States. Here is the calculation based on the world population in 2009.

$$\frac{6.8 \text{ billion people} \times 9.7 \text{ hectares/person} \times 1 \text{ planet}}{11 \text{ billion hectares}} = 6.0 \text{ planets}$$

Thus, to sustain this same standard of living for everybody on the planet we would need 5 more Earths in addition to the one we currently have!

"*Only One Earth.*" On this Earth, the number of people has risen dramatically in the last few hundred years. So has economic development. As a result, the estimated global ecological footprint is rising, as shown in Figure 0.3. In 2003, we estimate that humanity used the equivalent of 1.25 Earths. By 2040, the projection is that we will be using 2 Earths. Clearly this rate of consumption cannot be sustained.

Through your study of chemistry, we hope you will learn ways either to reduce your ecological footprint or to keep it low, if it already is. The next section describes how chemists can help make this process work in some ways you might not expect.

0.6 | Our Responsibilities as Citizens and Chemists

We humans have a special responsibility to take care of our planet. Living out this responsibility, however, has proven to be no easy task. Each chapter in *Chemistry in Context* highlights an issue of interest such as air quality, water quality, or nutrition.

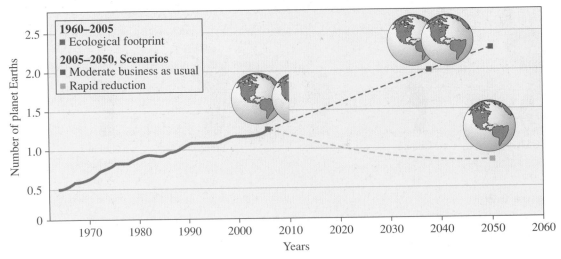

Figure 0.3

The ecological footprint of humans, current (solid line) and projected (dashed lines). Current estimates place the footprint above 1.0 Earth, that is, above what the Earth can sustain. Note: The upper projection assumes "moderate business as usual." The lower projection assumes an increase in sustainable practices.

Source: Adapted from data provided by the Global Footprint Network

These issues not only affect you personally, they also affect the health and well-being of the wider communities of which you are a part. For each issue, you will work on two related tasks: (1) learning about the issue and (2) finding ways to act constructively.

Chemists are learning and working right along with you. Recall the words of Bill Carroll that we quoted earlier in the chapter: "By 2015, the chemistry enterprise will be judged under a new paradigm of sustainability. Sustainable operations will become both economically and ethically essential."

How do chemists meet the challenges of sustainability? The answer lies in part with "green chemistry," a set of principles originally articulated by people at the U.S. Environmental Protection Agency (EPA) and now actively pursued by the American Chemical Society (ACS). **Green chemistry** is the design of chemical products and processes to use less energy, produce fewer hazardous materials, and use renewable resources whenever possible. The desired outcome is to produce less waste, especially toxic waste, and to use fewer resources.

Recognize that green chemistry is a tool in achieving sustainability, not an end in itself. As stated in the article "Color Me Green" from an issue of *Chemical & Engineering News* published in 2000: "Green chemistry represents the pillars that hold up our sustainable future. It is imperative to teach the value of green chemistry to tomorrow's chemists." Actually, we believe that it is imperative to teach the value of green chemistry to citizens as well. This is why so many applications of green chemistry are woven throughout *Chemistry in Context*.

To get you started, we list six big ideas about green chemistry (Table 0.2). They also are printed on the inside cover of this book.

Initiated under the EPA's *Design for the Environment Program*, green chemistry leads to cleaner air, water, and land, and the consumption of fewer resources. Chemists are now designing new processes (or retooling older ones) to make them more environmentally friendly. We call this "benign by design." Every green innovation does not necessarily have to be successful in achieving all six of these principles. But achieving several of them is an excellent step on the road to sustainability.

For example, an obvious way to reduce waste is to design chemical processes that don't produce it in the first place. One way is to have most or all of the atoms in the reactants end up as part of the desired product molecules. This "atom economy" approach, although not applicable to all reactions, has been used for the synthesis of many products, including pharmaceuticals, plastics, and pesticides. The approach saves

Look for green chemistry examples throughout this book, designated with this icon.

ACS Green Chemistry Institute

Table 0.2	Principles of Green Chemistry

1. It is better to **prevent waste** than to treat or clean up waste after it is formed.

2. It is better to **minimize the amount of materials used** in the production of a product.

3. It is better to **use and generate substances that are not toxic.**

4. It is better to **use less energy.**

5. It is better to **use renewable materials.**

6. It is better to **use materials that degrade** into innocuous products **at the end of their useful life.**

Source: Adapted from the *Twelve Principles of Green Chemistry*, by Paul Anastas and John Warner.

money, uses fewer starting materials, and minimizes waste. The connection between green chemistry and the Triple Bottom Line should be apparent!

Many chemical manufacturing processes now use innovative green chemical methods. For example, you will see applications of green chemistry that have led to cheaper, less wasteful, and less toxic production of low VOC paint (Chapter 1). You also will see the principles of green chemistry applied to processing raw cotton and to dry-cleaning methods (Chapter 5). There are more economical and healthier ways to process vegetable oils (Chapter 11).

VOC stands for volatile organic compounds. Look for more about VOCs in connection with air pollution in Chapter 1, "The Air You Breathe."

"Green chemistry and engineering hold the key for our sustainable future."

Robert Peoples, 2009
Director
Green Chemistry Institute

Green chemistry efforts have been rewarded! A select group of research chemists and chemical engineers has received the Presidential Green Chemistry Challenge Award. Initiated in 1996, this presidential-level award recognizes chemists and the chemical industry for their innovations aimed at reducing pollution. These awards recognize innovations in "cleaner, cheaper, and smarter chemistry."

0.7 | Back to the Blue Marble

Before we send you off to Chapter 1, we revisit the 1987 United Nations document *Our Common Future*. Earlier, we drew from this document our definition of sustainability: "Meeting the needs of the present without compromising the ability of future generations to meet their own needs." The foreword to this report was written by the chair, Gro Harlem Brundtland. She also wrote these words, ones that call us back to the image of Earth that opened this chapter:

> "In the middle of the 20th century, we saw our planet from space for the first time. Historians may eventually find that this vision had a greater impact on thought than did the Copernican revolution of the 16th century, which upset the human self-image by revealing that the Earth is not the centre of the universe. From space, we see a small and fragile ball dominated not by human activity and edifice but by a pattern of clouds, oceans, greenery, and soils. Humanity's inability to fit its activities into that pattern is changing planetary systems, fundamentally. Many such changes are accompanied by life-threatening hazards. This new reality, from which there is no escape, must be recognized—and managed."

We agree. The new reality must be recognized. There is no escape. And all of us—students and teachers alike—have important roles to play. With *Chemistry in Context,* we will strive to provide you with the chemical information that can make a difference in your life and in the lives of others. We hope that you will use it to meet the challenges of today and tomorrow with understanding and good spirit.

Questions

1. This chapter opened with a famous quote from a Saudi astronaut, Prince Sultan. Here is what he said in a 2005 interview: "Being an astronaut has had an enormous impact on me. Looking at the planet from the perspective of the blackness of space, it makes you wonder, . . ." Prince Sultan went on to describe what he wondered about. We did not reprint his words. Rather, we hoped that you would write your own.

   a. Write a three-paragraph essay. In the first, introduce yourself briefly. In the second, describe what is important to you as you begin your study of chemistry. And in the third, describe what you most wonder about that relates to planet Earth.

   b. Share your self-introduction with others in your class, as indicated by your instructor.

2. Read the full text of the Foreword to the Brundtland report. It is only a few pages and contains some of the most compelling language ever written. Pick a small section and write a short piece that connects it to something you care about or is of concern to you. You may choose a stance of agreement or disagreement. Your textbook's website contains a link to the document.

3. This chapter introduces the idea that the species on our planet are all connected, sometimes in ways that are not obvious. From your studies in other fields, give three examples of how organisms are linked or in some way depend on one another.

4. Figure 0.1 shows one possible representation of the Triple Bottom Line. Here is another. Comment on the similarities and differences between these two figures.

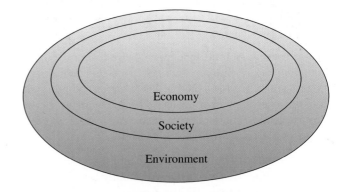

5. Calculate your ecological footprint. Feel free to use one of the websites suggested at your textbook's website.

6. A bus serving a university campus had this sign painted on its side: "Reduce your footprints. Take the bus." Explain the double meaning.

7. Select a corporation, visit its website, and look for the pages that describe its efforts at sustainability or corporate responsibility. Possibilities include Coca-Cola, WalMart, Exxon Mobil, and Burger King. Report on what you find.

8. Mel George, a retired mathematics professor and one of the architects of *Shaping the Future, New Expectations for Education in Undergraduate Science, Technology, Engineering, and Mathematics* (National Science Foundation), remarked: "Putting a man on the moon did not require me to do anything. In contrast, we all must do something to save the planet." Describe five things that you could do, given who you are and your field of study.

9. The principles of green chemistry are not just for chemists. Perhaps you are an economics major. Or you are planning to become a nurse or a teacher. Perhaps you spend time gardening or you like to commute by bike. Pick two of the principles of green chemistry and show how they connect to your life or intended profession.

10. Newspaper and magazine advertisements often proclaim how "green" a business or corporation is. Find one and read it closely. What do you think—is it a case of "greenwashing," in which a corporation is trying to use as a selling point one tiny green drop in an otherwise wasteful bucket? Or is it a case of a real improvement that significantly reduces the waste stream? Note: It may be difficult to tell. For example, removing 7 tons of waste may sound large unless you know that the actual waste stream is in the billions of tons.

11. The mission statement of the American Chemical Society, the world's largest scientific society, is "To advance the broader chemistry enterprise and its practitioners for the benefit of Earth and its people." Explain how the final six words of this statement connect to the definition of sustainability used in this chapter.

12. Thomas Berry, theologian and cultural historian (1914–2009), described in his book *The Great Works* (1999) how humanity is faced with the job of moving from the present geological age, the Cenozoic, to the coming age. He names this age as "Ecozoic," reflecting the tremendous power of humans to shape the face of the world.

   "The universities must decide whether they will continue training persons for temporary survival in the declining Cenozoic Era or whether they will begin educating students for the emerging Ecozoic. . . . While this is not the time for continued denial by the universities or for attributing blame to the universities, it is the time for universities to rethink themselves and what they are doing."

   a. What do you see as the coming age? Describe it. If you don't like the term *Ecozoic Age*, suggest one of your own.

   b. How have your studies to date been preparing you for the future world in which you will live?

   c. In what ways do you think your study of chemistry might prepare you?

1 The Air We Breathe

California blue skies, Lake Tahoe region.

"Ancient Greeks saw air as one of nature's basic elements, along with earth, fire, and water. Californians see it . . . Ah, perhaps those words offer clarification: Californians see too much of something that ought to be less visible. They also feel effects from breathing that air, which too often brings the routine act of respiration to their attention."

David Carle, *Introduction to Air in California*, 2006. page xiii.

People have always noticed—and been curious about—the air they breathe. Together with earth, fire, and water, the ancient Greeks named air as a basic element of nature. Hundreds of years later, chemists experimented to learn more about the composition of air. Today, we can view the Earth's atmosphere from outer space. And daily, just like the ancients, we can peer up through the night air to catch a glimpse of the twinkling stars.

Our atmosphere is the thin veil between us and outer space. This chapter describes the atmospheric gases that support the life on Earth. The next chapter describes the ozone in the stratosphere that protects us from harmful ultraviolet radiation emitted by the Sun. The third chapter describes the greenhouse gases in our atmosphere that protect us from the bitter cold of outer space. Truly, our atmosphere is a resource beyond price. Our goal is to convey its beauty, its magnitude, and its frailty.

As you will learn in this chapter, we humans have altered the composition of the atmosphere. This is not surprising, because about 6.9 billion humans currently live on the planet. In a few decades, the population may reach 9 or 10 billion. The next activity invites you to think about how our actions, both individually and collectively, can change the air we breathe.

Consider This 1.1   Footprints in the Air

Hiking boot treads, asphalt pavement, corn fields. Each of these is an example of a "ground print" left by humans because each one alters the lay of the land. Similarly, our activities leave "air prints" that alter the composition of our atmosphere.

   a. Name three things that leave an *indoor* air print.
   b. Does each (1) hurt the air quality, (2) improve the air quality, or (3) have some effect, but you don't know what it is? Explain.
   c. Repeat parts **a** and **b** for an *outdoor* air print.

Answers
   a. Indoor air prints are left by growing house plants, burning candles, and applying paint.
   b. Green plants remove carbon dioxide from the air and add oxygen, both a plus. Some plants emit pollen. For some people, this reduces the air quality.

As *Homo sapiens*, we have a special responsibility to guard the quality of the air on our planet. Living out this responsibility has proven to be no easy task. In fact, we have made some tragic errors that have killed people, animals, and vegetation. Ultimately, as we pointed out in Chapter 0, our responsibility is to live in ways today that will not compromise either our own health or the health of future generations. Keeping the air clean is part of this responsibility.

In this chapter, you will learn more about the air you breathe and its importance to your well-being. We hope you will also learn how important it is to make choices about air quality—both as an individual and as a member of a larger society—that demonstrate wisdom now and in the years to come.

In 1948, smog killed 20 and sickened between 5000 and 7000 residents of Donora, Pennsylvania. In 1952, the Great London Smog killed thousands.

1.1 | What's in a Breath?

Take a breath! Automatically and unconsciously, you do this thousands of times each day. Certainly you do not need us to tell you to breathe! Although a doctor or nurse may have encouraged your first breath, from then on nature took over. Even if you were to hold your breath in a moment of fear or suspense, you soon would involuntarily gasp a lungful of that invisible stuff we call air. Indeed, you could survive only minutes without a fresh supply.

Consider This 1.2   Take a Breath

What total volume of air do you inhale (and exhale) in a typical day? Figure this out. First determine how much air you exhale in a single "normal" breath. Then determine how many breaths you take per minute. Finally, calculate how much air you exhale per day. Describe how you made your estimate, provide your data, and list any factors you believe may have affected the accuracy of your answer.

Were you surprised at how much you breathe? For an adult, the value typically is more than 11,000 liters (about 3000 gallons) of air per day. The value would be even higher had you spent the day on a bike trail or paddling a kayak.

Although you cannot tell by looking, the air you are breathing is not a single pure substance. Rather, it is a **mixture;** that is, a physical combination of two or more pure substances present in variable amounts. Mixtures are one of the two forms of matter that we encounter on our planet (Figure 1.1). The other form is pure substances. In this section, we focus on the pure substances that are the major components of air: nitrogen, oxygen, argon, carbon dioxide, and water vapor. All are colorless gases, invisible to the eye.

Some mixtures are composed of gases. But gasoline is a mixture of liquids (Section 1.10), and soil is a mixture of solids and liquids.

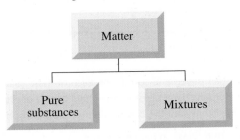

Figure 1.1
Matter can be classified either as a single pure substance or as a mixture.

The composition of the mixture that we call "air" depends on where you are. There's less oxygen in a stuffy room and more pollutants in an urban area. Exhaled air is a slightly different mixture than inhaled air. Trace amounts of substances may vary in the air as well. For example, the scent of lilacs may permeate the air outside. Indoors, the aroma of freshly brewed coffee may beckon you to the kitchen. In fact, the human nose is an extremely sensitive odor detector. In some cases, only a minute trace of a substance is needed to trigger the olfactory receptors responsible for detecting odors. Thus, tiny amounts of substances can have a powerful effect on our noses, as well as on our emotions.

Consider This 1.3   Your Nose Knows

The air is different in a pine forest, a bakery, an Italian restaurant, and a dairy barn. Blindfolded, you could smell the difference. Our noses alert us to the fact that air contains trace quantities of many substances.

  a. Name three indoor and three outdoor smells that indicate small quantities of chemicals in the air.
  b. Our noses warn us to avoid certain things. Give three examples of when a smell indicates a hazard.

The composition of our atmosphere has not been constant over the millennia. For example, the concentration of oxygen has varied.

Look for more about atoms and molecules in Section 1.7.

Using a pie chart and a bar graph, Figure 1.2 represents the composition of air. The pie chart emphasizes the fractions of the whole, whereas the bar graph emphasizes the relative amounts of each substance. Regardless of how we present the data, the air you breathe is primarily nitrogen and oxygen. More specifically, the composition of air by volume is about 78% nitrogen, 21% oxygen, and 1% other gases. **Percent (%)** means "parts per hundred." In this case, the parts are either molecules or atoms.

The percents shown in Figure 1.2 are for *dry* air. Water vapor is not included, as its concentration varies by location. In dry desert air, the concentration of water vapor can be close to 0%. In contrast, it can reach 5% by volume in a warm tropical rain forest.

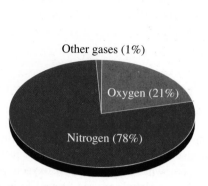

Figure 1.2

The composition of dry air, by volume.

 Figures Alive! Visit the textbook's website to learn more about the molecules and atoms that make up air. Watch for the Figures Alive! icon throughout this text.

Whether at high or low concentration, water vapor is a colorless gas that is invisible to the eye. Since you can see fog banks and clouds, they are not composed of water vapor. Rather, these consist of tiny droplets of liquid water or crystals of ice (Figure 1.3).

Nitrogen is the most abundant substance in the air and constitutes about 78% of what we breathe. This gas is colorless, odorless, and relatively unreactive, passing in and out of our lungs unchanged (Table 1.1). Although nitrogen is essential for life and part of all living things, most plants and animals obtain it from sources other than the nitrogen in our atmosphere.

Even though oxygen is less abundant than nitrogen in our atmosphere, it still plays a key role on our planet. Oxygen is absorbed into our blood via the lungs and reacts with the foods we eat to release the energy to power the chemical processes within our bodies. It is necessary for many other chemical reactions as well, including burning and

Water vapor is the gas that we think of as "humidity."

Section 6.9 describes the cycle by which atmospheric nitrogen becomes part of living plants and animals.

Chapter 11 provides more information about the energy content of foods.

Figure 1.3

Clouds consist of minuscule droplets of water that remain suspended because of upward air currents. Clouds can weigh millions of pounds.

Table 1.1	Typical Composition of Inhaled and Exhaled Air	
Substance	Inhaled Air (%)*	Exhaled Air (%)*
nitrogen	78.0	78.0
oxygen	21.0	16.0
argon	0.9	0.9
carbon dioxide	0.04	4.0
water vapor	variable	variable

*percents are by volume

We explain the term *element* in Section 1.6.

rusting. As the "O" in H_2O (water), oxygen is the most abundant element in the human body. Present in many rocks and minerals, it is also the most abundant element in the Earth's crust. Given this broad distribution, it is somewhat surprising that oxygen was not isolated as a pure substance until 1774. But once isolated, oxygen proved to be of great significance in establishing the young science of chemistry.

Consider This 1.4   More Oxygen . . . ?

We live in an atmosphere of 21% oxygen. A match burns in less than a minute, a fireplace consumes a small pine log in about 20 minutes, and we exhale about 15 times a minute. Life on Earth would be *very* different if the oxygen concentration were twice as high. List at least four differences.

Respiration also provides the energy to power chemical reactions in our bodies. Look for more about breathing and respiration in Chapter 4.

Every time we exhale, we add carbon dioxide to the atmosphere. Table 1.1 indicates the difference between inhaled dry air and exhaled air. Clearly some changes have taken place that use up some of the oxygen and give off carbon dioxide and water. In the process of **respiration,** the foods we eat are metabolized to produce carbon dioxide and water. With each breath, some of the water in our bodies evaporates from the moist tissue in our lungs.

Other gases are found in our atmosphere as well (see Figure 1.2). Argon, for example, is about 0.9% of the air. The name *argon,* meaning "lazy" in Greek, reflects the fact that argon is chemically inert. As you can see from Table 1.1, any argon that you inhale you simply exhale.

1 liter = 1.06 quart. Appendix 1 contains this and many other conversion factors.

The percentages we have been using to describe the composition of the atmosphere are based on volume; that is, the amount of space that each gas occupies. If we wanted to, we could closely approximate 100 liters (L) of dry air by mixing 78 L of nitrogen, 21 L oxygen, and 1 L argon (78% nitrogen, 21% oxygen, and 1% argon).

The composition of air also can be represented in terms of the numbers of molecules and atoms present. This works because equal volumes of gases contain equal numbers of molecules, providing the gases are at the same temperature and pressure. Thus, if you took a sample of 100 of the molecules and atoms in air (an unrealistically small amount), 78 would be nitrogen molecules, 21 would be oxygen molecules, and 1 would be an argon atom. In other words, when we say that air is 21% oxygen, we mean that there are 21 molecules of oxygen per 100 of the molecules and atoms in the air. A bit later we'll explain why nitrogen and oxygen are found as molecules. In contrast, argon is found as an atom.

Look for more about atoms and molecules in Section 1.7.

You now know that air contains nitrogen, oxygen, argon, carbon dioxide, and usually some water vapor. As you might imagine, there is more to the story.

1.2 | What Else Is in a Breath?

No matter where you live, each lungful of air you inhale contains tiny amounts of some substances. Many are present at concentrations of less than 1%, or one part per hundred. Such is the case with carbon dioxide, a gas you both inhale and exhale. In our atmosphere, the concentration of carbon dioxide currently is approximately 0.0390%. This value is slowly but steadily rising as we humans burn fossil fuels.

Although we could express 0.0390% as 0.0390 molecules of carbon dioxide per 100 molecules and atoms in the air, it sounds strange to refer to 0.0390 of a molecule. For low concentrations, it is more convenient to use **parts per million (ppm)**. One ppm is a unit of concentration 10,000 times smaller than 1% (one part per hundred). Here are the relationships.

Chapter 3 tells more about the rising CO_2 concentrations in the atmosphere.

0.0390% means		0.0390 parts per hundred
	means	0.390 parts per thousand
	means	3.90 parts per ten thousand
	means	39.0 parts per hundred thousand
	means	390 parts per million

Changing between % and ppm involves moving the decimal point four places to the right. You can practice using the activities in Figures Alive!

Out of a sample of air containing 1,000,000 molecules (and some atoms), we now can say that 390 of them will be carbon dioxide molecules. The carbon dioxide concentration is 390 ppm, or 0.0390%.

Skeptical Chemist 1.5   Really One Part per Million?

It has been said that a part per million is the same as one second in nearly 12 days. Is this a correct analogy? How about one step in a 568-mile journey? Check the validity of these analogies, explaining your reasoning. Then come up with an analogy of your own.

Your Turn 1.6   Practice with Parts per Million

a. In some countries, the limit for the average concentration of carbon monoxide in an 8-hour period is set at 9 ppm. Express this as a percentage.
b. Exhaled air typically contains about 78% nitrogen. Express this concentration in parts per million.

Answers
a. 0.0009%   b. 780,000 ppm

Even at tiny concentrations, some air pollutants are harmful and even life-threatening. Although officials enact air pollution laws, 100% clean air is not attainable. Our human activities always add some waste to the air. Furthermore, our air gets contaminated by natural events such as wild fires and volcanic eruptions. There is no such thing as pure air!

Nonetheless, dirty air definitely exists. Large metropolitan areas are one place you are likely to encounter it. For example, Figure 1.4 shows smog against the mountains near Santiago, Chile, a city of over 6 million inhabitants. Other large cities such as Los Angeles, Mexico City, Mumbai, and Beijing often have dirty air. Earlier, we noted how human activities leave "air prints," both indoors and out. When large numbers of people cook meals and drive vehicles, they tend to dirty the air.

Today, over half of the people in the world live in cities. In contrast, in 1900 this value was 10–15%.

In this chapter, we focus on four gases that contribute to air pollution at the surface of the Earth. One of these gases, carbon monoxide, is odorless; the other three—ozone, sulfur dioxide, and nitrogen dioxide—have characteristic odors. With sufficient exposure, all are hazardous to your health, even at concentrations well below 1 ppm. Together with particulate matter (PM), they represent the most serious air pollutants. Let's now examine the health effects of each.

Figure 1.4

Sunny November spring day in Santiago, Chile.

Figure 1.5

A propane camping stove.

In the stratosphere, ozone absorbs some wavelengths of UV radiation. Look for more about this in Chapter 2.

Figure 1.6

White pine needles damaged by ozone.

(Courtesy of Missouri Botanical Garden PlantFinder)

SO_2 and NO_2 dissolve in moist lung tissue to form sulfurous acid and nitric acid, respectively. Look for more about these acids in Chapter 6.

Chapter 6 provides more information about acid rain.

Carbon monoxide (CO) has earned the nickname "silent killer," as it has no color, taste, or smell. When you inhale carbon monoxide, it passes "silently" into your bloodstream. Once there, it interferes with the ability of your hemoglobin to carry oxygen. If exposed to carbon monoxide, at first you may feel dizzy, nauseous, or your head may hurt, symptoms that easily could be mistaken for other illnesses. Continued exposure, however, can lead to severe illness, even death. Automobile exhaust is one source of carbon monoxide. Charcoal grills and propane camping stoves are others. These can be used both outdoors and indoors. However, if used indoors, they need proper venting (Figure 1.5).

Ozone (O_3) has a sharp odor, one that you may have detected around a photocopier, an electric motor, or welding equipment. Even at very low concentrations, ozone can reduce lung function in healthy people. Symptoms of exposure include chest pain, coughing, sneezing, and lung congestion. Ozone also mottles the leaves of crops and yellows pine needles (Figure 1.6). At the Earth's surface, ozone is definitely a bad actor. At high altitudes, however, it plays an essential role in screening out ultraviolet radiation.

Sulfur dioxide (SO_2) has a sharp, unpleasant odor and dissolves in the moist tissue of your lungs to form an acid. The elderly, the young, and individuals with lung diseases such as emphysema or asthma are most susceptible to sulfur dioxide poisoning. At present, sulfur dioxide in the air comes primarily from the burning of coal. For example, the 1952 London smog that eventually killed over 10,000 people was in part caused by the emissions of coal-fired stoves. The causes of death included acute respiratory distress, heart failure (from preexisting conditions), and asphyxiation. Some who survived had permanent lung damage.

Nitrogen dioxide (NO_2) has a characteristic brown color and also damages lung tissue. Like sulfur dioxide, it can combine with fog, mist, snow, or rain to produce acid rain. Nitrogen dioxide is produced in our atmosphere from another pollutant (nitrogen monoxide, colorless) emitted both from the hot engines of vehicles and from the hot fires of coal-fired power plants. Oxides of nitrogen also can form naturally in grain silos, causing injury or death to the farmers who may inadvertently inhale the gases.

Particulate matter (PM) is the least understood of these five air pollutants. Particulate matter is a complex mixture of tiny solid particles and microscopic liquid droplets. It is classified by size rather than composition, because the particle size determines the health consequences. **PM_{10}** includes particles with an average diameter of

Figure 1.7

A 2004 wildfire near San Jose, California. This fire is releasing particulate matter, some of which is visible as soot.

10 μm or less, a length on the order of 0.0004 inches. **PM$_{2.5}$** is a subset of PM$_{10}$ and includes particles with an average diameter of less than 2.5 μm. These tinier and more deadly particles are sometimes called *fine particles*. Particulate matter originates from many sources, including truck and car engines, coal-burning power plants, fires, and blowing dust. Sometimes particulate matter is visible as soot or smoke (Figure 1.7). However, of more concern are the particles too tiny to see: PM$_{10}$ and PM$_{2.5}$. These particles, when inhaled, go deep into your lungs and cause irritation. The smallest particles pass from your lungs into your bloodstream and can cause heart disease or aggravate an existing condition.

> A **micrometer** (μm) is a millionth (10^{-6}) of a meter (m). The term *micron* also is used for this unit.

We end this section with a fact that may surprise you. All of the air pollutants that we just listed can occur naturally! For example, a wildfire (see Figure 1.7) produces particulate matter and carbon monoxide, lightning produces ozone and nitrogen oxides, and volcanoes release sulfur dioxide. The pollutants have the same hazards, whether released from natural or human sources. What are the risks to your health? We now turn to this topic.

1.3 | Air Pollutants and Risk Assessment

Risk is part of living. Although we cannot avoid risk, we still try to minimize it. For example, certain practices are illegal because they carry risks that are judged to be unacceptable. Other activities carry high risks and we label them as such. For example, cigarette packages carry a warning about lung cancer. Wine bottles carry warnings about birth defects and about operating machinery under the influence of alcohol. The absence of a warning, however, does not guarantee safety. The risk may be too low to label, it may be obvious or unavoidable, or it may be far outweighed by other benefits.

Warnings are just that. They do not mean that somebody *will* be affected. Rather, they report the likelihood of an adverse outcome. Let's say that the odds of dying from an accident are one in a million for each 30,000 miles traveled. On average, this means that one person out of every million traveling 30,000 miles would die in an accident. Such prediction is not simply a guess, but the result of **risk assessment,** the process of evaluating scientific data and making predictions in an organized manner about the probabilities of an outcome.

When is it risky to breathe the air? Fortunately, existing air quality standards can offer you guidance. We say *guidance,* because standards are set through a complex interaction of scientists, medical experts, governmental agencies, and politicians. People may not necessarily agree on which standards are reasonable and safe. Standards also change over time, as new scientific knowledge is generated.

In the United States, national air quality standards were established in 1970 as a result of the Clean Air Act. If pollutant levels fall below these standards, presumably the air is healthy to breathe. We say "presumably" because air quality standards change over time, usually becoming stricter. If you look worldwide, you will find that air quality regulations vary both in their strictness and in the degree to which they are enforced.

The risks presented by an air pollutant are a function of both **toxicity,** the intrinsic health hazard of a substance, and **exposure,** the amount of the substance encountered. Toxicities are difficult to accurately assess for many reasons, including that it is unethical to run experiments on people. Even if data were available, we still would have to determine the levels of risk that are acceptable for different groups of people. In spite of the complexities, government agencies have succeeded in establishing limits of exposure for the major air pollutants. Table 1.2 shows the National Ambient Air Quality Standards established by the U.S. Environmental Protection Agency (EPA). Here, **ambient air** refers to the air surrounding us, usually meaning the outside air. As our knowledge grows, we modify these standards. For example, in 2006 these standards were lowered for $PM_{2.5}$. Similarly, in 2008 they were lowered for ozone.

The U.S. EPA was formed in 1970 by President Richard Nixon. Senators from earlier years also played key roles in the legislation.

Table 1.2	U.S. National Ambient Air Quality Standards	
Pollutant	Standard (ppm)	Approximate Equivalent Concentration ($\mu g/m^3$)
Carbon monoxide		
8-hr average	9	10,000
1-hr average	35	40,000
Nitrogen dioxide		
Annual average	0.053	100
Ozone		
8-hr average	0.075	147
1-hr average	0.12	235
*Particulates**		
PM_{10}, annual average	—	50
PM_{10}, 24-hr average	—	150
$PM_{2.5}$, annual average	—	15
$PM_{2.5}$, 24-hr average[†]	—	35
Sulfur dioxide		
Annual average	0.03	80
24-hr average	0.14	365
3-hr average	0.50	1,300

*PM_{10} refers to all airborne particles 10 μm in diameter or less. $PM_{2.5}$ refers to particles 2.5 μm in diameter or less.

—The unit of ppm is not applicable to particulates.

[†]$PM_{2.5}$ standards are likely to be revised after 2011.

Source: U.S. Environmental Protection Agency. Standards also exist for lead, but are not included here.

Exposure is far more straightforward to assess than toxicity, because exposure depends on factors that we more easily can measure. These include:

- **Concentration in the air**
  The more toxic the pollutant, the lower the concentration must be set. Concentrations are expressed either as parts per million or as micrograms per cubic meter ($\mu g/m^3$), as shown in Table 1.2. Earlier, we used the prefix *micro-* with micrometers (μm), meaning a millionth of a meter (10^{-6} m). Similarly, one **microgram** (μg) is a millionth of a gram (g), or 10^{-6} g.

- **Length of time**
  Higher concentrations of a pollutant can be tolerated only briefly. A pollutant may have several standards, each for a different length of time.

- **Rate of breathing**
  Physically active people, such as athletes or laborers, breathe at a higher rate. If the air quality is poor, reducing activity is one way to reduce exposure.

1 μg is approximately the mass of a period printed on a page.

Suppose you collect an air sample on a city street. An analysis shows that it contains 5000 μg of carbon monoxide (CO) per cubic meter of air. Is this concentration of CO harmful to breathe? We can use Table 1.2 to answer this question. Two standards are reported for carbon monoxide, one for a 1-hour exposure and another for an 8-hour exposure. The 1-hour exposure is set at a higher level (4×10^4 μg CO/m^3) because a higher concentration can be tolerated for a short time.

Both concentrations are expressed in **scientific notation,** a system for writing numbers as the product of a number and 10 raised to the appropriate power. Scientific notation enables us to avoid writing strings of zeros either before or after the decimal point. For example, the value 1×10^4 is equivalent to 10,000. To understand this conversion, simply count the number of zeros to the right of the 1 in 10,000. There are four of them. The number 1 is then multiplied by 10^4 to obtain 1×10^4 μg CO/m^3. Similarly, 4×10^4 μg CO/m^3 is equivalent to 40,000 μg CO/m^3. Scientific notation is even more useful for very large numbers, such as the 20,000,000,000,000,000,000,000 molecules in a typical breath. In scientific notation, this value is written as 2×10^{22} molecules.

If you need help with exponents, consult Appendix 2.

Using scientific notation, we now can express the value of 5000 μg CO/m^3 as 5×10^3 μg CO/m^3. Clearly, this value is *less* than either standard. In the case of an 8-hour exposure, 5×10^3 is less than 1×10^4. Similarly, for a 1-hour period, 5×10^3 is less than 4×10^4. For all values, the units are μg CO/m^3.

Table 1.2 also allows us to assess the relative toxicities of pollutants. For example, we can compare the 8-hour average exposure standards for carbon monoxide and ozone: 9 ppm vs. 0.075 ppm. Doing the math, ozone is about 130 times more hazardous to breathe than carbon monoxide! Nonetheless, carbon monoxide still can be exceedingly dangerous. As the "silent killer," it may impair your judgment before you recognize the danger.

Your Turn 1.7   Estimating Toxicities

a. Which pollutant in Table 1.2 is likely to be the most toxic? Exclude particulate matter.

b. Examine the particulate matter standards. Earlier, we stated that "fine particles," $PM_{2.5}$, are more deadly than the coarser ones, PM_{10}. Do the values in Table 1.2 bear this out?

Answer

a. O_3. This is a hard call, as no common exposure period exists on which to base the comparison. Clearly CO is not the most toxic, as all the standards are higher. It is not NO_2, because SO_2 has a lower annual standard. Between NO_2 and O_3, ozone has the tighter standards.

Although the standards for air pollutants are expressed in parts per million, the concentrations of sulfur dioxide and nitrogen dioxide could conveniently be reported in **parts per billion (ppb),** meaning one part out of one billion, or 1000 times less concentrated than one part per million.

$$\text{sulfur dioxide } 0.030 \text{ ppm} = 30 \text{ ppb}$$
$$\text{nitrogen dioxide } 0.053 \text{ ppm} = 53 \text{ ppb}$$

As these values reveal, converting from parts per million to parts per billion involves moving the decimal point three places to the right.

Your Turn 1.8   Living Downwind

Copper metal can be recovered from copper ore by smelting, a process that releases sulfur dioxide (SO_2). Let's assume that a woman living downwind of a smelter inhaled 1050 µg of SO_2 in a day.

a. If she inhaled 15,000 liters (15 m³) of air per day, would she exceed the 24-hour average for the U.S. National Ambient Air Quality Standards for SO_2? Support your answer with a calculation.

b. If she were exposed at this rate every day, would she exceed the annual average?

To end this section, we note that our *perception* of a risk also plays an important role. For example, the risks of traveling by car far exceed those of flying. Each day in the United States, more than 100 people die in automobile accidents. Yet some people avoid taking a flight because of their fear of a plane crash. Similarly, some people fear living near a nuclear power plant. Yet as some notorious hurricanes have demonstrated, living in a coastal area can be a far riskier proposition. Whether perceived as a risk or not, *air pollution presents real hazards,* both to present and future generations. In the next section, we offer you the tools to assess these hazards.

1.4 | Air Quality and You

Depending on where you live, you will breathe air of different quality. Some locations always have good air; others have air of moderate quality, and still others have unhealthy air much of the time. As we will see, the differences arise because of the number of people living in a region, their activities, the geographical features of the region, the prevailing weather patterns, and the activities of people in neighboring regions.

To improve air quality, many nations have enacted legislation. For example, we already have cited the U.S. Clean Air Act (1970) that led to the establishment of air quality standards. Like many environmental laws, this one focused on limiting our exposure to hazardous substances. It has been named as a "command and control law" or an "end of the pipe solution" because it tries to limit the spread of hazardous substances or clean them up after the fact.

The Pollution Prevention Act (1990) was a significant piece of legislation that followed the Clean Air Act. It focused on *preventing* the formation of hazardous substances, stating that "pollution should be prevented or reduced at the source whenever feasible." The language shift is significant. Rather than attempting to regulate existing pollutants, people should not produce them in the first place! With the Pollution Prevention Act, it became national policy to employ practices that reduce pollutants at their source.

This act was the impetus for green chemistry, a topic introduced in Chapter 0.

Your Turn 1.9   The Logic of Prevention

Take off your muddy shoes at the door rather than cleaning up the carpet later! List three "common sense" examples that prevent air pollution rather than cleaning it up after the fact.

Hint: Revisit the first activity in this chapter on "air prints."

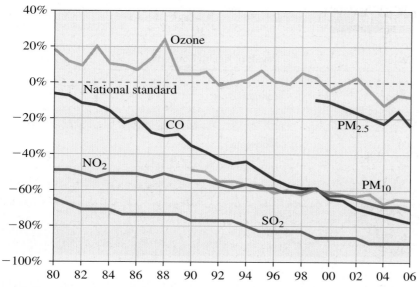

Figure 1.8

U.S. average levels of air pollutants (at selected sites) compared with national ambient air quality standards, 1980–2006. Note: Data for PM_{10} and $PM_{2.5}$ start in 1990 and 1999, respectively.

Source: U.S. Environmental Protection Agency, Latest Findings on National Air Quality, www.epa.gov/air/airtrends/2007

The decrease in the concentration of air pollutants in the United States has been dramatic (Figure 1.8). Some improvements occurred through a combination of laws and regulations, such as the ones we just mentioned. Others stemmed from local decisions. For example, a community may have built a new public transportation system or an industry may have installed more modern equipment. Still others occurred because of the ingenuity of chemists, most notably via a set of practices called "green chemistry." Look for green chemistry examples throughout this book that are designated with the Green Chemistry icon.

However, aggregate data such as those in Figure 1.8 hide the fact that people in some places still breathe dirty air. Although air quality may have improved on average, people in some metropolitan areas breathe air that contains unhealthy levels of pollutants. Check the data for the United States presented in Table 1.3.

The label "unhealthy" means just that. As we described earlier, air pollutants are the perpetrators of biological mischief. To help you more quickly assess the hazards,

The green chemistry examples in this text include some of the winners of the prestigious Presidential Green Chemistry Challenge Awards.

Table 1.3	Air Quality Data for Selected U.S. Metropolitan Areas	
	# of Unhealthy Days/Year*	
Metropolitan Area	**O₃**	**PM₂.₅**
Boston	10	5
Chicago	11	10
Cleveland	17	9
Houston	35	2
Los Angeles	62	38
Pittsburgh	14	42
Phoenix	10	2
Sacramento	28	10
Seattle	1	4
Washington, DC	19	8

Seattle has few bad air days because of its rainy climate. Section 1.12 describes the connections between climate and smog formation.

*Days for which the AQI for the pollutant exceeds 100, five year average (2001–2005).

Source: EPA, Latest Findings on National Air Quality.

Table 1.4	Levels for the Air Quality Index	
Air Quality Index (AQI) Values	**Levels of Health Concern**	**Colors**
When the AQI is in this range:	*...air quality conditions are:*	*...as symbolized by this color.*
0–50	Good	Green
51–100	Moderate	Yellow
101–150	Unhealthy for sensitive groups	Orange
151–200	Unhealthy	Red
201–300	Very unhealthy	Purple
301–500	Hazardous	Maroon

EPA air quality logo

The AQI for a particular day is set by the highest pollutant. To quote the EPA, "If a certain area had AQI values of 90 for ozone and 88 for sulfur dioxide, the AQI value would be 90 on that day."

the U.S. EPA developed the color-coded Air Quality Index (AQI) shown in Table 1.4. This Index is scaled from 1–500, with the value of 100 pegged to the national standard for the pollutant. Green or yellow ($\leq$100) indicates air of good or moderate quality. Orange indicates that the air has become unhealthy for some groups. Red, purple, or maroon (>150) indicates that the air is unhealthy for *everybody* to breathe.

Some newspapers provide only a general air quality report. For example, the air quality may be listed as "moderate" for a city. This means that at least one pollutant was moderate, but possibly others were as well. Sometimes the report is given in general terms, but numerical. For example, if two pollutants were present, one with a value of 85 and the other with a value of 91, the daily value would be reported as 91.

More and more, though, metropolitan areas are beginning to report each pollutant separately. This is helpful, because how you act depends on which pollutants are present. For example, Figure 1.9 shows the air quality forecast for carbon monoxide, ozone, and particulates on a sunny spring day in Phoenix, Arizona. Ozone clearly was the pollutant of concern.

FORECAST DATE	YESTERDAY WED 05/13/2009	TODAY THU 05/14/2009	TOMORROW FRI 05/15/2009	EXTENDED SAT 05/16/2009
AIR POLLUTANT	Highest AQI Reading/SITE			
O_3	61 CAVE CREEK & FOUNTAIN HILLS	80 MODERATE	97 MODERATE	104 UNHEALTHY FOR SENSITIVE GROUPS
CO	07 GREENWOOD	09 GOOD	11 GOOD	09 GOOD
PM_{10}	53 WEST FORTY THIRD	55 MODERATE	61 MODERATE	48 GOOD
$PM_{2.5}$	45 PHOENIX SUPERSITE	46 GOOD	49 GOOD	45 GOOD

Figure 1.9

Air quality forecast for Phoenix on May 14, 2009. Values >100 are unhealthy.

Source: Arizona Department of Environmental Quality.

1.5 | Where We Live: The Troposphere

As we pointed out earlier, our atmosphere is a mixture that varies slightly in composition at different locations. About 75% of the air is found within about 6 miles (~10 kilometers) of the surface of our planet. This is the **troposphere,** the lower region of the atmosphere in which we live that lies directly above the surface of the Earth. *Tropos* is Greek for "turning" or "changing." The troposphere contains the air currents and turbulent storms that turn and mix our air.

The warmest air in the troposphere usually lies at ground level because the Sun heats the ground, which in turn warms the air above it. Cooler air is found higher up, a phenomenon you may have observed as you hike or drive to higher elevations. However, air inversions occur when cooler air gets trapped beneath warmer air. Air pollutants can accumulate in an inversion layer, especially if the layer remains stationary for an extended period. This often occurs in cities ringed by mountains, such as Salt Lake City (Figure 1.10).

Air pollutants might better be termed "people fumes." One hundred years ago, our Earth was home to fewer than 2 billion people. Soon we will reach the 7-billion mark, with the majority of people living in urban regions. This growth in population has been accompanied by a massive growth in both the consumption of resources and the production of waste. The waste that we stash in our atmosphere is called air pollution.

Air pollution provides the first context in which we can discuss **sustainability,** the topic that set the stage for this book. As we pointed out in Chapter 0, we need to make decisions with an eye not only for today's outcomes but also for the needs of generations to come. It makes sense to avoid actions that produce pollutants that can compromise our health and well-being. Does this sound familiar? Again, this is the logic behind the Pollution Prevention Act of 1990, legislation that calls for preventing pollution rather than controlling our exposure to it.

The Pollution Prevention Act provided the impetus for **green chemistry,** a set of principles to guide all in the chemical community, including teachers and students. Green chemistry is "benign by design." It calls for designing chemical products and processes that reduce or eliminate the use or generation of hazardous substances. Begun under the EPA Design for the Environment Program, green chemistry reduces pollution through the design or redesign of chemical processes. The goal is to use less energy, create less waste, use fewer resources, and use renewable resources. Green chemistry is a tool for achieving sustainability, rather than an end in itself.

Innovative "green" chemical methods already have decreased or eliminated toxic substances used or created in chemical manufacturing processes. For example,

The depth of the troposphere varies from the mid-latitudes (~12 miles) to the poles (~5 miles).

Hurricanes in the troposphere clearly illustrate the meaning of the Greek word *tropos* (= turning or changing).

Principles of green chemistry are listed on the inside cover of this book.

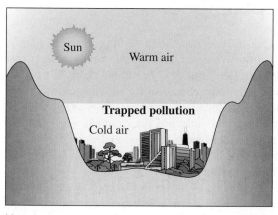

(a)

(b)

Figure 1.10

(a) An air inversion can trap pollution. **(b)** A air inversion, trapping a smoggy layer of air over Salt Lake City, Utah.

we now have cheaper and less wasteful ways to produce ibuprofen, pesticides, disposable diapers, and contact lenses. We have new dry cleaning methods and recyclable silicon wafers for integrated circuits. The research chemists and chemical engineers who developed these methods have received Presidential Green Chemistry Challenge Awards. Begun in 1995, these presidential-level awards recognize chemists for their innovations on behalf of a less polluted world. Each year since 1996, five awards have been given with the theme "Chemistry is not the problem, it's the solution."

The green chemistry awards are given in five areas: greener synthetic pathways, greener reaction conditions, designing greener chemicals, small business, and academia.

Consider This 1.10   Green Chemistry

Recall these two green chemistry principles:

It is better to *prevent waste* than to treat or clean up waste after it is formed.
It is better to *use less energy*.

a. Why is it better to use less energy? Give two examples that demonstrate the connection between using energy and putting waste in the air.
b. Now choose an air pollutant. Give two examples that demonstrate why it makes more sense to prevent its formation rather than to try to clean it up once in the air.

Bottom line: Nobody wants dirty air. It makes you sick, reduces the quality of your life, and may hasten your death. However, the problem is that many people have become so accustomed to breathing dirty air, that they don't notice it. Recall the concept of **shifting baselines** mentioned in Chapter 0. Haze is now so common that we have forgotten the clear days when it seemed we really could see forever. Burning eyes and breathing disorders have become so common, that we have forgotten that they once were not. We have become accustomed to living in **megacities,** urban areas with 10 million people or more. Tokyo, New York City, Mexico City, and Mumbai are examples. Pollutants such as wood smoke, car exhaust, and industrial emissions all concentrate in the troposphere around megacities.

Clearly, we have some problems! As promised, your knowledge of chemistry can lead you to making better choices to deal with these problems, both as an individual and in your local community. The next two sections give you a better grasp of the language of chemistry. We then use this language to approach the issues of air pollution in more detail.

1.6 | Classifying Matter: Pure Substances, Elements, and Compounds

In describing air and its quality, we already have employed several chemical names. For example, in Section 1.2 we listed nitrogen, oxygen, argon, and usually water vapor as four of the pure substances that make up the majority of our atmosphere. We named some pollutants found at very low concentrations: ozone, sulfur dioxide, carbon monoxide, and nitrogen dioxide. We included their chemical formulas as well: O_3, SO_2, CO, and NO_2. We also mentioned the terms *atom* and *molecule*. For example, we noted that air is a mixture that contains different molecules (and a few atoms). In this section, we explain more about elements and compounds. In the next, we will focus on atoms and molecules that these elements and compounds contain.

Matter consists of elements and compounds, as shown in Figure 1.11. An **element** is one of the 100 or so pure substances in our world from which compounds are formed. As we will see shortly, they contain only one type of atom. Nitrogen (N_2), oxygen (O_2), and argon (Ar) are examples of elements. So is ozone (O_3), another form of oxygen. Over 100 elements are known. In contrast, a **compound** is a pure substance made up of two or more different elements in a fixed, characteristic chemical combination. Compounds

The periodic table lists the known elements, as we will see momentarily.

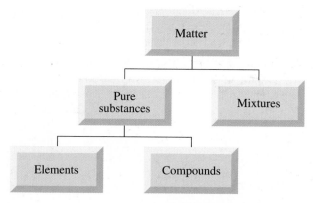

Figure 1.11

One way to classify matter.

contain two or more different types of atoms. For example, water (H_2O) is a compound of the elements oxygen and hydrogen. Similarly, carbon dioxide (CO_2) is a compound of the elements oxygen and carbon. In CO_2, the two elements are chemically combined and are no longer in their elemental forms. Sulfur dioxide (SO_2) and methane (CH_4) are other examples of compounds.

About 90 elements occur naturally on Earth and, as far as we know, elsewhere in the universe. The others have been created from existing elements through nuclear reactions. Plutonium is probably the best known of the human-made elements, although it does occur in trace amounts in nature. The vast majority of elements are solids. Nitrogen, oxygen, argon, and eight other elements are gases; and only bromine and mercury are liquids at room temperature.

An alphabetical list of the elements and their **chemical symbols,** one- or two-letter abbreviations for the elements, appears facing the inside back cover of the text. These symbols, established by international agreement, are used throughout the world. Some of them make immediate sense to those who speak English or related languages. For example, oxygen is O, nitrogen is N, sulfur is S, and nickel is Ni. Other symbols have their origin in other languages. For example, Fe is iron, Pb is lead, Au is gold, and Hg is mercury. These metals were known to the ancients and given Latin names long ago. For example, *ferrum* is iron, *plumbum* is lead, *aurum* is gold, and *hydrargyrum* is mercury.

Elements have been named for properties, planets, places, and people. Hydrogen (H) means "water former," because hydrogen gas (H_2) burns in oxygen (O_2) to form the compound water (H_2O). Neptunium (Np) and plutonium (Pu) were named after two planets in our solar system. Berkelium (Bk) and californium (Cf) honor the Berkeley lab in which a team of researchers first produced them. Darmstadtium (Ds) and roentgenium (Rg) were the most recently produced elements with names at the time this book went to press. The former was named after Darmstadt, the city in Germany in which it was discovered; the latter was named after Wilhelm Roentgen, Only a few atoms of each have been produced. A new element, #117, was reported in 2010 and currently is the heaviest element known.

Plutonium can fuel both nuclear reactors and nuclear bombs. See Chapter 7 for the details.

Chemical symbols sometimes also are referred to as atomic symbols.

In 2006, Pluto lost its status as a planet.

William Roentgen discovered X-rays and received the first Nobel Prize in physics.

Your Turn 1.11   **Pure Substances in Air**

a. Hydrogen (0.54 ppm), helium (5 ppm), and methane (17 ppm) are found in our atmosphere. Two of these are elements. Which ones?

b. List five other substances found in the air and classify them as elements or compounds.

c. Express each concentration in part **a** as a percent.

Lothar Meyer, a German chemist, also developed a periodic table at the same time as Mendeleev.

It is fitting that the 19th-century Russian chemist Dmitri Mendeleev has his own element (Md), because the most common way of arranging the elements—the

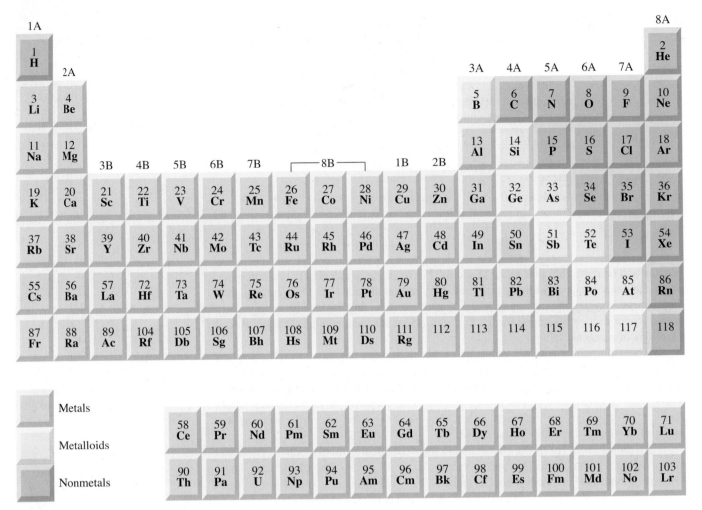

Figure 1.12

A simplified period table that shows the locations of metals, metalloids, and nonmetals.

Metals, nonmetals, and the ions they form are discussed in Section 5.6.

Semiconductors are explained in Section 8.7.

periodic table—reflects the system he developed. This is an orderly arrangement of all the elements based on similarities in their properties. The inside back cover has a copy for handy reference, and we will explain more about it in Chapter 2.

Figure 1.12 shows a simplified version of the periodic table that lists the elements by number but does not include their masses. The *light green* shading indicates the **metals,** elements that are shiny and conduct electricity and heat well. These include familiar substances such as iron, gold, and copper. Far fewer are the **nonmetals,** elements that do not conduct heat or electricity well and have no one characteristic appearance. These elements are indicated by the *light blue* shading and include sulfur, chlorine, and oxygen. A mere eight elements fall into a category known as **metalloids,** elements that lie between metals and nonmetals on the periodic table and do not fall cleanly into either category. Metalloids also are called semimetals and are indicated in *light red* shading. The semiconductors silicon and germanium are examples of metalloids.

The elements fall into vertical columns called **groups.** These organize elements according to important properties that they have in common and are numbered from left to right. Some groups are given names as well. For example, a **halogen** is one of the reactive nonmetals in Group 7A, such as fluorine (F), chlorine (Cl), bromine (Br), or iodine (I). Similarly, a **noble gas** is one of the inert elements in Group 8A that undergoes few, if any, chemical reactions. We already mentioned argon, a noble gas that is a constituent of our atmosphere. You may recognize helium as the noble gas used to make balloons float, as it is less dense than air. Radon is the one noble gas that is radioactive, a characteristic that distinguishes it from the other elements in Group 8A.

Although only 100 or so elements exist, over 20 million compounds have been isolated, identified, and characterized. Some are very familiar naturally occurring substances such as water, table salt, and sucrose (table sugar). Many known compounds were chemically synthesized by men and women across our planet. You might be wondering how 20 million compounds could possibly be formed from so few elements. In short, elements have the ability to combine in many different ways.

For example, carbon dioxide is a chemical combination of the elements carbon and oxygen. All pure samples of carbon dioxide contain 27% carbon and 73% oxygen by mass. Thus, a 100-g sample of carbon dioxide always consists of 27 g of carbon and 73 g of oxygen, chemically combined to form this particular compound. These values never vary, no matter the source of the carbon dioxide. This illustrates the fact that every compound exhibits a constant characteristic chemical composition.

Carbon monoxide (CO) is a different compound of carbon and oxygen. Pure samples of carbon monoxide contain 43% carbon and 57% oxygen by mass. Thus, 100 g of carbon monoxide contain 43 g of carbon and 57 g of oxygen, a composition different from that of carbon dioxide. This is not surprising, because carbon monoxide and carbon dioxide are two different compounds.

As we will see, each compound has its own set of properties. For example, water (H_2O) is a compound that is 11% hydrogen and 89% oxygen by weight. At room temperature, water is a colorless, tasteless liquid. At sea level, it boils at 100 °C and freezes at 0 °C. All samples of pure water have these same properties. Water is composed of molecules, a term that we now have used several times. The next section will help you to use the terms atom and molecule with more confidence.

1.7 | Atoms and Molecules

The definitions we just gave for elements and compounds made no assumptions about the nature of matter. We now know that elements are made up of **atoms,** the smallest unit of an element that can exist as a stable, independent entity. The word *atom* comes from the Greek for "uncuttable." Although today it is possible to "cut" atoms using specialized processes, atoms remain indivisible by ordinary chemical or mechanical means.

Atoms are extremely small. Because they are so tiny, we need huge numbers of atoms in order to see, touch, or weigh them. For example, the molecules in a single drop of water contain about 5×10^{21} atoms. This is roughly a trillion times greater than the 6.9 billion people on Earth, almost enough to give each person a trillion atoms.

As Figure 1.13 reveals, atoms now can be photographed. Using a scanning tunneling microscope, scientists at the IBM Almaden Research Center lined up iron atoms on a copper surface to create the kanji (Japanese character) for "atom." **Nanotechnology** refers to the creation of materials at the atomic and molecular (nanometer) scale: 1 nanometer (nm) $= 1 \times 10^{-9}$ m. This kanji is a few nanometers high and wide. At this size, about 250 million nanoletters could fit on a cross section of a human hair, equivalent to about 90,000 pages of text!

Using the concept of atoms, we can better explain the terms *element* and *compound*. Elements are made up of only one kind of atom. For example, the element carbon is made up only of carbon atoms. By contrast, compounds are made up of two or more different kinds of atoms. For example, the compound carbon dioxide contains carbon and oxygen atoms. Similarly, water is made up of hydrogen and oxygen atoms.

But we need to be careful with our language. The carbon and oxygen atoms in carbon dioxide are *not* present as such. Rather, the carbon and oxygen atoms are

Look for more about density in Chapter 5.

Radon affects the quality of indoor air, as we will see in Section 1.13. Chapter 7 tells more about other radioactive substances.

A small paper clip weighs about a gram.

Water is not tasteless if it contains dissolved gases or minerals. Look for more about water as a solvent in Chapter 5. As we will see, water is rarely "pure."

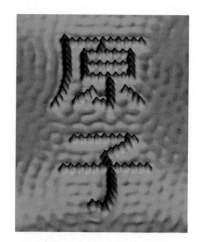

Figure 1.13

Iron atoms arranged on a copper surface, as imaged with a scanning tunneling microscope. This is the Japanese kanji for the word *genshi* (atom).

Chapter 1

Atoms are color-coded:

carbon ⬛

hydrogen

oxygen

nitrogen

sulfur

Section 3.3 explains why these two molecules have different shapes.

chemically combined to form a carbon dioxide **molecule,** two or more atoms held together by chemical bonds in a certain spatial arrangement. More specifically, two oxygen atoms (*red*) are combined with one carbon atom (*black*) to form a carbon dioxide molecule. Similarly, the water molecule contains two hydrogen atoms (*white*) combined with one oxygen atom (*red*).

water molecule   carbon dioxide molecule

A **chemical formula** is a symbolic way to represent the elementary composition of a substance. It reveals both the elements present (by chemical symbols) and the atom ratio of those elements (by the subscripts). For example, in the compound CO_2, the elements C and O are present in a ratio of one carbon atom for every two oxygen atoms. Similarly, H_2O indicates two hydrogen atoms for each oxygen atom. Note that when an atom occurs only once, such as the O in H_2O or the C in CO_2, the subscript of "1" is omitted.

Some elements exist as single atoms, such as helium or radon. We represent these as He and Rn, respectively. Other elements exist as molecules. For example, nitrogen and oxygen are found in our atmosphere as N_2 and O_2 molecules. Each is a **diatomic molecule,** meaning that it is a molecule consisting of two atoms. These representations clearly show the difference.

oxygen molecule   nitrogen molecule   helium atom

Table 1.5 summarizes our discussion of elements, compounds, and mixtures, listing both what we can observe experimentally and what exists at the atomic level that we cannot see.

We now apply these concepts to the mixture we call air. Some of its components, such as nitrogen, oxygen, and argon, are elements. Others, most notably water vapor and carbon dioxide, are compounds. All of the compounds discussed so far contain molecules (e.g., CO_2 and H_2O). But with the elements, it is not so simple. In the troposphere, the elements nitrogen and oxygen exist primarily as diatomic molecules (N_2 and O_2). In contrast, elements such as argon and helium exist as uncombined atoms.

Table 1.5	Types of Matter		
Substance	Example	Definition	Contains...
element	O_2, oxygen Ar, argon	cannot be broken down into simpler substances, although different forms of an element are possible	one type of atom only
compound	H_2O, water	can be broken down into elements and has a fixed composition	two or more different types of atoms
mixture	air	can be separated into two or more components and has a variable composition of elements, compounds, or both	many types of atoms, molecules, or both

Dry air is composed mainly of nitrogen and oxygen, that is, *molecules* of N_2 and O_2. If the air is humid, add in some water vapor in the form of H_2O *molecules*. Dry air contains just under 1% of Ar (argon) *atoms*, as well as tiny amounts of He (helium) *atoms*, Xe (xenon) *atoms*, and extremely small amounts of Rn (radon) *atoms*. Remember also to include the 390 ppm of carbon dioxide, meaning 390 CO_2 *molecules* per 1×10^6 of the *molecules and atoms* in the air.

Skeptical Chemist 1.12   **The Chemistry of Lawn Care**

News reports and advertisements should be viewed with a critical eye for their accuracy. For example, a lawn care service ad reports its fertilizers as "a balanced blend of nitrogen, phosphorus, and potassium. They have an organic nature, made up of carbon molecules. These fertilizers are biodegradable and turn into water." Edit the text of this ad to fix the chemical glitches.

1.8 | Names and Formulas: The Vocabulary of Chemistry

If chemical symbols are the alphabet of chemistry, then chemical formulas are the words. The language of chemistry, like any other language, has rules of spelling and syntax. In this section we help you to "speak chemistry" using chemical formulas and names. As you'll see, each name corresponds uniquely to one chemical formula. However, chemical formulas are *not* unique and may correspond to more than one name. In addition, some compounds are known by several names.

In this section, we follow a need-to-know philosophy. We help you learn what you *need to know* to understand the topic at hand and omit other naming rules until the need arises. Right now, you need to know the chemical names and formulas of compounds that relate to the air you breathe. So, for now, we work on these.

We already named some of the pure substances found in air, including carbon monoxide, carbon dioxide, sulfur dioxide, ozone, water vapor, and nitrogen dioxide. Although it may not be apparent, this list includes two types of names: systematic and common.

Systematic names for compounds follow a reasonably straightforward set of rules. Here are the rules for compounds composed of two nonmetals, such as carbon dioxide (CO_2) and carbon monoxide (CO):

- Name each element in the chemical formula, modifying the name of the second element to end in *-ide*. For example, oxygen becomes oxide, and sulfur becomes sulfide.
- Use prefixes to indicate the numbers of atoms in the chemical formula (Table 1.6). For example, *di-* means 2, and thus the name carbon *di*oxide means two oxygen atoms for each one carbon atom.
- Omit the prefix *mono-* if there is only one atom for the first element in the chemical formula. For example, CO is carbon monoxide, not monocarbon monoxide.

If instead you are writing a chemical formula from a name, keep in mind that the subscript of 1 is not used in chemical formulas. Thus the chemical formula for carbon dioxide is CO_2 and *not* C_1O_2. Similarly, carbon monoxide is CO and *not* C_1O_1. The next activity gives you a chance to practice.

Section 5.7 explains another set of naming rules, those for ionic compounds.

Remember that ozone, O_3, is an element, not a compound.

Table 1.6	Prefixes Used in Naming Compounds		
Prefix	**Meaning**	**Prefix**	**Meaning**
mono-	1	hexa-	6
di- or bi-	2	hepta-	7
tri-	3	octa-	8
tetra-	4	nona-	9
penta-	5	deca-	10

Your Turn 1.13   Oxides of Sulfur and Nitrogen

a. Write chemical formulas for nitrogen monoxide, nitrogen dioxide, dinitrogen monoxide, and dinitrogen tetraoxide.

b. Give chemical names for SO_2 and SO_3.

Answer

a. NO, NO_2, N_2O, and N_2O_4. *Note:* NO and N_2O also are called nitric oxide and nitrous oxide.

Some names ("common names") do not follow a set of rules. Water is one example. You might have expected H_2O to be called dihydrogen monoxide. This makes sense! But water was given its name long before anybody knew anything about hydrogen and oxygen. Chemists, being reasonable folks, did not rename water. Rather, they call the stuff they swim in and drink by the common name water, just like everybody else. Ozone (O_3) is another common name, as is ammonia (NH_3). Common names cannot be figured out; you have to know them or look them up.

In the next two sections, we explore the connection between air quality and the fuels we burn. Following our need-to-know philosophy, we need to introduce the names of several **hydrocarbons;** that is, compounds made up only of the elements hydrogen and carbon. Hydrocarbons follow a very different set of naming rules from the ones just presented.

Methane (CH_4) is the smallest hydrocarbon. Other small hydrocarbons include ethane, propane, and butane. Although methane may not appear to be a systematic name, it indeed is one if you are willing to accept that *meth-* means 1 carbon atom. Similarly, *eth-* means 2 carbon atoms, and C_2H_6 is ethane. *Prop-* means 3 carbon atoms, and *but-* means 4. So propane is C_3H_8, and butane is C_4H_{10}. Just as *mono-, di-, tri-,* and *tetra-* are used to count, so are *meth-, eth-, prop-,* and *but-*. In Chapter 4, we will explain the suffix *-ane* as well as explain the different numbers of C atoms and H atoms in the chemical formulas.

These new prefixes are very versatile. They can be used not only at the beginning of chemical names, but also within them, as the next activity shows.

Prefixes in hydrocarbon names

meth-	1 C atom
eth-	2 C atoms
prop-	3 C atoms
but-	4 C atoms

Look for more about hydrocarbons in Section 4.4.

Your Turn 1.14   Mother Eats Peanut Butter

Many generations of students have used the memory aid "<u>m</u>other <u>e</u>ats <u>p</u>eanut <u>b</u>utter" for *meth-, eth-, prop-, but-*. Use this or another memory aid of your choice to tell how many carbon atoms are in each of these compounds.

a. ethanol (a gasoline additive)

b. methylene chloride (a component of paint strippers and sometimes an indoor air pollutant)

c. propane (the major component in LPG, liquid petroleum gas)

Answer

b. The *meth-* in methylene indicates 1 C atom in the chemical formula.

As we'll see, hydrocarbon molecules can contain over 50 carbon atoms! For the smaller molecules, use the prefixes shown in Table 1.6. For example, the octane molecule contains 8 carbon atoms.

That's it for names and chemical formulas, at least for now. In the next section, we put this chemical vocabulary to work.

1.9 | Chemical Change: The Role of Oxygen in Burning

Life on Earth bears the stamp of oxygen. Compounds containing oxygen occur in the atmosphere, in your body, and in the rocks and soils of the planet. Why? The answer is that many different elements combine chemically with oxygen. One such element is carbon. You were already introduced to the compound carbon monoxide, CO, a pollutant listed in Table 1.2. Fortunately, CO is relatively rare in our atmosphere. In contrast, carbon dioxide, CO_2, is far more abundant, but still only a mere 390 ppm. Even so, at this concentration CO_2 plays an important role as a greenhouse gas. In this section, we explain how both CO_2 and CO are emitted into our atmosphere.

As you know, humans exhale CO_2 with each breath. Breathing is one natural source of CO_2 in our atmosphere. Carbon dioxide also is produced when humans burn fuels. **Combustion** is the chemical process of burning; that is, the rapid combination of fuel with oxygen to release energy in the form of heat and light. When carbon-containing compounds burn, the carbon combines with oxygen to produce carbon dioxide (CO_2). When the oxygen supply is limited, carbon monoxide (CO) is likely to form as well.

Combustion is a major type of **chemical reaction,** a process whereby substances described as reactants are transformed into different substances called products. A **chemical equation** is a representation of a chemical reaction using chemical formulas. To students, a chemical equation is probably better known as "the thing with an arrow in it." Chemical equations are the sentences in the language of chemistry. They are made up of chemical symbols (corresponding to letters) that often are combined in the formulas of compounds (the words of chemistry). Like a sentence, a chemical equation conveys information, in this case about the chemical change taking place. A chemical equation also must obey some of the same constraints that apply to a mathematical equation.

At the most fundamental level, a chemical equation is a qualitative description of this process:

$$\text{reactant(s)} \longrightarrow \text{product(s)}$$

By convention, the reactants are always written on the left and the products on the right. The arrow represents a chemical transformation and can be read as "is converted to."

The combustion of carbon (charcoal) to produce carbon dioxide as shown in Figure 1.14 can be represented in several ways. One is with chemical names.

$$\text{carbon} + \text{oxygen} \longrightarrow \text{carbon dioxide}$$

Another more common way is to use chemical formulas.

$$C + O_2 \longrightarrow CO_2 \qquad \textbf{[1.1]}$$

This compact symbolic statement conveys a good deal of information. It might sound something like this: "One atom of the element carbon reacts with one molecule of the element oxygen to yield one molecule of the compound carbon dioxide." Using black for carbon and red for oxygen, we also can represent the molecules and atoms involved.

The concentration of CO_2 in the atmosphere is increasing, with serious consequences for planet Earth, as we will describe in Chapter 3.

See Section 4.1 for more about combustion.

Figure 1.14

Charcoal burns in air.

The colors displayed here for atoms reflect the standard used in molecular modeling software and many model kits.

These equations are similar to a mathematical expression in that the number and kind of atom on the left side of the arrow *must* equal the number and kind of each atom on the right:

$$\text{Left side: 1 C and 2 O} \longrightarrow \text{Right side: 1 C and 2 O}$$

Here is an analogy. The building materials used to construct a warehouse (reactants) can be disassembled and rearranged to build three houses and a garage (products).

Atoms are neither created nor destroyed in a chemical reaction. The elements present do not change their identities when converted from reactants to products, although they may be bonded in different ways. This relationship is known as the **law of conservation of matter and mass:** In a chemical reaction, matter and mass are conserved. The mass of the reactants consumed equals the mass of the products formed.

Let's look at another example. Using yellow for sulfur, we can represent how sulfur burns in oxygen to produce the air pollutant sulfur dioxide.

$$S + O_2 \longrightarrow SO_2 \qquad [1.2]$$

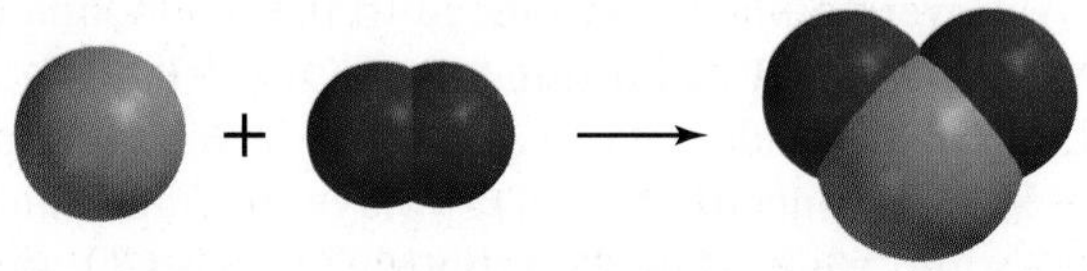

This equation is balanced: the same number and types of atoms are present on each side of the arrow. These atoms, however, were rearranged. This is what a chemical reaction is all about!

It is possible to pack even more information into a chemical equation by specifying the physical states of the reactants and products. A solid is designated by *(s)*, a liquid by *(l)*, and a gas by *(g)*. Because carbon and sulfur are solids, and oxygen, carbon dioxide, and sulfur dioxide are gases at ordinary temperatures and pressures, equations 1.1 and 1.2 become

$$C(s) + O_2(g) \longrightarrow CO_2(g)$$

$$S(s) + O_2(g) \longrightarrow SO_2(g)$$

We designate the physical states when this information is particularly important, but otherwise for simplicity we will omit them.

In a correctly balanced chemical equation, some things must be equal, others need not be. Table 1.7 summarizes our discussion so far.

Equation 1.1 describes the combustion of pure carbon in an ample supply of oxygen. But this is not always the case. If the oxygen supply is limited, CO may be one of the products. Let's take the extreme case in which carbon monoxide is the sole product.

$$C + O_2 \longrightarrow CO \text{ (unbalanced equation)}$$

This equation is not balanced because there are 2 oxygen atoms on the left but only 1 on the right. You might be tempted to balance the equation by simply adding an additional oxygen atom to the right side. But once we write the *correct* chemical formulas for the reactants and products, we cannot change them. We only can use whole-number coefficients

Table 1.7 Characteristics of Chemical Equations
Always Conserved
identity of atoms in reactants = identity of atoms in products
number of atoms of each element in reactants = number of atoms of each element in products
mass of all reactants = mass of all products
May Change
number of molecules in reactants may differ from those in products
physical states (*s*, *l*, or *g*) of reactants may differ from those in products

(or occasionally fractional ones) in front of the given chemical formulas. In simple cases like this, the coefficients can be found by trial and error. If we place a 2 in front of CO, it signifies two molecules of carbon monoxide. This balances the oxygen atoms.

$$C + O_2 \longrightarrow 2\,CO \text{ (still not balanced)}$$

But now the carbon atoms do not balance. Fortunately, this is easily corrected by placing a 2 in front of the C on the left side of the equation.

$$2\,C + O_2 \longrightarrow 2\,CO \text{ (balanced equation)} \qquad \textbf{[1.3]}$$

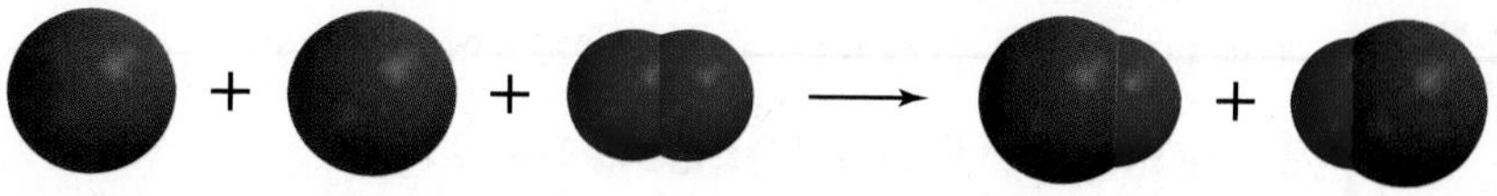

By comparing equations 1.1 and 1.3, you can see that more O_2 is required to produce CO_2 from carbon than is needed to produce CO. This matches the conditions we stated for the formation of carbon monoxide; namely, that the supply of oxygen was limited.

You may be surprised to learn the origin of the air pollutant nitrogen monoxide (also called nitric oxide). It comes from the nitrogen and oxygen found in the air! These two gases chemically combine in the presence of something very hot, such as an automobile engine or a forest fire.

$$N_2 + O_2 \xrightarrow{\text{high temperature}} NO \text{ (unbalanced equation)}$$

The equation is not balanced, as 2 oxygen atoms are on the left side, but only one is on the right. The same is true for nitrogen atoms. Placing a 2 in front of NO supplies 2 N and 2 O atoms on the right. The equation is now balanced.

$$N_2 + O_2 \xrightarrow{\text{high temperature}} 2\,NO \qquad \textbf{[1.4]}$$

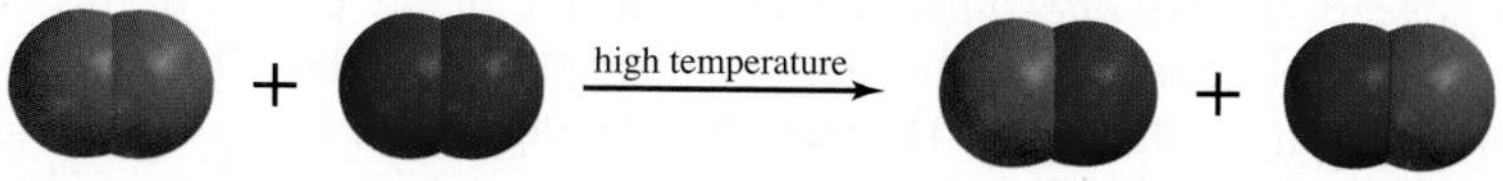

> A subscript follows a chemical symbol, as in O_2 or CO_2. A coefficient precedes a symbol or a formula, as in 2 C or 2 CO.

> Nitrogen and oxygen both are diatomic molecules.

Your Turn 1.15 Chemical Equations

Balance these chemical equations and draw representations of all reactants and products, analogous to equation 1.4. In H_2O and NO_2, O and N are the central atoms, respectively.

a. $H_2 + O_2 \longrightarrow H_2O$

b. $N_2 + O_2 \longrightarrow NO_2$

Note: Both H_2O and NO_2 are bent molecules. We will explain why in the next chapter.

Answer

a. $2\,H_2 + O_2 \longrightarrow 2\,H_2O$

Consider This 1.16 Advice from Grandma

A grandmother offered this advice to rid the garden of pesky caterpillars. "Hammer some iron nails about a foot up from the base of your trees, spacing them every 3 to 5 inches." According to this grandmother, the iron nails convert the sugary tree sap (a compound

(continued)

Consider This 1.16 **Advice from Grandma** *(continued)*

containing the elements carbon, hydrogen, and oxygen) into ammonia (NH_3), a substance the caterpillars cannot stand. Comment on the accuracy of grandma's chemistry (allowing for the possibility that the nails may still work, regardless of her explanation).

1.10 | Fire and Fuel: Air Quality and Burning Hydrocarbons

Look for other examples of burning fuels throughout Chapter 4.

As we mentioned earlier, hydrocarbons are compounds of hydrogen and carbon. The hydrocarbons that we use today are primarily obtained from crude oil. Methane (CH_4), the simplest hydrocarbon, is the primary component of natural gas. Both gasoline and kerosene are mixtures of many hydrocarbons.

Given an ample supply of oxygen, hydrocarbon fuels burn completely. You may hear this called "complete combustion." In essence, all of the carbon atoms in the hydrocarbon molecule combine with O_2 molecules from the air to form CO_2. Similarly, all the hydrogen atoms combine with O_2 to form H_2O. For example, here is the chemical equation for the complete combustion of methane. This equation is your first peek at why burning carbon-based fuels releases carbon dioxide into the atmosphere.

$$CH_4 + O_2 \longrightarrow CO_2 + H_2O \text{ (unbalanced equation)}$$

When balancing chemical equations, it is fine to use fractional coefficients.

Note that O appears in *both* products: CO_2 and H_2O. To balance it, start with an element that appears in *only one substance* on each side of the arrow. In this case, both H and C qualify. No coefficients need to be changed for carbon, because both sides contain one C atom. Balance the H atoms by placing a 2 in front of the H_2O.

$$CH_4 + O_2 \longrightarrow CO_2 + 2\,H_2O \text{ (still not balanced)}$$

Balance the oxygen atoms last. Four O atoms are on the right side and two O atoms are on the left, so we need 2 O_2 to balance the equation.

$$CH_4 + 2\,O_2 \longrightarrow CO_2 + 2\,H_2O \text{ (balanced equation)} \qquad \textbf{[1.5]}$$

A nice feature of chemical equations is that counting the atoms on both sides of the arrow tells you if it is balanced. Here, the equation is balanced because each side has 1 C atom, 4 H atoms, and 4 O atoms.

Most automobiles run on the complex mixture of hydrocarbons that we call gasoline. Octane, C_8H_{18}, is one of the pure substances in this mixture. With sufficient oxygen, octane burns to form carbon dioxide and water.

$$2\,C_8H_{18} + 25\,O_2 \longrightarrow 16\,CO_2 + 18\,H_2O \qquad \textbf{[1.6]}$$

Both products travel from the engine out the exhaust pipe and into the air. Are these combustion products visible? Usually not. Water, in the form of water vapor, and carbon dioxide are both colorless gases. But if you happen to be outside on a winter day, the water vapor condenses to form clouds of steam or tiny ice crystals that you can see. Occasionally, the frozen vapor gets trapped in an inversion layer and forms an ice fog (Figure 1.15).

With less oxygen, the hydrocarbon mixture we call gasoline burns incompletely ("incomplete combustion"). Water is still produced together with both CO_2 and CO. The extreme case occurs when only carbon monoxide is formed, as is shown here for the incomplete combustion of octane.

$$2\,C_8H_{18} + 17\,O_2 \longrightarrow 16\,CO + 18\,H_2O \qquad \textbf{[1.7]}$$

Compare the coefficient of 17 for O_2 in this equation with that of 25 for O_2 in equation 1.6. Less oxygen is needed for incomplete combustion, as CO contains less oxygen than CO_2.

Figure 1.15

A winter ice fog in Fairbanks, Alaska.

Your Turn 1.17   **Balancing Equations**

Demonstrate that equations 1.6 and 1.7 are balanced by counting the number of atoms of each element on both sides of the arrow.

Answer

Equation 1.6 contains 16 C, 36 H, and 50 O on each side.

What actual mixture of products is formed when gasoline is burned in your car? This is not a simple question, as the products vary with the fuel, the engine, and its operating conditions. It is safe to say that gasoline burns primarily to form H_2O and CO_2. However, some CO and soot also are produced. The amounts of soot, CO, and CO_2 that go out the tailpipe indicate how efficiently the car burns the fuel, which in turn indicates how well the engine is tuned. Some regions of the United States monitor auto emissions with a probe that detects CO (Figure 1.16). The CO concentrations in the exhaust are compared with established standards, for example, 1.20% in the state of Minnesota. If the vehicle fails the emissions test, it must be serviced.

Consider This 1.18   **Auto Emissions Report**

a. Figure 1.16 reports NO_x emissions in grams per mile. NO_x is a way to collectively represent the oxides of nitrogen. If $x = 1$ and $x = 2$, write the corresponding chemical formulas. Also give the chemical names.

b. NO is the primary oxide of nitrogen emitted. What is the source of this compound? *Hint:* Revisit equation 1.4.

c. The green line is missing on the CO_2 graph, but present on the others. Explain.

Answers

a. NO, nitrogen monoxide and NO_2, nitrogen dioxide

c. In the year that this graph was produced, CO_2 was not classified as an air pollutant in the United States. Accordingly, it has no green line indicating an acceptable range.

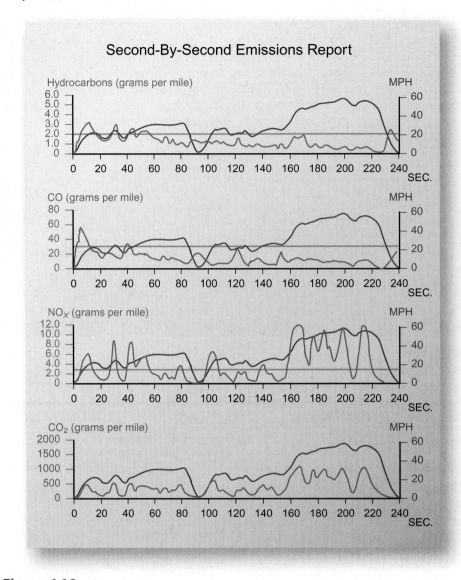

Figure 1.16

A U.S. auto emissions report. The blue line shows the change in engine speed; the red line shows the change in emissions. Any emissions below the green line are in the acceptable range.

CO_2 emissions are measured but not yet regulated. Look for more about CO_2 emissions in Chapter 3.

1.11 | Air Pollutants: Direct Sources

In this section, we examine two major sources of air pollutants: motor vehicles and coal-fired plants that generate electricity. These sources directly emit SO_2, CO, NO, and PM, and we will revisit each of these pollutants in turn. We also digress to discuss VOCs (volatile organic compounds), pollutants that are not regulated but are still intimately connected with the ones that are. We tackle ozone in the section that follows.

Your Turn 1.19   Tailpipe Gases

What comes out of the tailpipe of an automobile? Start your list now and build it as you work through this section.

Hint: Some of the air that enters the engine also comes out the tailpipe.

Sulfur dioxide emissions are linked to the coal that is burned to generate electric power. Although coal consists mostly of carbon, it may contain 1–3% sulfur together with small amounts of minerals. The sulfur burns to form SO_2, and the minerals end up as fine ash particles. If not contained, the SO_2 and ash go right up the smokestack. The hundreds of millions of tons of coal burned by the United States translates into millions of tons of waste in the air. As we will see in Chapter 6, the SO_2 produced by burning coal can dissolve in the water droplets of clouds and fall to the ground as acid rain.

See Section 4.6 for more about coal and its chemical composition.

The story does not end with SO_2. Once in the air, sulfur dioxide can react with oxygen to form sulfur trioxide, SO_3.

$$2\,SO_2 + O_2 \longrightarrow 2\,SO_3 \qquad \text{[1.8]}$$

Although normally quite slow, this reaction is faster in the presence of small ash particles. The particles also aid another process. If the humidity is high enough, they help condense water vapor into an aerosol of tiny water droplets. **Aerosols** are liquid and solid particles that remain suspended in the air rather than settling out. Smoke, such as from a campfire or a cigarette, is a familiar aerosol made up of tiny particles of solids and liquids.

Section 6.12 describes how sulfuric acid aerosols contribute to haze.

The aerosol of concern here is made up of tiny droplets of sulfuric acid, H_2SO_4. It forms because sulfur trioxide dissolves readily in water droplets to produce sulfuric acid.

$$H_2O + SO_3 \longrightarrow H_2SO_4 \qquad \text{[1.9]}$$

If inhaled, the droplets of the sulfuric acid aerosol are small enough to become trapped in the lung tissue and cause severe damage.

The good news? Sulfur dioxide emissions in the United States are declining (see Figure 1.8). For example, in 1985, approximately 20 million tons of SO_2 was emitted from the burning of coal. Today the value is closer to 9 million tons. This impressive decrease can be credited to the Clean Air Act of 1970 that mandated many reductions, including those from coal-fired electric power plants. More stringent regulations were established in the Clean Air Act Amendments and the Pollution Prevention Act of 1990. For example, gasoline and diesel fuel both once contained small amounts of sulfur, but the allowable amounts were drastically lowered in 1993 and in 2006, respectively. But continued progress has a price tag. Cleaning up the smaller older and dirty power plants will not come cheaply. Then again, allowing emissions to continue is costly in terms of human and environmental health.

Look for more about the economic and societal costs of atmospheric SO_2 in Chapter 6.

Your Turn 1.20   SO₂ from the Mining Industry

Burning coal is not the only source of sulfur dioxide. As you saw in Your Turn 1.8, smelting is another. For example, silver and copper metal can be produced from their sulfide ores. Write the balanced chemical equations.

a. Silver sulfide (Ag_2S) is heated in air to produce silver and sulfur dioxide.
b. Copper sulfide (CuS) is heated in air to produce copper and sulfur dioxide.

Answer
a. $Ag_2S + O_2 \longrightarrow 2\,Ag + SO_2$

With more than 250 million vehicles (and over 300 million people), the United States has more vehicles per capita than any other nation. Do these vehicles emit sulfur dioxide? Fortunately, the answer is no, because cars have internal combustion engines primarily fueled by gasoline. We already mentioned that the combustion of hydrocarbons in gasoline produces—at best—carbon dioxide and water vapor (see equation 1.6). Because gasoline contains little or no sulfur, burning it produces little or no sulfur dioxide. Nonetheless, each tailpipe puffs out its share of air pollutants. The ubiquitous motor car adds to the atmospheric concentrations of carbon monoxide, volatile organic compounds, nitrogen oxides, and particulate matter. We discuss each of these in turn.

Carbon monoxide pollution comes primarily from automobiles. But think in terms of *all* the tailpipes out there, not just those attached to cars. Some are attached to heavy trucks, SUVs, and the three m's: motorcycles, minibikes, and mopeds. Others are on equipment such as farm tractors, bulldozers, and motor boats. The tailpipes attached to *all* gasoline and diesel engines emit carbon monoxide.

Your Turn 1.21   Other Tailpipes

Visit "Nonroad Engines, Equipment, and Vehicles" courtesy of the EPA. A link is provided at the textbook's website.

a. The text mentioned tractors, bulldozers, and boats. Name five other engine-powered machines or vehicles that do not run on roads.
b. Select a machine or vehicle of interest to you. How are emissions from its engine being reduced? What is the time scale for the reduction?

Even though the number of cars has risen, a dramatic reduction in CO emissions has occurred. Based on measurements by the EPA at over 250 sites in the United States, since 1980 the average CO concentration has decreased almost 60% (see Figure 1.8). If wildfires are excluded, today's levels are the lowest reported in three decades. The decrease is due to several factors, including improved engine design, computerized sensors that better adjust the fuel–oxygen mixture, and most importantly, the requirement that all cars manufactured since the mid-1970s have catalytic converters (Figure 1.17). Catalytic converters reduce carbon monoxide in the exhaust stream by catalyzing the combustion of CO to CO_2. They also lower NO_x emissions by catalyzing the conversion of nitrogen oxides back to N_2 and O_2, the two atmospheric gases that formed them. In general, a **catalyst** is a chemical substance that participates in a chemical reaction and influences its rate without itself undergoing permanent change. Catalytic converters typically use metals such as platinum and rhodium as catalysts.

Wildfires contribute additional CO, increasing emissions as much as 10% each year.

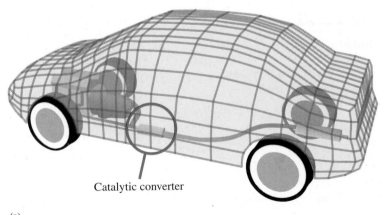

(a)

Catalytic converter

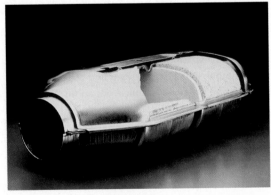

(b)

Figure 1.17

(a) Location of catalytic converter in car. **(b)** Cutaway view of a catalytic converter. Metals such as platinum and rhodium serve as catalysts and form a coating on the surface of ceramic beads.

Cars not only emit carbon in the form of carbon monoxide but also in the form of unburned and partially burned hydrocarbons. This leads us to the topic of VOCs, volatile organic compounds. A **volatile** substance readily passes into the vapor phase; that is, it evaporates easily. Gasoline and nail polish remover are both volatile. If you were to spill either of these, the puddle would soon evaporate. When you apply varnish to a surface, you can smell the volatile compounds that evaporate with each brush stroke. An **organic compound** always contains carbon, almost always contains hydrogen, and may contain elements such as oxygen and nitrogen. Organic compounds include methane and octane, hydrocarbons mentioned earlier. They also include alcohol and sugar, compounds that contain oxygen in addition to carbon and hydrogen.

Accordingly, **volatile organic compounds (VOCs)** are carbon-containing compounds that pass easily into the vapor phase. They originate from a variety of sources. For example, you can smell naturally occurring VOCs in a spruce or pine forest. VOCs from tailpipes are not so pleasant, as they are vapors of incompletely burned gasoline molecules or fragments of these molecules. The exhaust gas still contains oxygen, as not all of it is consumed in the engine. Catalytic converters utilize this oxygen to burn VOCs to form carbon dioxide and water. Section 1.12 describes the connection between VOCs and ozone formation. But right now, we want to connect VOCs with the formation of NO_2.

Nitrogen monoxide and nitrogen dioxide are collectively known as NO_x, as mentioned in Consider This 1.18. NO_2 is brown in color, giving smog its characteristic brownish tinge. Recall that N_2 and O_2 combine to produce NO which is a colorless gas (see equation 1.4). But what is the origin of NO_2? Here is a balanced equation that appears to be a likely candidate.

$$2\,NO + O_2 \longrightarrow 2\,NO_2 \qquad \textbf{[1.10]}$$

However, this is not what actually occurs. Instead, NO_2 is formed by other pathways that are more complex. Here is the one that predominates in urban settings where you are likely to find NO. In some cities, this actually lowers the ozone concentrations along highways congested with vehicles emitting NO.

$$NO + O_3 \longrightarrow NO_2 + O_2 \qquad \textbf{[1.11]}$$

To further complicate things, on a sunny day, some of the NO_2 converts back into NO, as we will see in the next section. Again, this is why people refer to NO_x, rather than to either NO or NO_2.

The conversion of NO to NO_2 connects to the breakdown of VOCs in the air. A new player is involved, the reactive hydroxyl radical ($\cdot$OH). This reactive species is present in tiny amounts in air, polluted or otherwise.

$$VOC + \cdot OH \longrightarrow A$$
$$A + O_2 \longrightarrow A'$$
$$A' + NO \longrightarrow A'' + NO_2 \qquad \textbf{[1.12]}$$

Here, A, A', and A'' represent reactive molecules that can form in the air from $\cdot$OH and VOCs. The bottom line? Atmospheric chemistry is complex and involves many players. You have met some of them, including NO, NO_2, O_2, O_3, VOCs, and $\cdot$OH.

The United States has had limited success in curbing NO_x emissions. In turn, as we'll see in the next section, this means limited success in curbing ozone. Nonetheless, given the increasing number of vehicles, *any* decrease in NO_x is impressive. Despite early claims from the auto industry that it would be impossible (or too costly) to meet new emissions standards, the industry is curbing emissions by improving catalytic converters, engine designs, and gasoline formulations.

Look for more about organic compounds in Chapter 4.

The dot in ($\cdot$OH) indicates an unpaired electron. In Chapter 2, you will encounter other reactive species with unpaired electrons.

Chapter 8 discusses fuel cells and
other alternatives to gasoline-
powered vehicles.

Consider This 1.22   Forget Road Rage

Burning less gasoline equates to fewer tailpipe emissions. Which driving practices conserve fuel? Which practices expend it more than necessary? Think about the behavior of motorists on highways, city streets, and in parking lots. For each of these venues, list at least three ways that drivers could burn less gasoline.

Hint: Consider how you accelerate, coast, idle, brake, and park.

Answer
Some possibilities relate to parking. For example, if you are able-bodied, take a spot further away and walk, rather than cruising around to find a closer spot. Find a parking space that you can pull through. This way, when you exit you don't need to back and turn, thus conserving fuel.

Particulate matter comes in a range of sizes, but only the tiny particles (PM_{10} and $PM_{2.5}$) are regulated as pollutants. This size can penetrate deeply into your lungs, pass into your bloodstream, and inflame your cardiovascular system. In terms of regulation, particles are the new pollutant on the block. Data collection in the United States for PM_{10} and $PM_{2.5}$ started in 1990 and 1999, respectively (see Figure 1.8). In 2006, the daily air quality standard for $PM_{2.5}$ was lowered from 65 to 35 $\mu g/m^3$ because these particles proved to be more hazardous than originally thought.

Particulate matter has many different sources. In the summer, wildfires may raise the concentration of particulate matter to a hazardous level. In the winter, wood stoves may produce exactly the same effect. At any time of the year in almost any urban environment, older diesel engines on trucks and buses emit clouds of black smoke. Diesel engines on tractors similarly can pollute. Construction sites, mining operations, and the unpaved roads that serve them also loft tiny particles of dust and dirt into the atmosphere. Particulate matter can even form right in the atmosphere. For example, the compound ammonia, used in agriculture, is a major player in forming ammonium sulfate and ammonium nitrate in the air, both $PM_{2.5}$.

Ammonia (NH_3) is a colorless gas with
a pungent odor. It is condensed to the
liquid phase and applied to soil as a
fertilizer. Look for more about
ammonia in Chapter 6.

Given all these sources, particulate matter has proven a tough pollutant to control. Even so, in 2007 the EPA reported a decrease of 14% in the annual $PM_{2.5}$ concentrations from 2000 to 2006. However, 10% of the sites monitored still showed an increase in particle pollution. Again, what you breathe very much depends on where you live.

Consider This 1.23   Particles Where You Live

Here is a map of the continental U.S. that shows $PM_{2.5}$ data for August 30, 2008.

a. In terms of air quality, what do the green, yellow, and orange colors indicate?

b. Which groups of people are most sensitive to particulate matter?

c. Visit *State of the Air*, a website posted by the American Lung Association. How many days a year does your state have "orange days" and "red days" for particle pollution? What is the difference?

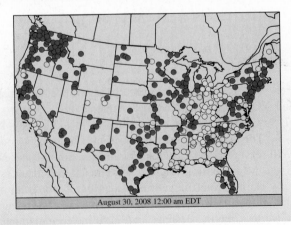

August 30, 2008 12:00 am EDT

1.12 | Ozone: A Secondary Pollutant

Ozone definitely is a bad actor in the troposphere. Even at very low concentrations, it reduces lung function in healthy people who are exercising outdoors. Ozone also damages crops and the leaves of trees. But ozone does *not* come out of a tailpipe and is *not* produced when coal is burned. How is it produced? Before we fill you in on the details, do this activity.

Today, background tropospheric levels for ozone are about 40 parts per billion (ppb). In pre-industrial times, the level was about 10 ppb.

Consider This 1.24   Ozone Around the Clock

Ozone concentrations vary during the day, as shown in Figure 1.18.

a. Near which cities is the air hazardous to one or more groups?

  Hint: Refer back to the color-coded AQI (see Table 1.4).

b. At about what time does the ozone level peak?

c. Can moderate levels (shown in yellow) of ozone exist in the absence of sunlight? Assume sunrise occurs around 6 AM and sunset at about 8 PM.

Answer

c. In the absence of sunlight, ozone does not persist for long. After sundown, the ozone levels drop.

The previous activity raises several related questions. Why is ozone more prevalent in some areas than others? What role does sunlight play in ozone production? We now address these.

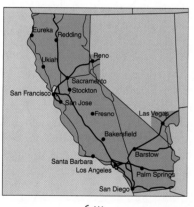

6 AM

10 AM

noon

4 PM

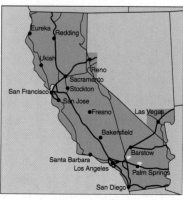

10 PM

Figure 1.18

Ozone level maps for a summer day in July 2006 in California.

Unlike the pollutants described in the previous section, ozone is a **secondary pollutant.** It is produced from chemical reactions involving one or more other pollutants. For ozone, these other pollutants are VOCs and NO_2. Recall from the previous section that NO, rather than NO_2, comes directly out of a tailpipe (or a smokestack). But over time and in the presence of VOCs and $\cdot OH$, NO in the atmosphere is converted to NO_2.

Recall from the previous section that ·OH is the hydroxyl radical.

Nitrogen dioxide meets several fates in the atmosphere. The one of most interest to us occurs when the Sun is high in the sky. The energy provided by sunlight splits one of the bonds in the NO_2 molecule:

$$NO_2 \xrightarrow{\text{sunlight}} NO + O \qquad \textbf{[1.13]}$$

The oxygen atoms produced then can react with oxygen molecules to produce ozone.

$$O + O_2 \longrightarrow O_3 \qquad \textbf{[1.14]}$$

This explains why ozone formation requires sunlight. Sunlight splits NO_2 to release O atoms. These in turn react with O_2 to form O_3. Thus once the Sun goes down, the ozone concentrations drop off sharply, as you can see in Figure 1.18. What happened to the ozone? In just a matter of hours, the ozone molecules react with many things, including animal and plant tissue.

Note that equation 1.14 contains three different forms of elemental oxygen: O, O_2, and O_3. All three are found in nature, but O_2 is the least reactive and by far the most abundant, constituting about one fifth of the air we breathe. Our atmosphere naturally contains tiny amounts of protective ozone up in the stratosphere. Oxygen atoms also exist in our upper atmosphere and are even more reactive than ozone.

"Good" ozone is in the stratosphere. "Bad" ozone is in the troposphere.

Consider This 1.25   O₃ Summary

Summarize what you have learned about ozone formation by developing your own way to arrange these chemicals sequentially and in relation to one another: O, O_2, O_3, VOCs, NO, NO_2. Chemicals may appear as many times as you like. You also may wish to include sunlight.

Because sunlight is involved in ozone formation, the concentration of ground-level ozone varies with weather, season, and latitude. High levels of O_3 are much more likely to occur on long sunny summer days, especially in congested urban areas. Stagnant air also favors the buildup of air pollution. For example, revisit the air quality data for cities shown in Table 1.3. Ozone was usually the culprit responsible for pollution in cities with sunny summer days. In contrast, windy and rainy cities have lower levels of ozone.

Consider This 1.26   Ozone and You

The AIRNOW website, courtesy of the EPA, provides a wealth of information about ground-level ozone levels in the United States.

a. Let's say that the ozone level is "orange," actually a common occurrence in many U.S. cities during the summer months. Does air of this quality affect you if you have no health concerns, but are actively exercising out of doors?
   Hint: Check the link to AIRNOW provided at the textbook's website.
b. Again assume you are active out of doors. How does the air quality in your state compare with others?

Canada also publishes daily ozone maps (Figure 1.19). Some of Canada's polluted air originates in the United States, blown northeastward from population centers in Ohio, Pennsylvania, and New York. Pollution knows no boundaries!

This is an example of the **tragedy of the commons.** The tragedy arises when a resource is common to all and used by many, but has no one in particular responsible

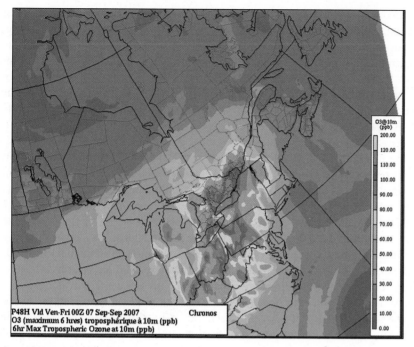

Figure 1.19

Tropospheric ozone map for September 7, 2007.

Source: Environment Canada.

for it. As a result, the resource may be destroyed by overuse to the detriment of all who use it. For example, we cannot lay individual claim to the air; it belongs to all of us. If the air we breathe has waste dumped into it, this leads to an unhealthy situation for everyone. Individuals whose activities have little or no effect on the air still suffer the same consequences as those who pollute. The costs are shared by all. In later chapters, we will see other examples of the tragedy of the commons that relate to water, energy, and food.

Air pollution, once primarily a local concern, is now a serious international issue. Many cities worldwide have high ozone levels. Couple vehicles with a sunny location anywhere on the planet, and you are likely to find unacceptable levels of ozone. Some places, however, are worse than others. London with its cool foggy days has low ozone levels. In contrast, ozone is a serious problem in Mexico City.

Ozone attacks rubber and so it damages the tires of the vehicles that led to its production in the first place. Should you park your car indoors in the garage to minimize possible rubber damage? In fact, should you park *yourself* indoors if the levels of ozone outside are unhealthy? The next section speaks to the quality of indoor air.

> Garrett Hardin is credited with coining the term "tragedy of the commons." In an article published in 1968, he pointed out how individuals using a common resource may destroy it such that ultimately nobody can use it.

1.13 | The Inside Story of Air Quality

In the Wizard of Oz, Dorothy hugged her dog Toto and exclaimed, "There's no place like home!" Of course she was right, but when it comes to air quality, your home may not always be the best place to be. Indoors, the levels of air pollution may far exceed those outside. Given that most of us sleep, work, study, and play indoors, we should learn about the air in the place we call home.

Indoor air may contain up to a thousand substances at low levels. If you are in a room where somebody is smoking, add another thousand or so. Indoor air contains some familiar culprits: VOCs, NO, NO_2, SO_2, CO, ozone, and PM. These pollutants are present either because they came in with the outside air or because they were generated right inside your dwelling.

Let's begin our discussion with the question posed in the previous section. Should you move indoors to escape the ozone present outside? In general, if a pollutant is

> Some copy machines and air cleaners generate ozone, which may increase indoor ozone levels.

Ponderosa pine bark emits compounds with a scent reminiscent of butterscotch.

Sick building syndrome has many causes. One way or another, most relate to air quality. The source of the bad air may come both from inside the building, outside the building, or both.

highly reactive, it does not persist long enough to be transported indoors. Thus, for highly reactive molecules such as O_3, NO_2, and SO_2, you expect lower levels indoors. Indeed, this is the case, as indoor air is typically 10–30% lower in ozone concentration than outdoor air. Similarly, sulfur dioxide and nitrogen dioxide levels are lower indoors, although the decrease is not as dramatic as that for ozone.

Carbon monoxide is a different story. As a relatively unreactive pollutant, CO has a long enough atmospheric lifetime to move freely in and out of buildings through doors, windows, or a ventilation system. The same is true for some VOCs, but not for the more reactive ones such as those that give pine forests their scent. If you want to inhale the delicious volatile compounds emitted from the bark of Ponderosa pines, it is best to remain outside near the trees.

Some pollutants are trapped by the filters in the heating or cooling system of the building. For example, many air-handling systems contain filters that remove larger-sized particulate matter and pollen. As a result, people who suffer from seasonal allergies can find relief indoors. Similarly, those near a wildfire can go inside to escape some of the irritating smoke particles. However, gas molecules such as O_3, CO, NO_2, and SO_2 are *not* trapped by the filters used in most ventilation systems.

These days, many buildings are constructed with an eye to increasing their energy efficiency. This is a win–win, lowering your heating bills and lowering the amounts of pollutants generated in producing the heat. But there can be a down side. A building that is air-tight with a limited intake of fresh air may have unhealthy levels of indoor air pollutants. Therefore what appeared initially to be a benefit (better energy efficiency) can turn into a higher risk (increased pollutant levels). In some cases, poor ventilation can cause indoor pollutants to reach hazardous levels, creating a condition known as "sick building syndrome." Clearly this is an undesirable outcome. Today, architects and builders are finding ways to make buildings more energy-efficient while maintaining a good air exchange.

Even with good ventilation, indoor activities can compromise air quality. For example, tobacco smoke is a serious indoor air pollutant, containing over a thousand chemical substances. Nicotine is one that you may recognize; others include benzene and formaldehyde. Taken as a whole, tobacco smoke is **carcinogenic,** meaning capable of causing cancer.

Combustion of carbon-containing fuels also can generate carbon monoxide and nitrogen oxides. For example, carbon monoxide from cigarette smoking in bars can reach 50 ppm, a value well within the unhealthy range. In cigarette smoke, NO_2 levels can exceed 50 ppb. Fortunately, smokers take puffs rather than constantly breathing cigarette smoke.

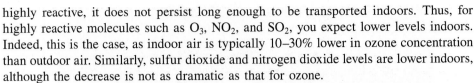

Your Turn 1.27   Indoor Cigar Party

In 2007, student researchers attended a cigar party and trade show in Times Square, New York. They carried concealed detectors and found levels of 1193 µg of particulate matter per cubic meter of air inside the ballroom.

a. The news article did not report whether the particulate matter was $PM_{2.5}$ or PM_{10}. For either, does the value they measured exceed the air quality standards in the United States?
Hint: See Table 1.2.
b. Assume that the students were measuring $PM_{2.5}$. What are the health implications?

People also burn candles, perhaps to soften the lighting or set a mood. However, candles deplete the oxygen in a room. They also can produce soot, carbon monoxide, and VOCs. Similarly, people may burn incense in their homes for one reason or another. Atmospheric scientist Stephen Weber, a researcher who studied incense burning in churches in Europe, found that "the pollutants in smoke from incense and candles may be more toxic than fine-particle pollution from sources such as vehicle engines."

Burning candles or incense can generate fumes more rapidly than these can be removed by your ventilation system or by the breezes that pass through open windows. The next activity gives you the opportunity to further investigate sources of indoor pollutants.

Your Turn 1.28   Indoor Activities

Name 10 activities that add pollutants or VOCs to indoor air. To get you started, two are pictured in Figure 1.20. Remember that some pollutants have no detectable odors.

As Figure 1.20 suggests, paints and varnishes are a source of VOCs. Your nose alerts you to these while you paint; so does your head if it starts aching. Although the amount of VOCs released per volume of paint applied varies widely, all oil-based paints and varnishes rank higher than water-based ones. Check out the VOC emission values printed right on the paint can. These range from zero for low-VOC paints to more than 600 grams VOCs per liter for some outdoor oil paints. You would be fortunate to find an oil paint with less than 350 grams VOCs per liter.

Consumers today can purchase high-quality paints that are nontoxic and odor-free. For example, check out the label on the paint can shown in Figure 1.21. This "zero-VOC" water-based paint emits less than 5 grams of VOCs per liter applied. This paint has earned a "Green Seal" certification, meaning that not only does it emit less than 50 grams of VOCs per liter, but also that it contains no toxic metals such as lead, mercury, or cadmium. Low- or zero-VOC paints are equally important to use outside. The paint applied to buildings, bridges, and railings in the United States once added up to over 7 million pounds of VOCs per year. In 2005, the value was reported at less than 4 million pounds, thanks to changes in the formulation of paints.

Before we can explain how volatile compounds have been removed from paint, we first need to address why paints contained them in the first place. Some VOCs are additives that evaporate as the paint dries. For example, you may recognize the two antifreeze additives listed in Table 1.8. Antifreeze ("glycols") allows the folks who live up north to store paint in their basements without worrying that it will be damaged by repeated freezing and thawing. Antifreeze also allows paint to be applied in colder weather and provides longer "open time," so that the paint does not dry out too quickly.

Coalescents are another additive used in latex paints. **Coalescents** are chemicals added to soften the latex particles in paint so that these particles spread to form a continuous film of uniform thickness. After all, you want your paint to brush on evenly! As the paint dries and hardens, the coalescents evaporate into the air. Between 2 and

(a)

(b)

Figure 1.20

Activities that can pollute indoor air.

Figure 1.21
All ingredients in YOLO paints are selected as "zero VOC."

3 gallons of volatile coalescents are used for every 100 gallons of paint. But do the math. In the United States this additive translates to over 100 million pounds of coalescents emitted each year, and approximately 3 times this amount worldwide.

Oil-based paints contain just that, oils. Most are oils derived from plants, such as linseed oil. These oils slowly react with the oxygen in the air and over time release a host of volatile compounds to the air. Oil paints also may contain solvents (thinners) that evaporate as the paint dries.

Look for more about the oils found in plants in Chapter 11.

Skeptical Chemist 1.29   **Varnish Fumes**

A can of clear satin floor varnish claims a maximum of 450 grams/liter of VOCs.
A consumer group computes this to just under 4 pounds of VOCs emitted per gallon as the varnish is applied. Did this group do the math correctly?

Table 1.8	VOCs Emitted by Some Paints

Glycols (antifreezes)

  ethylene glycol

  propylene glycol

Coalescents (used in latex paints)

  2,2,4-trimethyl-1,3-pentanediol monoisobutyrate (trade name: Texanol)

Hazardous air pollutants (solvents and preservatives)

  benzene

  formaldehyde

  ethylbenzene

  methylene chloride

  vinyl chloride

Stricter government regulations on VOC emissions have prompted paint manufacturers to devise new formulations for latex paints. In 2005, the Archer Daniels Midland Company won a Presidential Green Chemistry Challenge award for developing nonvolatile coalescents. The coalescents developed by the company react with the oxygen in the air, enabling them to chemically bond to the latex. Thus, the coalescents become part of the paint film rather than evaporating into the atmosphere.

Another advantage of these new coalescents is that they are produced from vegetable oils (a renewable resource) as opposed to crude oil (nonrenewable). Their production also creates less waste and requires less energy, a plus for both environmental and cost savings. There is no loss in quality, as the paints formulated with vegetable oil-based coalescents meet or exceed the performance of traditional paints. They have lower odors, better scrub resistance, and better opacity. Do these environmental, economic, and societal benefits sound reminiscent of the Triple Bottom Line? Again, this is the heart of sustainability.

If you have ever been to an auto body shop, you most likely have smelled the odors from the VOCs in the paints. More stringent laws and regulations in many countries have forced manufacturers to reformulate automotive primers (undercoats) and finishes. Most traditional primers are made by mixing two components that have a limited shelf life. The primer is then applied and cured in an oven that requires large amounts of energy to heat. BASF Corporation won a 2005 Presidential Green Chemistry Challenge award for the development of a one-component primer that reduces VOCs by 50% and that is rapidly cured in sunlight or with a UV-A lamp. This greatly reduces the time required to make and cure the primer. Here again, the reduced costs associated with less waste, reduced energy consumption, and greater throughput equate to an improved Triple Bottom Line.

UV-A lamps are longer-wave ultraviolet light, similar to tanning lamps. Look for more about UV light in Chapter 2.

Your Turn 1.30   **Bring Home the Green**

Revisit the principles of Green Chemistry listed on the inside front cover of this book. Which of the principles are met by the new coalescents developed by the Archer Daniels Midland Company? By the new undercoat developed by BASF? Prepare a list for each company.

We end our discussion of indoor air quality by turning to radon, a noble gas (Group 8A) that we mentioned earlier. Radon is a special case of indoor air pollution. It occurs naturally in tiny amounts and usually is no problem. But it may reach hazardous levels in basements, mines, and caves. Like all noble gases, radon is colorless, odorless, tasteless, and chemically unreactive. But unlike the others, it is radioactive. Radon is generated in the decay series of uranium, another naturally occurring radioactive element. Because uranium occurs at a concentration of about 4 ppm in the rocks of our planet, radon is ubiquitous. Depending on how your apartment or dorm is constructed, the radon produced from uranium-containing rocks may find entry into the basement. Radon causes lung cancer and is the second leading cause behind tobacco smoke. As is the case for other pollutants, the threshold for danger can be estimated but is not precisely known. Radon test kits, such as the one shown in Figure 1.22, are used to measure the radon concentration in living spaces.

Indoors or out, we need to breathe healthy air. And with each breath, we inhale a truly prodigious number of molecules and atoms. We end this chapter by revisiting these molecules and atoms.

Look for more about uranium and its natural decay series in Chapter 7.

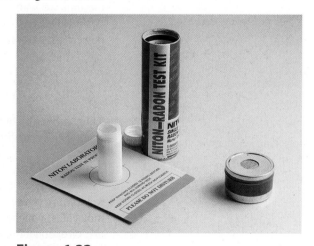

Figure 1.22

A home radon test kit.

1.14 | Back to the Breath—at the Molecular Level

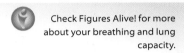 Check Figures Alive! for more about your breathing and lung capacity.

The maximum concentrations of pollutants allowed by the air quality standards seem very small (see Table 1.2). Indeed, an exposure to 9 ppm CO is a tiny amount! But even this low concentration of CO contains a staggering number of carbon monoxide molecules. This seeming contradiction is a consequence of the minuscule mass of molecules. Recall Consider This 1.2: Take a Breath. If you are an average-sized adult, the capacity of your lungs is between 5 and 6 L. You do not empty your lungs each time you take a breath. Rather, as you are at rest reading this, you are inhaling about 500 milliliters of air or approximately half a quart with each breath.

Accurately measuring the volume of air you inhale and exhale can be done with the help of a spirometer (Figure 1.23). Determining the number of molecules and atoms in this volume of air is a harder task, but it can be done. From experiments, we know that a typical breath of 500 mL contains about 2×10^{22} molecules and atoms. Remember that air is primarily N_2 and O_2 molecules together with a small amount of Ar atoms and a varying amount of H_2O molecules (humidity).

Using this number of molecules and atoms in the air (2×10^{22}), we now can calculate the number of CO molecules in the breath you just inhaled. We assume the breath contained 2×10^{22} molecules and atoms, and that the CO concentration in the air was at the air quality standard of 9 ppm. Thus, out of every million (1×10^6) molecules and atoms in the air, nine will be CO molecules. To compute the number of CO molecules in a breath, multiply the total number of molecules and atoms in the air by the fraction that are CO molecules.

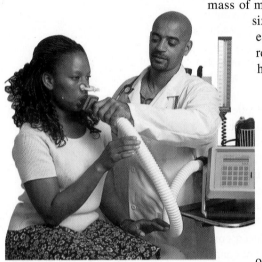

Figure 1.23

A spirometer is used for measuring an individual's breathing capacity.

$$\frac{\text{\# of CO molecules}}{1 \text{ breath of air}} = \frac{2 \times 10^{22} \text{ molecules and atoms in air}}{1 \text{ breath of air}} \times \frac{9 \text{ CO molecules}}{1 \times 10^6 \text{ molecules and atoms in air}}$$

$$= \frac{2 \times 9 \times 10^{22}}{1 \times 10^6} \frac{\text{CO molecules}}{\text{breath of air}}$$

$$= \frac{18}{1} \times \frac{10^{22}}{10^6} \frac{\text{CO molecules}}{\text{breath of air}}$$

In writing this out, we carefully retain the units on the numbers. Not only does this remind us of the physical entities involved, but also it guides us in setting up the problem correctly. The units "molecules and atoms in the air" cancel, and we are left with the unit we want: CO molecules per breath of air.

However, we need to divide 10^{22} by 10^6 to determine a final answer. To *divide* powers of 10, simply *subtract* the exponents. In this case,

$$\frac{10^{22}}{10^6} = 10^{(22-6)} = 10^{16}$$

Thus, a breath contains 18×10^{16} CO molecules.

The preceding answer is mathematically correct, but in scientific notation it is customary to have only one digit to the left of the decimal point. Here we have two: 1 and 8. Therefore, our last step is to rewrite 18×10^{16} as 1.8×10^{17}. We can make this conversion because $18 = 1.8 \times 10$, which is the same as 1.8×10^1. We *add* exponents to *multiply* powers of 10. Thus, 18×10^{16} CO molecules equals $(1.8 \times 10^1) \times 10^{16}$ CO molecules, which equals 1.8×10^{17} CO molecules in that last breath you inhaled. If this use of exponents is coming at you a little too fast, consult Appendix 2.

It may sound surprising, but it is more accurate to round off the answer and report it as 2×10^{17} CO molecules. Certainly 1.8×10^{17} looks more accurate, but the data that went into our calculation were not very exact. The breath contains *about* 2×10^{22}

molecules, but it might be 1.6×10^{22}, 2.3×10^{22}, or some other number. We say that 2×10^{22} expresses a physically based property to "one significant figure." A **significant figure** is a digit that is included (or excluded) to correctly represent the accuracy with which an experimental quantity is known. Only one digit, the 2 from the value 2.3 is used, and so 2×10^{22} has only one significant figure. Accordingly, the number of molecules in the breath is closer to 2×10^{22} than to 1×10^{22} or to 3×10^{22}, but anything beyond this we cannot say with certainty.

Similarly, the concentration of carbon monoxide is known to only one significant figure, 9 ppm. That 2×9 equals 18 is certainly correct mathematically, but our question about CO is based on physical data. The answer, 1.8×10^{17} CO molecules, includes two significant figures: the 1 and the 8. Two significant figures imply a level of knowledge that is not justified. The accuracy of a calculation is limited by the *least accurate* piece of data that goes into it. In this case, both the concentration of CO and the number of molecules and atoms in the breath were known only to one significant figure (9 and 2, respectively). Thus two significant figures in the answer are unjustified. The common-sense rule is that you cannot improve the accuracy of experimental measurements by manipulations like multiplying and dividing. Therefore, the answer must also contain only one significant figure and is 2×10^{17}.

Your Turn 1.31   Ozone Molecules

The local news reports that today's ground-level ozone readings are at the unacceptable level, 0.12 ppm. How many ozone molecules do you inhale in each breath? Assume that one breath contains 2×10^{22} molecules and atoms.

Answer
Start with the number of molecules and atoms in a breath. If the ozone concentration is 0.12 ppm, this gives the ratio 0.12 O_3 molecules per 10^6 molecules and atoms in air.

$$\frac{2 \times 10^{22} \text{ molecules and atoms in air}}{1 \text{ breath of air}} \times \frac{0.12 \, O_3 \text{ molecules}}{1 \times 10^6 \text{ molecules and atoms in air}}$$

$$= 2.4 \times 10^{15} \, O_3 \text{ molecules / breath}$$

$$= 2 \times 10^{15} \, O_3 \text{ molecules / breath (to one significant figure)}$$

You may well question the significance of all of this talk about significant figures. It has been observed that "figures don't lie, but liars can figure." Numbers often lend an air of authenticity to newspaper or television stories, so popular press accounts are full of numbers. Some are meaningful; others are not. Informed citizens can discriminate one from the other. For example, the assertion that the concentration of carbon dioxide in the atmosphere is 390.2537 ppm should be taken with a rather large grain of sodium chloride (salt). Values such as 390 ppm or 390.6 ppm (three or four significant figures) better represent what we actually can measure; any assertion with seven significant figures simply is not valid.

Your Turn 1.32   CO Monitors

Carbon monoxide monitors are available for homes and businesses. Figure 1.24 shows a convenient handheld CO detector that reads 35 ppm.

a. Would it be more helpful to have a meter that read 35.0388217 ppm? Explain.
b. Would 35.0388217 ppm be more valid? Explain.

Answer
a. No, it wouldn't be more helpful. The issue is whether the concentration of CO exceeds a certain value, such as 9 ppm over an 8-hour period or 35 ppm over a 1-hour period. The extra decimal places are of no use.

Figure 1.24
Carbon monoxide meter showing 35 ppm.

Absence of evidence is not the same
as evidence of absence. The
substance may be present but in
undetectable amounts.

Recall that we started with the concentration of CO in an air sample of 9 ppm. Even so, the number of CO molecules in a breath is enormous, about 2×10^{17}. From these numbers, you can see that it is *impossible* to completely remove all the CO molecules from the air. "Zero pollutants" is an unattainable goal. At present, our most sensitive methods of chemical analysis are capable of detecting one target molecule out of a trillion. One part per trillion is analogous to moving 6 inches in the 93 million-mile trip to the Sun, a single second in 320 centuries, or a pinch of salt in 10,000 tons of potato chips.

Again, air always has trace levels of contaminants, ones that we cannot detect. A breath of air contains molecules of hundreds, perhaps thousands of different compounds, most in minuscule concentrations. Their origin could be either natural or related to human activity. As with all chemicals, "natural" is not necessarily good, and "human-made" is not necessarily bad. As you learned earlier, exposure and toxicity are what matter.

Your Turn 1.33   CO Molecules in Perspective

To help you comprehend the magnitude of the 2×10^{17} CO molecules in one breath, assume that they were equally distributed among the 6.9 billion (6.9×10^9) inhabitants of the Earth. Calculate each person's share of the 2×10^{17} CO molecules you just inhaled.

Answer

You are trying to distribute the huge number of molecules in a breath among all the human inhabitants of the Earth. This can be found by dividing the total number of CO molecules by the total number of humans:

$$\text{Each person's share is } \frac{2 \times 10^{17} \text{ CO molecules}}{6.9 \times 10^9 \text{ people}}$$

Thus, to one significant figure, each person's share is 3×10^7 (or 30,000,000) molecules of CO.

In addition to being extremely small, the molecules and atoms you breathe possess other remarkable characteristics. They are in constant motion. At room temperature and pressure, a nitrogen molecule travels at about 1000 feet per second and experiences approximately 400 billion collisions with other molecules in that time interval. Nevertheless, relatively speaking, the molecules are quite far apart. The actual volume of the extremely tiny molecules making up the air is only about 1/1000 of the total volume of the gas. If the particles in your half-liter breath were squeezed together, their volume would be about 0.5 mL, less than a quarter teaspoon. Sometimes people mistakenly think that air is empty space. It is 99.9% empty space, but the matter it contains is literally a matter of life and death!

Moreover, it is matter that we continuously exchange with other living things. The carbon dioxide we exhale is used by plants to make the food we eat. The oxygen that plants release is essential for our existence. Our lives are linked by the elusive medium of air. With every breath, we exchange millions of molecules with one another. As you read this, your lungs contain 4×10^{19} molecules that have been previously breathed by other human beings, and 6×10^8 molecules that have been breathed by some *particular* person, say Julius Caesar, Mahatma Gandhi, or Joan of Arc. In fact, the odds are very good that right now your lungs contain one molecule that was in Caesar's *last* breath. The consequences are breathtaking!

Skeptical Chemist 1.34   Caesar's Last Breath

We just claimed that your lungs currently contain one molecule that was in Caesar's last breath. That assertion is based on some assumptions and a calculation. Are these assumptions reasonable? We are not asking you to reproduce the calculation, but rather to identify some of the assumptions and arguments we might have used.

Hint: The calculation assumes that all of the molecules in Caesar's last breath have been uniformly distributed throughout the atmosphere.

Consider This 1.35   **Air Quality Today**

The addition of waste to our atmosphere has not occurred overnight. Rather, air pollution has been a growing concern since at least the time of the Industrial Revolution. Why have nations and the larger world community become more concerned about air quality? Identify at least four factors that have brought air quality to the attention of citizens and voters.

CONCLUSION

The air we breathe affects both our health and the health of the planet. Our atmosphere contains the essentials for life, including two elements (oxygen and nitrogen) and two compounds (water and carbon dioxide). Our very existence on this planet depends on having a large supply of relatively clean, unpolluted air.

But the air you breathe may be polluted with carbon monoxide, ozone, sulfur dioxide, and the oxides of nitrogen. Polluted air is more common in large cities, the very places where most people now live. Emergency room visits correlate with bad air quality. So do shortness of breath, scratchy throats, and stinging eyes. The pollutants that cause us harm are, for the most part, relatively simple chemical substances. They largely are produced as consequences of our dependence on coal for electricity production in power plants, gasoline in internal combustion engines, and the fuels we burn to heat and cook.

Over the past 30 years, governmental regulations, industry initiatives, and modern technology have reduced pollutant levels. Both catalytic converters on cars and emissions controls on smokestacks have been important players. But it makes more sense not to generate "people fumes" in the first place. Here is where green chemistry plays an important role. By designing new processes that do not produce air pollutants, we do not later have to clean them up.

Indoors or out, the oxygen-laden air we breathe is very close to the surface of the Earth. However, the Earth's atmosphere extends upward for considerable distance and contains other gases that also are essential for life on this planet. Chapters 2 and 3 will describe two of these: stratospheric ozone and carbon dioxide. We will see that our human footprints and "air prints" on planet Earth connect in surprising ways to both of these gases.

> Recall the first green chemistry principle listed on the inside cover: "It is better to prevent waste than to treat or clean up waste after it is formed."

Chapter Summary

The numbers in parentheses indicate the sections in which the topics are introduced. Having studied this chapter, you should be able to:

- Explain the connection between your health and what you breathe (entire chapter)
- Describe air in terms of its major components, their relative amounts, and the local and regional variations in the composition of air (1.1, 1.5)
- List the major air pollutants and describe the health effects of each (entire chapter)
- Compare and contrast indoor and outdoor air in terms of which pollutants are likely to be present and their sources (1.3, 1.13)
- Interpret local air quality data, including why air quality standards are set separately for each pollutant (1.3)
- Evaluate the risks and benefits of a particular activity (1.3)
- Discuss the green chemistry initiative and why it makes sense to prevent pollution rather than to clean it up afterward (1.5)
- Relate these terms: matter, pure substances, mixtures, elements, compounds, metals, nonmetals (1.6)
- Discuss the features of the periodic table, including the groups it contains (1.6)

- Explain the difference between atoms and molecules, giving examples of each (1.7)
- Name chemical elements and compounds that relate to air quality (1.7)
- Write and interpret chemical formulas that relate to air quality (1.8)
- Balance and interpret chemical equations that relate to air quality (1.9–1.10)
- Understand oxygen's role in combustion, including how hydrocarbons burn to form carbon dioxide, carbon monoxide, and soot (1.9–1.10)

- Describe how ozone forms, including how sunlight, NO, NO_2, and VOCs are involved (1.12)
- Identify the sources and nature of indoor air pollution (1.13)
- Explain the unreasonableness of "pollution-free" air (1.14)
- Use scientific notation and significant figures in performing basic calculations (1.4, 1.14)
- Apply what you know about air pollution to ways of living that result in cleaner air (entire chapter)

Questions

The end-of-chapter questions are grouped in three ways:

- **Emphasizing Essentials** questions give you the opportunity to practice fundamental skills. They are similar to the *Your Turn* exercises in the chapter.
- **Concentrating on Concepts** questions are more difficult and may relate to societal issues. They are similar to the *Consider This* activities in the chapter.
- **Exploring Extensions** questions challenge you to go beyond the information presented in the text.

Appendix 5 contains the answers to questions with numbers in **blue**.

 Questions marked with this icon require the resources of the Internet.

 Questions marked with this icon relate to green chemistry.

Emphasizing Essentials

1. a. Calculate the volume of air in liters that you might inhale (and exhale) in an 8-hour working day. Assume that each breath has a volume of about 0.5 L, and that you are breathing 15 times a minute.

    b. From this calculation, you can see that breathing exposes you to large volumes of air. Name five things that you can do to improve the quality of the air that you and others breathe.

2. Our atmosphere can be characterized both as a thin veil that supports life and as a few vertical miles of chemicals. Explain what makes each description accurate. Also state which feature(s) of our atmosphere each description emphasizes and which ones it omits.

3. These gases are found in the troposphere: Rn, CO_2, CO, O_2, Ar, N_2.

    a. Rank them in order of their abundance in the troposphere.

    b. For which of these gases is it convenient to express its concentration in parts per million?

    c. Which of these gases is/are currently regulated as an air pollutant where you live?

    d. Which of these gases are found in Group 8A of the periodic table, the noble gases?

4. Give three examples of particulate matter found in air. Explain the difference between $PM_{2.5}$ and PM_{10} in terms of size and health effects.

5. Radon is one of the noble gases, found in Group 8A on the periodic table. Which properties does it share with the other inert gases? In which way is it distinctly different?

6. a. The concentration of argon in air is approximately 0.9%. Express this value in ppm.

    b. The air exhaled from the lungs of a smoker has a concentration of 20–50 ppm CO. In contrast, air exhaled by nonsmokers is 0–2 ppm CO. Express each concentration as a percent.

    c. In a tropical rain forest, the water vapor concentration may reach 50,000 ppm. Express this as a percent.

    d. In the dry polar regions, water vapor may be a mere 10 ppm. Express this as a percent.

7. In these diagrams, two different types of atoms are represented by color and size. Characterize each sample as an element, a compound, or a mixture. Explain your reasoning.

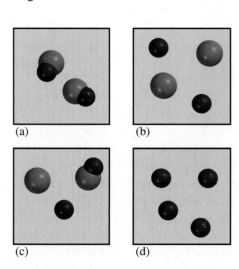

8. Consider this representation of the reaction between nitrogen and hydrogen to form ammonia (NH_3).

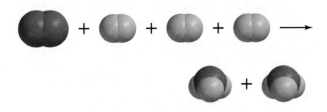

   a. Are the masses of reactants and products the same? Explain.

   b. Are the numbers of molecules of reactants and of products the same? Explain.

   c. Are the total number of atoms in the reactants and the total number of atoms in the products the same? Explain.

9. Express each of these numbers in scientific notation.

   a. 1500 m, the distance of a foot race

   b. 0.0000000000958 m, the distance between O and H atoms in a water molecule

   c. 0.0000075 m, the diameter of a red blood cell

   d. 150,000 mg of CO, the approximate amount breathed daily

10. Write each of these values as a "regular" number.

    a. 8.5×10^4 g, the mass of air in an average room

    b. 1.0×10^7 gallons, the volume of crude oil spilled by the *Exxon Valdez*

    c. 5.0×10^{-3}%, the concentration of CO in the air on a city street

    d. 1×10^{-5} g, the recommended daily allowance of vitamin D

11. The threshold for detecting NO_2 by smell is 0.00022 g/m^3 of air.

    a. Express this value in scientific notation.

    b. Would you expect a similar value for the threshold of CO?

    c. Name another pollutant that has a sharp, easily detected odor.

12. Wildfires occur all across our planet. The one shown here was photographed on a commercial flight north of Phoenix, AZ.

   a. What combustion products would you expect from the burning of wood?

   b. This fire is emitting at least three pollutants. Which are visible and which are not?

13. Consider this portion of the periodic table and the groups shaded on it.

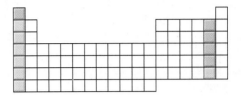

   a. What is the group number for each shaded region?

   b. Name the elements that make up each group.

   c. Give a general characteristic of the elements in each of these groups.

14. Consider the following blank periodic table.

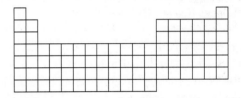

   a. Shade the region of the periodic table in which metals are found.

   b. Common metals include iron, magnesium, aluminum, sodium, potassium, and silver. Give the chemical symbol for each.

   c. Give the name and chemical symbol for five nonmetals (elements that are not in your shaded region).

15. Classify each of these substances as an element, a compound, or a mixture.

    a. a sample of "laughing gas" (dinitrogen monoxide, also called nitrous oxide)

    b. steam coming from a pan of boiling water

    c. a bar of deodorant soap

    d. a sample of copper

    e. a cup of mayonnaise

    f. the helium filling a balloon

16. These gases are found in the atmosphere in small amounts: CH_4, SO_2 and O_3.

    a. What information does each chemical formula convey about the number and types of atoms present?

    b. Give the names of these gases.

17. Hydrocarbons are important fuels that we burn for many different reasons.

    a. What is a hydrocarbon?

    b. Rank these hydrocarbons by the number of carbons they contain: propane, methane, butane, octane, ethane.

c. We suggested "mother eats peanut butter" as a memory aid for the names of the first four hydrocarbons. Propose a new one that includes *pent-*, the prefix that indicates five carbon atoms.

18. Write balanced chemical equations to represent these reactions. *Hint:* Nitrogen and oxygen are both diatomic molecules.

    a. Nitrogen reacts with oxygen to form nitrogen monoxide.

    b. Ozone decomposes into oxygen and atomic oxygen (O).

    c. Sulfur reacts with oxygen to form sulfur trioxide.

19. Analogous to equation 1.8, draw models to represent the chemical equations from the previous question.

20. These questions relate to combustion of hydrocarbons.

    a. LPG (liquid petroleum gas) is mostly propane, C_3H_8. Balance this equation.

    $$C_3H_8(g) + O_2(g) \longrightarrow CO_2(g) + H_2O(g)$$

    b. Cigarette lighters burn butane, C_4H_{10}. Write a balanced equation, assuming complete combustion, that is, plenty of oxygen.

    c. With a limited supply of oxygen, both propane and butane can burn incompletely to form carbon monoxide. Write balanced equations for both reactions.

21. Balance these equations in which ethane (C_2H_4) burns in oxygen.

    a. $C_2H_4(g) + O_2(g) \longrightarrow C(s) + H_2O(g)$

    b. $C_2H_4(g) + O_2(g) \longrightarrow CO(g) + H_2O(g)$

    c. $C_2H_4(g) + O_2(g) \longrightarrow CO_2(g) + H_2O(g)$

22. Examine the coefficients for oxygen in the balanced equations from the previous question. Explain why they vary, depending on whether C, CO, or CO_2 is formed.

23. Count the atoms on both sides of the arrow to demonstrate that these equations are balanced.

    a. $2\, C_3H_8(g) + 7\, O_2(g) \longrightarrow 6\, CO(g) + 8\, H_2O(l)$

    b. $2\, C_8H_{18}(g) + 25\, O_2(g) \longrightarrow 16\, CO_2(g) + 18\, H_2O(l)$

24. Platinum, palladium, and rhodium are used in the catalytic converters of cars.

    a. Give the chemical symbol for each metal.

    b. Locate each metal on the periodic table.

    c. What can you infer about the properties of these metals, given that they are useful in this application?

25. Nail polish remover containing acetone was spilled in a room 6 m × 5 m × 3 m. Measurements indicated that 3600 mg of acetone evaporated. Calculate the acetone concentration in micrograms per cubic meter.

Concentrating on Concepts

26. "Air prints" were mentioned in the opening activity of this chapter. Examine these two photographs. The first is a beautiful view from a lodging on the Hilo coast of the island Hawaii. The second shows the tarmac on a hazy summer day at the Narita International Airport in Tokyo. List three ways in which each photo shows the air print of humans. *Hint:* Some may not be visible but rather implied by the photograph.

27. The AIRNOW website (EPA) states that "Quality of air means quality of life." Demonstrate the wisdom of this statement for two air pollutants of your choice.

28. In Consider This 1.2, you calculated the volume of air exhaled in a day. How does this volume compare with the volume of air in your chemistry classroom? Show your calculations. *Hint:* Think ahead about the most convenient unit to use for measuring or estimating the dimensions of your classroom.

29. According to Table 1.1, the percentage of carbon dioxide in inhaled air is *lower* than it is in exhaled air, but the percentage of oxygen in inhaled air is *higher* than in exhaled air. How can you account for these relationships?

30. Cars don't inhale and exhale like humans do. Nonetheless, the air that goes into a car is different from what comes out. In Your Turn 1.19 you listed what comes out of a tailpipe. Now comment on the *differences* between the air that goes into the car engine and what comes out the tailpipe. For which chemicals have the concentrations noticeably increased or decreased?

31. A headline from the *Anchorage Daily News* in Alaska (January 17, 2008): "Family in car overcome by carbon monoxide. Fire department saves five after slide into snow bank."

a. If your car is in a snow bank and the engine is running, CO may accumulate inside the car. Normally, however, CO does not accumulate in the car. Explain.

b. Why didn't the occupants detect the CO?

32. A headline from the *Pioneer Press* in St. Paul, Minnesota (January 8, 2008): "Man dies after exposure to gas; carbon monoxide sickens five others."

a. Name two possible sources of CO inside a home.

b. The level measured was 4700 ppm. Express this value as a percent.

c. How does this level compare with the U.S. ambient air quality standards set by the EPA?

d. Name three symptoms that the survivors most likely experienced.

e. Where in a home should you install CO detectors? Note: Adjacent to a furnace is *not* usually recommended.

33. In Consider This 1.4, you considered how life on Earth would change if the concentration of oxygen were twice as high. Now consider how life would change if the concentration of O_2 were cut in half. Give two examples of things that what would be affected.

34. Explain why CO is named the "silent killer." Select two other pollutants for which this name would not apply and explain why not.

35. Undiluted cigarette smoke may contain 2–3% carbon monoxide.

a. How many parts per million is this?

b. How does this value compare with the National Ambient Air Quality Standards for CO in both a 1-hour and an 8-hour period?

c. Propose a reason why smokers do not die from carbon monoxide poisoning.

36. For many states, the ozone season runs from May 1 to October 1. Why are ozone levels typically not reported in the winter months?

37. The EPA characterizes ozone as "good up high, bad nearby." Explain.

38. Here are ozone air quality data for London, Ontario, from June 8–20, 2005.

a. In general, which groups of people are the most sensitive to ozone?

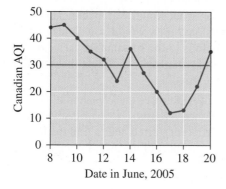

Date in June, 2005

b. Air rated above 30 is hazardous for some or all groups by the Canadian Environmental Assessment Agency. For the data shown, how many days was the air hazardous?

c. Ozone levels drop off sharply at night. Explain why.

d. During the daytime, the ozone dropped off sharply after June 10. Propose two different reasons that could account for this observation.

e. Data for the month of December is not shown. Would you expect the ozone levels to be higher or lower than those in June? Explain.

39. Here are air quality data for April 1–10, 2005 in Beijing. The primary pollutant was PM_{10}.

a. In general, which groups of people are the most sensitive to particulate matter?

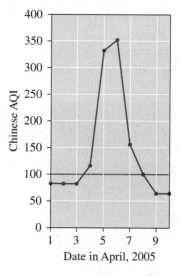

Date in April, 2005

b. Air rated above 100 is hazardous for some or all groups by the National Environmental Monitoring Centre in China. For the data shown, how many days was the air hazardous?

c. The levels of PM do not necessarily drop off at night the way they do for ozone. Explain.

d. The levels of particulate matter increased sharply on April 5. Propose two different reasons that could account for this observation.

40. Prior to 1990, diesel fuel could contain as much as 2% sulfur. New regulations have changed this, and today most diesel fuel is ultra-low sulfur diesel (ULSD) containing a maximum of 15 ppm sulfur.

a. Express 15 ppm as a percent. Likewise, express 2% in terms of ppm. How many times lower is the ULSD than the older formulation of diesel fuel?

b. Write a chemical equation that shows how burning diesel fuel containing sulfur contributes to air pollution.

c. Diesel fuel contains the hydrocarbon $C_{12}H_{26}$. Write a chemical equation that shows how burning diesel adds carbon dioxide to the atmosphere.

d. Comment on burning diesel fuel as a sustainable practice, both in terms of how things have improved and in terms of where they still need to go.

41. A certain city has an ozone reading of 0.13 ppm for 1 hour, and the permissible limit is 0.12 for that time. You have the choice of reporting that the city has exceeded the ozone limit by 0.01 ppm or saying that it has exceeded the limit by 8%. Compare these two methods of reporting.

42. Here is the air quality outlook for the United States for early August, 2005. *Source:* www.airnow.gov

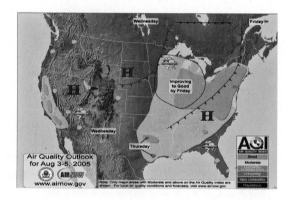

a. This forecast is typical in that most of the ozone pollution is expected in California, Denver, Texas, the Midwest, and the East Coast. Why these regions of the country?

b. Phoenix typically has high ozone levels in the summer, but not on this particular day. Offer a possible explanation.

c. The air quality is forecasted to improve in Illinois and Wisconsin. Offer a possible explanation.

d. Why are inland areas in California, such as the Sacramento Valley which is shown as unhealthy for sensitive groups, likely to have worse air quality than the California coast?

43. Look up the ozone air quality data for two states, one in the Sun Belt and one not. Account for any difference you find. *Hint:* Use *State of the Air*, a website posted by the American Lung Association.

44. At certain times of the year, inhabitants of the beautiful city of Santiago, Chile, breathe some of the worst air on the planet.

a. Driving private cars has been severely restricted in Santiago. How specifically does this improve air quality?

b. Although the population of Santiago is comparable to that in other cities, its air quality is much worse. Suggest geographical features that might be responsible.

45. a. Explain why jogging outdoors (as opposed to sitting outdoors) increases your exposure to pollutants.

b. Jogging indoors at home can decrease your exposure to some pollutants, but may increase your exposure to others. Explain.

46. Consumers now can purchase paints that emit only low amounts of VOCs. However, these consumers may not know why it matters to purchase this paint.

a. What would you print on the label of a paint can to make the point that a low-VOC paint is a good idea?

b. We apply paint to many outdoor surfaces, such as buildings, bridges, and fence posts. Comment on the environmental effects of the VOCs that these paints emit.

c. Explain how producing low-VOC paint meets the Triple Bottom Line.

47. One can purchase a carbon monoxide monitor that immediately sounds an alarm if the concentration of CO reaches a threshold. In contrast, most radon detection systems sample the air over a period of time before an alarm sounds. Why the difference?

48. Select a profession of your choice, possibly the one you intend to pursue. Name at least one way that a person in this profession could have a positive effect on air quality.

Exploring Extensions

49. "Air pollution is a diffuse problem, the shared fault of many emitters. It is a classic example of the tragedy of the commons." *Source: Introduction to Air in California*, by David Carle, 2006. Explain the phrase "tragedy of the commons" and how air pollution is a classic example.

50. Mercury, another serious air pollutant, is not described in this chapter. If you were a textbook author, what would you include about mercury emissions? How would you connect mercury emissions to the sustainable use of resources? Write several paragraphs in a style that would match that of this textbook.

51. The EPA oversees the Presidential Green Chemistry Challenge Awards. Use the EPA website to find when the program started and to find the list of the most recent winners of the award. Pick one winner and

summarize in your own words the green chemistry advance that merited the award.

52. Recreational scuba divers usually use compressed air that has the same composition as normal air. A mixture being used is called Nitrox. What is its composition, and why is it being used?

53. Here are two scanning electron micrograph images of particulate matter, courtesy of the National Science Foundation and researchers at Arizona State University. The first is of a soil particle and the second of a rubber particle, and each is about 10 μm in diameter.

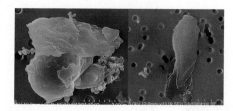

   a. What is a likely source of the rubber particle? Name two other substances that might contribute PM to the air.

   b. The soil particle is composed mainly of silicon and oxygen. What other elements are commonly present in the rocks and minerals in Earth's crust?

   c. What about these photographs suggests that these particles would inflame your blood vessels?

54. Ultrafine particles have diameters less than 0.1 μm. In terms of their sources and health effects, how do these particles compare with $PM_{2.5}$ and PM_{10}? Use the Internet to locate the most up-to-date information.

55. Most lawnmowers do not have catalytic converters (at least as this book went to press). What comes out of the tailpipe of a gasoline-powered lawn mower? Why has adding a catalytic converter been so controversial? What are the immediate benefits to curbing these emissions, as well as the longer term ones?

56. Consider this graph that shows the effects of carbon monoxide inhalation on humans.

   a. Both the amount of exposure and the duration of exposure have an effect on CO toxicity in humans. Use the graph to explain why.

   b. Use the information in this graph to prepare a statement to include with a home carbon monoxide detection kit about the health hazards of carbon monoxide gas.

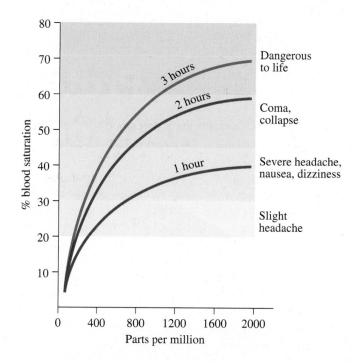

57. Consider This 1.4 asks you to consider how our world would be different if the oxygen content of the atmosphere were doubled. Develop your answer into an essay. Title your essay "An Hour in the Life of . . ." and describe how things would be different for a person of your choice. If an hour is too short to make your point, substitute "A Morning . . ." or "A Day . . .".

58. You may have admired the beauty of hardwood floors. Polyurethane is the finish of choice for floors because it is more durable than varnishes and shellacs. Until recently, polyurethane was always an oil-based paint. But recently, the Bayer Corporation developed a water-based polyurethane that reduces the amount of VOCs by 50–90%. In 2000, Bayer was awarded a Presidential Green Chemistry Challenge award for this development. Prepare a summary of this work. Also check stores to see if any water-based polyurethanes are available in your area.

59. Composite wood is made by gluing smaller pieces of wood (often waste scraps of wood) together. Examples include plywood, particle board, and fiber board.

   a. Many glues release formaldehyde, a volatile compound. What are its hazards?

   b. Professor Kaichang Li of Oregon State University and Columbia Forest Products developed a new soy-based adhesive glue, winning a Presidential Green Chemistry Challenge Award in 2007. Prepare a summary of his accomplishments.

2 Protecting the Ozone Layer

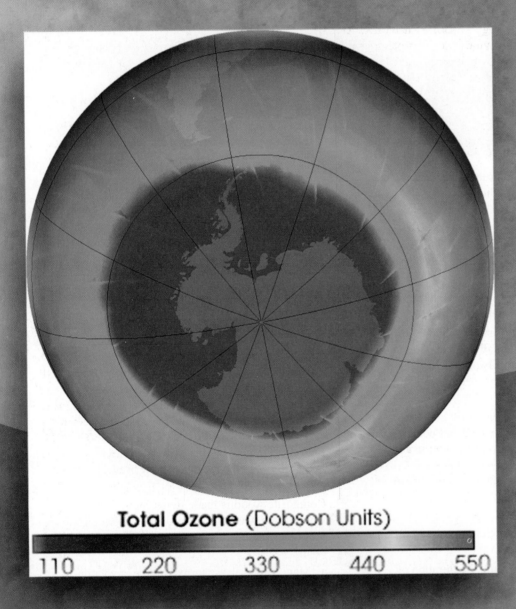

Total Ozone (Dobson Units)

110   220   330   440   550

The stratospheric ozone "hole" in 2009 (the purple and dark blue areas) over Antarctica reached a maximum area of 24.0 million km² on September 17, 2009. The record 2000 hole area peaked at 28.4 million km².

Source: **NASA.**

Good up high; bad nearby. To understand ozone, think *location*. Down here in the troposphere, where we live, ozone is a pollutant that forms in the presence of sunlight from other pollutants in our atmosphere. Once the Sun goes down, the generation of ozone ceases. Any ozone that is present quickly reacts with something else, and the concentrations drop in the twilight hours. Were the Sun not to rise again, we wouldn't have to worry about breathing ozone (but we'd face a few other problems).

But up in the stratosphere, the story of ozone is altogether different. Unlike ground-level ozone, the ozone up high is formed naturally. Also unlike ground-level ozone, stratospheric ozone plays a vital role in protecting us from damaging solar radiation. One might say it acts like the Earth's sunglasses.

In the 1970s, chemists discovered that certain chemicals could make their way into the upper atmosphere and partially destroy the protective ozone found there. Ever since, scientists, policy makers, and indeed concerned citizens worldwide have participated in efforts to control and reverse ozone destruction. Somewhat surprisingly, the most severe depletion has been over Antarctica, and the yearly images of the ozone hole have become some of the most widely recognized scientific graphics. Later in this chapter, you will have the opportunity to examine past trends and update the Antarctic ozone-hole story.

You may be wondering what this story has to do with you, because last time we checked, not many college students were living in Antarctica. Even though ozone depletion was first documented in that faraway region, it also has been observed to a lesser extent in many other locations on Earth, including over North America. Where you live and the season of the year both influence the amount of stratospheric ozone overhead and how well it provides its protective effects. Take a look at some of the important data for yourself.

Consider This 2.1   Ozone Levels Above You

As you read this, an instrument onboard a satellite is measuring ozone levels in the stratosphere. The textbook's website provides a link to the data.

a. What is the current total column ozone (in Dobson units) at your location? Request data for three different years and compute an average.

b. Now retrieve data for Antarctica for the same dates. How does the ozone measurement where you live compare? *Note:* If you are making the comparison in September or October, you are likely to see the biggest difference.

One DU (Dobson unit) corresponds to about one ozone molecule for every billion molecules and atoms present in air.

What caused this stratospheric ozone depletion? Why is this depletion so serious? Look for the answers to these questions and more in the sections that follow.

As we explore the topic of ozone depletion, we also highlight the **precautionary principle.** This principle stresses the wisdom of acting, even in the absence of complete scientific data, before the adverse affects on human health or the environment become significant or irrevocable. As you will learn, the world community did act. The wisdom of the collective actions is evident, as the measures taken to protect the ozone layer appear to be working. Even so, another warning bell will sound in the final section of this chapter. Listen for it in regards to global climate change—the topic of Chapter 3.

2.1 | Ozone: What and Where Is It?

If you have ever been near a sparking electric motor or in a severe lightning storm, most likely you have smelled ozone. Its odor is unmistakable but difficult to describe. Some compare it to that of chlorine gas; others think the odor reminds them of newly mown grass. It is possible for humans to detect concentrations as low as 10 parts per billion (ppb), that is, 10 molecules out of 1 billion. Appropriately enough, the name *ozone* comes from a Greek word meaning "to smell."

The ozone and oxygen molecules differ by only one atom.

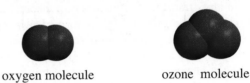

oxygen molecule   ozone molecule

As we will see, this difference in molecular structure translates to significant differences in chemical properties. One difference is that ozone is far more chemically reactive than O_2. As you will learn in Chapter 5, ozone can be used to kill microorganisms in water. Ozone is also used to bleach paper pulp and fabrics. At one time, ozone was even advocated as a deodorizer for air. This use only makes sense, however, if nobody breathes the air during the deodorizing process. In contrast, you safely can (and must) breathe oxygen day in and day out. Although oxygen is still quite chemically reactive, it is not reactive enough to bleach paper or purify water.

Ozone forms both naturally and as a result of human activities. Given the high reactivity of ozone, it does not usually persist very long. If it were not for the fact that ozone is formed anew on our planet, you would not find it except as a curiosity in the chemistry lab.

Ozone can be formed from oxygen, but the process requires energy. A simple chemical equation summarizes the process:

$$\text{energy} + 3\,O_2 \longrightarrow 2\,O_3 \qquad \text{[2.1]}$$

This chemical equation helps to explain why ozone is formed from oxygen in the presence of an electrical discharge, whether from an electric spark or lightning.

Ozone is reasonably rare in the troposphere, the region of the atmosphere closest to the Earth's surface (Figure 2.1). Only somewhere between 20 and 100 ozone molecules typically occur for each billion molecules and atoms that make up the air. Unhealthy concentrations are sometimes found near Earth's surface, the result of chemical reactions that produce it as a component of photochemical smog. But these concentrations are very low. As was noted in the previous chapter, the air quality standard in the United States for ground-level ozone was set at 0.075 ppm for an 8-hr average as of 2009.

But what is detrimental in one region of the atmosphere, even at very low concentrations, can be essential in another. The stratosphere, at an altitude of 15 to 30 km, is where ozone does most of its filtering of some types of ultraviolet light from the Sun. The concentration of ozone in this region is several orders of magnitude greater than in the troposphere, but still very low. As an upper limit, about 12,000 ozone molecules are present for every billion molecules and atoms of gases that make up the atmosphere at this level.

Most of the ozone on our planet, about 90% of the total, is found in the stratosphere. The term **ozone layer** refers to a designated region in the stratosphere of maximum ozone concentration. Figure 2.2 shows the ozone concentrations in the troposphere and stratosphere.

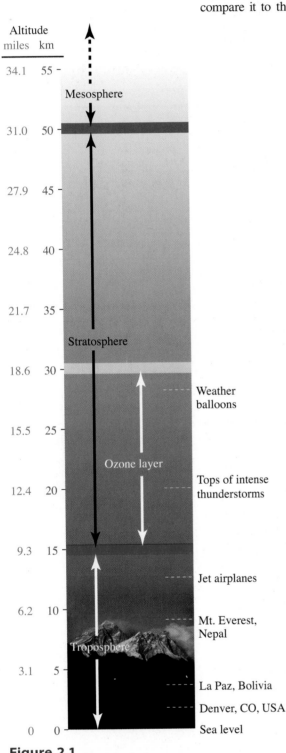

Figure 2.1

The regions of the atmosphere.

0.075 parts per *million* is equivalent to 75 ozone molecules for every *billion* molecules and atoms found in air.

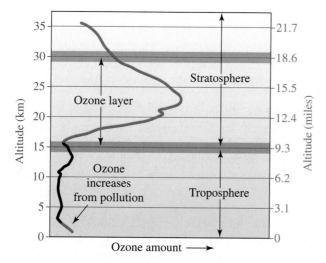

Figure 2.2

The ozone layer is a region of maximum ozone concentration in the stratosphere.

Source: Global Ozone Research and Monitoring Project Report No. 44, 1998. Reprinted with permission of World Meteorological Organization.

Your Turn 2.2   The Ozone Layer

Use Figure 2.2 and values given in the text to answer these questions.

a. What is the approximate altitude of maximum ozone concentration?
b. What is the maximum number of ozone molecules per billion molecules and atoms of all types found in the stratosphere?
c. What is the maximum number of ozone molecules per billion molecules and atoms of all types found in ambient air just meeting the EPA limit for an 8-hr average?

Answer
a. About 23 km (14 miles).

Because the range of altitudes is so broad, the concept of an "ozone layer" can be a little misleading. No thick, fluffy blanket of ozone exists in the stratosphere. At the altitudes of the maximum ozone concentration, the atmosphere is very thin, so the total amount of ozone is surprisingly small. If all the O_3 in the atmosphere could be isolated and brought to the average pressure and temperature at Earth's surface (1.0 atm and 15 °C), the resulting layer of gas would have a thickness of less than 0.5 cm, or about 0.25 inch. On a global scale, this is a minute amount of matter. Yet, this ozone shield protects the surface of the Earth and its inhabitants from the harmful effects of ultraviolet radiation.

The total amount of ozone in a vertical column of air of known volume can be determined fairly easily. The determination can be done from Earth's surface by measuring the amount of UV radiation reaching a detector; the lower the intensity of the radiation, the greater the amount of ozone in the column. G. M. B. Dobson, a scientist at Oxford University, pioneered this measurement method. In 1920, he invented the first instrument to quantitatively measure the concentration of ozone in a column of the Earth's atmosphere. Therefore, it is fitting that the unit of such measurements is named for him.

One Dobson unit (DU) is equivalent to about 3×10^{16} O_3 molecules in a vertical column of air with a cross section of 1 cm^2.

Consider This 2.3   Interpreting Ozone Values

A classmate used the National Aeronautics and Space Administration (NASA) website to check the atmospheric ozone above her hometown in Ohio. She found the readings to be 417 DU (Dobson units) on April 10 and 386 DU on May 10. The student, reassured by these findings, concluded there had been an improvement in protection from damaging UV radiation. Do you agree? Explain.

NASA's EOS *Aura* mission also is collecting data about tropospheric air quality (Chapter 1) and key climate parameters (Chapter 3).

Scientists continue to measure and evaluate ozone levels using ground observations, weather balloons, and high-flying aircraft. However, since the 1970s, measurements of total column ozone have also been made from the top of the atmosphere. Satellite-mounted detectors record the intensity of the UV radiation scattered by the upper atmosphere. The results are then related to the amount of O_3 present.

The Space Shuttle, *Columbia,* tested a new approach for monitoring ozone. Rather than looking directly downward toward Earth from a satellite, the equipment aboard the Shuttle looked sideways through the thin blue haze (see photo) that rises above the denser regions of the troposphere and follows the curve of the Earth. This region is known as the Earth's "limb" and is responsible for the name of this new technique, "limb viewing." Reliable information can be gathered at each level of the atmosphere, particularly allowing scientists to better understand chemistry taking place in the lower regions of the stratosphere. In January 2004, NASA launched a new mission called Earth Observing System (EOS) *Aura* that also uses a variety of viewing geometries, including limb viewing, to gather additional data about changes in Earth's stratospheric ozone layer.

The process by which ozone protects us from damaging solar radiation involves the interaction of matter and energy from the Sun. To help you to understand this, we turn first to a submicroscopic view of matter.

2.2 | Atomic Structure and Periodicity

Both the O_2 and O_3 molecules are composed of oxygen atoms. What do we know about these atoms? During the 20th century, scientists probed the inner workings of the atom. The physicists were almost too successful in their endeavors, finding more than 200 subatomic particles. Fortunately, most chemical behavior can be explained with only three.

Every atom has a **nucleus,** a minuscule and highly dense center of an atom composed of protons and neutrons. **Protons** are positively charged particles, and **neutrons** are electrically neutral particles. Both have almost exactly the same mass. Indeed, the protons and neutrons in the nucleus account for almost all the mass of an atom. Outside the nucleus are the electrons that define the boundary of the atom. An **electron** has a mass much smaller than that of a proton or neutron and a negative electric charge equal in magnitude to that of a proton, but opposite in sign. Therefore, in any electrically neutral atom, the number of electrons equals the number of protons. The properties of these particles are summarized in Table 2.1.

The number of protons in the nucleus determines the identity of the atom. The term **atomic number** refers to the number of protons in the nucleus of an atom. For example, all hydrogen (H) nuclei contain 1 proton; hydrogen has an atomic number of 1. Similarly, all helium (He) nuclei contain 2 protons and have an atomic number of 2. With each successive element in the periodic table, the atomic number increases. For example, the nucleus of element #92 (U, uranium) contains 92 protons.

Table 2.1	Properties of Subatomic Particles		
Particle	Relative Charge	Relative Mass	Actual Mass, kg
proton	+1	1	1.67×10^{-27}
neutron	0	1	1.67×10^{-27}
electron	−1	0*	9.11×10^{-31}

*This value is zero when rounded to the nearest whole number. The electron does indeed have mass, though very small!

Your Turn 2.4   Atomic Bookkeeping

Using the periodic table as a guide, specify the number of protons and electrons in a neutral atom of each of these elements.

a. carbon (C)   **b.** calcium (Ca)   **c.** chlorine (Cl)   **d.** chromium (Cr)

Answers
a. 6 protons, 6 electrons
b. 20 protons, 20 electrons

We wish we could show you a picture of a typical atom. However, atoms defy easy representation, and depictions in textbooks are at best oversimplifications. Electrons are sometimes pictured as moving in "orbits" about the nucleus, but the modern view of electrons is a good deal more complicated and abstract. For one thing, the relative size of the nucleus and the atom creates serious problems for the illustrator. If the nucleus of a hydrogen atom were the size of a period on this page, the atom's single electron would most likely be found at a distance of about 10 feet from that period. Moreover, electrons do not follow specific circular orbits. In spite of what you may have learned early in your education, an atom is really not very much like a miniature solar system. Rather, the distribution of electrons in an atom is described best using concepts of probability and statistics.

If this sounds rather vague to you, you are not alone. Common sense and our experience of ordinary things are not particularly helpful in our efforts to visualize the interior of an atom. Instead, we are forced to resort to mathematics and metaphors. The mathematics required (a field called quantum mechanics) can be formidable. Chemistry majors do not normally encounter this field until rather late in their undergraduate study. Although we cannot fully share with you the strange beauties of the peculiar quantum world of the atom, we can provide some useful generalizations.

The periodic table lists elements in order of increasing atomic number. The table also has elements arranged so that those with similar chemical properties fall in the same column (group). For example, lithium (Li, atomic number 3), sodium (Na, 11), potassium (K, 19), rubidium (Rb, 37), and cesium (Cs, 55) all fall in the same column and all are highly reactive metals. What fundamental feature accounts for this?

Today we know that the chemical properties of elements are the consequence of the distribution of electrons in the atoms of these elements. When chemical properties repeat themselves, this signals a repeat in electronic arrangement. As we will see, the electrons farthest from the nucleus are the main determinant of chemical properties.

Both experiment and calculation demonstrate that the electrons are arranged in certain energy levels about the nucleus. The electrons in the innermost level are the most strongly attracted by the positively charged protons in the nucleus. The greater the distance between an electron and the nucleus, the weaker the attraction between them. We say that the more distant electron is in a higher energy level, which means that the electron itself possesses more potential energy.

Each energy level has a maximum number of electrons that can be accommodated and is particularly stable when fully occupied. The innermost level, corresponding to the lowest energy, can hold only two electrons. The second level has a maximum capacity of eight, and the higher levels are also particularly stable when they contain eight electrons.

Table 2.2 shows some important information about electrons in neutral atoms of the first 18 elements. The total number of electrons in each atom is printed in blue and the number of outer electrons is printed in maroon. **Outer (valence) electrons** are found in the highest energy level and help to account for many of the observed trends in chemical properties. Observe that the group designation (1A, 2A, etc.) corresponds to the number of *outer* electrons for the A group elements, one of the most useful organizing features of the periodic table.

What we call "levels" used to be referred to as "shells," using the earlier solar system model of atomic structure.

Look for more about potential energy in Chapter 4.

The periodic table also contains B group elements. Table 2.2 does not show these, as these elements start with the fourth row.

The smallest noble gas, helium, has 2 valence electrons rather than 8.

Lithium (stored in oil)

Sodium (removed from oil, being cut)

Potassium (in a sealed glass tube)

Rubidium (in a sealed glass tube)

Figure 2.3
Selected Group 1A elements.

The group number does not *necessarily* indicate the number of outer electrons for elements in B groups.

Table 2.2	Atoms of the First 18 Elements (Total and Outer Electrons)							
Group 1A	**2A**	**3A**	**4A**	**5A**	**6A**	**7A**	**8A**
1							2
H							He
1							2
3	4	5	6	7	8	9	10
Li	Be	B	C	N	O	F	Ne
1	2	3	4	5	6	7	8
11	12	13	14	15	16	17	18
Na	Mg	Al	Si	P	S	Cl	Ar
1	2	3	4	5	6	7	8

• *Above* the atomic symbol is the atomic number, the number of protons in the nucleus. For a neutral atom, this also is the number of electrons.
• *Below* the atomic symbol is the number of **outer** electrons in a neutral atom.

Take another look at the first column in Table 2.2. Lithium and sodium atoms both have one *outer* electron per atom, despite having different *total* numbers of electrons. This fact explains much of the chemistry that these two alkali metals have in common. It places them in Group 1A of the periodic table (the 1 indicates one outer electron). Moreover, we would be correct in assuming that potassium, rubidium, and the other elements in column 1A of the periodic table also have a single outer electron in each of their atoms. They are all metals that react readily with oxygen, water, and a wide range of other chemicals. Figure 2.3 shows photographs of some Group 1A elements.

The periodic table is a useful guide to electron arrangement. In the families (another name for groups) of elements marked "A," the number that heads the column indicates the number of outer electrons in each atom. We introduced the terms *alkali metal, alkaline earth metal, halogen,* and *noble gas* in Chapter 1. We now connect these terms with their group number.

- Alkali metals (Group 1A)—highly reactive metals with one outer electron
- Alkaline earth metals (Group 2A)—reactive metals with two outer electrons
- Halogens (Group 7A)—reactive nonmetals with 7 outer electrons
- Noble gases (Group 8A)—unreactive nonmetals with 8 outer electrons

Your Turn 2.5   Outer Electrons

Using the periodic table as a guide, specify the group number and number of outer electrons in a neutral atom of each element.

a. sulfur (S)   **b.** silicon (Si)   **c.** nitrogen (N)   **d.** krypton (Kr)

Answers
a. Group 6A; 6 outer electrons
b. Group 4A; 4 outer electrons

Your Turn 2.6   Family Features

a. What electronic feature do fluorine (F), chlorine (Cl), bromine (Br), and iodine (I) have in common?
b. The element beryllium (Be), like the other elements in Group 2A, has two outer electrons. Give the names and symbols for the other Group 2A elements.

Answer
a. All have seven outer electrons. They belong to Group 7A, the halogens.

Table 2.3	Isotopes of Hydrogen			
Name	Isotope	# of Protons (atomic number)	# of Neutrons	#n + #p (mass number)
hydrogen	H-1 or $_1^1$H	1	0	1
deuterium	H-2 or $_1^2$H	1	1	2
tritium	H-3 or $_1^3$H	1	2	3

Elements have both stable and radioactive isotopes. For example, H-3 (tritium) is radioactive, but H-1 and H-2 are not. Look for more about radioisotopes in Chapter 7.

In addition to electrons and protons, atoms also contain neutrons. The one (and only) exception is an atom of the most common form of hydrogen, which consists of only one proton in its nucleus. But the nucleus of 1 out of every 6700 hydrogen atoms also contains a neutron. This naturally occurring form of hydrogen is called deuterium. Tritium, a radioactive form of hydrogen that is quite rare in nature, has two neutrons in its nucleus. Hydrogen, deuterium, and tritium are examples of **isotopes,** two or more forms of the same element (same number of protons) whose atoms differ in number of neutrons, and hence in mass.

An isotope is identified by its **mass number,** the sum of the number of protons and neutrons in the nucleus of an atom. The mass number can vary for the same element. In contrast, the atomic number cannot vary for the same element. For example, the full atomic symbol $_1^1$H represents the most common isotope of hydrogen. Because the atomic number of 1 for hydrogen is invariant, the subscript is sometimes omitted. Thus you also may see ^{1}H, hydrogen-1, or H-1. Table 2.3 summarizes information about the isotopes of hydrogen.

Your Turn 2.7   Protons and Neutrons

Specify the number of protons and electrons in the nucleus of each of these.

a. carbon-14 ($_6^{14}$C)   **b.** uranium-235 ($_{92}^{235}$U)   **c.** iodine-131 ($_{53}^{131}$I)

Answers
a. 6 protons, 8 neutrons
b. 92 protons, 143 neutrons

All elements have more than one isotope, but the number of stable ones varies considerably. Each element's atomic mass, the number you see on every periodic table, takes the relative natural abundance of isotopes, as well as their masses, into account. Following our general rule of introducing information on a need-to-know basis, we will return to a discussion of atomic masses in Chapter 3.

Mass number is the total number of protons and neutrons in a specific isotope. Atomic mass refers to a weighted average of all naturally occurring isotopes of that element.

2.3 | Molecules and Models

Having taken a short excursion into the atomic realm, we now move to the topic of bonding in molecules so that in turn, we can understand the ozone hole.

Let's begin with the simplest molecule, H_2. Each hydrogen atom has one electron. If two hydrogen atoms bond, the two electrons become common property. If we represent each electron by a dot, the two hydrogen atoms might look something like this:

$$H\cdot \quad \text{and} \quad \cdot H$$

Bringing the two atoms together yields a molecule that can be represented this way.

$$H : H$$

Each atom effectively has a share in both electrons. The resulting H_2 molecule has a lower energy than the sum of the energy in the two individual H atoms, and consequently the molecule with its bonded atoms is more stable than the separate atoms. The two electrons that are shared constitute a **covalent bond.** Appropriately, the name *covalent* implies "shared strength."

A **Lewis structure** is a representation of an atom or molecule that shows its outer electrons. The name honors Gilbert Newton Lewis (1875–1946), an American chemist who pioneered its use. Lewis structures, also called dot structures, can be predicted for many simple molecules by following a set of straightforward steps. We first illustrate the procedure with hydrogen fluoride, HF, another simple molecule.

1. Note the number of outer electrons contributed by each of the atoms.
   Hint: The periodic table is a useful guide for Group A elements.

   > 1 H atom (H·) × 1 outer electron per atom = 1 outer electron
   >
   > 1 F atom (:F̈·) × 7 outer electrons per atom = 7 outer electrons

2. Add the outer electrons contributed by the individual atoms to obtain the total number of outer electrons available.

   > 1 + 7 = 8 outer electrons

3. Arrange the outer electrons in pairs. Then distribute them so as to maximize stability by giving each atom a share in enough electrons to fully fill its outer level: 2 electrons in the case of hydrogen, 8 electrons for most other atoms.

   > H:F̈:

We surrounded the F atom with 8 dots, organized into 4 pairs. The pair of dots between the H and the F represents the electron pair that forms the bond uniting the hydrogen and fluorine atoms. The other 3 pairs of electrons are not shared with other atoms. As such, they are called nonbonding electrons, or "lone pairs."

A **single covalent bond** is formed when two electrons (one pair) are shared between two atoms. A line may be used to represent the two electrons in the bond.

> H—F̈:

Sometimes the nonbonding electrons are removed from a Lewis structure, simplifying it still more. The result is called a **structural formula.**

> H—F

Remember that the single line represents one pair of shared electrons. These 2 electrons plus the 6 electrons in the 3 nonbonding pairs mean that the fluorine atom is associated with 8 outer electrons, whether or not the electrons are specifically shown. Remember that the hydrogen atom has no additional electrons other than the single pair shared with fluorine. It is at maximum capacity with two electrons.

The fact that electrons in many molecules are arranged so that every atom (except hydrogen) shares in eight electrons is called the **octet rule.** This generalization is useful for predicting Lewis structures and the formulas of compounds. Consider the Cl_2 molecule, the diatomic form of elemental chlorine. From the periodic table, we can see that chlorine, like fluorine, is in Group 7A, which means that its atoms each have 7 outer electrons. Using the scheme given for HF earlier, we first count and add up the outer electrons for Cl_2.

> 2 Cl atoms (:C̈l·) × 7 outer electrons per atom = 14 outer electrons

For the Cl_2 molecule to exist, a bond must connect the two atoms. The remaining 12 electrons constitute 6 nonbonding pairs, distributed in such a way as to give each chlorine atom a share in 8 electrons (2 bonding and 6 nonbonding). Here is the Lewis structure.

> :C̈l—C̈l:

You are unlikely to run into HF in your chemistry laboratory. It is a highly reactive compound, and in aqueous solution it is used to etch glass.

Your Turn 2.8   **Lewis Structures for Diatomic Molecules**

Draw the Lewis structure for each molecule.

a. HBr   **b.** Br_2

Answer

a.   1 H atom (H•) $\times$ 1 outer electron per atom = 1 outer electron

1 Br atom ($\cdot\overset{\cdot\cdot}{\underset{\cdot\cdot}{Br}}$:) $\times$ 7 outer electrons per atom = 7 outer electrons

Total = 8 outer electrons

Here is the Lewis structure: H:$\overset{\cdot\cdot}{\underset{\cdot\cdot}{Br}}$:   or   H—$\overset{\cdot\cdot}{\underset{\cdot\cdot}{Br}}$:

So far we have dealt only with molecules having just two atoms. But the octet rule applies to larger molecules as well. Let's use a water molecule, H_2O, as an example. Just as with two-atom molecules, first tally the outer electrons.

2 H atoms (H•) $\times$ 1 outer electron per atom = 2 outer electrons

1 O atom ($\cdot\overset{\cdot\cdot}{\underset{\cdot\cdot}{O}}\cdot$) $\times$ 6 outer electrons per atom = 6 outer electrons

Total = 8 outer electrons

In molecules like water that have a single atom bonded to two or more atoms of a different element (or elements), *the single atom is the central one.* You'll encounter exceptions, but this is a useful rule. Since oxygen is the "single atom" in H_2O, we place it in the center of the Lewis structure. Each of the H atoms bonds to the O atom, using 4 electrons. The remaining 4 electrons go on the O atom as 2 nonbonding pairs.

H:$\overset{\cdot\cdot}{\underset{\cdot\cdot}{O}}$:H

A quick count confirms that the O atom is surrounded by 8 electrons, as predicted by the octet rule. Alternatively, we could use lines for the single bonds.

H—$\overset{\cdot\cdot}{\underset{\cdot\cdot}{O}}$—H

Chemical formulas show the types and ratio of atoms present. In contrast, Lewis structures also indicate how the atoms are connected and show the nonbonding pairs of electrons, if present. Note that Lewis structures do *not* directly reveal the shape of a molecule. For example, from the Lewis structure we drew it might appear that the water molecule is linear. In fact, the molecule is bent.

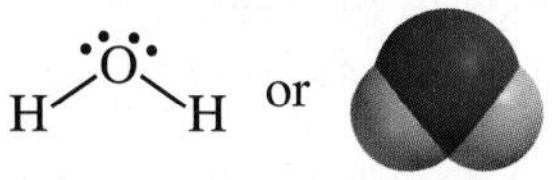

Another molecule to consider is methane, CH_4. Again, we begin by tallying the valence electrons.

4 H atoms (H•) $\times$ 1 outer electron per atom = 4 outer electrons

1 C atom ($\cdot\overset{\cdot}{\underset{\cdot}{C}}\cdot$) $\times$ 4 outer electrons per atom = 4 outer electrons

Total = 8 outer electrons

The central carbon atom is surrounded by the 8 electrons, giving carbon an octet of electrons. In the Lewis structure, each H atom uses 2 of the electrons to bond with the C atom, for a total of 4 single covalent bonds.

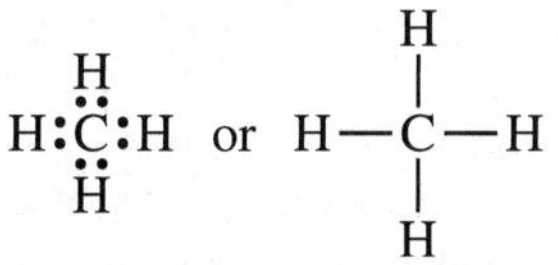

$$\begin{array}{c} H \\ H:\overset{\cdot\cdot}{\underset{\cdot\cdot}{C}}:H \\ H \end{array} \quad \text{or} \quad \begin{array}{c} H \\ | \\ H-C-H \\ | \\ H \end{array}$$

Remember that H can only accommodate a pair of electrons. The next activity gives you the opportunity to practice with other molecules.

Each hydrogen atom forms only one bond (two shared electrons). Oxygen can form two bonds and is the central atom in H_2O.

The space-filling model of water was shown in Section 1.7. We will explain why the water molecule is bent in Chapter 3.

The combustion of methane was discussed in Section 1.10. Look in Chapter 3 for an explanation of the shape of the methane molecule.

In some structures, single covalent bonds do not allow the atoms to follow the octet rule. Consider, for example, the O_2 molecule. Here we have 12 outer electrons to distribute, 6 from each of the oxygen atoms. There are not enough electrons to give each of the atoms a share in eight electrons if only one pair is held in common. However, the octet rule can be satisfied if the two atoms share four electrons (two pairs). A covalent bond consisting of two pairs of shared electrons is called a **double bond.** This bond is represented by four dots or by two lines.

$$\ddot{O}::\ddot{O} \quad \text{or} \quad \ddot{O}=\ddot{O}$$

Double bonds are shorter, stronger, and require more energy to break than single bonds involving the same atoms. The experimentally measured length and strength of the bond in the O_2 molecule correspond to a double bond. However, oxygen has a property that is not fully consistent with the Lewis structure just drawn. When liquid oxygen is poured between the poles of a strong magnet, it sticks there like iron filings. Such magnetic behavior implies the presence of unpaired electrons rather than the paired arrangement shown in the preceding Lewis structures. But this is hardly a reason to discard the useful generalizations of the octet rule. After all, simple scientific models seldom if ever explain all phenomena, but they can be helpful approximations. There are other common examples in which the straightforward application of the octet rule leads to discrepancies in interpreting experimental evidence. Coming across data that do not seem to fit has led to the development of more sophisticated models.

A **triple bond** is a covalent linkage made up of three pairs of shared electrons. For the same atoms, triple bonds are even shorter, stronger, and harder to break than double bonds. For example, the nitrogen molecule, N_2, contains a triple bond. Each Group 5A nitrogen atom contributes 5 outer electrons for a total of 10. These 10 electrons can be distributed in accordance with the octet rule if 6 of them (three pairs) are shared between the two atoms, leaving 4 of them to form two nonbonding pairs, one on each nitrogen atom.

$$:N:::N: \quad \text{or} \quad :N\equiv N:$$

The ozone molecule introduces another structural feature. We again start with the octet rule. Each of the three oxygen atoms contributes 6 outer electrons for a total of 18. These 18 electrons can be arranged in two ways; each way gives a share in 8 outer electrons to each atom.

$$\ddot{O}::\ddot{O}:\ddot{O}: \qquad :\ddot{O}:\ddot{O}::\ddot{O}$$

         a   **b**

The stability of the triple bond linking N atoms in N_2 gas helps explain nitrogen's relative inertness in the troposphere.

Structures **a** and **b** predict that the molecule should contain one single bond and one double bond. In structure **a,** the double bond is shown to the left of the central atom; in **b** it is shown to the right. But experiments reveal that the two bonds in the O_3 molecule are identical, being intermediate between the length and strength of a single and double bond. Structures **a** and **b** are called **resonance forms,** Lewis structures that represent hypothetical extremes of electron arrangements in a molecule. For example, no single resonance form represents the electron arrangement in the ozone molecule. Rather, the actual structure is something like a hybrid of the two resonance forms. A double-headed arrow linking the different forms is used to represent the resonance phenomenon.

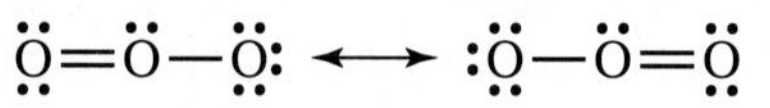

$$\ddot{O}{=}\ddot{O}{-}\ddot{\underset{..}{O}}{:} \longleftrightarrow {:}\ddot{\underset{..}{O}}{-}\ddot{O}{=}\ddot{O}$$

Resonance is just another modeling concept invented by chemists to represent the complex microworld of molecules. It is not intended to be "the truth" but rather just a way to describe the structures of molecules that do not exactly fit the octet rule model. Figure 2.4 compares the Lewis structures of several different oxygen-containing species relevant to the chemistry in this and other chapters.

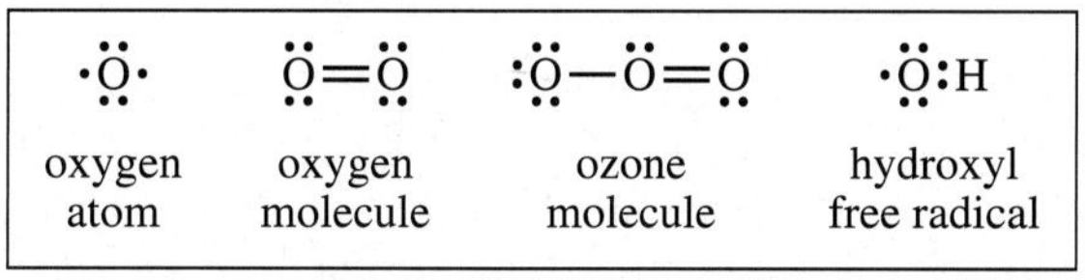

Figure 2.4

Lewis structures for several oxygen-containing species. Only one resonance form of ozone is shown.

A closer experimental inspection of that microworld reveals that the O_3 molecule is not linear as the simple Lewis structures just drawn seems to indicate. Remember that Lewis structures tell us only what is connected to what and do not necessarily show the shape of the molecule. The O_3 molecule is actually bent, as in this representation.

$$\ddot{O}{=}\overset{\ddot{O}}{\diagdown}\ddot{\underset{..}{O}}{:} \longleftrightarrow {:}\ddot{\underset{..}{O}}\diagup\overset{\ddot{O}}{=}\ddot{O}$$

An explanation of why the O_3 molecule is bent must wait until Chapter 3. At this point, we only need to know how the bonding in the O_2 and O_3 molecules relates to their interaction with sunlight.

The hydroxyl radical was mentioned in Chapter 1 in connection with smog formation.

Observe that both H_2O and O_3 are bent molecules, with O as the central atom.

Your Turn 2.10   **Lewis Structures with Multiple Bonds**

Draw the Lewis structure for each compound. Both follow the octet rule.

a. carbon monoxide (CO)   **b.** sulfur dioxide (SO_2)

Answer

a. 1 C atom ($\cdot\ddot{C}\cdot$) × 4 outer electrons per atom = 4 outer electrons

1 O atom ($\cdot\ddot{\underset{..}{O}}\cdot$) × 6 outer electrons per atom = 6 outer electrons

Total = 10 outer electrons

The Lewis structure is $:C:::O:$ or $:C{\equiv}O:$ and has 10 outer electrons. The N_2 molecule also has 10 outer electrons and similarly forms a triple bond.

2.4 | Waves of Light

Every second, light is emitted by the Sun and, after some time, reaches our planet. Some of this light we can see; some we cannot. Prisms and raindrops break the light we see into a spectrum of colors. Sometimes we name these colors simply as violet, indigo, blue, green, yellow, orange, and red. Other times, we distinguish between the hues with more descriptive names such as cherry red or forest green.

Another way to describe a color is with a numerical value that corresponds to its wavelength. The word *wavelength* correctly suggests that light behaves something like a wave in a body of water. **Wavelength** is the distance between successive peaks. It is expressed in units of length and symbolized by the Greek letter lambda (λ). Waves are also characterized by a certain **frequency,** the number of waves passing a fixed point in 1 second. Frequency is symbolized by the Greek letter nu (ν). Figure 2.5 shows two waves of different wavelength and frequency.

The relationship between frequency and wavelength can be summarized in a simple equation in which ν is the frequency and c is the constant speed at which visible light and other forms of electromagnetic radiation travel, 3.00×10^8 m·s^{-1}.

$$\text{frequency } (\nu) = \frac{\text{speed of light } (c)}{\text{wavelength } (\lambda)} \qquad [2.2]$$

Equation 2.2 indicates that wavelength and frequency are *inversely* related. As the λ decreases, the ν increases, and vice versa.

As wavelength ↑, frequency ↓.

It is both interesting and humbling to realize that out of the vast array of radiant energies, our eyes are sensitive only to the tiny portion between roughly 700×10^{-9} meters (red light) and 400×10^{-9} meters (violet light). These wavelengths are very short, so we typically express them in nanometers. One **nanometer (nm)** is defined as one billionth of a meter (m).

$$1 \text{ nm} = \frac{1}{1,000,000,000} \text{ m} = \frac{1}{1 \times 10^9} \text{ m} = 1 \times 10^{-9} \text{ m}$$

We can use this equivalence to convert meters to nanometers. For example, this calculation shows how many nanometers are in 700×10^{-9} m.

$$\text{wavelength } (\lambda) = 700 \times 10^{-9} \text{ m} \times \frac{1 \text{ nm}}{1 \times 10^{-9} \text{ m}} = 700 \text{ nm}$$

The units of meters cancel, leaving nanometers.

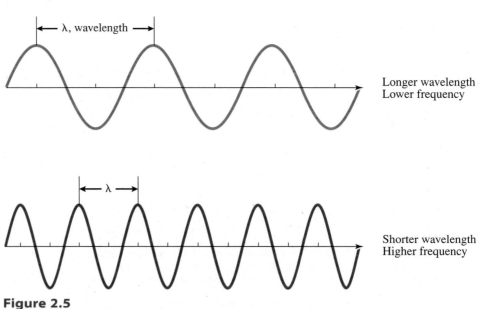

λ, wavelength

Longer wavelength
Lower frequency

λ

Shorter wavelength
Higher frequency

Figure 2.5

Comparison of two different waves.

Consider This 2.11   Analyzing a Rainbow

Water droplets in a rainbow act as prisms to separate visible light into its colors.

   a. In Figure 2.6, which color has the longest wavelength? The highest frequency?
   b. Green light has a wavelength of 500 nm. Express this value in meters.

Answer
   b. 500×10^{-9} m. In scientific notation, this is 5×10^{-7} m.

The **electromagnetic spectrum** is a continuum of waves that ranges from short, high-energy X-rays and gamma rays to long, low-energy radio waves. Visible light is only a narrow band in this spectrum. The term **radiant energy** refers to the entire collection of different wavelengths, each with its own energy. Figure 2.7 shows the electromagnetic spectrum, the relative wavelengths (not drawn to scale), and some examples to help you develop perspective on the range of wavelengths represented.

In this chapter, we consider the **ultraviolet (UV)** region that lies adjacent to the violet end of the visible region of the electromagnetic spectrum, but at shorter wavelengths. At still shorter wavelengths are the X-rays used in medical diagnosis and the determination of crystal structures, and gamma rays that are given off in processes of nuclear decay. At wavelengths longer than those of red visible light lies the **infrared (IR)** region. We cannot see these wavelengths, but can feel their heating effect. The microwaves used in radar and to cook food quickly have wavelengths on the order of centimeters. At still longer wavelengths are the regions of the spectrum used to transmit your favorite AM and FM radio and television programs.

Figure 2.6

A rainbow of color.

We will consider the IR region of the spectrum in Chapter 3.

Your Turn 2.12   Relative Wavelengths

Consider these four types of radiant energy from the electromagnetic spectrum: infrared, microwave, ultraviolet, visible.

   a. Arrange them in order of *increasing* wavelength.
   b. Approximately how many times longer is a wavelength associated with a radio wave than one associated with an X-ray?
   Hint: See Figure 2.7.

Answer
   a. ultraviolet < visible < infrared < microwave

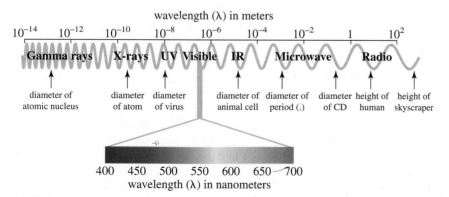

Figure 2.7

The electromagnetic spectrum. The wavelength variation from gamma rays to radio waves is not drawn to scale.

 Figures Alive! Visit the textbook's website to learn more about relationships in the electromagnetic spectrum.

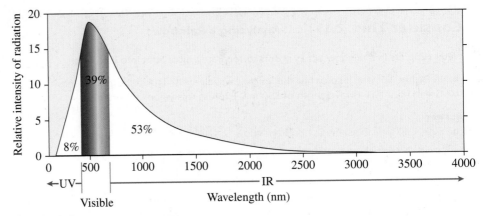

Figure 2.8

Wavelength distribution of solar radiation above Earth's atmosphere.

Source: From *An Introduction to Solar Radiation* by Muhammad Iqbal, Academic Press, 1983. Copyright Elsevier 1983.

Our local star, the Sun, emits many types of radiant energy but not with equal intensity. This is evident from Figure 2.8, a plot of the relative intensity of solar radiation as a function of wavelength. The curve represents the spectrum as measured *above* the atmosphere, before there has been opportunity for interaction of radiation with the molecules found in air. The peak indicating the greatest intensity is in the visible region. However, 53% of the total energy emitted by the Sun is radiated to Earth as infrared radiation. This is the major source of heat for the planet. Approximately 39% of the energy comes to us as visible light and only about 8% as ultraviolet. (The areas under the curve give an indication of these percentages.) But in spite of its small percentage, the Sun's UV radiation can be the most damaging to living things. To understand why, we need to look at electromagnetic radiation in terms of its energy.

2.5 | Radiation and Matter

The idea that radiation can be described in terms of wave-like character is well established and very useful. However, around the beginning of the 20th century, scientists found several phenomena that seemed to contradict this model. In 1909, a German physicist named Max Planck (1858–1947) argued that the shape of the energy distribution curve pictured in Figure 2.8 could only be explained if the energy of the radiating body were the sum of many energy levels of minute but discrete size. In other words, the energy distribution is not really continuous, but consists of many individual steps. Such an energy distribution is called **quantized.** An often-used analogy is that the quantized energy of a radiating body is like steps on a staircase, which are also quantized (no partial steps allowed), not like a ramp that allows any size stride. Albert Einstein (1879–1955), in the work that won him the 1921 Nobel Prize in physics, suggested that radiation itself should be viewed as constituted of individual bundles of energy called **photons.** One can regard these photons as "particles of light," but they are definitely not particles in the usual sense. For example, they have no mass. These ideas form the basis of modern quantum theory.

The wave model remains useful, even with the development of the quantum theory. Both are valid descriptions of radiation. The dual nature of radiant energy seems to defy common sense. How can light be described in two different ways at the same time, both waves and particles? There is no obvious answer to that very reasonable question—that's just the way nature is. The two views are linked in a simple relationship that is one of the most important equations in modern science. It is also an equation relevant to the role of ozone in the atmosphere.

Planck and Einstein were both amateur violinists who played duets together.

$$\text{energy } (E) = \frac{hc}{\lambda} \qquad \textbf{[2.3]}$$

Here E represents the energy of a single photon. Both symbols h and c represent constants. The symbol h is called Planck's constant and c is the speed of light. This equation therefore shows that energy, E, is *inversely* proportional to the wavelength, λ. Consequently, as the wavelength of radiation gets shorter, its energy increases. This qualitative relationship is important in the story of ozone depletion.

As wavelength ↓, energy ↑.

Your Turn 2.13   **Color and Energy Relationships**

Arrange these colors of the visible spectrum in order of *increasing* energy per photon:

green, red, yellow, violet.

Answer

red < yellow < green < violet

Using equation 2.3, one can calculate that the energy associated with a photon of UV radiation is approximately 10 million times larger than the energy of a photon emitted by your favorite radio station. A consequence of this large difference in energy is that you can damage your skin with exposure to UV radiation, but not by listening to the radio—unless you happen to be listening to it outside in the sunlight. Whether or not your radio is turned on, you are continuously bombarded by radio waves. Your body cannot detect them, but your radio can. The energy associated with each of the radio photons is very low and not sufficient to produce a local increase in the concentration of the skin pigment, melanin, as happens with exposure to UV. Producing melanin involves a quantum jump, an electronic transition between energy levels that requires far more energy than radio wave photons can supply.

The Sun bombards Earth with countless photons—indivisible packages of energy. The atmosphere, the planet's surface, and Earth's living things all absorb these photons. Radiation in the infrared region of the spectrum warms Earth and its oceans. The cells of our retinas are tuned to the wavelengths of visible light. Photons associated with different wavelengths are absorbed, and the energy is used to "excite" electrons in biological molecules. Some electrons jump to higher energy levels, triggering a series of complex chemical reactions that ultimately lead to sight. Compared with animals, green plants capture photons in an even narrower region of the visible spectrum (corresponding to red light) and use the energy to convert carbon dioxide and water into food, fuel, and oxygen in the process of photosynthesis.

Remember that as the wavelength of light *decreases,* the energy carried by each photon *increases.* Photons in the UV region of the spectrum are sufficiently energetic to displace electrons from neutral molecules, converting them into positively charged species. Even shorter UV wavelength photons break bonds, causing molecules to come apart. In living things, such changes disrupt cells and create the potential for genetic defects and cancer. The interaction of UV radiation with chemical bonds is shown schematically in Figure 2.9.

It is part of the fascinating symmetry of nature that this interaction of radiation with matter explains both the damage ultraviolet radiation can cause and the atmospheric mechanism that protects us from it. We turn next to understanding the ultraviolet shield provided by oxygen and ozone in our stratosphere.

From Figure 2.8 you can see that the photons from the Sun with energies in the visible light range are the most intense.

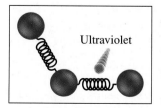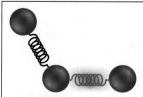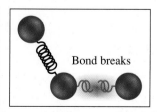

Figure 2.9

Ultraviolet radiation can break some chemical bonds. Bonds are represented as springs that hold the atoms together but allow the atoms to move relative to each other.

2.6 | The Oxygen–Ozone Screen

We know the colors of visible light by their names, red, blue, yellow and so on. Similarly, we call ultraviolet light by different names. Admittedly, however, these names are not as colorful: UV-A, UV-B, and UV-C. UV-A lies closest to the violet region of visible light and is the lowest in energy; you may know it as "black light." In contrast, UV-C has the highest energy and lies next to the X-ray region of the electromagnetic spectrum. Table 2.4 shows the characteristics of the different types of UV light.

Your Turn 2.14　　　The ABCs of Solar UV

a. Arrange UV-A, UV-B, and UV-C in order of increasing wavelength.
b. Is the order for increasing energy the same as for wavelength? Explain.
c. Should you use a sunscreen that claims to protect against UV-C? Explain.

Answer
c. No, you should not. No protection is needed for UV-C, because this set of UV wavelengths is absorbed up in the stratosphere.

As you saw from the previous activity, UV-C radiation from the Sun is absorbed in the upper atmosphere before it ever reaches the ground. Both oxygen and ozone absorb light of these wavelengths. As we noted in Chapter 1, about 21% of the atmosphere consists of oxygen, O_2. Photons with energy corresponding to 242 nm or less have sufficient energy to break the bonds in an O_2 molecule. These wavelengths are found in the UV-C region.

$$O_2 \xrightarrow[\lambda \leq 242\,nm]{UV\ photon} 2\,O \qquad\qquad [2.4]$$

If O_2 were the only molecule absorbing UV light from the Sun, Earth's surface and the creatures that live on it would still be subjected to damaging radiation in the range of 242–320 nm. It is here that O_3 plays its important protective role. The O_3 molecule is more easily broken apart than O_2. Recall that the atoms in the O_2 molecule are connected with a double bond, but each of the bonds in O_3 is somewhere between a single and double bond in length and in strength. Accordingly, the bonds in O_3 are weaker than the double bonds in O_2. Therefore, photons of a lower energy (longer wavelength) are sufficient to separate the atoms in O_3. Indeed, photons of wavelength 320 nm or less break the O-to-O bond in ozone.

$$O_3 \xrightarrow[\lambda \leq 320\,nm]{UV\ photon} O_2 + O \qquad\qquad [2.5]$$

Table 2.4	Types of UV Radiation		
Type	**Wavelength**	**Relative Energy**	**Comments**
UV-A	320–400 nm	Lowest energy	Least damaging and reaches the Earth's surface in greatest amount
UV-B	280–320 nm	Higher energy than UV-A but less energetic than UV-C	More damaging than UV-A but less damaging than UV-C. Most UV-B is absorbed by O_3 in the stratosphere
UV-C	200–280 nm	Highest energy	Most damaging but not a problem because it is totally absorbed by O_2 and O_3 in the stratosphere

Consider This 2.15   Energy and Wavelength

We just stated that it takes photons in the UV-C range ($\leq$ 242 nm) to break the double bond in O_2. The bonds in O_3 are somewhat weaker than those in O_2, so lower energy photons ($\leq$ 320 nm) can break those bonds. Just how much greater is the energy of a 242-nm photon than that of a 320-nm photon?

Hint: One approach could be to calculate the ratio of the energies for a 242-nm photon and that of a 320-nm photon and then to compare that with the ratio of their wavelengths. Look for values of Planck's constant and the speed of light in Appendix 1.

Equations 2.4 and 2.5, together with earlier equation 2.1 (that showed the formation of O_3 from O_2), are part of a set of chemical reactions in the stratosphere. Every day, 300,000,000 (3×10^8) tons of stratospheric O_3 forms, and an equal mass decomposes. New matter is neither created nor destroyed but merely changes its chemical form. So the overall concentration of ozone remains constant in this natural cycle. The process is an example of a **steady state,** a condition in which a dynamic system is in balance so that there is no net change in concentration of the major species involved. A steady state arises when a number of chemical reactions, typically competing reactions, balance each other. The **Chapman cycle,** as shown Figure 2.10, represents the first set of natural steady-state reactions proposed for stratospheric ozone. This natural cycle includes chemical reactions for both ozone formation and decomposition.

This set of reactions is named after Sydney Chapman, a physicist who first proposed it in 1929.

Your Turn 2.16   The Ozone Layer

a. Ozone is formed by the reaction of oxygen atoms with oxygen molecules. Write the chemical reaction.
b. Up in the stratosphere, what is the source of the oxygen atoms?
c. Up in the stratosphere, the lifetime of a given ozone molecule ranges from days to years. For example, in the ozone layer, an O_3 molecule can persist for several months. What does stratospheric ozone break down into?
d. In contrast, at ground level, ozone molecules react in a matter of minutes rather than months. Why the difference?

Answer
d. Down in the troposphere, the air is much denser ("thicker"). Ozone molecules quickly bump into other molecules and react with them. For example, if you happen to breathe air containing ozone pollution, the O_3 quickly reacts with your lung tissue, potentially damaging it.

In a later section, we consider what happens when something disturbs the steady state of the Chapman cycle, leading to destruction of the protective stratospheric ozone.

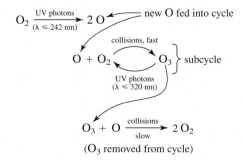

Figure 2.10
The Chapman cycle.

Because of the presence of O_2 and O_3 in the stratosphere, only certain UV wavelengths reach the surface of the Earth. However, these wavelengths still can cause harm, which is the topic of the next section.

2.7 | Biological Effects of Ultraviolet Radiation

Look for more about DNA in Chapter 12.

For animals and plants, the effects of UV radiation depend primarily on the energy of the radiation and the sensitivity of the organism. These relationships are evident from Figure 2.11, where biological sensitivity is plotted versus wavelength. As defined here, biological sensitivity is based on experiments in which the damage to deoxyribonucleic acid (DNA), the chemical basis of heredity, is measured at various wavelengths. In the figure, the biological sensitivity is expressed in relative units using a logarithmic scale. On this scale, each mark on the y-axis represents a biological sensitivity that is 10 times the value corresponding to the mark immediately below it on the vertical axis.

Biological sensitivity at 320 nm is about 1×10^{-5}, or 0.00001 units. But at 280 nm, the sensitivity is 1×10^{0}, or 1 unit. This means that radiation at 280 nm is 100,000 times more damaging than radiation at 320 nm. However, shorter wavelength solar radiation (<320 nm) is absorbed by O_3 in the stratosphere. This is most fortunate, because radiation in this region of the spectrum is particularly damaging to living things.

Consider This 2.17   Relative Biological Sensitivity

DNA sensitivity decreases with increasing UV wavelength, as shown by Figure 2.11.

a. Propose an explanation for this.
b. Why is there no need to include the UV-C region in this figure?

As we will see in the next section, around 1980 the average concentrations of stratospheric ozone unexpectedly began to decrease. Although this has happened to varying extents in different regions of the globe, generally speaking living things are

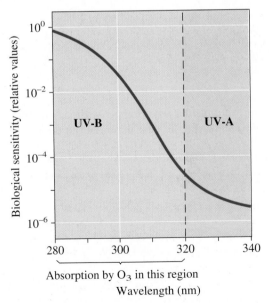

Figure 2.11

Variation of biological sensitivity of DNA with UV wavelength. The UV-C region is below 280 nm.

Source: John E. Fredrick, University of Chicago Reprinted by permission.

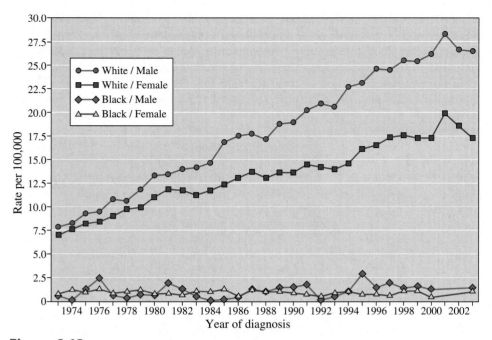

Figure 2.12

Increase in incidence of melanoma skin cancer in the United States, 1973–2003.

Source: Surveillance, Epidemiology, and End Results (SEER) Program of the National Cancer Institute.

now exposed to greater intensities of damaging radiation than in the past. According to calculations, a given percent decrease in stratospheric ozone is expected to increase the biological damage done by UV radiation by twice that percentage. For example, from a 6% decrease in stratospheric ozone we predict a 12% rise in skin cancer, especially the more easily treated forms such as basal cell and squamous cell carcinomas. These conditions are considerably more common among people with light-colored skin than among people whose skin is more heavily pigmented (Figure 2.12).

Skeptical Chemist 2.18   Skin Cancer in Men and Women

Did Figure 2.12 mislabel the cancer rates for white men and women? That is, should the rate be higher for women than for men? Argue the case either way and then check out the data, either from the source given or from one of your choice. Who or what did you consult and what did you learn?

At least to date, geography seems to affect skin cancer more than ozone depletion. For example, evidence links the incidence of melanomas, the most deadly form of skin cancer, with the latitude at which you live. In general, the disease becomes more prevalent as one moves south in the Northern Hemisphere. Those who endure the long nights and short days of winter are compensated by a level of skin cancer that is only about half that of those who live near the equator.

Skin cancer rates are high in the Southern Hemisphere as well. For example, two of every three Australians will develop skin cancer sometime during their lifetime; as of 2005, 1600 Australians died from this form of cancer per year. The Australian government has acted in several ways to reverse this trend, including banning tanned models from all advertising media. National Skin Cancer Action Week is held each year at the start of their summer season in November, with the goals of raising awareness and urging the use of clothing and lotions for sun protection.

However, geographic location is not the only factor influencing development of skin cancer. Skin cancer rates generally continue to rise in *all* countries, despite

According to the Australian Department of Health and Aging (2007), Australia has the highest skin cancer rates in the world.

Figure 2.13
Blue Lizard Suncream.

Nanotechnology was defined in
Section 1.7. Look for more about this
topic in Chapter 8.

increased awareness of the dangers of exposure to UV radiation. Changes in the natural protection afforded by the stratospheric ozone layer are only partially responsible for higher rates of skin cancer. About a million new cases of skin cancers occur each year in the United States, almost as many as the total number of cases of all other cancers. Although those with lighter skin tones are about 40 times more likely to develop melanoma, those with darker skin have a significantly lower long-term survival rate, according to a 2005 report issued by the American Academy of Dermatology. Skin cancers can develop many years after repeated, excessive exposure has stopped. Skin cancers even have been linked to a single episode of extreme sunburn in adolescence with the effects showing up many years later. Public service campaigns center on early detection and promote regular inspection of suspicious moles for people of all ethnic backgrounds. Other possible causes need to be considered. One of these is tanning, either naturally or in a tanning bed. This practice presents a risk–benefit activity for everyone, because skin cancer can strike people of all skin tones.

Consider This 2.19   Tanning Spas

The indoor tanning industry runs a public relations campaign that highlights positive findings about indoor tanning, promoting it as part of a healthy lifestyle. Countering these claims are the studies published in scientific journals that support the view of dermatologists that there is no such thing as a "safe tan" for any skin type. Investigate at least two websites that present different points of view. List the arguments on both sides. What do you conclude?

Wearing protective sunscreen is one way to reduce the risk of skin cancer. Such products contain compounds that absorb UV-B to some extent together with others for absorbing UV-A. The American Academy of Dermatology recommends a sunscreen with a skin protection factor (SPF) of 15 to 30. But wearing a sunscreen does not mean that you are without risk from UV rays. Because sunscreens allow you to be exposed for a longer time without burning, they may ultimately cause greater skin damage. The Australian product Blue Lizard Suncream (Figure 2.13) uses "smart bottle" technology for containing and marketing their product. The bottle itself changes color from white to blue in UV light, sending an extra reminder that the dangers of UV light are still present, even if a sunscreen is being used.

Wearing protective sunblock is another. These products physically block the light from reaching your skin, much as tightly woven clothing would. Sunblocks reflect the incident light; some absorb UV as well. A familiar example may be the white opaque cream used by lifeguards ("lifeguard nose") at a pool or beach. This cream contains small white particles of ZnO (zinc oxide) or TiO_2 (titanium dioxide) (or both) and has a long-term track record of safety.

However, the "see-through" formulations of ZnO and TiO_2 that contain these compounds in nanoparticle form are more controversial. Because the particles of ZnO and TiO_2 are so microscopically tiny, they do not scatter light. As a result, the cream is clear rather than opaque, definitely a plus to those who wear it. These nanoparticle products spread more evenly, are cost-effective, and are extremely effective at both absorbing and reflecting UV radiation. However, nanoparticles may present a risk if they penetrate the skin; the verdict is not yet in. As of 2009, both consumers and government agencies were calling for further studies to better quantify the risks.

Controversial or not, sunscreens play an important role in protecting us from UV radiation. For example, the U.S. National Weather Service issues an Ultraviolet Index forecast that appears on the web. UV Index values range from 0 to 15 and are based on how long it takes for skin damage to occur (Table 2.5). The UV Index is color-coded for ease of interpretation. Specific steps are suggested to protect eyes and skin from

Table 2.5		The UV Index
Exposure Category	**Index**	**Tips to Avoid Harmful Exposure to UV**
LOW	< 2	Wear sunglasses on bright days. Be aware that snow and water can reflect the Sun's rays. If you burn easily, cover up and use sunscreen.
MODERATE	3–5	Take precautions, such as covering up, if you will be outside. Stay in shade near midday when the Sun is strongest.
HIGH	6–7	Protect against sunburn. Reduce time in the Sun between 10 AM and 4 PM. Cover up, wear a hat and sunglasses, and use sunscreen.
VERY HIGH	8–10	Take extra precautions against sunburn, as unprotected skin will be damaged and can burn quickly. Minimize Sun exposure between 10 AM and 4 PM. Otherwise, seek shade, cover up, wear a hat and sunglasses, and use sunscreen.
EXTREME	11+	Take all precautions against sunburn, as unprotected skin can burn in minutes. Be aware that white sand and other bright surfaces reflect UV and will increase UV exposure. Avoid the Sun between 10 AM and 4 PM. Seek shade, cover up, wear a hat and sunglasses, and use sunscreen.

Source: U.S. EPA, 2009.

Sun damage. As you might guess, one is to not expose yourself in the first place. Stay in the shade and wear protective clothing!

Consider This 2.20   UV Index Forecasts

The UV Index indicates the amount of UV radiation reaching Earth's surface at solar noon, the time when the Sun is highest in the sky.

a. The UV Index depends on the latitude, the day of the year, time of day, amount of ozone above the city, elevation, and the predicted cloud cover. How is the UV Index affected by each of these?

b. The U.S. EPA provides a UV Index forecast. The textbook's website provides a direct link. Suggest reasons for values that you see on today's map.

Although the UV Index focuses on skin damage, this is not the only biological effect of UV radiation. Your eyes can be damaged as well. For example, all people, no matter what the pigmentation level of their skin, are susceptible to retinal damage caused by UV exposure. Another effect is cataracts, a clouding of the lens of the eye caused by excessive exposure to UV-B radiation. It has been estimated that a 10% decrease in the ozone layer could create up to 2 million new cataract cases globally. However, just as proper clothing and sunscreen can cut down on skin damage, wearing optical-quality sunglasses capable of blocking at least 99% of UV-A and UV-B can protect your eyes. Learn more about sunglasses in this next activity.

Consider This 2.21   Protecting Your Eyes

Sunglasses make far more than a fashion statement. They offer protection from harmful UV rays. Check out several manufacturers to learn the virtues of their products. What activities especially require good UV eye protection? Report on your findings.

Other species on the globe are affected by UV radiation as well. For example, increases in UV radiation can harm young marine life, such as floating fish eggs, fish larvae, juvenile fish, and shrimp larvae. Experimental evidence also exists for damage to the DNA of the eggs of Antarctic ice fish. Plant growth is suppressed by increased UV radiation, and experiments have measured the negative effect that increased UV-B radiation has on phytoplankton. These photosynthetic microorganisms live in the oceans, where they occupy a fundamental niche in the food chain. Any significant decrease in their number could have a major effect globally. In 1999, an international panel of scientists confirmed that exposure to elevated levels of UV-B affected both the movement of phytoplankton up and down in water as well as their ability to move through the water. If these microorganisms are not able to achieve proper position in the water, they cannot carry out photosynthesis effectively. Moreover, these tiny plant-like organisms play an important role in the carbon dioxide balance of the planet by absorbing atmospheric carbon dioxide.

Given the harmful effects of too much UV radiation, you now can see why the decreasing stratospheric ozone concentrations observed in the 1980s set off planetary alarm bells. The next two sections tell the tale of the ozone hole and how it unexpectedly appeared on our planet.

2.8 | Stratospheric Ozone Destruction: Global Observations and Causes

Switzerland holds the record for the longest continuous set of ozone level measurements. Since 1926, stratospheric O_3 concentrations have been measured at the Swiss Meteorological Institute. More recently, starting in 1979, satellite-mounted detectors have been beaming down data on ozone levels at many points. These measurements show both that the natural concentration of stratospheric O_3 is not uniform across the globe and that the levels have changed over time.

On average, the total O_3 concentration is higher the closer one gets to either pole, with the exception of the seasonal ozone "hole" over the Antarctic. The formation of ozone in the Chapman cycle is triggered when an O_2 molecule absorbs a photon of UV-C light, splitting into two O atoms. These in turn react with O_2 to form O_3.

$$O_2 \xrightarrow[(\lambda \leq 242 \text{ nm})]{\text{UV-C}} 2\,O \qquad \text{[2.6a]}$$

$$O + O_2 \longrightarrow O_3 \qquad \text{[2.6b]}$$

Therefore, ozone production increases with the intensity of the radiation striking the stratosphere, which in turn depends primarily on the angle of the Earth with respect to the Sun and the distance between the Sun and the Earth. At the equator, the period of highest intensity occurs at the equinox (March and October) when the Sun is directly overhead. Outside the tropics, the Sun is never directly overhead, so the maximum intensity occurs at the summer solstice (June in the Northern Hemisphere, December in the Southern). The angle of Earth with respect to Sun dominates both ozone production and the seasons. There is a slight (~7%) increase, however, in solar power reaching Earth in early January, when the Earth is nearest the Sun, compared with July, when the Earth is farthest away. In addition, the amount of radiation emitted by the Sun changes over an 11- to 12-year cycle related to sunspot activity. This variation also influences O_3 concentrations, but only by a percent or two. The wind patterns in the stratosphere cause other variations in ozone concentrations, some on a seasonal basis and others over a longer cycle. To further complicate matters, seemingly random fluctuations often occur.

Extraordinary images of the Earth, such as the one that opens this chapter, are color-coded to show stratospheric ozone concentrations. The dark blue and purple regions indicate where the lowest concentrations of O_3 are observed. Total ozone levels above Earth's surface are expressed in Dobson units (DU). A value of 250–270 DU is

In conjunction with the Earth's energy balance and global warming, look for information about solar irradiance in Section 3.9.

Recall that a Dobson unit corresponds to about one ozone molecule for every billion molecules and atoms of air.

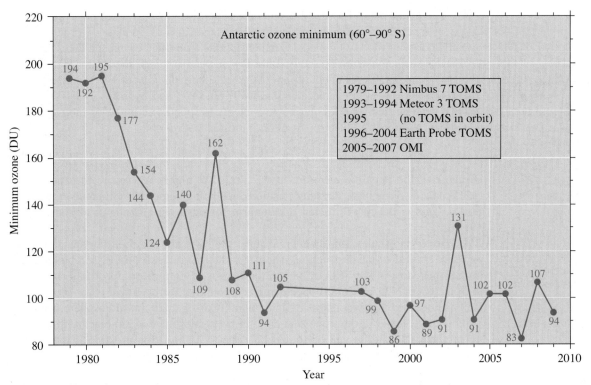

Figure 2.14

Minima in spring stratospheric ozone in Antarctica, 1979–2009. The number is the minimum reading in Dobson units. TOMS (Total Ozone-Measuring Spectrometer) and OMI (Ozone Monitoring Instrument) are analytical instruments.

Source: NASA.

typical at the equator. As one moves away from the equator, values range between 300 and 350 DU, with seasonal variations. At the highest northern latitudes, values can be as high as 400 DU.

Of special interest is the thinning of ozone (the "ozone hole") that occurs seasonally over the South Pole. Indeed, these changes were so pronounced that, when the British monitoring team at Halley Bay in Antarctica first observed it in 1985, they thought their instruments were malfunctioning! The area over which ozone levels are reported to be less than 220 DU is usually considered to be the "hole." From the mid-1990s on, the annual size of the ozone hole has nearly equaled the total area of the North American continent, and in some cases exceeded it.

Check out the dramatic decline in stratospheric ozone levels observed near the South Pole shown in Figure 2.14. In recent years, the minimum has been around 100 DU. Keep in mind that seasonal variation has always occurred in ozone concentration over the South Pole, with a minimum in late September or early October—the Antarctic spring. Unprecedented, however, is the striking decrease in this minimum that has been observed in recent decades.

Consider This 2.22   This Year's Ozone Hole

Starting in September, citizens and scientists alike examine the data for the ozone hole over Antarctica. What is happening this year? NASA posts the data, and the textbook's website provides a link. For the most recent year:

a. What is the area of the hole? How does this compare with recent years?

b. What is the lowest reading observed for ozone? Again, how does this compare?

The major natural cause of ozone destruction, wherever it takes place around the globe, is a series of reactions involving water vapor and its breakdown products. The great majority of the H_2O molecules that evaporate from the oceans and lakes fall back to Earth's surface as rain or snow. But a few molecules reach the stratosphere, where the H_2O concentration is about 5 ppm. At this altitude, photons of UV radiation trigger the dissociation of water molecules into hydrogen (H·) and hydroxyl (·OH) free radicals. A **free radical** is a highly reactive chemical species with one or more unpaired electrons. An unpaired electron is often indicated with a dot:

$$H_2O \xrightarrow{\text{photon}} H\cdot + \cdot OH \qquad [2.7]$$

Because of the unpaired electron, free radicals are highly reactive. Thus, the H· and ·OH radicals participate in many reactions, including some that ultimately convert O_3 to O_2. This is the most efficient mechanism for destroying ozone at altitudes above 50 km.

Water molecules and their breakdown products are not the only agents responsible for natural ozone destruction. Another is the free radical ·NO, also called nitrogen monoxide. Most of the ·NO in the stratosphere is of natural origin. It is formed from nitrous oxide, N_2O, a naturally occurring compound of nitrogen that is produced in the soil and oceans by microorganisms and gradually drifts up to the stratosphere. Although N_2O is quite stable in the troposphere, up in the stratosphere it can react with O atoms to produce ·NO. Little can or should be done to control this process. It is part of a natural cycle involving compounds of nitrogen, as we will see in Chapter 6.

Free radicals appear in several other contexts.

Chapter 1: ·OH, formation of NO_2 and then tropospheric ozone
Chapter 6: ·OH, formation of SO_3 in acid rain
Chapter 7: ·OH and $H_2O\cdot^+$, damage to cells by nuclear radiation
Chapter 9: R·, polymerization of ethylene

The air pollutant NO (nitrogen monoxide or nitric oxide) is highly reactive, which is one of the reasons it causes harm to your lungs. In contrast, N_2O (nitrous oxide or "laughing gas") is a very stable molecule that persists for decades.

Your Turn 2.23   Free Radicals

a. Draw the Lewis structure for the ·OH free radical. The unpaired electron goes on the O atom.
b. Draw the Lewis structure for the ·NO free radical. The unpaired electron goes on the N atom. *Note:* In Chapter 1, in the context of air pollution, we wrote simply NO.
c. In contrast, N_2O has no unpaired electrons. Draw its Lewis structure. *Hint:* Place one of the N atoms in the middle.

Human activities also can alter steady-state concentrations of NO. In the 1970s, people became concerned about the increase in NO concentration that would result from building a fleet of Concorde SSTs (supersonic transports). These planes were designed to fly at an altitude of 15–20 km, the region of the ozone layer. As you learned in Chapter 1, hot engines produce NO as part of the exhaust gas stream. The NO is produced by the hot engines of a jet plane on takeoff, landing, and during flight.

$$N_2 + O_2 \xrightarrow{\text{high temperature}} 2\,NO \qquad [2.8]$$

Scientists carried out experiments and calculations to predict the effects of a fleet of SSTs. They concluded that the risks would outweigh the benefits. So people decided, partly on scientific grounds, not to build an American fleet of SSTs. Until 2003, the Anglo–French Concorde was the only commercial plane that operated at this altitude. The Concorde took its last flight on October 24, 2003. Both safety concerns and economic factors played roles in ending the flights of these remarkable jets.

Even when the effects of water, nitrogen oxides, and other naturally occurring compounds are included in stratospheric models, the measured ozone concentration is still lower than predicted. Measurements worldwide indicate that the ozone concentration has been decreasing over the past 20 years. There is a good deal of fluctuation in the data, but the trend is clear. The stratospheric ozone concentration at midlatitudes (60° south to 60° north) has decreased by more than 8% in some cases. These changes cannot be correlated with changes in the intensity of solar radiation, so we must look elsewhere for an explanation. Thus it is time to turn our attention to chlorofluorocarbons.

2.9 | Chlorofluorocarbons: Properties, Uses, and Interactions with Ozone

A major cause of stratospheric ozone depletion was uncovered through the masterful scientific sleuthing of F. Sherwood Rowland, Mario Molina, and Paul Crutzen. For their work, the trio jointly won the 1995 Nobel Prize in chemistry. They analyzed vast quantities of atmospheric data and studied hundreds of chemical reactions. As with most scientific investigations, some uncertainties remained. Nonetheless, their evidence all pointed to an unlikely group of compounds: the chlorofluorocarbons.

As the name implies, **chlorofluorocarbons** (CFCs) are compounds composed of the elements chlorine, fluorine, and carbon (but do not contain the element hydrogen). Fluorine and chlorine are members of the same chemical group, the halogens (Figure 2.15). In their elemental state, all of the halogens are diatomic molecules, but only fluorine and chlorine are gases. Fluorine is not shown in the figure because it is so reactive that it would react with the glass vessel. In contrast, CFCs are highly unreactive.

To get started with CFCs, let's examine two examples.

CCl_3F
trichlorofluoromethane
Freon-11

and

CCl_2F_2
dichlorodifluoromethane
Freon-12

Note how the names show the connection of CFCs to methane, CH_4. The prefixes *di-* and *tri-* specify the number of chlorine and fluorine atoms that substitute for hydrogen atoms of methane. These two CFCs also are known by their trade names, Freon-11 and Freon-12. You may also hear them called CFC-11 and CFC-12, respectively, following a naming scheme developed in the 1930s by chemists at DuPont.

CFCs do not occur in nature; we humans synthesized them for a variety of uses. This is an important verification point in the debate over the role of CFCs in stratospheric ozone depletion. As we saw in the previous section, other contributors to the destruction of ozone, such as the ·OH and ·NO free radicals, are formed in the atmosphere both from natural sources and human activities.

Rightly, the introduction of CCl_2F_2 as a refrigerant gas in the 1930s was hailed as a great triumph of chemistry and an important advance in consumer safety. It replaced ammonia or sulfur dioxide, two toxic and corrosive refrigerant gases. In many respects, CCl_2F_2 was (and still is) an ideal substitute. It is nontoxic, odorless, colorless, and does not burn. In fact, the CCl_2F_2 molecule is so stable that it does not react with much of anything!

Given the desirable nontoxic properties of CFCs, they soon were put to other uses. For example, CCl_3F was often blown into polymer mixtures to make foams for cushions and foamed insulation. Other CFCs served as propellants in aerosol spray cans and as nontoxic solvents for oil and grease.

Halons are close cousins of CFCs. Like them, halons are inert, nontoxic compounds that contain chlorine or fluorine (or both, but no hydrogen). But in addition, they contain bromine. For example, here is the Lewis structure for bromotrifluoromethane, $CBrF_3$, also known as Halon-1301.

Chlorine

Bromine

Iodine

Figure 2.15
Selected elements from Group 7A, the halogen family.

Methane, the smallest hydrocarbon, was described in Chapter 1.

For more about Freons and how they are named, see end-of-chapter question #53.

Polymers and plastics are the topic of Chapter 9. Gases that "puff up" the plastic, making it into a foam, are called blowing agents.

You will find different definitions for the chemical composition of halons, depending on where you look. Sometimes halons are defined by their use (in suppressing fires) rather than by their chemical composition.

Halons are used as fire suppressants. They are especially helpful when a fire hose or sprinkler system would be inappropriate, for instance in libraries (especially rare book rooms), grease fires (where water might spread the fire), chemical stockrooms (where some chemicals react with water), and aircraft (where hosing down the cockpit would definitely be a bad idea).

For better or worse, the synthesis of CFCs has had a major effect on our lives. Because CFCs are nontoxic, nonflammable, cheap, and widely available, they revolutionized air conditioning, making it readily accessible for homes, office buildings, shops, schools, and automobiles. Beginning in the 1960s and 1970s, CFCs helped to spur the growth of cities in hot and humid parts of the world. In effect, a major demographic shift occurred because of CFC-based technology that transformed the economy and business potential of entire regions of the globe.

In the United States, CFCs helped spur the growth of cities with hot, humid weather, including Atlanta, Houston, Tampa, and Memphis.

Ironically, the very property that makes CFCs ideal for so many applications—their chemical inertness—ended up doing harm to our atmosphere. The C–Cl and C–F bonds in the CFCs are so strong that the molecules are virtually indestructible. For example, it has been estimated that an average CCl_2F_2 molecule can persist in the atmosphere for 120 years before it meets some fate that decomposes it. In contrast, it only takes about five years for atmospheric wind currents to bring molecules up to the stratosphere, which is exactly where some of the CFC molecules ended up.

In 1973, Rowland and Molina, motivated largely by intellectual curiosity, set out to study the fate of stratospheric CFC molecules. They understood that with increasing altitude, the concentrations of oxygen and ozone decrease, but the intensity of UV radiation increases. They reasoned photons of high-energy UV-C light (<220 nm) would break C–Cl bonds. Here is the chemical reaction that releases chlorine atoms from dichlorodifluoromethane.

$$F\!-\!\underset{\underset{\displaystyle F}{|}}{\overset{\overset{\displaystyle Cl}{|}}{C}}\!-\!Cl \quad \xrightarrow[\lambda \le 220\,\text{nm}]{\text{UV photon}} \quad F\!-\!\underset{\underset{\displaystyle F}{|}}{\overset{\overset{\displaystyle Cl}{|}}{C}}\!\cdot \;+\; \cdot Cl \qquad \textbf{[2.9]}$$

A chlorine atom has seven outer electrons, one of them unpaired. We depict it as Cl· or ·Cl to emphasize this unpaired electron. The chlorine atom exhibits a strong tendency to achieve a stable octet by combining and sharing electrons with another atom. Rowland and Molina and subsequent researchers hypothesized that this reactivity would result in a series of reactions. Although CFCs are known to destroy stratospheric ozone via several pathways, we illustrate with a typical one known to take place in polar regions.

First, the Cl· free radical pulls an oxygen atom away from an O_3 molecule to form ClO·, chlorine monoxide and an O_2 molecule. The coefficient 2 is not canceled because we anticipate using it the next step.

The Br· free radical undergoes a comparable reaction, starting another cycle of ozone destruction. Br· is up to 10 times more effective than Cl· in destroying O_3.

$$2\,Cl\cdot + 2\,O_3 \longrightarrow 2\,ClO\cdot + 2\,O_2 \qquad \textbf{[2.10]}$$

The ClO· species is another free radical; it has 13 outer electrons (7 + 6). Recent experimental evidence indicates that 75–80% of stratospheric ozone depletion involves joining two ClO· radicals to form ClOOCl.

$$2\,ClO\cdot \longrightarrow ClOOCl \qquad \textbf{[2.11]}$$

In turn, ClOOCl decomposes in a two-step sequence.

$$ClOOCl \xrightarrow{\text{UV photon}} ClOO\cdot + \cdot Cl \qquad \textbf{[2.12a]}$$

$$ClOO\cdot \longrightarrow Cl\cdot + O_2 \qquad \textbf{[2.12b]}$$

We can treat this set of chemical equations as if they were mathematical equations. If we add them together, here is the result.

$$2\,\cancel{Cl\cdot} + 2\,O_3 + 2\,\cancel{ClO\cdot} + ClOOCl + \cancel{ClOO\cdot} \longrightarrow$$
$$2\,\cancel{ClO\cdot} + 2\,O_2 + \cancel{ClOOCl} + \cancel{ClOO\cdot} + \cancel{Cl\cdot} + \cancel{Cl\cdot} + O_2 \qquad \textbf{[2.13]}$$

Just as is done with mathematical equations, we can eliminate the duplicate Cl•, ClO•, and ClOOCl species from both sides of the chemical equation. The terms for O_2 on the right side of the equation, $2\,O_2$ and O_2, can be combined into $3\,O_2$. What remains is the net equation showing the conversion of ozone into oxygen gas.

$$2\,O_3 \longrightarrow 3\,O_2 \qquad\qquad [2.14]$$

Thus, the complex interaction of ozone with atomic chlorine provides a pathway for the destruction of ozone.

Notice that Cl• appears both as a reactant in equation 2.13 and as a product in equations 2.12a and 2.12b. This means that Cl• is both consumed *and* regenerated in the cycle, with no net change in its concentration. Such behavior is characteristic of a catalyst, a chemical substance that participates in a chemical reaction and influences its speed without undergoing permanent change. Atomic chlorine acts catalytically by being regenerated and recycled to remove more ozone molecules. On average, a single atom can catalyze the destruction of as many as 1×10^5 ozone molecules before it is carried back to the lower atmosphere by winds.

Interestingly, the mechanism just described for ozone destruction by CFCs in the stratosphere was not the one first proposed by Rowland and Molina. Their initial hypothesis was that Cl• reacted with O_3 to form ClO• and O_2. The second step proposed was that ClO• reacted with oxygen atoms to form O_2 and regenerate radicals.

> The term catalyst also was mentioned in Section 1.11 in connection with catalytic converters.

$$Cl\bullet + O_3 \longrightarrow ClO\bullet + O_2 \qquad\qquad [2.15]$$

$$ClO\bullet + O \longrightarrow Cl\bullet + O_2 \qquad\qquad [2.16]$$

Although this mechanism did not prove to be the major one in the formation of the ozone hole, it did provide a reasonable explanation for why recycling a limited number of chlorine atoms could be responsible for the destruction of a large number of ozone molecules. But it is the cycle that primarily accounts for the destruction of ozone in the tropical and midlatitudes, regions in which the incident sunlight is more intense. As is often true in science, hypotheses need to be recast in light of experimental evidence.

Thankfully, almost all of the chlorine in the stratosphere is not in the active form of Cl• or ClO•. Rather, chlorine is incorporated into stable compounds that do not destroy ozone. Hydrogen chloride (HCl) and chlorine nitrate ($ClONO_2$) are two such compounds. These form quite readily at altitudes below 30 km. Thus, chlorine atoms are fairly effectively removed from the region of highest ozone concentration (about 20–25 km). These gases, as we will see in Chapter 5, are water-soluble. So in the troposphere, they are removed from the air when they wash out in the rain.

> Although HCl and $ClONO_2$ do not destroy ozone, they still are potential sources of Cl•. For example, HCl can react with the hydroxyl radical (•OH) to produce Cl•.

Your Turn 2.24   Bromine, too!

Although we have been casting the discussion in terms of chlorine atoms, bromine atoms also play a role.

a. Write chemical reactions involving bromine analogous to equations 2.10 and 2.15.

b. Bromine concentrations are much lower than those of chlorine. Propose a reason why.

Answer

b. Fewer of the substances that deplete ozone contain the element bromine. These compounds, such as $CBrF_3$ (Halon-1301) have been manufactured in smaller amounts.

Rowland, a professor at the University of California at Irvine, and Molina, then a postdoctoral fellow in Rowland's laboratory, published their first paper on CFCs and ozone depletion in 1974 in the scientific journal *Nature*. At about the same time, other scientists were obtaining the first experimental evidence of stratospheric ozone depletion and CFCs in the stratosphere. The conclusions were troublesome; the implications were that the use of CFCs should be discontinued. These initial reports were met with

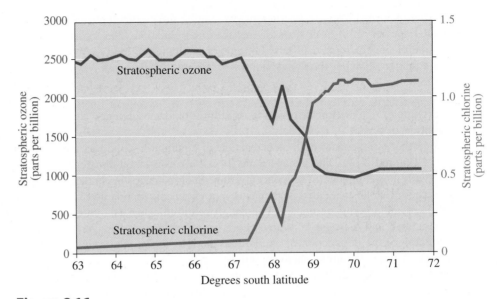

Dr. Susan Solomon, a chemist, headed the team that first gathered stratospheric ClO• and ozone data over Antarctica. The data solidified the connection between CFCs and the ozone hole. She was just 30 years old at the time.

Figure 2.16

Antarctic stratospheric concentrations of ozone and reactive chlorine (from a flight into the Antarctic ozone hole, 1987).

Source: United Nations Environment Programme.

skepticism, as might be expected when much was at stake economically. But the precautionary principle ultimately prevailed. Action was taken to mitigate the loss of ozone before even worse ozone destruction took place.

Over the years, the correctness of the Rowland–Molina hypothesis has been well established. Perhaps the most compelling evidence for the involvement of chlorine and chlorine monoxide is presented in Figure 2.16. It shows two sets of data from the Antarctic, one for O_3 concentration and the other for ClO•. Both are plotted versus the latitude at which samples were measured. As stratospheric O_3 concentration decreases, the concentration increases; the two curves mirror each other almost perfectly. The major effect is a decrease in ozone and an increase in chlorine monoxide as the South Pole is approached. Because equation 2.10 links ClO•, Cl•, and O_3, the conclusion is compelling. Figure 2.16 is sometimes described as the "smoking gun" for stratospheric ozone depletion.

Not all of the chlorine implicated in stratospheric ozone destruction comes from CFCs. Other chlorinated carbon compounds come from natural sources, such as sea water and volcanoes. However, most chlorine from natural sources is in water-soluble forms. Therefore, any natural chlorine-containing substances are washed out of the atmosphere by rainfall long before they can reach the stratosphere. Of particular significance are the data gathered by NASA and by international researchers that establish that high concentrations of HCl (hydrogen chloride) and HF (hydrogen fluoride) always occur together. Although some of the HCl might conceivably arise from a variety of natural sources, the only reasonable origin of stratospheric concentrations of HF is CFCs.

Consider This 2.25   Radio Talk Show Opinions

"And if prehistoric man merely got a sunburn, how is it that we are going to destroy the ozone layer with our air conditioners and underarm deodorants and cause everybody to get cancer? Obviously we're not . . . and we can't . . . and it's a hoax. Evidence is mounting all the time that ozone depletion, if occurring at all, is not doing so at an alarming rate."*

Consider the first thing you would ask this talk-show host about these statements. Remember that you need to formulate a short and focused question to get any airtime!

*Source: Limbaugh, R. 1993. *See, I Told You So.* New York: Pocket Books.

2.10 | The Antarctic Ozone Hole: A Closer Look

Ozone-depleting gases are present throughout the stratosphere. Furthermore, as a result of global wind patterns, CFCs are present in comparable abundance in lower parts of the atmosphere over *both* hemispheres. Why, then, have the greatest losses of stratospheric ozone occurred over Antarctica? And given that more ozone-depleting gases are emitted in the Northern Hemisphere, why are their effects felt most strongly in the Southern Hemisphere?

A special set of conditions exist in Antarctica, ones that relate to the fact that the lower stratosphere over the South Pole is the coldest spot on Earth. From June to September (Antarctic winter), the winds that circulate around the South Pole form a vortex that prevents warmer air from entering the region. As a result, the temperature may drop as low as $-90\,°C$. Under these conditions, **polar stratospheric clouds (PSCs)** can form. These thin clouds are composed of tiny ice crystals formed from the small amount of water vapor present in the stratosphere. The chemical reactions that occur on the surface of these ice crystals convert molecules that do not deplete ozone, such as $ClONO_2$ and HCl that we mentioned previously, to the more reactive species that do: $HOCl$ and Cl_2.

Neither $HOCl$ nor Cl_2 causes any harm in the dark of winter. But when sunlight returns to the South Pole in late September, the light splits $HOCl$ and Cl_2 to release chlorine atoms. Given this increase in $Cl\cdot$, a species that destroys vast quantities of ozone, the hole starts to form. Notice the conditions required: extreme cold, a circular wind pattern (vortex), enough time for ice crystals to form and provide a surface for the reactions, and darkness followed by rapidly increasing levels of sunlight. Figure 2.17 shows the seasonal variation and compares the minimum temperatures above the Arctic and the Antarctic. As you can see, the necessary conditions more often are found in Antarctica.

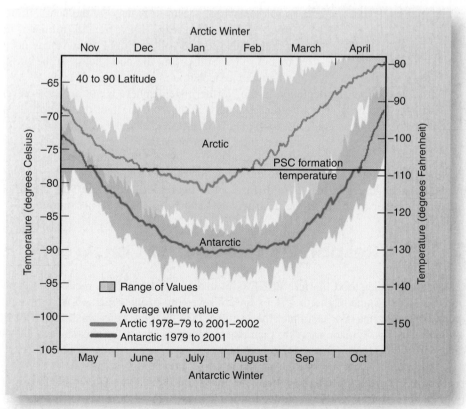

Figure 2.17

Minimum air temperatures in the polar lower stratosphere. Polar stratospheric clouds (PSCs) are thin clouds of ice crystals that form at very low temperatures.

Source: Scientific Assessment of Ozone Depletion: 2002, World Meteorological Organization, United Nations Environment Programme.

Figure 2.18

Arctic polar stratospheric clouds in the northern part of Sweden.

Photo credit: Ross J. Salawitch, University of Maryland.

In the context of air quality, the tragedy of the commons was first mentioned in Chapter 1.

The colors in PSCs are caused by diffraction of light by the ice particles in the clouds.

Changes in ozone above Antarctica closely follow the seasonal temperatures. Typically a rapid ozone decline takes place during spring at the South Pole, that is, from September to early November. As the sunlight warms the stratosphere, the polar stratospheric clouds dissipate, halting the chemistry that occurs on the surfaces of their ice crystals. Then air from lower latitudes flows into the polar region, replenishing the depleted ozone levels. By the end of November, the hole is largely refilled. Although the deepest decrease in the ozone layer over Antarctica occurs during the spring, recent discoveries by British Antarctic Survey researchers indicate that the ozone depletion may begin earlier, as early as midwinter at the edges of the Antarctic, including over populated southern areas of South America.

This situation presents us with another example of the tragedy of the commons—a resource that is common to all and used by many, but has no one in particular responsible for it. As a result, the resource may be harmed to the detriment of all. Here, the resource is our protective ozone layer. Decreased stratospheric ozone over the South Pole leads to increased UV-B levels reaching the Earth. In turn, skin cancer rates increase in Australia and southern Chile. Australian scientists believe that wheat, sorghum, and pea crop yields have decreased a result of increased UV radiation. Similar effects are also being felt in southern Chile in the area around Punta Arenas, and on the island of Tierra del Fuego at the southernmost tip of South America. Chile's health minister has warned the 120,000 residents of Punta Arenas not to be out in the Sun during the noon hours in the spring, when ozone depletion is greatest.

It turns out that the depletion in the Northern Hemisphere is not nearly as severe as it is in the Southern. The difference stems mainly from the fact that the air above the North Pole is not as cold. Even so, polar stratospheric clouds have been repeatedly observed in the Arctic. For example, the "mother-of-pearl" polar stratospheric clouds shown in Figure 2.18 were photographed above Porjus, a village in Swedish Lapland. However, these clouds do not lead to the formation of an ozone hole, as the air trapped over the Arctic generally begins to diffuse out of the region before the Sun gets bright enough to trigger as much ozone destruction as has been observed in Antarctica.

2.11 | Responses to a Global Concern

Once people understood the role of CFCs in ozone destruction, the response was surprisingly rapid. Some of the first steps toward reversing ozone depletion were taken by individual countries. For example, the use of CFCs in spray cans was banned in the United States and Canada in 1978; their use as foaming agents for plastics was discontinued in 1990. The problem of CFC production and subsequent release, however, was a global one. As such, it required international cooperation.

In 1977, in response to growing experimental evidence, the United Nations Environment Programme (UNEP) convened a meeting. Those in attendance adopted a World Plan of Action on the Ozone Layer and established a coordinating committee to guide future international actions. In 1985, world leaders participated in the Vienna Convention on the Protection of the Ozone Layer. Through action taken at the convention, the nations represented committed themselves to protecting the ozone layer and to conducting scientific research to better understand atmospheric processes. A major breakthrough came in 1987 with the signing of a treaty: the Montreal Protocol on Substances That Deplete the Ozone Layer. Each nation that signed the treaty then needed to ratify it. Many did so immediately, but it was not until 2009 until the last four nations joined the other 182 nations who had ratified the treaty.

Consider This 2.26   Graffiti with a Message

a. This cartoon dates from the mid-1970s. Explain the basis of its humor.

b. Is this cartoon still relevant to the problem of ozone depletion today? Explain.

c. Create a cartoon of your own that deals with ozone depletion. Be sure that the chemistry is correct!

The key initial strategy for reducing chlorine in the stratosphere was to stop production of CFCs. The United States and 140 other countries agreed to a complete halt in CFC manufacture after December 31, 1995. Figure 2.19 indicates that the decline in global CFC production has been dramatic. By 1996, production of CFCs had fallen to 1960 levels. Without the international action required by the Montreal Protocol, stratospheric abundances of chlorine found in the stratosphere could have tripled by the middle of the 21st century.

The Protocol included a provision to hold future meetings to revise goals as new scientific knowledge evolved. These meetings turned out to be of key importance, because all soon agreed—atmospheric scientists, environmentalists, chemical manufacturers, and government officials—that the Montreal Protocol was not stringent enough. Subsequent meetings have been held in locations worldwide. Currently, the goal is to phase out not just CFCs, but also 96 other substances that deplete ozone!

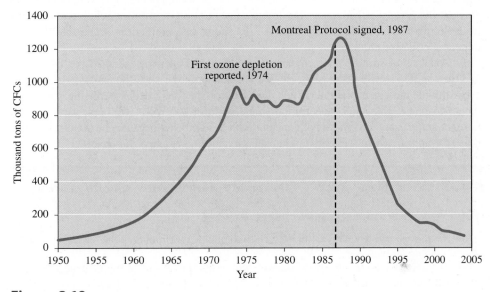

Figure 2.19

Global production of CFCs, 1950–2004.

Source: United Nations Environment Programme (UNEP).

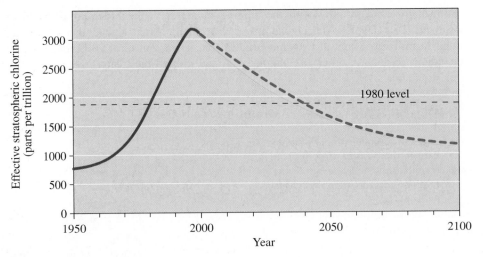

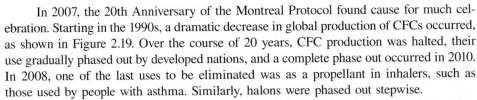

Figure 2.20

Concentrations of effective chlorine, 1950–2100. The height of the yellow band for any year is an estimate of the uncertainty in the prediction.

Source: Scientific Assessment of Ozone Depletion: 2002, World Meteorological Organization, United Nations Environment Programme.

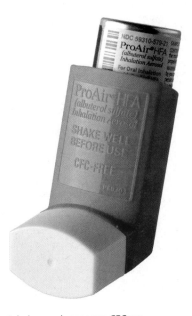

Inhalers no longer use CFCs as propellants.

In 2007, the 20th Anniversary of the Montreal Protocol found cause for much celebration. Starting in the 1990s, a dramatic decrease in global production of CFCs occurred, as shown in Figure 2.19. Over the course of 20 years, CFC production was halted, their use gradually phased out by developed nations, and a complete phase out occurred in 2010. In 2008, one of the last uses to be eliminated was as a propellant in inhalers, such as those used by people with asthma. Similarly, halons were phased out stepwise.

Stopping production of CFCs and restricting their uses did not immediately translate into a drop in the stratospheric concentration of chlorine. Most Earth systems, including those in the atmosphere, are complex and slow to respond to changes. In fact, the atmospheric concentrations of ozone-depleting gases continued to rise steadily through the 1990s, despite the restrictions of the Montreal Protocol and its subsequent amendments.

One reason for the slow rate of change is that many of the CFCs have long atmospheric lifetimes, some estimated to be over 100 years. Even so, the signs are encouraging that the Montreal Protocol has had a beneficial effect. Decreases are now being observed in the amount of **effective stratospheric chlorine,** a term reflecting both chlorine- and bromine-containing gases in the stratosphere. The values take into account the greater effectiveness but lower concentration of bromine relative to chlorine in depleting stratospheric ozone. Figure 2.20 shows one prediction of the future abundance of effective chlorine.

Analysis of stratospheric chlorine levels indicates that it peaked in the late 1990s and then started slowly decreasing. But we are not yet in the clear. Scientists estimate that even with the most stringent international controls on the use of ozone-depleting chemicals, the stratospheric chlorine concentration would not drop to 2 ppb (2000 ppt, parts per trillion) for some years to come. That concentration is significant because the Antarctic ozone hole first appeared when effective stratospheric chlorine reached that level.

Consider This 2.27   Past and Future Effective Chlorine Levels

Use Figures 2.19 and 2.20 to answer these questions.

a. In approximately what year did effective chlorine concentration peak? What was the reading in that year?

b. Is the peak year for effective chlorine concentration the same as the peak year for CFC production? Why or why not?

c. According to predictions, in approximately what year will the effective chlorine level return to 1980 levels? What will the reading be in that year?

Although the Montreal Protocol and its subsequent adjustments set dates for the halt of all CFC production, the sale of existing stockpiles and recycled materials will remain legal for some time into the future. This is necessary because appliances that were designed to operate using CFCs still remain in use. For example, air conditioning units produced in the United States for cars and homes before 1996 were designed to run on CFCs. But as you might surmise, both the paperwork and the price of legally obtained CFCs have risen sharply. As of 2010, the production of CFCs is banned for all signatories of the Montreal Protocol, with the possible exception of a small number of essential uses, such as those in medicine.

2.12 | Replacements for CFCs

No one seriously advocates returning to ammonia and sulfur dioxide in home refrigeration units or giving up air conditioning as solutions to necessary restrictions on CFCs. Instead, chemists are synthesizing new compounds, concentrating on those similar to the CFCs but without their problematic long-term effects on stratospheric ozone. Substitute molecules might include one or two carbon atoms, at least one hydrogen atom, and fluorine or chlorine atoms. The rules of molecular structure limit the options.

In synthesizing substitutes for CFCs, chemists initially weighed three undesirable properties—toxicity, flammability, and extreme stability—and attempted to achieve the most suitable compromise. For example, introducing hydrogen atoms in place of one or more of the halogen atoms reduces molecular stability and promotes destruction of the compounds at low altitudes, long before they enter the ozone-rich regions of the atmosphere. However, too many hydrogen atoms increase flammability.

Moreover, if a hydrogen atom replaces a halogen atom, the total mass of the molecule is decreased. This results in a decrease in boiling point, making the compounds less than ideal for use as refrigerants. A boiling point in the −10 to −30 °C range is an important property for a refrigerant gas. Too many chlorine atoms seem to increase toxicity. Chloroform, $CHCl_3$, would therefore not be a good CFC substitute both because of its toxicity and its higher boiling point (61 °C). The relationships among composition, molecular structure, boiling point, and proposed use must all be considered along with toxicity, flammability, and stability for any substitute.

Fortunately, chemists were able to use their knowledge to synthesize some promising replacements for CFCs. Here are two examples of **hydrochlorofluorocarbons, (HCFCs),** compounds of hydrogen, chlorine, fluorine, and carbon.

CHClF$_2$
chlorodifluoromethane
HCFC-22

$C_2H_3Cl_2F$
dichlorofluoroethane
HCFC-141b

Starting with the 1996 model year, automobiles in the United States use HCFCs in their air conditioning units rather than CFCs. Even so, older cars are still on the roads.

The first of these, $CHClF_2$ (also called R-22) was once the most widely used HCFC. Its ozone-depleting potential is about 5% that of CFC-12 and its estimated atmospheric lifetime is only 20 years, compared with 111 years for CFC-12. It is suitable both for air conditioners and as a blowing agent to make fast-food containers. The second, $C_2H_3Cl_2F$, is also used as a blowing agent to make foam insulation.

Look for more about foam containers for fast food and blowing agents in Section 9.4.

To their credit, HCFCs decompose in the troposphere more readily than CFCs and hence do not accumulate to the same extent in the stratosphere. However, HCFCs still contain chlorine. As a result, they still have adverse effects on the ozone layer and thus can be regarded only as an interim solution.

The Montreal Protocol and its amendments are tackling the problems associated with HCFCs. This situation is complex in that developing nations want and need to use these compounds as refrigerant gases. In 2007, an adjustment was made concerning the production and consumption of HCFCs. Little change was made for the developed nations that are Parties to the Montreal Protocol, as they already were subject to caps on their consumption and production, ultimately leading to a phase out by 2030. For example, by 2015, the production or importation of *all* HCFCs will end in the United States. Any continued demand to service refrigerating equipment manufactured prior to those deadlines must be met with recovered HCFCs. But with the 2007 adjustment, a timetable was launched for the developing nations, one that also would lead to a 100% reduction by 2030.

During the phaseout period, the price of available HCFCs will undoubtedly rise dramatically. One factor will be the limited supply of HCFCs for existing equipment but another is change in air-conditioning energy-efficiency regulations mandated by the Department of Energy's new Seasonal Energy Efficiency Rating (SEER) standards for residential air conditioner manufacturers. Starting in January 2006, SEER ratings of new air-conditioning units were required to show a 30% increase in energy efficiency, a performance that HCFCs helped manufacturers achieve.

If HCFCs are being phased out, what options do we have? At first glance, **hydrofluorocarbons, (HFCs),** compounds of hydrogen, fluorine, and carbon, would seem to be likely candidates. Here are two examples.

C_2HF_5
pentafluoroethane
HFC-125

and

CH_2F_2
difluoromethane
HFC-32

From these structures, you can see that HFCs have no chlorine atoms to deplete ozone. In addition, their hydrogen atoms facilitate decomposition in the lower atmosphere, so they do not have excessively long atmospheric lifetimes. Furthermore, compounds containing only C and F (fluorocarbons) are nontoxic and nonflammable under normal conditions.

The two HFCs for which we just drew Lewis structures, C_2HF_5 and CH_2F_2, often are blended to produce the refrigerant known as R-410A. Newer designs for air conditioners will be engineered to use this blend as a replacement for HCFC-22.

Consider This 2.28   Blended HFCs

R-407c is a blend of HFC-125, HFC-32, and HFC-134a. The formula, name, and Lewis structure for the first two components of R-407c were given earlier. The formula for HFC-134a is $C_2H_2F_4$ and its name is tetrafluoroethane.

a. In what ways are HFC-125, HFC-32, and HFC-134a different from CFCs?
b. Draw the Lewis structure for HFC-134a.
   Hint: Place two F atoms on each C atom.

At the risk of getting ahead of the story of global warming, which we will tell in Chapter 3, we now mention that HFCs are greenhouse gases. Like CO_2, HFCs absorb infrared radiation, trap heat in the atmosphere, and contribute to global warming. Actually, the same is true for CFCs, as you can see from Figure 2.21. Thus, over the long haul, HFCs cannot serve as replacements either.

Also to be phased out under the Montreal Protocol are the bromine-containing halons, compounds discussed earlier in Section 2.9. Although very effective in fighting fires, molecule for molecule, halons are even more effective than CFCs in harming the ozone layer. In addition and just like HFCs, halons contribute to global warming as you can see in Figure 2.21.

In 1998, Pyrocool Technologies of Monroe, Virginia, won a Presidential Green Chemistry Challenge Award for its development of foam that is environmentally benign and yet more effective than the halons it replaces. The product, Pyrocool

Global climate change is the topic of Chapter 3.

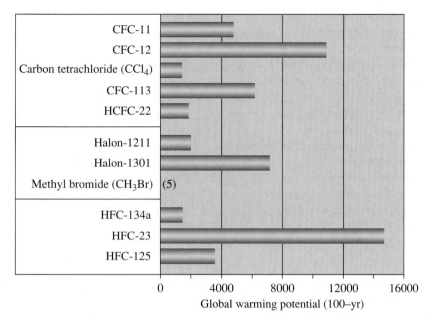

Figure 2.21

Relative importance of equal amounts (by mass) of selected CFCs, HCFCs, HFCs and halons in terms of their global warming potential. The values are for a 100-year time interval after emission.

Source: Taken from D.W. Fahey, 2006, Twenty Questions and Answers about the Ozone Layer—2006 Update, a supplement to Scientific Assessment of Ozone Depletion: 2006, the World Meteorological Organization Global Ozone Research and Monitoring Project—Report No. 50, released 2007, and reproduced here with the kind permission of the United Nations Environment Programme.

The x-axis of Figure 2.21 is global warming potential. Look for an explanation of the term in Section 3.8.

Fire-Extinguishing Foam (FEF), can replace halons in fighting even large-scale fires such as those on oil tankers or jet aircraft. A 0.4% solution of Pyrocool FEF dissolved in water to produce a foam was used to control the spread of fires in the sublevels beneath the collapsed towers of the World Trade Center towers following the terrorist attack of September 11, 2001 (Figure 2.22). Many of the hot spots buried in the debris were dangerous to any rescue operations. The foam also helped protect the huge tanks that stored Freon for the air-conditioning systems. The Pyrocool FEF foam also has a cooling effect that helps firefighters, a useful feature when fighting brush fires. Other companies across the globe are working on the challenge of replacing halons.

Figure 2.22

An aqueous foam of Pyrocool FEF is being applied to subterranean fires of the north tower of the World Trade Center, September 30, 2001.

The phaseout of CFCs and the continuing development of alternative materials are not without major economic concerns. At its peak, the annual worldwide market for CFCs reached $2 billion, but that was only the tip of a very large financial iceberg. In the United States alone, CFCs were used in or used to produce goods valued at about $28 billion per year. Although the conversion to CFC replacements has had some additional costs associated with it, the overall effect on the U.S. economy actually was minimal. Companies that produce refrigerators, air conditioners, insulating plastics, and other goods have adapted to using new compounds. Some substitutes for CFC refrigerants are less energy-efficient, hence increasing energy consumption somewhat. But the conversions provide a market opportunity for innovative syntheses based on the principles of green chemistry to produce environmentally benign substances, a win for both present and future generations.

Developing countries face a different set of economic problems and priorities. CFCs have played an important role in improving the quality of life in the industrialized nations. Few of their citizens would be willing to give up the convenience and health benefits of refrigeration or the comfort of air conditioning. It is understandable that millions of people over the globe aspire to the lifestyle of the industrialized nations. But, if the developing nations are banned from using the relatively inexpensive CFC-based technology, they may not be able to afford alternatives. "Our development strategies cannot be sacrificed for the destruction of the environment caused by the West," asserts Ashish Kothari, a member of an Indian environmental group. Both India and China originally refused to sign the Montreal Protocol because they felt that it discriminated against developing countries. To gain the participation of these highly populated nations, the industrially developed nations created a special fund that is administered through the World Bank. The goal of the fund is to help countries phase out their use of ozone-depleting materials without hurting their economic development.

Consider This 2.30   History Still Being Written

Each year, nations gather to continue conversations about the Montreal Protocol and its amendments. The textbook's website links you to the recent conferences. Where was this year's gathering held? Summarize the accomplishments.

Clearly, an understanding of chemistry is necessary to protect the ozone layer, but it is not sufficient. Chemists can help unravel the causes of ozone depletion and develop alternative materials to replace CFCs. Ultimately, though, the debate among governments and their citizens about how best to protect the stratospheric ozone layer continues in the global political arena.

Conclusion

Chemistry is intimately entwined with the story of ozone depletion. Chemists created the chlorofluorocarbons, the near-perfect properties of which only later revealed their dark side as predators of stratospheric ozone. Chemists worked internationally to discover the mechanism by which CFCs destroy stratospheric ozone and warned of the dangers of increased ultraviolet radiation reaching the Earth. And chemists will continue to synthesize the substitutes necessary to replace CFCs and other related compounds.

Although chemistry is a necessary part of the solution, it was only part of the solution. At the 2005 meeting in Dakar, Senegal, of parties to the Montreal Protocol on Substances That Deplete the Ozone Layer, Executive Secretary Marco González reminded delegates that the final 20% of any global cooperative effort can be the hardest. Fundamental differences in domestic regulatory approaches have the potential to deplete the stockpiles of goodwill, placing long-term goals in jeopardy. In their complexity, the economic, social, and political issues rival that of the scientific and technological ones.

Thus the problem of ozone depletion brought together an array of different participants in pursuit of a common end. Chemists provided the information on the causes and effects of ozone depletion. People in industry, responding to the stimulus provided by the control measures, developed alternatives far more rapidly and more cheaply than anyone initially thought possible, participating fully in the debates over further reductions. Nongovernmental organizations (NGOs) and the media served as essential channels of communication with the peoples of the world in whose name the measures had been taken. Governments worked well together in patiently negotiating agreements acceptable to a range of countries with widely varying circumstances, aims, and resources—and showed courage and foresight in applying the **precautionary principle** before the scientific evidence was entirely clear.

Clearly our problems are global. Ultimately, our solutions must be global as well. As we mentioned earlier in this chapter, in 2007 a symposium was held on the 20th Anniversary of the Montreal Protocol. Georgios Souflias from Greece gave the opening address, pointing out that we cannot remain indifferent toward the environment, as the environment is our home. *"The environment does not only provide people a better quality of life, but life itself."*

Souflias also pointed out the connection between CFCs and their replacements and global climate change. His call to action was unequivocal: *"We all, breathing on this planet today and having the potential, must guarantee its future, rapidly and decisively. We have no right to delay; we have no luxury of losing time."*

We urge you to carry these words with you as we turn to our next topic, the chemistry of climate change. As we mentioned at the outset of this chapter, CFCs, HCFCs, and HFCs all are greenhouse gases.

Chapter Summary

Having studied this chapter, you should be able to:

- Differentiate between harmful ground-level ozone and beneficial stratospheric ozone layer (2.1)

- Describe the chemistry of ozone, including how it is formed in our atmosphere. (2.1, 2.6, 2.8–2.10)

- Describe the ozone layer, characterizing it in several different ways (2.1, 2.6, 2.8–2.10)

- Apply the basics of atomic structure to atoms of certain elements (2.2)

- Understand what it means when elements fall into the same group of the periodic table (2.2)

- Differentiate atomic number from mass number and apply the latter to isotopes (2.2)

- Write Lewis structures for small molecules with single, double, and triple covalent bonds (2.3)

- Describe the electromagnetic spectrum in terms of frequency, wavelength, and energy (2.4, 2.5)

- Interpret graphs related to wavelength and energy, radiation and biological damage, and ozone depletion (2.4–2.8)

- Understand the natural Chapman cycle of stratospheric ozone depletion (2.6)

- Understand how the stratospheric ozone layer protects against harmful ultraviolet radiation (2.6, 2.7)
- Compare and contrast UV-A, UV-B, and UV-C radiation along several different lines (2.6, 2.7)
- Discuss the interaction of radiation with matter and changes caused by such interactions, including biological sensitivity (2.6, 2.7)
- Relate the meaning and the use of the UV Index (2.7)
- Write Lewis structures for chlorine and bromine atoms, as well as for some other free radicals. Be able to explain why these free radicals are so reactive (2.8)

- Recognize the complexities of collecting accurate data for stratospheric ozone depletion and interpreting them correctly (2.8, 2.9)
- Understand the chemical nature and role of CFCs in stratospheric ozone depletion (2.9, 2.10)
- Explain the unique circumstances responsible for seasonal ozone depletion in the Antarctic (2.10)
- Summarize the outcomes of the Montreal Protocol and its amendments (2.11, 2.12)
- Evaluate articles on green chemistry alternatives to stratospheric ozone-depleting compounds (2.12)
- Discuss factors that will help lead to the recovery of the ozone layer (2.11, 2.12)

Questions

Emphasizing Essentials

1. How does ozone differ from oxygen in its chemical formula? In its properties?

2. Explain why it is possible to detect the pungent odor of ozone after a lightning storm or around electrical transformers.

3. The text states that the odor of ozone can be detected in concentrations as low as 10 ppb. Would you be able to smell ozone in either of these air samples?

   a. 0.118 ppm of ozone, a concentration reached in an urban area

   b. 25 ppm of ozone, a concentration measured in the stratosphere

4. A journalist wrote "Hovering 10 miles above the South Pole is a sprawling patch of stratosphere with disturbingly low levels of radiation-absorbing ozone."

   a. How big is this sprawling patch?

   b. Is the figure of 10 miles correct? Express this value in kilometers.

   c. What type of radiation does ozone absorb?

5. It has been suggested that the term *ozone screen* would be a better descriptor than *ozone layer* to describe ozone in the stratosphere. What are the advantages and disadvantages to each term?

6. Assume there are 2×10^{20} CO molecules per cubic meter in a sample of tropospheric air. Furthermore, assume there are 1×10^{19} O_3 molecules per cubic meter at the point of maximum concentration of the ozone layer in the stratosphere.

   a. Which cubic meter of air contains the larger number of molecules?

   b. What is the ratio of CO to O_3 molecules in a cubic meter?

7. a. What is a Dobson unit?

   b. Does a reading of 320 DU or 275 DU indicate more total column ozone overhead?

8. Using the periodic table as a guide, specify the number of protons and electrons in a neutral atom of each of these elements.

   a. oxygen (O)   b. nitrogen (N)

   c. magnesium (Mg)   d. sulfur (S)

9. Consider this representation of a periodic table.

   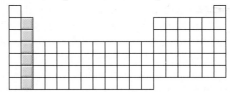

   a. What is the group number of the shaded column?

   b. Which elements make up this group?

   c. What is the number of electrons for a neutral atom of each element in this group?

   d. What is the number of outer electrons for a neutral atom of each element of this group?

10. Give the name and symbol for the element with this number of protons.

    a. 2   b. 19   c. 29

11. Give the number of protons, neutrons, and electrons in each of these neutral atoms.

    a. oxygen-18 ($^{18}_{8}O$)   b. sulfur-35 ($^{35}_{16}S$)

    c. uranium-238 ($^{238}_{92}U$)   d. bromine-82 ($^{82}_{35}Br$)

    e. neon-19 ($^{19}_{10}Ne$)   f. radium-226 ($^{226}_{88}Ra$)

12. Give the symbol showing the atomic number and the mass number for the isotope that has:

    a. 9 protons and 10 neutrons (used in nuclear medicine).

    b. 26 protons and 30 neutrons (the most stable isotope of this element).

    c. 86 protons and 136 neutrons (the radioactive gas found in some homes).

13. Draw the Lewis structure for each of these atoms.

    a. calcium   b. nitrogen

    c. chlorine   d. helium

14. Assuming that the octet rule applies, draw the Lewis structure for each of these molecules.

    a. CCl_4 (carbon tetrachloride, a substance formerly used as a cleaning agent)

    b. H_2O_2 (hydrogen peroxide, a mild disinfectant; the atoms are bonded in this order: H–O–O–H)

    c. H_2S (hydrogen sulfide, a gas with the unpleasant odor of rotten eggs)

    d. N_2 (nitrogen gas, the major component of the atmosphere)

    e. HCN (hydrogen cyanide, a molecule found in space and a poisonous gas)

    f. N_2O (nitrous oxide, "laughing gas"; the atoms are bonded N–N–O)

    g. CS_2 (carbon disulfide, used to kill rodents; the atoms are bonded S–C–S)

15. Several oxygen species play important chemical roles in the stratosphere, including oxygen atoms, oxygen molecules, ozone molecules, and hydroxyl radicals. Draw Lewis structures for each.

16. Consider these two waves representing different parts of the electromagnetic spectrum. How do they compare in terms of:

Wave 1   Wave 2

    a. wavelength   b. frequency   c. speed of travel

17. Use Figure 2.7 to specify the region of the electromagnetic spectrum where radiation of each wavelength is found. *Hint:* Change each wavelength to meters before making the comparison.

    a. 2.0 cm   b. 400 nm

    c. 50 μm   d. 150 mm

18. Arrange the wavelengths in question 17 in order of *increasing* energy. Which wavelength possesses the most energetic photons?

19. Arrange these types of radiation in order of *increasing* energy per photon: gamma rays, infrared radiation, radio waves, visible light.

20. The microwaves in home microwave ovens have a frequency of 2.45×10^9 s^{-1}. Is this radiation more or less energetic than radio waves? Than X-rays?

21. Ultraviolet radiation is categorized as UV-A, UV-B, or UV-C. Arrange these types in order of increasing:

    a. wavelength

    b. energy

    c. potential for biological damage

22. Draw Lewis structures for any three different CFCs.

23. CFCs were used in hair sprays, refrigerators, air conditioners, and plastic foams. Which properties of CFCs made them desirable for these uses?

24. a. Can a molecule that contains hydrogen be classified as a CFC?

    b. What is the difference between an HCFC and an HFC?

25. a. Most CFCs are based either on methane, CH_4, or ethane, C_2H_6. Use structural formulas to represent these two compounds.

    b. Substituting chlorine atoms, fluorine atoms, or both for all of the hydrogen atoms on a methane molecule, you obtain CFCs. How many possibilities exist?

    c. Which of the substituted CFC compounds in part **b** has been the most successful?

    d. Why weren't all of these compounds equally successful?

26. These free radicals all play a role in catalyzing ozone depletion reactions: Cl·, ·NO_2, ClO·, and ·OH.

    a. Count the number of outer electrons available and then draw a Lewis structure for each free radical.

    b. What characteristic is shared by these species that makes them so reactive?

27. a. How were the original measurements of increases in chlorine monoxide and the stratospheric ozone depletion over the Antarctic obtained?

    b. How are these measurements made today?

28. Which graph shows how measured increases in UV-B radiation correlate with percent reduction in the concentration of ozone in the stratosphere over the South Pole?

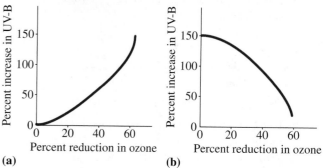

(a)   (b)

Concentrating on Concepts

29. The EPA has used the slogan *"Ozone: Good Up High, Bad Nearby"* in some of its publications for the general public. Explain the message.

30. Nobel laureate W. Sherwood Rowland referred to the ozone layer as the Achilles heel of our atmosphere. Explain the metaphor.

31. In the abstract of a talk he gave in 2007, Nobel laureate W. Sherwood Rowland wrote "Solar UV radiation creates an ozone layer in the atmosphere which in turn completely absorbs the most energetic fraction of this radiation."

    a. What is the most energetic fraction? *Hint:* See Figure 2.8.

    b. How does solar UV radiation "create an ozone layer"?

32. In the conclusion of this chapter, we reported the words that Georgios Souflias spoke at the Symposium for the 20th Anniversary of the Montreal Protocol: "*We all, breathing on this planet today and having the potential, must guarantee its future, rapidly and decisively. We have no right to delay; we have no luxury of losing time.*"

    a. What danger in delaying was he referring to?

    b. Look back at the definitions of sustainability in the prologue. How do his words connect to these definitions?

33. Consider the Chapman cycle in Figure 2.10.

    a. Explain the source of the oxygen atoms.

    b. Can this cycle take place in the troposphere as well? Explain.

34. What are some of the reasons that the solution to ozone depletion proposed in this Sydney Harris cartoon will not work?

"OH, FOR PETE'S SAKE, LET'S JUST GET SOME OZONE AND SEND IT BACK UP THERE!"

Source: ScienceCartoonsPlus.com. Reprinted with permission.

35. "*We risk solving one global environmental problem while possibly exacerbating another unless other alternatives can be found.*" The date of this quote by a U.S. official is 2009, and the context is phasing out the use of HCFCs.

    a. What compounds were HCFCs being replaced with in 2009?

    b. What is the risk of this replacement?

36. It is possible to write three resonance structures for ozone, not just the two shown in the text. Verify that all three structures satisfy the octet rule and offer an explanation of why the triangular structure is not reasonable.

$$\ddot{O}=\ddot{O}-\ddot{O}: \longleftrightarrow :\ddot{O}-\ddot{O}=\ddot{O} \longleftrightarrow :\ddot{O}-\overset{\ddot{O}}{\underset{}{}}-\ddot{O}:$$

37. The average length of an O–O single bond is 132 pm. The average length of an O–O double bond is 121 pm. What do you predict the O–O bond lengths will be in ozone? Will they all be the same? Explain your predictions.

38. Consider the Lewis structures for SO_2. How do they compare with the Lewis structures for ozone?

39. Even if you have skin with little pigment, you cannot get a tan from standing in front of a radio. Why?

40. The morning newspaper reports a UV Index Forecast of 6.5. Given the amount of pigment in your skin, how might this affect how you plan your daily activities?

41. All the reports of the damage caused by UV radiation focus on UV-A and UV-B radiation. Why is there no attention on the damaging effects of UV-C radiation?

42. If all 3×10^8 tons of stratospheric ozone that are formed every day are also destroyed every day, how is it possible for stratospheric ozone to offer any protection from UV radiation?

43. How does the chemical inertness of CCl_2F_2 (Freon-12) relate both the usefulness and the problems associated with this compound?

44. Explain how the small changes in concentrations (measured in parts per billion) can cause the much larger changes in O_3 concentrations (measured in parts per million).

45. Development of the stratospheric ozone hole has been most dramatic over Antarctica. What set of conditions exist over Antarctica that help to explain why this area is well-suited to studying changes in stratospheric ozone concentration? Are these same conditions not operating in the Arctic? Explain.

46. The free radical $CF_3O \cdot$ is produced during the decomposition of HFC-134a.

    a. Propose a Lewis structure for this free radical.

    b. Offer a possible reason why $CF_3O \cdot$ does not cause ozone depletion.

47. One mechanism that helps break down ozone in the Antarctic region involves the $BrO \cdot$ free radical. Once formed, it reacts with $ClO \cdot$ to form BrCl and O_2. BrCl, in turn, reacts with sunlight to break into $Cl \cdot$ and $Br \cdot$, both of which react with O_3 and form O_2.

    a. Represent this information with a set of equations similar to those shown for the Chapman cycle.

    b. What is the net equation for this cycle?

48. Polar stratospheric clouds (PSCs) play an important role in stratospheric ozone depletion.

    a. Why do PSCs form more often over Antarctica than in the Arctic?

    b. Reactions occur more quickly on the surface of PSCs than in the atmosphere. One such reaction is the reaction of hydrogen chloride and chlorine nitrate ($ClONO_2$), two species that do not deplete ozone, to produce a chlorine molecule and nitric acid (HNO_3). Write the chemical equation.

    c. The chlorine molecule produced does not deplete ozone either. However, when the Sun returns to the Antarctic in the springtime, it is converted to a species that does. Show how with a chemical equation.

49. Consider this graph that shows the atmospheric abundance of bromine-containing gases from 1950 to 2100.

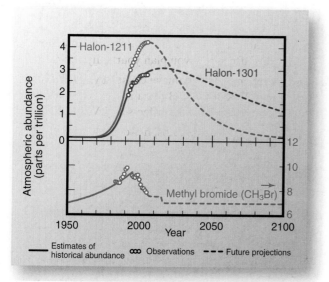

Source: Taken from D.W. Fahey, 2006, Twenty Questions and Answers about the Ozone Layer—2006 Update, a supplement to Scientific Assessment of Ozone Depletion: 2006, the World Meteorological Organization Global Ozone Research and Monitoring Project—Report No. 50, released 2007, and reproduced here with the kind permission of the United Nations Environment Programme.

a. Halon-1301 is $CBrF_3$ and Halon-1211 is $CClBrF_2$. Why were these compounds once manufactured?

b. Compare the patterns for Halon-1211 and Halon-1301. Why doesn't Halon-1301 drop off as quickly?

c. In 2005, methyl bromide was phased out in the United States except for critical uses. Why is its future use predicted as a straight line, rather than tailing off?

Exploring Extensions

50. Chapter 1 discussed the role of nitrogen monoxide (NO) in forming photochemical smog. What role, if any, does NO play in stratospheric ozone depletion? Are NO sources the same in the stratosphere as in the troposphere?

51. Resonance structures can be used to explain the bonding in charged groups of atoms as well as in neutral molecules, such as ozone. The nitrate ion, NO_3^-, has one additional electron plus the outer electrons contributed by nitrogen and oxygen atoms. That extra electron gives the ion its charge. Draw the resonance structures, verifying that each obeys the octet rule.

52. Although oxygen exists as O_2 and O_3, nitrogen exists only as N_2. Propose an explanation for these facts. *Hint:* Try drawing a Lewis structure for N_3.

53. The chemical formulas for a CFC, such as CFC-11 (CCl_3F), can be figured out from its code number by adding 90 to it to get a three-digit number. For example, with CFC-11 you get $90 + 11 = 101$. The first digit is the # of C atoms, the second is the # of H atoms, and the third is the # of F atoms. Accordingly, CCl_3F has 1 C atom, no H atoms, and 1 F atom. All remaining bonds are assumed to be chlorine.

a. What is the chemical formula for CFC-12?

b. What is the code number for CCl_4?

c. Does this "90" method work for HCFCs? Use HCFC-22 ($CHClF_2$) in explaining your answer.

d. Does this method work for halons? Use Halon-1301 (CF_3Br) in explaining your answer.

54. Many different types of ozone generators ("ozonators") are on the market for sanitizing air, water, and even food. They are often sold with a slogan such as this one from a pool store. "Ozone, the world's most powerful sanitizer!"

a. What claims are made for ozonators intended to purify air?

b. What risks are associated with these devices?

55. The effect a chemical substance has on the ozone layer is measured by a value called its *ozone-depleting potential,* ODP. This is a numerical scale that estimates the lifetime potential stratospheric ozone that could be destroyed by a given mass of the substance. All values are relative to CFC-11, which has an ODP defined as equal to 1.0. Use those facts to answer these questions.

a. Name two factors that affect the ODP value of a compound and explain the reason for each one.

b. Most CFCs have ODP values ranging from 0.6 to 1.0. What range do you expect for HCFCs? Explain your reasoning.

c. What ODP values do you expect for HFCs? Explain your reasoning.

56. Recent experimental evidence indicates that $ClO\cdot$ initially reacts to form Cl_2O_2.

a. Predict a reasonable Lewis structure for this molecule. Assume the order of atom linkage is Cl–O–O–Cl.

b. What effect does this evidence have on understanding the mechanism for the catalytic destruction of ozone by $ClO\cdot$?

3

The Chemistry of Global Climate Change

"Global warming is a misnomer, because it implies something that is gradual, something that is uniform, something that is quite possibly benign. What we are experiencing with climate change is none of those things."

John Holdren, *Meeting the Climate Change Challenge,* **National Council for Science and the Environment, 2008.**

Is global warming really a misnomer? Could those two little words misrepresent three large aspects of the issue we are facing? Let's examine each of the assertions that Holdren has made.

First, he asserts that global warming isn't gradual. By this he means that in comparison with the past, the climate changes we are seeing today are occurring much more rapidly. Natural climate changes are part of our planet's history. Glaciers, for example, have advanced and retreated numerous times, and global temperatures have been both much higher and much lower than the temperatures we currently experience. But the geologic evidence indicates these past changes occurred over millennia, not decades as they are today. So Holdren is correct. Global warming is not gradual, at least not in comparison with the geologic time frames of the past.

Second, he asserts that global warming does not occur uniformly across the globe. Holdren is right again. To date, the most dramatic effects have been observed at the poles. These include quickly receding glaciers, shrinking sea ice, and melting permafrost. So far, the more densely populated lower latitudes have experienced far smaller effects from climate change.

His third assertion, that global warming might not be benign, is the most difficult to assess. The issue is complicated in part because we cannot predict with certainty which aspects of our planet global warming will affect and to what degree. It is further complicated because we cannot easily understand why only a couple of degrees of warming might be catastrophic.

It is important not to take one person's word regarding a topic as complex as climate change. Therefore, in this chapter, we delve into Holdren's assertions in detail but do so by taking things one step at a time. The first step is to examine how our Earth maintains its energy balance. We consider incoming and outgoing solar radiation and explain how our atmosphere functions much like a greenhouse. The second step in our discussion of climate change examines scientific data on both the current and past state of the planet. In particular, we examine the concentrations of atmospheric greenhouse gases, both past and present. Carbon dioxide resides at the center of this discussion, yet the reason is far from obvious. After all, CO_2 is an essential component of the atmosphere, a gas that all animals exhale and green plants absorb. Key to understanding climate change is studying the molecular mechanism by which CO_2 and other compounds absorb the infrared radiation emitted by the planet, helping to keep it warm. Some knowledge of molecular structure and shape is necessary to understand this mechanism. Climate change also has a significant quantitative component; we need numbers to help grasp the magnitude of the problem. Therefore, our third step is to introduce the way that chemists count the unimaginably numerous (and unimaginably small) particles we call atoms and molecules.

Lastly, we examine the limits to which climate scientists can predict what the future may hold and describe several of the more dangerous consequences of a warmer planet Earth. As with many of the issues addressed in this book, we must consider not only what is happening today, but also the effects our actions (and our inactions) will have on future generations. Developing an understanding of these issues will lead us on a journey into the realm of chemical knowledge and its connections with public policy around the globe.

The terms climate change and global warming are both used, both in the popular press and by scientists. Although they are not the same, they are closely related. We will use both in this chapter. See Consider This 3.1.

Consider This 3.1   What's in a Name?

Sometimes people, including scientists, talk about global warming. Other times, folks refer to global climate change.

a. Interview two friends or family members and ask them to list what comes to mind when they hear the term "global warming." Do the same for the term "global climate change." Comment on the two lists.

b. Do you prefer one term over the other? Explain.

Figure 3.1

Venus, as photographed by the Galileo spacecraft.

These and other types of electromagnetic radiation were introduced in Section 2.4

3.1 | In the Greenhouse: Earth's Energy Balance

The brightest and most beautiful body in the night sky, after our own Moon, is considered by many to be Venus (Figure 3.1). It is ironic that the planet named for the goddess of love is a most unlovely place by earthly standards. The Venetian atmosphere has a pressure 90 times greater than that of Earth, and it is 96% carbon dioxide, with clouds of sulfuric acid. It makes the worst smog-bound day anywhere on Earth seem like a breath of fresh country air. Spacecraft have revealed a desolate, eroded surface with an average temperature of about 450 °C (840 °F). In contrast, the beautiful blue-green ball we inhabit has an average annual temperature of 15 °C (59 °F).

The point of this little astronomical digression is that both Venus and Earth are warmer than one would expect based solely on their distances from the Sun and the amount of solar radiation they receive. If distance were the *only* determining factor, the temperature of Venus would average approximately 100 °C, the boiling point of water. Earth, on the other hand, would have an average temperature of −18 °C (0 °F), and the oceans would be frozen year-round.

Processes that keep the energy of our Earth in balance are shown in Figure 3.2. The Earth receives nearly all of its energy from the Sun (orange arrows), primarily in the form of ultraviolet, visible, and infrared radiation. Accounting for what happens to the incoming radiation is relatively straightforward. Some of it is reflected back to space (blue arrows), either by the molecules, dust, and aerosol particles that are suspended in our atmosphere (25%) or by the surface of the Earth itself, especially those regions white with snow (6%). But most of the incoming radiation warms the Earth, either by being absorbed by the atmosphere (23%) or by being absorbed by the landmasses and oceans (46%). The numbers tally quite nicely: 25% reflected + 6% reflected + 23% absorbed + 46% absorbed = 100% of the incoming radiation.

Your Turn 3.2   **Light from the Sun**

Consider these three types of radiant energy, all emitted by the Sun: infrared (IR), ultraviolet (UV), and visible.

a. Arrange them in order of *increasing* wavelength.
b. Arrange them in order of *increasing* energy.

Answer
a. ultraviolet, visible, infrared

Accounting for what happens to the outgoing radiation is more complicated. Observe that our Earth, like the Sun, gives off radiation. If this were not the case, our planet would quickly become unbearably hot! The energy absorbed by the Earth must be reemitted to maintain the energy balance. But unlike the Sun, the Earth emits primarily in the infrared region (red arrows). A small amount of this IR radiation (9%) passes directly from the surface of the Earth out into space. However, most of the heat radiated by the Earth (37%) is absorbed by the atmosphere and returned to Earth rather than being lost to space. Heat is transferred by collisions between neighboring molecules, and these molecules are found in greater abundance in the denser regions of the lower atmosphere. Check the math: 46% absorbed = 9% + 37% emitted. The percent of solar radiation striking the Earth that remains in the atmosphere is about 80% (37% ÷ 46%). The **greenhouse effect** is the natural process by which atmospheric gases trap a major portion (about 80%) of the infrared radiation radiated by the Earth. Again, Earth's average annual temperature of 15 °C (59 °F) is a result of the heat trapping gases in our atmosphere.

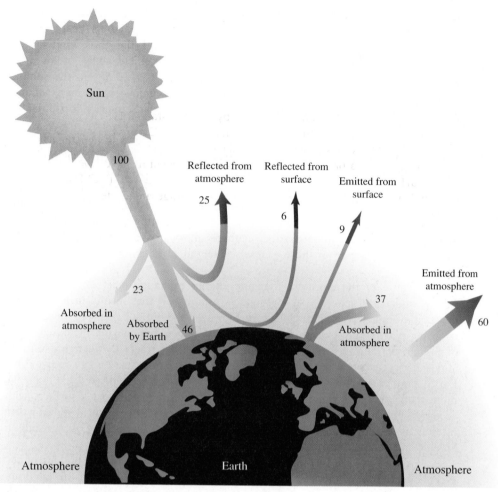

Figure 3.2

The Earth's energy balance. Orange represents a mixture of wavelengths. Shorter wavelengths of radiation are shown in blue, longer ones in red. The values are given in percentages of the total incoming radiation.

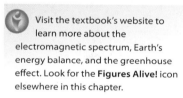
Visit the textbook's website to learn more about the electromagnetic spectrum, Earth's energy balance, and the greenhouse effect. Look for the **Figures Alive!** icon elsewhere in this chapter.

Your Turn 3.3   **Earth's Energy Balance**

Refer to Figure 3.2 to answer these questions.

a. Incoming solar radiation (100%) is either absorbed or reflected. Outgoing radiation from the Earth into space also can be accounted for (100%), as required for energy balance. Show how.

b. What percent of the outgoing energy is absorbed in the Earth's atmosphere? Calculate this by adding the percentage of incoming solar energy absorbed in the atmosphere to that absorbed in the atmosphere after being radiated from Earth's surface. How does this value compare with the percentage emitted from the atmosphere?

c. Suggest reasons why the different colors were used for incoming and outgoing radiation.

If you have ever parked a car with its windows closed on a sunny day, you probably have experienced firsthand how a greenhouse can trap heat. The car, with its glass windows, operates much the same way as does a greenhouse for growing plants. The glass windows transmit visible and a small amount of UV light from the Sun. This energy is absorbed by the car's interior, particularly by any dark surfaces. Some of this radiant energy is then reemitted as longer wavelength IR radiation (heat). Unlike visible light, infrared light is not easily transmitted through the glass windows and so becomes "trapped" inside the car. When you reenter the vehicle, a blast of hot air greets you. The temperature inside of a car can exceed 49 °C (120 °F) in the summer in certain climates! Although the physical barrier of the windows is not an exact analogy to the Earth's atmosphere, the effect of warming the car's interior is similar to the warming of the Earth.

Although the ability of the atmosphere to trap heat was hypothesized by the French mathematician Jean-Baptiste Joseph Fourier (1768–1830) around 1800, it took another 60 years for scientists to identify the molecules that were responsible. Irish physicist John Tyndall (1820–1893) first demonstrated that both carbon dioxide and water vapor absorb infrared radiation. **Greenhouse gases** are those gases capable of absorbing and trapping infrared radiation, thereby warming the atmosphere. Examples include water vapor, carbon dioxide, methane, nitrous oxide, ozone, and chlorofluorocarbons. The presence of those gases is essential in keeping our planet at habitable temperatures.

Because of the constant, dynamic energy exchange between Earth, its atmosphere, and outer space, a steady state is established that results in a more or less constant average terrestrial temperature. However, the buildup of greenhouse gases that is taking place today is changing the energy balance and causing increased warming of the planet. The term **enhanced greenhouse effect** refers to the process in which atmospheric gases trap and return *more than* 80% of the heat energy radiated by the Earth. An increase in the concentration of greenhouse gases will very likely mean that more than 80% of the radiated energy will be returned to Earth's surface, with an accompanying increase in average global temperature. The popular term **global warming** often is used to describe the increase in average global temperatures that results from an enhanced greenhouse effect.

We need look no further than ourselves to find the cause of the buildup of certain greenhouse gases in our atmosphere. **Anthropogenic** influences on the environment stem from human activities, such as industry, transportation, mining, and agriculture. In the late 19th century, Swedish scientist Svante Arrhenius (1859–1927) considered the problems that increased industrialization might cause by building up CO_2 in the atmosphere. He calculated that doubling the concentration of CO_2 would result in an increase of 5–6 °C in the average temperature of the planet's surface. Writing in the *London, Edinburgh, and Dublin Philosophical Magazine*, Arrhenius dramatically described the phenomenon: "We are evaporating our coal mines into the air." At the end of the 19th century, the Industrial Revolution was already well under way in Europe and America, and it was "picking up steam" as well as generating it.

Water vapor is the most abundant greenhouse gas in our atmosphere. However, contributions of H_2O from human activity are negligible compared with those from natural sources.

Another steady-state process, the Chapman cycle, was discussed in Section 2.6.

Section 1.9 described the chemistry of combustion. Look for more about coal (a fossil fuel) in Chapter 4.

Consider This 3.4   Evaporating Coal Mines

Although the Arrhenius statement about "evaporating our coal mines into the air" certainly was effective in grabbing attention in 1898, what process do you think he really was referring to in discussing the amount of CO_2 being added to the air? Explain your reasoning.

To further investigate global climate change, we need answers to several important questions. For example, how have the atmospheric concentrations of greenhouse gases changed over time? Similarly, how has the average global temperature changed and how did we measure the changes? Can we determine if the changes in greenhouse gases and temperature are correlated? Can we distinguish natural climate variability from human influences? In the following section, we provide some data to help answer these questions.

3.2 | Gathering Evidence: The Testimony of Time

Over the past 4.5 billion years, both Earth's climate and its atmosphere have varied widely as a result of astronomical, chemical, biological, and geological processes. Earth's climate has been directly affected by periodic astronomical changes in the shape of Earth's orbit and the tilt of Earth's axis. Such changes are thought to be responsible for the ice ages that have occurred regularly during the past million years. Even the Sun itself has changed. Its energy output half a billion years ago was 25–30% less than it is today. In addition, changes in atmospheric greenhouse gas concentrations affect the Earth's energy balance, and hence its climate. Carbon dioxide was once 20 times more prevalent in the atmosphere than it is today. Chemical processes lowered that level by dissolving much of the CO_2 in the oceans, or incorporating it in rocks such as limestone. The biological process of photosynthesis also radically altered the composition of our atmosphere by removing CO_2 and producing oxygen. Certain geological events like volcanic eruptions add millions of tons of CO_2 and other gases to the atmosphere.

The ionic compounds calcium carbonate ($CaCO_3$) and magnesium carbonate ($MgCO_3$) are both insoluble. Look for more about solubility in Chapter 5.

Although these natural phenomena will continue to influence Earth's atmosphere and its climate in the coming years, we must also assess the role that human activities are playing. With the development of modern industry and transportation, humans have moved huge quantities of carbon from terrestrial sources like coal, oil, and natural gas into the atmosphere in the form of CO_2. To evaluate the influence humans are having on the atmosphere, and hence on any enhanced greenhouse effect, it is important to investigate the fate of this large unnatural influx of carbon dioxide. Indeed, CO_2 concentrations in the atmosphere have increased significantly in the past half century. The best direct measurements are taken from the Mauna Loa Observatory in Hawaii (Figure 3.3). The red zigzag line shows the average monthly concentrations, with a small increase each April followed by a small decrease in October. The black line is a 12-month moving average. Notice the steady increase in average annual values from 315 ppm in 1960 to about 388 ppm in 2009. Later in this chapter, we will examine the evidence linking much of the added carbon dioxide to the burning of **fossil fuels,** combustible substances derived from the remnants of prehistoric organisms, the most common of which are coal, petroleum, and natural gas.

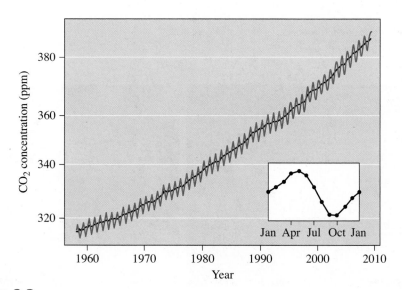

Figure 3.3

Carbon dioxide concentrations from 1958 to 2009, as measured at Mauna Loa, Hawaii. Inset: One year of the monthly variations.

Source: Scripps Institution of Oceanography, NOAA Earth System Research Laboratory, 2009.

Your Turn 3.5   The Cycles of Mauna Loa

a. Calculate the percent increase in CO_2 concentration during the last 50 years.
b. Estimate the variation in parts per million (ppm) CO_2 within any given year.
c. On average, the CO_2 concentrations are higher each April than each October. Explain.

Answer
c. Photosynthesis removes CO_2 from the atmosphere. Spring begins in the northern latitudes in April; October is the start of spring in the southern latitudes. But the landmasses (and number of green plants) are greater in the northern hemisphere so the seasons in the northern hemisphere control the fluctuations.

In the history of human civilization, 50 years of direct measurements are not much to rely on. How can we obtain data about the composition of our atmosphere farther back in time? Much relevant information comes from the analysis of ice core samples. Regions on the planet that have permanent snow cover contain preserved histories of the atmosphere, buried in layers of ice. Figure 3.4a shows a dramatic example of annual ice layers from the Peruvian Andes. The oldest ice on the planet is located in Antarctica, and scientists have been drilling and collecting ice core samples there for over 50 years (Figure 3.4b). Air bubbles trapped in the ice (Figure 3.4c) provide a vertical timeline of the history of the atmosphere; the deeper you drill, the farther back in time you go.

Relatively shallow ice core data show that for the first 800 years of the last millennium the CO_2 concentration was relatively constant at about 280 ppm. Figure 3.5 combines the Mauna Loa data (*red dots*) with data from a 200-meter ice core from the Siple station in Antarctica (*green triangles*), and a deeper core from the Law Dome, also in Antarctica (*blue squares*). Beginning about 1800, CO_2 began accumulating in the atmosphere at an ever-increasing rate, corresponding to the beginning of the Industrial Revolution and the accompanying combustion of fossil fuels that powered that transformation.

Skeptical Chemist 3.6   Checking the Facts on CO_2 Increases

a. A recent government report states that the atmospheric level of CO_2 has increased 30% since 1860. Use the data in Figure 3.5 to evaluate this statement.
b. A global warming skeptic states that the percent increase in the atmospheric level of CO_2 since 1957 has been only about half as great as the percent increase from 1860 to the present. Comment on the accuracy of that statement and how it could affect potential greenhouse gas emissions policy.

(a)

(b)

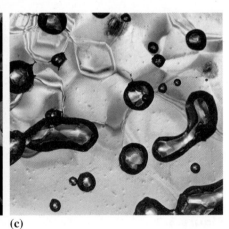
(c)

Figure 3.4

(a) Quelccaya ice cap (Peruvian Andes) showing the annual layers. (b) Ice core that can be used to determine changes in concentrations of greenhouse gases over time. (c) Microscopic air bubbles in ice.

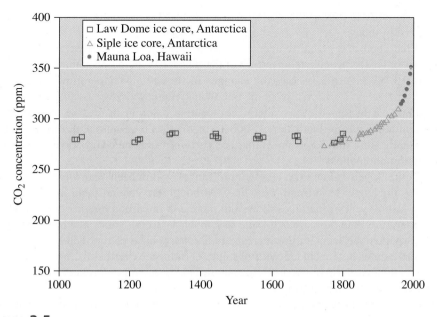

Figure 3.5

Carbon dioxide concentrations over the last millennium as measured from Antarctic ice cores (*blue squares and green triangles*) and the Mauna Loa observatory (*red dots*).

Source: "Climatic Feedbacks on the Global Carbon Cycle," in *The Science of Global Change: The Impact of Human Activities on the Environment*, American Chemical Society Symposium Series, 1992.

What about further back in time? Drilling by a team of Russian, French, and U.S. scientists at the Vostok Station in Antarctica yielded over a mile of ice cores taken from the snows of 400 millennia. The atmospheric carbon dioxide concentrations going back over 400,000 years are shown in Figure 3.6, with the data from Figure 3.5 in the inset.

Most obvious from the graph are the periodic cycles of high and low carbon dioxide concentrations, which occur roughly in 100,000-year intervals. Although not shown on the graph, analysis of other ice cores indicate these regular cycles go back at least 1 million years. Two important conclusions can be drawn from these data. First, the current atmospheric CO_2 concentration is 100 ppm *higher* than any time in the last

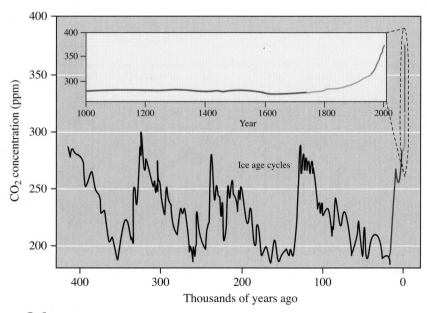

Figure 3.6

Carbon dioxide concentrations for the last 400,000 years. *Inset:* Data from Figure 3.5 for comparison.

million years. Also during that time, never has the CO_2 concentration risen as rapidly as it is rising today.

What about the global temperature? Measurements indicate that during the past 120 years or so, the average temperature of the planet has increased somewhere between 0.4 and 0.8 °C. Figure 3.7 shows the changes in surface air temperature from 1880 to 2006. Some scientists correctly point out that a century or two is an instant in the 4.5-billion-year history of our planet. They caution restraint in reading too much into short-term temperature fluctuations. Short-term changes in atmospheric circulation patterns like El Niño and La Niña events are certainly implicated in some of observed temperature anomalies.

Figure 3.7 also shows the temperature ranges within each year (*black error bars*) as well as the longer term trend (*blue line*). Although the general trend in temperatures over the last 50 years generally follows the increases in carbon dioxide concentrations, the temperature data from year to year are much less consistent. Furthermore, determining whether the temperature increase is a *consequence* of the increased CO_2 concentration cannot be concluded with absolute certainty. Nevertheless, as we shall see, experimental evidence implicates carbon dioxide from human-related sources as a cause of recent global warming.

It is important to realize that an increase in global average temperature does not mean that across the globe every day is now 0.6 °C warmer than it was in 1970. A map of the temperatures for 2006 compared to the average temperature between 1951 and 1980 is displayed in Figure 3.8. Many regions have experienced just a little warming, and some others have even cooled (*blue areas*). Yet there are other regions (*dark red areas*), particularly in the higher latitudes, that have experienced much more than the average warming. The increases are most drastic in the Arctic, where not surprisingly, much of the tangible effects of climate change have already been observed.

Ice cores also can provide data for estimating temperatures further back in time because of the hydrogen isotopes found in the frozen water. Water molecules containing the most abundant form of hydrogen atoms, ^1H, are lighter than those that contain deuterium, ^2H. The lighter H_2O molecules evaporate just a bit more readily than the heavier ones. As a result, there is more ^1H than ^2H in the water vapor of the atmosphere than in the oceans. Likewise, the heavier H_2O molecules in the atmosphere condense just a bit more readily than the lighter ones. Therefore, snow that condenses from atmospheric water vapor is enriched in ^2H. The degree of enrichment depends on temperature. The ratio of ^2H to ^1H in the ice core can be measured and used to estimate the temperature at the time the snow fell.

When we look back into the past, we see that the global temperature has undergone fairly regular cycles, matching the highs and lows in CO_2 concentration quite

El Niño and La Niña are names given to natural cyclical changes in the ocean–atmosphere system in the tropical Pacific. El Niño events lead to warmer ocean temperatures in the middle latitudes, and La Niña cycles produce cooler ocean temperatures.

Isotopes of hydrogen (and other elements) were discussed in Section 2.2.

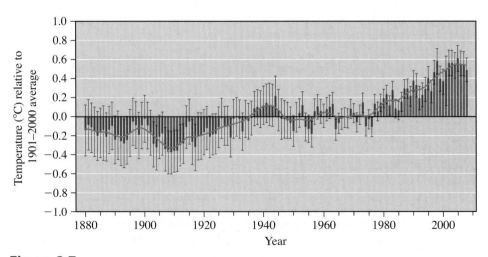

Figure 3.7

Global surface temperatures (1880–2006). The red bars indicate the average temperature for each year, and the ranges for each year are shown as the black error bars. The blue line shows the 5-year moving average.

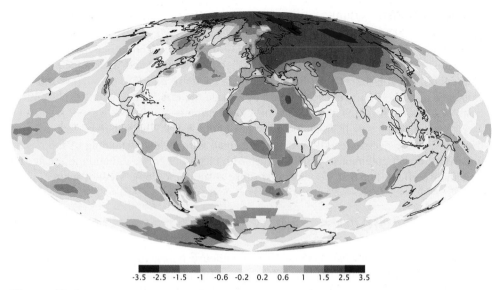

Figure 3.8

Global temperatures for 2006 (in °C) relative to the 1951–1980 average.

Source: NASA

remarkably (Figure 3.9). Other data show that periods of high temperature also have been characterized by high atmospheric concentrations of methane, another significant greenhouse gas. The precision of these data do not allow an assignment of cause and effect. It is difficult to conclude whether increasing greenhouse gases caused the temperature increases, or vice versa. What is clear, however, is that the current CO_2 and methane levels are much higher than any time in the last million years. Notice that the variation from hottest to coldest is only about 10 °C, yet that is the difference between the moderate climate we have today, and ice covering much of northern North America and Eurasia, as was the case during the last glacial maximum 20,000 years ago.

Over the past million years, Earth has experienced 10 major periods of glacier activity and 40 minor ones. Without question, mechanisms other than greenhouse gas concentrations are involved in the periodic fluctuations of global climate. Some of this temperature variation is caused by minor changes in Earth's orbit that affect the distance from Earth to the Sun and the angle at which sunlight strikes the planet. However,

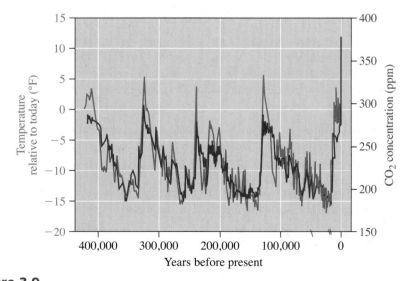

Figure 3.9

Carbon dioxide concentration (*blue*) and global temperatures (*red*) over the last 400,000 years from ice core data.

this hypothesis cannot fully explain the observed temperature fluctuations. Orbital effects most likely are coupled with terrestrial events such as changes in reflectivity, cloud cover, and airborne dust, as well as CO_2 and CH_4 concentrations. The feedback mechanisms that couple these effects together are complicated and not completely understood, but it is likely that the effects from each are *additive*. In other words, the existence of natural climate cycles doesn't preclude the effect that increased concentrations of greenhouse gases would have on global climate.

We are a long way from the out-of-control hothouse of Venus, but we face difficult decisions. These decisions will be better informed with an understanding of the mechanisms by which greenhouse gases interact with electromagnetic radiation to create the greenhouse effect. For that we must again take a submicroscopic view of matter.

3.3 | Molecules: How They Shape Up

Carbon dioxide, water, and methane are greenhouse gases; in contrast, nitrogen and oxygen are not. Why the difference? The answer relates in part to molecular shape. In this section, we'll help you to put your knowledge of Lewis structures to work to predict shapes of molecules. In the next, we connect these shapes to molecular vibrations, which can help us to explain the difference between greenhouse gases and nongreenhouse gases.

In Chapter 2 you used Lewis structures to predict how electrons are arranged in atoms and molecules. Shape was not the primary consideration. Even so, in a few cases the Lewis structure did dictate the shape of the molecule. One example is for diatomic molecules such as O_2 and N_2. Here, the shape is unambiguous because the molecule must be linear.

$$:N:::N: \quad \text{or} \quad :N\equiv N: \quad \text{or} \quad N\equiv N$$

$$\ddot{O}::\ddot{O} \quad \text{or} \quad \ddot{O}=\ddot{O} \quad \text{or} \quad O=O$$

> Remember, the atmosphere is composed of 78% nitrogen and 21% oxygen.

Even though different geometries are possible with larger molecules, Lewis structures still can help us with the process of predicting the shape. Therefore, the first step in predicting the shape of a molecule is to draw its Lewis structure. If the octet rule is obeyed throughout the molecule, each atom (except hydrogen) will be associated with four pairs of electrons. Some molecules include nonbonding lone-pair electrons, but all molecules must contain some bonding electrons or they would not be molecules!

A basic rule of physics is that opposite charges attract and like charges repel. Negatively charged electrons are attracted to a positively charged nucleus in every case. However, the electrons all have the same charge and therefore are found as far from each other in space as possible while still maintaining their attraction to the positively charged nucleus. Groups of negatively charged electrons repel one another. *The most stable arrangement is the one in which the mutually repelling electron groups are as far apart as possible.* In turn, this determines the atomic arrangement and the shape of the molecule.

We illustrate the procedure for predicting the shape of a molecule with methane, a greenhouse gas.

1. **Determine the number of outer electrons associated with each atom in the molecule.** The carbon atom (Group 4A) has four outer electrons; each of the four hydrogen atoms contributes one electron. This gives $4 + (4 \times 1)$, or 8 outer electrons.

2. **Arrange the outer electrons in pairs to satisfy the octet rule.** This may require single, double, or triple bonds. For the methane molecule, use the eight outer electrons to form four single bonds (four electron pairs) around the central carbon atom. This is the Lewis structure.

$$
\begin{array}{c}
\quad\quad\quad\quad H \\
\quad\quad H \quad\quad\quad | \\
H:\overset{\cdot\cdot}{\underset{\cdot\cdot}{C}}:H \quad \text{or} \quad H-C-H \\
\quad\quad H \quad\quad\quad | \\
\quad\quad\quad\quad H
\end{array}
$$

Although this structure seems to imply that the CH_4 molecule is flat, it is not. In fact, the methane molecule is tetrahedral, as we will see in the next step.

3. **Assume that the most stable molecular shape has the bonding electron pairs as far apart as possible.** (*Note:* In other molecules we need to consider nonbonding electrons as well, but CH_4 has none.) The four bonding electron pairs around the carbon atom in CH_4 repel one another, and in their most stable arrangement they are as far from one another as possible. As a result, the four hydrogen atoms also are as far from one another as possible. This shape is *tetrahedral,* because the hydrogen atoms correspond to the corners of a **tetrahedron,** a four-cornered geometric shape with four equal triangular sides, sometimes called a triangular pyramid.

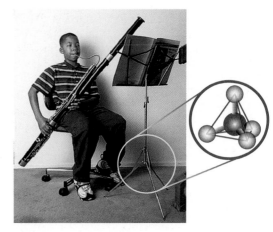

Figure 3.10

The legs and the shaft of a music stand approximate the geometry of the bonds in a tetrahedral molecule such as methane.

One way to describe the shape of a CH_4 molecule is by analogy to the base of a folding music stand. The four C–H bonds correspond to the three evenly spaced legs and the vertical shaft of the stand (Figure 3.10). The angle between each pair of bonds is 109.5°. The tetrahedral shape of a CH_4 molecule has been experimentally confirmed. Indeed, it is one of the most common atomic arrangements in nature, particularly in carbon-containing molecules.

Consider This 3.7   Methane: Flat or Tetrahedral?

a. If the methane molecule really were flat, as the two-dimensional Lewis structure seems to indicate, what would the H–C–H bond angle be?

b. Offer a reason why the tetrahedral shape, not the two-dimensional flat shape, is more advantageous for this molecule.

c. Consider the music stand shown in Figure 3.10. In the analogy of shape using a music stand, where would the carbon atom be located? Where would each of the hydrogen atoms lie?

Answer

a. 90° (at right angles). The two H atoms across from each other would be at 180°.

Chemists represent molecules in several different ways. The simplest, of course, is the chemical formula itself. In the case of methane, this is simply CH_4. Another is the Lewis structure, but again this is only a two-dimensional representation that gives information about the outer electrons. Figure 3.11 shows these two representations as well as two others that are three-dimensional in appearance. One has a wedge-shaped line that represents a bond coming out of the paper in a direction generally toward the reader. The dashed wedge in the same structural formula represents a bond pointing away from the reader. The two solid lines lie in the plane of the paper. The other, a space-filling model, was drawn with the help of a molecular modeling program. Space-filling models enclose the volume occupied by electrons in an atom or molecule. Seeing and manipulating physical models, either in the classroom or laboratory, can also help you visualize the structure of molecules.

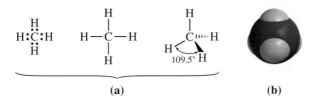

(a)   (b)

Figure 3.11

Representations of CH_4.

(a) Lewis structures and structural formula; (b) Space-filling model.

$$\text{H:\ddot{N}:H} \qquad \text{H—\ddot{N}—H} \qquad \underset{107.3°}{\text{H}\overset{\text{N}\cdots\text{H}}{<}\text{H}}$$

(a)   (b)

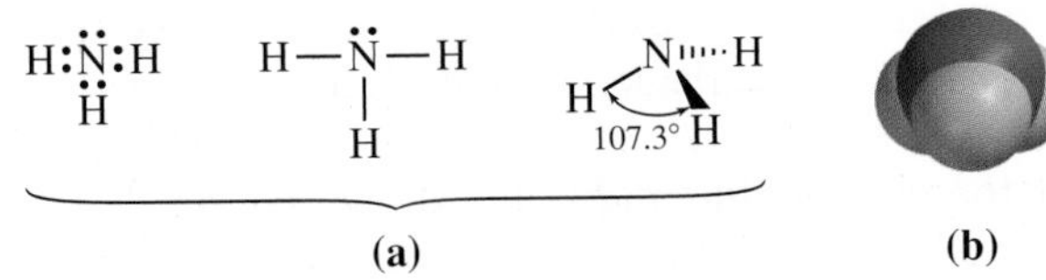

Figure 3.12

Representations of NH_3.
(a) Lewis structures and structural formula; **(b)** Space-filling model.

Section 2.9 discussed replacement of NH_3 as a refrigerant gas by CFCs. The role of ammonia in the nitrogen cycle is the subject of Section 6.9 and Section 11.8 describes the importance of NH_3 in agriculture.

Not all outer electrons reside in bonding pairs. In some molecules, the central atom has nonbonding electron pairs, also called lone pairs. For example, Figure 3.12 shows the ammonia molecule in which nitrogen completes its octet with three bonding pairs and one nonbonding pair.

A nonbonding electron pair effectively occupies greater space than a bonding pair of electrons. Consequently, the nonbonding pair repels the bonding pairs somewhat more strongly than the bonding pairs repel one another. This stronger repulsion forces the bonding pairs closer to one another, creating an H–N–H angle slightly less than the predicted 109.5° associated with a regular tetrahedron. The experimental value of 107.3° is close to the tetrahedral angle, again indicating that our model is reasonably reliable.

The shape of a molecule is described in terms of its arrangement of atoms, not electrons. The hydrogen atoms of NH_3 form a triangle with the nitrogen atom above them at the top of the pyramid. Thus, ammonia is said to have a *trigonal pyramidal* shape. Going back to the analogy of the folding music stand (see Figure 3.10), you could expect to find hydrogen atoms at the tip of each leg of the music stand. This places the nitrogen atom at the intersection of the legs with the shaft, with the non-bonded electron pair forming around the shaft of the stand.

The water molecule is *bent*, illustrating yet another shape. There are eight outer electrons on the central oxygen atom: one from each of the two hydrogen atoms plus six from the oxygen atom (Group 6A). These eight electrons are distributed in two bonding and two lone pairs of electrons (Figure 3.13a).

If these four pairs of electrons were arranged as far apart as possible, we might predict the H–O–H bond angle to be 109.5°, the same as the H–C–H bond angle in methane. However, unlike methane, water has two nonbonding pairs of electrons. The repulsion between the two nonbonding pairs causes the bond angle to be less than 109.5°. Experiments indicate a value of approximately 104.5°.

Your Turn 3.8   Predicting Molecular Shapes, Part 1

Using the strategies just described, sketch the shape for each of these molecules.

a. CCl_4 (carbon tetrachloride)
b. CCl_2F_2 (Freon-12; dichlorodifluoromethane)
c. H_2S (hydrogen sulfide)

Answer

a. Total outer electrons: 4 + 4(7) = 32. Eight of these electrons form 4 single bonds around the central C atom, one to each Cl atom. The other 24 are in 12 nonbonding pairs on the 4 Cl atoms. The bonding electron pairs on C arrange themselves to maximize the separation, and the molecule is tetrahedral.

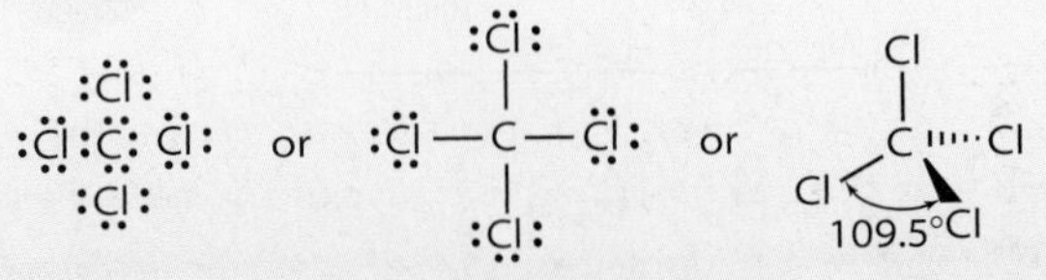

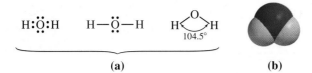

Figure 3.13

Representations of H_2O.
(a) Lewis structures and structural formula; **(b)** Space-filling model.

We already looked at the structures of several molecules important for understanding the chemistry of climate change. What about the structure of the carbon dioxide molecule? With 16 outer electrons, the C atom contributes 4 electrons and 6 come from each of the 2 oxygen atoms. If only single bonds were involved, each atom would not have an octet. But the octet rule still can be obeyed if the central carbon atom forms a double bond with each of the 2 oxygen atoms, thus sharing 4 electrons.

What is the shape of the CO_2 molecule? Again, groups of electrons repel one another, and the most stable configuration provides the furthest separation of the negative charges. In this case, the groups of electrons are the double bonds, and these are furthest apart with an O–C–O bond angle of 180°. The model predicts that all three atoms in a CO_2 molecule will be in a straight line and that the molecule will be *linear*. This is, in fact, the case as shown in Figure 3.14.

> Revisit Section 2.3 for more about drawing Lewis structures for molecules with double bonds.

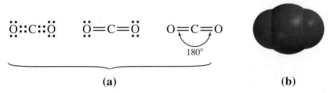

Figure 3.14

Representations of CO_2.
(a) Lewis structures and structural formula; **(b)** Space-filling model.

We applied the idea of electron pair repulsion to molecules in which there are four groups of electrons (CH_4, NH_3, and H_2O) and two groups of electrons (CO_2). Electron pair repulsion also applies reasonably well to molecules that include three, five, or six groups of electrons. In most molecules, the electrons and atoms are still arranged to keep the separation of the electrons at a maximum. This logic accounts for the bent shape we associated with the ozone molecule.

The Lewis structure for the ozone (O_3) molecule with its 18 outer electrons contains a single bond and a double bond, and the central oxygen atom carries a nonbonding lone pair of electrons. Thus, the central O atom has three groups of electrons: the pair that makes up the single bond, the two pairs that constitute the double bond, and the lone pair. These three groups of electrons repel one another, and the minimum energy of the molecule corresponds to their furthest separation. This occurs when the electron groups are all in the same plane and at an angle of about 120° from one another. We predict, therefore, that the O_3 molecule should be bent, and the angle made by the three atoms should be approximately 120°. Experiments show the bond angle to be 117°, just slightly smaller than the prediction (Figure 3.15). The nonbonding electron pair on the central oxygen atom occupies an effectively greater volume than bonding pairs of electrons, causing a greater repulsion force responsible for the slightly smaller bond angle.

> The O_3 molecule is best represented by two equivalent resonance structures. Again see Section 2.3.

Your Turn 3.9   Predicting Molecular Shapes, Part 2

Using the strategies just described, predict and sketch the shapes of SO_2 (sulfur dioxide) and SO_3 (sulfur trioxide).

Hint: Because S and O are in the same group on the periodic table, the structures for SO_2 and O_3 will be closely related.

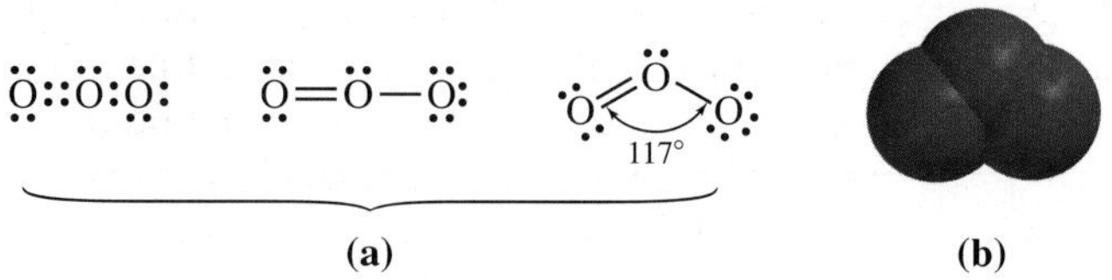

$$\ddot{\text{O}}::\ddot{\text{O}}:\ddot{\text{O}}: \qquad \ddot{\text{O}}=\ddot{\text{O}}-\ddot{\text{O}}: \qquad :\ddot{\text{O}}\overset{\ddot{\text{O}}}{\underset{117°}{\diagdown}}\ddot{\text{O}}:$$

$$\underbrace{\phantom{\ddot{\text{O}}::\ddot{\text{O}}:\ddot{\text{O}}: \qquad \ddot{\text{O}}=\ddot{\text{O}}-\ddot{\text{O}}: \qquad :\ddot{\text{O}}\overset{\ddot{\text{O}}}{\diagdown}\ddot{\text{O}}:}}$$

(a)   **(b)**

Figure 3.15

Representations of O_3.

(a) Lewis structures and structural formula for one resonance form; **(b)** Space-filling model.

As promised, in this section we helped you see that molecules have different shapes, ones that can be predicted. In the next, we return to our story of greenhouse gases, putting your knowledge of shapes to work to help you understand why not all gases are greenhouse gases.

3.4 | Vibrating Molecules and the Greenhouse Effect

How do greenhouse gases trap heat, keeping our planet at more or less comfortable temperatures? In part, the answer lies in how molecules respond to photons of energy. This topic is complex, but even so, we can give you enough basics so you can understand how the greenhouse gases in our atmosphere function. At the same time, we'll reveal why some gases do *not* trap heat.

We begin this topic by revisiting the interaction of ultraviolet (UV) light with molecules, something we discussed earlier in Chapter 2 in connection with the ozone layer. You saw that a photon in the UV region of the electromagnetic spectrum had sufficient energy to break some covalent bonds. In particular, you saw that UV-C could break the bonds in O_2 and that photons of lower energy (UV-B) could break the bonds in O_3. Put another way, both the ozone and the oxygen molecule can absorb UV radiation. When this absorption occurs, an oxygen-to-oxygen bond is broken.

Fortunately, IR photons do not contain enough energy to cause chemical bonds to break. Instead, a photon of IR radiation can add energy to the vibrations in a molecule. Depending on the molecular structure, only certain vibrations are possible. The energy of the incoming photon must correspond exactly to the vibrational energy of the molecule for the photon to be absorbed. This means that different molecules absorb IR radiation at different wavelengths and thus vibrate at different energies.

We illustrate these ideas with the CO_2 molecule, representing the atoms as balls and the covalent bonds as springs. Every CO_2 molecule is constantly vibrating in the four ways pictured in Figure 3.16. The arrows indicate the direction of motion of each atom during each vibration. The atoms move forward and backward along the arrows. Vibrations **a** and **b** are stretching vibrations. In vibration **a**, the central carbon atom is

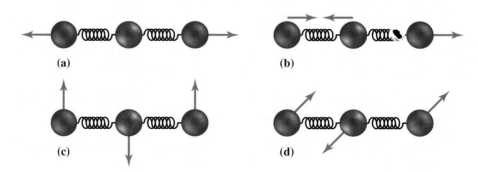

Figure 3.16

Molecular vibrations in CO_2. Each spring represents a C=O double bond. Vibrations **a** and **b** are stretching vibrations; **c** and **d** are bending vibrations.

stationary and the oxygen atoms move back and forth (stretch) in opposite directions away from the central atom. Alternatively, the oxygen atoms can move in the same direction and the carbon atom in the opposite direction (vibration **b**). Vibrations **c** and **d** look very much alike. In both cases, the molecule bends from its normal linear shape. The bending counts as two vibrations because it occurs in either of two possible planes. In vibration **c** the molecule is shown bending up and down in the plane of the paper on which the diagram is printed, whereas in vibration **d** the molecule is shown bending out of the plane of the paper.

If you ever examined a spring, you probably noticed that more energy is required to stretch it than to bend it. Similarly, more energy is required to stretch a CO_2 molecule than to bend it. This means that more energetic photons, those with shorter wavelengths, are needed to add energy to stretching vibrations **a** or **b** than to add energy to bending vibrations **c** or **d**. For example, absorption of IR radiation with a wavelength of 15.0 micrometers (μm) adds energy to the bending vibrations (**c** and **d**). When that occurs, the atoms move farther from their equilibrium positions and move faster (on average) than they do normally. For the same thing to happen with vibration **b**, higher energy radiation having a wavelength of 4.3 μm is required. Together, vibrations **b**, **c**, and **d** account for the greenhouse properties of carbon dioxide.

In contrast, direct absorption of IR radiation does not add energy to vibration **a**. In a CO_2 molecule, the average concentration of electrons is greater on the oxygen atoms than on the carbon atom. This means that the oxygen atoms carry a partial negative charge relative to the carbon atom. As the bonds stretch, the positions of the electrons change, thereby changing the charge distribution in the molecule. Because of the linear shape and symmetry of CO_2, the changes in charge distribution during vibration **a** cancel and no infrared absorption occurs.

The infrared (heat) energy that molecules absorb can be measured with an instrument called an infrared spectrometer. IR radiation from a glowing filament is passed through a sample of the compound to be studied, in this case gaseous CO_2. A detector measures the amount of radiation, at various wavelengths, transmitted by the sample. High transmission means low absorbance, and vice versa. This information is displayed graphically, where the relative intensity of the transmitted radiation is plotted versus wavelength. The result is the *infrared spectrum* of the compound. Figure 3.17 shows the infrared spectrum of CO_2.

The infrared spectrum shown in Figure 3.17 was acquired using a laboratory sample of CO_2, but the same absorption takes place in the atmosphere. Molecules of CO_2 that absorb specific wavelengths of infrared energy experience different fates. Some hold that extra energy for a brief time, and then reemit it in all directions as heat. Others collide with atmospheric molecules like N_2 and O_2 and can transfer some

A micrometer is equal to one-millionth of a meter:
$1 \ \mu m = 1 \times 10^{-6} \ m = 1000 \ nm$.

The property of electronegativity, a measure of an atom's ability to attract bonded electrons, is discussed in Section 5.5.

Spectroscopy is the field of study that examines matter by passing electromagnetic energy through a sample.

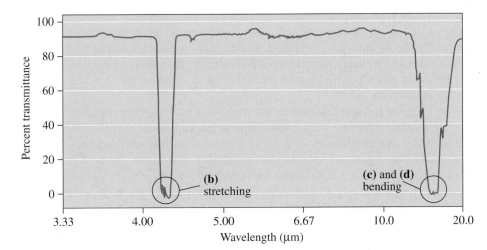

Figure 3.17

Infrared spectrum of carbon dioxide. The letters (**b**), (**c**), and (**d**) refer to the molecular vibrations shown in Figure 3.16.

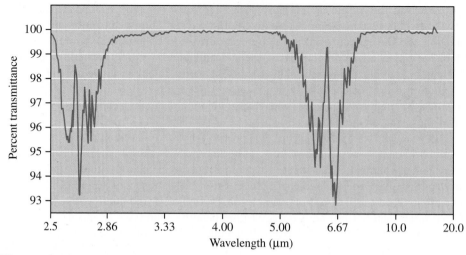

Figure 3.18

Infrared spectrum of water vapor.

of the absorbed energy to those molecules, also as heat. Through both of these processes, CO_2 "traps" some of the infrared radiation emitted by the Earth, keeping our planet comfortably warm. This is what makes CO_2 a greenhouse gas.

Any molecule that can absorb photons of IR radiation can behave as a greenhouse gas. There are many such molecules. Water is by far the most important gas in maintaining Earth's temperature, followed by carbon dioxide. Figure 3.18 shows the IR spectrum of H_2O molecules absorbing IR radiation. However, methane, nitrous oxide, ozone, and chlorofluorocarbons (such as CCl_3F) are among the other substances that help retain planetary heat.

> You already learned about the role of CFCs in ozone depletion in Chapter 2. Nitrous oxide, N_2O, is also called dinitrogen monoxide. You will encounter this gas again in Chapter 6.

Consider This 3.10   Bending and Stretching Water Molecules

a. Use Figure 3.18 to estimate the wavelengths corresponding to the strongest IR absorbencies for water vapor.

b. Which wavelength do you predict represents bending vibrations and which represents stretching? Explain the basis of your predictions.

Hint: Compare the IR spectrum of H_2O with that of CO_2.

Diatomic gases, such as N_2 and O_2, are not greenhouse gases. Although molecules consisting of two identical atoms do vibrate, the overall electric charge distribution does not change during these vibrations. Hence, these molecules cannot be greenhouse gases. Earlier we discussed this lack of overall electric charge distribution as the reason why stretching vibration **a** in Figure 3.17 was not responsible for the greenhouse gas behavior of CO_2.

So far, you have encountered two ways that molecules respond to radiation. Highly energetic photons with high frequencies and short wavelengths (such as UV radiation) can break bonds within molecules. The less energetic photons (such as IR radiation) cause an increase in molecular vibrations. Both processes are depicted in Figure 3.19, but the figure also includes another response of molecules to radiant energy that is probably a good deal more familiar to you. Longer wavelengths than those in the IR range have only enough energy to cause molecules to rotate faster.

For example, microwave ovens generate electromagnetic radiation that causes water molecules to spin faster. The radiation generated in such a device is of relatively long wavelength, about a centimeter. Thus the energy per photon is quite low. As the H_2O molecules absorb the photons and spin more rapidly, the resulting friction cooks

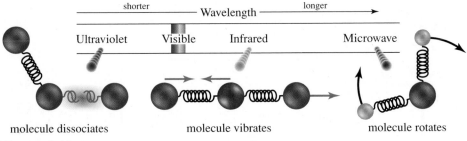

Figure 3.19

Molecular response to types of radiation.

your food, warms up the leftovers, or heats your coffee. The same region of the spectrum is used for radar. Beams of microwave radiation are sent out from a generator. When the beams strike an object such as an airplane, the microwaves bounce back and are detected by a sensor.

3.5 | The Carbon Cycle: Contributions from Nature and Humans

In his book *The Periodic Table,* the late chemist, author, and World War II concentration camp survivor Primo Levi wrote eloquently about CO_2.

> *"This gas which constitutes the raw material of life, the permanent store upon which all that grows draws, and the ultimate destiny of all flesh, is not one of the principal components of air but rather a ridiculous remnant, an 'impurity' thirty times less abundant than argon, which nobody even notices. . . . [F]rom this ever renewed impurity of the air we come, we animals and we plants, and we the human species, with our four billion discordant opinions, our millenniums of history, our wars and shames, nobility and pride."*

Levi's book, *The Periodic Table* was written in 1975. Over 6.9 billion people now inhabit Earth.

In the essay from which this quotation is taken, Levi traces a brief portion of the life history of a carbon atom from a piece of limestone (calcium carbonate, $CaCO_3$), where it lies "congealed in an eternal present," to a CO_2 molecule, to a molecule of glucose in a leaf, and ultimately to the brain of the author. And yet that is not the final destination. "The death of atoms, unlike our own," writes Levi, "is never irrevocable." That carbon atom, already billions of years old, will continue to persist into the unimagined future.

This marvelous continuity of matter, a consequence of its conservation, is beautifully illustrated by the carbon cycle (Figure 3.20). Even without Primo Levi's poetic description, the cycle is important to our understanding of the human effects on the global ecosystem. It is certain that without the proper functioning of the carbon cycle, every aspect of life on Earth could undergo dramatic change.

Your Turn 3.11   **Understanding the Carbon Cycle**

  a. Which processes add carbon (in the form of CO_2) to the atmosphere?
  b. Which processes remove carbon from the atmosphere?
  c. What are the two largest reservoirs of carbon?
  d. Which parts of the carbon cycle are most influenced by human activities?

The carbon cycle is a dynamic system, with all the processes illustrated happening simultaneously but at far different rates. Michael B. McElroy of Harvard University estimated, "The average carbon atom has made the cycle from sediments through the

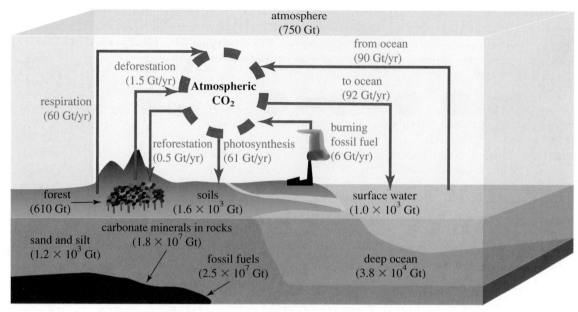

Figure 3.20

The global carbon cycle. The numbers show the quantity of carbon, expressed in gigatonnes (Gt), that is stored in various carbon reservoirs (*black numbers*) or moving through the system per year (*red numbers*).

A gigatonne (Gt) is a billion metric tons, or about 2200 billion pounds. For comparison, a fully loaded 747 jet weighs about 800,000 lb. It would take nearly 3 million 747s to have a total mass of 1 Gt.

more mobile compartments of the Earth back to sediments some 20 times over the course of Earth's history." CO_2 in the air today may have been released from campfires burning more than a thousand years ago.

Notice the presence of both natural emission and removal mechanisms. Respiration adds carbon dioxide to the atmosphere, and photosynthesis removes it. Similarly, the oceans both absorb and emit carbon dioxide. As members of the animal kingdom, we *Homo sapiens* participate in the carbon cycle along with our fellow creatures. As is true for any animal, we inhale and exhale, ingest and excrete, live and die. In addition though, human civilization relies on processes that put much more carbon into the atmosphere than they remove (Figure 3.21). Widespread burning of coal for electricity production, of petroleum products for transportation, and of natural gas for home heating all transfer carbon from the largest underground carbon reservoir into the atmosphere.

Another human influence on CO_2 emissions is deforestation by burning, a practice that releases about 1.5 Gt of carbon to the atmosphere each year. It is estimated

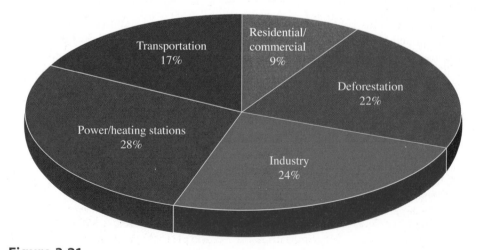

Figure 3.21

Global carbon dioxide emissions by end use.

Source: IPCC Fourth Assessment Report, Working Group III, 2007.

that forested land the size of two football fields is lost every second of every day from the rain forests of the world. Although firm numbers are rather elusive, Brazil continues as the country with the greatest annual loss of rain forest acreage; over 5.4 million acres of Amazon rain forest is vanishing each year. Trees, those very efficient absorbers of carbon dioxide, are removed from the cycle through deforestation. If the wood is burned, vast quantities of CO_2 are generated; if it is left to decay, that process also releases carbon dioxide, but more slowly. Even if the lumber is harvested for construction purposes and the land is replanted in cultivated crops, the loss in CO_2-absorbing capacity may approach 80%.

The total quantity of carbon released by the human activities of deforestation and burning fossil fuels is about 7.5 Gt per year. About half of this is eventually recycled into the oceans and the biosphere, which serve as **carbon sinks,** natural processes that remove CO_2 from the atmosphere. These processes do not always remove CO_2 with the speed required by the ever-increasing concentrations of CO_2. Much of the CO_2 emitted stays in the atmosphere, adding between 3.1 and 3.5 Gt of carbon per year to the existing base of 750 Gt noted in Figure 3.20. We are concerned primarily with the relatively rapid *increase* in atmospheric carbon dioxide, because the *excess* CO_2 is implicated in global warming. Therefore, it would be useful to know the mass (Gt) of CO_2 added to the atmosphere each year. In other words, what mass of CO_2 contains 3.3 Gt of carbon, the midpoint between 3.1 Gt and 3.5 Gt? Answering this question requires that we return to some more quantitative aspects of chemistry.

Remember that the natural "greenhouse effect" makes life on Earth possible. Problems occur when the amount of greenhouse gases *increases* faster than the sinks can accommodate the increases. The result is the *enhanced* greenhouse effect.

3.6 | Quantitative Concepts: Mass

To solve the problem just posed, we need to know how the mass of C is related to the mass of CO_2. Regardless of the source of CO_2, its chemical formula is stubbornly the same. The mass percent of C in CO_2 is also unwavering, and therefore we must calculate the mass percent of C in CO_2, based on the formula of the compound. As you work through this and the next section, keep in mind that we are seeking a value for that percentage.

The approach requires the use of the masses of the elements involved. But this raises an important question: How much does an individual atom weigh? The mass of an atom is mainly attributable to the neutrons and protons in the nucleus. Thus, elements differ in mass because their atoms differ in composition. Rather than using absolute masses of individual atoms, chemists have found it convenient to employ relative masses—in other words, to relate the masses of all atoms to some convenient standard. The internationally accepted mass standard is carbon-12, the isotope that makes up 98.90% of all carbon atoms. C-12 has a mass number of 12 because each atom has a nucleus consisting of 6 protons and 6 neutrons plus 6 electrons outside the nucleus.

The periodic table in the text shows that the atomic mass of carbon is 12.01, not 12.00. This is not an error; it reflects the fact that carbon exists naturally as three isotopes. Although C-12 predominates, 1.10% of carbon is C-13, the isotope with six protons and *seven* neutrons. In addition, natural carbon contains a trace of C-14, the isotope with six protons and *eight* neutrons. The tabulated mass value of 12.01 is often called by the name atomic weight, an average that takes into consideration the masses and percent natural abundance of all naturally occurring isotopes of carbon. This isotopic distribution and average mass of 12.01 characterize carbon obtained from any chemical source—a graphite ("lead") pencil, a tank of gasoline, a loaf of bread, a lump of limestone, or your body.

Isotopes and the relative masses of subatomic particles were discussed in Section 2.2.

The term *atomic weight* is a familiar (but not technically correct) term used for the relative scale of atomic masses.

The radioactive isotope carbon-14, although present only in trace amounts, provides direct evidence that the combustions of fossil fuels is the *predominant* cause of the rise in atmospheric CO_2 concentrations over the past 150 years. In all living things, only 1 out of 10^{12} carbon atoms is a C-14 atom. A plant or animal constantly exchanges CO_2 with the environment, and this maintains a constant C-14 concentration in the organism. However, when the organism dies, the biochemical processes that exchange

You will learn to write equations for nuclear reactions in Section 7.2.

carbon stop functioning and the C-14 is no longer replenished. This means that after the death of the organism, the concentration of C-14 decreases with time because it undergoes radioactive decay to form N-14. Coal, oil, and natural gas are remnants of plant life that died hundreds of millions of years ago. Hence, in fossil fuels, and in the carbon dioxide released when fossil fuels burn, the level of C-14 is essentially zero. Careful measurement show that the concentration of C-14 in atmospheric CO_2 has recently decreased. This strongly suggests that the origin of the added CO_2 is indeed the burning of fossil fuels, a decidedly human activity.

Your Turn 3.12   Isotopes of Nitrogen

Nitrogen (N) is an important element in the atmosphere and in biological systems. It has two naturally occurring isotopes: N-14 and N-15.

a. Use the periodic table to find the atomic number and atomic mass of nitrogen.
b. What is the number of protons, neutrons, and electrons in a neutral atom of N-14?
c. Compare your answers for part b with those for a neutral atom of N-15.
d. Given the atomic mass of nitrogen, which isotope has the greatest natural abundance?

Having reviewed the meaning of isotopes, we return to the matter at hand—the masses of atoms and particularly the atoms in CO_2. Not surprisingly, it is impossible to weigh a single atom because of its extremely small mass. A typical laboratory balance can detect a minimum mass of 0.1 mg; this corresponds to 5×10^{18} carbon atoms, or 5,000,000,000,000,000,000 carbon atoms. An atomic mass unit is far too small to measure in a conventional chemistry laboratory. Rather, the gram is the chemist's mass unit of choice. Therefore, scientists use exactly 12 g of carbon-12 as the reference for the atomic masses of all the elements. We define **atomic mass** as the mass (in grams) of the same number of atoms that are found in exactly 12 g of carbon-12. This number of atoms is, of course, *very* large. This important chemical number is named after an Italian scientist with the impressive name of Count Lorenzo Romano Amadeo Carlo Avogadro di Quaregna e di Ceretto (1776–1856). (His friends called him Amadeo.) **Avogadro's number** is the number of atoms in exactly 12 g of C-12. Avogadro's number, if written out, is 602,000,000,000,000,000,000,000. It is more compactly written in scientific notation as 6.02×10^{23}. This is the incredible number of atoms in 12 g of carbon, no more than a tablespoonful of soot!

Avogadro's number counts a large collection of atoms, much like the term *dozen* counts a collection of eggs. It does not matter if the eggs are large or small, brown or white, "organic" or not. No matter, for if there are 12 eggs, they are still counted as a dozen. A dozen ostrich eggs has a greater mass than a dozen quail eggs. Figure 3.22 illustrates this point with a half-dozen tennis and a half-dozen golf balls. Like atoms of different elements, the masses of a tennis ball and a golf ball differ. The number of balls is the same—six in each bag, a half dozen.

Figure 3.22

Six tennis balls have a greater mass than six golf balls.

Skeptical Chemist 3.13   Marshmallows and Pennies

Avogadro's number is so large that about the only way to hope to comprehend it is through analogies. For example, one Avogadro's number of regular-sized marshmallows, 6.02×10^{23} of them, would cover the surface of the United States to a depth of 650 miles. Or, if you are more impressed by money than marshmallows, assume 6.02×10^{23} pennies were distributed evenly among the approximately 7 billion inhabitants of the Earth. Every man, woman, and child could spend $1 million every hour, day and night, and half of the pennies would still be left unspent at death.

Can these fantastic claims be correct? Check one or both, showing your reasoning. Come up with an analogy of your own.

Knowledge of Avogadro's number and the atomic mass of any element permit us to calculate the average mass of an individual atom of that element. Thus, the mass of 6.02×10^{23} oxygen atoms is 16.00 g, the atomic mass from the periodic table. To find the average mass of just one oxygen atom, we must divide the mass of the large collection of atoms by the size of the collection. In chemist's terms, this means dividing the atomic mass by Avogadro's number. Fortunately, calculators help make this job quick and easy.

$$\frac{16.00 \text{ g oxygen}}{6.02 \times 10^{23} \text{ oxygen atoms}} = 2.66 \times 10^{-23} \text{ g oxygen/oxygen atom}$$

This very small mass confirms once again why chemists do not generally work with small numbers of atoms. We manipulate trillions at a time. Therefore, practitioners of this art need to measure matter with a sort of chemist's dozen—a very large one, indeed. To learn about it, read on . . . but only after stopping to practice your new skill.

Your Turn 3.14   **Calculating Mass of Atoms**

a. Calculate the average mass in grams of an individual atom of nitrogen.
b. Calculate the mass in grams of 5 trillion nitrogen atoms.
c. Calculate the mass in grams of 6×10^{15} nitrogen atoms.

Answer
a. $\dfrac{14.01 \text{ g nitrogen}}{6.02 \times 10^{23} \text{ nitrogen atoms}} = 2.34 \times 10^{-23}$ g nitrogen/nitrogen atom

Calculation Tip
Predict:
Will the answer be a large or a small number?
Check:
Does your answer match your prediction and is it reasonable?

3.7 | Quantitative Concepts: Molecules and Moles

Chemists have another way of communicating the number of atoms, molecules, or other small particles present. This is to use the term **mole (mol),** defined as containing an Avogadro's number of objects. The term is derived from the Latin word to "heap," or "pile up." Thus, 1 mol of carbon atoms consists of 6.02×10^{23} C atoms, 1 mol of oxygen gas is made up of 6.02×10^{23} oxygen molecules, and 1 mol of carbon dioxide molecules corresponds to 6.02×10^{23} carbon dioxide molecules.

As you already know from previous chapters, chemical formulas and equations are written in terms of atoms and molecules. For example, reconsider the equation for the complete combustion of carbon in oxygen.

$$C + O_2 \longrightarrow CO_2 \qquad \qquad [3.1]$$

This equation tells us that one atom of carbon combines with one molecule of oxygen to yield one molecule of carbon dioxide. Thus it reflects the *ratio* in which the particles interact. It is equally correct to say that 10 C atoms react with 10 O_2 molecules (20 O atoms) to form 10 CO_2 molecules. Or, putting the reaction on a grander scale, we can say 6.02×10^{23} C atoms combine with 6.02×10^{23} O_2 molecules (12.0×10^{23} O atoms) to yield 6.02×10^{23} CO_2 molecules. The last statement is equivalent to saying: "one *mole* of carbon plus one *mole* of oxygen yields one *mole* of carbon dioxide." The point is that the numbers of *atoms and molecules* taking part in a reaction are proportional to the numbers of *moles* of the same substances. The ratio of two oxygen atoms to one carbon atom remains the same regardless of the number of carbon dioxide molecules, as summarized in Table 3.1.

When used together with a number, *mol* is an abbreviation for *mole*.

There are 2 mol of oxygen atoms, O, in every mole of oxygen molecules, O_2.

Table 3.1	Ways to Interpret a Chemical Equation		
C	+	O₂ →	CO₂
1 atom		1 molecule	1 molecule
6.02×10^{23} atoms		6.02×10^{23} molecules	6.02×10^{23} molecules
1 mol		1 mol	1 mol

In the laboratory and the factory, the quantity of matter required for a reaction is often measured by mass. The mole is a practical way to relate number of particles to the more easily measured mass. The **molar mass** is the mass of one Avogadro's number, or *mole*, of whatever particles are specified. For example, the mass of a mole of carbon atoms, rounded to the nearest tenth of a gram, is 12.0 g. A mole of oxygen atoms has a mass of 16.0 g. But we can also speak of a mole of O_2 molecules. Because there are two oxygen atoms in each oxygen molecule, there are two moles of oxygen atoms in each mole of molecular oxygen, O_2. Consequently, the molar mass of O_2 is 32.0 g, twice the molar mass of O. Some refer to this as the molecular mass or molecular weight of O_2, emphasizing its similarity to atomic mass or atomic weight.

The same logic for the molar mass of the element O_2 applies to compounds, which brings us, at last, to the composition of carbon dioxide. The formula, CO_2, reveals that each molecule contains one carbon atom and two oxygen atoms. Scaling up by 6.02×10^{23}, we can say that each mole of CO_2 consists of 1 mol of C and 2 mol of O atoms (see Table 3.1). But remember that we are interested in the mass composition of carbon dioxide—the number of grams of carbon per gram of CO_2. This requires the molar mass of carbon dioxide, which we obtain by adding the molar mass of carbon to twice the molar mass of oxygen:

$$1 \text{ mol CO}_2 = 1 \text{ mol C} + 2 \text{ mol O}$$

$$= \left(1 \text{ mol C} \times \frac{12.0 \text{ g C}}{1 \text{ mol C}}\right) + \left(2 \text{ mol O} \times \frac{16.0 \text{ g O}}{1 \text{ mol O}}\right)$$

$$= 12.0 \text{ g C} + 32.0 \text{ g O}$$

$$1 \text{ mol CO}_2 = 44.0 \text{ g CO}_2$$

This procedure is routinely used in chemical calculations, where molar mass is an important property. Some examples are included in the next activity. In every case, you multiply the number of moles of each element by the corresponding atomic mass in grams and add the result.

Your Turn 3.15   Molecular Molar Mass

Calculate the molar mass of each of these greenhouse gases.

a. O_3 (ozone)
b. N_2O (dinitrogen monoxide or nitrous oxide)
c. CCl_3F (Freon-11; trichlorofluoromethane)

Answer

a. $1 \text{ mol O}_3 = 3 \text{ mol O}$

$$= 3 \text{ mol O} \times \frac{16.0 \text{ g O}}{1 \text{ mol O}}$$

$$= 48.0 \text{ g O}_3$$

We started out on this mathematical excursion so that we could calculate the mass of CO_2 produced from burning 3.3 Gt of carbon. We now have all the pieces assembled. Out of every 44.0 g of CO_2, 12.0 g is C. This mass ratio holds for all samples of CO_2, and we can use it to calculate the mass of C in any known mass of CO_2. More to the point, we can use it to calculate the mass of CO_2 released by any known mass of carbon. It only depends on how we arrange the ratio. The C-to-CO_2 ratio is $\dfrac{12.0 \text{ g C}}{44.0 \text{ g CO}_2}$, but it is equally true that the CO_2-to-C ratio is $\dfrac{44.0 \text{ g CO}_2}{12.0 \text{ g C}}$.

For example, we could compute the number of grams of C in 100.0 g CO_2 by setting up the relationship in this manner.

$$100.0 \text{ g } \cancel{CO_2} \times \frac{12.0 \text{ g C}}{44.0 \text{ g } \cancel{CO_2}} = 27.3 \text{ g C}$$

The fact that there is 27.3 g of carbon in 100.0 g of carbon dioxide is equivalent to saying that the mass percent of C in CO_2 is 27.3%. Note that carrying along the units "g CO_2" and "g C" helps you do the calculation correctly. The unit "g CO_2" can be canceled, and you are left with "g C." Keeping track of the units and canceling where appropriate are useful strategies in solving many problems. This method is sometimes called "unit analysis."

Your Turn 3.16   Mass Ratios and Percents

a. Calculate the mass ratio of S in SO_2.
b. Find the mass percent of S in SO_2.
c. Calculate the mass ratio and the mass percent of N in N_2O.

Answers

a. The mass ratio is found by comparing the molar mass of S with the molar mass of SO_2.

$$\frac{32.1 \text{ g S}}{64.1 \text{ g SO}_2} = \frac{0.501 \text{ g S}}{1.00 \text{ g SO}_2}$$

b. To find the mass percent of S in SO_2, multiply the mass ratio by 100.

$$\frac{0.501 \text{ g S}}{1.00 \text{ g SO}_2} \times 100 = 50.1\% \text{ S in SO}_2$$

Calculation Tip

Predict:

Will the answer be larger or smaller than the given value? What are the units?

Check:

Does the answer match your prediction? Have units canceled, leaving the one needed for the answer?

To find the mass of CO_2 that contains 3.3 gigatons (Gt) of C, we use a similar approach. We could convert 3.3 Gt to grams, but it is not necessary. As long as we use the same mass unit for C and CO_2, the same numerical ratio holds. Compared with our last calculation, this problem has one important difference in how we use the ratio. We are solving for the mass of CO_2, not the mass of C. Look carefully at the units this time.

The question to be answered is: What mass of CO_2 contains 3.3 Gt of carbon?

$$3.3 \text{ } \cancel{\text{Gt C}} \times \frac{44.0 \text{ Gt CO}_2}{12.0 \text{ } \cancel{\text{Gt C}}} = 12 \text{ Gt CO}_2$$

Once again the units cancel and we are left with Gt of CO_2.

Our burning question, "What is the mass of CO_2 added to the atmosphere each year from the combustion of fossil fuels?" has finally been answered: 12 gigatons. Of course, we also managed to demonstrate the problem-solving power of chemistry and to introduce five of its most important ideas: atomic mass, molecular mass, Avogadro's number, mole, and molar mass. The next few activities provide opportunities to practice your skill with these concepts.

Your Turn 3.17   SO$_2$ from Volcanoes

a. It is estimated that volcanoes globally release about 19×10^6 t (19 million metric tons) of SO$_2$ per year. Calculate the mass of sulfur in this amount of SO$_2$.

b. If 142×10^6 t of SO$_2$ is released per year by fossil-fuel combustion, calculate the mass of sulfur in this amount of SO$_2$.

Answer

a. The mass ratio of S to SO$_2$ is known from Your Turn 3.16.

$$19 \times 10^6 \text{ t SO}_2 \times \frac{32.1 \times 10^6 \text{ t S}}{64.1 \times 10^6 \text{ t SO}_2} = 9.5 \times 10^6 \text{ t S}$$

If you know how to apply these ideas, you have gained the ability to critically evaluate media reports about releases of C or CO$_2$ (and other substances as well) and judge their accuracy. One can either take such statements on faith or check their accuracy by applying mathematics to the relevant chemical concepts. Obviously, there is insufficient time to check every assertion, but we hope that readers develop questioning and critical attitudes toward all statements about chemistry and society, including those found in this book.

Skeptical Chemist 3.18   Checking Carbon from Cars

A clean-burning automobile engine emits about 5 pounds of C in the form of CO$_2$ for every gallon of gasoline it consumes. The average American car is driven about 12,000 miles per year. Using this information, check the statement that the average American car releases its own weight in carbon into the atmosphere each year. List the assumptions you make in solving this problem. Compare your list and your answer with those of your classmates.

3.8 | Methane and Other Greenhouse Gases

Concerns about an enhanced greenhouse effect are based primarily on increases in concentrations of atmospheric CO$_2$. However, other gases also play a role. Methane, nitrous oxide, chlorofluorocarbons, and even our friend ozone all take part in trapping heat in the atmosphere.

Our level of concern regarding each of these gases is related to their concentration in the atmosphere but also to other important characteristics. The **global atmospheric lifetime** characterizes the time required for a gas added to the atmosphere to be removed. It is also referred to as the "turnover time." Greenhouse gases also vary in their effectiveness in absorbing infrared radiation. This is quantified by the **global warming potential (GWP),** a number that represents the relative contribution of a molecule of the atmospheric gas to global warming. The GWP of carbon dioxide is assigned the reference value of 1; all other greenhouse gases are indexed with respect to it. Gases with relatively short lifetimes, such as water vapor, tropospheric ozone, tropospheric aerosols, and other ambient air pollutants, are distributed unevenly around the world. It is difficult to quantify their effect, and therefore GWP values are not usually assigned. Table 3.2 lists four greenhouse gases, their main sources, and their important properties in the climate change conversation.

Table 3.2	Examples of Greenhouse Gases				
Name and Chemical Formula	Preindustrial Concentration (1750)	Concentration in 2008	Atmospheric Lifetime (years)	Anthropogenic Sources	Global Warming Potential
carbon dioxide CO_2	270 ppm	388 ppm	50-200*	Fossil fuel combustion, deforestation, cement production	1
methane CH_4	700 ppb	1760 ppb	12	Rice paddies, waste dumps, livestock	21
nitrous oxide N_2O	275 ppb	322 ppb	120	Fertilizers, industrial production, combustion	310
CFC-12 CCl_2F_2	0	0.56 ppb	102	Liquid coolants, foams	8100

*A single value for the atmospheric lifetime of CO_2 is not possible. Removal mechanisms take place at different rates. The range given is an estimate based on several removal mechanisms.

Your Turn 3.19   Greenhouse Gases on the Rise

Using the data in Table 3.2, calculate the percentage increases for CO_2, CH_4, and N_2O since the beginning of the Industrial Revolution. Rank the three in order of their percentage increase.

Global atmospheric lifetime values, although useful for comparison, are best thought of as approximations.

The current atmospheric concentration of CH_4 is about 50 times lower than that of CO_2, but as an infrared absorber, methane is about 20 times more efficient than carbon dioxide. Fortunately, CH_4 is quite readily converted to other chemical species by interaction with tropospheric free radicals, and therefore has a relatively short lifetime. By comparison, carbon dioxide is much less reactive. The primary removal mechanisms for CO_2 are dissolution in oceans, and the much longer process of mineralization into carbonate rocks.

Methane emissions arise from both natural and human sources. About 40% of total CH_4 emissions come from natural sources, of which emanations from wetlands are by far the largest contributor. These marshy habitats are perfectly suited for **anaerobic bacteria,** those that can function without the use of molecular oxygen. As they decompose organic matter, many types of anaerobic bacteria produce methane, which then escapes into the atmosphere. In Alaska, Canada, and Siberia, however, much of the methane produced from thousands of years of decomposition has remained trapped underground by the permafrost. There is concern that melting of the surface in the Northern latitudes might trigger a massive release of methane into the atmosphere. There is geological evidence that such a release has occurred in the past, and led to higher global temperatures.

Methane is also released from the oceans, where a substantial amount of it appears to be trapped in "cages" made of water molecules. Such deposits are referred to as methane hydrates. Australia's Commonwealth Scientific and Industrial Research Organization (CSIRO) has been taking a series of ocean core drillings to gather evidence about methane hydrates and their role in global warming (Figure 3.23). There is concern that if some of these hydrates become unstable then large amounts of methane might rapidly be released to the atmosphere.

Termites are another natural source of methane. These ubiquitous insects have special bacteria in their guts that allow them to metabolize cellulose, the main component of wood. But instead of making water and CO_2, termites produce methane and CO_2. Not only can they inflict direct damage to homes, but they also add to greenhouse gas concentrations. The sheer number of termites is staggering, estimated to be more than half a metric ton for every man, woman, and child on the planet!

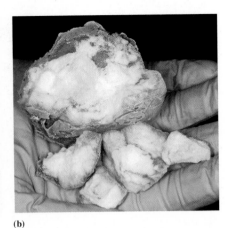

(a)   (b)

Figure 3.23

(a) A floating drilling platform used by the CSIRO. **(b)** A sample of methane hydrate from the continental shelf off the coast of Florida.

The major human source of CH_4 is agriculture, with the biggest culprits being rice cultivation and the raising of livestock. Rice is grown with its roots under water, where, again, anaerobic bacteria produce methane. Most of the methane is released to the atmosphere. Additional agricultural CH_4 comes from an increasing number of cattle and sheep. The digestive systems of these ruminants (animals that chew their cud) contain bacteria that break down cellulose. In the process, methane is formed and released through belching and flatulence—about 500 liters of CH_4 per cow per day! The ruminants of the Earth release a staggering 73 million metric tons of CH_4 each year.

Landfills add another large quantity of methane to the atmosphere. The chemistry occurring within our buried garbage is controlled by the same anaerobic bacteria found in wetlands and produces the same result. Some of this methane is captured (biogas) and burned as a fuel, but the vast majority is released into the atmosphere.

For more information on using biogas as a fuel, check Section 4.10.

The other main anthropogenic source of methane originates from our extraction of fossil fuels. Methane is often found with oil and coal deposits, and drilling and mining procedures release most of that methane to the atmosphere while recovering the liquid or solid products. There are also significant losses from transporting, purifying, and using natural gas.

Consider This 3.20   Methane Concentrations Stabilizing?

Concentrations of atmospheric methane have stabilized in recent years. Use the resources of the web to find some hypotheses about why the concentrations have leveled out.

The role of N_2O in destroying stratospheric ozone was discussed in Section 2.8.

Another gas that contributes to global warming is nitrous oxide, also known as "laughing gas." It has been used as an inhaled anesthetic for dental and medical purposes. Its sources and sinks are not as well established as are those for carbon dioxide and methane. The majority of N_2O molecules in the atmosphere come from the bacterial removal of nitrate ion (NO_3^-) from soils, followed by removal of oxygen. Agricultural practices, again linked to population pressures, can speed up the removal of reactive compounds of nitrogen from soils. Other sources include ocean upwelling, and stratospheric interactions of nitrogen compounds with high-energy oxygen atoms. Major anthropogenic sources of N_2O are automobile catalytic converters, ammonia fertilizers, biomass burning, and certain industrial processes (nylon and nitric acid production). In the atmosphere, a typical N_2O molecule persists for about 120 years,

Table 3.3	Climate Change and Ozone Depletion: A Comparison	
	Climate Change	**Ozone Depletion**
region of atmosphere	primarily the troposphere	the stratosphere
major players	H_2O, CO_2, CH_4, and N_2O	O_3, O_2, and CFCs
interaction with radiation	Molecules absorb IR radiation. This causes them to vibrate and return heat energy to the Earth.	Molecules absorb UV radiation. This causes one or more bonds in the molecule to break.
nature of problem	Greenhouse gases are increasing in concentration. In turn this is trapping more heat, causing an increase in the average global temperatures.	CFCs are causing a decrease in concentrations of O_3 in the stratosphere. In turn, this is causing an increase in the UV radiation at the surface of the Earth.

absorbing and emitting infrared radiation. Over the past decade, global atmospheric concentrations of N_2O have shown a slow but steady rise.

A few comments need to be made about ozone, a gas we encountered in Chapter 2. Often there is confusion between the phenomena of climate change and ozone depletion. Both are often in the news, both involve complex atmospheric processes, and both have anthropogenic as well as natural sources. In fact, ozone itself can act like a greenhouse gas, but its efficiency depends very much on its altitude. It appears to have its maximum warming effect in the upper troposphere, around 10 km above the Earth. Therefore, depletion of ozone has a *slight cooling effect* in the stratosphere, and it may also promote slight cooling at Earth's surface. Other differences are summarized in Table 3.3.

Depletion of the stratospheric ozone layer is clearly *not* a principal cause of climate change. However, stratospheric ozone depletion and climate change are linked in an important way, through ozone-destroying substances. CFCs, HCFCs, and halons, all implicated in the destruction of stratospheric ozone, also absorb infrared radiation and are all greenhouse gases. Emissions of these synthetic gases have risen by 58% from 1990–2005, although their concentrations are still very low.

HCFCs were discussed in Section 2.12.

Your Turn 3.21   Comparing Greenhouse Gas Effectiveness

Multiplying GWP by tropospheric abundance provides a number that can be used to compare the warming effectiveness of a greenhouse gas.

a. Compare the effectiveness of CFC-12 (CCl_2F_2) as a greenhouse gas relative to that of CO_2.

b. HFC-134a (CF_3CH_2F lifetime = 13.8 years) is a replacement for Freon-12 and has a GWP value of 1300 and a tropospheric abundance of 7.5 parts per trillion (1998 data). Compare its effectiveness as a greenhouse gas to that of CO_2.
Hint: Use the same unit of tropospheric abundance for both gases.

c. How do their global atmospheric lifetimes affect their overall ability to function as greenhouse gases?

3.9 | How Warm Will the Planet Get?

"Prediction is very difficult, especially about the future." Niels Bohr, one of the foremost contributors to our modern view of the atom, spoke these words years ago. His words still hold true today!

The unique properties of water, including its unusually large specific heat, will be described in Chapter 5.

Figure 3.24

Climate scientists use computer simulations to understand future climate change.

Although admittedly a difficult task, we still need to make predictions. To this end, in 1988, the United Nations Environment Programme and the World Meteorological Organization teamed up to establish the UN Intergovernmental Panel on Climate Change (IPCC). The IPCC was charged with assessing the data on climate change, not just the scientific data, but the socioeconomic information as well. Thousands of international scientists were involved in this review. In their fourth and most recent report published in 2007, the vast majority of scientists agreed on several key points: (1) The Earth is getting warmer; (2) Human activities (primarily the combustion of fossil fuels and deforestation) are responsible for much of the recent warming; and (3) If the rate of greenhouse gas emissions is not curtailed, our water resources, food supply, and even our health will suffer.

The challenge, however, is to understand current climate change well enough to *predict* future changes and by doing so, to determine the decrease in emissions required to minimize harmful changes. To make predictions, scientists work with models. They design computer models of the oceans and the atmosphere that take into account the ability of each to absorb heat as well as to circulate and transport matter (Figure 3.24). If that weren't difficult enough, the models must also include astronomical, meteorological, geological, and biological factors, ones that are often incompletely understood. Human influences, such as population, industrialization levels, and pollution emissions must also be included. Dr. Michael Schlesinger, who directs climate research at the University of Illinois, remarked: "If you were going to pick a planet to model, this is the *last* planet you would choose."

Climate scientists call the factors (both natural and anthropogenic) that influence the balance of Earth's incoming and outgoing radiation by the term **radiative forcings.** Negative forcings have a cooling effect; positive forcings a warming effect. The primary forcings used in climate models are solar irradiance, greenhouse gas concentrations, land use, and aerosols. The effects of these forcings on the Earth's energy balance are summarized in Figure 3.25. Red, orange, and yellow bars represent positive forcings, and blue bars indicate negative ones. Each forcing has an error bar associated with it; the larger the error bar, the more uncertain the value.

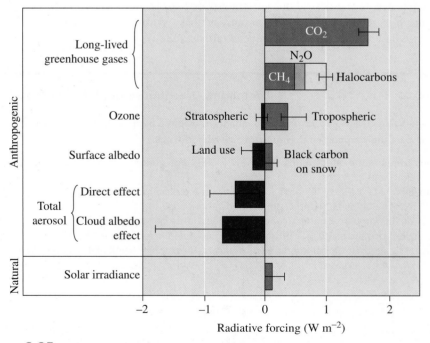

Figure 3.25

Selected radiative forcings of climate from 1750 to 2005. The units are in watts per square meter (W m^{-2}), the light energy hitting a square meter of the Earth's surface every second.

Source: Adapted from *Climate Change 2007: The Physical Science Basis. Contribution of Working Group I to the Fourth Assessment Report of the Intergovernmental Panel on Climate Change.*

Solar Irradiance ("solar brightness")

We can directly observe the natural seasonal variations in sunlight intensity. In the higher latitudes, temperatures are warmer in the summer. Compared to winter months, the Sun is higher in the sky and stays up longer. Across the globe, these variations essentially cancel, because when it is winter in the Northern Hemisphere it is summer in the Southern Hemisphere.

Subtle periodic changes occur in the brightness of the Sun. The Earth's orbit oscillates slightly over a 100,000-year period, changing its shape. In addition, the magnitude of the tilt of the Earth's axis and the direction of that tilt both change over the course of several tens of thousands of years, affecting the amount of solar radiation hitting the Earth. Neither of these occurs on a time scale short enough to explain the recent warming, however.

Additionally, sunspots occur in large numbers about every 11 years. You might think that dark spots on the Sun would mean a smaller amount of radiation hitting the Earth, but exactly the opposite is true. Sunspots occur when there is increased magnetic activity in the outer layers of the Sun, and the stronger magnetic fields stir up a larger amount of charged particles that emit radiation. Notably, the 17th and 18th centuries, sometimes called the "Little Ice Age" because of the below average temperatures in Europe, were preceded by a period of almost no sunspot activity. However, the solar brightness over those 11-year cycles varies only by about 0.1%. As you can see from Figure 3.25, this natural variability is the *smallest* of any positive forcing listed.

> Periodic orbital eccentricities are a possible cause of the ice age oscillations shown in Figure 3.9.

> During periods of high sunspot activity, the aurora borealis ("northern lights") is more spectacular because of the greater number of charged particles striking the Earth's atmosphere.

Your Turn 3.22   Radiation from the Sun

Sunlight strikes the Earth continually. Which types of light are emitted by the Sun? Which one makes up the largest percentage of sunlight?
Hint: Refer back to Figure 2.7.

Greenhouse Gases

These are the dominant anthropogenic forcings. Largest among these is CO_2, constituting about two thirds of the warming from all greenhouse gases. However, as we explained in the previous section, methane, nitrous oxide, and other gases do contribute. Notice the relatively small contribution from "halocarbons" (CFCs and HCFCs) as shown in Figure 3.25. It has been estimated that without the ban on CFC production imposed by the Montreal Protocol, by 1990 the forcings from CFCs would have outweighed those from CO_2. In sum, the positive forcings from greenhouse gases are more than 30 times greater than the natural changes in solar irradiance.

> The Montreal Protocol was discussed in Section 2.11.

Land Use

Changes in land use drive climate change because these changes alter the amount of incoming solar radiation that is absorbed by the surface of the Earth. The ratio of electromagnetic radiation *reflected* from a surface relative to the amount of radiation *incident* on it is called the **albedo.** In short, albedo is a measure of the reflectivity of a surface. The albedo of the Earth's surface varies between about 0.1 and 0.9, as you can see from the values listed in Table 3.4. The higher the number, the more reflective the surface.

As the seasons change, so does the albedo of the Earth. When a snow-covered area melts, the albedo decreases and more sunlight is absorbed, creating a positive feedback loop and additional warming. This effect helps to explain the greater increases in average temperature observed in the Arctic, where the amount of sea ice and permanent snow cover is decreasing. Similarly, when glaciers retreat and expose darker rock, the albedo decreases, causing further warming.

Human activity also can change the Earth's albedo, most notably through deforestation in the tropics. The crops we plant reflect more incoming light than does the

> Earth has an average albedo of 0.39. In contrast, that of the Moon is about 0.12.

Table 3.4	Albedo Values for Different Ground Covers
Surface	Range of Albedo
fresh snow	0.80–0.90
old/melting snow	0.40–0.80
desert sand	0.40
grassland	0.25
deciduous trees	0.15–0.18
coniferous forest	0.08–0.15
tundra	0.20
ocean	0.07–0.10

dark green foliage of the rainforests, causing an increase in the albedo and hence resulting in *cooling*. In addition, sunlight is more consistent in the tropics, so changes in land use at low latitudes produce greater effects than changes in the polar regions. The conversion of tropical rainforest to crop and pastureland has more than offset the decrease in the amount of sea ice and snow cover near the poles. Therefore, the changes in the Earth's albedo have caused a net *cooling* effect.

Consider This 3.23   White Roofs, Green Roofs

a. In 2009, U.S. Energy Secretary Steven Chu suggested that painting roofs white would be one way to combat global warming. Explain the reasoning behind this course of action.
b. The idea of "green roofs" is also attracting attention. Planting gardens on rooftops has benefits in addition to those of white roofs. But such gardens also have limitations. Explain.

The term *aerosol* was defined in Section 1.11. The role of aerosols in acid rain will be discussed in Section 6.6.

Mt. Pinatubo eruption, 1991.

Aerosols

A complex class of materials, aerosols have a correspondingly complex effect on climate. Many natural sources of aerosols exist, including dust storms, ocean spray, forest fires, and volcanic eruptions. Human activity can also release aerosols into the environment in the form of smoke, soot, and sulfate aerosols from coal combustion.

The effect of aerosols on climate is probably the least well understood of the forcings listed in Figure 3.25. Tiny aerosol particles (<4 μm) are efficient at scattering incoming solar radiation. Other aerosols absorb incoming radiation, and still other particles both scatter and absorb. Both processes decrease the amount of radiation available for greenhouse gases to absorb and therefore have a cooling effect (negative forcing). In a dramatic example, the 1991 eruption of Mt. Pinatubo in the Philippines spewed over 20 million tons of SO_2 into the atmosphere. In addition to providing spectacular sunsets for several months, the sulfur dioxide caused temperatures around the world to drop slightly. The results provided the climate modelers a mini-control experiment. The most reliable models were able to reproduce the cooling effect caused by the eruption.

In addition to that direct cooling effect, aerosol particles can serve as nuclei for the condensation of water droplets and hence promote cloud formation. Clouds reflect incoming solar radiation, although the effects of increased cloud cover are more complex than this. Therefore, in both direct and indirect ways, aerosols *counter* the warming effects of greenhouse gases.

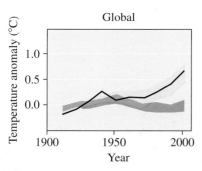
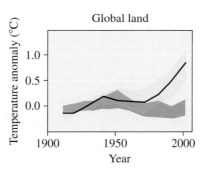
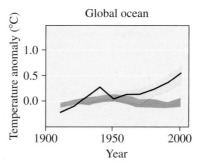

Figure 3.26

Climate model predictions of annual global mean surface temperatures for the 20th century. Black lines display temperature data relative to the average temperature for the years 1901–1950. The blue bands indicate the predicted temperature range using natural forcings only. The pink bands indicate the predicted temperature range using *both* natural and anthropogenic forcings.

Source: Adapted from Climate Change 2007: The Physical Science Basis. Contribution of Working Group I to the Fourth Assessment Report of the Intergovernmental Panel on Climate Change.

Given the complexity inherent in all the forcings that we have just described, you can appreciate that assembling these forcings into a climate model is no easy task. Furthermore, once a model has been built, scientists have difficulty assessing its validity. However, scientists do have one trick in their back pockets. They can test climate models with known data sets as a means to tease apart the contributions of different forcings. For example, we know the temperature data of the 20th century. In Figure 3.26, the black lines represent the known data. Next examine the blue bands. These represent temperature ranges that were predicted by the climate model using *only* natural forcings. As you can see, the natural forcings do not map well onto the actual temperatures. Finally, examine the pink bands to see that when anthropogenic forcings are included, the temperature increases of the 20th century can be accurately reproduced. So although the last 30 years of warming were *influenced* by natural factors, the actual temperatures cannot be accounted for without including the effects of human activities.

Your Turn 3.24   **Assessing Climate Models**

Between 1950 and 2000, the climate models that used natural forcings only (blue bands in Figure 3.26) showed an overall cooling effect and thus did not match the observed temperatures.

a. Name the forcings included in the models that only included natural forcings.

b. List two additional forcings included in the models that more accurately recreate the temperatures of the 20th century (pink bands in Figure 3.26).

Answer

a. Aerosols (such as those from volcanic eruptions), solar irradiance.

The magnitude of future emissions, and hence the magnitude of future warming, depends on many factors. As you might expect, one is population. As of 2010, the global population stood at about 6.9 billion. Assuming that there will be more feet on the planet in the future, we humans are likely to have a larger **carbon footprint,** an estimate of the amount of CO_2 and other greenhouse gas emissions in a given time frame, usually a year. Having more people to feed, clothe, house, and transport will require the consumption of more energy. In turn, this translates to more CO_2 emissions, at least if using current energy sources. In addition, scientists who create climate models have to include values for two factors: (1) the rate of economic growth and (2) the rate of development of "green" (less carbon-intensive) energy sources. Again as you might expect, both are difficult to predict.

Chapter 0 introduced the concept of an ecological footprint. Carbon footprints are a subset of the more general term.

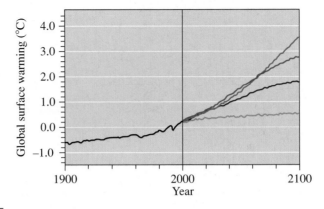

Figure 3.27

Four model projections for temperature scenarios in the 21st century based on different socioeconomic assumptions. The black line is the data for the 20th century with the gray regions indicating the uncertainty in those values. The four dark lines represent projected 21st-century temperatures, with the wider lighter colored bands representing the uncertainty range for each scenario.

Source: Adapted from Climate Change 2007: The Physical Science Basis. Contribution of Working Group I to the Fourth Assessment Report of the Intergovernmental Panel on Climate Change.

So what, if anything, can computer models tell us about the Earth's future climate? Given the uncertainties that we have listed, hundreds of different projected temperature scenarios for the 21st century are possible. Figure 3.27 shows four of these, together with the actual temperature data for the 20th century.

The four scenarios for 21st-century temperatures are based on different assumptions. The orange line assumes that emissions levels are kept at 2000 levels, admittedly an unrealistic target given the increases that already have occurred since 2000. Even with this most optimistic scenario, some additional warming will take place due to the persistence of CO_2 in the atmosphere for years to come. Both the blue and green lines assume that the global population will increase to 9 billion by 2050 but then gradually decrease. However, the blue line includes the more rapid development of energy-efficient technologies, leading to lower CO_2 emissions. The red line assumes a continually increasing population combined with a slower and less globally integrated transition to new, cleaner technologies.

All of the lines point in the same direction—up. With some amount of future warming virtually ensured, we now turn our discussion to the consequences of climate change.

3.10 | The Consequences of Climate Change

Considering even the most extreme predictions of warming described in the last section, you may be thinking, "So what?" After all, the temperature changes predicted in Figure 3.27 are only a few degrees. At any single spot on the planet, the temperature fluctuates several times that amount daily.

An important distinction needs to be made between the terms *climate* and *weather*. **Weather** includes the daily highs and lows, the drizzles and downpours, the blizzards and heat waves, and the fall breezes and hot summer winds, all of which have relatively short durations. In contrast, **climate** describes regional temperatures, humidity, winds, rain, and snowfall over decades, not days. And while the weather varies on a daily basis, our climate has stayed relatively uniform over the last 10,000 years. The values quoted for the "average global temperature" are but one measure of climate phenomena. The key point is that relatively small changes in average global temperature can have huge effects on many aspects of our climate.

In addition to modeling various future temperature scenarios (see Figure 3.27), the 2007 IPCC report estimated the likelihood of various consequences. The report employed descriptive terms ("judgmental estimates of confidence") to help both policy

Table 3.5	Judgmental Estimates of Confidence
Term	**Probability That a Result Is True (%)**
virtually certain	>99
very likely	90–99
likely	66–90
medium likelihood	33–66
unlikely	10–33
very unlikely	1–10

Source: Adapted from Climate Change 2007: The Physical Science Basis. Contribution of Working Group I to the Fourth Assessment Report of the Intergovernmental Panel on Climate Change.

makers and the general public better understand the inherent uncertainty of the data. These terms are listed in Table 3.5 together with the definitions that will continue to be used in all subsequent updates to the IPCC report.

Conclusions from the 2007 IPCC report are listed in Table 3.6. For example, it was judged *very unlikely* that all of the observed global warming was due to natural climate variability. Rather, the scientific evidence strongly supports the position that human activity is a significant factor causing the increase in average global temperature observed over the last century. Furthermore, from the scientific evidence for global warming, it was judged *virtually certain* that human activities were the main drivers of recent warming. Check Table 3.6 for other conclusions relevant to any discussion of global climate change.

Many scientific organizations, including the American Association for the Advancement of Science and the American Chemical Society, also have recognized the threats posed by climate change. In an open letter to United States senators, the organizations cited sea level rises, more extreme weather events, increased water scarcity, and disturbances of local ecosystems as likely eventualities of a warmer planet. To conclude this section, we describe these and other outcomes we can expect, including sea ice disappearance, more extreme weather, changes in ocean chemistry, loss of biodiversity, and harm to human health.

Each of the potential consequences can be considered in the context of the tragedy of the commons, which we encountered in Chapters 1 and 2.

Table 3.6	IPCC Conclusions, 2007

Very Likely

- Human-caused emissions are the main factor causing warming since 1950.
- Higher maximum temperatures are observed over nearly all land areas.
- Snow cover decreased about 10% since the 1960s (satellite data); lake and river ice cover in the middle and high latitudes of the Northern Hemisphere was reduced by 2 weeks per year in the 20th century (independent ground-based observations).
- In most of the areas in the Northern Hemisphere, precipitation has increased.

Likely

- Temperatures in the Northern Hemisphere during the 20th century have been the highest of any century during the past 1000 years.
- Arctic sea ice thickness declined about 40% during late summer to early autumn in recent decades.
- An increase in rainfall, similar to that in the Northern Hemisphere, has been observed in tropical land areas falling between 108° North and 108° South.
- Summer droughts have increased.

Very Unlikely

- The observed warming over the past 100 years is due to climate variability alone, providing new and even stronger evidence that changes must be made to stem the influence of human activities.

Figure 3.28

The extent of Arctic ice in 1979 (*left*) and in 2003 (*right*), composite images.

Source: Earth Observatory, NASA.

Sea Ice Disappearance

As shown in Figure 3.8, the temperatures in the Arctic are rising faster than anywhere else on Earth. One result is that sea ice is shrinking (Figure 3.28). In 2007, the ice cover at the end of the Arctic summer was 23% lower than the previous record. A new analysis that uses both computer models and data from actual conditions in the Arctic region forecasts that most of the Arctic sea ice will be gone in 30 years. Not only would significant populations of wildlife be endangered, but the accompanying decrease in albedo would lead to even more warming.

Sea-Level Rise

Warmer temperatures result in an increase in sea level. This increase occurs primarily because as water warms, it expands. A smaller effect is caused by the influx of fresh-water into the ocean from glacier runoff. According to a 2008 study published in the journal *Nature*, the increase was about 1.5 millimeters each year between 1961 and 2003. However, the increases are not seen uniformly across the globe. In addition, they are influenced by regional weather patterns. Even so, these small increases in sea levels can cause erosion in coastal areas and the stronger storm surges associated with hurricanes and cyclones.

Consider This 3.25   External Costs

The consequences described earlier and later on are examples of what are known as external costs. These costs are not reflected in the price of a commodity, such as the price of a gallon of gasoline or a ton of coal, but nevertheless take a toll on the environment. The external costs of burning fossil fuels often are shared by those who emit very little carbon dioxide, such as the people of the island nation of Maldives. Although a rise of sea level of just a few millimeters may not seem like much, the effects could be catastrophic for nations that lie close to sea level. Use the resources of the web to investigate how the people of Maldives are preparing for rising sea levels.

More Extreme Weather

An increase in the average global temperature could cause more extreme weather, including storms, floods, and droughts. In the Northern Hemisphere, the summers are predicted to be drier and the winters wetter. Over the past several decades, more frequent wildfires and floods have occurred on every continent. The severity (although not the frequency) of cyclones and hurricanes also may be increasing. These tropical storms extract their energy from the oceans; a warmer ocean provides more energy to feed the storms.

Changes in Ocean Chemistry

"Over the past 200 years, the oceans have absorbed approximately 550 billion tons of CO_2 from the atmosphere, or about a third of the total amount of anthropogenic emissions over that period," reports Richard A. Feely, a senior scientist with the National Pacific Marine Environmental Laboratory in Seattle. Scientists estimate that a million tons of CO_2 is absorbed into the oceans every hour of every day! In their role as carbon sinks, the world's oceans have mitigated some of the warming that carbon dioxide would have caused had it remained in the atmosphere. However, this absorption has come with a cost. Critical changes are already occurring in the oceans, as we will further explore in Chapter 6. For example, carbon dioxide is slightly soluble in water and dissolves to form carbonic acid. In turn, this is affecting marine organisms that rely on a constant level of acidity in the ocean to maintain the integrity of their shells and skeletons. The increase in carbon dioxide concentrations in the atmosphere (and hence the corresponding concentration in the oceans) is putting entire marine ecosystems at risk.

Look for more about carbon dioxide and ocean acidification in Chapter 6.

 Consider This 3.26   **Plankton and You**

Plankton are microscopic plant- and animal-like creatures found in both salt and freshwater systems. Many plankton species have shells made of calcium carbonate that could be weakened by more acidic environments. Although humans do not eat plankton, many other marine organisms do. Construct a food chain to show the link between plankton and humans.

Loss of Biodiversity

Climate change already is affecting plant, insect, and animal species around the world. Species as diverse as the California starfish, Alpine herbs, and checkerspot butterflies all have exhibited changes in either their ranges or their habits. Dr. Richard P. Alley, a Pennsylvania State University expert on past climate shifts, sees particular significance in the fact that animals and plants that rely on each other will not necessarily change ranges or habits at the same rate. Referring to affected species, he said, "You'll have to change what you eat, or rely on fewer things to eat, or travel farther to eat, all of which have costs." In extreme cases, those costs can cause the extinction of species. Currently, the rate of extinction worldwide is nearly 1000 times greater than at any time during the last 65 million years! A 2004 report in the journal *Nature* projects that about 20% of the plants and animals considered will face extinction by 2050, even under the most optimistic climate forecasts.

Many different species of checkerspot butterflies exist. This one is found in parts of Wisconsin.

Vulnerability of Freshwater Resources

Like polar and sea ice, glaciers in many parts of the world are shrinking (Figure 3.29). Billions of people rely on glacier runoff for both drinking water and crop irrigation. The 2007 report of the IPCC predicts that a 1 °C increase in global temperature corresponds to more than half a billion people experiencing water shortages that they have not known before. The redistribution of freshwater also has implications in food production. Drought and high temperatures could reduce crop yields in the American Midwest, but the growing range might extend farther into Canada. It is also possible that some desert regions could get sufficient rain to become arable. One region's loss may well become another locale's gain, but it is too early to tell.

For more on the chemistry of water availability and use, see Section 5.3.

Figure 3.29
A view of the Exit Glacier in Kenji Fjords National Park, Alaska in 2008. The sign in the foreground marks the extent of the ice flow in 1978.

Human Health

We may all be losers in a warmer world. In 2000, the WHO attributed over 150,000 premature deaths worldwide to the effects of climate change. Those effects included more frequent and severe heat waves, increased droughts in already water-stressed regions, and infectious diseases in regions where they had not occurred before. Further increases in average temperatures are expected to expand the geographical range of mosquitoes, tsetse flies, and other disease-carrying insects. The result could be a significant upturn in illnesses such as malaria, yellow and dengue fevers, and sleeping sickness in new areas, including Asia, Europe, and the United States.

3.11 | What Can (or Should) We Do About Climate Change?

The debate over climate change has shifted in the last 15 years. Today's scientific data leave little room for doubt about whether it is occurring. For example, measurements of higher surface and ocean temperatures, retreating glaciers and sea ice, and rising sea levels are unequivocal. In addition, the carbon isotopic ratio found in atmospheric CO_2 (discussed in Section 3.4) leaves little doubt that human activity is responsible for much of the observed warming. However, at issue is what we *can* do and what we *should* do about the changes that are occurring.

Consider This 3.27   Carbon Footprint Calculations

Investigate three websites that calculate your carbon footprint. To save you time, the textbook's website provides a list.

a. For each site, list the name, the sponsor, and the information requested in order to calculate the carbon footprint.

b. Does the information requested differ from site to site? If so, report the differences.

c. List two advantages and two disadvantages of doing a carbon footprint calculation.

Energy is essential for every human endeavor. Personally, you obtain the energy you need by eating and then metabolizing food. As a community or nation, we meet our energy needs in a variety of ways, including by burning coal, petroleum, and natural gas. The combustion of these carbon-based fuels produces several waste products, including carbon dioxide. The countries with large populations and those that are highly industrialized tend to burn the largest quantities of fuels and as a result emit the most CO_2. According to the Carbon Dioxide Information Analysis Center (CDIAC) of Oak Ridge National Laboratory, in 2006, the top CO_2 emitters were China, the United States, the Russian Federation, India, and Japan. Which other nations rank high on the list? The next activity shows you how to find out.

For more on food metabolism as a source of energy, see Section 11.9.

 Consider This 3.28   **Carbon Emissions by Nation**

CDIAC publishes a list of the top 20 nations for CO_2 emissions.

   a. From what you already know, predict any five of the nations (in addition to those listed in the previous paragraph) that are on this list. Then use the link provided at the textbook's website to check how accurate your predictions were.
   b. How would these rankings change if they were listed per capita?

Recall the quotation that opened this chapter. In that same 2008 address, John Holdren summarized our options in dealing with climate change with three words: mitigation, adaptation, and suffering. "Basically, if we do less mitigation and adaptation, we're going to do a lot more suffering," he concluded. But who will be responsible for the mitigation? Who will be forced to adapt? Who will bear the brunt of the suffering? It is likely that significant disagreements will arise regarding answers to these questions. But we can agree that any practical solution must be global in nature and include a complicated mix of risk perception, societal values, politics, and economics.

Climate mitigation is any action taken to permanently eliminate or reduce the long-term risk and hazards of climate change to human life, property, or the environment. The most obvious strategy for minimizing anthropogenic climate change is to reduce the amount of CO_2 emitted into the atmosphere in the first place. Take a look back at Figure 3.21. It is difficult to imagine curtailing any of these "necessities" to any great extent. Therefore decreasing our energy consumption will not be easy, at least in the short term. The simplest and least expensive approach is to improve energy efficiency. Due to the inefficiencies associated with energy production, saving energy on the consumer end multiplies its effect on the production end three to five times. However, relying on the individual consumers worldwide to buy the right goods and do the right things will not be sufficient to hold CO_2 emissions below dangerous levels.

Chapter 4 focuses on energy from fossil fuels, Chapter 7 on nuclear energy, and Chapter 8 on some alternative energy sources such as wind and solar power.

A developing technology aimed at slowing the rate of carbon dioxide emissions is to capture and isolate the gas after combustion. **Carbon capture and storage (or CCS)** involves separating CO_2 from other combustion products and storing (sequestering) it in a variety of geologic locations. If the CO_2 is properly immobilized, it cannot reach the atmosphere and contribute to global warming. In addition to the large technological challenges posed by CCS, high start-up costs, usually in excess of $1 billion per plant, so far are limiting this approach as a mitigation strategy.

Although at least two dozen projects are in development worldwide, as of 2009, only four industrial-scale CCS projects were in operation. Three remove CO_2 from natural gas reservoirs and store it in various underground geologic formations (Figure 3.30). The fourth and largest project, located in Saskatchewan, Canada, takes CO_2 captured from a coal-fired power plant in North Dakota and injects it into a depleted oil field. By doing so, additional oil is forced up through the existing wells for recovery. The benefits of enhancing oil recovery combined with CO_2 sequestration could become a model for other types of projects. Combined, these CCS efforts store about 5 million metric tons of carbon dioxide annually.

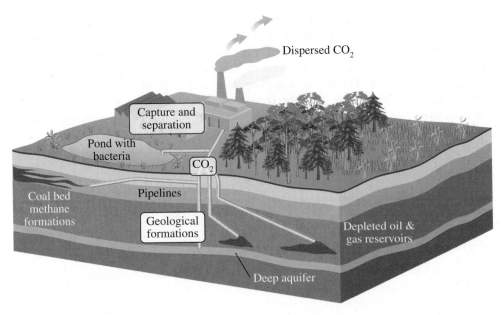

Figure 3.30
Methods for carbon dioxide sequestration.

Your Turn 3.29   Carbon Capture Limitations

Refer back to the global carbon cycle in Figure 3.20. What percent of global carbon dioxide emissions from fossil fuel burning is captured by current CCS technology?

Critics of CCS technology dispute its ultimate efficacy for slowing atmospheric CO_2 buildup, citing high costs as well as the long time frame for commercial implementation. Others contend that pursuing CCS simply delays and distracts attention from developing carbon-free energy sources. Finally, there is the sheer magnitude of the problem. According to the International Energy Agency in order for CCS to make a meaningful contribution to mitigation efforts by 2050, it would require nearly 6000 installations *each* injecting a million metric tons of CO_2 per year into the ground.

A low-tech sequestration strategy is to reverse the extensive deforestation activities that are occurring predominantly in the world's tropical rainforests. Started in 2006 by the United Nations Development Programme and the World Agroforestry Center, the "Billion Trees Campaign" seeks to slow climate change by planting trees in depleted forests. During the first 18 months of the program, over 2 billion trees were planted, mostly in Africa. The organizers have now expanded their goals to 7 billion plantings by 2010. If 7 billion trees seem like a lot, remember the scale of deforestation; in 2005, forest area equal to about 35,000 football fields was cleared *every day*!

Your Turn 3.30   Trees as Carbon Sinks

An average-sized tree absorbs 25 to 50 pounds of carbon dioxide each year. In the United States, the average annual per capital CO_2 emission is 19 tons.

a. How many trees would be required to absorb the annual CO_2 emissions for an average U.S. citizen?

b. What percentage of annual global emissions could be absorbed by 7 billion trees? *Hint:* Refer back to Figure 3.20.

Regardless of any potential decreases in future emissions, some effects of climate change are unavoidable. As mentioned previously, many of the CO_2 molecules emitted today will remain in the atmosphere for centuries. **Climate adaptation** refers to the ability of a system to adjust to climate change (including climate variability and extremes) to moderate potential damage, to take advantage of opportunities, or to cope with the consequences. Some adaptive methods include developing new crop varieties and shoring up or constructing new coastline defense systems for low-lying countries and islands. The further spread of infectious diseases could be minimized by enhanced public health systems. Many of these strategies are win–win situations that would benefit societies even in the absence of climate change challenges.

Compared with the scientific consensus on understanding the role greenhouse gases play in the Earth's climate, there is much less agreement among governments regarding what actions should be taken to limit greenhouse gas emissions. One outcome from the Earth Summit held in 1992 in Rio de Janeiro was the Framework Convention on Climate Change. The goal of this international treaty was "to achieve stabilization of greenhouse gas concentrations in the atmosphere at a low enough level to prevent dangerous anthropogenic interference with the climate system." Not only was this treaty nonbinding, but also there was no agreement about what "dangerous anthropogenic interference" meant, or what level of greenhouse gas emissions would be necessary to avoid it.

In 1997, the first international treaty imposing legally binding limits on greenhouse gas emissions was written by nearly 10,000 participants from 161 countries gathered in Kyoto, Japan. The result has come to be known as the Kyoto Protocol. Binding emission targets based on 1990 levels were set for 38 developed nations to reduce their emissions of six greenhouse gases. The gases regulated include carbon dioxide, methane, nitrous oxide, hydrofluorocarbons (HFCs), perfluorocarbons (PFCs), and sulfur hexafluoride. The United States was expected to reduce emissions to 7% below its 1990 levels, the European Union (EU) nations 8%, and Canada and Japan 6% by 2012.

Genetically engineered foods, specifically corn, are described in Section 12.6.

 Consider This 3.31   **The British Experience**

The British Labour Party in 1997, under the leadership of Tony Blair, boldly committed to cut British greenhouse gas emissions 20% by 2010. This is significantly more than the 12.5% required by the Kyoto treaty. Did Britain meet its goal? Research this question and write a short report on the British experience in reducing greenhouse gases. Have other countries been able to reduce their emissions significantly since 1997?

Although the treaty went into effect in 2005 (when ratified by the Russian Federation), the United States never opted to participate. One reason was the belief that meeting the reduction requirements set by the protocol would cause serious harm to the U.S. economy. Another reason for not ratifying the protocol was concern about the lack of emissions limitations on developing nations, mainly China and India; those countries are expected to show the most dramatic increases in carbon dioxide emissions in the coming years. The administration of President George W. Bush argued that such unequal burdens between developed and developing countries would be economically disastrous to the United States.

The United States has also resisted domestic legislation to restrict CO_2 emissions on similar economic grounds. Voluntary reduction programs implemented during the early 2000s proved insufficient to reduce emissions for a variety of reasons. One "problem" is that fossil fuels are too cheap. A second problem is that any mitigation measures entail significant up-front costs, and just as importantly, the cost of mitigation is not known with certainty, making it difficult for corporations to plan effectively. The world's current energy infrastructure cost $15 trillion to develop and distribute, and reducing carbon dioxide emissions will mean replacing much of that infrastructure. A final problem lies in the fact that the benefits of emissions reductions will not be felt for decades because of the long residence time of CO_2 molecules in the atmosphere.

Now, 15 years after the Earth Summit, scientific consensus is beginning to focus on determining what levels of CO_2 are considered "dangerous." At the United Nations Climate Conference in 2007, participating scientists concluded that greenhouse gas emissions need to peak by about 2020, and then be reduced to well below half of current levels by 2050. In absolute terms, that means that annual global emissions must be decreased by about 9 billion tons. To give you a scale of the magnitude of this goal, reducing emissions by 1 billion tons requires one of the following changes.

- Cutting energy usage in the world's buildings by 20–25% below business-as-usual.
- Having *all* cars get 60 mpg instead of 30 mpg.
- Capturing and sequestering carbon dioxide at 800 coal-burning power plants.
- Replacing 700 large coal-burning power plants with nuclear, wind, or solar power.

Clearly, implementation of any one of those (and the projected goal is 9 billion tons) will not be accomplished on a purely voluntary basis. In the United States and elsewhere, there is a burgeoning realization that laws and regulations are needed to reduce greenhouse gas emissions. One example is a "cap-and-trade" system, such as the one that has been successful in reducing the emission of oxides of both sulfur and nitrogen in the United States. The "trade" part of the cap-and-trade system works through a system of allowances. Companies are assigned allowances that authorize the emission of a certain quantity of CO_2, either during the current year or any year thereafter. At the end of a year, each company must have sufficient allowances to cover its actual emissions. If it has extra allowances, it can trade or sell them to another company that might have exceeded their emissions limit. If a company has insufficient allowances, it must purchase them. The "cap" is enforced by creating only a certain number of allowances each year.

Here's an example of how cap-and-trade works. Without emission restrictions, Plant A emits 600 tons of CO_2 and Plant B emits 400 tons. To get under the imposed cap, they are required to reduce their combined emissions by 300 tons (30%). One way to accomplish this is for each to reduce their own emissions by 30%, each accruing the associated costs. It is likely, however, that one of the plants (Plant B in Figure 3.31)

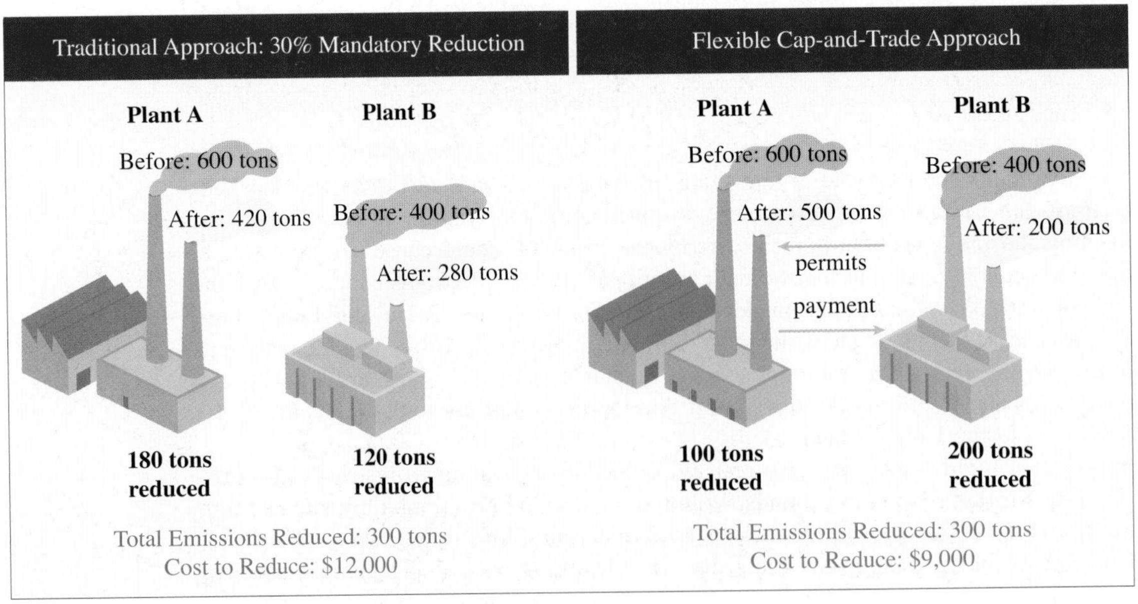

Section 6.11 describes in detail the damage caused by the oxides of nitrogen and sulfur.

Figure 3.31

The emissions cap-and-trade concept.

Source: EPA, *Clearing the Air, The Facts About Capping and Trading Emissions*, 2002, page 3.

would be more efficient in their emissions reductions, and lower their emissions below the proscribed 30%. In that case, Plant A can purchase some unused emissions permits from Plant B, at a cost less than that required for Plant A to comply with the 30% emissions reduction. The *overall* emissions reductions are then arrived at in the most financially beneficial way for both plants.

The cap-and-trade system has some possible disadvantages, including a potentially volatile market for the emissions permits. Energy providers might experience wide, often unpredictable swings in their energy costs. Those swings would result in large fluctuations in consumer costs. As an alternative to cap-and-trade, some advocate a carbon tax instead of a cap-and-trade program. Instead of limiting emissions and letting the market decide how "best" to comply, a carbon tax simply increases the cost of burning fossil fuels. Placing an additional cost based on the amount of carbon contained in a certain quality of fuel is intended to make alternative energy sources more competitive in the near term. Of course, levying a tax on carbon fuels or emissions will mean higher prices for consumers as well.

Consider This 3.32   Climate Change Insurance?

Mitigation of climate change can be seen as a risk–benefit scenario. As such, uncertainty about future effects may discourage governments from taking financially costly actions. Another way of tackling climate change is to view it as a risk-management problem, analogous to the reasons we buy insurance. Having car insurance doesn't reduce the likelihood of being involved in an accident, but it can limit the costs if an accident should occur. How might the insurance analogy fit in with climate change actions and policies?

Although the U.S. federal government has been slow to produce binding climate change legislation, individual states have taken matters into their own hands. The 10 northeastern states that make up the Regional Greenhouse Gas Initiative (RGGI) signed the first U.S. cap-and-trade program for carbon dioxide. The program began by capping emissions at current levels in 2009 and then reducing emissions 10% by 2019. The Midwestern Regional Greenhouse Gas Reduction Accord states developed a multisector cap-and-trade system to help meet a long-term target of 60–80% below current emissions levels. Western Climate Initiative states, as well as British Columbia and Manitoba (the first participating jurisdictions outside of the United States), agreed to mandatory emissions reporting, as well as regional efforts to accelerate development of renewable energy technologies.

More locally, the U.S. Mayors Climate Protection Agreement included 227 cities committed to cutting emissions to meet the targets of the Kyoto Protocol. The cities represented include some of the largest in the Northeast, the Great Lakes region, and West Coast, and their mayors represent some 44 million people.

Skeptical Chemist 3.33   Drop in the Bucket?

Critics suggest that actions made by individual states or countries, even if successful, cannot possibly have a significant effect on global emissions of greenhouse gases. Proponents for immediate action, such as NASA climate scientist James Hansen, take a different approach. "China and India have the most to lose from uncontrolled climate change because they have huge populations living near sea level. Conversely then, they also have the most to gain from reduced local air pollution. They must be a part of the solution to global warming, and I believe they will be if developed nations such as the United States take the appropriate first steps." After studying this chapter, which side do you fall on? Explain.

Conclusion

Climate change is not the first environmental challenge created by our highly technological world, but two aspects make it the most daunting. The first is the timescale. We have recognized only recently the harm that air pollution, stratospheric ozone depletion, and acid rain can bring. Yet when we realized that human activities were responsible for the creation of those crises, we did not shrink from the challenges they presented. Combining regulation, new technologies, and adaptation strategies, we quickly began to restore air quality, retard the growth of the Antarctic ozone hole, and reduce emissions that caused acid rain.

Our ability to address and correct those problems relied on the relatively short lifetimes of the pollutants involved; we have no such advantage with carbon dioxide and climate change. In 2008, United Nations Resident Coordinator Khalid Malik spoke to this very issue:

> *"What we do today about climate change has consequences that will last a century or more. The part of that change that is due to greenhouse gas emissions is not reversible in the foreseeable future. The heat trapping gases we send into the atmosphere in 2008 will stay there until 2108 and beyond. We are therefore making choices today that will affect our own lives, but even more so the lives of our children and grandchildren. This makes climate change different and more difficult than other policy challenges."*

The words bear striking resemblance to the definition of sustainability in Chapter 0. If we wait until our children experience the effects of climate change first-hand, it will likely be too late to avoid disastrous consequences. Yet our current political institutions are ill-suited to prepare for and respond to such long-term threats, rendering the possibility of corrective global action slim at best.

The second daunting aspect is the immense scale of global greenhouse gas emissions. In 2009, over 80% of the energy produced worldwide was supplied by burning fossil fuels and was responsible for the majority of greenhouse gas emissions. The next chapter explores in detail how that energy is derived from combustion, mainly of fossil fuels, but also of biofuels. Given the projected doubling of energy demand by 2050, the need for developing large-scale, carbon-free energy sources becomes clear. Toward that end, the advantages and disadvantages associated with harnessing the energy of splitting atomic nuclei is the subject of Chapter 7. Technological challenges of converting radiation from our most abundant energy source, the Sun, directly to electricity are discussed in Chapter 8.

Like it or not, we are in the midst of conducting a planet-wide experiment, one that will test our ability to sustain both our economic development and our environment. It is vital that we as individuals and as a society respond with wisdom, compassion, commitment, and courage. We may get only one chance.

Chapter Summary

Having completed this chapter you should be able to:

- Understand the different processes that take part in Earth's energy balance (3.1)

- Compare and contrast the Earth's natural greenhouse effect and the enhanced greenhouse effect (3.1)

- Understand the major role that certain atmospheric gases play in the greenhouse effect (3.1–3.2)

- Explain the methods used to gather past evidence of greenhouse gas concentrations and global temperatures (3.2)

- Use Lewis structures to determine molecular geometry and bond angles (3.3)

- Relate molecular geometry to absorption of infrared radiation (3.4)

- List the major greenhouse gases and explain why each has the appropriate molecular geometry to be a greenhouse gas (3.4)

- Explain the roles that natural processes play in the carbon cycle and climate change (3.5)

- Evaluate how human activities contribute to the carbon cycle and climate change (3.5)

- Understand how molar mass is defined and used (3.6)

- Calculate the average mass of an atom using Avogadro's number (3.6)

- Demonstrate the usefulness of the chemical mole (3.7)

- Assess the sources, relative emission quantities, and effectiveness of greenhouse gases other than CO_2 (3.8)

- Evaluate the roles of natural and anthropogenic climate forcings (3.9)

- Recognize the successes and limitations of computer-based models in predicting climate change (3.9)

- Correlate some of the major consequences of climate change with their likelihood (3.10)

- Evaluate the advantages and disadvantages of proposed greenhouse gas regulations (3.11)

- Provide examples of climate mitigation and climate adaptation strategies (3.11)

- Analyze, interpret, evaluate, and critique news stories on climate change (3.1–3.12)

- Take an informed position with respect to issues surrounding climate change (3.1–3.12)

Questions

Emphasizing Essentials

1. The chapter opened with a quote from John Holdren: *"Global warming is a misnomer, because it implies something that is gradual, something that is uniform, something that is quite possibly benign. What we are experiencing with climate change is none of those things."* Use examples to:

    a. explain why climate change is not uniform.

    b. explain why it is not gradual, at least in comparison to how quickly social and environmental systems can adjust.

    c. explain why it probably will not be benign.

2. The surface temperatures of both Venus and Earth are warmer than would be expected on the basis of their respective distances from the Sun. Explain.

3. Using the analogy of a greenhouse to understand the energy radiated by Earth, of what are the "windows" of Earth's greenhouse made? In what ways is the analogy not precisely correct?

4. Consider the photosynthetic conversion of CO_2 and H_2O to form glucose, $C_6H_{12}O_6$, and O_2.

    a. Write the balanced equation.

    b. Is the number of each type of atom on either side of the equation the same?

    c. Is the number of molecules on either side of the equation the same? Explain.

5. Describe the difference between climate and weather.

6. a. It is estimated that 29 megajoules per square meter (MJ/m^2) of energy comes to the top of our atmosphere from the Sun each day, but only 17 MJ/m^2 reaches the surface. What happens to the rest?

    b. Under steady-state conditions, how much energy would leave the top of the atmosphere?

7. Consider Figure 3.9.

    a. How does the present concentration of CO_2 in the atmosphere compare with its concentration 20,000 years ago? With its concentration 120,000 years ago?

    b. How does the present temperature of the atmosphere compare with the 1950–1980 mean temperature? With the temperature 20,000 years ago? How does each of these values compare with the average temperature 120,000 years ago?

    c. Do your answers to parts **a** and **b** indicate causation, correlation, or no relation? Explain.

8. Understanding Earth's energy balance is essential to understanding the issue of global warming. For example, the solar energy striking the Earth's surface averages 168 watts per square meter (W/m^2), but the energy leaving Earth's surface averages 390 W/m^2. Why isn't the Earth cooling rapidly?

9. Explain each of these observations.

    a. A car parked in a sunny location may become hot enough to endanger the lives of pets or small children left in it.

    b. Clear winter nights tend to be colder than cloudy ones.

    c. A desert shows much wider daily temperature variation than a moist environment.

    d. People wearing dark clothing in the summertime put themselves at a greater risk of heatstroke than those wearing white clothing.

10. Construct a methane molecule (CH_4) from a molecular model kit (or use Styrofoam balls or gumdrops to represent the atoms and toothpicks to represent the bonds). Demonstrate that the hydrogen atoms would be farther from one another in a tetrahedral arrangement than if they all were in the same plane (square planar arrangement).

11. Draw the Lewis structure and name the molecular geometry for each molecule.

    a. H_2S

    b. OCl_2 (oxygen is the central atom)

    c. N_2O (nitrogen is the central atom)

12. Draw the Lewis structure and name the molecular geometry for these molecules.

  a. PF_3

  b. HCN (carbon is the central atom)

  c. CF_2Cl_2

13. a. Draw the Lewis structure for methanol (wood alcohol), H_3COH.

  b. Based on this structure, predict the H–C–H bond angle. Explain your reasoning.

  c. Based on this structure, predict the H–O–C bond angle. Explain your reasoning.

14. a. Draw the Lewis structure for ethene (ethylene), H_2CCH_2, a small hydrocarbon with a C=C double bond.

  b. Based on this structure, predict the H–C–H bond angle. Explain your reasoning.

  c. Sketch the molecule showing the predicted bond angles.

15. Three different modes of vibration of a water molecule are shown. Which of these modes of vibration contributes to the greenhouse effect? Explain.

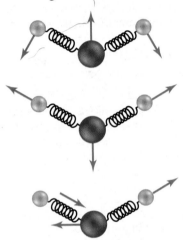

16. If a carbon dioxide molecule interacts with certain photons in the IR region, the vibrational motions of the atoms are increased. For CO_2, the major wavelengths of absorption occur at 4.26 μm and 15.00 μm.

  a. What is the energy corresponding to each of these IR photons?

  b. What happens to the energy in the vibrating CO_2 species?

17. Water vapor and carbon dioxide are greenhouse gases, but N_2 and O_2 are not. Explain.

18. Explain how each of these relates to global climate change.

  a. volcanic eruptions

  b. CFCs in the stratosphere

19. Termites possess enzymes that allow them to break down cellulose into glucose, $C_6H_{12}O_6$, and then metabolize the glucose into CO_2 and CH_4.

  a. Write a balanced equation for the metabolism of glucose into CO_2 and CH_4.

  b. What mass of CO_2, in grams, could one termite produce in one year if it metabolized 1.0 mg glucose in one day?

20. Consider Figure 3.21.

  a. Which sector has the highest CO_2 emission from fossil-fuel combustion?

  b. What alternatives exist for each of the major sectors of CO_2 emissions?

21. Silver has an atomic number of 47.

  a. Give the number of protons, neutrons, and electrons in a neutral atom of the most common isotope, Ag-107.

  b. How do the numbers of protons, neutrons, and electrons in a neutral atom of Ag-109 compare with those of Ag-107?

22. Silver only has two naturally occurring isotopes: Ag-107 and Ag-109. Why isn't the average atomic mass of silver given on the periodic table simply 108?

23. a. Calculate the average mass in grams of an individual atom of silver.

  b. Calculate the mass in grams of 10 trillion silver atoms.

  c. Calculate the mass in grams of 5.00×10^{45} silver atoms.

24. Calculate the molar mass of these compounds. Each plays a role in atmospheric chemistry.

  a. H_2O

  b. CCl_2F_2 (Freon-12)

  c. N_2O

25. a. Calculate the mass percent of chlorine in CCl_3F (Freon-11).

  b. Calculate the mass percent of chlorine in CCl_2F_2 (Freon-12).

  c. What is the maximum mass of chlorine that could be released in the stratosphere by 100 g of each compound?

  d. How many atoms of chlorine correspond to the masses calculated in part **c**?

26. The total mass of carbon in living systems is estimated to be 7.5×10^{17} g. Given that the total mass of carbon on Earth is estimated to be 7.5×10^{22} g, what is the ratio of carbon atoms in living systems to the total carbon atoms on Earth? Report your answer in percent and in ppm.

27. Consider the information presented in Table 3.2.

  a. Calculate the percent increase in CO_2 when comparing 1998 concentrations with preindustrial concentrations.

  b. Considering CO_2, CH_4, and N_2O, which has shown the greatest percentage increase when comparing 1998 concentrations with preindustrial concentrations?

28. Other than atmospheric concentration, what two other properties are included in the calculation of the global warming potential for a substance?

29. Total greenhouse gas emissions in the United States rose 16% from 1990 to 2005, growing at a rate of 1.3% a year since 2000. How is this possible when CO_2 emissions grew by 20% in the same time period? *Hint:* See Table 3.2.

Concentrating on Concepts

30. John Holdren, quoted at the opening of the chapter, suggests that we use the term *global climatic disruption* rather than *global warming.* After studying this chapter, do you agree with his suggestion? Explain.

31. The Arctic has been called "our canary in the coal mine for climate impacts that will affect us all."

    a. What does the phrase "canary in the coal mine" mean?

    b. Explain why the Arctic serves as a canary in a coal mine.

    c. The melting of the tundra accelerates changes elsewhere. Give one reason.

32. Do you think the comment made in the cartoon is justified? Explain.

Pepper . . . and Salt

"This winter has lowered my concerns about global warming . . ."

Source: From *The Wall Street Journal.* Permission by Cartoon Features Syndicate.

33. Given that direct measurements of Earth's atmospheric temperature over the last several thousands of years are not available, how can scientists estimate past fluctuations in the temperature?

34. A friend tells you about a newspaper story that stated, "The greenhouse effect poses a serious threat to humanity." What is your reaction to that statement? What would you tell your friend?

35. Over the last 20 years, about 120 billion tons of CO_2 has been emitted from the burning of fossil fuels, yet the amount of CO_2 has risen only by about 80 billion tons. Explain.

36. Carbon dioxide gas and water vapor both absorb IR radiation. Do they also absorb visible radiation? Offer some evidence based on your everyday experiences to help explain your answer.

37. How would the energy required to cause IR-absorbing vibrations in CO_2 change if the carbon and oxygen atoms were connected with single rather than with double bonds?

38. Explain why water in a glass cup is quickly warmed in a microwave oven, but the glass cup warms much more slowly, if at all.

39. Ethanol, C_2H_5OH, can be produced from sugars and starches in crops such as corn or sugarcane. The ethanol is used as a gasoline additive and when burned, it combines with O_2 to form H_2O and CO_2.

    a. Write a balanced equation for the complete combustion of C_2H_5OH.

    b. How many moles of CO_2 are produced from each mole of C_2H_5OH completely burned?

    c. How many moles of O_2 are required to burn 10 mol of C_2H_5OH?

40. Explain whether each of the radiative forcings described in Section 3.9 is positive or negative and rank them in terms of importance to overall climate change predictions.

41. Why is the atmospheric lifetime of a greenhouse gas important?

42. Compare and contrast stratospheric ozone depletion and climate change in terms of the chemical species involved, the type of radiation involved, and the predicted environmental consequences.

43. Explain the term *radiative forcings* to someone unfamiliar with climate modeling.

44. It is estimated that Earth's ruminants, such as cattle and sheep, produce 73 million metric tons of CH_4 each year. How many metric tons of carbon are present in this mass of CH_4?

45. The 10 warmest years since 1880 all occurred between 1997 and 2009. Does this *prove* that the enhanced greenhouse effect (global warming) is taking place? Explain.

46. A possible replacement for CFCs is HFC-152a, with a lifetime of 1.4 years and a GWP of 120. Another is HFC-23, with a lifetime of 260 years and a GWP of 12,000. Both of these possible replacements have a significant effect as greenhouse gases and are regulated under the Kyoto Protocol.

    a. Based on the given information, which appears to be the better replacement? Consider only the potential for global warming.

    b. What other considerations are there in choosing a replacement?

47. This figure shows global emissions of CO_2 in metric tons per person per year for the selected countries.

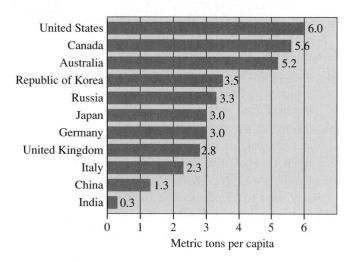

Metric tons per capita

a. The values in this figure are reported as metric tons of CO_2, not as metric tons of C in CO_2. How are these values related? *Hint:* Think about the mass relationships developed in Section 3.6.

b. Find the value for U.S. CO_2 emissions from a source other than shown in this figure. Do the values agree? Explain.

48. Compare and contrast a cap-and-trade system with a carbon tax.

49. When Arrhenius first theorized the role of atmospheric greenhouses, he calculated that doubling the concentration of CO_2 would result in an increase of 5–6 °C in the average global temperature. How far off was he from the current IPCC modeling?

50. Now that you have studied air quality (Chapter 1), stratospheric ozone depletion (Chapter 2), and global warming (Chapter 3), which do you believe poses the most serious problem for you in the short run? In the long run? Discuss your reasons with others and draft a short report on this question.

Exploring Extensions

51. Former Vice President Al Gore writes in his 2006 book and film, *An Inconvenient Truth*: "We can no longer afford to view global warming as a political issue—rather, it is the biggest moral challenge facing our global civilization."

a. Do you believe that global warming is a moral issue? If so, why?

b. Do you believe that global warming is a political issue? If so, why?

52. China's growing economy is fueled largely by its dependence on coal, described as China's "double-edged sword." Coal is both the new economy's "black gold" and the "fragile environment's dark cloud."

a. What are some of the consequences of dependence on high-sulfur coal?

b. Sulfur pollution from China may slow global warming, but only temporarily. Explain.

c. What other country is rapidly stepping up its construction of coal-fired power plants and is expected to have a larger population than China by the year 2030?

53. The quino checkerspot butterfly is an endangered species with a small range in northern Mexico and southern California. Evidence reported in 2003 indicates that the range of this species is even smaller than previously thought.

a. Propose an explanation why this species is being pushed north, out of Mexico.

b. Propose an explanation why this species is being pushed south, out of southern California.

c. Propose a plan to prevent further harm to this endangered species.

54. Data taken over time reveal an increase in CO_2 in the atmosphere. The large increase in the combustion of hydrocarbons since the Industrial Revolution is often cited as a reason for the increasing levels of CO_2. However, an increase in water vapor has *not* been observed during the same period. Remembering the general equation for the combustion of a hydrocarbon, does the difference in these two trends *disprove* any connection between human activities and global warming? Explain your reasoning.

55. In the energy industry, 1 standard cubic foot (SCF) of natural gas contains 1196 mol of methane (CH_4) at 15.6 °C (60 °F). *Hint:* See Appendix 1 for conversion factors.

a. How many moles of CO_2 could be produced by the complete combustion of 1 SCF of natural gas?

b. How many kilograms of CO_2 could be produced?

56. An international conference on climate change was held in Copenhagen in December 2009. Write a brief summary of the outcomes of this conference.

57. A solar oven is a low-tech, low-cost device for focusing sunlight to cook food. How might solar ovens help mitigate global warming? Which regions of the world would benefit most from using this technology?

58. In 2005, the European Union adopted a cap-and-trade policy for carbon dioxide. Write a short report on the outcomes of this policy, both in terms of the economic result and the effect it has had on European greenhouse gas emissions.

59. The world community responded differently to the atmospheric problems described in Chapters 2 and 3. The evidence of ozone depletion was met with the Montreal Protocol, a schedule for decreasing the production of ozone-depleting chemicals. The evidence of global warming was met with the Kyoto Protocol, a plan calling for targeted reduction of greenhouse gases.

a. Suggest reasons why the world community dealt with the issue of ozone depletion *before* that of global warming.

b. Compare the current status of the two responses. When was the latest amendment to the Montreal Protocol? How many nations have ratified it? Has the level of chlorine in the stratosphere dropped as a result of the Montreal Protocol? How many nations have ratified the Kyoto Protocol? What has happened since it went into effect? Have any other initiatives been proposed? Have levels of greenhouse gases dropped as a result of the Kyoto Protocol?

4 Energy from Combustion

Since the beginning of recorded history, fire has been a source of heat, light, and security.

Ever since fire was harnessed by our early ancestors, combustion has been central to society. Our modern fuels—the substances we burn—come in many different forms. We use coal in power plants to generate electricity. We use gasoline to run our cars. We use natural gas or heating oil to warm our homes. We use propane, charcoal, or wood to cook our food at a summer barbeque. We might even use wax to provide light for a romantic candlelight dinner. In each of these cases, *using* fuels means *burning* them. The process of combustion releases the energy stored in the molecules that these substances contain.

However, the rate at which we are burning fuels is not sustainable. Perhaps you are somewhat skeptical of this claim. The supply of coal, petroleum, and natural gas may appear to be adequate, because new deposits are always being found and extraction technologies are continually improving. But even if the supply of fossil fuels were infinite (it is not), sustainability involves more than just availability. In Chapter 0, we mentioned the need to consider how our actions today will affect those who live tomorrow. A Lakota Sioux proverb emphasizes the same idea: *"We don't inherit this land from our ancestors, we borrow it from our children."* The effects of our current fossil fuel use will be felt for many decades to come.

Consider This 4.1   Fuels in the News!

a. Locate two recent news articles concerning a fuel of your choice. Cite the title, author, date, and source. According to what you read, what is the intended use of the fuel?

b. Interpret each article in terms of the Lakota Sioux proverb. What, if anything, are we borrowing from our children?

Burning fossil fuels for energy fails to meet the criteria of sustainability in two ways. First, the fuels themselves—taking hundreds of millions of years to produce—are nonrenewable. Once gone, they cannot be replaced. Although our modern economies, which are based on fossil fuels, have been functioning for nearly two centuries, we need new long-term solutions. Second, the waste products of combustion have adverse effects on our environment. The previous chapter described how atmospheric carbon dioxide concentrations have risen dramatically since the beginning of the Industrial Revolution. Burning coal also releases soot, carbon monoxide, mercury, and the oxides of sulfur and nitrogen. These emissions have undeniable links to serious environmental concerns such as global warming, acid rain, and the deterioration of air and water quality.

In this chapter, we will describe fuels and their characteristics. We begin with what happens inside a power plant. In the context of energy transformation, we introduce a law that tells us that energy is never created or destroyed; rather, it just changes forms. We also consider the efficiency (actually the inefficiency) of energy transformations, a factor in our ability to harness energy in convenient forms. But fuels differ, and so we need to describe how all fuels are not created equal; that is, how they have different heat contents and release different amounts of carbon dioxide. To do this, we take a closer look at coal and petroleum, describing their chemical composition, physical properties, the structures of the molecules they contain, and the ways we manipulate them for use. We learn how these molecules store energy and how to write the chemical reactions that describe energy release. We then move to biofuels, exploring the advantages and disadvantages of these renewable resources. This chapter closes with a discussion of the scale of global energy production and usage, revisiting the challenges of meeting our future energy needs.

Chapter 1 described the connection between combustion and poor air quality. Chapter 3 dealt with carbon dioxide as a greenhouse gas. Look for acid rain in Chapter 6.

4.1 | Fossil Fuels and Electricity

About 70% of the electricity generated in the United States comes from burning fossil fuels—primarily coal. How do electrical power plants "produce" electricity and what really goes on inside them? Our task in this section is to take a closer look at the energy transformations in a power plant. In Section 4.3, we discuss the chemistry of coal.

The first step in producing electricity from coal is to burn it. Examine the photographs in Figure 4.1. You can almost feel the heat from the burning coal! In the coal beds of the boilers, the temperature can reach 650 °C. To generate this heat, this small power plant burns a train car load of coal every few hours. As we pointed out in Chapter 1, **combustion** is the chemical process of burning; that is, the rapid combination of fuel with oxygen to release energy in the form of heat and light. Note that the two most common combustion products, CO_2 and H_2O, both contain oxygen.

The second step in producing electricity is to use the heat released from combustion to boil water, usually in a closed, high-pressure system (Figure 4.2). The elevated pressure serves two purposes: it raises the boiling point of the water and it compresses the resulting water vapor. The hot high-pressure steam is then directed at a steam turbine.

The third and final step generates electricity. As the steam expands and cools, it rushes past the turbine, causing it to spin. The shaft of the turbine is connected to a large coil of wire that rotates within a magnetic field. The turning of this coil generates an electric current. Meanwhile, the water vapor leaves the turbine and continues to cycle through the system. It passes through a condenser where a stream of cooling water carries away the remainder of the heat energy originally acquired from the fuel. The condensed water then reenters the boiler, ready to resume the energy transfer cycle.

To help you better understand these different steps, we define two types of energy. **Potential energy,** as the name suggests, is stored energy or the energy of position. For example, energy can be stored in the position of a book lifted against the force of gravity. The heavier the book and the higher you lift it, the more potential energy it has. The potential energy of a reactant or product is often referred to as its "chemical

Operating at full capacity, a large power plant can burn up to 10,000 tons of coal a day!

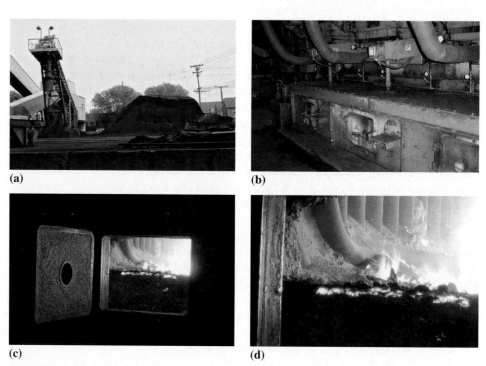

(a)

(b)

(c)

(d)

Figure 4.1

Photos from a small coal-fired electric power plant.
(a) Piles of coal outside the plant.
(b) A row of boilers into which the coal is fed.
(c) Behind the blue door in photograph **(b)**.
(d) A close-up of coal burning on the boiler bed.

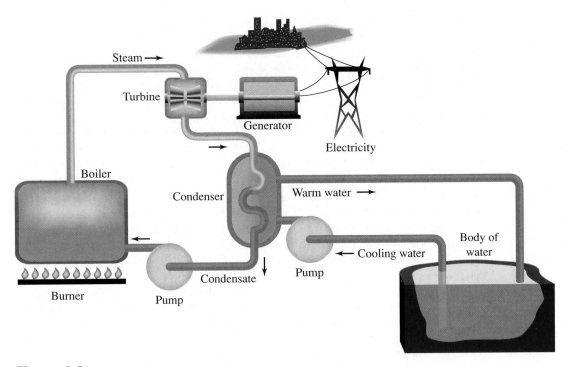

Figure 4.2

Diagram of an electric power plant illustrating the conversion of energy from the combustion of fuels to electricity.

energy." We discuss the chemical energy stored in fuel molecules in Section 4.5. In contrast, **kinetic energy** is the energy of motion. The heavier an object is and the faster it is moving, the more kinetic energy it possesses. Would you rather be hit by a baseball traveling at 90 mph or a Ping-Pong ball traveling at 90 mph? The baseball has considerably more kinetic energy because of its larger mass.

Molecules that have high potential energy make good fuels. The process of combustion converts some of the potential energy of the fuel molecules into heat, which in turn is absorbed by the water in the boiler. As the water molecules absorb the heat, they move faster and faster in all directions; their kinetic energy increases. The temperature we observe is simply a measure of the average speed of that molecular motion. Hence, the temperature increases as the amount of kinetic energy of the molecules increases. When the water is vaporized to steam, the water molecules acquire a tremendous amount of kinetic energy. That energy is transformed into the *mechanical energy* of the spinning turbine that then turns the generator converting the mechanical energy into *electrical energy*. These energy transformation steps are summarized in Figure 4.3.

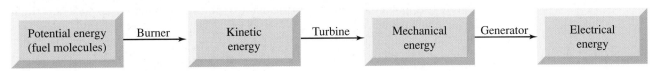

Figure 4.3

Energy transformations in a fossil fuel electric power plant.

Consider This 4.2   Energy Conversion

Although power plants require several steps to transform potential energy into electrical energy, other devices do this more simply. For example, a battery converts chemical energy to electrical energy in one step. List three other devices that convert energy from one form to another. For each one, name the types of energy involved.

Section 3.2 introduced three fossil fuels: coal, petroleum, and natural gas.

Although coal is usable as a fuel, its combustion products are not. We stated that good fuels have high potential energy, but from what source does this energy come? A clue lies in the name "fossil fuel" itself. The creation of fossil fuels began when sunlight was captured by green plants that flourished on our primeval planet. Also, as mentioned earlier, photosynthesis is the process by which green plants capture the energy of sunlight to produce glucose and oxygen from carbon dioxide and water. In essence, the energy from sunlight is converted into the potential energy of glucose and oxygen.

The Sun's energy is nuclear in origin; more specifically, nuclear fusion.

$$6\,CO_2 + 6\,H_2O \xrightarrow{\text{chlorophyll}} \underset{\text{glucose}}{C_6H_{12}O_6} + 6\,O_2 \qquad [4.1]$$

When living organisms die and decay, they release energy and reverse this process, producing CO_2 and H_2O. Under certain conditions, however, the carbon-containing compounds that make up the organism only *partially* decompose. This happened in the prehistoric past when vast quantities of plant and animal life became buried beneath layers of sediment in swamps or at the bottom of the oceans. Oxygen failed to reach the decaying material, thus retarding the decomposition process. The temperature and pressure increased as additional layers of mud and rock covered the buried remnants, causing additional chemical reactions to occur. Over time, the plants that once captured the Sun's rays were transformed into the substances we call coal, petroleum, and natural gas. In a very real sense, these fossils are ancient solar energy (sunshine) stored in the solid, liquid, and gaseous state.

Plants contain nitrogen as well. Section 6.9 tells more about the nitrogen cycle.

Yes, today's plants will become tomorrow's fossil fuels. But this will not occur in a time frame useful to humans. It is staggering to realize that we will consume in a few centuries what it took nature hundreds of millions of years to produce. We discuss the details of fuels at the molecular level in Section 4.6.

Your Turn 4.3   Steamy Compost

Want to recycle and reuse plant and animal material? Start a compost pile. Under the right weather conditions, steam can be seen rising from a pile of compost. Explain this observation.

Revisit the processes of combustion and photosynthesis. Energy is released in combustion, but is required for photosynthesis. The relationship between the two hints at a cycle, as shown in Figure 4.4. The **first law of thermodynamics,** also called the law of conservation of energy, states that energy is neither created nor destroyed. It implies that although the *forms* of energy change, the total amount of energy before and after any transformation remains the same. The solar energy that is stored as potential energy during photosynthesis is released as heat and light during combustion.

The law of conservation of matter and mass was introduced in Section 1.9.

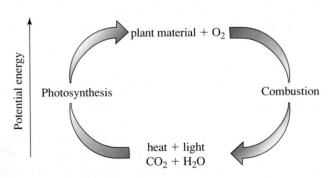

Figure 4.4

The energy relationship between photosynthesis and combustion.

4.2 | Efficiency of Energy Transformation

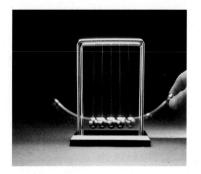

Figure 4.5
A Newton's cradle.

By the first law of thermodynamics, we are assured that the total energy of the universe is conserved. If this is true, how can we ever experience an energy crisis? To be sure, no new energy is created during combustion, but none is destroyed, either. Although we may not be able to win, can we at least break even? The question is not as facetious as it might sound. In fact, we *cannot* break even. In burning coal, natural gas, and petroleum, we always convert at least some of the energy in the fuels into forms that we cannot easily use.

You probably have seen the desktop toy known as a Newton's cradle (Figure 4.5). This device transforms energy but is much simpler than a power plant! Here's how it works.

- A ball is lifted at one end. This gives potential energy to this ball.
- The ball is released and falls back toward its starting point. The potential energy (energy of position) is converted into kinetic energy (energy of motion).
- The ball hits the row of stationary balls. Kinetic energy is transferred along the row to the ball at the other end.
- The ball at the other end swings up. Kinetic energy is gradually converted into potential energy as the ball rises and slows.

The second ball begins to fall and the process repeats. With each successive cycle, however, each ball does not rise quite as high as the previous one. Eventually, the balls all come to rest at their original positions.

Why do they stop moving? Where did their energy go? Is this a violation of the first law of thermodynamics? Fortunately, it isn't. In each collision, some of the energy is used to make sound and some is used to generate heat. If we could measure precisely enough, we would observe the balls heating up slightly. This heat is then transferred to the surrounding atoms and molecules in the air, thus increasing their kinetic energy. In keeping with the law of conservation of energy, neither kinetic energy nor potential energy is conserved independently, but the sum of the two is. Therefore, when all the balls finally come to rest, the energy of the universe has been conserved. All the energy that you initially put into the system has been dissipated as random motion of the atoms and molecules in the surrounding air. In essence, the device is a fun way of dissipating a little bit of potential energy into heat (the kinetic energy of the atoms and molecules in the air).

These same principles can be used to explain why no electric power plant, no matter how well designed, can completely convert one type of energy into another. In spite of the best engineers and the most competent green chemists, inefficiency is inevitable. It is caused by the transformation of energy into useless heat. Overall, the net efficiency is given by the ratio of the electrical energy produced to the energy supplied by the fuel.

$$\text{Net efficiency} = \frac{\text{electrical energy produced}}{\text{heat from fuel}} \times 100 \qquad \textbf{[4.2]}$$

Newer boiler systems and advanced turbine technologies have pushed the efficiencies of each step in Figure 4.3 to 90% or better. Efficiencies are multiplicative, so you might be surprised to learn that the net efficiencies of most fossil fuel power plants are between 35 and 50%. Why so low?

The problem is that not all of the heat energy from the fuel combustion in the boilers can be converted into electricity. Consider, for example, the high-temperature steam that initially spins the turbines. As the steam transfers energy to the turbines, the kinetic energy of the steam decreases, it cools, and its pressure drops. It isn't long before the steam does not have enough energy to spin the turbines anymore. Yet, the production of this "unused" steam still required a significant amount of energy; energy that is not converted into electricity.

Power plants using very high temperature steam (600 °C) have efficiencies at the high end of the range. In fact, the efficiency goes up as the difference between the steam temperature and the temperature outside the plant increases. Of course there is a limit. Higher temperature steam means higher pressures and improved construction materials that need to be able to withstand such extreme conditions.

Before we discuss a specific example, we need to say a word or two about energy units. The **calorie** was introduced with the metric system in the late 18th century and was defined as the amount of heat necessary to raise the temperature of one gram of water by one degree Celsius. When Calorie is capitalized, it generally means kilocalorie. The values tabulated on package labels and in cookbooks are, in fact, kilocalories.

One potato (148 g) has about 100 food Calories (100 kilocalories).

$$1 \text{ kilocalorie (kcal)} = 1000 \text{ calories (cal)} = 1 \text{ Calorie (Cal)}$$

The modern system of units uses the **joule (J),** a unit of energy equal to 0.239 cal. One joule (1 J) is approximately equal to the energy required to raise a 1-kg book 10 cm against the force of gravity. On a more personal basis, each beat of the human heart requires about 1 J of energy.

Now consider the case of electrical home heating, sometimes advertised as being clean and efficient. Assume that electricity from a coal-burning power plant (efficiency of 37%) is used to heat a house. If the house requires 3.5×10^7 kJ of energy for heat each day, a typical value for a northern city in January, how much coal would be burned?

To answer this question, we need a value for the energy content of the coal. Let's assume that the combustion of 1 gram of this particular coal releases about 29 kJ. Remember that only 37% of the energy released by burning the coal is available to heat the house. We now can calculate the total quantity of heat that we need to generate by burning coal at the power plant.

$$\text{energy generated at plant} \times \text{efficiency} = \text{energy required to heat house}$$

$$\text{energy generated at plant} \times 0.37 = 3.5 \times 10^7 \text{ kJ} = \text{energy required to heat house}$$

$$\text{energy generated at plant} = \frac{3.5 \times 10^7 \text{ kJ}}{0.37} = 9.5 \times 10^7 \text{ kJ} = \text{energy required to heat house}$$

In these calculations, note that we expressed the % efficiency in decimal form. We now take into account that each gram of coal burned yields 29 kJ.

$$9.5 \times 10^7 \text{ kJ} \times \frac{1 \text{ g coal}}{29 \text{ kJ}} = 3.3 \times 10^6 \text{ g coal} = \text{coal required to heat house}$$

This shows that 3.3×10^6 g of coal must be burned in order to furnish the needed 3.5×10^7 kJ of energy to heat the house.

This calculation assumed an efficiency of 37% at the coal plant. Higher efficiencies would mean that less fuel would have to be burned to generate the same amount of energy and that less carbon dioxide and other pollutants would be emitted. The next activity explores these connections.

Your Turn 4.4   Comparing Power Plants

Consider two coal-fired power plants that generate 5.0×10^{12} J of electricity daily. Plant A has an overall net efficiency of 38%. Plant B, a proposed replacement, would operate at higher temperatures with an overall net efficiency of 46%. The grade of coal used releases 30 kJ of heat per gram. Assume that coal is pure carbon.

a. If 1000 kg of coal costs $30, what is the difference in daily fuel costs for the two plants?
b. How many fewer grams of CO_2 are emitted daily by Plant B, assuming complete combustion?

Answer
a. Coal costs for Plant A = $13,150/day. Coal costs for Plant B = $10,900/day.

Cars and trucks also convert energy from one form to another. The internal combustion engine uses the gaseous combustion products (CO_2 and H_2O) to push a series of pistons, thus converting the potential energy of the gasoline or diesel fuel into mechanical energy. Other mechanisms transform that mechanical energy eventually into the kinetic energy of the vehicle's motion. Internal combustion engines are even less efficient than coal-fired power plants. Only about 15% of the energy released by the combustion of the gasoline actually is used to move the vehicle. Much of the energy is dissipated as waste heat, including about 60% lost from the internal combustion engine alone.

Sections 8.4 and 8.6 discuss more efficient hybrid and fuel cell vehicles.

Consider This 4.5   Transportation Inefficiency

a. List some of the energy losses that take place when driving a car. Use the web to verify and expand your list if necessary. The textbook's website provides a direct link.

b. Given the assumption that only 15% of the energy from fuel combustion is used to move the vehicle, estimate the percent used to move the passengers.

To bring this section to a close, we ask you to revisit the Newton's cradle. You would never expect the balls at rest to start knocking into one another on their own, right? For this to occur, all the heat energy dissipated when the balls were colliding would have to be gathered back together. The inability of a Newton's cradle to start up on its own relates to another concept—entropy. **Entropy** is a measure of how much energy gets dispersed in a given process. The **second law of thermodynamics** has many versions, the most general of which is that the entropy of the universe is constantly increasing. The Newton's cradle provides an example of the second law of thermodynamics. When we lift one of the balls of the Newton's cradle, we add potential energy. After the balls knock for awhile and come to rest, this potential energy has become transformed into the chaotic (and hence more random and dispersed) motion of heat energy and never the other way around. The entropy of the universe has increased.

Do you find it difficult to visualize how energy can disperse? If so, here is an analogy that might help. Imagine that you were sitting in the middle of a large auditorium and someone down in the front broke a bottle of perfume. You don't smell anything at first, because it takes time for the molecules of the perfume to diffuse to where you're sitting. This process of diffusion is predicted by the second law of thermodynamics. When the perfume molecules disperse into a larger volume (from the smaller volume of the bottle), the energy of the molecules gets dispersed as well. As with the Newton's cradle, the end result is an increase in the entropy of the universe. It is extremely unlikely that all of the perfume molecules would suddenly gather in one corner of the room. Rather, once dispersed they stay dispersed unless energy is expended to recollect them.

In the same way, it is essentially impossible for the Newton's cradle to begin to move on its own after the energy originally added was dissipated as heat. Though it may not be as obvious, the second law of thermodynamics also explains the inability of a power plant or an auto engine to convert energy from one type to another with 100% efficiency.

Consider This 4.6   Can Entropy Decrease?

Processes that result in a decrease in "local" entropy require an input of energy. For example, it requires energy to arrange the socks in a bureau drawer. Identify another process in which entropy appears to decrease but is actually coupled with an increase in entropy elsewhere in the universe.

Hint: Consider the entropy associated with burning coal.

4.3 | The Chemistry of Coal

About two centuries ago, the Industrial Revolution began the great exploitation of fossil fuels that continues today. In the early 1800s, wood was the major energy source in the United States. Coal turned out to be an even better energy source than wood, because it yielded more heat per gram. Coal continued to provide more than 50% of the nation's energy until about 1940.

162   Chapter 4

By the 1960s, most coal was used for generating electricity, and today the electrical power sector accounts for 92% of all U.S. coal consumption. Figure 4.6 displays the history of U.S. energy consumption.

Consider This 4.7   Evolution of Fuels

a. In the period from 1950 to 2005, which sources of energy have shown significant growth and which have not? Propose reasons for the observed trends.
Hint: See Figure 4.6.
b. Estimate the percent total energy consumption now supplied by coal.

In Your Turn 4.4, we assumed that coal was pure carbon. In fact, coal contains small amounts of other elements as well. Although not a single compound, coal can be approximated by the chemical formula $C_{135}H_{96}O_9NS$. This formula corresponds to a carbon content of about 85% by mass. The smaller amounts of hydrogen, oxygen, nitrogen, and sulfur come from the ancient plant material and other substances present when the plants were buried. In addition, some samples of coal typically contain trace amounts of silicon, sodium, calcium, aluminum, nickel, copper, zinc, arsenic, lead, and mercury.

Your Turn 4.8   Coal Calculations

a. Assuming the composition of coal can be approximated by the formula $C_{135}H_{96}O_9NS$, calculate the mass of carbon (in tons) in 1.5 million tons of coal. This quantity of coal might be burned by a typical power plant in 1 year.
b. Compute the amount of energy (in kilojoules) released by burning this mass of coal. Assume the process releases 30 kJ/g of coal. Recall that 1 ton = 2000 lb and that 1 pound = 454 g.
c. What mass of CO_2 would be formed by the complete combustion of 1.5 million tons of this coal?
d. How many molecules of CO_2 would be formed by the complete combustion of 1.5 million tons of this coal?

Answers
a. Calculate the approximate molar mass of coal. The subscripts for each element give the number of moles:

$$135 \text{ mol C} \times \frac{12.0 \text{ g C}}{1 \text{ mol C}} = 1620 \text{ g C}$$

$$96 \text{ mol H} \times \frac{1.0 \text{ g H}}{1 \text{ mol H}} = 96 \text{ g H}$$

$$9 \text{ mol O} \times \frac{16.0 \text{ g O}}{1 \text{ mol O}} = 144 \text{ g O}$$

$$1 \text{ mol N} \times \frac{14.0 \text{ g N}}{1 \text{ mol N}} = 14.0 \text{ g N}$$

$$1 \text{ mol S} \times \frac{32.1 \text{ g S}}{1 \text{ mol S}} = 32.1 \text{ g S}$$

The sum of these elemental contributions for $C_{135}H_{96}O_9NS$ is 1906 g/mol. Therefore, every 1906 g of coal contains 1620 g C. Similarly, 1906 tons of coal contains 1620 tons of carbon.

$$\text{Mass of carbon} = 1.5 \times 10^6 \text{ tons C}_{135}\text{H}_{96}\text{O}_9\text{NS} \times \frac{1620 \text{ tons C}}{1906 \text{ tons C}_{135}\text{H}_{96}\text{NS}}$$

b. 4.1×10^{13} kJ   **c.** 4.8 million tons

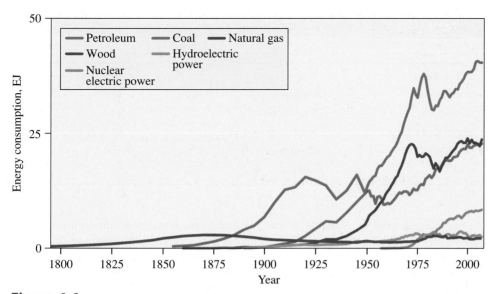

Figure 4.6

History of U.S. energy consumption by source, 1800–2008. *Note:* 1 EJ = 10^{18} J.

Source: Annual Energy Review, 2008. DOE/EIA-0384, June 2009.

Alternative energy sources, such as solar, wind, geothermal, and biofuels are barely visible in Figure 4.6 because they appeared only in the most recent decade.

Coal occurs in varying grades, but all grades are better fuels than wood because they contain a higher percentage of carbon and a lower percentage of oxygen. Generally speaking, the more oxygen a fuel contains, the less energy per gram it releases on combustion. In other words, oxygen-containing fuels lie lower on the potential energy scale. For example, burning 1 mole of C to produce CO_2 yields about 40% more energy than is obtained from burning 1 mole of CO to produce CO_2.

Soft lignite, or brown coal, is the lowest grade (Figure 4.7). The plant matter from which it originated underwent the least amount of change, and its chemical composition is similar to that of wood or peat (Table 4.1). Consequently, the amount of energy released when lignite is burned is only slightly greater than that of wood. The higher grades of coal, bituminous and anthracite, have been exposed to higher pressures and temperatures for longer periods of time in the earth. During that process, they lost more oxygen and moisture and became a good deal harder, more mineral than vegetable (see Figure 4.7). These grades of coal contain a higher percentage of carbon than lignite. Anthracite has a relatively high carbon content and a low sulfur content, both of which make it the most desirable grade of coal. Unfortunately, the deposits of anthracite are relatively small, and the supply of it in the United States is almost exhausted.

Figure 4.7

Samples of anthracite (*left*) and lignite coal (*right*).

Table 4.1	Energy Content of U.S. Coals	
Type of Coal	State of Origin	Energy Content (kJ/g)
anthracite	Pennsylvania	30.5
bituminous	Maryland	30.7
sub-bituminous	Washington	24.0
lignite (brown coal)	North Dakota	16.2
peat	Mississippi	13.0
wood	various	10.4–14.1

Although coal is available across the globe and remains a widely used fuel, it has serious drawbacks, the first of which relates to underground mining, which is both dangerous and expensive. Although mine safety has dramatically improved in the United States, since 1900 more than 100,000 workers have been killed by accidents, cave-ins, fires, explosions, and poisonous gases. Many more have been incapacitated by respiratory diseases. Worldwide, the picture is far worse.

A second drawback is the environmental harm caused by coal mining. Many streams and rivers in Appalachia suffer from the effects of decades of mining operations. When groundwater floods abandoned mine shafts, or comes in contact with sulfur-rich rock often associated with coal deposits, it becomes acidified. This acid mine drainage also dissolves excessive amounts of iron and aluminum, making the water uninhabitable for many fish species and placing drinking water sources at risk for many communities.

In 2009, a U.S. district court in West Virginia issued an injunction against new mountaintop removal projects in the southern part of that state. Further legislation in other parts of the country is likely.

When coal deposits lie sufficiently close to the surface, mining techniques safer for miners are possible, but they still have environmental costs. One technique, called mountaintop mining, is most common in West Virginia and eastern Kentucky. The process calls for scraping away the overlying vegetation and then blasting off the top several hundred feet of a mountain to reveal the underlying coal seam. Mountaintop mining creates massive quantities of rubble ("overburden") that often is disposed of by dumping the debris into nearby river valleys. In 2005, the U.S. EPA estimated that over 700 miles of Appalachian streams were completely buried as a result of mountaintop mining between 1985 and 2001. Furthermore, increased sediments and mineral content in the surrounding water systems has adversely affected many aquatic ecosystems.

A third drawback is that coal is a dirty fuel. It is, of course, physically dirty, but the issue here is the dirty combustion products. Soot from countless coal fires in cities in the 19th and early 20th century blackened both buildings and lungs. The oxides of nitrogen and sulfur are less visible but equally damaging. Although coal contains only minor amounts of mercury (50–200 ppb), mercury is concentrated in the fly ash that escapes as particulate matter into the atmosphere. In the United States, coal-fired power plants emit roughly 48 metric tons of mercury to the environment each year. The "bottom" ash left on site also presents a storage hazard. For example, Figure 4.8 shows the devastation caused by millions of gallons of fly ash sludge that spilled down a valley when the retaining walls of a storage pond failed.

Mercury, a contaminant in soils and in drinking water, will be discussed in Section 5.5.

Your Turn 4.9   Coal Emissions

In the United States, coal-burning power plants are responsible for two thirds of the sulfur dioxide emissions and one fifth of the nitrogen monoxide emissions.

a. Why does burning coal produce SO_2? Name another source of SO_2 in the atmosphere.
b. Why does burning coal produce nitrogen oxides? Name two other sources of NO.

Hint: Revisit Chapter 1.

Figure 4.8

In December 2008, 300 million gallons of coal sludge buried homes near Knoxville, Tennessee.

A fourth drawback may ultimately be the most serious, that burning coal produces carbon dioxide, a greenhouse gas. Coal combustion produces more CO_2 per kilojoule of heat released than either petroleum or natural gas. As of 2009, coal combustion accounted for about 40% of global CO_2 emissions.

Because of these drawbacks, and given that coal reserves are relatively plentiful in the United States, significant research efforts are underway to develop new coal technologies. Though it may sound like an oxymoron, "clean coal" is promoted by its supporters as an important step toward decreasing our reliance on petroleum imports and reducing air pollution. The term "clean coal technology" actually encompasses a variety of methods that aim to increase the efficiency of coal-fired power plants while decreasing harmful emissions. Here we list several technologies already implemented in selected power plants.

- "Coal washing" to remove sulfur and other mineral impurities from the coal before it is burned.
- "Gasification" to convert coal to a mixture of carbon monoxide and hydrogen (equation 4.10). The resulting gas burns at a lower temperature, thus reducing the generation of nitrogen oxides.
- "Wet scrubbing" to chemically remove SO_2 before it goes up the smokestack. This is accomplished by reacting the SO_2 with a mixture of ground limestone and water.

Of course, none of these address greenhouse gas emissions. This requires the most ambitious clean coal technology: carbon capture and storage. Serious questions remain about the viability of the technology involved.

What does the future hold for the dirtiest of the fossil fuels? The answer may depend on where you live. Figure 4.9 compares coal consumption in different regions of the globe for 1999 and 2009. Though most regions showed modest increases, the use of coal in Asia is skyrocketing. On one hand, this makes sense, as China has enormous coal reserves. But on the other, coal burning (by any nation) clearly does not meet the criteria for sustainability. In contrast, Europe and Eurasia have actually decreased their coal consumption.

The United States owns more than one quarter of the world's coal reserves, and coal combustion accounts for more than half of all U.S. electricity generation. According to the U.S. Department of Energy, 214 new coal-fired power plants were proposed in the United States between 2000 and 2009. This upturn was met with significant public, political, and financial opposition; one third have been refused construction

Carbon dioxide capture and storage, also known as sequestration, was discussed in Section 3.11.

China has about 12% of the world's coal reserves. Only the United States (28%) and Russia (21%) have more.

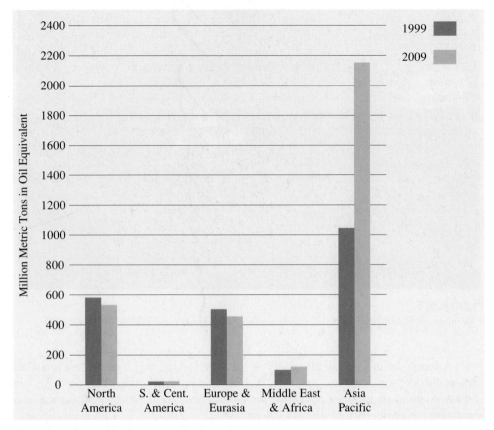

Figure 4.9

Global coal consumption by region for 1999 (*green*) and 2009 (*gold*). The unit is million metric tons oil equivalent, the approximate energy released in burning a million metric tons of oil.

Source: British Petroleum International, 2010.

licenses or are not being pursued. Another one third face on-going court battles, and many banks are refusing to finance traditional coal-fired power plants because of potential new emissions limits.

Consider This 4.10   How Clean Is Coal?

The merits of clean coal technology are widely debated. In 2004, the Sierra Club's Dan Becker proclaimed, "There is no such thing as 'clean coal' and there never will be. It's an oxymoron." That same year, David Hawkins, director of the Natural Resources Defense Council's Climate Center took a different stand. "Coal is an inevitable and substantial part of the global energy mix [for the foreseeable future], so to the extent that we are going to use it, we believe coal-based generation should be . . . with carbon capture and storage." Use the resources available to you to develop your own conclusions about clean coal. Then write an entry for an energy blog that states your position.

4.4 | Petroleum

Although you may never have set eye on a lump of coal, undoubtedly you have seen gasoline. Around 1950, petroleum surpassed coal as the major energy source in the United States. The reasons are relatively easy to understand. Petroleum, like coal, is partially decomposed organic matter. However, it has the distinct advantage of being a liquid, making it easily pumped to the surface from its natural, underground reservoirs, transported via pipelines, refined, and fed to its point of use. Moreover, petroleum yields approximately 40–60% more energy per gram than does coal. Typical figures are 48 kJ/g for petroleum versus 30 kJ/g for a high grade of coal.

Petroleum ("crude oil") is a mixture of thousands of different compounds. The great majority are **hydrocarbons,** compounds that consist only of the elements hydrogen and carbon. The hydrocarbons in petroleum can contain from 1 to as many as 60 carbon atoms per molecule. Many are **alkanes,** hydrocarbons with only single bonds between carbon atoms (Table 4.2). Compared with coal, the concentrations of sulfur and other contaminants in petroleum generally are quite low, which minimizes the emissions of pollutants such as SO_2.

The oil refinery is the icon of the petroleum industry (Figure 4.10). During the initial step in the refining process, the crude oil is separated into fractions that consist of compounds with similar properties. **Distillation** is a separation process in which a solution is heated to its boiling point and the vapors are condensed and collected. The distillation of crude oil takes place in this manner.

The simplest hydrocarbon, methane, contains only one carbon atom, and therefore no carbon–carbon bonds.

Petroleum is classified as "sweet crude" when it contains less than 0.5% sulfur. "Sour crude" has more than 1% sulfur.

Table 4.2	Selected Alkanes	
Name and Chemical Formula	**Structural Formula**	**Condensed Structural Formula**
methane CH_4	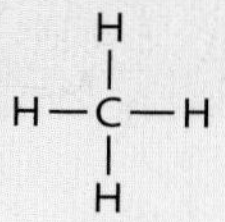	CH_4
ethane C_2H_6	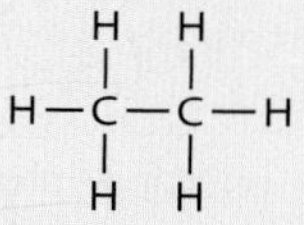	CH_3CH_3
propane C_3H_8	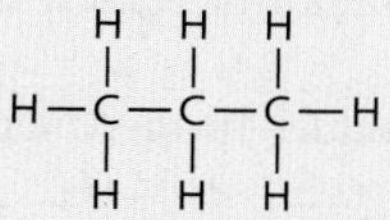	$CH_3CH_2CH_3$
n-butane C_4H_{10}	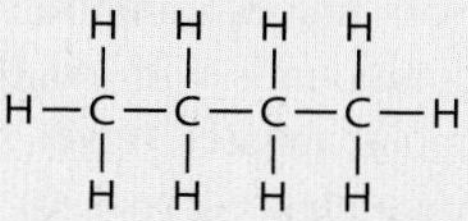	$CH_3CH_2CH_2CH_3$
n-pentane C_5H_{12}	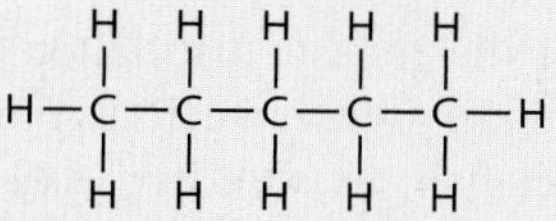	$CH_3CH_2CH_2CH_2CH_3$
n-hexane C_6H_{14}	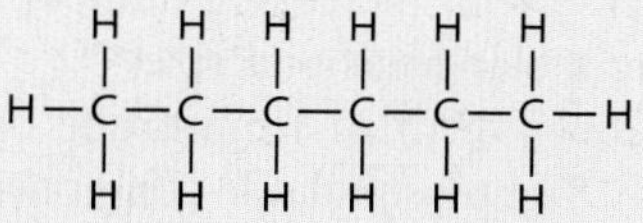	$CH_3CH_2CH_2CH_2CH_2CH_3$
n-heptane C_7H_{16}	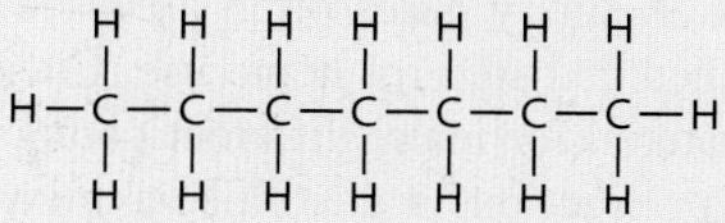	$CH_3CH_2CH_2CH_2CH_2CH_2CH_3$
n-octane C_8H_{18}	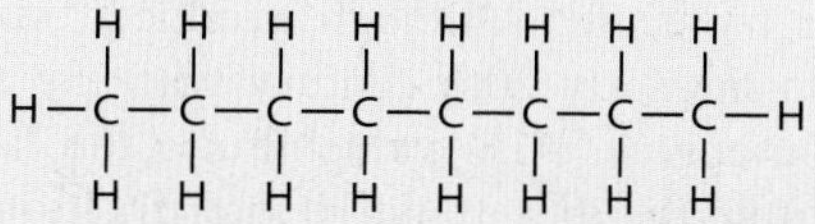	$CH_3CH_2CH_2CH_2CH_2CH_2CH_2CH_3$

Note: Butane, pentane, hexane, heptane, and octane all have other isomers. We explain names such as *n*-butane and *n*-pentane in Section 4.7.

Figure 4.10
An oil refinery showing the tall distillation towers.

As of 2008, 143 oil refineries were operating in the United States. No new refineries have been built since 1976.

- The crude oil is pumped into a large vessel and heated.
- As the temperature increases, the compounds with the lowest boiling points vaporize.
- As the temperature further increases, compounds with higher boiling points begin to vaporize.
- Once in the gas phase, compounds travel up the tall distillation or fractionation tower. The compounds with smaller molar masses travel higher; the compounds with larger molar masses travel shorter distances up the column.
- Fractions are condensed back to the liquid state at different levels in the tower.

Figure 4.11 illustrates a distillation tower and lists some of the fractions obtained. These include refinery gases such as methane and propane, liquids such as gasoline, kerosene, and jet fuel, as well as waxy solids and asphalt. Note that boiling points generally increase with increasing number of carbon atoms in the molecule and hence with increasing molar mass and size.

The fractions distilled from crude oil differ in the hydrocarbon molecules that constitute them. The most volatile components, "refinery gases," have 1–4 carbon atoms per molecule. Refinery gases often are used to fuel the distillation towers. They also can be liquefied and sold for home use. Chemical manufacturers use refinery gases as a starting material for many different compounds.

A major fraction is gasoline, a mixture of hydrocarbons with 5–12 carbon atoms per molecule. Although produced since the mid-1800s, gasoline became valuable in the early 20th century with the advent of the automobile and internal combustion engine. Also valuable are the larger molecules (12–16 carbon atoms) that constitute the diesel and jet fuel fractions. The highest boiling fractions, containing molecules with over 20 carbon atoms, are useful as industrial oils, lubricants, and asphalt.

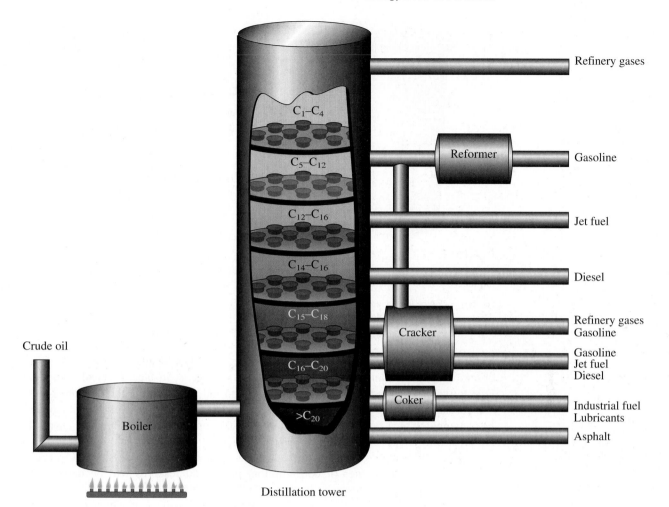

Figure 4.11

Diagram of a distillation tower for crude oil that shows the fractions and typical uses.

Figure 4.12 shows the range of products from a barrel (42 gallons) of crude oil. Most are burned for heating and transportation. The remaining 7 or so gallons are used primarily for nonfuel purposes, including the gallon or two that serve as starting materials, or "feedstocks," to produce a myriad of plastics, pharmaceuticals, fabrics, and other carbon-based products. As these hydrocarbon feedstocks are nonrenewable resources, one easily can predict how one day petroleum products might become too valuable to burn.

Consider This 4.11   **Products from a Barrel of Crude**

Through research, chemists have increased the amount of gasoline that can be derived from a barrel of crude oil. For example, in 1904 a barrel of crude oil produced 4.3 gal of gasoline, 20 gal of kerosene, 5.5 gal of fuel oil, 4.9 gal of lubricants, and 7.1 gal of miscellaneous products. By 1954, the products were 18.4 gal of gasoline, 2.0 gal of kerosene, 16.6 gal of fuel oil, 0.9 gal of lubricants, and 4.1 gal of miscellaneous products.

a. Compare these values with those shown in Figure 4.12.
b. Offer two reasons why the distribution of products has changed over time.

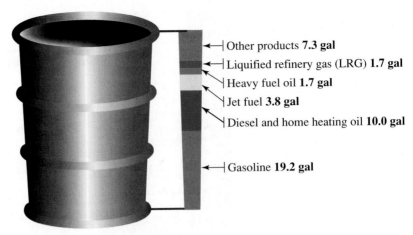

Other products **7.3 gal**
Liquified refinery gas (LRG) **1.7 gal**
Heavy fuel oil **1.7 gal**
Jet fuel **3.8 gal**
Diesel and home heating oil **10.0 gal**

Gasoline **19.2 gal**

Figure 4.12

Products (in gallons) from the refining of a barrel of crude oil. *Note:* A barrel is 42 gallons. The refinery products occupy a volume of about 44 gallons.

Source: U.S. Energy Information Administration, 2009.

In his 2006 State of the Union address, former President George W. Bush observed that the United States is "addicted to oil." It is, in fact, a global addiction. Due to increases in population, urbanization, and industrialization, the demand for oil has reached unprecedented levels. The largest increases are in developing countries. For example, China put nearly 10 million new cars on the road in 2008 and currently is the second largest car market in the world.

> The United States has about 770 cars for every 1000 people. In China, there are only 10 cars per 1000 people.

The 2004 update to the United Nations World Energy Assessment Overview claims that at current rates of consumption, proven oil reserves will be depleted in 50–100 years and coal reserves in about 250 years. Nontraditional sources and new extraction technologies may extend these limits, but regardless of which scenario we choose to believe, the supply of fossil fuels is finite.

The real issue here is not the quantity of fossil fuels remaining on Earth, but rather the rate at which we are able to extract them (and, of course, whether it makes sense to burn what we extract). In the mid-1950s, yearly global oil consumption was 4 billion barrels, and over 30 billion barrels of new deposits were found annually. Today, those numbers are nearly reversed. Experts therefore predict that sometime in the near future, oil production will peak and then decline as we deplete the most easily recoverable deposits. In fact, this has already happened in the United States, where oil production from the lower 48 states has slowly but steadily declined since 1970. Oil company executives voice doubts that production could ever keep pace with future demand. In 2008, Royal Dutch Shell's CEO, Jeroen van der Veer, commented that "after 2015, supplies of easy-to-access oil and gas will no longer keep up with demand."

> Oil production from Alaskan oil fields peaked in 1988.

Assuming petroleum consumption increases 2% per year, the U.S. Energy Information Agency has developed three "peak oil" scenarios (Figure 4.13); one low, one average, and one high. The differences lie in the degree to which oil reserves worldwide are recoverable, and even the most optimistic scenario gives a maximum before 2050. In none of the scenarios do we abruptly "run out" of oil. Rather, dramatically higher prices and increasing scarcity will characterize the era when oil production peaks. Thus at some point in this century, we will no longer be able to sustain our addiction to this "black gold."

> Other estimates of the amount of recoverable petroleum are more pessimistic, with some experts claiming that peak oil may have already occurred.

We now turn our attention to natural gas, also a fossil fuel. Although the natural gas that enters your home is practically pure methane, the "raw" gas from oil and gas wells usually includes ethane (2–6%), other small hydrocarbons, and varying quantities of water vapor, carbon dioxide, hydrogen sulfide, and helium. Before natural gas can be transported by pipeline, it must be processed to remove these impurities.

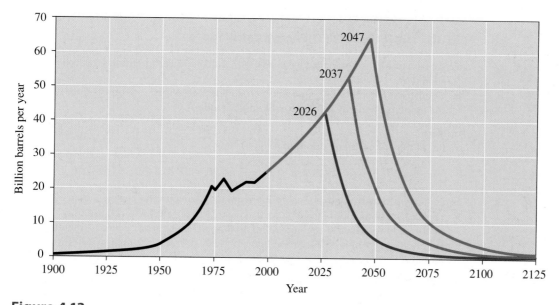

Figure 4.13

Worldwide oil production since 1900 (*black*) and three peak oil scenarios. The blue, red, and green lines correspond to low, average, and high estimates of recoverable oil, respectively.

Source: Energy Information Agency, U.S. Department of Energy.

Natural gas provides heat for about two thirds of the single-family homes and apartment buildings in the United States. Increasingly, though, natural gas is burned to generate electricity and to power vehicles. The reason is that natural gas burns more cleanly than other fossil fuels. Natural gas releases essentially no sulfur dioxide when burned because the sulfur-containing compounds are removed at the refinery. In addition, the levels of unburned volatile hydrocarbons, particulates, carbon monoxide, and nitrogen oxides are relatively low. No ash residue containing toxic metals remains after combustion. Although burning natural gas does produce the greenhouse gas carbon dioxide, the amount is less per unit of energy released than for the other fossil fuels. Check the numbers yourself in the next activity.

Your Turn 4.12   Coal Versus Natural Gas

The combustion of one gram of natural gas releases 50.1 kJ of heat.

a. Calculate the mass of CO_2 released when natural gas is burned to produce 1500 kJ of heat. Assume that natural gas is pure methane, CH_4.

b. Select one of the grades of coal from Table 4.1. Compare the mass of CO_2 produced when enough of this coal is burned to produce the same 1500 kJ of heat.
Hint: Assume coal is $C_{135}H_{96}O_9NS$. You calculated its molar mass in Your Turn 4.8.

4.5 | Measuring Energy Changes

The ability of a substance to release energy makes it a good fuel. As you have seen, both calories and joules can be used to express the energy contained in a food or fuel. In this section, you learn to quantify energy changes in chemical reactions. The next activity gives you practice with energy units.

Donuts contain lipids and carbohydrates, which are discussed in Sections 11.3 and 11.5, respectively.

Your Turn 4.13   Energy Calculations

a. When a donut is metabolized, 425 kcal (425 Cal) are released. Express this value in kilojoules.

b. Calculate the number of 1-kg books you could lift to a shelf 2 m off the floor with the amount of energy from metabolizing one donut.

Answers

a. Recall that 1 kcal is equivalent to 4.184 kJ.

$$425 \ \text{kcal} \times \frac{4.184 \ \text{kJ}}{1 \ \text{kcal}} = 1.78 \times 10^3 \ \text{kJ}$$

b. Earlier, we stated that one joule is approximately equal to the energy required to raise a 1-kg book a distance of 10 cm against Earth's gravity. We can use this information to calculate the number of 1-kg books that could be lifted 2 meters. First note that 2 m is equivalent to 200 cm. Next, calculate the energy (joules) required to lift one 1-kg book the entire 2 m.

$$200 \ \text{cm} \times \frac{1 \ \text{J}}{10 \ \text{cm}} = 20 \ \text{J}$$

Then, express this value in kilojoules.

$$20 \ \text{J} \times \frac{1 \ \text{kJ}}{10^3 \ \text{J}} = 0.020 \ \text{kJ}$$

Use this value to make the final calculation.

$$1.78 \times 10^3 \ \text{kJ} \times \frac{1 \ \text{book}}{0.020 \ \text{kJ}} = 8.9 \times 10^4 \ \text{books}$$

To work off one donut requires lifting almost 90,000 books!

Skeptical Chemist 4.14   Checking Assumptions

A simplifying (and erroneous) assumption was made in doing the calculations in part **b** of the preceding activity. What was the assumption and is it reasonable? Based on this assumption, is your answer too high or too low? Explain your reasoning.

The metabolism of foods, including minimally healthy ones such as donuts, helps keep our bodies at a constant temperature. **Temperature** is a measure of the average kinetic energy of the atoms and/or molecules present in a substance. Everything around us is at some temperature—hot, cold, or lukewarm. When we perceive a particular object as "cold," this means its atoms and molecules are moving more slowly on average relative to an object we perceive as "hot." Therefore, for the temperature of an object to increase, the kinetic energy of its atoms and molecules must increase. Where does that energy come from? **Heat** is the kinetic energy that flows from a hotter object to a colder one. When two bodies are in contact, heat always flows from the object at the higher temperature to one at a lower temperature.

Although the concepts of temperature and heat are related, they are not identical. Your bottle of water and the Pacific Ocean may be at the same temperature, but the ocean contains and can transfer far more heat than the bottle of water. Indeed, bodies of water can affect the climate of an entire region as a consequence of their ability to absorb and transfer heat.

The connections between oceans and climate are discussed further in Sections 3.10 and 6.5.

The **calorimeter** is a device used to experimentally measure the quantity of heat energy released in a combustion reaction. Figure 4.14 shows a schematic representation of a calorimeter. To use it, you introduce a known mass of fuel and an excess of oxygen into the heavy-walled stainless steel container. The container is then sealed and submerged in a bucket of water. The reaction is initiated with a spark. The heat evolved by the reaction flows from the container to the water and the rest of the apparatus. As a consequence, the temperature of the entire calorimeter system

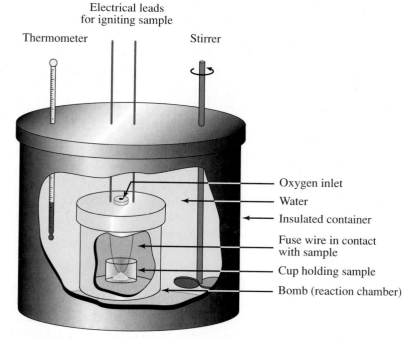

Electrical leads
for igniting sample

Thermometer   Stirrer

Oxygen inlet
Water
Insulated container
Fuse wire in contact
with sample
Cup holding sample
Bomb (reaction chamber)

Figure 4.14

Schematic drawing of a calorimeter.

increases. The quantity of heat given off by the reaction can be calculated from this temperature rise and the known heat-absorbing properties of the calorimeter and the water it contains. The greater the temperature increase, the greater the quantity of energy evolved from the reaction.

Experimental measurements of this sort are the source of most of the tabulated values of heats of combustion. As the name suggests, the **heat of combustion** is the quantity of heat energy given off when a specified amount of a substance burns in oxygen. Heats of combustion are typically reported as positive values in kilojoules per mole (kJ/mol), kilojoules per gram (kJ/g), kilocalories per mole (kcal/mol), or kilocalories per gram (kcal/g). For example, the experimentally determined heat of combustion of methane is 802.3 kJ/mol. This means that 802.3 kJ of heat is given off when 1 mole of $CH_4(g)$ reacts with 2 moles of $O_2(g)$ to form 1 mole of $CO_2(g)$ and 2 moles of $H_2O(g)$.

$$CH_4(g) + 2\ O_2(g) \longrightarrow CO_2(g) + 2\ H_2O(g) + 802.3\ kJ \qquad \textbf{[4.3]}$$

We can use this value to calculate the number of kilojoules released for a gram, rather than for a mole. The molar mass of CH_4, calculated from the atomic masses of carbon and hydrogen, is 16.0 g/mol. We then can calculate the heat of combustion per gram of methane.

$$\frac{802.3\ kJ}{1\ mol\ CH_4} \times \frac{1\ mol\ CH_4}{16.0\ g\ CH_4} = 50.1\ kJ/g\ CH_4$$

As fuels go, this is a high heat of combustion! Look ahead to Figure 4.16 to see how this value compares with those for other fuels.

Burning methane is analogous to water tumbling down from the top of a waterfall. Initially in a state of higher potential energy, the water drops down to one of lower potential energy. The potential energy is converted into kinetic energy, which is then released when the water hits the rocks below. Similarly, when methane is burned, energy is released when the atoms in the reactants "fall" to a state of lower potential energy as the products are formed. Figure 4.15 is a schematic representation of this process. The downward arrow indicates that the energy associated with 1 mole of $CO_2(g)$ and 2 moles of $H_2O(g)$ is less than the energy associated with 1 mole of $CH_4(g)$

Heats of combustion, by convention, are tabulated as positive values even though all combustion reactions *release* heat.

The mole was defined in Section 3.7.

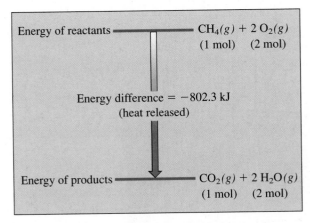

Figure 4.15

Energy difference in the combustion of methane—an exothermic reaction.

For an exothermic reaction,

$E_{products} - E_{reactants} < 0$.

and 2 moles of $O_2(g)$. The combustion of methane is **exothermic,** a term applied to any chemical or physical change accompanied by the release of heat. In this reaction, the energy difference is -802.3 kJ. The negative sign attached to the energy change for all exothermic reactions signifies the decrease in potential energy going from reactants to products. Not surprisingly, the amount of energy released depends on the amount of fuel burned.

By now, it is probably clear that good fuels have high potential energies. The higher the potential energy of a fuel, the more heat it releases when it is burned to produce CO_2 and H_2O. Figure 4.16 compares the energy difference (in kJ/g) of several different fuels. We can make some interesting generalizations based on the chemical formulas of the fuels. First, the fuels with the highest heats of combustion are hydrocarbons. Second, as the ratio of hydrogen-to-carbon decreases, the heat of combustion decreases. And third, as the amount of oxygen in the fuel molecule increases, the heat of combustion decreases.

Consider This 4.15   Coal Versus Ethanol

On the basis of their chemical composition, explain why ethanol and coal have very different chemical formulas but similar heats of combustion.

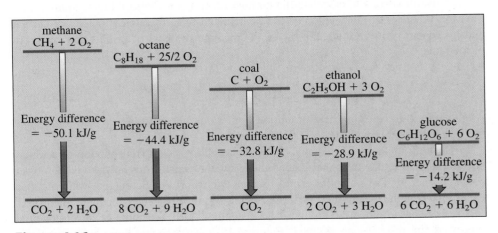

Figure 4.16

Energy differences (in kJ/g) for the combustion of methane (CH_4), n-octane (C_8H_{18}), coal (assumed to be pure carbon), ethanol (C_2H_5OH), and wood (assumed to be glucose).

Table 4.3	Endothermic Versus Exothermic Reactions	
Endothermic Reaction	**Exothermic Reaction**	
$Energy_{products} > Energy_{reactants}$	$Energy_{products} < Energy_{reactants}$	
Net energy change is positive.	Net energy change is negative.	
Energy is absorbed.	Energy is released.	

Many naturally occurring reactions are not exothermic but rather *absorb* energy as they occur. We discussed two important examples in earlier chapters. One is the decomposition of O_3 to yield O_2 and O, and the other is the combination of N_2 and O_2 to yield two molecules of NO. Both reactions require energy in the form of an electrical discharge, a high-energy photon, or a high temperature. These reactions are **endothermic,** the term applied to any chemical or physical change that absorbs energy. A chemical reaction is endothermic when the potential energy of the products is *higher* than that of the reactants. The energy change for an endothermic reaction is always positive. Table 4.3 compares the energy changes in endothermic and exothermic reactions.

Photosynthesis also is endothermic. This process requires the absorption of 2800 kJ of sunlight per mole of $C_6H_{12}O_6$, or 15.5 kJ per gram of glucose formed. The complete process involves many steps, but the overall reaction can be described with this equation.

$$2800 \text{ kJ} + 6\,CO_2(g) + 6\,H_2O(l) \xrightarrow{\text{chlorophyll}} \underset{\text{glucose}}{C_6H_{12}O_6(s)} + 6\,O_2(g) \qquad \textbf{[4.4]}$$

The reaction requires the participation of the green pigment chlorophyll. The chlorophyll molecule absorbs energy from the photons of visible sunlight and uses this energy to drive the photosynthetic process, an energetically uphill reaction. Photosynthesis plays an essential role in the carbon cycle, as it removes CO_2 from the atmosphere.

The potential energy of any specific chemical species, and therefore the amount of energy released on combustion, is related to the chemical bonds in the molecules of the fuel. In the following section, we illustrate how knowledge of molecular structure can be used to calculate heats of combustion and can allow us to pinpoint the differences between fuels.

4.6 | Energy Changes at the Molecular Level

Chemical reactions involve the breaking and forming of chemical bonds. Energy is required to break bonds, just as energy is required to break chains or to tear paper. In contrast, forming chemical bonds releases energy. The overall energy change associated with a chemical reaction depends on the net effect of the bond breaking and bond forming. If the energy required to break the bonds in the reactants is greater than the energy released when the products form, the overall reaction is *endothermic;* energy is absorbed. If, on the other hand, the bond-making energy of the products is greater than the bond breaking in the reactants, then the net energy change is *exothermic;* energy is released by the reaction.

For example, consider the combustion of hydrogen. Hydrogen is desirable as a fuel because, compared with other fuels, it releases a large amount of energy per gram when it burns.

$$2\,H_2(g) + O_2(g) \longrightarrow 2\,H_2O(g) + \text{energy} \qquad \textbf{[4.5]}$$

To calculate the energy change associated with the combustion of hydrogen to form water vapor, let us assume that all the bonds in the reactant molecules are broken, and then the individual atoms are reassembled to form the products. In fact, the reaction does not occur this way, but we are interested in only the overall (net) change, not

Section 2.6 described the decomposition of O_3 with the absorption of UV light. Section 1.9 described the formation of NO at high temperature.

Look for more about hydrogen as a fuel in Sections 8.5 and 8.6.

Table 4.4		Covalent Bond Energies (in kJ/mol)							
	H	C	N	O	S	F	Cl	Br	I
Single Bonds									
H	436								
C	416	356							
N	391	285	160						
O	467	336	201	146					
S	347	272	—	—	226				
F	566	485	272	190	326	158			
Cl	431	327	193	205	255	255	242		
Br	366	285	—	234	213	—	217	193	
I	299	213	—	201	—	—	209	180	151
Multiple Bonds									
C=C	598			C=N	616		C=O*	803	
C≡C	813			C≡N	866		C≡O	1073	
N=N	418			O=O	498				
N≡N	946								

*in CO_2.

the details. Therefore, we proceed with our convenient plan and see how well our calculated result agrees with the experimental value.

The covalent bond energies given in Table 4.4 provide the numbers needed for our computation. **Bond energy** is the amount of energy that must be absorbed to break a specific chemical bond. Thus, because energy must be absorbed, breaking bonds is an endothermic process, and all the bond energies in Table 4.4 are positive. The values are expressed in kilojoules per mole of bonds broken. Note that the atoms appear both across the top and down the left side of the table. The number at the intersection of any row and column is the energy (in kilojoules) needed to break a mole of the bonds between the two atoms.

The amount of energy required depends on the number of bonds broken: more bonds take more energy. Each value in Table 4.4 is for one mole of bonds. For example, the energy required to break 1 mole of H–H bonds, as in the H_2 molecule, is 436 kJ. Similarly, the energy required to break 1 mole of O=O double bonds, as in the O_2 molecule, is 498 kJ.

We need to keep track of whether energy is absorbed or released. To do this, we indicate the energy absorbed with a positive sign. This is the energy absorbed when the bond is broken. Forming a bond releases energy, and the sign is negative. For example, the bond energy for the O=O double bond is 498 kJ/mol. Accordingly, when 1 mole of O=O double bonds is broken, the energy change is +498 kJ, and when 1 mole of O=O double bonds is formed, the energy change is −498 kJ.

Now we are finally ready to apply these concepts and conventions to the burning of hydrogen gas, H_2. The next equation shows the Lewis structures of the species involved so that we can count the bonds that need to be broken and formed:

$$2\,H\!-\!H \;+\; \ddot{\underset{\cdot\cdot}{O}}\!=\!\ddot{\underset{\cdot\cdot}{O}} \longrightarrow 2\; {}_{H}\!\!\nearrow\!\!\overset{\cdot\cdot\overset{\cdot\cdot}{O}}{}\!\!\searrow_{H} \qquad\qquad [4.6]$$

Remember that chemical equations can be read in terms of moles. Both equation 4.5 and 4.6 indicate "2 moles of H_2 plus 1 mole of O_2 yields 2 moles of H_2O." To use bond energies, we need to count the number of moles of *bonds* involved. Here is a summary.

Molecule	Bonds per Molecule	Moles in Reaction	Moles of Bonds	Bond Process	Energy per Bond	Total Energy
H–H	1	2	$1 \times 2 = 2$	breaking	+436 kJ	$2 \times (+436) = +872$ kJ
O=O	1	1	$1 \times 1 = 1$	breaking	+498 kJ	$1 \times (+498) = +498$ kJ
H–O–H	2	2	$2 \times 2 = 4$	forming	−467 kJ	$4 \times (−467) = −1868$ kJ

From the last column, we can see that the overall energy change in breaking bonds (872 kJ + 498 kJ = 1370 kJ) and forming new ones (−1868 kJ) results in a net energy change of −498 kJ.

This calculation is diagrammed in Figure 4.17. The energy of the reactants, 2 H_2 and O_2, is set at zero, an arbitrary but convenient value. The green arrows pointing upward signify energy absorbed to break the bonds in the reactant molecules and form 4 H atoms and 2 O atoms. The red arrow on the right pointing downward represents energy released as these atoms bond to form the product molecules: 2 H_2O. The shorter red arrow corresponds to the net energy change of −498 kJ signifying that the overall combustion reaction is strongly exothermic. The products are lower in energy than the reactants, so the energy change is negative. The net result is the release of energy, mostly in the form of heat. Another way to look at such exothermic reactions is as a conversion of reactants involving weaker bonds to products involving stronger ones. In general, the products are more stable (lower potential energy) and less reactive than the starting substances.

The energy change we just calculated from bond energies, −498 kJ for burning 2 mol of hydrogen, compares favorably with the experimentally determined value when all of the species are gases. This agreement justifies our rather unrealistic assumption that all the bonds in the reactant molecules are first broken and then all the bonds in

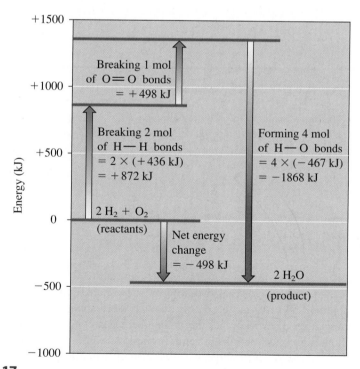

Figure 4.17

The energy changes during the combustion of hydrogen to form water vapor.

 Figures Alive! Visit the textbook's website to learn more about the energy changes of this reaction.

the product molecules are formed. The energy change that accompanies a chemical reaction depends on the energy *difference* between the products and the reactants, not on the particular process, mechanism, or individual steps that connect the two. This is an extremely powerful idea when doing calculations related to energy changes in reactions.

Not all calculations come out as easily and well as this one did. For one thing, the bond energies in Table 4.4 apply only to gases, so calculations using these values agree with experiment only if all the reactants and products are in the gaseous state. Moreover, tabulated bond energies are average values. The strength of a bond depends on the overall structure of the molecule in which it is found; in other words, on what else the atoms are bonded to. Thus, the strength of an O–H bond is slightly different in HOH, HOOH, and CH_3OH. Nevertheless, the procedure illustrated here is a useful way of estimating energy changes in a wide range of reactions. The approach also helps illustrate the relationship between bond strength and chemical energy.

This analysis also helps clarify why products of combustion reactions (such as H_2O or CO_2) cannot be used as fuels. There are no substances into which these compounds can be converted that have stronger bonds and that are lower in energy. Bottom line: You cannot run a car on its exhaust fumes!

> Experimental values differ somewhat from those calculated using bond energies.

Your Turn 4.16   Heat of Combustion for Ethyne

Use the bond energies in Table 4.4 to calculate the heat of combustion for ethyne, C_2H_2, also called acetylene. Report your answer both in kilojoules per mole (kJ/mol) C_2H_2 and kilojoules per gram (kJ/g) C_2H_2. Here is the balanced chemical equation.

$$2\,H\!-\!C\!\equiv\!C\!-\!H \;+\; 5\,\ddot{\underset{\cdot\cdot}{O}}\!=\!\ddot{\underset{\cdot\cdot}{O}} \;\longrightarrow\; 4\,\ddot{\underset{\cdot\cdot}{O}}\!=\!C\!=\!\ddot{\underset{\cdot\cdot}{O}} \;+\; 2\;{}_{H}\!\!\overset{\ddot{\cdot\cdot}O}{\diagdown}{}_{H}$$

Answer

Energy change = −1256 kJ/mol C_2H_2, or −48.3 kJ/g C_2H_2
Heat of combustion = 1256 kJ/mol C_2H_2, or 48.3 kJ/g C_2H_2

Your Turn 4.17 O_2 Versus O_3

As noted in Chapter 2, ozone absorbs UV radiation having wavelengths less than 320 nm, and oxygen absorbs electromagnetic radiation with wavelengths less than 242 nm. Use the bond energies in Table 4.4 plus information about the resonance structures of O_3 from Chapter 2 to explain why.

4.7 | The Chemistry of Gasoline

Equipped with our understanding of the molecular nature of fuels and the energy changes associated with combustion, we now return to petroleum. The distribution of compounds obtained by distilling crude oil does not correspond to the prevailing pattern of commercial use. For example, the demand for gasoline is considerably greater than that for higher boiling fractions. Chemists employ several processes to change the natural distribution and to obtain more gasoline of higher quality. These include cracking, combining, and reforming (see Figure 4.11).

Thermal cracking, a process that breaks large hydrocarbon molecules into smaller ones by heating them to a high temperature, was developed first. In this procedure, the heaviest crude oil fractions are heated between 400 and 450 °C. This heat "cracks" the heaviest tarry crude oil molecules into smaller ones useful for gasoline and diesel fuel. For example, at high temperature, one molecule of $C_{16}H_{34}$ can be cracked into two nearly identical molecules.

$$C_{16}H_{34} \longrightarrow C_8H_{18} + C_8H_{16} \qquad \text{[4.7]}$$

Thermal cracking also can produce different-sized molecules.

$$C_{16}H_{34} \longrightarrow C_{11}H_{22} + C_5H_{12} \qquad \text{[4.8a]}$$

In either case, the total number of carbon and hydrogen atoms is unchanged from reactants to products. The larger reactant molecules simply have been fragmented into smaller, more economically important molecules. We can use space-filling models to show the size difference more clearly.

The space-filling model of $C_{11}H_{22}$ shows a "bend" because of the geometry of the atoms at the C=C double bond.

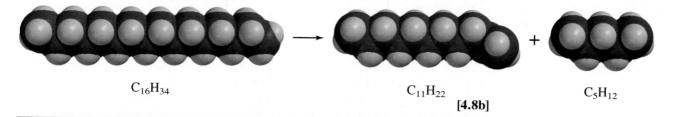

$C_{16}H_{34}$ $C_{11}H_{22}$ C_5H_{12}

[4.8b]

Your Turn 4.18 More Practice with Cracking

a. Draw structural formulas for one pair of products formed when $C_{16}H_{34}$ is thermally cracked (see equation 4.8a).
Hint: Draw the atoms of the product molecules in an unbranched chain and include one double bond (only in one product).

b. Revisit the alkanes shown in Table 4.2. Look closely to find a pattern for the number of H atoms per C atom. Use this pattern to write the generic chemical formula.

c. Write the generic chemical formula for a hydrocarbon with one C=C double bond.

Answer

b. C_nH_{2n+2} (where n is an integer)

The problem with thermal cracking is the energy required to produce the high temperature. **Catalytic cracking** is a process in which catalysts are used to crack larger hydrocarbon molecules into smaller ones at relatively low temperatures, thus reducing energy use. Chemists at all major oil companies have developed important cracking catalysts and continue to find more selective and inexpensive processes. We discuss how catalysts affect the rates of chemical reactions in Section 4.8.

Sometimes chemists want to combine molecules, rather than split them apart. To produce more of the intermediate-sized molecules needed for gasoline, catalytic combination can be used. In this process, smaller molecules are joined.

$$4\ C_2H_4 \xrightarrow{\text{catalyst}} C_8H_{16} \qquad \text{[4.9]}$$

Another important chemical process is **catalytic reforming.** Here, the atoms within a molecule are rearranged, usually starting with linear molecules and producing ones with more branches. As we will see, the more highly branched molecules burn more smoothly in automobile engines.

It turns out that molecules with the same molecular formula are not necessarily identical. For example, octane has the formula C_8H_{18}. Careful analysis reveals 18 different compounds with this formula. Molecules with the same molecular formula but with different chemical structures and different properties are called **isomers.** In *n*-octane (normal octane) the carbon atoms are all in a continuous chain (Figure 4.18a). In iso-octane, the carbon chain has several branch points (Figure 4.18b). The chemical and physical properties of these two isomers are similar, but they are not identical. For example, the boiling point of *n*-octane is 125 °C, compared with 99 °C for iso-octane.

Although the heats of combustion for *n*-octane and iso-octane are nearly identical, they burn differently in an auto engine. The more compact shape of the latter compound imparts a "smoother" burn. In a well-tuned car engine, gasoline vapor and

Using catalysts to produce *very* large molecules (polymers) from smaller ones (monomers) is discussed in Section 9.3.

Combustion of branched hydrocarbons releases 2–4% more energy than combustion of their straight-chain isomers.

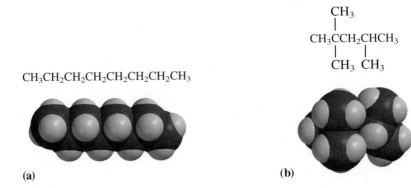

$$CH_3CH_2CH_2CH_2CH_2CH_2CH_2CH_3$$

(a)   (b)

Figure 4.18

Condensed structural formula and space-filling model for **(a)** *n*-octane and **(b)** iso-octane.

air are drawn into a cylinder, compressed by a piston, and ignited by a spark. Normal combustion occurs when the spark plug ignites the fuel–air mixture and the flame front travels rapidly across the combustion chamber, consuming the fuel. Sometimes, however, compression alone is enough to ignite the fuel before the spark occurs. This premature firing is called preignition. It results in lower engine efficiency and higher fuel consumption because the piston is not in its optimal location when the burned gases expand. "Knocking," a violent and uncontrolled reaction, occurs after the spark ignites the fuel, causing the unburned mixture to burn at supersonic speed with an abnormal rise in pressure. Knocking produces an objectionable metallic sound, loss of power, overheating, and engine damage when severe.

In the 1920s, knocking was shown to depend on the chemical composition of the gasoline. The "octane rating" was developed to designate a particular gasoline's resistance to knocking. Iso-octane performs exceptionally well in automobile engines and arbitrarily has been assigned an octane rating of 100. Like *n*-octane, *n*-heptane is a straight-chain hydrocarbon, but with one fewer –CH₂ group. It also has a high tendency to knock and is assigned an octane rating of 0 (Table 4.5). When you go to the gasoline pump and fill up with 87 octane, you are buying gasoline that has the same knocking characteristics as a mixture of 87% iso-octane (octane number 100) and 13% *n*-heptane (octane number 0). Higher grades of gasoline also are available: 89 octane (regular plus) and 92 octane (premium). These blends contain a higher percent of compounds with higher octane ratings (Figure 4.19).

Although *n*-octane has a poor octane rating, it is possible to catalytically re-form *n*-octane to iso-octane, thus greatly improving its performance. This rearrangement is accomplished by passing *n*-octane over a catalyst consisting of rare and expensive elements such as platinum (Pt), palladium (Pd), rhodium (Rh), and iridium (Ir). Re-forming isomers to improve their octane rating became important starting in the late 1970s because of the nationwide efforts to ban the use of tetraethyl lead (TEL) as an antiknock additive.

Younger people today may never have heard the sound of an engine "knocking," because it rarely occurs with current engine technology and gasoline blends.

Figure 4.19

Gasoline is available in a variety of octane ratings.

Table 4.5	Octane Ratings of Several Compounds
Compound	**Octane Rating**
n-octane	−20
n-heptane	0
iso-octane	100
methanol	107
ethanol	108
MTBE	116

MTBE, *methyl tertiary-butyl ether.*

In the United States, TEL was phased out during the 1970s because of the toxic effects of lead, especially to children. See Section 5.10 for more information on lead in the environment.

![web] **Consider This 4.19**   **Getting the Lead Out**

The United States completed the ban on leaded gasoline in 1996 because of the hazards associated with lead exposure. But other sources of lead still exist. Be a detective on the web to identify:

  a. an occupational source of lead exposure.
  b. a hobby that is a source of lead exposure.
  c. a source of lead exposure that particularly affects children.

The textbook's website provides helpful links to aid your search.

Elimination of TEL as an octane enhancer necessitated finding substitutes that were inexpensive, easy to produce, and environmentally benign. Several were tried, including ethanol and MTBE (*m*ethyl *t*ertiary-*b*utyl *e*ther), each with an octane rating greater than 100 (see Table 4.5). As we will see, however, MTBE did not turn out as well as expected.

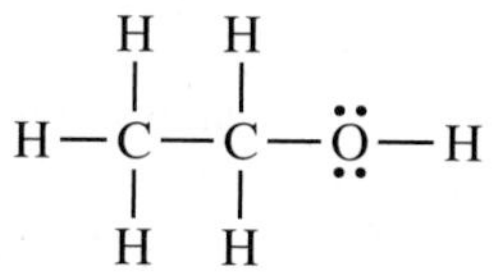

ethanol 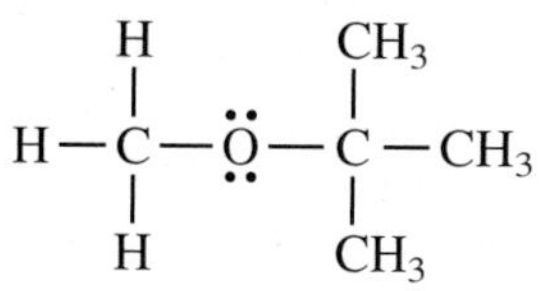   MTBE

Fuels containing these additives are referred to as **oxygenated gasolines,** blends of petroleum-derived hydrocarbons with added oxygen-containing compounds such as MTBE, ethanol, or methanol (CH_3OH). Because they contain oxygen, these gasoline blends burn more cleanly and produce less carbon monoxide than their nonoxygenated counterparts.

Since 1995, about 90 cities and metropolitan areas with the highest ground-level ozone levels have adopted the Year-Round Reformulated Gasoline Program mandated by the Clean Air Act Amendments of 1990. This program requires the use of **reformulated gasolines (RFGs),** oxygenated gasolines that also contain a lower percentage of certain more volatile hydrocarbons found in nonoxygenated conventional gasoline. RFGs cannot contain more than 1% benzene (C_6H_6) and must be at least 2% oxygen. Because of their composition, reformulated gasolines evaporate less readily than conventional gasolines and produce less carbon monoxide emissions.

As pointed out earlier in Chapter 1, the volatile organic compounds (VOCs) in conventional gasoline play a role in tropospheric ozone formation, especially in high-traffic areas. When RFGs were introduced in the 1990s, MTBE was the oxygenate of choice. However, concerns over its toxicity and its ability to leach from gasoline storage tanks into the groundwater have led many states to ban MTBE and switch to ethanol. As an additive and a fuel in its own right, ethanol is described more fully in Section 4.9.

Benzene, C_6H_6, is a known carcinogen. Its molecular structure is introduced in Section 9.6 and further discussed in Section 10.2.

Volatile organic compounds and their role in ozone formation were explained in Section 1.11.

4.8 | New Uses for an Old Fuel

World supplies of coal are predicted to last for hundreds of years, much longer than current estimates of remaining available oil reserves. Unfortunately, the fact that coal is a solid makes it inconvenient for many applications, especially as a fuel for vehicles. Therefore, research and development projects are underway aimed at converting solid coal into fuels that possess characteristics similar to petroleum products.

Before large supplies of natural gas were discovered and exploited, cities were lighted with water gas, a mixture of carbon monoxide and hydrogen. Water gas is formed by blowing steam over hot coke, the impure carbon that remains after volatile components have been distilled from coal.

$$C(s) + H_2O(g) \longrightarrow CO(g) + H_2(g) \qquad \text{[4.10]}$$
$$\text{coke} \qquad\qquad\qquad \text{water gas}$$

This same reaction is the starting point for the Fischer–Tropsch process for producing synthetic gasoline from coal. German chemists Emil Fischer (1852–1919) and Hans Tropsch (1889–1935) developed the process during the 1920s. At that time, Germany had abundant coal reserves, but little petroleum.

The Fischer–Tropsch process can be described by this general equation.

$$n\,CO(g) + (2n + 1)\,H_2(g) \xrightarrow{\text{catalyst}} C_nH_{2n+2}(g,l) + n\,H_2O(g) \qquad \text{[4.11]}$$

The hydrocarbon products can range from small molecules like methane, CH_4 ($n = 1$), to the medium-sized molecules ($n = 5$–8) typically found in gasoline. This chemical reaction proceeds when the carbon monoxide and hydrogen are passed over a catalyst containing iron or cobalt.

To better understand the role of the catalyst, consider a typical exothermic reaction, as shown in Figure 4.20. The potential energy of the reactants (left side) is higher than the potential energy of the products (right side) because it is an exothermic reaction. Now examine the pathways that connect the reactants and products. The green line indicates the energy changes during a reaction in the absence of a catalyst. Overall, this reaction gives off energy, but the energy initially *increases* because some bonds break (or start to break) first. The energy necessary to initiate a chemical reaction is called its **activation energy** and is indicated by the green arrow. Although energy must be expended to get the reaction started, energy is given off as the process proceeds to a lower potential energy state. Generally, reactions that occur rapidly have low activation energies; slower reactions have higher activation energies. However, there is no direct relationship between the height of the activation barrier and the net energy change in the reaction. In other words, a highly exothermic reaction can have a large or a small activation energy.

Increasing the temperature often results in increased reaction rates; when molecules have extra energy, a greater fraction of collisions can overcome the required activation energy. Sometimes, however, increasing the temperature isn't a practical solution. The blue line shows how a catalyst can provide an alternative reaction pathway and thus a lower activation energy (represented by the blue arrow), without raising the temperature.

In the Fischer–Tropsch process, strong $C{\equiv}O$ triple bonds must be broken for the reaction to proceed. Breaking this bond corresponds to an activation energy so large the reaction simply does not proceed. This is the point at which the metal catalyst

Catalysts were introduced in the context of automobile catalytic converters in Section 1.11.

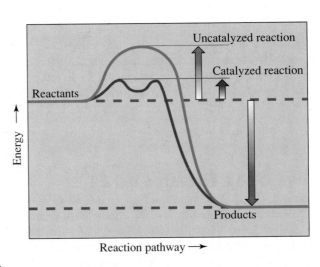

Figure 4.20

Energy–reaction pathway for the same reaction with (*blue line*) and without (*green line*) a catalyst. The green and blue arrows represent the activation energies. The red arrow represents the overall energy change for either pathway.

enters the reaction. Molecules of CO can form bonds with the metal surface, and when this happens, the C≡O bonds weaken. The hydrogen molecules also attach to the metal surface, completely breaking the H–H single bonds. The rest of the reaction proceeds quickly, producing the higher molecular weight hydrocarbons. The beauty of a catalyst is that it is not consumed and thus only small amounts of it must be used. Green chemists value catalytic reactions not only because small amounts of catalysts are employed, but also because the reaction often can be carried out at lower temperatures.

Historically, commercialization of Fischer–Tropsch technology has been limited. South Africa, a coal-rich and oil-poor nation, is the only country that synthesizes a majority of its gasoline and diesel fuel from coal. Any spike in oil prices, coupled with a plentiful domestic coal supply, may spark increased use of the Fischer–Tropsch process in other energy-hungry countries. As of 2008, China is constructing a coal-to-liquid fuels plant in Inner Mongolia. In the United States, an Australian energy corporation announced plans to build a $7 billion coal-to-liquids plant in Big Horn County, Montana, home to the Crow Tribe. The deposits there are estimated to contain almost 9 billion tons of coal.

Coal, whether solid or converted to liquid fuels, still burns to produce CO_2. Recent work by the National Renewable Energy Laboratory indicates that greenhouse gas emissions over the entire fuel cycle for producing coal-based liquid fuels are nearly twice as high as their petroleum-based equivalent. Clearly, we need to search for fuels to replace coal.

4.9 | Biofuels I—Ethanol

As we mentioned in the opening of this chapter, the current rate of fossil fuel combustion is *not* sustainable. In Chapter 0, we also quoted Donella Meadows, a biophysicist and the founder of the Sustainability Institute: "A sustainable society is one that is far-seeing enough, flexible enough, and wise enough not to undermine either its physical or social systems of support." Clearly, burning fossil fuels today is undermining our support systems.

What are our options? Many people today believe that moving to a more sustainable energy future may be possible with the increased use of **biofuels,** the generic term for renewable fuels derived from plant matter such as trees, grasses, agricultural crops, or other biological material. Compared with fossil fuels, burning biofuels should release less net CO_2, because the plants from which biofuels are grown had absorbed CO_2 from the atmosphere through photosynthesis. This statement assumes, however, that the fuels burned to grow the crop did not cancel out this net benefit.

The most common biofuel, wood, is in insufficient supply to meet our energy demands. Cutting down trees for fuel also destroys effective absorbers of carbon dioxide. Instead of relying on direct combustion of wood or other biomass, scientists currently are eyeing liquid fuels, including ethanol made via fermentation of grains, and biodiesel made from different plant oils (Figure 4.21).

Since ancient times, people have known how to produce ethanol from the fermentation of the starches and sugars found in grains. Admittedly, though, this fermentation was to brew alcoholic beverages rather than to fuel auto engines. In early human history, corn was not the grain of choice. As we will see in Chapter 12 on genetic engineering, the people of the New World bred corn from a wild strain. Those living in other continents brewed ethanol from other grains, including rice and barley.

Today, however, we ferment corn in the United States to produce fuel grade ethanol. The first step involves making a "soup" of corn kernels and water. The second step uses enzymes to catalyze the breakdown of the large starch molecules into individual glucose molecules. In the third step, yeast cells take over by releasing different enzymes that catalyze the conversion of glucose to ethanol.

Donella Meadows 1941–2001

In many regions of the planet, entire ecosystems are being lost because people burn wood for cooking and heat. Section 3.5 discussed how deforestation releases carbon dioxide, a greenhouse gas.

Biofuels for automobiles are not new. Henry Ford envisioned ethanol as the fuel of choice for the Model T, and Rudolph Diesel ran the first of his engines on peanut oil.

Enzymes (biological catalysts) are introduced in Section 10.4. Further examples are given in Chapters 11 and 12.

$$C_6H_{12}O_6 \longrightarrow 2\ C_2H_5OH + 2\ CO_2 \qquad \textbf{[4.12]}$$
$$\text{glucose} \qquad\qquad \text{ethanol}$$

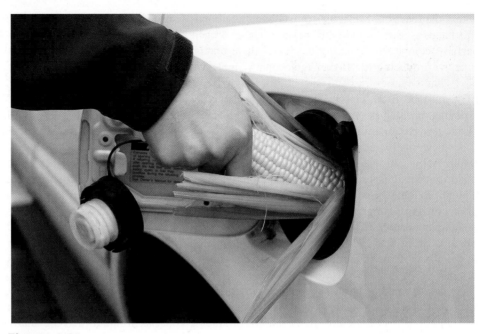

Figure 4.21

An advertisement for renewable fuels.

The distillation of ethanol works on the same principles as that of crude oil. Both distillations separate components by their boiling points. The more volatile ethanol boils first, allowing it to be separated.

The result is essentially "beer minus the hops," as the yeast cells die when the alcohol content reaches about 10%. The final step is to distill the mixture to separate, concentrate, and purify the ethanol.

Most of the ethanol produced in the United States is blended with gasoline to make "gasohol." Usually the mixture contains 10% ethanol, one that can be used in standard automobile engines (Figure 4.22a). As mentioned earlier, the oxygenated blends reduce emissions that produce ozone in urban areas. Significant interest (and investment) also exists in producing cars and trucks that run on a higher percent of ethanol. For example, of the 13 million vehicles in Brazil, more than 4 million use the pure ethanol produced from fermented sugarcane. The remainder operate on a mixture of ethanol and gasoline. As of 2007, more than 6 million flexible fuel vehicles (FFVs) on the road in the United States can use E85 (85% ethanol and 15% gasoline), gasoline, or any mixture of gasoline and E85 (Figure 4.22b).

(a)   (b)

Figure 4.22

Ethanol can be blended with gasoline to make **(a)** Gasohol, 10% ethanol, 90% gasoline or **(b)** E85, 85% ethanol, 15% gasoline.

Now let's do the energy analysis. First, recall that ethanol releases less energy per gram when burned than does a hydrocarbon mixture such as gasoline: 29.7 kJ/g (C_2H_5OH) versus 47.8 kJ/g (C_8H_{18}). Ethanol releases less energy because it contains oxygen (see Figure 4.16).

$$C_2H_5OH(l) + 3\ O_2(g) \longrightarrow 2\ CO_2(g) + 3\ H_2O(l) + 1367\ kJ \qquad \textbf{[4.13]}$$

Don't confuse gas mileage with the octane rating. The octane rating relates to how smoothly the fuel burns in the engine rather than to the energy content of the fuel. The octane rating of gasohol is higher than that of gasoline; the gas mileage is typically lower with an alcohol blend.

Your Turn 4.20   Ethanol and Your Gas Mileage

Even though E85 (85% ethanol and 15% gasoline) has a significantly higher octane rating than regular or even premium gasoline, the energy content per gallon is less. This translates into fewer miles per gallon. Use the values in Figure 4.16 to estimate the percent decrease in gas mileage when using E85 instead of "regular" gasoline.

Consider This 4.21   Ethanol in Brazil

Brazil derives almost all of its automotive fuel from fermented sugarcane. Use the resources of the web to learn why ethanol is so much more cost-effective in Brazil than in other countries.

Growing our fuel is far from a magic bullet. Scientists and citizens alike are questioning the sustainability of ethanol production, both short term and long. Recall the Triple Bottom Line: Healthy economies, healthy communities, and healthy ecosystems. To give you a better sense of how sustainable ethanol production might be, we now dive into the details.

We begin with the economic bottom line. With the crude oil prices of 2009, it was still more expensive to produce a gallon of ethanol than a gallon of gasoline. So why the booming corn ethanol market? The answer varies by region. For example, in 2007, the U.S. government provided more than $3 billion dollars in tax credits to ethanol producers. This is one factor that encouraged the use of this fuel.

In terms of other energy costs, the good news is that the Sun freely provides the energy for plants to grow. The bad news, however, is that growing corn requires additional energy inputs. Planting, cultivating, and harvesting all require energy. The same is true for producing and applying the fertilizers, manufacturing and maintaining the necessary farm equipment, and distilling the alcohol from the fermented grains. Currently, this energy is supplied by burning fossil fuels at a significant monetary cost and with significant carbon dioxide emissions. The overall energy cost for corn ethanol is difficult to quantify. Some studies estimate that for every joule put into ethanol production, 1.2 J is recovered. Others contend that the combined energy inputs outweigh the energy content of the ethanol produced.

Next, we consider the social bottom line. Many rural communities in the Midwest have benefited greatly from booming ethanol demand. Construction of a distillery not only provides jobs to local workers, but also a buyer for locally grown corn. Several communities, hit hard by the depressed viability of family farms, have experienced a resurgence thanks to demand for ethanol. There are certain drawbacks as well. Increased demand for corn leads to increased prices on many other products (especially foods); prices that everyone in that community (and others around the world) must pay.

The Triple Bottom Line was first introduced in Chapter 0.

Can genetic engineering reduce the pesticides used to grow corn? Section 12.1 explores this question.

Lastly, we turn to the environmental bottom line. Growing corn relies on the heavy use of fertilizers, herbicides, and insecticides. The manufacturing and transport of these chemicals requires the burning of fossil fuels, which in turn releases carbon dioxide. Furthermore, these chemicals, once introduced to the field, degrade the soil and water quality. Although corn growers can and do follow responsible practices, they certainly face challenges when called on to produce more corn. Increased demand for corn has also directly affected deforestation rates. In some countries, entire sections of rainforests have been cut to provide increased acreage on which to grow corn for fuel.

There are also ethical considerations, including whether valuable farmland, normally used to grow crops that feed people and animals, should be used to produce ethanol. Historically, the United States has produced a significant surplus of corn and other grains for export. That surplus is dwindling as corn is diverted to fuel production. For example, in 2007, over 25% of the U.S. corn crop was used for ethanol. Increasing nonfood demand for corn is part of the cause of the recent spike in corn prices. In turn, food prices are up worldwide, and several countries are facing possible food shortages. People around the world who count on imports of American grain are paying the price for our increasing use of ethanol.

Even with these issues, it appears likely that ethanol production will increase in the coming years. A record 6.5 billion gallons of ethanol was produced in the United States in 2007, more than double the output in 2002. In 2007, then President George W. Bush signed the Energy Independence and Security Act. A major part of this legislation was the Renewable Fuels Standard, a law that requires the use of 36 billion gallons of renewable fuels by 2022, to be made up predominantly of ethanol and biodiesel. The Department of Energy estimated that only about 12 billion gallons (10% of U.S. gasoline demand) could be met by corn ethanol.

The topic of genetic modification is explored in Section 12.6. Genetically modified bacteria may be able to efficiently convert agricultural wastes into ethanol.

To address the problems with corn ethanol, researchers are looking for ways to produce ethanol from agricultural waste products, and nonfood-based crops, and the use of marginal lands to produce these crops. **Cellulosic ethanol** is the ethanol produced from corn stalks, switchgrass, wood chips, and other materials that are nonedible by humans. Agricultural wastes are more chemically complex than the starch from corn kernels and are more difficult to break down into glucose. As of 2009, there were no commercial cellulosic ethanol plants in the United States, but the Department of Energy has commissioned six commercial-scale plants to test the technological viability of new conversion processes. For cellulosic ethanol, the supply of raw materials is significantly greater. Without taking additional food from anyone's plate, we have enough raw materials to make 60 billion gallons of ethanol a year. This would satisfy about 30% of the current U.S. gasoline needs.

As we pointed out at the start of this section, ethanol is not the only biofuel in town. The next section is devoted to one of the newcomers: biodiesel.

Consider This 4.24   Crops for Fuel

List some desirable characteristics of nonfood crops designed for ethanol production. Consideration of the Triple Bottom Line might help in your deliberations.

4.10 | Biofuels II—Biodiesel, Garbage, and Biogas

The production of biodiesel has grown dramatically during the last few years. It is unique among transportation fuels in that it can be produced economically in small batches by individual consumers (Figure 4.23). Biodiesel is primarily made from vegetable oils, but animal fats work as well. Both vegetable oils and animal fats contain compounds called triglycerides. Because of their high molar masses, triglycerides are not suitable for direct use in an auto engine. In contrast, biodiesel is. Reacting a triglyceride molecule with methanol and sodium hydroxide (as a catalyst) produces three molecules of biodiesel and one molecule of glycerol ($C_3H_8O_3$).

> Section 11.3 discusses triglycerides in the context of the fats and oils found in foods.

$$\text{a triglyceride} + 3\ CH_3OH \xrightarrow{\text{NaOH}} 3\ CH_3CH_2CH_2CH_2CH_2CH_2CH_2CH_2\overset{\overset{\textstyle O}{\|}}{C}OCH_3 + C_3H_8O_3$$
$$\text{a biodiesel molecule} \qquad\qquad \text{glycerol}$$

$$[4.14]$$

Consider This 4.25   Heat of Combustion for Biodiesel

Examine the structural formula of a biodiesel molecule. Then estimate the heat of combustion per gram of biodiesel in comparison to that for gasoline and to that for ethanol. Explain your reasoning.

Notably, biodiesel releases much more energy when burned than it costs to produce. In 1998, the U.S. Department of Energy and the U.S. Department of Agriculture undertook a life cycle study of the energy balance (energy in vs. energy out) of biodiesel.

(a)

(b)

Figure 4.23
(a) Biodiesel formulated from recycled restaurant vegetable oil by a regional vendor.
(b) B20 is a mixture of 80% petroleum diesel and 20% biodiesel.

The study concluded that for each unit of fossil fuel energy used in the entire biodiesel production cycle, 3.2 units of energy are gained when the fuel is burned; a positive energy balance of 320%.

Consider This 4.26   Origin of Biodiesel

The U.S. Department of Energy has an extensive website dedicated to biofuels. Use the link provided at the textbook's website to answer these questions.

a. How many biodiesel plants are located in your state?
b. The feedstocks for the synthesis of biodiesel differ by region. What might account for this?

Like fossil fuels, ethanol and biodiesel contain carbon and thus produce carbon dioxide when burned. But because they are derived from plants, they are more carbon-neutral fuels. By this we mean that the carbon released in combustion is at least partially offset by the carbon absorbed as the plants grew. Some of the oxygen produced by the plants is then used to burn the fuel. The production of the fuels from the raw plants does require energy (usually from fossil fuels). Thus there is a net release of CO_2 even with biofuels, though less than that of coal, petroleum, or natural gas.

One interesting facet of biodiesel synthesis is the glycerol by-product. Glycerol is an important compound used in many different consumer products. However, for every 9 pounds of biodiesel, 1 pound of glycerol is produced, which has resulted in a glut of glycerol on the market. In 2006, Galen Suppes and coworkers at the University of Missouri earned a Presidential Green Chemistry Challenge Award for a process to convert glycerol to propylene glycol.

$$
\begin{array}{c}
\underset{\text{glycerol}}{
\begin{array}{c}
\text{H}\ \ \text{H}\ \ \text{H} \\
|\ \ \ |\ \ \ | \\
\text{H}-\text{C}-\text{C}-\text{C}-\text{H} \\
|\ \ \ |\ \ \ | \\
\text{OH}\ \text{OH}\ \text{OH}
\end{array}}
\xrightarrow{\text{copper catalyst}}
\underset{\text{propylene glycol}}{
\begin{array}{c}
\text{H}\ \ \text{H}\ \ \text{H} \\
|\ \ \ |\ \ \ | \\
\text{H}-\text{C}-\text{C}-\text{C}-\text{H} \\
|\ \ \ |\ \ \ | \\
\text{OH}\ \text{OH}\ \text{H}
\end{array}}
\end{array}
\qquad [4.15]
$$

The FDA has approved propylene glycol as a food additive in alcoholic beverages, ice cream, seasonings, and flavorings. The compound also finds uses in cosmetics, pharmaceuticals, paints, and detergents. Conversion of the glycerol into value-added products lowers the cost of biodiesel production, making it more competitive with petroleum-derived diesel fuel. Here, the production of propylene glycol is from a renewable resource, whereas other methods of production require petroleum as a feedstock. Biomass will become an increasingly important source of organic compounds as the supply of crude oil begins to decline.

Section 1.13 cited propylene glycol as an antifreeze used in water-based interior paints. However, since the compound is volatile, the new low VOC paints no longer contain this.

Another energy source that is cheap, always present in abundant supply, and always being renewed is garbage. Other than in a movie, no one is likely to design a car that will run on orange peels and coffee grounds. However, approximately 90 waste-to-energy power plants in the United States generate electricity from garbage. One of these, pictured in Figure 4.24, is the Hennepin Energy Resource Recovery Facility in Minneapolis, Minnesota. Hennepin County produces about 1 million tons of solid waste each year. One truckload of garbage, or about 27,000 pounds, generates the same quantity of energy as 21 barrels of oil. In 2008, the plant converted 365,000 tons of garbage into electricity to provide power to the equivalent of about 25,000 homes. In addition, over 11,000 tons of iron-containing metals are recovered from the garbage and recycled. Elk River Resource Recovery Facility, the second in Hennepin County, converts another 235,000 tons of garbage to electricity.

This "resource recovery" approach, as it is sometimes called, simultaneously addresses two major problems: our growing need for energy and our growing mountains

Figure 4.24
Hennepin County Resource Recovery Facility, a garbage-burning power plant.

of garbage. When used as a power plant fuel, the great majority of the trash is converted to carbon dioxide and water; no supplementary fuel is needed. The unburned residue is disposed of in landfills, but it represents only about 10% of the volume of the original refuse. Although some citizens have expressed concern about gaseous emissions from garbage incinerators, the incinerator's stack effluent is carefully monitored and must be maintained within established limits. The energy needs of over 500 million people worldwide, mostly in Europe and Japan, are supplied by power plants running on garbage. In the United States, waste-to-energy plants supply power to about 2.3 million households.

Consider This 4.27   Building a Waste-Burning Plant

Imagine you were the administrator of a city of a million residents charged with drafting a proposal to your city council outlining the pros and cons of a waste-burning plant. Use the link provided at the textbook's website for the Hennepin Energy Recovery Center in Minnesota as a model to collect information and examples. Issues such as resource recovery, pollution, and the concerns of local residents might be helpful starting points.

Biogas generators provide another good example of using waste as an energy source. In the absence of oxygen, certain strains of bacteria are able to decompose organic matter. The bacteria produce a fuel composed of about 60% CH_4, 35% CO_2 (by volume), and small amounts of water, hydrogen sulfide, and carbon monoxide. The biogas can be used for cooking, heating, lighting, refrigeration, and generating electricity. Both sewage and manure can be used as the source material. The leftover waste also can be used as excellent compost. The technology lends itself very well to small-scale applications. The daily manure from one or two cows can generate enough biogas to meet most of the cooking and lighting needs of a farm family. For example, more than 25 million rural households in China were using biogas in 2008. Within a decade, this number is expected to more than double.

Section 3.8 discussed methane as a potent greenhouse gas. Methane capture from landfills equates to less methane in the atmosphere.

4.11 | The Way Forward

One thing is certain. Future generations are going to need *more* energy than we do today. Figure 4.25 displays the trends in energy consumption for the past several decades together with projections for the next 15 years. The data on this figure are split into three different economies. The mature market economies include the United States, Canada, and Western Europe. The emerging economies include Central and South America, India, China, and Africa. And the transitional economies include Eastern Europe and the former Soviet Union. One piece of the energy puzzle *not* shown in Figure 4.25 is how energy use varies across the globe. Roughly 25% of the world's energy supply is consumed by the 5% of the world's population that lives in North America.

Why do all three future projections slope upward? One reason is increasing population. More people will require more energy to go about their daily lives. Another reason is that energy is a primary driver of industrial and economic progress. Although the future energy use in each type of economy is projected to increase, the 4.5% rate of increase in the emerging economies—dominated by the tremendous growth in China and India—far exceeds that of either the established (1.2%) or transitional economies (1.7%).

Energy consumption and gross domestic product (GDP) strongly correlate. However, the GDP does not tell the full story of the quality of life for people. A more complete picture is given by the Human Development Index (HDI), one that includes GDP but also takes into account two other factors. One is life expectancy, a general measure of the health of a population. The other relates to education and includes the adult literacy rate as well as amount of schooling. For convenience, the HDI values are scaled to fall between 0 and 1.0. Norway and Iceland both fall close to the upper bound of 1.0 and thus have very high qualities of life.

Examine Figure 4.26 to see the HDI is the *y*-axis value. Note that nations are color-coded:

- very low quality of life (Haiti, Pakistan, and many African nations)
- average quality of life (Egypt, most of Central America, states of the former Soviet Union)
- very high quality of life (much of Europe, Japan, Israel, Saudi Arabia, Australia)

These categories roughly correlate with a nation's being underdeveloped, developing, or developed.

In Figure 4.26, the *x*-axis is energy consumption. The unit is the kilograms of oil equivalent (kgoe) per year, analogous to the million metric tons of oil equivalent that

> GDP is a measure of the total values of all goods as services produced by a country during a specific period of time.

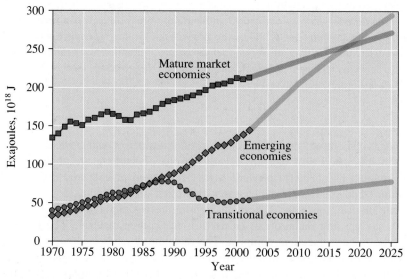

Figure 4.25

The history (data points) and projected future (solid lines) of energy consumption worldwide.

Source: Annual Energy Review 2005, Energy Information Agency, U.S. Department of Energy.

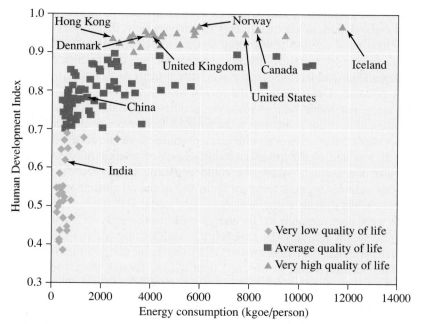

Figure 4.26

The relationship between Human Development Index and energy consumption. Energy consumption is reported in kilograms of oil equivalent (kgoe) per person. 1 kgoe = 4.4×10^{10} J or the approximate energy released in burning a kilogram of oil.

Source: United Nations Human Development Report 2007/2008.

we encountered earlier with Figure 4.9. As you might expect, energy consumption correlates to some degree with the quality of life. All of the countries whose people have a low quality of life and many of those with an average quality of life have low per capita energy consumptions, many less than 1000 kgoe per person.

However, the story of energy consumption does not end here. Look again at Figure 4.26 and you will see that countries with high standards of living vary greatly in the amount of energy they consume per person. For example, Norway has the highest HDI (tied with Iceland), but a person in Norway consumes considerably less energy than does one in the United States. What does this mean? Most simply, it means that some countries are much more energy-efficient than others.

A related question concerns the minimum amount of energy per capita required to produce a high standard of living. Although there is no definitive value, from Figure 4.26 we can ballpark it in the range of 2000–2500 kgoe per person. Hong Kong currently falls in this range. Energy consumption beyond the value of 2500 kgoe per person results in relatively small and possibly insignificant changes in the quality of living. To those living in countries with higher per capita energy consumption, this is good news. It brings some measure of hope that many in the world can maintain a certain quality of life while decreasing energy use.

Consider This 4.28   Not So Fast!

It is one thing to say that we could simply produce more energy, but it is another to say that we could do so sustainably. In this chapter, you learned about the combustion of fossil fuels.

a. Why is burning coal not sustainable? List three reasons.
b. In terms of sustainability, do natural gas and petroleum stack up any better?

Do we have enough energy available so that everybody on the planet can achieve a high standard of living, say, to the 0.9 HDI level? This question may be on your mind; actually, many people are asking it. If you do the math, our world energy production falls short of the math. Furthermore, as the world population increases, so does the total amount of energy needed.

Given that global energy resources are limited, nations face a dual challenge. First, they must make choices to use their energy resources wisely; that is, they must conserve. Secondly, they must develop new energy sources, especially ones that come closer to meeting the criteria for sustainability. For example, China's construction of the immense Three Gorges Dam and hydroelectric power station helps to meet skyrocketing electrical energy demands. Denmark leads the way in wind power. Other countries are exploring other clean, renewable sources including geothermal, tidal, and solar energy.

Conservation is not as easy as it may sound. Remember from our earlier discussions in Section 4.2 that, even with efficient technologies, about two thirds of the energy produced is lost due to the inherent inefficiencies of energy transformations. Let's consider just one example where we could conserve: transportation. In the United States, motor vehicles account for more than 70% of the oil consumption and for about one fifth of carbon dioxide emissions.

One legislative response to the energy crisis in the early 1970s was the Energy Policy Conservation Act of 1975. One lasting result of that act was the establishment of Corporate Average Fuel Economy (CAFE) standards that set fuel economy limits for each manufacturer's fleet of cars and light trucks. Though the United States is the leader in many aspects of energy, fuel economy is not one of them. The average European car gets about 40 miles per gallon (mpg), and those in Japan get about 45 mpg. In contrast, automobiles in the United States average around 25 mpg.

Section 8.6 describes photovoltaic cells, a means of capturing the energy from the Sun.

Skeptical Chemist 4.29   Cars and CO_2

A news reporter asserts that the average car in the United States emits 6 to 9 tons of CO_2 yearly. What do you think, is this a reasonable estimate? Do a calculation to evaluate the reporter's statement.

The era of stagnant CAFE standards for U.S. cars and trucks is coming to an end. In addition to mandating significant increases in biofuel production (as noted earlier), the Energy Independence and Security Act of 2007 also raises CAFE standards to 35.5 mpg by 2020. In his first four months in office, President Obama fast-tracked the transition by requiring the new standards be in place nationwide by 2016. Critics contend that to meet the stringent standards, vehicles will cost more and will need to be significantly lighter and therefore more dangerous in accidents. Improvements in safety technology, as well as development of lighter alloys and composites, will likely mean that the vehicles of the next decade will be both economical and safe.

The fact remains that the automobile is an energy-intensive means of transportation and an inefficient one at that. A mass transit system is far more economical, provided it is heavily used. In Japan, 47% of travel is by public transportation, compared with only 6% in the United States. Of course, Japan is a compact country with a high population density. Although the entirety of North America is not ideally suited to mass transit, such systems could be employed in population-dense regions. However, one also must reckon with the long love affair between Americans and their automobiles.

Improving end-use efficiency is another way to save energy and energy resources. In a *Scientific American* article published in September 2005, Amory B. Lovins asserted, "With the help of efficiency improvements and competitive renewable energy sources, the U.S. can phase out oil use by 2050." Driving this push for efficiency is simple economics, as individuals and corporations realize that it is much cheaper to conserve fossil fuels than it is to burn them.

To be sure, the future is filled with more questions than answers. Without doubt, making sustainable development a reality will be one of the greatest challenges we humans have ever faced. Often we feel pessimistic about our ability to affect change, especially on a global scale; recycling an aluminum can here and there will not lead to a sustainable future. The fact is, our actions don't just leave our footprints in our neighborhood anymore, they leave global impressions. A few slightly smaller shoes won't make a difference, but a few billion most certainly will. And as always, with challenge comes opportunity. To end the chapter on a note of optimism, we invite you to complete the following activity.

Consider This 4.30   **A Sustainable Future**

In 2002, then Secretary General of the United Nations Kofi Annan called sustainability, "...an exceptional opportunity—economically to build markets, socially to bring people in from the margins, and politically to reduce tensions over resources that could give every man and woman a voice and a choice in deciding their own future." Expand on the Secretary General's remarks, giving some specifics for each area mentioned.

Conclusion

Fire! To early humans, fire was a source of security to ward off animals and brought the ability to cook food and minimize the spread of certain diseases. But it did more than that. Fire allowed people to venture into colder regions of the planet. The nightly campfire also became an important social vehicle, a place to come together and share stories as a community.

Today, combustion still is central to our human community. We use it daily to cook; heat or cool our dwellings; transport goods and crops; and to travel the roads, rails, and skies of our planet. As we have seen in this chapter, the process of combustion converts matter and energy into less useful forms. For example, when we burn a hydrocarbon, we dissipate the potential energy it contains in the form of heat. The products of combustion—carbon dioxide and water—are not usable as fuels. One of these products, CO_2, is a greenhouse gas linked to global warming. Few chemical reactions have as far-sweeping consequences as does our combustion of fuels.

Given the realities of combustion, we have no option. We must expand and diversify our energy sources. Biofuels like ethanol and biodiesel likely will be a part of a sustainable energy future, but they will not be enough. We discuss nuclear energy in Chapter 7 and solar energy in Chapter 8 as other possible ways of satisfying our ever-increasing appetite for energy. The next energy crisis, if not already here, will be fundamentally different from those of the past. There is little doubt that overcoming the challenges it will bring will require fundamental changes. As individuals and as a society, we must decide what sacrifices we are willing to make in speed, comfort, and convenience for the sake of our dwindling fuel supplies and the good of the planet.

Recall the definition of sustainability from Chapter 0, "meeting the needs of the present without compromising the ability of future generations to meet their own needs." We must continue to ask if we are borrowing too much from future generations. One thing is clear: the sooner we honestly examine our options, our priorities, and our will, the better. Sustainable energy, chemistry, and society are complexly intertwined, and this chapter has been an attempt to untangle them.

Chapter Summary

Having studied this chapter, you should be able to:

- Name the fossil fuels, describe the characteristics of each, and compare them in terms of how cleanly they burn and how much energy they produce (4.1–4.7)

- Evaluate fossil fuels as a sustainable source of energy (4.1–4.7)

- Correlate the process of electricity generation from fossil fuels with the steps in energy transformation (4.1)

- Compare and contrast kinetic energy and potential energy, both on the macroscopic and molecular level (4.1)

- Apply the concept of entropy to explain the second law of thermodynamics (4.2)

- Describe "clean coal technologies" and comment on their viability, long-term and short (4.3)

- Explain how and why petroleum is refined (4.4)

- List the different fractions obtained by distilling petroleum. Compare and contrast these in terms of their chemical composition, chemical properties, boiling points, and end uses (4.4)
- Apply the terms *endothermic* and *exothermic* to chemical reactions based on calculations or chemical intuition (4.5)
- Calculate energy changes in reactions using bond energies (4.6)
- Assess how gasoline additives affect automobile efficiencies, tailpipe emissions, human health, and the environment (4.7)
- Understand activation energy and how it relates to the rates of reaction (4.8)

- Compare and contrast the production and uses of ethanol and biodiesel as fuels (4.9–4.10)
- Compare and contrast biofuels with gasoline in terms of chemical composition, energy released on combustion, and energy required to produce (4.9–4.10)
- Correlate energy use to population, environmental pollution, and economic expansion (4.10)
- Take an informed stand on various energy conservation measures, including to what extent they are likely to produce energy savings (4.11)
- Evaluate news articles on energy sustainability measures and judge their accuracy (4.11)

Questions

Emphasizing Essentials

1. a. List three fossil fuels.

    b. What is the origin of fossil fuels?

    c. Are fossil fuels renewable resources?

2. Explain each energy transformation step that takes place when coal is burned in a power plant.

3. Compare the processes of combustion and photosynthesis.

4. Describe how grades of coal differ and the significance of these differences.

5. A coal-burning power plant generates electrical power at a rate of 500 megawatts (MW) or 5.00×10^8 J/s. The plant has an overall efficiency of 37.5% (0.375) for the conversion of heat to electricity.

    a. Calculate the electrical energy (in joules) generated in 1 year of operation and the heat energy used for that purpose.

    b. Assuming the power plant burns coal that releases 30 kJ/g, calculate the mass of coal (in grams and metric tons) that is burned in 1 year of operation. *Hint:* 1 metric ton = 1×10^3 kg = 1×10^6 g.

6. The complete combustion of methane is given in equation 4.3.

    a. By analogy, write a chemical equation for the combustion of ethane, C_2H_6.

    b. Rewrite this equation using Lewis structures.

    c. The heat of combustion for ethane, C_2H_6, is 47.8 kJ/g. How much heat is produced if 1.0 mol of ethane undergoes complete combustion?

7. List three major drawbacks of using coal as a fuel.

8. Mercury (Hg) is a contaminant of coal, ranging from 50–200 ppb. Consider the amount of coal burned by the power plant in Your Turn 4.8. Calculate tons of mercury in the coal based on the lower (50 ppb) and higher (200 ppb) concentrations.

9. An energy consumption of 650,000 kcal per person per day is equivalent to an annual consumption of 65 barrels of oil or 16 tons of coal. Calculate the amount of energy available in kilocalories for each of these.

    a. one barrel of oil

    b. 1 gallon of oil (42 gallons per barrel)

    c. 1 ton of coal

    d. 1 pound of coal (2000 pounds per ton)

10. Use the information in the previous question to find the ratio of the quantity of energy available in 1 pound of coal to that in 1 pound of oil. *Hint:* One pound of oil has a volume of 0.56 quart.

11. Consider the data for these three hydrocarbons.

Compound, Formula	Melting Point (°C)	Boiling Point (°C)
pentane, C_5H_{12}	−130	36
triacontane, $C_{30}H_{62}$	66	450
propane, C_3H_8	−188	−42

Predict the physical state (solid, liquid, or gas) of each at room temperature.

12. a. Write the chemical equation for the complete combustion of *n*-heptane, C_7H_{16}.

    b. The heat of combustion for *n*-heptane is 4817 kJ/mol. How much heat is released if 250 kg of *n*-heptane burns completely?

13. Figure 4.17 shows energy differences for the combustion of hydrogen, an exothermic chemical reaction. The combination of nitrogen gas and oxygen gas to form nitrogen monoxide is an example of an endothermic reaction:

$$N_2(g) + O_2(g) \longrightarrow 2\ NO(g)$$

The bond energy in NO is 607 kJ/mol. Sketch an energy diagram for this reaction, and calculate the overall energy change.

14. A single serving bag of Granny Goose Hawaiian Style Potato Chips has 70 Cal. Assuming that all of the energy

from eating these chips goes toward keeping your heart beating, how long can these chips sustain a heartbeat of 80 beats per minute? *Note:* 1 kcal = 4.184 kJ, and each human heart beat requires approximately 1 J of energy.

15. A 12-oz serving of a soft drink has an energy equivalent of 92 kcal.

 a. In kilojoules, what is the energy released when metabolizing this beverage?

 b. Assume that you use this energy to lift concrete blocks that weigh 10 kg each. How many blocks could you lift to a height of 3.0 m with the energy calculated in part **a**?

16. One way to produce ethanol for use as a gasoline additive is the reaction of water vapor with ethylene:

$$CH_2CH_2(g) + H_2O(g) \longrightarrow CH_3CH_2OH(l)$$

 a. Rewrite this equation using Lewis structures.

 b. Use the bond energies in Table 4.4 to calculate the energy change for this reaction. Is the reaction endothermic or exothermic?

17. From personal experience, state whether these processes are endothermic or exothermic.

 a. A charcoal briquette burns.

 b. Water evaporates from your skin.

 c. Ice melts.

18. Use the bond energies in Table 4.4 to explain why:

 a. chlorofluorocarbons, CFCs, are so stable.

 b. it takes less energy to release Cl atoms than F atoms from CFCs.

19. Use the bond energies in Table 4.4 to calculate the energy changes associated with each of these reactions. Label each reaction as endothermic or exothermic. *Hint:* Draw Lewis structures of the reactants and products to determine the number and kinds of bonds.

 a. $N_2(g) + 3\ H_2(g) \longrightarrow 2\ NH_3(g)$

 b. $2\ C_5H_{12}(g) + 11\ O_2(g) \longrightarrow 10\ CO(g) + 12\ H_2O(l)$

 c. $H_2(g) + Cl_2(g) \longrightarrow 2\ HCl(g)$

20. Use the bond energies in Table 4.4 to calculate the energy changes associated with each of these reactions. Label each reaction as endothermic or exothermic.

 a. $2\ H_2(g) + CO(g) \longrightarrow CH_3OH(g)$

 b. $H_2(g) + O_2(g) \longrightarrow H_2O_2(g)$

 c. $2\ BrCl(g) \longrightarrow Br_2(g) + Cl_2(g)$

21. Use Figure 4.6 to compare the sources of U.S. energy consumption. Arrange the sources in order of decreasing percentage and comment on the relative rankings.

22. The structural formulas of straight-chain (normal) alkanes containing 1 to 8 carbon atoms are given in Table 4.2.

 a. Draw the structural formula for *n*-decane, $C_{10}H_{22}$.

 b. Predict the chemical formula for *n*-nonane (9 carbon atoms) and for *n*-dodecane (12 carbon atoms).

 c. The structural formulas shown are two-dimensional. Use the bond angle information in Chapter 3 to predict the C–C–C and H–C–H bond angles in *n*-decane.

23. Consider this equation representing the process of cracking.

$$C_{16}H_{34} \longrightarrow C_5H_{12} + C_{11}H_{22}$$

 a. Which bonds are broken and which bonds are formed in this reaction? Use Lewis structures to help answer this question.

 b. Use the information from part **a** and Table 4.4 to calculate the energy change during this cracking reaction.

24. Here is a ball-and-stick representation for one isomer of butane (C_4H_{10}).

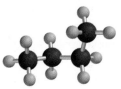

 a. Draw the Lewis structure for this isomer.

 b. Draw Lewis structures for all other isomers. *Hint:* Watch for duplications!

25. A premium gasoline available at most stations has an octane rating of 92. What does that tell you about:

 a. the knocking characteristics of this gasoline?

 b. whether the fuel contains oxygenates?

26. Figure 4.16 gives the energy content of several fuels in kilojoules per gram (kJ/g). Calculate the energy content in kilojoules per mole (kJ/mol) for each. How does the chemical composition of a fuel relate to its energy content? Visit *Figures Alive!* at the textbook's website for related activities.

Concentrating on Concepts

27. How might you explain the difference between temperature and heat to a friend? Use some practical, everyday examples.

28. Write a response to this statement: "Because of the first law of thermodynamics, there can never be an energy crisis."

29. A friend tells you that hydrocarbon fuels containing larger molecules liberate more heat than those with smaller ones.

 a. Use these data, together with appropriate calculations, to discuss the merits of this statement.

Hydrocarbon	Heat of Combustion
octane, C_8H_{18}	5070 kJ/mol
butane, C_4H_{10}	2658 kJ/mol

 b. Based on your answer to part **a**, do you expect the heat of combustion per gram of candle wax, $C_{25}H_{52}$, to be more or less than that of octane? Do you expect the molar heat of combustion of candle wax to be more or less than that of octane? Justify your predictions.

30. Halons are synthetic chemicals similar to CFCs but include bromine. Although halons are excellent materials for fire fighting, they more effectively deplete ozone than CFCs. Here is the Lewis structure for halon-1211.

   a. Which bond in this compound is broken most easily? How is that related to the ability of this compound to deplete ozone?

   b. The compound C_2HClF_4 is being considered as a replacement for halons in fire extinguishers. Draw its Lewis structure and identify the bond broken most easily.

31. The Fischer–Tropsch conversion of hydrogen and carbon monoxide into hydrocarbons and water was given in equation 4.11:

$$n\ CO + (2n + 1)\ H_2 \longrightarrow C_nH_{2n+2} + n\ H_2O$$

   a. Determine the heat evolved by this reaction when $n = 1$.

   b. Without doing a calculation, do you think that more or less energy is given off per mole in the formation of larger hydrocarbons ($n > 1$)? Explain your reasoning.

32. During petroleum distillation, kerosene and hydrocarbons with 12–18 carbons used for diesel fuel condense at position C marked on this diagram.

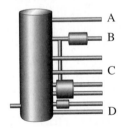

   a. Separating hydrocarbons by distillation depends on differences in a specific physical property. Which one?

   b. How does the number of carbon atoms in the hydrocarbon molecules separated at A, B, and D compare with those separated at position C? Explain your prediction.

   c. How do the uses of the hydrocarbons separated at A, B, and D differ from those separated at position C? Explain your reasoning.

33. Explain why cracking is necessary in the refinement of crude oil.

34. Consider equation 4.15. Are glycerol and propylene glycol isomers? Explain.

35. Catalysts speed up cracking reactions in oil refining and allow them to be carried out at lower temperatures. What other examples of catalysts were given in the first three chapters of this text?

36. Octane ratings of several substances are listed in Table 4.5.

   a. What evidence can you give that the octane rating is or is not a measure of the energy content of a gasoline?

   b. Octane ratings are measures of a fuel's ability to minimize or prevent engine knocking. Why is the prevention of knocking important?

   c. Why are higher octane gasolines more expensive than lower octane gasoline?

37. Section 4.8 states that both n-octane and iso-octane have essentially the same heat of combustion. How is that possible if they have different structures?

38. At present, the United States is dependent on foreign oil. One possible consequence is periodic gasoline shortages. These shortages affect more than individual motorists. List two ways in which a gasoline shortage could affect your life.

39. It was stated in the text that emissions of some pollutants are lower using biodiesel than using petroleum diesel. Based on the methods of production for each fuel, explain the lower amounts of

   a. sulfur dioxide emissions.   b. CO emissions.

40. These three structures have the chemical formula C_8H_{18}. The hydrogen atoms and C–H bonds have been omitted for simplicity.

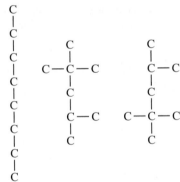

structure 1   structure 2   structure 3

   a. Redraw the structures to show the missing hydrogen atoms. *Hint:* Check that all structures have 18 H atoms.

   b. Which (if any) of these structures are identical?

   c. Obtain a model kit and build one of these molecules. What are the C–C–C bond angles?

   d. Draw the structural formulas of two additional isomers of C_8H_{18}.

   e. If you were to build models of these two isomers, would the C–C–C bond angles be the same as that in part c? Explain.

41. Here is a ball-and-stick model of ethanol, C_2H_6O. Another compound, dimethyl ether, has this same chemical formula. Draw the Lewis structure of dimethyl ether. *Hint:* Remember to follow the octet rule.

42. Describe how the growth in oxygenated gasolines relates to each of these.

    a. restrictions on the use of lead in gasoline

    b. federal and state air quality regulations

43. Compare the energy released on combustion of 1 gallon of ethanol and 1 gallon of gasoline. Assume gasoline is pure octane (C_8H_{18}). Explain the difference.

44. Your neighbor is shopping for a new family vehicle. The salesperson identified a van of interest as a flexible fuel vehicle (FFV).

    a. Explain what is meant by FFV to your neighbor.

    b. What is E85 fuel?

    c. Would your neighbor and his family be particularly interested in using E85 fuel depending on what region of the country they live?

45. The concept of entropy and probability is used in games like poker. Describe how the rank of hands (from a simple high card to a royal flush) is related to entropy and probability.

46. Bond energies such as those in Table 4.4 are sometimes found by "working backward" from heats of reaction. A reaction is carried out, and the heat absorbed or evolved is measured. From this value and known bond energies, other bond energies can be calculated. For example, the energy change associated with the combustion of formaldehyde (H_2CO) is -465 kJ/mol.

$$H_2CO(g) + O_2(g) \longrightarrow CO_2(g) + H_2O(g)$$

Use this information and the values found in Table 4.4 to calculate the energy of the C=O double bond in formaldehyde. Compare your answer with the C=O bond energy in CO_2 and speculate on why there is a difference.

Exploring Extensions

47. Revisit the Six Principles of Green Chemistry found on the inside of the front cover. Which of these are met by the synthesis by Suppes of propylene glycol from glycerol? *Hint:* See equation 4.15.

48. Another claim in the *Scientific American* article by Lovins referenced in Section 4.11 was that replacing an incandescent bulb (75 W) with a compact fluorescent bulb (18 W) would save about 75% in the cost of electricity. Electricity is generally priced per kilowatt-hour (kWh). Using the price of electricity where you live, calculate how much money you would save over the life of one compact fluorescent bulb (about 10,000 hr).

49. Section 4.7 states that RFGs burn more cleanly by producing less carbon monoxide than nonoxygenated fuels. At the molecular level, what evidence supports this statement?

50. Another type of catalyst used in the combustion of fossil fuels is the catalytic converter that was discussed in Chapter 1. One of the reactions that these catalysts speed up is the conversion of NO(g) to N$_2$(g) and O$_2$(g).

    a. Draw a diagram of the energy of this reaction similar to the one shown in Figure 4.20.

    b. Why is this reaction important? *Hint:* See Sections 1.9 and 1.11.

51. Chemical explosions are *very* exothermic reactions. Describe the relative bond strengths in the reactants and products that would make for a good explosion.

52. Because the United States has large natural gas reserves, there is significant interest in developing uses for this fuel. List two advantages and two disadvantages of using natural gas to fuel vehicles.

53. You may have seen some General Motors advertisements using the slogan "Live Green by Going Yellow" for their FlexFuel vehicles that can use E85 gasoline. To what do the colors in this slogan refer?

54. China's large population has increased energy consumption as the standard of living increases.

    a. Report on China's increasing number of automobiles over the last 10 years.

    b. What evidence suggests that the increase in the number of vehicles has affected air quality? What interventions, if any, does the Chinese government have underway?

55. Quality of life and energy consumption are related as shown in Figure 4.26. What ethical considerations (if any) about their lifestyle do citizens of a country having a per capita consumption of 8000 kgoe (kilograms oil equivalent) have to the rest of the world?

56. What are the advantages and disadvantages of replacing gasoline with renewable fuels such as ethanol? Indicate your personal position on the issue and state your reasoning.

57. According to the EPA, driving a car is "a typical citizen's most polluting daily activity."

    a. Do you agree? Explain.

    b. What pollutants do cars emit? *Hint:* Information on automobile emissions provided by the EPA (together with the information in this text) can help you fully answer this question.

    c. RFGs play a role in reducing emissions. Where in the country are RFGs required? Check the current list published on the web by the EPA.

    d. Explain which emissions RFGs are supposed to lower.

58. Research the Three Gorges Dam in China. Investigate some of the major issues concerning this dam. Present your findings in a format of your choice.

59. C. P. Snow, a noted scientist and author, wrote an influential book called *The Two Cultures,* in which he stated: "The question, 'Do you know the second law of thermodynamics?' is the cultural equivalent of 'Have you read a work by Shakespeare?'" How do you react to this comparison? Discuss his remark in light of your own educational experiences.

5 Water for Life

"Water has never lost its mystery. After at least two and a half millennia of philosophical and scientific inquiry, the most vital of the world's substances remains surrounded by deep uncertainties. Without too much poetic license, we can reduce these questions to a single bare essential: What exactly is water?"

Philip Ball, in *Life's Matrix: A Biography of Water*, University of California Press, Berkeley, CA, 2001, p. 115.

Neeru, shouei, maima, aqua. In any language, water is the most abundant compound on the surface of the Earth. Look at any map and you will see that oceans, rivers, lakes, and ice cover more than 70% of our planet's surface. Recognizing the importance of water, the United Nations General Assembly proclaimed 2005–2015 as a decade for international action on water. This "Water for Life" decade addresses many themes, including the scarcity of water, sanitation, food, agriculture, and water pollution.

Indeed, *water is for life*. It plays a key role in the cycling of nutrients on our planet. The water cycle drives weather and climate and helps to shape the contours of land masses. Water is incredibly versatile, dissolving many substances and suspending others. It is the essential medium for the biochemical reactions in the cells of all living species, including humans. Your body can go weeks without food but only days without water. If the water content in your body were reduced by 2%, you would get thirsty. With a 5% water loss, you would feel fatigue and have a headache. At a 10–15% loss your muscles would become spastic and you would feel delirious. And with greater than 15% dehydration, you would die.

Water has unique properties that make life on Earth possible. It is the only common substance that you can find as a solid, a liquid, and a gas. Most solids are denser than their liquid counterparts, but ice is an exception. It is less dense than liquid water, and so it floats, allowing ecosystems in lakes to survive under the ice during winter. And because water absorbs more heat per gram than many other substances, bodies of water act as heat reservoirs. Oceans and lakes moderate extreme temperature swings. Because water has properties that are unique in supporting life, when scientists search for life on other planets, they search for water.

Take a sip from a tap, bottle, or can. Steam some vegetables. Wash some laundry. Flush a toilet. Water is part of your daily routine. You also depend on water in ways that may be less obvious. For example, it takes water to irrigate crops and to prepare them for consumption. Industrial processes also require water to produce our vast array of consumer goods.

Individually and collectively, we get water, use it for purposes that most likely dirty it, and then dispose of it without thinking where that dirty water ends up. This has been called the "flush and forget" syndrome. Natural cycles can clean water, but these processes occur over long periods of time. As a resource, fresh water is neither unlimited nor renewable fast enough to meet our burgeoning needs. We are creating dirty water faster than nature can clean it.

Consider also that water is distributed unevenly on our planet. Just as we fight wars over oil, in the future we could be fighting wars over a much more basic necessity—water. Water is a strategic resource, and its scarcity brews conflicts and raises questions of who has the right to access and use it.

In this chapter, we explore many facets of water, including how we use it, the issues related to its use, and how we might arrive at local and global water solutions. As we do so, we will keep a close eye on water and its properties. The behavior of the water molecule drives many of the phenomena we observe on this, our wet planet.

WATER FOR LIFE
2005–2015

The amount of water you need daily depends on your size, age, health, and physical activity. To stay hydrated, a rule of thumb is to drink when you feel thirsty.

A renewable natural resource can be replenished over short periods of time.

About 20% of the people in the world lack access to safe drinking water.

"Many of the wars of the 20th century were about oil, but wars of the 21st century will be over water." Ismail Serageldin, former Vice-President for Environmentally and Socially Sustainable Development at the World Bank.

Consider This 5.1   Keep a Water Log

Pick a 12-hour waking segment of your day. Log all of your activities that involve water by time and activity. Also log:

a. The role the water played in your life. For example, are you consuming it? Are you using it in some process? Is it part of your outdoor experience?
b. The source of the water, the quantity involved, and where it went afterward.
c. The extent to which you got the water dirty.

🕸 **Consider This 5.2**   **Beyond Toilets**

Flushing a toilet is just one part your daily water routine. Learn more about your daily indoor water use. A link is provided at the textbook's website.

a. What surprised you about your water use?
b. How does this information relate to your water log from the previous activity?

5.1 | The Unique Properties of Water

Clearly, water is essential to our lives. What may not be as apparent is that water has a number of unusual properties. In fact, these properties are quite peculiar and we are *very* fortunate that they are. If water were a more conventional compound, life as we know it could not exist.

Let us begin with its physical state. Water is a liquid at room temperature (about 25 °C, or 77 °F) and normal atmospheric pressure. This is surprising, because almost all other compounds with similar molar masses are gases under those conditions. Consider these three gases found in air: N_2, O_2, and CO_2. Their molar masses are 28, 32, and 44 g/mol, respectively, all greater than that of water (18 g/mol). Yet none of these are liquids!

Not only is water a liquid under these conditions, but also it has an anomalously high boiling point of 100 °C (212 °F). When water freezes, it exhibits another somewhat bizarre property—it expands. Most liquids contract when they solidify.

These and other unusual properties derive from the molecular structure of water. First, recall the chemical formula of water, H_2O. This is probably the world's most widely known bit of chemical trivia. Next, recall that water is a covalently bonded molecule with a bent shape. Figure 5.1 shows the same representations of the water molecule that we used in Chapter 3.

New to our discussion in this chapter is the fact that the electrons are not shared equally in the O–H covalent bond. Experimental evidence indicates that the O atom attracts the shared electron pair more strongly than does the H atom. In chemical language, oxygen is said to have a higher electronegativity than hydrogen. **Electronegativity** is a measure of the attraction of an atom for an electron in a chemical bond. The scale runs from about 0.7 to 4.0. The values have no units and are set relative to each other. The greater the electronegativity, the more an atom attracts the electrons in a chemical bond toward itself.

Table 5.1 shows electronegativity values for the first 18 elements. Examine it to see that:

- Fluorine and oxygen have the highest values.
- Metals such as lithium and sodium have low values.
- Values *increase* from left to right in a row of the periodic table (from metals to nonmetals) and *decrease* going down a group.

The greater the difference in electronegativity between two bonded atoms, the more polar the bond is. Accordingly, we can use electronegativity values to estimate bond polarities. For example, the electronegativity difference between oxygen and

For covalent substances, as the molar mass increases, the boiling point generally increases as well.

Revisit Sections 2.3 and 3.3 for more information about the water molecule.

Electronegativity values were developed by the quantum chemist and biochemist Linus Pauling (1901–1994).

If the electronegativity difference between two atoms is more than 1.0, the bond is considered polar. If it is greater than 2.0, the bond is considered ionic. Use this information as a guideline rather than as a rule.

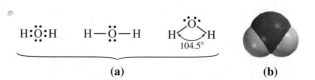

(a)   (b)

Figure 5.1

Representations of H_2O. **(a)** Lewis structures and structural formula; **(b)** Space-filling model.

Table 5.1	Electronegativity Values for Selected Elements							
1A	**2A**	**3A**	**4A**	**5A**	**6A**	**7A**	**8A**
H 2.1							He *
Li 1.0	Be 1.5	B 2.0	C 2.5	N 3.0	O 3.5	F 4.0	Ne *
Na 0.9	Mg 1.2	Al 1.5	Si 1.8	P 2.1	S 2.5	Cl 3.0	Ar *

*Noble gases rarely (if ever) bond to other elements

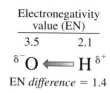

Electronegativity
value (EN)

3.5   2.1

$\delta^- O \longleftarrow H \delta^+$

EN *difference* = 1.4

Figure 5.2

Representation of the polar covalent bond between a hydrogen and oxygen atom. The electrons are pulled toward the more electronegative oxygen atom.

hydrogen is 1.4. The electrons in an O–H bond are pulled closer to the more electronegative oxygen atom. This unequal sharing results in a partial negative charge (δ^-) on the O atom and a partial positive charge (δ^+) on the H atom, as shown in Figure 5.2. An arrow is used to indicate the direction in which the electron pair is displaced. The result is a **polar covalent bond,** a covalent bond in which the electrons are not equally shared but rather are closer to the more electronegative atom. A polar covalent bond is an example of an **intramolecular force,** a force that exists within a molecule.

Compare:

■ *Intramolecular* forces are *within* molecules.

■ *Intramural* sports are played *within* a college.

Your Turn 5.3   Polar Bonds

For each pair, which is the more polar bond? In the bond you select, the electron pair is more strongly attracted to one of the atoms. Which one?
Hint: Use Table 5.1.

 a. H–F or H–Cl
 b. N–H or O–H
 c. N–O or O–S

Answer

 a. The H–F bond is more polar. The electron pair is more strongly attracted to the F atom.

We have made the case that bonds can be polar, some more than others. What about molecules? To help you predict if a molecule is polar, we offer two useful generalizations:

■ A molecule that contains only nonpolar bonds *must be* nonpolar. For example, the Cl_2 and H_2 molecules are nonpolar.
■ A molecule that contains polar covalent bonds *may or may not be* polar. The polarity depends on the geometry of the molecule.

For example, the water molecule contains two polar bonds and the molecule is polar (Figure 5.3). Each H atom carries a partial positive charge (δ^+), and the oxygen atom carries a partial negative charge (δ^-). Because the molecule is bent, overall it is polar.

Many of the unique properties of water are a consequence of its polarity. But before we continue the story of water, take a moment to complete this activity.

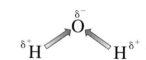

Figure 5.3

H_2O, a polar covalent molecule with polar covalent bonds.

Consider This 5.4   The Carbon Dioxide Molecule

Revisit the carbon dioxide molecule. You can find its Lewis structure in Figure 3.14.

 a. Are the covalent bonds in CO_2 polar or nonpolar? Use Table 5.1.
 b. Analogous to Figure 5.3, draw a representation for CO_2.
 c. In contrast to the H_2O molecule, the CO_2 molecule is *not* polar. Explain.

5.2 | The Role of Hydrogen Bonding

Consider what happens when two water molecules approach each other. Because opposite charges attract, a H atom (δ^+) on one of the water molecules is attracted the O atom (δ^-) on the neighboring water molecule. This is an example of an **intermolecular force,** that is, a force that occurs between molecules.

But with more than two water molecules, the story gets more complicated. Examine each H_2O molecule in Figure 5.4 and note the two H atoms and two nonbonding pairs of electrons on the O atom. These allow for multiple intermolecular attractions. This phenomenon of attracting between molecules is called "hydrogen bonding." A **hydrogen bond** is an electrostatic attraction between a H atom bonded to a highly electronegative atom (O, N, or F) and a neighboring O, N, or F atom, either in another molecule or in a different part of the same molecule. Hydrogen bonds typically are only about one tenth as strong as the covalent bonds connecting atoms *within* molecules. Also, the atoms involved in hydrogen bonding are farther apart than they are in covalent bonds. In liquid water there may be three or four hydrogen bonds per water molecule, as shown in Figure 5.4.

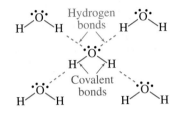

Figure 5.4

Hydrogen bonding in water (distances not to scale).

 Figures Alive! Visit the textbook's website to learn more about hydrogen bonding.

Compare:

- *Intermolecular* forces are *between* molecules.
- *Intercollegiate* sports are played *between* colleges.

Your Turn 5.5   Bonding in Water

a. Explain the dashed lines between water molecules in Figure 5.4.
b. In the same figure, label the atoms on two adjacent water molecules with δ^+ or δ^-. How do these partial charges help to explain the orientation of the molecules?
c. Are hydrogen bonds intermolecular or intramolecular forces? Explain.

Although hydrogen bonds are not as strong as covalent bonds, hydrogen bonds still are quite strong compared with other types of intermolecular forces. The boiling point of water gives us evidence for this assertion. For example, consider H_2S, a molecule that is analogous to water but does not hydrogen bond. H_2S boils at about $-60\ °C$ and so is a gas at room temperature. In contrast, water boils at $100\ °C$. Because of hydrogen bonding, water is a liquid at room temperature as well as at body temperature (about $37\ °C$). Life's very existence on our planet depends on this fact.

Sulfur is less electronegative than oxygen and nitrogen. Although H atoms bonded to N or O atoms can form hydrogen bonds, H atoms bonded to S atoms cannot.

Consider This 5.6   Bonds Within and Between Water Molecules

Are any covalent bonds broken when water boils? Explain with drawings.
Hint: Start with molecules of water in the liquid state as shown in Figure 5.4. Make a second drawing to show water in the vapor phase.

Hydrogen bonding also can help you understand why ice cubes and icebergs float. Ice is a regular array of water molecules in which every H_2O molecule is hydrogen-bonded to four others. The pattern is shown in Figure 5.5. Note the empty space in the form of hexagonal channels. When ice melts, the pattern is lost, and individual H_2O molecules can enter the open channels. As a result, the molecules in the liquid state are more closely packed than in the solid state. Thus, a volume of one cubic centimeter ($1\ cm^3$) of liquid water contains more molecules than $1\ cm^3$ of ice. Consequently, liquid water has a greater mass per cubic centimeter than ice. This is simply another way of saying that the **density,** the mass per unit volume, of liquid water is greater than that of ice.

People often confuse density with mass. For example, popcorn has a low density, and people say that a bag of popcorn feels "light." Similarly, you may hear someone say that lead is "heavy." Large pieces of lead are indeed often quite heavy, but it is more accurate to say that lead has a high density ($11.3\ g/cm^3$).

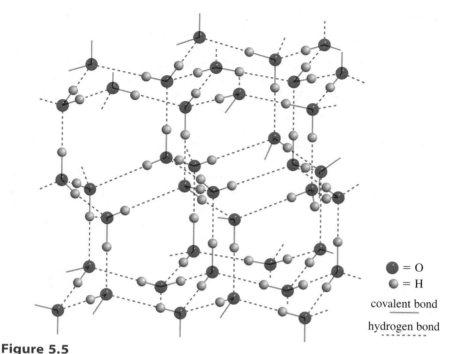

= O

= H

covalent bond

hydrogen bond

Figure 5.5

The hydrogen-bonded lattice structure of the common form of ice. Note the open channels between "layers" of water molecules that cause ice to be less dense than water.

We usually express the mass of water in grams. Expressing its volume is a bit trickier. We use either cubic centimeters or milliliters (mL)—the two units are equivalent. The density of water is 1.00 g/cm^3 at 4 °C and varies only slightly with temperature. So for convenience, we sometimes say that 1 cm^3 of water has a mass of 1 g. On the other hand, 1.00 cm^3 of ice has a mass of 0.92 g, so its density is 0.92 g/cm^3. The bottom line? The ice cubes in your favorite beverage float rather than sink.

Unlike water, most substances are denser as solids. The fact that water shows the reverse behavior means that in the winter, ice floats on lakes rather than sinking. This topsy-turvy behavior means that surface ice, often covered by snow, can act as an insulator and keep the lake water beneath from freezing solid. Aquatic plants and fish thus can live in a freshwater lake during winter. And when the ice melts in spring, the water formed sinks, helping to mix the nutrients in the freshwater ecosystem. Needless to say, water's unique behavior has implications both for the biological sciences and for life itself.

The phenomenon of hydrogen bonding is not restricted to water. It can occur in other molecules that contain covalent O–H or N–H bonds. The H bonds help stabilize the shape of large biological molecules, such as proteins and nucleic acids. For example, the double-helix structure of the DNA molecule is stabilized by hydrogen bonds between the two DNA strands. When DNA undergoes transcription, it "unzips" as the hydrogen bonds across the two strands break. Again, hydrogen bonding plays an essential role in the processes of life.

We end this section by examining one last unusual property of water, its uncommonly high capacity to absorb and release heat. **Specific heat** is the quantity of heat energy that must be absorbed to increase the temperature of 1 gram of a substance by 1 °C. The specific heat of water is 4.18 J/g · °C. This means that 4.18 J of energy is needed to raise the temperature of 1 g of liquid water by 1 °C. Conversely, 4.18 J of heat must be removed in order to cool 1 g of water by 1 °C. Water has one of the highest specific heats of any substance and is said to have a high heat capacity. Because of this, it is an exceptional coolant. When water evaporates, it can be used to carry away the excess heat in a car radiator, in a power plant, or in the human body.

For any liquid at any temperature: 1 cm^3 = 1 mL

To reiterate, water is most dense at 4 °C. At 0 °C, it is slightly less dense.

DNA molecules form hydrogen bonds between *different* strands of DNA. In contrast, proteins can form hydrogen bonds within different regions within the *same* molecule. Look for more about the structures of proteins and DNA in Chapter 12.

The joule and the calorie, units of energy, were defined in Section 4.2. The specific heat of water can also be expressed (using calories) as 1.00 cal/g · °C.

Consider This 5.7   **A Barefoot Excursion**

Have you ever walked barefoot across a carpeted floor and then onto a tile or stone floor? If not, try it and see what you notice. Based on your observation, does carpet or tile have the higher heat capacity?

    Because of water's high specific heat, large bodies of water influence regional climate. When water evaporates from seas, rivers, and lakes, heat is absorbed. By absorbing vast quantities of heat, the oceans and the droplets of water in clouds help mediate global temperatures. Since water has a higher capacity to "store" heat than the ground does, when the weather turns cold, the ground cools more quickly. Water retains more heat and is able to provide more warmth for a longer time to the areas bordering it. Such properties should be familiar to anyone who has ever lived near a large body of water.

    We have just examined some of the critical properties of water that influence life on our planet. Before we explore its ability to dissolve many different substances, we seek a broader picture of what water is used for and what issues are related to its use.

5.3 | Water Use

Just as we need clean, unpolluted air to breathe, we also need **potable water;** that is, water that is safe to drink and to cook with. Nonpotable water may contain toxic metals such as arsenic, or it may be contaminated with bacteria such as those that cause cholera. Nonetheless, water that is not safe to drink still has its uses. For example, untreated water from nearby rivers and lakes can be transported for street washing, keeping down dust, or irrigation, as shown in Figure 5.6.

Figure 5.6
Water truck at the University of Alaska, Fairbanks, with a warning that the water is not fit to drink.

Our need for potable water is but a small part of the larger picture of water use. Over 390 liters (~ 100 gallons) of water per day are required to support the lifestyle of the average U.S. citizen. In addition, it can take 4 L to process one can of fruit or vegetables, and a whopping 240,000 L is needed to produce 1 ton of steel.

Globally, we consume about 10% of water for domestic or household use, 20% for industrial needs, and 70% for agriculture. For a given region of the globe, these percentages vary. For example, water availability depends on climate, and so more water is needed for agriculture in the arid parts of our planet. Water use also can change over time. If a country shifts its economic base, its water use may increase or decrease. For example, as China shifts from agriculture to industry, its need for water will change.

Agriculture takes the biggest gulp of water. Worldwide, we grow wheat, rice, corn, soybeans, and other crops that we have come to depend on. In feedlots, water is needed to raise the beef, pork, and chicken sold in grocery stores. Table 5.2 shows the relative amounts of water needed to produce some of the foods we eat. Although it takes a smaller sip, industry also uses its share. Industrial water use includes chemical processing (such as dying textiles), cooling (running power plants), and washing (such as cleaning fibers).

Our discussion of water use falls in a larger context. A **water footprint** is an estimate (for an individual or a nation) of the amount of water required to sustain the consumption of goods and services. The total global water footprint is about 7×10^{15} liters per year (L/yr). Doing the math, this translates to 1×10^6 L/yr for every person on Earth. Alternatively, this is about half the water required to fill an Olympic-sized swimming pool. The numbers listed in Table 5.2 are examples of water footprints.

> A liter (L) contains 1000 milliliters (mL). One gallon is about 3.8 L.

Consider This 5.8   Your Own Water Footprint

Thanks to several organizations, you can now calculate your water footprint. A link is provided at the textbook's website.

a. Calculate your personal water footprint. What did you discover about your water usage?
b. Do you feel that the survey was fair in assessing the water you use? Explain.
c. Name ways you now might use water differently, knowing what you know.

Figure 5.7 ranks nations according to their water footprints. The United States has the largest per capita water footprint at 2×10^6 L/person/yr. In part, this is due to the high consumption of meat and industrial goods. Although India has the largest total consumption of any nation at 1×10^{15} L/yr, this value stems from its large population. As you can see from Figure 5.7, each U.S. citizen uses nearly two and a half times more water than does a citizen in India.

Table 5.2	Fresh Water Needed to Produce Food		
Food (1 kg)	Water (L)	Food (1 kg)	Water (L)
beef	15,500	rice	3,400
pork	4,800	soybeans	1,800
chicken	3,900	wheat	1,300
sheep	6,100	corn	900

Source: Water Footprint Network. www.waterfootprint.org.

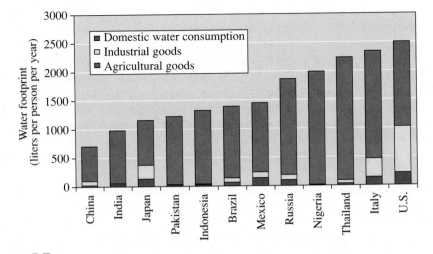

Figure 5.7

The national water footprint per capita and the contribution of different consumption categories for some selected countries.

Source: Data from A. Y. Hoekstra and A. K. Chapagain. Water footprints of nations, *Water Resource Management,* (2007) 21:35–48.

On a more personal level, our water use can be viewed through products we may encounter each day. Consider a 200-mL glass of milk. The volume of water used to *produce* this glass of milk is 2000 L, or ten thousand times the volume in one glass of milk! This includes the water to care for the cow and the water used to grow the food that it eats. It also includes the water used at a dairy farm to collect the milk and clean the equipment. You can check out the water footprints for some of your favorite beverages and consumer goods in Table 5.3.

Water footprint values are controversial and far from exact. Our intent in providing them is not to label items as either good or bad for consumption. Rather, these values are meant to increase your awareness of *how* we use water to produce goods and to provide you with a more-inclusive picture of water use. For example, on first inspection of Table 5.3, you might be tempted to forego cotton T-shirts. Cotton is indeed a thirsty crop and has been grown in arid climates in which farmers have had to import water, often using a canal system. Huge amounts of water are required to process cotton fibers into the shirts that we wear. This large water footprint for cotton can encourage us to irrigate more efficiently and to design industrial practices that conserve water.

A green chemistry solution applied to the processing of cotton is discussed in Section 5.12.

Table 5.3	Product Water Footprints	
Product	**Water Footprint (L)**
1 cup of coffee (125 mL)	140
1 apple (100 g)	70
1 orange (100 g)	50
1 glass of orange juice (200 mL)	170
1 egg (60 g)	200
1 hamburger (150 g)	2400
1 cotton T-shirt (250 g)	2700
1 computer chip (2 g)	32

Source: Adapted from A. Y. Hoekstra and A. K. Chapagain, Water footprints of nations, *Water Resource Management* (2007) 21: 35–48.

5.4 | Water Issues

Of all the water on Earth, amazingly only 3% is fresh water, as shown in Figure 5.8. About 68% of this fresh water is locked up in glaciers, ice caps, and snowfields, although with global warming, the ice is decreasing. About 30% of our fresh water is underground and must be pumped for our use. Less than 1% is tied up in our atmosphere, and finally about 0.3% is the most easily accessible on Earth's surfaces in lakes, swamps, and rivers. If all the water on Earth were represented by the contents of a 2-L bottle, then 60 mL of this would be fresh water, and the amount of this fresh water that is accessible is about four drops!

Skeptical Chemist 5.9   A Drop to Drink

The previous section states that four drops in 2 L corresponds to the amount of fresh water available for our use. Is this accurate? Make a determination of your own.
Hint: Use both the relationships shown in Figure 5.8 and assume 20 drops/mL.

The previous activity makes water look precious, and arguably it is, depending on where you live. We strive to use the most convenient source of fresh water for human activities, which in many cases comes from **surface water,** the fresh water found in lakes, rivers, and streams (Figure 5.9). Less convenient to access is **groundwater,** fresh water found in underground reservoirs also known as aquifers. However, by pumping groundwater from wells drilled into these underground reservoirs, people worldwide come to depend on this source of fresh water. A third plentiful source, sea water, only works if we remove its salt through a process called desalination.

In the United States, the average household spends about $2.00 for 1000 gallons (3800 L) of water for home use. Compared with other commodities such as gasoline, this is relatively inexpensive. Most citizens in the United States obtain their water from a faucet or drinking fountain (Figure 5.10a). Municipal tap water may be consumed with or without further home filtering. Some people draw water from their own wells. If out on a hiking trail, water may be purified from nearby streams. Water also can be purchased in plastic, aluminum, or glass containers (Figure 5.10b). Each of these options most certainly has advantages and disadvantages. However, the point is that an

Look for more about desalination in Section 5.12.

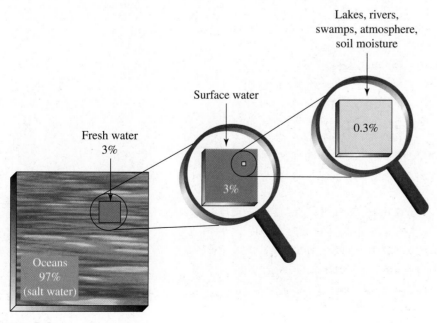

Figure 5.8

Distribution of fresh water on Earth.

Figure 5.9

Lakes and reservoirs provide much of our drinking water. This one, Hetch Hetchy, provides water to San Francisco, California.

(a)

(b)

Figure 5.10

(a) Some people (but not all) take the safety of drinking water for granted. (b) Across the planet, people drink bottled water for many different reasons.

infrastructure supports the availability of water. Those who live in economically advantaged nations with a strong infrastructure can make decisions about *how frequently* they drink, *how much* they drink, and about the *source* of the water they drink.

What if you live where you cannot turn on a tap or buy bottled water? You might need to walk for miles to reach a water source, fill a container, and carry it home to your family (Figure 5.11a). You might need to regularly depend on a water truck to stop by your house and deliver water (Figure 5.11b). On a larger scale, you might have to depend on governments working with engineers to design mega-structures to help move water from one region of the country to another. Aqueducts in the United States move water from the Colorado River to the Southwest. In China, a massive south-to-north

(a)   (b)

Figure 5.11

(a) Young girls walking home with water buckets. (b) A truck provides water for a community.

water transfer project costing billions of dollars promises to funnel more than 45 trillion liters of water per year from the Yangtze River basin in southern China to the water-thirsty north. Such major diversions of water usually come with a considerable cost, as we will explore in a later section.

Unfortunately, a mismatch exists between where water is found on our planet and where people need to use it. The phrase "location, location, location" applies well to water. Several issues, including global climate change, overconsumption and inefficient use of water, and contamination further complicate the availability of water. We now discuss each in turn.

Global Climate Change

Violent storms and floods bring water in ferocious abundance, as witnessed by recent flooding in the midwestern United States, Europe, and China. At the other extreme, drought or desertification creates crippling shortages. Either way, climate affects the supply and demand for water.

When regions become hotter and drier, the demand increases for water, especially water for irrigation. For example, consider the Great Lakes Basin Water Resource Compact that was first signed in 2005 by the governors of the eight states and the leaders of the two Canadian provinces that border the Great Lakes. In part, this pact was driven by fears that drought-stricken regions might try to raid the water supply of the Great Lakes (Figure 5.12).

> Weather describes the physical conditions of the atmosphere, such as its temperature, pressure, moisture, and wind. In contrast, climate describes the long-term weather patterns in a region.

> Lake Superior holds over 10% of the world's fresh water supply.

Figure 5.12

Political cartoon of the evolution of the Great Lakes Compact.

Source: Joe Heller

Your Turn 5.10   Great Lakes, Great News!

A long-debated agreement to regulate the withdrawal of water from the Great Lakes passed the U.S. House of Representatives on September 23, 2008. A local newspaper in Wisconsin offered its readers these quiz questions.

a. How long does it take a single drop of rain to cycle through Lake Superior?
b. How much water is permanently lost from the Great Lakes drainage basin every year?
c. How many national parks and lakeshores are there on the Great Lakes?

Note: As a reference point, the Great Lakes contain about 220 quadrillion liters of water.

Answers
a. About 200 years
b. About 1 trillion liters, or more than 15,000 L for each of the basin's 37 million residents
c. Ten national parks plus hundreds of state and provincial parks with more than 70 million visitors annually

In Chapter 3, you learned about the carbon cycle.

This newspaper quiz brings up an important point: A drop of water that enters Lake Superior will one day leave it. Just as our planet has a carbon cycle, it also has a water cycle. The water that falls on land either evaporates or eventually finds its way to the ocean. The water cycle not only includes lakes and rivers, but also their frozen cousins—glaciers and sea ice.

Climate plays an important role in the *timing* of the water cycle. For example, glaciers stabilize water flow over time as they accumulate snow pack during winter months and then release a regular stream of water during summer months. The great glaciers of the Himalayas feed seven of the largest rivers in Asia, ensuring a reliable water supply for 2 billion people—almost one third of the world's population. If climate is altered and these glaciers are not replenished on an annual cycle to a point where they cannot melt and sustain the rivers in the region, the effect on the people in Asia who have come to rely on glaciers as water reservoirs could be devastating.

Climate change also affects the *timing* of events in ecosystems. For example, insects, birds, and plants need to appear in the right order so that the birds can feed, the insects can pollinate, and the plants can grow. If birds migrate earlier in the spring, they may arrive before enough insects have hatched for food. Conversely, if too many insects hatch before the birds are present to eat them, the insects may devastate crops. Either way, water is a key variable.

Overconsumption and Inefficient Use

In many places, water is being pumped out of the ground faster than it can be replenished by the natural water cycle. For example, much of the bountiful grain harvest from the central United States is a result of using water from the Ogallala Aquifer (also called the High Plains Aquifer). This vast aquifer trapped water from the last ice age and runs from South Dakota to Texas (Figure 5.13). Clearly, it is not a sustainable practice to continuously pump water from all aquifers. Some aquifers do recharge more quickly by precipitation and runoff. Others may take hundreds or even thousands of years to recharge naturally. Continuous pumping can bring other harmful outcomes as well. For example, if water is removed from a geologically unstable area near the coast, salt water may intrude into a freshwater aquifer.

In the context of air quality, the tragedy of the commons was first mentioned in Section 1.12.

This situation presents us with another example of the tragedy of the commons. The water from aquifers is a resource used in common, yet no one in particular is responsible for this resource. If water is overdrawn for agriculture or some other purpose, this act can be to the detriment of all.

An overdraw of surface water can create other problems. Consider Kazakhstan and Uzbekistan, countries that border the Aral Sea. Until recently, this sea was the world's fourth largest inland body of fresh water. In the 1960s, workers in the former Soviet Union built a network of canals that diverted this water from the rivers that fed the Aral Sea in order to grow cotton in the arid climate. Consequently, the Aral Sea dried up, as shown in Figure 5.14. Although the ecosystem once was rich as a fishery, today only three very salty pools of water remain. The United Nations has called this the greatest environmental disaster of the 20th century. Dust that is laden with toxins, pesticides, and salt now blows in the region, causing health problems and contributing to poverty.

Not only were the rivers feeding the Aral Sea diverted, but the river water taken was also used inefficiently. For example, the water used to irrigate cotton was transported in open canals. Given the arid climate, much of the water was lost through evaporation. Other wasteful practices in areas where water is scarce include using water sprinklers to irrigate fields, cultivating lush green lawns in residential areas, and not fixing leaky pipes in aging water distribution systems. Many factors influence this unsustainable use of water, including a lack of knowledge of other irrigation options, subsidies to keep the cost of water low, and the high cost of repairing a water distribution system.

Contamination

We expect our water to be safe; that is, devoid of harmful chemicals and microbes. Access to clean water varies worldwide, as shown in Figure 5.15. More than a billion people (1 in 6), principally in developing nations, lack access to safe drinking water.

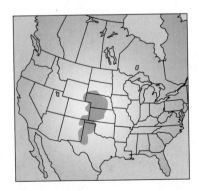

Figure 5.13

One of the world's largest aquifers, the High Plains Aquifer, is show in dark blue on this map.

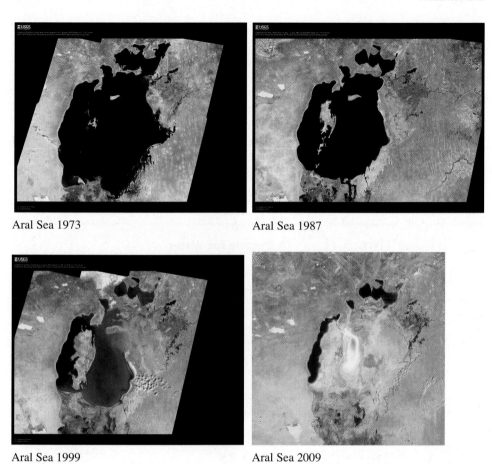

Aral Sea 1973   Aral Sea 1987

Aral Sea 1999   Aral Sea 2009

Figure 5.14

The Aral Sea has lost more than 80% of its water over a period of 30 years. The rivers that fed it were diverted to irrigate crops.

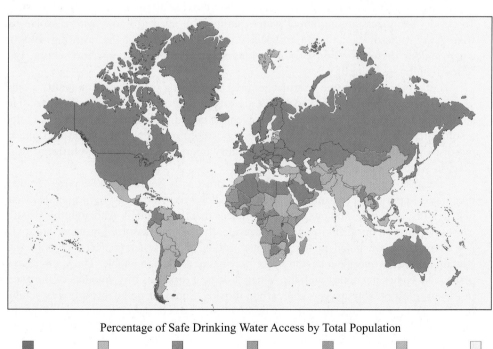

Percentage of Safe Drinking Water Access by Total Population

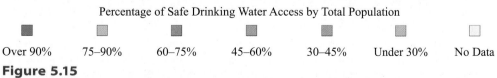

Over 90%   75–90%   60–75%   45–60%   30–45%   Under 30%   No Data

Figure 5.15

Access to safe drinking water varies widely across the world.

Source: © Compare Infobase.

Another 2.6 billion lack basic sanitation; that is, garbage disposal and treatment of industrial and other wastewater. Each day, more than 5000 deaths worldwide occur in infants and young children because of infectious disease agents borne by unsafe water.

In response to the global water needs in the 1980s, the United Nations Children's Fund (UNICEF) funded pumps and the sinking of wells to tap into underground sources of water. Although water from aquifers is generally potable, tragically, the water pumped to the surface in India and Bangladesh was not. It contained arsenic and fluoride ions, both naturally occurring in minerals found in the bedrock of the aquifers. Arsenic and fluoride ions are cumulative poisons, and it may take years before the amount ingested reaches a harmful level. Thousands of people have been irreversibly poisoned, because the arsenic and fluoride ions were detected too late in their drinking water.

Arsenic and fluoride ions occur in water in the form of a cation and an anion, respectively. Look for more about ions in Section 5.6.

WATER FOR LIFE
2005–2015

Consider This 5.11   A Decade for Water

As pointed out earlier in this chapter, the United Nations has designated 2005–2015 as the Decade for Action regarding water. The textbook's website provides a link.

a. Find two significant issues related to water that were not discussed in this section. Make a case for the importance of each.
b. Summarize the theme for this year's World Water Day.

Let us now turn to topics that help us better understand why water is able to dissolve and mix with so many substances, including essential nutrients and contaminants.

5.5 | Aqueous Solutions

Water dissolves a remarkable variety of substances. As we will see, some of them, including salt, sugar, ethanol, and the air pollutant SO_2, are *very* soluble in water. In comparison, limestone rock, oxygen, and carbon dioxide dissolve only in tiny amounts. To build your understanding about water quality, you need to know *what* dissolves in water, *why* it dissolves, and *how* to specify the concentration of the resulting solution when it does dissolve. This section tackles solution concentrations; the section that follows addresses solubility.

Let's begin with some useful chemical terminology. Water is a **solvent,** a substance, often a liquid, that is capable of dissolving one or more pure substances. The solid, liquid, or gas that dissolves in a solvent is called the **solute.** The result is called a **solution,** a homogeneous (of uniform composition) mixture of a solvent and one or more solutes. In this section, we are particularly interested in **aqueous solutions,** solutions in which water is the solvent.

Because water is such a good solvent, it practically never is "100% pure." Rather, it contains impurities. For example, when water flows over the rocks and minerals of our planet, it dissolves tiny amounts of the substances that they contain. Although this usually causes no harm to our drinking water, occasionally the ions dissolved in water are toxic. For example, as we noted in the previous section, if the water contacts minerals that contain arsenic or fluoride ions, the water may be rendered nonpotable. The water on our planet also comes in contact with air. When it does, it dissolves tiny amounts of the gases in the air, most notably oxygen and carbon dioxide. Some air pollutants are *very* soluble in water. So when it rains, the water actually cleans some of the pollutants out of the air, including SO_2 and NO_2. As we will see in Chapter 6, the acidic solutions that form can have serious consequences for the environment.

Humans also contribute to the number of substances dissolved in water. When we wash clothes, we add not only the spent detergent, but also whatever made our clothing dirty in the first place. When we flush a toilet, we add liquid and solid wastes. Our urban streets add solutes to rain water during the process of storm run-off. And our agricultural practices add fertilizers and other soluble compounds to water.

What does water's being a good solvent mean for our drinking water? In order to assess water quality, you need to know several things. One is a way to specify *how much* of a substance has dissolved, so that you can compare the value with a known standard. In other words, you need to understand the concept of concentration. This was first introduced in Chapter 1 in relation to the composition of air. For example, O_2 and N_2 are about 21% and 78% of dry air, respectively. We revisited concentration again in Chapters 2 and 3, exploring the concentrations of chlorine compounds in the stratosphere and greenhouse gases in the troposphere. For example, carbon dioxide has a concentration of about 390 ppm in the air. Now we examine this concept in terms of substances dissolved in water. As we will see, percent and parts per million are valid ways of expressing concentrations for aqueous solutions as well.

To get started with solution concentrations, let's use a familiar analogy—sweetening a cup of tea. If 1 teaspoon of sugar is dissolved in a cup of tea, the resulting solution has a concentration of 1 teaspoon per cup. Note that you would have this same concentration if you were to dissolve 3 teaspoons of sugar in 3 cups of tea, or half a teaspoon in half a cup of tea. If your recipe is tripled or halved, the sugar and tea are adjusted proportionally. Therefore, the **concentration,** the ratio of the amount of solute to the amount of solution—or in this case the ratio of sugar dissolved to make the solution—is the same in each case.

Solute concentrations in aqueous solution follow the same pattern but are expressed with different units. We use four ways to express concentration: percent, parts per million, parts per billion, and molarity. Three of these should already be familiar to you. The fourth, molarity, uses the mole concept introduced in Chapter 3.

Percent (%) means parts per hundred. For example, an aqueous solution containing 0.9 grams of sodium chloride (NaCl) in 100 grams of solution is a 0.9% solution by mass. This concentration of sodium chloride is referred to as "normal saline" in medical settings when given intravenously. You may find the antiseptic hydrogen peroxide (H_2O_2) in your medicine cabinet as a 3% aqueous solution by volume. It contains 3 milliliters of H_2O_2 in every 100 milliliters of aqueous solution. Percent is used to express the concentration of a wide range of solutions.

But when the concentration is very low, as is the case for many substances dissolved in drinking water, **parts per million (ppm)** is more commonly used. For example, water that contains 1 ppm of calcium ion contains the equivalent of 1 gram of calcium (in the form of the calcium ion) dissolved in 1 million grams of water. The water we drink contains substances naturally present in the parts per million range. For example, the acceptable limit for nitrate ion, found in well water in some agricultural areas, is 10 ppm; the limit for the fluoride ion is 4 ppm.

Although parts per million is a useful concentration unit, measuring 1 million grams of water is not very convenient. We can do things more easily by switching to the unit of a liter. One ppm of any substance in water is equivalent to 1 mg of that substance dissolved in a liter of solution. Here is the math:

$$1 \text{ ppm} = \frac{1 \text{ g solute}}{1 \times 10^6 \text{ g water}} \times \frac{1000 \text{ mg solute}}{1 \text{ g solute}} \times \frac{1000 \text{ g water}}{1 \text{ L water}} = \frac{1 \text{ mg solute}}{1 \text{ L water}}$$

Municipal water utilities typically use the unit milligrams per liter (mg/L) to report the minerals and other substances dissolved in tap water. For example, Table 5.4 shows a tap water analysis from an aquifer that supplies a midwestern community in the United States.

For solutions at low concentration, the mass of the solution is approximately the mass of the solvent.

These limits for fluoride and nitrate ions reflect the U.S. standards. See Section 5.10.

1000 grams (1×10^3 g) of H_2O can be taken to have a volume of 1 liter. Strictly speaking, this is true only at 4 °C.

Table 5.4	Tap Water Mineral Report		
Cation	**mg/L**	**Anion**	**mg/L**
calcium ion	97	sulfate ion	45
magnesium ion	51	chloride ion	75
sodium ion	27	nitrate ion	4
		fluoride ion	1

Some contaminants are of concern at concentrations much lower than parts per million and are reported as **parts per billion (ppb).** Assuming that 1 ppm corresponds to 1 second in nearly 12 days, then 1 ppb corresponds to 1 second in 33 years. Another way of looking at this is that one part per billion corresponds to a few centimeters on the circumference of the Earth!

One contaminant found in the range of parts per billion is mercury. For humans, the primary source of exposure to mercury is food, mainly fish and fish products. Even so, the concentration of mercury in water needs to be monitored. One part per billion of mercury (Hg) in water is equivalent to 1 gram of Hg dissolved in 1 billion grams of water. In more convenient terms, this means 1 microgram (1×10^{-6} g, or 1 μg) of Hg dissolved in 1 liter of water. The acceptable limit for mercury in drinking water is 2 ppb:

Mercury in water is present in a soluble form (Hg^{2+}) rather than as elemental Hg ("quicksilver").

In aqueous solutions,
1 ppb = 1 μg/L
1 ppm = 1 mg/L

$$2 \text{ ppb Hg} = \frac{2 \text{ g Hg}}{1 \times 10^9 \text{ g H}_2\text{O}} \times \frac{1 \times 10^6 \text{ μg Hg}}{1 \text{ g Hg}} \times \frac{1000 \text{ g H}_2\text{O}}{1 \text{ L H}_2\text{O}} = \frac{2 \text{ μg Hg}}{1 \text{ L H}_2\text{O}}$$

Convince yourself that the units cancel, as in the previous example.

Your Turn 5.12   Mercury Ion Concentrations

a. A 5-L sample of water contains 80 μg of dissolved mercury ion. Express the concentration of the solution in ppm and ppb.

b. Would your answer in part **a.** be in compliance with a federal maximum mercury concentration of 2 ppb? Explain.

Molarity (M), another useful concentration unit, is defined as a unit of concentration represented by the number of moles of solute present in 1 liter of solution.

$$\text{Molarity (M)} = \frac{\text{moles of solute}}{\text{liter of solution}}$$

The great advantage of molarity is that solutions of the same molarity contain exactly the same number of moles of solute and hence the same number of molecules (ions or atoms) of solute. The mass of a solute varies depending on its identity. For example, 1 mole of sugar has a different mass than 1 mole of sodium chloride. But if you take the same volume, all 1 M solutions (read as "one molar") contain the same number of moles.

The molar mass of NaCl (58.5 g) is calculated by adding the molar mass of sodium (23.0 g) plus the molar mass of chlorine (35.5 g). Section 3.7 explains molar mass calculations.

(aq) is short for aqueous, indicating that the solvent is water.

As an example, consider a solution of NaCl in water. The molar mass of NaCl is 58.5 g; therefore, 1 mol of NaCl has a mass of 58.5 g. By dissolving 58.5 g of NaCl in some water and then adding enough water to make exactly 1.00 L of solution, we would have a 1.00 M NaCl aqueous solution (Figure 5.16). We have prepared a one-molar solution of sodium chloride. Note the use of a **volumetric flask,** a type of glassware that contains a precise amount of solution when filled to the mark on its neck. But because concentrations are simply ratios of solute to solvent, there are many ways to make a 1.00 M NaCl(*aq*) solution. Another possibility is to use 0.500 mol NaCl (29.2 g) in 0.500 L of solution. This requires the use of a 500-mL volumetric flask, rather than the 1-L flask shown in Figure 5.16.

$$1 \text{ M NaCl}(aq) = \frac{1 \text{ mol NaCl}}{1 \text{ L solution}} \text{ or } \frac{0.500 \text{ mol NaCl}}{0.500 \text{ L solution}}, \text{ etc.}$$

Remember that 1 ppm = 1 mg/L and that the molar mass of Hg is 200.6 g/mol.

Let's say you have a water sample with 150 ppm of dissolved mercury. What is this concentration expressed in molarity? You might do the calculation this way:

$$150 \text{ ppm Hg} = \frac{150 \text{ mg Hg}}{1 \text{ L H}_2\text{O}} \times \frac{1 \text{ g Hg}}{1000 \text{ mg Hg}} \times \frac{1 \text{ mol Hg}}{200.6 \text{ g Hg}} = \frac{7.5 \times 10^{-4} \text{ mol Hg}}{1 \text{ L H}_2\text{O}} = 7.5 \times 10^{-4} \text{ M Hg}$$

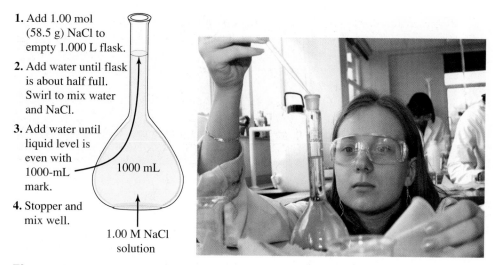

1. Add 1.00 mol (58.5 g) NaCl to empty 1.000 L flask.

2. Add water until flask is about half full. Swirl to mix water and NaCl.

3. Add water until liquid level is even with 1000-mL mark.

4. Stopper and mix well.

1000 mL

1.00 M NaCl solution

Figure 5.16

Preparing a 1.00 M NaCl aqueous solution.

Thus, a sample of water containing 150 ppm of mercury also can be expressed as 7.5×10^{-4} M Hg.

Your Turn 5.13   Moles and Molarity

a. Express a concentration of 16 ppb Hg in units of molarity.

b. For 1.5 M and 0.15 M NaCl(*aq*), how many moles of solute are present in 500 mL of each?

c. A solution is prepared by dissolving 0.50 mol NaCl in enough water to form 250 mL of solution. A second solution is prepared by dissolving 0.60 mol NaCl to form 200 mL of solution. Which solution is more concentrated? Explain.

d. A student was asked to prepare 1.0 L of a 2.0 M $CuSO_4$(*aq*) solution. The student placed 40.0 g of $CuSO_4$ crystals in a volumetric flask and filled it with water to the 1000-mL mark. Was the resulting solution 2.0 M? Explain.

In this section, we made the case that water is an excellent solvent for a wide variety of substances and that we can express the concentration of these substances numerically. As promised, the next section helps you to build an understanding of how and why substances dissolve in water.

5.6 | A Closer Look at Solutes

Sugar and salt both dissolve in water. However, the solutions of these two solutes are quite different in nature—the former conducts electricity and the latter does not. Experimentally, we demonstrate this difference using a **conductivity meter,** an apparatus that produces a signal to indicate that electricity is being conducted. The conductivity meter shown in Figure 5.17 is built from wires, a battery, and a light bulb. As long as the electrical circuit is not completed, the bulb does not glow. For example, if the two wires are placed into distilled water or a sugar solution in distilled water, the bulb does not light. However, if the separate wires are placed into an aqueous solution of salt, the bulb turns on. Perhaps the light also has gone on in the mind of the experimenter!

Distilled water does not conduct electricity. The same is true for an aqueous solution of sugar. Sugar is a **nonelectrolyte,** a solute that is nonconducting in aqueous solutions. But an aqueous solution of common table salt, NaCl, conducts electricity and the light bulb glows. Sodium chloride is classified as an **electrolyte,** a solute that conducts electricity in aqueous solution.

Distillation, a process to purify water, is discussed in Section 5.12.

The term *electrolyte* is used in connection with sports drinks. Some taste slightly salty because they contain sodium salts.

(a)   **(b)**   **(c)**

Figure 5.17

Conductivity experiments. **(a)** Distilled water (nonconducting). **(b)** Sugar dissolved in distilled water (nonconducting). **(c)** Salt dissolved in distilled water (conducting).

Remember to indicate ions in aqueous solution using *(aq)*.

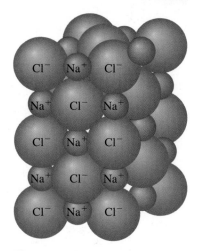

Figure 5.18

The arrangement of Na⁺ and Cl⁻ in a crystal of sodium chloride.

Valence (outer) electrons were described in Section 2.2.

What makes salt solutions behave differently from sugar solutions or from pure water? The observed flow of electric current through a solution involves the transport of electric charge. The ability of aqueous NaCl solutions to conduct electricity suggests that they contain some charged species capable of moving electrons through the solution. When solid NaCl dissolves in water, it separates into $Na^+(aq)$ and $Cl^-(aq)$. An **ion** is an atom or group of atoms that has acquired a net electric charge as a result of gaining or losing one or more electrons. The term is derived from the Greek for "wanderer." Na^+ is an example of a **cation,** a positively charged ion. Similarly, Cl^- is an example of an **anion,** a negatively charged ion. No such separation occurs with covalently bonded sugar or water molecules, making these liquids unable to carry electric charge.

It may surprise you to learn that Na^+ and Cl^- exist both in crystals of salt (such as those in a salt shaker) and in an aqueous solution of NaCl. Solid sodium chloride is a three-dimensional cubic arrangement of sodium and chloride ions. An **ionic bond** is the chemical bond formed when oppositely charged ions attract. In the case of NaCl, ionic bonds hold the crystal together; there are no covalently bonded atoms, only positively charged cations and negatively charged anions held together by electrical attractions. An **ionic compound** is composed of ions that are present in fixed proportions and arranged in a regular, geometric structure. In the case of NaCl, each Na^+ is surrounded by six oppositely charged Cl^- ions. Likewise, each Cl^- is surrounded by six positively charged Na^+ ions. A single, tiny crystal of sodium chloride consists of many trillions of sodium ions and chloride ions in the arrangement shown in Figure 5.18.

We've described ionic compounds, but we still need to explain *why* certain atoms lose or gain electrons to form ions. Not surprisingly, the answer involves the distribution of electrons within atoms. For example, recall that a neutral sodium atom has 11 electrons and 11 protons. Sodium, like all metals in Group 1A, has one valence electron. This electron is rather loosely attracted to the nucleus and can be easily lost. When this happens, the Na atom forms Na^+, a cation.

$$Na \longrightarrow Na^+ + e^- \qquad\qquad [5.1]$$

The Na^+ has a 1+ charge because it contains 11 protons but only 10 electrons. It also has a complete octet, just like the neon atom. Table 5.5 shows the comparison between Na, Na^+, and Ne.

Table 5.5	Electronic Bookkeeping for Cation Formation	
Sodium Atom	**Sodium Ion**	**Neon Atom**
Na	Na⁺	Ne
11 protons	11 protons	10 protons
11 electrons	10 electrons	10 electrons
net charge of 0	net charge of 1+	net charge of 0

Table 5.6	Electronic Bookkeeping for Anion Formation		
	Chlorine Atom	**Chloride Ion**	**Argon Atom**
	Cl	Cl⁻	Ar
	17 protons	17 protons	18 protons
	17 electrons	18 electrons	18 electrons
	net charge of 0	net charge of 1−	net charge of 0

Unlike sodium, chlorine is a nonmetal. Recall that a neutral chlorine atom has 17 electrons and 17 protons. Chlorine, like all nonmetals in Group 7A, has seven valence electrons. Because of the stability associated with eight outer electrons, it is energetically favorable for chlorine to gain one electron.

Revisit Section 1.6 for more about metals and nonmetals.

$$Cl + e^- \longrightarrow Cl^- \qquad\qquad [5.2]$$

The chloride ion (Cl^-) has 18 electrons and 17 protons; thus the net charge is 1− (Table 5.6).

Sodium metal and chlorine gas react vigorously when they come in contact. The result is the aggregate of Na^+ and Cl^-, known as sodium chloride. In the formation of an ionic compound such as sodium chloride, the electrons are actually transferred from one atom to another, not simply shared as they would be in a covalent compound.

Is there evidence for electrically charged ions in pure sodium chloride? Experimental tests show that crystals of sodium chloride do not conduct electricity. This makes sense, because in the crystal, the ions are fixed in place and so are unable to move and transport charge. However, when these crystals are melted, the ions are free to move, and the hot liquid conducts electricity. This provides evidence that ions are present.

Like other ionic compounds, NaCl crystals are hard yet brittle. When hit sharply, they shatter rather than being flattened. This suggests the existence of strong forces that extend throughout the ionic crystal. Strictly speaking, there is no such thing as a specific, localized "ionic bond" analogous to a covalent bond in a molecule. Rather, ionic bonding holds together a large assembly of ions; in this case, Na^+ and Cl^-.

Generally speaking, electron transfer to form cations and anions occurs between metallic elements and nonmetallic elements. Sodium, lithium, magnesium, and other metals have a strong tendency to give up electrons and form positive ions. They have very low electronegativity values. On the other hand (or the other side of the periodic table), chlorine, fluorine, oxygen, and other nonmetals have a strong attraction for electrons and readily gain them to form negative ions. Nonmetals have relatively high electronegativity values. Potassium chloride (KCl) and sodium iodide (NaI) are two of many such compounds.

Your Turn 5.14   Predicting Ionic Charge

a. Predict whether these atoms will form an anion or a cation based on their electronegativity values.

Li   S   K   N

b. Predict the ion that each of these will form. Then draw a Lewis structure for the atom and the ion, clearly labeling the charge on the ion.

Br   Mg   O   Al

Hint: Use the periodic table as a guide to the number of outer electrons. Then determine how many electrons must be lost or gained to achieve stability with an octet of electrons.

Answer

b. Bromine (Group 7A) gains one electron. The resulting ion has a charge of 1−, just as was the case for chlorine. Here are the Lewis structures.

:B̈r· and [:B̈r:]⁻

This section opened with a discussion of sugar and salt. The names *salt* and *sugar* are both in common use, and you knew what we were talking about. Salt is the stuff you sprinkle on French fries. And sugar is the stuff that some people use to sweeten coffee. In fact, ordinary table salt (NaCl) is such an important example of an ionic compound that chemists frequently refer to others simply as "salts," meaning crystalline ionic solids. As you will see in Chapter 11, sugars are another important class of compounds, and what we call "sugar" is really the compound sucrose.

To delve further into the issues of water quality, you need to know the names of other salts; that is, ionic compounds. As you might guess, chemists name them using a long and careful set of rules. Luckily, we won't discuss all of these rules here. Rather, we follow the "need-to-know" philosophy, helping you learn the ones that you need for understanding water quality.

5.7 | Names and Formulas of Ionic Compounds

In this section, we work on the "vocabulary" you need in order to work with ionic compounds. As we pointed out in Chapter 1, chemical symbols are the alphabet of chemistry and chemical formulas are the words. Earlier, we helped you to "speak chemistry" by correctly using chemical formulas and names for the substances in the air you breathe. Now we do the same for the substances in the water you drink.

Let's begin with the ionic compound formed from the elements calcium and chlorine: $CaCl_2$. The explanation for the 1:2 ratio of Ca to Cl lies in the charges of the two ions. Calcium, a member of Group 2A, loses its two outer electrons to form Ca^{2+}.

$$Ca \longrightarrow Ca^{2+} + 2\,e^- \qquad\qquad [5.3]$$

Chlorine, as we saw in equation 5.2, gains an outer electron to form Cl^-. In an ionic compound, the sum of the positive charges equals the sum of the negative charges. Hence, the formula for this compound is $CaCl_2$.

The logic is the same with MgO and Al_2O_3, two other ionic compounds. These both contain oxygen, but in different ratios. Recall that oxygen, Group 6A, has six outer electrons. Thus a neutral oxygen atom can gain two electrons to form O^{2-}. The magnesium atom loses two electrons to form Mg^{2+}. These two ions must then combine in a 1:1 ratio so the overall charge is zero; the chemical formula is MgO. Note that although the charge *always* must be written on an individual ion, we omit the charges in the chemical formulas of ionic compounds. Thus, it is *not* correct to write the chemical formula as $Mg^{2+}O^{2-}$. The charges are implied by the chemical formula.

Here is another example. Armed with the knowledge that aluminum tends to lose three electrons to form Al^{3+}, you can write the chemical formula of the ionic compound formed from Al^{3+} and O^{2-} as Al_2O_3. Here, a 2:3 ratio of ions is needed so that the overall electric charge on the compound is zero. Again, it is *not* correct to write the chemical formula as $Al_2^{3+}O_3^{2-}$.

Earlier in the chapter, we referred to several ionic compounds by their names, including sodium chloride, sodium iodide, and potassium chloride. Observe the pattern: name the cation first, then the anion, modified to end in the suffix *-ide*. Thus, $CaCl_2$ is calcium chloride, with each ion named for its element and with chlorine modified to read chloride. Similarly, NaI is sodium iodide and KCl is potassium chloride.

The elements presented thus far formed only one type of ion. Group 1A and 2A elements only form 1+ and 2+ ions, respectively. The halogens form only 1− ions. Lithium bromide is LiBr. The ratio of 1:1 is understood because lithium only forms Li^+ and bromine only forms Br^-. The prefixes *mono-, di-, tri-,* and *tetra-* are *not* used when naming ionic compounds such as these. There is no need to call it monolithium monobromide. $MgBr_2$ is magnesium bromide, not magnesium dibromide. Magnesium *only* forms Mg^{2+}, and the ratio of 1:2 is understood and so does not need to be stated.

Halogens were described in Sections 1.6 and 2.9.

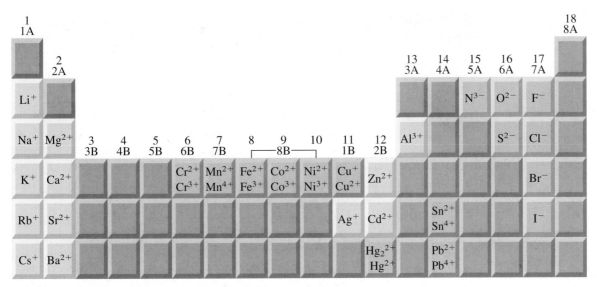

Figure 5.19

Common ions formed from their elements. Ions in green (cations) or blue (anions) have only one charge. Ions in red (cations) have more than one possible ionic charge.

But some elements do form more than one ion, as you can see in Figure 5.19. Prefixes still are not used, but rather the charge on the ion must be specified using a Roman numeral. Take copper for example. If your instructor asks you to head down to the stockroom and grab some copper oxide, what do you do? You ask if what is wanted is copper(I) oxide or copper(II) oxide, right? Similarly, iron can form different oxides. Two forms are FeO (formed from Fe^{2+}) and Fe_2O_3 (commonly called rust and formed from Fe^{3+}). The names for FeO and Fe_2O_3 are iron(II) oxide and iron(III) oxide, respectively. Note the space after but not before the parenthesis enclosing the Roman numeral.

Again compare. The name $CuCl_2$ is copper(II) chloride, but the name of $CaCl_2$ is calcium chloride. Calcium only forms one ion (Ca^{2+}), whereas copper can form two ions: Cu^+ and Cu^{2+}.

> Prefixes such as *di-* and *tri-* generally are not used in naming ionic compounds. Roman numerals are used with the name of the cation if it has more than one possible charge.

Your Turn 5.15   Ionic Compounds

Each pair of elements forms one or more ionic compounds. For each, write the chemical formulas and names.

a. Ca and S   **b.** F and K   **c.** Mn and O
d. Cl and Al   **e.** Co and Br

Answer
e. From Figure 5.19, Co can form Co^{2+} and Co^{3+}. Br forms only Br^-. The possible chemical formulas are $CoBr_2$, cobalt(II) bromide, and $CoBr_3$, cobalt(III) bromide.

One or both of the ions in an ionic compound can be a **polyatomic ion**, two or more atoms covalently bound together that have an overall positive or negative charge. An example is the hydroxide ion, OH^-, with an oxygen atom covalently bonded to a hydrogen atom. The Lewis structure shown in Figure 5.20 reveals that there are 8 electrons, 1 more than the 6 valence electrons provided by one O atom and one H atom. The "extra" electron gives the hydroxide ion a charge of 1−. Table 5.7 lists common polyatomic ions. Most are anions, but polyatomic cations also are possible, as in the case of the ammonium ion, NH_4^+. Note that some elements (carbon, sulfur, and nitrogen) form more than one polyatomic anion with oxygen.

The rules for naming ionic compounds containing polyatomic ions are similar to those for ionic compounds of two elements. Consider, for example, aluminum sulfate,

> Brackets in a Lewis structure, such as with the hydroxide ion, call attention to an ion that fulfilled its octet by losing or gaining one or more electrons.

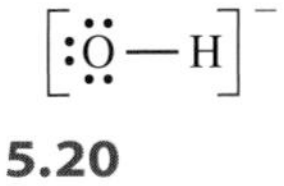

$$\left[:\overset{..}{\underset{..}{O}} - H \right]^-$$

Figure 5.20

The Lewis structure for the hydroxide ion, OH^-.

Table 5.7	Common Polyatomic Ions		
Name	**Formula**	**Name**	**Formula**
acetate	$C_2H_3O_2^-$	nitrite	NO_2^-
bicarbonate*	HCO_3^-	phosphate	PO_4^{3-}
carbonate	CO_3^{2-}	sulfate	SO_4^{2-}
hydroxide	OH^-	sulfite	SO_3^{2-}
hypochlorite	ClO^-	ammonium	NH_4^+
nitrate	NO_3^-		

*Also called the hydrogen carbonate ion.

an ionic compound that is used in many water purification plants. The compound is formed from Al^{3+} and SO_4^{2-}. When you see $Al_2(SO_4)_3$, mentally read this chemical formula as a compound that contains two ions: aluminum and sulfate. These ions are in a 2:3 ratio. As is true for all ionic compounds, the name of the cation is given first.

The parentheses in $Al_2(SO_4)_3$ are meant to help you. The subscript 3 applies to the *entire* SO_4^{2-} ion that is enclosed in parentheses. Accordingly, "read" this as three sulfate ions, not as one larger unit composed of three sulfate ions. Similarly, in the ionic compound ammonium sulfide (see Table 5.8), the NH_4^+ is enclosed in parentheses. The subscript of 2 indicates that there are two ammonium ions for each sulfide ion. Note that the charges are not shown in the chemical formula; they are assumed to be there. In some cases, though, the polyatomic ion is *not* enclosed in parentheses. Table 5.8 shows two examples. The PO_4^{3-} ion in aluminum phosphate has no parentheses; similarly, the NH_4^+ ion in ammonium chloride has no parentheses. Parentheses are omitted when the subscript of the polyatomic ion is 1. Nonetheless, you still have to "read" the chemical formula of $AlPO_4$ as containing the phosphate ion, and you have to "read" NH_4Cl as containing the ammonium ion.

These activities will help you practice using polyatomic ions.

Your Turn 5.16   Polyatomic Ions I

Write the chemical formula for the ionic compound formed from each pair of ions.

a. Na^+ and SO_4^{2-} **b.** Mg^{2+} and OH^- **c.** Al^{3+} and $C_2H_3O_2^-$ **d.** CO_3^{2-} and K^+

Answers
a. Na_2SO_4 **b.** $Mg(OH)_2$

Your Turn 5.17   Polyatomic Ions II

Name each of these compounds.

a. KNO_3 **b.** $(NH_4)_2SO_4$ **c.** $NaHCO_3$ **d.** $CaCO_3$ **e.** $Mg_3(PO_4)_2$

Answers
a. potassium nitrate   **b.** ammonium sulfate

Table 5.8	Ionic Compounds Containing Polyatomic Ions			
Chemical formula	$Al_2(SO_4)_3$	$(NH_4)_2S$	$AlPO_4$	NH_4Cl
Cation(s)	Al^{3+} Al^{3+}	NH_4^+ NH_4^+	Al^{3+}	NH_4^+
Anion(s)	SO_4^{2-} SO_4^{2-} SO_4^{2-}	S^{2-}	PO_4^{3-}	Cl^-

Your Turn 5.18   Polyatomic Ions III

Write the chemical formula for each of these compounds.

a. sodium hypochlorite (used to disinfect water)
b. magnesium carbonate (found in some limestone rocks, makes water "hard")
c. ammonium nitrate (fertilizer, runoff can contaminate groundwater)
d. calcium hydroxide (an agent used to remove impurities from water)

Answer

d. $Ca(OH)_2$. Two hydroxide ions (OH^-) are needed for each calcium ion (Ca^{2+}).

5.8 | The Ocean—An Aqueous Solution with Many Ions

Salt water! As we pointed out earlier, about 97% of the water on our planet is found in the oceans. This source of water contains much more than simple table salt (NaCl) dissolved in water. You are now in a position to understand why so many other ionic compounds can be found dissolved in our oceans.

Recall from Section 5.1 that water molecules are polar. When you take salt crystals and dissolve them in water, the polar H_2O molecules are attracted to the Na^+ and Cl^- ions contained in these crystals. The partial negative charge (δ^-) on the O atom of a water molecule is attracted to the positively charged Na^+ cations of the salt crystal. At the same time, the H atoms in H_2O, with their partial positive charges (δ^+), are attracted to the negatively charged Cl^- anions. Over time, the ions are separated and then surrounded by water molecules. Equation 5.4 and Figure 5.21 represent the process of forming an aqueous solution.

$$NaCl(s) \xrightarrow{H_2O} Na^+(aq) + Cl^-(aq) \qquad [5.4]$$

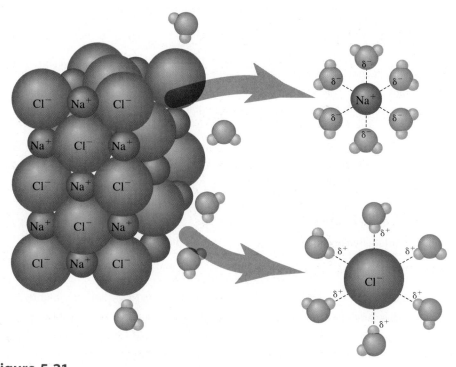

Figure 5.21

Sodium chloride dissolving in water.

 Figures Alive! Visit the textbook's website to learn more about sodium chloride dissolving in water.

The process is similar for forming solutions of compounds containing polyatomic ions. For example, when solid sodium sulfate dissolves in water, the sodium ions and sulfate ions simply separate. Note that the sulfate ion stays together as a unit.

$$Na_2SO_4(s) \xrightarrow{H_2O} 2\,Na^+(aq) + SO_4^{2-}(aq) \qquad [5.5]$$

Many ionic compounds dissolve in this manner. This explains why almost all naturally occurring water samples contain various amounts of ions. The same is also true for our bodily fluids, as these also contain significant concentrations of electrolytes.

Consider This 5.19   Electricity and Water Don't Mix

Small electric appliances such as hair dryers and curling irons carry prominent warning labels advising the consumer not to use the appliance near water. Why is water a problem since it does not conduct electricity? What is the best course of action if a plugged-in hair dryer accidentally falls into a sink full of water?

If the principles we just described applied to *all* ionic compounds, our planet would be in trouble. When it rained, ionic compounds such as calcium carbonate (limestone) would dissolve and end up in the ocean! Fortunately, many ionic compounds are only slightly soluble or have extremely low solubilities. The differences arise because of the sizes and charges of the ions, how strongly they attract one another, and how strongly the ions are attracted to water molecules.

Table 5.9 is your guide to solubility. For example, calcium nitrate, $Ca(NO_3)_2$, is soluble in water, as are all compounds containing the nitrate ion. Calcium carbonate, $CaCO_3$, is insoluble, as are most carbonates. By similar reasoning, copper(II) hydroxide, $Cu(OH)_2$, is insoluble, but copper(II) sulfate, $CuSO_4$, is soluble.

Your Turn 5.20   Solubility of Ionic Compounds

Which of these compounds are soluble in water? Use Table 5.9 as your guide.

a. ammonium nitrate, NH_4NO_3, a component of fertilizers
b. sodium sulfate, Na_2SO_4, an additive in laundry detergents
c. mercury(II) sulfide, HgS, known as the mineral cinnabar
d. aluminum hydroxide, $Al(OH)_3$, used in water purification processes

Answer
a. Soluble. All ammonium compounds and all nitrate compounds are soluble.

The landmasses on Earth are largely composed of minerals, that is, ionic compounds. Most have extremely low solubility in water as we mentioned earlier. Table 5.10 summarizes some environmental consequences of solubility.

Table 5.9	Water Solubility of Ionic Compounds		
Ions	Solubility of Compounds	Solubility Exceptions	Examples
sodium, potassium, and ammonium	all soluble	none	$NaNO_3$ and KBr. Both are soluble.
nitrates	all soluble	none	$LiNO_3$ and $Mg(NO_3)_2$. Both are soluble.
chlorides	most soluble	silver and some mercury chlorides	$MgCl_2$ is soluble. AgCl is insoluble.
sulfates	most soluble	strontium, barium, and lead sulfate	K_2SO_4 is soluble. $BaSO_4$ is insoluble.
carbonates	mostly insoluble*	Group 1A and NH_4^+ carbonates are soluble.	Na_2CO_3 is soluble. $CaCO_3$ is insoluble.
hydroxides and sulfides	mostly insoluble*	Group 1A and NH_4^+ hydroxides and sulfides are soluble.	KOH is soluble. $Al(OH)_3$ is insoluble.

*Insoluble means that the compounds have extremely low solubilities in water (less than 0.01 M). All compounds have at least a very small solubility in water.

Table 5.10	Environmental Consequences of Solubility	
Source	Ions	Solubility and Consequences
salt deposits	sodium and potassium halides*	These salts are soluble. Over time, they dissolve and wash into the sea. Thus, oceans are salty and sea water cannot be used for drinking without expensive purification.
agricultural fertilizers	nitrates	All nitrates are soluble. The runoff from fertilized fields carries nitrates into surface and groundwater. Nitrates can be toxic, especially for infants.
metal ores	sulfides and oxides	Most sulfides and oxides are insoluble. Minerals containing iron, copper, and zinc are often sulfides and oxides. If these minerals had been soluble in water, they would have washed out to sea long ago.
mining waste	mercury, lead	Most mercury and lead compounds are insoluble. However, they may leach slowly from mining waste piles and contaminate water supplies.

*Halides are the anions in Group 7A, such as Cl^- and I^-.

5.9 | Covalent Compounds and Their Solutions

From the previous discussion, you might have gotten the impression that only ionic compounds dissolve in water. But remember that sugar dissolves in water as well. The white granules of "table sugar" that you use to sweeten your coffee or tea are sucrose, a polar covalent compound with the chemical formula $C_{12}H_{22}O_{11}$ (Figure 5.22).

When sucrose dissolves in water, the sucrose molecules disperse uniformly among the H_2O molecules. The sucrose molecules remain intact and do *not* separate into ions. Evidence for this includes the fact that aqueous sucrose solutions do not conduct electricity (see Figure 5.17b). However, the sugar molecules do interact with the water molecules, as they are both polar and are attracted to one another. Furthermore, the sucrose molecule contains eight –OH groups and three additional O atoms that can participate in hydrogen bonding (see Figure 5.22). Solubility is always promoted when an attraction exists between the solvent molecules and the solute molecules or ions. This suggests a general solubility rule: *Like dissolves like.*

Let's also consider two other familiar polar covalent compounds, both of which are highly soluble in water. One is ethylene glycol, the main ingredient in antifreeze; and the other is ethanol, or ethyl alcohol, found in beer and wine. These molecules both contain the polar –OH group and are classified as alcohols (Figure 5.23).

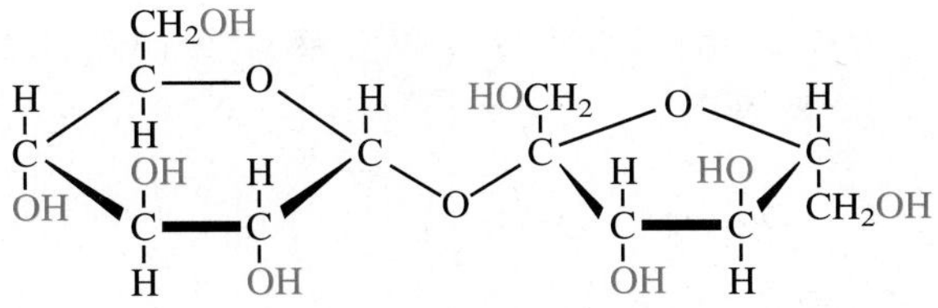

Look for more about sucrose and other sugars in Chapter 11.

"Like dissolves like" is a useful generalization for solubility.

Figure 5.22
Structural formula of sucrose. The –OH groups are shown in red.

Your Turn 5.21   Alcohols and Hydrogen Bonds

Some of the H and O atoms in the ethanol and the ethylene glycol molecule (see Figure 5.23) bear partial charges. Label these δ^+ and δ^-, respectively.
Hint: If the electronegativity difference between two atoms is more than 1.0, the bond is considered polar.

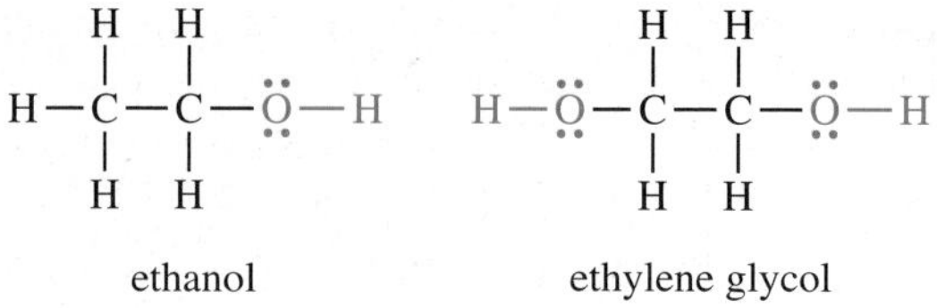

$$\begin{array}{ccc}
& H & H \\
& | & | \\
H - & C - C - & \ddot{O} - H \\
& | & | \\
& H & H
\end{array}
\qquad
\begin{array}{cccc}
& & H & H \\
& & | & | \\
H - \ddot{O} - & C - & C - & \ddot{O} - H \\
& & | & | \\
& & H & H
\end{array}$$

ethanol   ethylene glycol

Figure 5.23

Lewis structures of ethanol and ethylene glycol. The –OH groups are shown in red.

The H in the –OH group can hydrogen bond, just as was the case for water (Figure 5.24). This is why water and ethanol have a great affinity for each other. Any bartender can tell you that alcohol and water form solutions in all proportions. Again, both molecules are polar and *like dissolves like.*

Ethylene glycol is another example of an alcohol, sometimes called a "glycol." Ethylene glycol is added to water, such as the water in the radiator of your car, to keep it from freezing. It also is one of the VOCs that some water-based paints emit when drying, an additive to keep the paint from freezing. Examine its structural formula in Figure 5.23 to see that it has two –OH groups available for hydrogen bonding. These intermolecular attractions give high water-solubility to ethylene glycol, a necessary property for any antifreeze.

It often has been observed that "oil and water don't mix." Water molecules are polar, and the hydrocarbon molecules in oil are nonpolar. When in contact, water molecules tend to attract to other water molecules; in contrast, hydrocarbon molecules stick with their own. Since oil is less dense than water, oil slicks float on top (Figure 5.25).

> In connection with indoor air quality, Chapter 1 mentioned propylene glycol, a "glycol" used as an antifreeze in paints.

Your Turn 5.22   More About Hydrocarbons

Hydrocarbon molecules such as pentane and hexane contain C–H and C–C bonds. Use the electronegativity values in Table 5.1 to determine whether these bonds are polar or nonpolar.

Since water is a poor solvent for grease and oil, we cannot use water to wash these off. Instead, we wash our hands (and clothes) with the aid of soaps and detergents. These compounds are **surfactants,** compounds that help polar and nonpolar compounds to mix, sometimes called "wetting agents." The molecules of surfactants contain both polar and nonpolar groups. The polar groups allow the surfactant to dissolve in water while the nonpolar ones are able to dissolve in the grease.

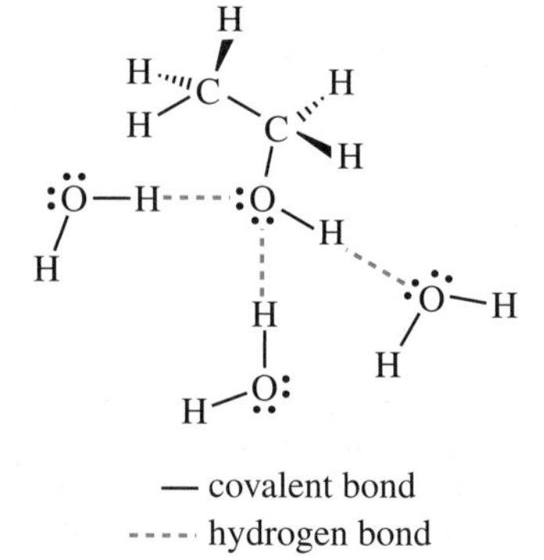

— covalent bond
---- hydrogen bond

Figure 5.24

Hydrogen bonding between an ethanol molecule and three water molecules.

Figure 5.25

Oil and water do not dissolve in each other.

Another way to dissolve nonpolar molecules is to use nonpolar solvents. Like dissolves like! Worldwide, the production of nonpolar solvents (sometimes called "organic solvents") is estimated to be 15 billion kilograms. These solvents are widely used, including in the production of drugs, plastics, paints, cosmetics, and cleaning agents. For example, dry cleaning solvents typically are chlorinated hydrocarbons. One example, "perc," is a cousin of ethene. Take ethene (sometimes called ethylene), a compound with a C=C double bond, and replace all the H atoms with Cl atoms. The result is tetrachloroethylene, also called perchloroethylene. Its nickname is perc.

Look for more about ethylene (ethene) in Chapter 9 on polymers.

ethylene   tetrachloroethylene ("perc")

Perc and other chlorinated hydrocarbons like it are carcinogens or suspected carcinogens. They have serious health consequences whether we are exposed to them in the workplace or as contaminants of our air, water, or soil.

Green chemists aim to redesign processes so that they don't require solvents. But if this is not possible, they try to replace harmful solvents like perc with ones that are friendly to the environment. One possibility is liquid carbon dioxide. Under conditions of high pressure, the gas you know as CO_2 can condense to form a liquid! Compared with organic solvents, $CO_2(l)$ offers many advantages. It is nontoxic, nonflammable, chemically benign, non-ozone-depleting, and it does not contribute to the formation of smog. Although you may be concerned with the fact that it is a greenhouse gas, carbon dioxide that is used as a solvent is a recovered waste product from industrial processes and it is generally recycled.

Adapting liquid CO_2 to dry cleaning posed a challenge, as it is not very good at dissolving oils, waxes, and greases found in soiled fabrics. To make carbon dioxide a better solvent, Dr. Joe DeSimone, a chemist and chemical engineer at the University of North Carolina–Chapel Hill, developed a surfactant to use with $CO_2(l)$. For his work, DeSimone received a 1997 Presidential Green Chemistry Challenge Award. His breakthrough process paves the way for designing environmentally benign, inexpensive, and easily recyclable replacements for conventional organic and water solvents currently in use. DeSimone was instrumental in the beginnings of Hangers Cleaners, a dry cleaning chain that uses the process that he developed.

Consider This 5.23   Liquid CO₂ as a Solvent

a. Which of the six principles of Green Chemistry (see inside the front cover) are met by the use of liquid carbon dioxide as a solvent to replace organic solvents? Explain.
b. Comment on this statement: "Using carbon dioxide as a replacement for organic solvents simply replaces one set of environmental problems with another."
c. If a local dry cleaning business switched from "perc" to carbon dioxide, how might this business report a different Triple Bottom Line?

The tendency of nonpolar compounds to dissolve in other nonpolar substances explains how fish and animals accumulate nonpolar substances such as PCBs (polychlorinated biphenyls) or the pesticide DDT (dichlorodiphenyltrichloroethane) in their fatty tissues. When fish ingest these, the molecules are stored in body fat (nonpolar) rather than in the blood (polar). PCBs can interfere with the normal growth and development of a variety of animals, including humans, in some cases at concentrations of trillionths of a gram per liter.

The higher you go on the food chain, the greater concentrations of harmful nonpolar compounds like DDT you find. This is called **biomagnification,** the increase in concentration of certain persistent chemicals in successively higher levels

PCBs (mixtures of highly chlorinated compounds) were widely used as coolants in electrical transformers until banned in 1977. Like CFCs, they do not burn easily. They were released into the environment during manufacture, use, and disposal.

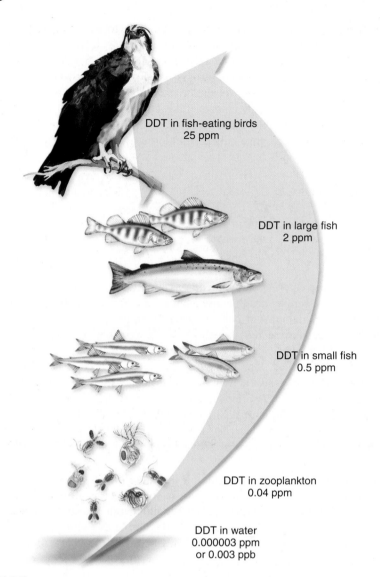

DDT in fish-eating birds
25 ppm

DDT in large fish
2 ppm

DDT in small fish
0.5 ppm

DDT in zooplankton
0.04 ppm

DDT in water
0.000003 ppm
or 0.003 ppb

Figure 5.26

Organisms in the water take up and store DDT. They are eaten by larger creatures that in turn are eaten by still larger ones. Creatures highest on the food chain have the highest concentration of DDT.

Source: From William and Mary Ann Cunningham. *Environmental Science: A Global Concern, 10th ed.,* 2008. Reprinted with permission of the McGraw-Hill Companies, Inc.

of a food chain. Figure 5.26 shows a biomagnification process that was studied extensively in the 1960s. At that time, DDT was shown to interfere with the reproduction of peregrine falcons and other predatory birds at the top of their food chain.

5.10 | Protecting Our Drinking Water: Federal Legislation

One way or another, many different substances get added to fresh water. Is this water safe to drink? The answer depends on *what* is present in the water, *how much* of it is present, and *how much* of it you drink in a day. In this section, we address issues relating to water quality.

The need to keep public water supplies safe to drink has long been recognized. In 1974, the U.S. Congress passed the Safe Drinking Water Act (SDWA) in response to public concern about harmful substances in the water supply. The aim of this act, as amended in 1996, was to ensure potable water to those who depend on community water supplies. As required by the SDWA, contaminants that may be health risks are regulated by the

The SDWA does not apply to the 10% of the people in the United States whose water comes from private wells.

Environmental Protection Agency (EPA). The EPA sets legal limits for contaminants according to their toxicities (Table 5.11). These limits also take into account the practical realities that water utilities face in trying to remove the contaminants with available technology.

For each water-soluble contaminant, the EPA established a **maximum contaminant level goal (MCLG),** the maximum level of a contaminant in drinking water at which no known or anticipated adverse effect on human health would occur. Expressed in parts per million or parts per billion, this level has no ill effects for a person weighing 70 kilograms (154 pounds) who consumes 2 liters of water every day for 70 years. Each MCLG allows for uncertainties in data collection and for how different people might react to each contaminant. An MCLG is *not* a legal limit to which water systems must comply; rather, it is a goal based on human health considerations. For known carcinogens, the EPA has set the health goal at zero under the assumption that *any* exposure presents a cancer risk.

Before regulatory action can be taken against a water utility, the concentration of an impurity must exceed the **maximum contaminant level (MCL),** the legal limit for the concentration of a contaminant expressed in parts per million or parts per billion. The EPA sets legal limits for each impurity as close to the MCLG as possible, keeping in mind any practical realities that may make it difficult to achieve the goals. Except for contaminants regulated as carcinogens (for which the MCLG is set at zero), most legal limits and health goals are the same. Even when less strict than the MCLGs, the MCLs still provide substantial public health protection.

 Consider This 5.24 **What's in Drinking Water?**

Table 5.11 is merely a starting point for information available about contaminants in drinking water. The EPA Office of Ground Water and Drinking Water offers a consumer fact sheet on dozens of contaminants. Both general summaries and technical fact sheets are available; the latter is recommended.

a. Select a contaminant listed in Table 5.11. How does it get into the water supply?
b. How would you know if it were in your drinking water? Is your state one of the top states that releases it? The textbook's website contains useful links as well as hints on how to locate the data.

Consider This 5.25 **Understanding MCLGs and MCLs**

Most people are unfamiliar with the terms MCLG and MCL from the Safe Drinking Water Act. How would you explain these abbreviations to the general public? Prepare an outline for a presentation. Be prepared to answer questions from the audience, including why MCLs are not set to zero for all carcinogens.

Water legislation continually needs to be updated. In part, this need arises because water chemists keep improving their ability to detect what is in the water. But the need

Table 5.11	MCLGs and MCLs for Drinking Water	
Contaminant	**MCLG (ppm)**	**MCL (ppm)**
cadmium (Cd^{2+})	0.005	0.005
chromium (Cr^{3+}, CrO_4^{2-})	0.1	0.1
lead (Pb^{2+})	0	0.015
mercury (Hg^{2+})	0.002	0.002
nitrates (NO_3^-)	10	10
benzene (C_6H_6)	0	0.005
trihalomethanes ($CHCl_3$ and others)	0	0.080

Look for more examples of trihalomethanes in the next section.

also arises because our knowledge base is growing. MCL limits should be raised or lowered as we learn more about toxicity. Currently, more than 80 contaminants are regulated:

- metal ions such as Cd^{2+}, Cr^{3+}, CrO_4^{2-}, Hg^{2+}, Cu^{2+}, and Pb^{2+}
- nonmetal ions such as NO_3^-, F^-, and various arsenic-containing ions
- miscellaneous compounds, including pesticides, industrial solvents, and compounds associated with plastics manufacturing
- radioisotopes, including radon and uranium
- biological agents, including *Cryptosporidium* and intestinal viruses

Depending on the particular contaminant, MCLs range from about 10 ppm to less than 1 ppb. Some contaminants interfere with liver or kidney function. Others can affect the nervous system if ingested over a long period at levels consistently above the legal limit (MCL). For example, unlike many contaminants, lead is a cumulative poison. Lead pipes and solder were once commonly used in water distribution systems. When ingested by humans and animals, lead accumulates in bones and the brain, causing severe and permanent neurological problems. Severe exposure in adults causes symptoms such as irritability, sleeplessness, and irrational behavior. Lead is a particular problem for children because Pb^{2+} can be incorporated rapidly into bone along with Ca^{2+}. Since children have less bone mass than adults, some Pb^{2+} may remain in the blood where it can damage cells, especially in the brain. Children may suffer mental retardation and hyperactivity as a result of lead exposure, even at relatively low concentrations.

Fortunately, very little lead is present in most public water supplies. Amounts exceeding allowable limits are estimated to be present in less than 1% of public water supply systems, and these serve less than 3% of the U.S. population. Most of this lead comes from corrosion of plumbing systems, not from the source water itself. When lead is reported, consumers are advised to take simple steps to minimize exposure, such as letting water run before using and using only cold water for cooking. Both actions minimize the chances of ingesting dissolved Pb^{2+}.

Until recently, the MCL for Pb^{2+} in drinking water was 15 ppb. In 1992, the EPA converted this to an "action level," meaning that the EPA will take legal action if 10% of tap water samples exceed 15 ppb. The hazard from lead is so great that the EPA has established an MCLG of 0, even though lead is not a carcinogen.

The chemical symbol Pb comes from the Latin name for lead, *plumbum*. The word *plumbing* comes from this Latin word as well and harkens back to the time when most water pipes were made of lead.

Cold water is recommended because some lead compounds, notably $PbCl_2$, are more soluble in hot water than in cold.

Your Turn 5.26   Comparing Lead Content

Two samples of drinking water contained lead ion. One had a concentration of 20 ppb; the other a concentration of 0.003 mg/L.

  a. Which sample has higher concentration of lead ion? Explain.
  b. How does each sample compare with the current acceptable limit?

Whereas contaminants such as lead cause chronic long-term health problems, other substances in drinking water present more immediate and acute effects. For example, in infants, nitrate ion (NO_3^-) may be converted into nitrite ion (NO_2^-), a substance that limits blood's ability to carry oxygen. Infants who drink formula made from water containing high levels of nitrate ion may experience difficulty breathing and possibly permanent brain damage from lack of oxygen. Although a maximum contaminant level (MCL) is set for nitrate ion in drinking water, this level may be exceeded for a variety of reasons, including fertilizer or manure runoff that gets into well water. Figure 5.27 shows water quality data for nitrate ion in California. As you can see, some water sources exceeded the MCL of 10 ppm. Because nitrate is toxic to infants, monitoring nitrate levels *and* informing communities of any violations are important.

Water can also be contaminated by biological agents such as bacteria, viruses, and protozoa. Examples include *Cryptosporidium* and *Giardia*. News media warnings announcing a "boil-water emergency" are typically the result of a "total coliform" violation. Coliforms are a broad class of bacteria that live in the digestive tracts of humans and other animals. Most are harmless. The presence of a high coliform concentration in water

In the United States, about 1000 nitrate violations are reported yearly. These violations usually affect a small number of people and only for a short period of time. In contrast, the number of microbial violations is up to 10 times this value.

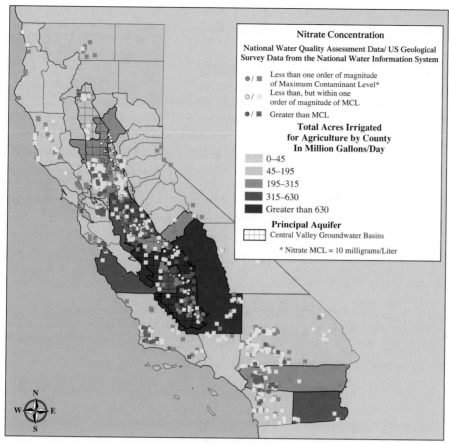

Figure 5.27

Map showing nitrate concentrations from California domestic groundwater wells and agricultural irrigation.

Source: Environmental Waikato, 2000.

usually indicates that the water-treatment or distribution system is allowing fecal contamination to enter the drinking water supply. Diarrhea, cramps, nausea, and vomiting—the symptoms of microbial-related illness—are generally not serious for a healthy adult but can be life-threatening for the very young, the elderly, or those with weakened immune systems.

Consider This 5.27   A Cryptic Microbe

The EPA surface water-treatment rules require systems using surface water or groundwater under the direct influence of surface water to remove or deactivate 99% of *Cryptosporidium*.

a. What is *Cryptosporidium* and how does it get into drinking water? What are its potential health effects?

b. Regarding *Cryptosporidium*, what is the LT2 Rule (Long Term 2 Surface Water Treatment Rule) and when did it take effect?

In addition to the Safe Drinking Water Act, other federal legislation controls pollution of lakes, rivers, and coastal areas. The Clean Water Act (CWA), passed in 1974 by Congress and amended several times, provided the foundation for reducing surface water pollution. The CWA established limits on the amounts of pollutants that industry can discharge, removing over a billion pounds of toxic pollutants from U.S. waters every year. In keeping with the new trend toward green chemistry, industries are finding ways both to convert waste materials into useful products and to design processes that neither use toxic substances nor harm water quality. Improvements in surface water quality have at least two major beneficial effects. First, they reduce the amount of cleanup needed for public drinking water supplies. And second, they result in a more healthful natural environment for aquatic organisms. In turn, more healthy aquatic ecosystems have many indirect benefits for humans.

5.11 | Water Treatment

This section explores the chemistry that takes place at a local drinking water treatment plant. We assume that the plant treats lake water, which is true in many municipalities. For example, if you live in Chicago, nearby Lake Michigan is the source. And if you live in San Francisco, the water comes from a reservoir in the Hetch Hetchy valley, over a hundred miles away (see Figure 5.9).

In a typical water treatment plant (Figure 5.28), the first step is to pass the water through a screen that physically removes items such as weeds, sticks, and beverage bottles. The next step is to add aluminum sulfate and calcium hydroxide. Take a moment to review these two chemicals.

Your Turn 5.28   Water Treatment Chemicals

a. Write chemical formulas for these ions: sulfate, hydroxide, calcium, and aluminum.
b. What compounds can be formed from these four ions? Write their chemical formulas.
c. The hypochlorite ion plays a role in water purification. Write chemical formulas for sodium hypochlorite and calcium hypochlorite.

Aluminum sulfate and calcium hydroxide are flocculating agents; that is, they react in water to form a sticky floc (gel) of aluminum hydroxide. This gel collects suspended clay and dirt particles on its surface.

$$Al_2(SO_4)_3(aq) + 3\ Ca(OH)_2(s) \longrightarrow 2\ Al(OH)_3(s) + 3\ CaSO_4(aq) \qquad \text{[5.7]}$$

As the $Al(OH)_3$ gel slowly settles, it carries particles with it that were suspended in the water (see Figure 5.28). Any remaining particles are removed as the water is filtered through charcoal or gravel and then sand.

The crucial step comes next—disinfecting the water to kill disease-causing microbes. In the United States, this is most commonly done with chlorine-containing compounds. Chlorination is accomplished by adding chlorine gas (Cl_2), sodium hypochlorite

Chlorine only can kill the microorganisms with which it comes in contact. Chlorine does not kill bacteria or viruses that hide inside particles of silt or clay. This is one reason why particles need to be removed before the chlorination step.

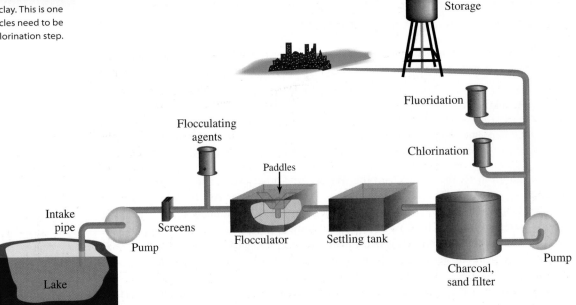

Figure 5.28
Typical municipal water treatment facility.

(NaClO), or calcium hypochlorite (Ca(ClO)$_2$). All of these generate the antibacterial agent hypochlorous acid, HClO. A very low concentration of HClO, 0.075 to 0.600 ppm, remains to protect the water against further bacterial contamination as it passes through pipes to the user. **Residual chlorine** refers to the chlorine-containing chemicals that remain in the water after the chlorination step. These include hypochlorous acid (HClO), the hypochlorite ion (ClO$^-$), and dissolved elemental chlorine (Cl$_2$).

Before chlorination, thousands died in epidemics spread via polluted water. In a classic study, John Snow, an English physician, was able to trace a mid-1800s cholera epidemic in London to water contaminated with the excrement of cholera victims. A more recent example occurred in 2007 in war-torn Iraq. After extremists had put chlorine tanks on suicide truck bombs earlier that year, authorities kept tight controls on chlorine. The chlorine killed two dozen people in several attacks, sending up noxious clouds that left hundreds of people panicked and gasping for breath. At one point, a shipment of 100,000 tons of chlorine was held up for a week at the Jordanian border amid fears for its safe passage through Iraq. With the water infrastructure disrupted and the quality of water and sanitation poor, levels of fecal coliform bacteria increased dramatically, resulting in thousands of Iraqis contracting cholera.

Even in peacetime when the transportation of chlorine is relatively safe, chlorination has its drawbacks. The taste and odor of residual chlorine can be objectionable and is commonly cited as a reason why people drink bottled water or use filters to remove residual chlorine. A more serious drawback is the reaction of residual chlorine with other substances in the water to form by-products at concentrations that may be toxic. The most widely publicized, **trihalomethanes (THMs),** are compounds such as CHCl$_3$ (chloroform), CHBr$_3$ (bromoform), CHBrCl$_2$ (bromodichloromethane), and CHBr$_2$Cl (dibromochloromethane) that form from the reaction of chlorine or bromine with organic matter in drinking water. Like HClO, hypobromous acid (HBrO) used to disinfect spa tubs can generate trihalomethanes.

NaClO is present in Clorox and other brands of laundry bleach. Ca(ClO)$_2$ is commonly used to disinfect swimming pools.

HClO is sometimes written as HOCl. The second shows the order in which the atoms are bonded.

Your Turn 5.29   THMs at a Glance

a. Draw Lewis structures for any two THM molecules.
b. THMs differ from CFCs in their chemical composition. How?
c. THMs differ from CFCs in their chemical properties. How?

Answer

c. CFCs are chemically inert and nontoxic. In contrast, THMs are chemically reactive and quite toxic.

Many European and a few U.S. cities use ozone to disinfect their water supplies. In Chapter 1, we discussed tropospheric O$_3$ as a serious air pollutant. But in water treatment, the toxicity of O$_3$ serves a beneficial purpose. One advantage is that a lower concentration of ozone than chlorine is required to kill bacteria. Furthermore, ozone is more effective than chlorine against water-borne viruses.

But ozonation also comes with disadvantages. One is cost. Ozonation only becomes economical for large water-treatment plants. Another is that ozone decomposes quickly and hence does not protect water from possible contamination as it is piped through the municipal distribution system. Consequently, a low dose of chlorine must be added to ozonated water as it leaves the treatment plant.

Disinfecting water using ultraviolet (UV) light is gaining in popularity. By UV, we mean UV-C, the high-energy UV radiation that can break down DNA in microorganisms, including bacteria. Disinfection with UV-C is fast, leaves no residual by-products, and is economical for small installations, including rural homes with unsafe well water. Like ozone, however, UV-C does not protect the water after it leaves the treatment site. Again, a low dose of chlorine must be added.

Depending on local needs, one or more additional purification steps may be taken after disinfection at the water-treatment facility. Sometimes the water is sprayed into the air to remove volatile chemicals that create objectionable odors and taste. If little natural

See Chapter 2 for more about UV light.
See Chapter 12 for more about DNA.

In water, sodium fluoride dissolves to form Na$^+$(aq) and F$^-$(aq).

fluoride ion is present in the water supply, some municipalities add fluoride ion (~1 ppm NaF) to protect against tooth decay. Learn more about fluoridation in the next activity.

 Consider This 5.30   **The Fluoride Ion Controversy**

Community water fluoridation is cited as one of 10 great public health achievements of the 20th century by the U.S. Centers for Disease Control and Prevention. In 2005, the American Dental Association celebrated "20 years of community water fluoridation." But in 2006, the U.S. National Academies published a report claiming that this practice can damage bones and teeth, and that Federal standards may put children at risk. List the arguments on both sides. Your class might use this topic for a lively debate.

We just described how water is treated before it is ready to drink out of the tap. But once we turn on this tap, we start the process of getting the water dirty again. We add waste to the water each time it leaves our bathrooms in a toilet flush, runs down the drain after a soapy shower, or goes down the sink after we wash the dishes. Clearly it makes sense to use as little water as possible because if we dirty it, it has to be cleaned again before it is released back to the environment. Remember Green Chemistry! It is better to prevent waste than to treat or clean up waste after it is formed.

How do we clean sewage and other wastes from water? If the drains in your home are connected to a municipal sewage system, then the wastewater flows to a sewage treatment plant. Once there, it undergoes similar cleaning processes to those for water treatment, with the exception of end-stage chlorination before it is released back to the environment.

Cleaning sewage is more complicated, though, because it contains waste in the form of organic compounds and nitrate ions. To many aquatic organisms, this waste is a source of food! As these organisms feed, they deplete oxygen from surface waters. **Biological oxygen demand (BOD)** is a measure of the amount of dissolved oxygen microorganisms use up as they decompose organic wastes found in water. A low BOD is one indicator of good quality water, whereas a high BOD indicates polluted water.

Typical BOD values: A pristine river, 1 mg/L; effluent from a municipal sewage treatment plant, 20 mg/L; and untreated sewage, 200 mg/L.

Nitrates and phosphates both contribute to BOD, as these ions are important nutrients for aquatic life. An overabundance of either can disrupt the normal flow of nutrients and lead to algal blooms that clog waterways and deplete oxygen from the water. In turn, this reduced oxygen can lead to massive fish kills. The problem of reduced oxygen in water is compounded by the fact that the solubility of oxygen in water is so very low in the first place.

Flood waters and agricultural runoff containing nitrates and phosphates disrupt ecosystems in the Mississippi Delta. Chapter 11 (Figure 11.15) shows a stunning photo.

Some treatment plants are using wetland areas to capture nutrients such as nitrates and phosphates before the water is returned to the surface water or recharges the groundwater. Plants and soil microorganisms in these wetland areas (marshes and bogs) facilitate nutrient recycling, thus reducing the nutrient load in the water.

If treated sewage water is clean enough, why not just use it as a source of drinking water? Los Angeles, California, has been considering doing this very thing. Los Angeles has an arid climate, and the Colorado River cannot supply enough water to meet the current needs of those who live there. Water conservation helps, but Los Angeles still is challenged by not having enough water. The next activity can help you to explore the issues that the citizens of metropolitan areas face.

Consider This 5.31   **Uses for Treated Sewage Water**

Use resources of the textbook's website to answer these questions.

a. Name the pros and cons of using treated sewage in agricultural practices.
b. If the quality of the water produced from the sewage treatment process matched the quality of the water in our current drinking water system, would you accept treated sewage water as drinking water? Comment either way.

5.12 | Water Solutions for Global Challenges

As we saw at the opening of this chapter, 2005 began the United Nations International Decade for Action called Water for Life. These words frame the U.N. call to action: "Water is crucial for sustainable development, including the preservation of our natural environment and the alleviation of poverty and hunger. Water is indispensable for human health and well-being."

In this final section, we showcase three efforts that demonstrate how water connects to sustainable development. The first relates to the production of fresh water from salt water. The second describes how individuals in developing nations can purify their own drinking water. And the third relates to industry—a green chemistry solution for cotton production.

Fresh Water from Salt Water

"Water, water everywhere, nor any drop to drink." These words from *The Rime of the Ancient Mariner* are as true today as they were in 1798 when written by Samuel Taylor Coleridge. The high salt content (3.5%) of sea water makes it unfit for human consumption. Neither the ancient mariner nor any who live by the seacoast today can survive drinking sea water.

Today, in certain parts of the world where fresh water is not available, we are able to tap the sea as a source of water for both agriculture and drinking. **Desalination** is any process that removes ions from salt water, thus producing potable water. According to the International Desalination Association, in 2008, over 13,000 desalination plants worldwide together produced over 60 billion L/day of water. The world's largest single desalination plant is in the United Arab Emirates and produces over 450 million L/day (Figure 5.29). Besides the Middle East, desalination facilities have been built in Europe, Australia, and North America.

One means of desalination is **distillation,** a separation process in which a liquid solution is heated and the vapors are condensed and collected. Impure water is heated. As the water vaporizes, it leaves behind most of its dissolved impurities. Distillation requires energy! Figure 5.30 shows this energy being provided by a Bunsen burner in one case and by the Sun in the other. Recall from Section 5.2 that water has an unusually high specific heat and requires an unusually large amount of energy to convert to a vapor. Both result from the extensive hydrogen bonding in water. The so-called distilled

> Recall from Section 5.3 that the average U.S. citizen uses about 390 L water/day.

> A process similar to distillation takes place in the natural water cycle. Water evaporates, condenses, and then falls as rain or snow.

Figure 5.29

Desalination plant (right) at Jebel Ali in the United Arab Emirates.

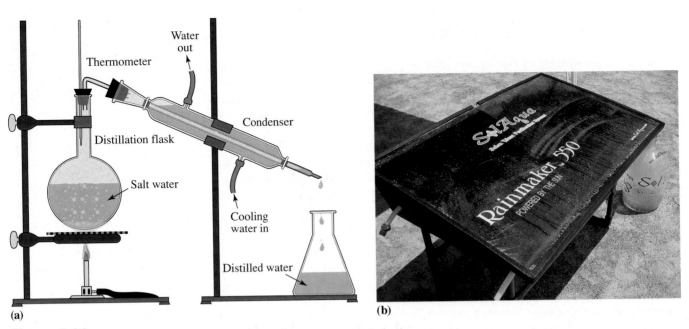

Figure 5.30

(a) Laboratory distillation apparatus. (b) Table-top solar still.

water that you use in introductory laboratory experiments most likely was "deionized" rather than distilled. Again, distillation requires too much energy.

Large-scale distillation operations used today often involve either multiple-effect distillation (MED) or multistage flash (MSF) evaporation. These newer technologies have increased energy efficiency over the basic distillation process shown in Figure 5.30a. However, because fossil fuels currently supply the majority of energy necessary to distill fresh water, these new technologies are not sustainable. An alternative to using fossil fuels to heat water is to use smaller solar distillation units (Figure 5.30b).

> ### Consider This 5.32   Other Forms of Distillation
>
> Both MED and MSF evaporation separate salt from water by changing the pressure of heated water as it passes through different compartments. List two ways in which these newer technologies are improvements over the basic process of distillation. The textbook's website provides some helpful links.

Reverse osmosis is another desalination technique. **Osmosis** is the passage of water through a semipermeable membrane from a solution that is less concentrated to a solution that is more concentrated. The water diffuses through the membrane and the soluble does not. This is why the membrane is called "semipermeable." However, with an input of energy, osmosis can be reversed. **Reverse osmosis** uses pressure to force the movement of water through a semipermeable membrane from a solution that is more concentrated to a solution that is less concentrated. When using this process to purify water, pressure is applied to the saltwater side, forcing water through the membrane to leave the ions behind (Figure 5.31).

Reverse osmosis desalination requires a tremendous amount of energy input to supply the pressure needed to make this process work. Despite this high requirement of energy, almost 60% of desalination plants use reverse osmosis technology due to the high quality of water produced. It can be used to produce some bottled water. Portable units are suitable for use on sailboats (Figure 5.32).

Point-of-Use LifeStraws

Imagine that your water source was contaminated with bacteria. A billion people are sickened or die each year due to cholera, diarrhea, typhoid, and other ailments caused by microbes in untreated water. A European company, Vestergaard Frandsen,

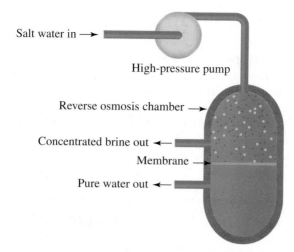

Figure 5.31

Water purification by reverse osmosis.

Figure 5.32

A small reverse osmosis apparatus for converting sea water to potable water.

The LifeStraw disinfects using iodine and possibly other halogens as well.

has developed the LifeStraw, a type of pipe filter that removes virtually all bacteria, viruses, and parasites from water. Aptly named, this device is used to suck water through filters and a disinfection unit (Figure 5.33). This unit can be used to drink water from a stream, river, or lake. It costs about $3, can purify about 700 liters of water, and lasts about a year. For household use, the LifeStraw Family unit filters a minimum of 18,000 liters of water per family for up to 3 years. The units cost about $25 each and are distributed with the help of charitable organizations.

The LifeStraw has limitations, however. It is not a long-term solution to the lack of potable water. In addition, it filters neither heavy metals such as arsenic and lead nor some of the protozoan parasites responsible for diarrhea. However, it does provide an interim stopgap in regions where there is plenty of fresh water contaminated with microbes. Although it is not known how many people have used the LifeStraw, charitable organizations such as the Carter Center purchased 23 million units to be distributed in developing nations.

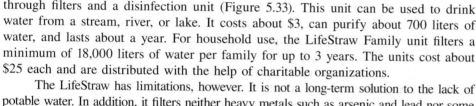

Skeptical Chemist 5.33   LifeStraw

The company that produces LifeStraw has a set of FAQs on the web. One reads: "Does LifeStraw filter heavy metals like arsenic, iron, and fluoride?" What might the Skeptical Chemist say in response to such a question? The textbook's website provides a link to the answer provided (which we hope isn't what you would provide).

(a)   **(b)**

Figure 5.33

(a) Boys using personal LifeStraw to drink. **(b)** The portable LifeStraw unit is easy to carry.

The Cotton Industry

A cotton T-shirt, anybody? Or perhaps pants, socks, or a baseball cap? Another solution has targeted a problem associated with cotton clothing. Globally, over 40 billion pounds of this very popular material is produced each year. Even with the growing trend toward synthetic fibers like nylon and polyester, cotton still accounts for over half the market share for apparel in the United States. Currently, the United States is the world's largest exporter of cotton and exports close to 50% of the world's supply. Cotton serves not only as the fiber for clothing, but also for many other consumer products. Explore these in the next activity.

Recall the water requirement for a cotton T-shirt, listed in Table 5.2.

Figure 5.34
Raw cotton has a waxy cuticle (outer layer) that must be removed.

Consider This 5.34   **Water Footprint Content of a T-Shirt**

In an earlier section of this chapter, we pointed out that the water footprint of a 250-g cotton T-shirt was 2700 L (see Table 5.3).

   a. List four ways that water is used in the production of a T-shirt.
   b. Other than for clothing, cotton has many other uses. Name three.
   c. For these uses, estimate whether the water footprint is larger, smaller, or roughly the same as for cotton clothing.

Answers
   b. Cotton uses include curtains, upholstery, netting, gauze bandages, and swabs.
   c. These all can be expected to have a large water footprint because of the water used both to grow and clean the cotton.

Even though cotton is a natural fiber, its production leaves a significant footprint on the environment. In Section 5.3, we mentioned that cotton was a thirsty crop that drank large amounts of water as it grew. Once produced, raw cotton must be treated before it can be bleached and dyed (Figure 5.34). In particular, the cuticle, or outermost layer of cotton must be removed. The process ("scouring") requires copious quantities of caustic chemicals, water, and energy. The wastewater produced has a BOD equal to that of raw sewage! In addition, the wastewater is contaminated with significant amounts of caustic chemicals, ones that weaken the cotton fibers.

In 2001, Novozymes won a Presidential Green Chemistry Challenge Award for developing an alternative process to remove the waxy cotton cuticle. The milder process ("biopreparation") uses an enzyme that breaks down the cuticle. As a result, cotton's environmental footprint improved. The BOD of wastewater dropped by more than 20%, the caustic chemicals were eliminated, and the amount of water, energy, and time required was significantly reduced. Recall one of the principles of Green Chemistry: *It is better to use and generate substances that are not toxic.*

Your Turn 5.35   **Green Chemistry in Action**

We just named one of the six principles of Green Chemistry (see inside front cover) that is met by through the Novozyme "biopreparation" process. Which others apply as well?

This is another example of how green chemistry improves the Triple Bottom Line (economic, societal, environmental) and helps to move us toward more sustainable practices. Reduced and less toxic waste benefits the environment. Using less energy and materials reduces the cost. And society gains a more robust cotton fiber and conserves water.

Consider This 5.36　　The Future of Water

a. Select two of the Green Chemistry principles listed on the inside front cover. For each, brainstorm an idea that might help us to use water more efficiently.
b. Identify an important global water issue. Suggest two factors that make it important. Name two ways in which people currently are addressing this issue.

Conclusion

Like the air we breathe, water is essential to our lives. It bathes our cells, transports nutrients through our bodies, provides most of our body mass, and cools us when it evaporates. Water also is central to our *way of life*. We drink it, cook with it, clean things in it, use it to irrigate our crops, and manufacture goods with it. As we do these things, we add waste to the water. Although fresh water purifies itself through a cycle of evaporation and condensation, we humans are dirtying water faster than nature can regenerate clean water.

Remember the first principle of Green Chemistry: *It is better to prevent waste than to treat or clean up waste after it is formed.* So catch the rain water and use it on a garden, rather than letting it run off and join the streams of runoff that pick up pollutants. Instead of using the garbage disposal to grind up food wastes, put the scraps in a compost pile and save the tap water. Turn off the faucet when you are brushing your teeth, limit your time in the shower, and fix that dripping faucet and running toilet!

You may feel like your efforts are a mere drop in a much larger bucket. Indeed they are. But remember that like the raindrop shown in the winning Earth Day poster below, your efforts are part of the bigger water picture on this planet (Figure 5.35).

Although fresh water is a renewable resource, the demands of population growth, rising affluence, and other global issues are amplifying shortages of this essential commodity. If we are to achieve sustainability we must think water! Think water! Your life and the lives of other creatures depend on it.

One little raindrop
Falling from the darkened sky
Landing in the lake

Figure 5.35
An Earth Day Haiku Poster Winner, 2008.

Chapter Summary

Having studied this chapter you should be able to:

- Describe how water is linked to life on this planet (introduction)
- Connect the electronegativity of atoms with the polarity of the bonds formed from these atoms (5.1)
- Describe hydrogen bonding and relate it to the properties of water (5.2)
- Compare the densities of ice and water and be able to account for the difference (5.2)
- Relate the specific heat of water to the roles played by water on the planet (5.2)
- Discuss the relationship between the properties of water and its molecular structure (5.2)
- Describe the major ways people use water on our planet (5.3)
- Discuss how the concept of water footprint shapes our view of water use (5.3)
- Connect global climate change with the supply and demand of water (5.4)
- Use concentration units: percent, ppm, ppb, and molarity (5.5)
- Discuss why water is such an excellent solvent for many (but not all) ionic and covalent compounds (5.5)
- Relate these terms: *cation, anion,* and *ionic compound* (5.6)

- Write the names and chemical formulas for ionic compounds, including those with common polyatomic ions (5.7)
- Describe what occurs when an ionic compound dissolves in water (5.8)
- Explain why some solutions conduct electricity and others do not (5.9)
- Describe the role of surfactants as solubility agents (5.9)
- Explain the saying "like dissolves like" and relate this to biomagnification (5.9)
- Understand the role of federal legislation in protecting safe drinking water (5.10)
- Contrast the maximum contaminant level goal (MCLG) and the maximum contaminant level (MCL) established by the EPA to ensure water quality (5.10)
- Discuss how drinking water can be made safe to drink (5.11)
- Understand the processes of distillation and reverse osmosis for producing potable water (5.12)
- Describe how green chemistry and its applications can contribute to clean water (5.9, 5.12)
- Summarize at least two possible solutions to our global water challenges (5.12)

Questions

Emphasizing Essentials

1. The chapter opens with these words: *"Neeru, shouei, maima, aqua.* In any language, water is the most abundant compound on the surface of the Earth."

   a. Explain the term *compound* and also why water is *not* an element.

   b. Draw the Lewis structure for water and explain why its shape is bent.

2. Today we are creating dirty water faster than nature can clean it for us.

   a. Name five daily activities that dirty the water.

   b. Name two ways in which polluting substances naturally are removed from water.

   c. Name five steps you could take to keep water cleaner in the first place.

3. Life on our planet depends on water. Explain each of these.

   a. Bodies of water act as heat reservoirs, moderating climate.

   b. Ice protects ecosystems in lakes because it floats rather than sinks.

4. Why might a water pipe break if left full of water during extended frigid weather?

5. Here are four pairs of atoms. Consult Table 5.1 to answer these questions.

N and C	N and H
S and O	S and F

   a. What is the electronegativity difference between the atoms?

   b. Assume that a single covalent bond forms between each pair of atoms. Which atom attracts the electron pair in the bond more strongly?

   c. Arrange the bonds in order of increasing polarity.

6. Consider a molecule of ammonia, NH_3.

   a. Draw its Lewis structure.

   b. Does the NH_3 molecule contain polar bonds? Explain.

   c. Is the NH_3 molecule polar? *Hint:* Consider its geometry.

   d. Would you predict NH_3 to be soluble in water? Explain.

7. In some cases, the boiling point of a substance increases with its molar mass.

   a. Does this hold true for hydrocarbons? Explain with examples. *Hint:* See Section 4.4.

b. Based on the molar masses of H_2O, N_2, O_2, and CO_2, which would you expect to have the lowest boiling point?

c. Unlike N_2, O_2, and CO_2, water is a liquid at room temperature. Explain.

8. Both methane (CH_4) and water are compounds of hydrogen and another nonmetal.

   a. Give four examples of nonmetals. In general, how do the electronegativity values of nonmetals compare with those of metals?

   b. How do the electronegativity values of carbon, oxygen, and hydrogen compare?

   c. Which bond is more polar, the C–H bond or the O–H bond?

   d. Methane is a gas at room temperature, but water is a liquid. Explain.

9. This diagram represents two water molecules in a liquid state. What kind of bonding force does the arrow indicate? Is this an *inter*molecular or *intra*molecular force?

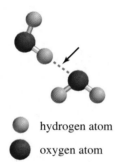

   ⬤ hydrogen atom

   ⬤ oxygen atom

10. The density of water at 0 °C is 0.9987 g/cm^3; the density of ice at this same temperature is 0.917 g/cm^3.

   a. Calculate the volume occupied at 0 °C by 100.0 g of liquid water and by 100.0 g of ice.

   b. Calculate the percentage increase in volume when 100.0 g of water freezes at 0 °C.

11. Consider these liquids.

Liquid	Density, g/mL
dishwashing detergent	1.03
maple syrup	1.37
vegetable oil	0.91

   a. If you pour equal volumes of these three liquids into a 250-mL graduated cylinder, in what order should you add the liquids to create three separate layers? Explain.

   b. Predict what would happen if a volume of water equal to the other liquids were poured into the cylinder in part **a** and the contents then were mixed vigorously.

12. Let's say the water in a 500-L drum represents the world's total supply. How many liters would be suitable for drinking? *Hint:* See Figure 5.8.

13. Based on your experience, how soluble is each of these substances in water? Use terms such as *very soluble, partially soluble,* or *not soluble.* Cite supporting evidence.

   a. orange juice concentrate

   b. household ammonia

   c. chicken fat

   d. liquid laundry detergent

   e. chicken broth

14. a. Bottled water consumption was reported to be 29 gallons per person in the United States in 2007. The 2000 U.S. census reported the population as 3.0×10^8 people. Given this, estimate the total bottled water consumption.

   b. Convert your answer in part **a** to liters.

15. NaCl is an ionic compound, but $SiCl_4$ is a covalent compound.

   a. Use Table 5.1 to determine the electronegativity difference between chlorine and sodium, and between chlorine and silicon.

   b. What correlations can be drawn about the difference in electronegativity between bonded atoms and their tendency to form ionic or covalent bonds?

   c. How can you explain on the molecular level the conclusion reached in part **b**?

16. For each of these atoms, draw a Lewis structure. Also draw the Lewis structure for the corresponding ion. *Hint:* Consult Tables 5.5 and 5.6.

   a. Cl   b. S

   c. Ne   **d.** Ba

   e. Li

17. Give the chemical formula and name of the ionic compound that can be formed from each pair of elements.

   a. Na and Br   b. Cd and S

   c. Ba and Cl   **d.** Al and O

   e. Rb and I

18. Write the chemical formula for each compound.

   a. calcium bicarbonate   **b.** calcium carbonate

   c. magnesium chloride   d. magnesium sulfate

19. Name each compound.

   a. $KC_2H_3O_2$ b. LiOH

   c. CoO   d. ZnS

   e. $Ca(ClO)_2$ f. Na_2SO_4

   g. $MnCl_2$ h. K_2O

20. Explain why $CoCl_2$ is named cobalt(II) chloride, whereas $CaCl_2$ is named calcium chloride.

21. The MCL for mercury in drinking water is 0.002 mg/L.

   a. Does this correspond to 2 ppm or 2 ppb mercury?

   b. Is this mercury in the form of elemental mercury ("quicksilver") or the mercury ion (Hg^{2+})?

22. The acceptable limit for nitrate, often found in well water in agricultural areas, is 10 ppm. If a water sample is found to contain 350 mg/L, does it meet the acceptable limit?

23. A student weighs out 5.85 g of NaCl to make a 0.10 M solution. What size volumetric flask does he or she need? *Hint:* See Figure 5.16.

24. Solutions can be tested for conductivity using this type of apparatus.

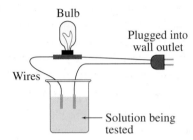

Predict what will happen when each of these dilute solutions is tested for conductivity. Explain your predictions briefly.

   a. $CaCl_2(aq)$

   b. $C_2H_5OH(aq)$

   c. $H_2SO_4(aq)$

25. An aqueous solution of KCl conducts electricity, but a solution of sucrose does not. Explain.

26. Based on the generalizations in Table 5.9, which compounds are likely to be water-soluble?

   a. $KC_2H_3O_2$

   b. LiOH

   c. $Ca(NO_3)_2$

   d. Na_2SO_4

27. For a 2.5 M solution of $Mg(NO_3)_2$, what is the concentration of each ion present?

28. Explain how you would prepare these solutions using powdered reagents and any necessary glassware.

   a. Two liters of 1.50 M KOH

   b. One liter of 0.050 M NaBr

   c. 0.10 L of 1.2 M $Mg(OH)_2$

29. a. A 5-minute shower requires about 90 L of water. How much water would you save for each minute that you shorten your shower?

   b. Running the water while you brush your teeth can consume another liter. How much water can you save in a week by turning it off?

30. Rank these in order of water footprint: producing the meat for a hamburger, growing an orange, growing a potato, and producing a pint of beer. Explain your reasoning.

Concentrating on Concepts

31. Explain why water is often called the *universal solvent*.

32. Is there any such thing as "pure" drinking water? Discuss what is implied by this term, and how the meaning of this term might change in different parts of the world.

33. Some vitamins are water-soluble, whereas others are fat-soluble. Would you expect either or both to be polar compounds? Explain.

34. A new sign is posted at the edge of a favorite fishing hole that says "Caution: Fish from this lake may contain over 1.5 ppb Hg." Explain to a fishing buddy what this unit of concentration means, and why the caution sign should be heeded.

35. This periodic table contains four elements identified by numbers.

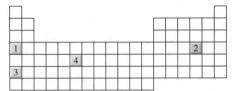

   a. Based on trends within the periodic table, which of the four elements would you expect to have the highest electronegativity value? Explain.

   b. Based on trends within the periodic table, rank the other three elements in order of decreasing electronegativity values. Explain your ranking.

36. A diatomic molecule XY that contains a polar bond *must* be a polar molecule. However, a triatomic molecule XY_2 that contains a polar bond *does not necessarily* form a polar molecule. Use some examples of real molecules to help explain this difference.

37. Imagine you are at the molecular level, watching water vapor condense.

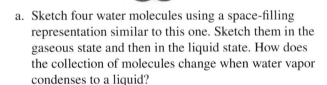

   a. Sketch four water molecules using a space-filling representation similar to this one. Sketch them in the gaseous state and then in the liquid state. How does the collection of molecules change when water vapor condenses to a liquid?

   b. What happens at the molecular level when water changes from a liquid to a solid?

38. Propose an explanation for the fact that NH_3, like H_2O, has an unexpectedly high specific heat. *Hint:* See Figure 3.12 for the Lewis structure and bond geometry in NH_3.

39. a. What type of bond holds together the two hydrogen atoms in the hydrogen molecule, H_2?

   b. Explain why the term *hydrogen bonding* does *not* apply to the bond within H_2.

40. Consider ethanol, an alcohol with the chemical formula of C_2H_5OH.

   a. Draw the Lewis structure for ethanol.

   b. A cube of solid ethanol sinks rather than floats in liquid ethanol. Explain this behavior.

41. The unusually high specific heat of water helps keep our body temperature within a normal range despite age, activity, and environmental factors. Consider some of the ways the body produces and loses heat. How would these differ if water had a low specific heat?

42. Health goals for contaminants in drinking water are expressed as MCLG, or maximum contaminant level goals. Legal limits are given as MCL, or maximum contaminant levels. How are MCLG and MCL related for a given contaminant?

43. Some areas have a higher than normal amount of THMs (trihalomethanes) in the drinking water. Suppose that you are considering moving to such an area. Write a letter to the local water district asking relevant questions about the drinking water.

44. Infants are highly susceptible to elevated nitrate levels because bacteria in their digestive tract convert nitrate ion into nitrite ion, a much more toxic substance.

    a. Give chemical formulas for both the nitrate ion and nitrite ion.

    b. Nitrite ion can interfere with the ability of blood to carry oxygen. Explain the role of oxygen in respiration. *Hint:* Review Sections 1.1 and 3.5 for more about respiration.

    c. Boiling nitrate-containing water will not remove nitrate ion. Explain.

45. Water quality in a chemistry building on campus was continuously monitored because testing indicated water from drinking fountains in the building had dissolved lead levels above those established by the Safe Drinking Water Act.

    a. What is the likely major source of the lead in the drinking water?

    b. Do the research activities carried out in this chemistry building account for the elevated lead levels found in the drinking water? Explain.

46. Explain why desalination techniques, despite proven technological effectiveness, are not used more widely to produce potable drinking water.

Exploring Extensions

47. In 2001, four countries in South America reached a historic agreement to share the immense Guarani Aquifer. This is significant because underground sources are routinely not considered by international law.

    a. Where is the Guarani Aquifer, and which four countries joined in this international "underground concordat"?

    b. What concerns did each country have about this aquifer that led to the agreement?

48. Liquid CO_2 has been used successfully for many years to decaffeinate coffee. Explain how and why this works.

49. How can you purify your water when you are hiking? Name two or three possibilities. Compare these

methods in terms of cost and effectiveness. Are any of these methods similar to those used to purify municipal water supplies? Explain.

50. Hydrogen bonds vary in strength from about 4 to 40 kJ/mol. Given that the hydrogen bonds between water molecules are at the high end of this range, how does the strength of a hydrogen bond *between* water molecules compare with the strength of a H–O covalent bond *within* a water molecule? Do your values bear out the assertion made in Section 5.2 that hydrogen bonds are about one tenth as strong as covalent bonds? *Hint:* Consult Table 4.4 for covalent bond energies.

51. Levels of naturally occurring mercury in surface water are usually less than 0.5 μg/L. The average intake of mercury from food is 2–20 μg daily, but may be much higher in regions where fish is a staple of the diet.

    a. Name three human activities that add Hg^{2+} ("inorganic mercury") to water.

    b. What is "organic mercury"? This chemical form of mercury tends to accumulate in the fatty tissues of fish. Explain why.

52. We all have the amino acid glycine in our bodies. Here is the structural formula.

$$H-\overset{\overset{\displaystyle H}{|}}{\underset{\underset{\displaystyle H}{|}}{N}}-\overset{\overset{\displaystyle H}{|}}{\underset{\underset{\displaystyle H}{|}}{C}}-\overset{\overset{\displaystyle :O:}{\|}}{C}-\overset{\cdot\cdot}{\underset{\cdot\cdot}{O}}-H$$

    a. Is glycine a polar or nonpolar molecule? Explain.

    b. Can glycine exhibit hydrogen bonding? Explain.

    c. Is glycine soluble in water? Explain.

53. Hard water may contain Mg^{2+} and Ca^{2+} ions. The process of water softening involves removing these ions.

    a. How hard is the water in your local area? One way to answer this question is to determine the number of water-softening companies in your area. Use the resources of the web, as well as ads in your local newspapers and yellow pages, to find out if your area is targeted for marketing water-softening devices.

    b. If you chose to treat your hard water, what are the options?

54. Suppose that you are in charge of regulating an industry in your area that manufactures agricultural pesticides. How will you decide if this plant is obeying necessary environmental controls? Which criteria affect the success of this plant?

55. Before the U.S. EPA banned their manufacture in 1979, PCBs were regarded as useful chemicals. What properties made them desirable? Besides being persistent in the environment, they bioaccumulate in the fatty tissues of animals. Use the electronegativity concept to show why PCB molecules are nonpolar and thus fat-soluble.

6 Neutralizing the Threat of Acid Rain

"I tell my son, go to see the corals now because soon it will be too late." James Orr, *Laboratory of Sciences of the Climate and Environment, France.*

Source: New Scientist, August 5, 2006. Ocean Acidification: the Other CO$_2$ Problem.

Coral reefs are massive structures found in shallow ocean waters. Often called "the rainforests of the ocean," reefs provide habitat for countless species of marine creatures. Attached to a reef or in the water nearby you may find sponges, mollusks, clams, crab, shrimp, sea urchins, sea worms, jellyfish, and many species of fish. Of benefit to humans, reefs protect fragile coastlines from powerful ocean waves. In some countries, they single-handedly support the tourism industry.

Coral reefs are alive. With lifetimes of hundreds of thousands of years, reefs can grow thousands of kilometers long, millimeter by millimeter. The tiny animals that form the reef grow and develop using nutrients and other essential chemical compounds dissolved in the ocean water. Why does James Orr want his son to go see the corals? Simply put, today's reefs may not exist tomorrow. Scientists estimate that at least a quarter of the reefs are now lost; very few are pristine. As energy use worldwide has climbed, changes in both the temperature and chemical composition of our atmosphere have occurred. In turn, changes have occurred in our oceans. At best, these changes slow the growth of coral reefs; at worst they damage the ecosystems established within the reefs. Here we are witness to another example of the tragedy of the commons.

> Over time, healthy reefs repair themselves. So in part, the question becomes how healthy the reefs were to begin with.

As you learned in earlier chapters, we burn fuels to harness their energy. Emissions from combustion include carbon dioxide, nitrogen oxides, and sulfur dioxide. These gases, especially the oxides of nitrogen and sulfur, are soluble in water, including the salt water of our oceans. They dissolve to produce acids; as a result, the water becomes more acidic. For example, over time oceans absorb 25–40% of the carbon dioxide that is emitted as a result of human activities. As more carbon dioxide is emitted, more carbon dioxide dissolves in the oceans. The resulting changes in the seawater have significant effects on the ocean ecosystems. For example, the increase in acidity leads to a reduction in the amount of carbonate ion available to build and maintain coral reefs.

> The tragedy of the commons was discussed in Chapter 1.

> Ocean acidification is one of several factors that harm coral reefs. Physical damage occurs with storms and warming ocean temperature; chemical damage through waste products dumped in the ocean.

Not only do the emissions from combustion dissolve in the oceans, but they also dissolve in water anywhere on the planet, including the rain, snow, and mist of our atmosphere. For example, when SO_2 and NO_x dissolve in rainwater, they fall back to the Earth in the form of acid rain. Just as added acidity damages ocean ecosystems, it damages the ecosystems of rivers and lakes as well.

> Recall from Chapter 5 that water is a polar compound. Nonpolar compounds such as CO_2 dissolve in water only to a small extent. In contrast, SO_2 is a polar compound and very soluble.

In Chapter 5 you learned about the special properties of water, what water is used for, how it tends to get dirty, and how we clean it. In this chapter, you will learn how certain compounds dissolve in water to produce acidic and basic solutions. In the right context, acids and bases are extremely useful compounds. We depend on them in many agricultural and manufacturing processes. Acids also impart flavors to the foods we eat. However, in the wrong place at the wrong time, acids can have devastating effects. Ocean acidification and acid rain are two examples. This chapter tells the story of both with an eye to helping you understand what is happening and why. But this story would not be complete without a discussion of bases and of pH, so we touch on these topics as well.

> Recall from Chapter 1 that NO_x is a shorthand notation for NO and NO_2.

Since people tend to be more familiar with acids than with bases, we begin with a discussion of acids and their properties. In what contexts have you already encountered acids? Before you read about acids in the next section, take a moment to do this activity.

Your Turn 6.1   Acids You Have Encountered

a. List the names for any three compounds that are acids.

b. In what context do you know of these acids? For example, is it from reading ingredient labels on foods? Did you run across these acids in some sport or activity? Did you read about these acids in a news article?

Figure 6.1

Citrus fruit contains both citric acid and ascorbic acid.

The plant dye litmus changes from blue to pink in acid. The term *litmus test* has also come to refer to something that quickly reveals a politician's point of view.

An antacid tablet dissolved in water generates a "fizz" that results from the carbonate in the tablet reacting with citric acid, also present in the tablet.

The notation *(aq)* is short for *aqueous*. Revisit equations 5.4 and 5.5 that show the formation of ions as a solute dissolves in water.

You will see the term *proton* again in Chapter 8 (proton exchange membranes) and Chapter 10 (the protonated form of drug molecules).

Here is the Lewis structure for the hydronium ion:

$$\left[\begin{array}{c} \overset{H}{H:\overset{..}{\underset{..}{O}}:H} \end{array} \right]^{+}$$

It obeys the octet rule.

6.1 | What Is an Acid?

We can approach acids either by listing their observable properties or by describing their behavior at the molecular level. Either way, the information is useful to our discussion, so we employ both approaches.

Historically, chemists identified acids by properties such as their sour taste. Although tasting is not a smart way to identify chemicals, you undoubtedly know the sour taste of acetic acid in vinegar. The sour taste of lemons comes from acids as well (Figure 6.1). Acids also show a characteristic color change with indicators such as litmus.

Another way to identify an acid is by its chemical properties. For example, under certain conditions acids can react with marble, eggshell, or the shells of marine creatures, causing them to dissolve. These materials all contain the carbonate ion (CO_3^{2-}) either as calcium carbonate or magnesium carbonate. An acid reacts with a carbonate to produce carbon dioxide. This gas is the "burp" when carbonate-containing stomach antacid tablets react with acids in your stomach. This chemical reaction also explains the dissolution of the skeletons of carbonate-based sea creatures such as coral in acidified oceans, as we will see in a later section.

At the molecular level, an **acid** is a compound that releases hydrogen ions, H^+, in aqueous solution. Remember that a hydrogen atom is electrically neutral and consists of one electron and one proton. If the electron is lost, the atom becomes a positively charged ion (H^+). Because only a proton remains, sometimes H^+ is referred to as a proton.

For example, consider hydrogen chloride (HCl), a compound that is a gas at room temperature. Hydrogen chloride is composed of HCl molecules. These dissolve readily in water to produce a solution that we name hydrochloric acid. As the polar HCl molecules dissolve, they become surrounded by polar water molecules. Once dissolved, these molecules break apart into two ions: $H^+(aq)$ and $Cl^-(aq)$. This equation represents the two steps of the reaction.

$$HCl(g) \xrightarrow{H_2O} HCl(aq) \longrightarrow H^+(aq) + Cl^-(aq) \qquad \textbf{[6.1]}$$

We also could say that HCl *dissociates* into H^+ and Cl^-. No HCl molecules remain in solution because they dissociate completely. Hydrochloric acid is a **strong acid;** that is, an acid that dissociates completely in aqueous solution.

There is a slight complication with the definition of acids as substances that release H^+ ions (protons) in aqueous solutions. By themselves, H^+ ions are much too reactive to exist as such. Rather, they attach to something else, such as water molecules. When dissolved in water, each HCl molecule donates a proton (H^+) to an H_2O molecule, forming H_3O^+, a hydronium ion. Here is a representation of the overall reaction.

$$HCl(aq) + H_2O(l) \longrightarrow H_3O^+(aq) + Cl^-(aq) \qquad \textbf{[6.2]}$$

The solution represented on the product side in *both* equations 6.1 and 6.2 is called hydrochloric acid. It has the characteristic properties of an acid because of the presence of H_3O^+ ions. Chemists often simply write H^+ when referring to acids (for example, in equation 6.1), but understand this to mean H_3O^+ (hydronium ion) in aqueous solutions.

Your Turn 6.2   **Acidic Solutions**

For each of these strong acids dissolved in water, write a chemical equation that shows the release of a hydrogen ion, H^+.

Hint: Remember to include the charges on the ions. The net charge on both sides of the equation should be the same.

a. HI*(aq)*, hydroiodic acid   **b.** HNO_3*(aq)*, nitric acid   **c.** H_2SO_4*(aq)*, sulfuric acid

Answer

c. $H_2SO_4(aq) \longrightarrow H^+(aq) + HSO_4^-(aq)$

Consider This 6.3   Are All Acids Harmful?

Although the word *acid* may conjure up all sorts of pictures in your mind, every day you eat or drink various acids. Check the labels of foods or beverages and make a list of the acids you find. Speculate on the purpose of each acid.

Hydrogen chloride is but one of several gases that dissolves in water to produce an acidic solution. Sulfur dioxide and nitrogen dioxide are two others. These two gases are emitted during the combustion of certain fuels (particularly coal) to produce heat and electricity. As we mentioned in the introduction to this chapter, SO_2 and NO_2 both dissolve in rain and mist. When they do so, they form acids that in turn fall back to the Earth's surface in rain and snow. The increased acidity of the Earth's water due to anthropogenic emissions is the central focus of this chapter.

But before delving into the acidity in rain caused by nitrogen oxides and sulfur dioxide, let's focus on carbon dioxide. With an atmospheric concentration of about 390 ppm in 2010, and rising, carbon dioxide is at a far higher concentration than either sulfur dioxide or nitrogen dioxide. Just as solids vary in their solubility in water, so do gases. Compared with more polar compounds such as SO_2 and NO_2, carbon dioxide is far less soluble in water. Even so, it dissolves to produce a weakly acidic solution.

At this stage, you as a Skeptical Chemist should be raising an important question. Given that an acid is defined as a substance that releases hydrogen ions in water, how can carbon dioxide act as an acid? There are no hydrogen atoms in carbon dioxide! The explanation is that when CO_2 dissolves in water, it produces carbonic acid, $H_2CO_3(aq)$. Here is a way to represent the process.

Refer back to Chapter 5 to review solubility. In general, "like dissolves like."

$$CO_2(g) \xrightarrow{\text{H}_2\text{O}} CO_2(aq) \qquad \text{[6.3a]}$$

$$CO_2(aq) + H_2O(l) \longrightarrow H_2CO_3(aq) \qquad \text{[6.3b]}$$

The carbonic acid dissociates to produce the H^+ ion and the hydrogen carbonate ion.

$$H_2CO_3(aq) \longrightarrow H^+(aq) + HCO_3^-(aq) \qquad \text{[6.3c]}$$

This reaction occurs only to a limited extent, producing only tiny amounts of H^+ and HCO_3^-. Accordingly, we say that carbonic acid is a **weak acid**; that is, an acid that dissociates only to a small extent in aqueous solution.

Although carbon dioxide is only slightly soluble in water and then only a tiny amount of the dissolved carbonic acid dissociates to produce H^+, these reactions are happening on a large scale across the planet. The carbon dioxide can dissolve in water in the upper atmosphere (making acids that may fall as acidic rain) or in the planet's oceans, lakes, and streams. We will return to this topic after we introduce bases.

6.2 | What Is a Base?

No discussion of acids would be complete without discussing their chemical counterparts—bases. For our purposes, a **base** is a compound that releases hydroxide ions, OH^-, in aqueous solution. Aqueous solutions of bases have their own characteristic properties attributable to the presence of $OH^-(aq)$. Unlike acids, bases generally taste bitter and do not lend an appealing flavor to foods. Aqueous solutions of bases have a slippery, soapy feel. Common examples of bases include household ammonia (an aqueous solution of NH_3) and NaOH, sometimes called lye. The cautions on oven cleaners (Figure 6.2) warn that lye can cause severe damage to eyes, skin, and clothing.

Dilute basic solutions have a soapy feel because bases can react with the oils of your skin to produce a tiny bit of soap.

Many common bases are compounds containing the hydroxide ion. For example, sodium hydroxide (NaOH), a water-soluble ionic compound, dissolves in water to produce sodium ions (Na^+) and hydroxide ions (OH^-).

$$NaOH(s) \xrightarrow{\text{H}_2\text{O}} Na^+(aq) + OH^-(aq) \qquad \text{[6.4]}$$

Figure 6.2
Oven cleaning products may contain NaOH, commonly called lye.

Although sodium hydroxide is very soluble in water, most compounds containing the hydroxide ion are not. Table 5.9 summarized the solubility trends for compounds containing specific types of anions. Bases that dissociate completely in water, such as NaOH, are called **strong bases.**

Your Turn 6.4   Basic Solutions

These solids dissolve in water to release hydroxide ions. For each, write a balanced chemical equation.

  a. KOH(s), potassium hydroxide
  b. LiOH(s), lithium hydroxide
  c. Ca(OH)₂(s), calcium hydroxide

Some bases, however, do not contain the hydroxide ion, OH^-, but rather react with water to form it. One example is ammonia, a gas with a distinctive sharp odor. Unlike carbon dioxide, ammonia is very soluble in water. It rapidly dissolves in water to form an aqueous solution.

$$NH_3(g) \xrightarrow{\text{H}_2\text{O}} NH_3(aq) \qquad \text{[6.5a]}$$

On a supermarket shelf, you may see a 5% (by mass) aqueous solution of ammonia with the name of "household ammonia." This cleaning agent has an unpleasant odor; if it gets on your skin, you should wash it off with plenty of water.

The chemical behavior of aqueous ammonia is difficult to simplify, but we will do our best to represent it for you with a chemical equation. Think in terms of an ammonia molecule reacting with a water molecule. In essence, the water molecule transfers an H^+ ion to the aqueous NH_3 molecule to form an aqueous ammonium ion, $NH_4^+(aq)$, and a hydroxide ion. However, this reaction only occurs to a small extent; that is, only a tiny amount of $OH^-(aq)$ is produced.

$$NH_3(aq) + H_2O(l) \xrightarrow{\text{only to a small extent}} NH_4^+(aq) + OH^-(aq) \qquad \text{[6.5b]}$$

The source of the hydroxide ion in household ammonia now should be apparent. When ammonia dissolves in water, it releases small amounts of the hydroxide ion and the ammonium ion. Aqueous ammonia is an example of a **weak base,** a base that dissociates only to a small extent in aqueous solution.

In order to more clearly indicate that aqueous ammonia is a base, some people use the representation NH₄OH(aq). If you add up the atoms (and their charges), you

In some industrial applications, ammonia (rather than HCFCs) is used as a refrigerant gas. Great care needs to be taken to prevent the exposure of workers to ammonia, as the gas can dissolve in moist lung tissue and injure or kill.

The ammonium ion, NH_4^+, is analogous to the hydronium ion, H_3O^+, in that each was formed by the addition of a proton (H^+) to a neutral compound.

will see that NH$_4$OH(aq) is equivalent to the left-hand side of equation 6.5b. It is unlikely, however, that this species exists as such in an aqueous solution of ammonia.

6.3 | Neutralization: Bases Are Antacids

Acids and bases react with each other—often very rapidly. Not only does this happen in laboratory test tubes, but also in your home and in almost every ecological niche of our planet. For example, if you put lemon juice on fish, an acid–base reaction occurs. The acids found in lemons neutralize the ammonia-like compounds that produce the "fishy smell." Similarly, if the ammonia fertilizer on a corn field comes in contact with the acidic emissions of a power plant nearby, an acid–base reaction occurs.

Let us first examine the acid–base reaction of solutions of hydrochloric acid and sodium hydroxide. When the two are mixed, the products are sodium chloride and water.

$$HCl(aq) + NaOH(aq) \longrightarrow NaCl(aq) + H_2O(l) \qquad \textbf{[6.6]}$$

This is an example of a **neutralization reaction,** a chemical reaction in which the hydrogen ions from an acid combine with the hydroxide ions from a base to form water molecules. The formation of water can be represented like this.

$$H^+(aq) + OH^-(aq) \longrightarrow H_2O(l) \qquad \textbf{[6.7]}$$

What about the sodium and chloride ions? Recall from equations 6.1 and 6.4 that the HCl(g) and NaOH(s), when dissolved in water, completely dissociate into ions. We can rewrite equation 6.6 to show this.

$$H^+(aq) + Cl^-(aq) + Na^+(aq) + OH^-(aq) \longrightarrow Na^+(aq) + Cl^-(aq) + H_2O(l) \quad \textbf{[6.8]}$$

Neither Na$^+(aq)$ nor Cl$^-(aq)$ take part in the neutralization reaction; they remain unchanged. Canceling these ions from both sides again gives us equation 6.7, which summarizes the chemical changes taking place in an acid–base neutralization reaction.

> Recall from Section 5.8 that NaCl is an ionic compound that dissolves in water to produce Na$^+(aq)$ and Cl$^-(aq)$.

Your Turn 6.5   Neutralization Reactions

For each acid–base pair, write a neutralization reaction. Then rewrite the equation in ionic form and eliminate ions common to both sides. What is the relevance of the final simplified step in each case?

a. HNO$_3$$(aq)$ and KOH(aq)
b. H$_2$SO$_4$$(aq)$ and NH$_4$OH(aq)
c. HBr(aq) and Ba(OH)$_2$$(aq)$

Answer

c. 2 HBr(aq) + Ba(OH)$_2$$(aq)$ $\longrightarrow$ BaBr$_2$$(aq)$ + 2 H$_2$O(l)

2 H$^+(aq)$ + 2 Br$^-(aq)$ + Ba$^{2+}(aq)$ + 2 OH$^-(aq)$ $\longrightarrow$ Ba$^{2+}(aq)$ + 2 Br$^-(aq)$ + 2 H$_2$O(l)

2 H$^+(aq)$ + 2 OH$^-(aq)$ $\longrightarrow$ 2 H$_2$O(l)

Divide by 2 to simplify this last equation.

H$^+(aq)$ + OH$^-(aq)$ $\longrightarrow$ H$_2$O(l)

In each case, the final step summarizes the reaction; that is, it shows that the hydrogen ion from the acid and the hydroxide ion from the base react with each other to form water.

Neutral solutions are neither acidic nor basic; that is, they have equal concentrations of H$^+$ and OH$^-$ ions. Pure water is a neutral solution. Some salt solutions also are neutral, such as the one formed by dissolving solid NaCl in water. In contrast, acidic solutions contain a higher concentration of H$^+$ than OH$^-$ ions, and basic solutions contain a higher concentration of OH$^-$ than H$^+$ ions.

> There is no such thing as "pure" water. Remember from Chapter 5 that water always has some kind of impurities in it.

It may seem strange that acidic and basic solutions contain both hydroxide ions *and* hydrogen ions. But when water is involved, it is not possible to have H^+ without OH^- (or vice versa). A simple, useful, and very important relationship exists between the concentration of hydrogen ion and hydroxide ion in any aqueous solution.

$$[H^+][OH^-] = 1 \times 10^{-14} \qquad \text{[6.9]}$$

The square brackets indicate that the ion concentrations are expressed in molarity, and $[H^+]$ is read as "the hydrogen ion concentration." When $[H^+]$ and $[OH^-]$ are multiplied together, the product is a constant with a value of 1×10^{-14} as shown in mathematical equation 6.9. This shows that the concentrations of H^+ and OH^- depend on each other. When $[H^+]$ increases, $[OH^-]$ decreases, and when $[H^+]$ decreases, $[OH^-]$ increases. Both ions are always present in aqueous solutions.

Knowing the concentration of H^+, we can use equation 6.9 to calculate the concentration of OH^- (or vice versa). For example, if a rain sample has a H^+ concentration of 1×10^{-5} M, we can calculate the OH^- concentration by substituting in 1×10^{-5} M for $[H^+]$.

$$(1 \times 10^{-5}) \times [OH^-] = 1 \times 10^{-14}$$

$$[OH^-] = \frac{1 \times 10^{-14}}{1 \times 10^{-5}}$$

$$[OH^-] = 1 \times 10^{-9}$$

Since the hydroxide ion concentration (1×10^{-9} M) is smaller than the hydrogen ion concentration (1×10^{-5} M), the solution is acidic.

In pure water or in a neutral solution, the molarities of the hydrogen and hydroxide ions both equal 1×10^{-7} M. Applying mathematical expression 6.9, we can see that $[H^+][OH^-] = (1 \times 10^{-7})(1 \times 10^{-7}) = 1 \times 10^{-14}$.

Your Turn 6.6   Acidic and Basic Solutions

Classify these solutions as acidic, neutral, or basic. Then, for parts **a** and **c**, calculate $[OH^-]$. For **b**, calculate $[H^+]$.

a. $[H^+] = 1 \times 10^{-4}$ M   **b.** $[OH^-] = 1 \times 10^{-6}$ M   **c.** $[H^+] = 1 \times 10^{-10}$ M

Answer
a. The solution is acidic because $[H^+] > [OH^-]$.
$[H^+][OH^-] = 1 \times 10^{-14}$. Solving, $[OH^-] = 1 \times 10^{-10}$ M.

Your Turn 6.7   Ions in Acidic and Basic Solutions

Classify each solution as acidic, basic, or neutral. All of the compounds are strong acids or strong bases. Then list all of the ions present in order of decreasing relative amounts in each solution.

a. $KOH(aq)$ **b.** $HNO_2(aq)$ **c.** $H_2SO_3(aq)$ **d.** $Ca(OH)_2(aq)$

Answer
d. When calcium hydroxide dissociates, two hydroxide ions are released for each calcium ion. The basic solution contains much more OH^- than H^+.

$$OH^-(aq) > Ca^{2+}(aq) > H^+(aq)$$

How can we know if the acidity of seawater and of rain is cause for concern? To make a judgment, we need a convenient way of reporting how acidic or basic a solution is. The pH scale is just such a tool because it relates the acidity of a solution to its H^+ concentration.

Margin notes:

The product $[H^+][OH^-]$ is dependent on temperature. The value 1×10^{-14} is valid at 25 °C.

Refer to Section 5.5 for the definition of molarity.

By definition, the product of the two concentrations is unitless.

Acidic solution

$[H^+] > [OH^-]$

Neutral solution

$[H^+] = [OH^-]$

Basic solution

$[H^+] < [OH^-]$

6.4 | Introducing pH

The term *pH* already may be familiar to you. For example, test kits for soils and for the water in aquariums and swimming pools report the acidity in terms of pH. Deodorants and shampoos claim to be pH-balanced (Figure 6.3). And, of course, articles about acid rain make reference to pH. The notation pH is always written with a small p and a capital H and stands for "power of hydrogen." In the simplest terms, **pH is a number, usually between 0 and 14, that indicates the acidity (or basicity) of a solution.**

As the midpoint on the scale, pH 7 separates acidic from basic solutions. Solutions with a pH less than 7 are acidic, and those with a pH greater than 7 are basic (alkaline). Solutions of pH 7 (such as pure water) have equal concentrations of H^+ and OH^- and are said to be neutral.

The pH values of common substances are displayed in Figure 6.4. You may be surprised that you eat and drink so many acids. Acids naturally occur in foods and contribute distinctive tastes. For example, the tangy taste of McIntosh apples comes from malic acid. Yogurt gets its sour taste from lactic acid, and cola soft drinks contain several acids, including phosphoric acid. Tomatoes are well known for their acidity, but with a pH of about 4.5, they are in fact less acidic than many other fruits.

 Consider This 6.8   **Acidity of Foods**

> **a.** Rank tomato juice, lemon juice, milk, cola, and pure water in order of increasing acidity. Check your order against Figure 6.4.
> **b.** Pick any other five foods and make a similar ranking. Look up the actual pH values. A helpful link is provided at the textbook's website.

Is the water on the planet acidic, basic, or neutral? Water would be expected to have a pH of 7.0, but Figure 6.4 shows that the pH of water depends on where it is found. "Normal" rain is slightly acidic, with a pH value between 5 and 6. Even though the acid formed by dissolved carbon dioxide is a weak acid, enough H^+ is produced to lower the pH of rain. Seawater is slightly basic, with a pH of approximately 8.2.

As you might have guessed, pH values are related to the hydrogen ion concentration. For solutions in which $[H^+]$ is 10 raised to some power, the pH value is this power (the exponent) with its sign changed. For example, if $[H^+] = 1 \times 10^{-3}$ M, then the pH is 3. Similarly, for $[H^+] = 1 \times 10^{-9}$ M, the pH is 9. Appendix 3 describes the relation between pH and $[H^+]$ in more detail.

Equation 6.9 shows that the hydrogen ion concentration multiplied by the hydroxide ion concentration is a constant, 1×10^{-14}. When the concentration of H^+ is high (and

For highly acidic or basic solutions, the pH may lie outside the 0-to-14 range.

Universal indicator paper is a quick way to estimate the pH of a solution. For more accurate results, pH meters are used.

Figure 6.3

This shampoo claims to be "pH-balanced," that is, adjusted to be closer to neutral. Soaps tend to be basic, which can be irritating to the skin.

Carbonated water is more acidic than rainwater, because more CO_2 gas is forced under pressure to dissolve. The pH is about 4.7.

The mathematical relationship is

$$pH = -\log[H^+]$$

More information can be found in Appendix 3.

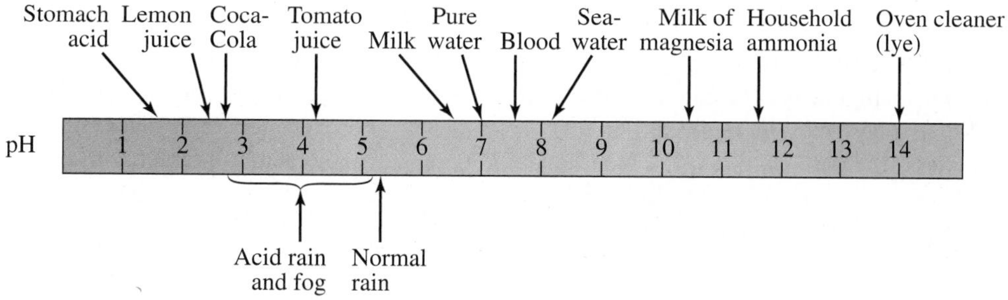

| Stomach acid | Lemon juice | Coca-Cola | Tomato juice | Milk | Pure water | Blood | Sea-water | Milk of magnesia | Household ammonia | Oven cleaner (lye) |

pH scale: 1 2 3 4 5 6 7 8 9 10 11 12 13 14

Acid rain and fog   Normal rain

Figure 6.4

Common substances and their pH values.

 Figures Alive! Visit the textbook's website to learn more about acids, bases, and the pH scale.

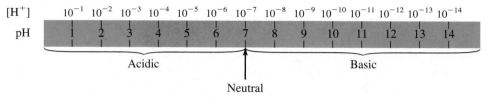

Figure 6.5

The relationship between pH and the concentration of H⁺ in moles per liter (M). As pH increases, [H⁺] decreases.

the pH is low), the concentration of OH⁻ is low. Likewise, as pH values rise above 7.0, the concentration of hydrogen ions decreases and the concentration of hydroxide ions increases. As the pH value *decreases,* the acidity *increases.* For example, a sample of water with a pH of 5.0 is 10 times *less* acidic than one with a pH of 4.0. This is because a pH of 4 means that the [H⁺] is 0.0001 M. By contrast, a solution with a pH of 5 is more dilute, with a [H⁺] = 0.00001 M. This second solution is *less* acidic with only 1/10 the hydrogen ion concentration of a solution of pH 4. Figure 6.5 shows the relationship between pH and the hydrogen ion concentration.

Your Turn 6.9   Small Changes, Big Effects

Compare the pairs of samples below. For each, which one is more acidic? Include the relative difference in hydrogen ion concentration between the two pH values.

a. Rain sample, pH = 5, and lake water sample, pH = 4.
b. Ocean water sample, pH = 8.3, and tap water sample, pH = 5.3.
c. Tomato juice sample, pH = 4.5, and milk sample, pH = 6.5.

Answer
c. Although the pH values differ only by 2, the tomato juice sample is 100 times more acidic and (for an equal volume) would have 100 times more H⁺ than the milk sample.

Consider This 6.10   On the Record

A legislator from the Midwest is on record with an impassioned speech in which he argued that the environmental policy of the state should be to bring the pH of rain all the way down to zero. Assume that you are an aide to this legislator. Draft a tactful memo to your boss to save him from additional public embarrassment.

6.5 | Ocean Acidification

How can seawater be basic when rain is naturally acidic? Indeed this is the case, as shown in Figure 6.4. Ocean water contains small amounts of three chemical species that all play roles in maintaining the ocean pH at approximately 8.2. These three species—the carbonate ion, the bicarbonate ion, and carbonic acid—interact with each other as well. All arise from dissolved carbon dioxide (equations 6.3a, b, and c). These species also help maintain your blood at a pH of about 7.4.

Ocean pH can vary by ±0.3 pH units, depending on latitude and region.

Only one resonance form is shown for the bicarbonate ion and for the carbonate ion. For more about resonance, see Section 2.3.

carbonate ion
$CO_3^{2-}(aq)$

bicarbonate ion
$HCO_3^-(aq)$

carbonic acid
$H_2CO_3(aq)$

Many organisms, such as mollusks, sea urchins, and coral, have connections to this bit of ocean chemistry because they build their shells out of calcium carbonate, $CaCO_3$. Changing the amount of one chemical species in the ocean (such as carbonic acid) can affect the concentration of the others, in turn affecting marine life.

Human industrial activity has rapidly increased the amount of carbon dioxide released into the atmosphere over the past 200 years. As a result, more carbon dioxide is dissolving into the oceans and forming carbonic acid. In turn, the pH of seawater has dropped by roughly 0.1 pH unit since the early 1800s. This may sound like a small number. Remember, though, that each full unit of pH represents a 10-fold difference in the concentration of H^+ ions. A decrease of 0.1 pH unit corresponds to a 26% increase in the amount of H^+ in seawater. The lowering of the ocean pH due to increased atmospheric carbon dioxide is called **ocean acidification.**

How can such a seemingly small change in pH pose a danger to marine organisms? Part of the answer lies in the chemical interactions between $CO_3^{2-}(aq)$, $HCO_3^-(aq)$, and $H_2CO_3(aq)$. The H^+ produced by the dissociation of carbonic acid reacts with carbonate ion in seawater to form the bicarbonate ion.

$$H^+(aq) + CO_3^{2-}(aq) \longrightarrow HCO_3^-(aq) \qquad \text{[6.10]}$$

The net effect is to reduce the concentration of carbonate ion in seawater. The calcium carbonate in the shells of sea creatures begins to dissolve in order to maintain the concentration of carbonate ions in seawater.

$$CaCO_3(s) \xrightarrow{H_2O} Ca^{2+}(aq) + CO_3^{2-}(aq) \qquad \text{[6.11]}$$

The interaction of carbonic acid, bicarbonate ion, and carbonate ion are summarized in Figure 6.6. As carbon dioxide dissolves in ocean water, it forms carbonic acid. This in turn dissociates to produce "extra" acidity in the chemical form of H^+. The H^+ reacts with the carbonate ion, depleting it and producing more bicarbonate ion. Calcium carbonate then dissolves to replace the carbonate that was depleted.

Ocean scientists predict that within the next 40 years, the carbonate ion concentration will reach a low enough level that the shells of sea creatures near the ocean surface will begin to dissolve. In fact, one study has shown that the Great Barrier Reef off the coast of Australia is already growing at slower and slower rates. However, other factors could be to blame, including ocean warming. One can examine growth rings in a slice of coral, much as one can view tree rings (Figure 6.7).

To date, only a small number of researchers have focused on the effects of thinning shells on sea creatures. However, negative effects on whole ecosystems have been

Recall (Section 3.2) that in recent decades, the concentration of CO_2 in the atmosphere has risen a few parts per million each year. The increase has been steady and relentless.

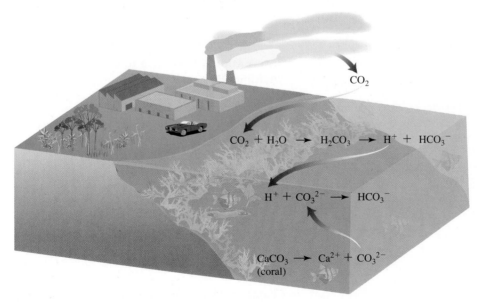

Figure 6.6

Chemistry of CO_2 in the ocean.

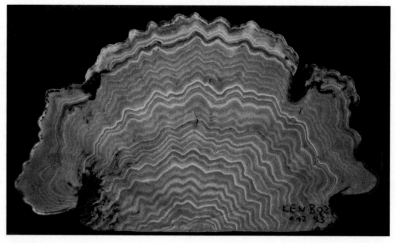

Figure 6.7

A thin slice of coral. Special lighting reveals annual growth rings. A recent study has shown that some corals have seen a dramatic decrease in their growth rate over the past 20 years.

projected. For example, weaker (or missing) coral reefs could fail to protect coastlines from harsh ocean waves. Coral reefs also provide fish species with their habitat, and damage to the reefs would translate into losses of marine life. Finally, a weakening of the reefs would make them more susceptible to further damage from storms and predators.

Can the ocean heal itself? Although we don't know the answer for sure, nonetheless we can speculate from what we know of past events. When changes in ocean pH have occurred over a very long period of time, the ocean has been able to compensate. This happens because large collections of sediment at the bottom of the ocean contain massive amounts of calcium carbonate, mostly from the shells of long-deceased marine creatures. Over long time periods, these sediments dissolve to replenish the carbonate lost to reaction with excess H^+. But today's changes in the pH of the oceans have happened rapidly on the geologic time scale. In just 200 years, the pH of the ocean has dropped to a level not seen in the past 400 million years. Because the acidification is occurring over a relatively short time and in water close to the surface, the sediment reserve has not had time to dissolve and counteract the effects of the added acidity.

Even if the amount of carbon dioxide in the atmosphere were to immediately level off, the oceans would take thousands of years to return to the pH measured in preindustrial times. Coral reefs would take even longer to regenerate, and any species lost to extinction, of course, would not return.

Consider This 6.11   **International Response to Ocean Acidification**

In October, 2008, a group of scientists met in Monaco to raise awareness about ocean acidification. They issued the Monaco Declaration, calling on the countries of the world to reverse carbon dioxide emissions trends by 2020. Have more recent gatherings of scientists and negotiators created a worldwide policy to address ocean acidification? Do research of your own and summarize your findings.

6.6 | The Challenges of Measuring the pH of Rain

Ours is a wet planet, as we saw in Chapter 5. Carbon dioxide not only has the opportunity to dissolve in the ocean but also can dissolve in rain and fresh water everywhere. The result is the same—when CO_2 dissolves, the pH drops slightly due to the formation

Figure 6.8

A pH meter with a digital display.

of carbonic acid. In the case of rain, the resulting pH is between 5 and 6. But at times the pH of rain can be even lower than this. We refer to this type of precipitation as **acid rain,** that is, rain with a pH below 5 that is more acidic than "normal" rain.

What are the levels of acidity in rain across the mainland United States, Alaska, Hawaii, and Puerto Rico? To measure acidity levels anywhere in the world, we need an analytical tool, the pH meter. Many types of pH meters are available, depending both on the conditions under which you wish to use them and how much you are willing to pay. The pH meter that you are most likely to encounter has a special probe capped with a membrane that is sensitive to H^+. When the probe is immersed in a sample, the difference in H^+ ion concentrations between the solution and the probe creates a voltage across the membrane. The meter measures this voltage and converts it to a pH value (Figure 6.8).

Although it is straightforward to measure the pH of a rain sample, certain procedures are necessary to ensure accurate results. For example, the electrode of the pH meter needs to be carefully calibrated. Another challenge is to collect and measure rain samples without contaminating them. The collection containers must be scrupulously clean and free of minerals from the water in which they were washed. When a container is placed on site, it must be set high enough to prevent splash contamination either from the ground or surrounding objects. Even if elevated, contamination may still occur from the pollen of nearby plants, insects, bird droppings, leaves, soil dust, or even the ash of a fire.

One way to minimize contamination is to fit a rain collection bucket with a lid and a moisture sensor that opens this lid when it begins to rain. This is how samples are collected at the approximately 250 sites of the National Atmospheric Deposition Program/National Trends Network (NADP/NTN). Figure 6.9a shows the sensor and two buckets at a NADP/NTN monitoring station in Illinois that has been in operation for over 30 years. One bucket is for dry deposition (open when it is not raining) and the other is covered. A sensor opens this bucket (closing the other) when it rains.

Deciding where to locate the collection sites also is a challenge. Due to budgetary constraints, the test sites cannot go in as many places as researchers might like. The relative advantages of widely dispersing the sites versus putting several near each other in a special ecosystem (such as a national park) need to be weighed. Currently there are more collection sites in the eastern United States, as historically the acidity levels have been higher there due in large part to coal-fired electric power plants.

Rain samples have been collected routinely in the United States and Canada since about 1970. Since 1978, NADP/NTN has collected over 250,000 samples, analyzing them for pH and for these ions: SO_4^{2-}, NO_3^-, Cl^-, NH_4^+, Ca^{2+}, Mg^{2+}, K^+, and Na^+. Figure 6.9b shows the five active NADP/NTN sites in the state of Illinois. To discover how many sites are in your state, complete the following activity.

Section 8.5 describes how hydrogen fuel cells also create a voltage across a membrane.

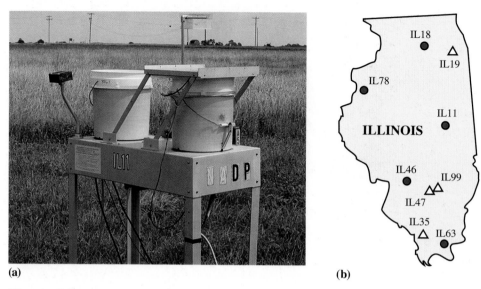

(a)   (b)

Figure 6.9

(a) The Bondville Monitoring Station in central Illinois (IL11) has been in operation since 1979. The black moisture sensor connected to the left of the table controls which bucket is open. It is not raining, so the right bucket for wet deposition is closed. **(b)** The five active NTN precipitation monitoring sites in Illinois, including IL11 in Bondville. The sites marked with triangles are inactive.

Source: National Atmospheric Deposition Program 2009. NADP Program Office, Illinois State Water Survey. http://nadp.sws.uiuc.edu/sites/sitemap.asp?state=il

---

 Consider This 6.12   **The Rain in Maine . . . Oregon or Florida**

Thanks to the NADP/NTN, almost every state plus Puerto Rico and the Virgin Islands has one or more precipitation monitoring sites.

a. In Figure 6.9a, name the precautions you see being taken to preserve the integrity of the rain samples.

b. How many monitoring sites are in your state? A map with links is provided at the textbook's website.

c. Do you think the number and placement of collection sites in your state fairly represent the acidic deposition?

d. On the web, select a collection site in your state (or in a neighboring one) that provides a photograph. Compare the picture with Figure 6.9a. What additional ways of minimizing contamination (for example, a fence or signage) can you spot, if any?

Answer

a. The collection buckets are located up off the ground, one is fitted with a lid that opens when it rains, and the area around the site is mowed. Also, the location is far from people and roads.

Each week, researchers at the Central Analytical Laboratory in Champaign, Illinois, receive hundreds of rain samples. The photographs in Figure 6.10 indicate the magnitude of the operation. At the left are sample collection buckets waiting to be cleaned prior to being shipped back to the collection sites. The center photo shows a set of rain samples in the queue to be analyzed, each assigned an alphanumeric label. A small portion of each sample is saved after analysis and stored under refrigeration. The photo on the right shows Chris Lehmann, director of the laboratory, standing inside a cold room in which samples are archived.

Each year, researchers at the Central Analytical Laboratory use the analytical data to construct maps such as the one shown in Figure 6.11. From these maps, we observe that all rain is slightly acidic. Remember that rain contains a small amount of dissolved carbon dioxide, and that the carbon dioxide dissolved in rainwater produces a weakly acidic solution.

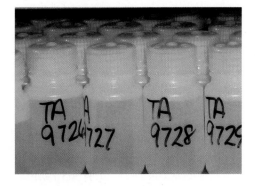

Figure 6.10

Photographs from the Central Analytical Laboratory (CAL), Champaign, Illinois. *Left:* Sample collection buckets waiting to be cleaned. *Center:* Rain samples in the queue to be analyzed. *Right:* Chris Lehmann, Director of the CAL.

Recall that rain naturally has a pH between 5 and 6. Figure 6.11 shows that the pH of rain samples is well below normal in the eastern third of the United States, especially in the Ohio River valley. Carbon dioxide is not the only source of H^+ in rain. Chemical analysis of rainwater confirms the presence of other substances that result in the formation of hydrogen ion: sulfur dioxide (SO_2), sulfur trioxide (SO_3), nitrogen monoxide (NO), and nitrogen dioxide (NO_2). These compounds are affectionately known as "Nox and Sox." Chemically, we write SO_x and NO_x, where $x = 2$ and 3 for SO_x, and $x = 1$ and 2 for NO_x.

Let's examine SO_x and NO_x in turn. First, here is a way to represent the process by which sulfur trioxide dissolves in water to form sulfuric acid.

$$SO_3(g) + H_2O(l) \longrightarrow H_2SO_4(aq) \qquad \textbf{[6.12]}$$
$$\text{sulfuric acid}$$

Equation 6.12 is analogous to the reaction of CO_2 with water.

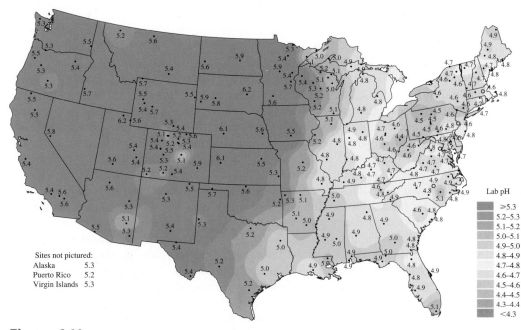

Figure 6.11

The pH of rain samples. Measurements made at the Central Analytical Laboratory, 2008. Values at stations in Alaska, Puerto Rico, and the Virgin Islands are given at the lower left. Hawaii data not available.

Source: National Atmospheric Deposition Program Illinois State Water Survey.

In water, sulfuric acid is a source of the hydrogen ion.

$$H_2SO_4(aq) \longrightarrow H^+(aq) + HSO_4^-(aq) \qquad \text{[6.13a]}$$
hydrogen sulfate ion

The hydrogen sulfate ion also can dissociate to yield another hydrogen ion.

Equations 6.13b and 6.13c actually are more complex than shown here.

$$HSO_4^-(aq) \longrightarrow H^+(aq) + SO_4^{2-}(aq) \qquad \text{[6.13b]}$$
sulfate ion

Adding equations 6.13a and 6.13b shows that sulfuric acid dissociates to yield two hydrogen ions and a sulfate ion (SO_4^{2-}).

$$H_2SO_4(aq) \longrightarrow 2\,H^+(aq) + SO_4^{2-}(aq) \qquad \text{[6.13c]}$$

The sulfate ion can be detected in rainwater, providing a clue to the source of the additional acidity in the rain.

Your Turn 6.13   Sulfurous Acid

Sulfur dioxide dissolves in water to form sulfurous acid, H_2SO_3. Write equations for the formation of $2\,H^+(aq)$ from sulfurous acid, analogous to chemical equations 6.13a, 6.13b, and 6.13c for sulfuric acid. Visit Figures Alive! at the textbook's website for a set of interactive activities relating to acids and bases.

Now let's turn to NO_x. Nitrogen oxides also dissolve in water to form acids, but the chemical reactions are more complex because the element O_2 also is a reactant. For example, NO_2 reacts in moist air to form nitric acid. This reaction is a simplification of the atmospheric chemistry that takes place.

$$4\,NO_2(g) + 2\,H_2O(l) + O_2(g) \longrightarrow 4\,HNO_3(aq) \qquad \text{[6.14]}$$
nitric acid

In water, nitric acid dissociates to release H^+.

$$HNO_3(aq) \longrightarrow H^+(aq) + NO_3^-(aq) \qquad \text{[6.15]}$$
nitrate ion

The nitrate ion that is produced in this reaction can be detected in rainwater.

Rain is only one of several ways that acids can be delivered to Earth's surface waters. Snow and fog obviously are others. The term **acid deposition** refers to both wet and dry forms of delivery of acids from the upper atmosphere to the surface of the Earth. Examples of wet deposition include rain, snow, and fog. Mountaintops are particularly susceptible to wet deposition resulting from direct contact with clouds containing microscopic water droplets. Because these droplets contain acids that are more concentrated than those found in larger raindrops, they are often more acidic and damaging than acid rain. If SO_x and NO_x are indeed responsible for the increased acidity of rain falling in the eastern part of the United States, these regions should show elevated levels of sulfate ion and nitrate ion in rainwater, from SO_x and NO_x, respectively. Indeed, this is the case. Figure 6.12 shows the nitrate and sulfate ion content for wet deposition.

Acid deposition also includes the "dry" forms of acids that deposit on land and water. For example, during dry weather, tiny solid particles (aerosols) of the acidic compounds ammonium nitrate, NH_4NO_3, and ammonium sulfate, $(NH_4)_2SO_4$, are deposited. Dry deposition can be just as significant as the wet deposition of the acids in rain, snow, and fog. These aerosols also contribute to haze, as we will see in Section 6.12.

Aerosols consist of tiny particles that remain suspended in our atmosphere (Section 1.11) and have a cooling effect on the Earth (Section 3.9).

Armed with the knowledge that oxides of sulfur and nitrogen contribute to acid rain, we now need to look more closely at how these oxides come to be released into the atmosphere. In the next section, we examine the chemistry of SO_2 and its connection to the burning of coal. In the two sections after, we turn our attention to nitrogen chemistry, including that of NO and NO_2.

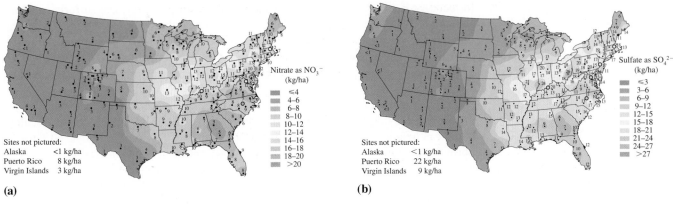

(a)   **(b)**

Figure 6.12

(a) 2008 wet deposition of nitrate ion in kilograms per hectare.
(b) 2008 wet deposition of sulfate ion in kilograms per hectare.

Source: National Atmospheric Deposition Program 2009. NADP Program Office, Illinois State Water Survey, Champaign, IL.

6.7 | Sulfur Dioxide and the Combustion of Coal

Let's look more closely at coal and its combustion products. At first glance, coal may not appear much different from charcoal or black soot, both of which are essentially pure carbon. When carbon is burned with plenty of oxygen, it forms carbon dioxide and liberates large amounts of energy, which of course is the reason for burning it.

$$C(in\ coal) + O_2(g) \longrightarrow CO_2(g) + energy \qquad \textbf{[6.16]}$$

Coal also contains varying amounts of sulfur. How did the sulfur wind up in the coal? Several hundred million years ago, coal formed from decaying vegetation such as that found in swamps or peat bogs. Plants, like other living things, contain sulfur, so some of the sulfur in coal can be traced to this ancient vegetation. However, most of the sulfur in coal came from the sulfate ion (SO_4^{2-}) naturally present in seawater. Bacteria on the sea floors use the sulfate ion as an oxygen source, removing the oxygen and releasing the sulfide ion (S^{2-}). In the past, the sulfide ion became incorporated into the ancient rocks (including coal) that were in contact with seawater. In contrast, the coal formed in freshwater peat bogs has a lower sulfur content. Thus, the percent of sulfur in coal can vary from less than 1% to as much as 6% by mass.

We can approximate its composition with the chemical formula $C_{135}H_{96}O_9NS$. Burning sulfur in air produces sulfur dioxide, a poisonous gas with an unmistakable choking odor (Figure 6.13).

$$S(s) + O_2(g) \longrightarrow SO_2(g) \qquad \textbf{[6.17]}$$

Because the sulfur content of coal varies, burning coal produces sulfur dioxide in varying amounts. This fact is central to the acid rain story. When coal is burned, the sulfur dioxide produced goes up the smokestack along with the carbon dioxide, water vapor, and small amounts of metal oxide ash. Emission control measures can reduce the amount of SO_2. Thus the levels of SO_2 emissions vary depending on how coal-burning electric utility plants are equipped.

Once in the atmosphere, SO_2 can react with oxygen to form sulfur trioxide, SO_3.

$$2\ SO_2(g) + O_2(g) \longrightarrow 2\ SO_3(g) \qquad \textbf{[6.18]}$$

This reaction, although extremely slow, is accelerated by the presence of finely divided solid particles such as the ash that goes up the stack along with the SO_2. Once SO_3 is

The movement of sulfur through the biosphere should remind you of the carbon cycle described in Section 3.5.

In ancient times sulfur was known as brimstone, thus the biblical admonition about "fire and brimstone."

Figure 6.13

Sulfur burns in air to produce SO_2. This gas produces an acidic solution when dissolved in water.

Sulfur trioxide plays a role in aerosol formation, as we will see in Section 6.12.

formed, it reacts rapidly with water vapor in the atmosphere to form sulfuric acid (see equation 6.12). A second important pathway involves the hydroxyl radical (•OH) that is formed from ozone and water in the presence of sunlight. The hydroxyl radical reacts with SO_2 and then the product reacts with oxygen to form SO_3 and regenerate •OH. The reaction goes faster in intense sunlight and thus is more important in summer and at midday.

Although pathways exist for the reaction of SO_2 with oxygen to form SO_3, the majority of SO_2 in the atmosphere directly contributes to acid rain. A chemical calculation can help us better appreciate the vast quantities of SO_2 produced by coal-burning power plants. Such utilities typically burn 1 million metric tons of coal a year, where a metric ton is equivalent to 1000 kg.

Emissions data are given both in metric tons (1000 kg, 2200 lb) and in short tons (2000 lb). To add to the confusion, short tons usually are simply called tons.

$$\frac{1 \times 10^6 \text{ metric ton coal}}{\text{yr}} \times \frac{1000 \text{ kg coal}}{\text{metric ton coal}} \times \frac{1000 \text{ g coal}}{\text{kg coal}} = 1 \times 10^{12} \text{ g coal/yr}$$

For this discussion, we assume a coal that contains 2.0% sulfur; that is, 2.0 g sulfur per 100 g coal. First we can calculate the grams of sulfur released each year from 1 million metric tons (1×10^{12} g) of coal.

$$\frac{1 \times 10^{12} \text{ g coal}}{\text{yr}} \times \frac{2.0 \text{ g S}}{100 \text{ g coal}} = \frac{2.0 \times 10^{10} \text{ g S}}{\text{yr}}$$

See Section 3.7 to review mole calculations.

Next, we use the fact that 1 mole of sulfur reacts with oxygen (O_2) to form 1 mole of SO_2 (see equation 6.17). The molar mass of sulfur is 32.1 g, and the molar mass of SO_2 is 64.1 g, that is, 32.1 g + 2(16.0 g). Therefore, 32.1 g of sulfur burn to produce 64.1 g of SO_2.

$$\frac{2.0 \times 10^{10} \text{ g S}}{\text{yr}} \times \frac{1 \text{ mol S}}{32.1 \text{ g S}} \times \frac{1 \text{ mol SO}_2}{1 \text{ mol S}} \times \frac{64.1 \text{ g SO}_2}{1 \text{ mol SO}_2} = \frac{4.0 \times 10^{10} \text{ g SO}_2}{\text{yr}}$$

This mass of SO_2 is equivalent to 40,000 metric tons, or 88 million pounds, of SO_2 per year. Power plants burning high-sulfur coal may emit more than twice this!

The connection between burning coal and sulfur dioxide emissions in the United States is evident in Figure 6.14. Most of the emissions arise from power plants ("fuel combustion") in which coal or other fossil fuels are burned to generate electricity for public or industrial consumption. Transportation is responsible only for a small percent of the emissions because gasoline and diesel fuel contain relatively low amounts of sulfur. Industrial processes, such as the producing of metals from their ores, account for the remainder of the emissions. For example, the ores of both copper and nickel are sulfides. When nickel sulfide is heated to a high temperature in a smelter, the ore decomposes and sulfur dioxide is released. Similarly, smelting copper sulfide releases SO_2. Although the large-scale production of nickel and copper contributes only a few percent to the total emissions, huge quantities of SO_2 are generated in some regions.

One of the world's largest smelters in Sudbury, Ontario, produces nickel from an ore that contains sulfur. The bleak, lifeless landscape in the immediate vicinity of the plant stands in mute testimony to past uncontrolled releases of SO_2. After a major renovation in 1993, the two major smelters in the area have decreased their sulfur dioxide emissions substantially. The smokestack is tall so that the emissions are carried away from Sudbury by the prevailing winds (Figure 6.15). Lest we point any fingers, Canadians report that more than half of acid deposition in the eastern portion of their country originates in the United States. The quantity of sulfur dioxide that drifts northward over the border into Canada is estimated at 4 million metric tons per year.

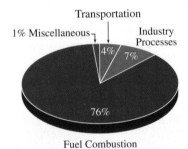

Figure 6.14

U.S. sulfur dioxide emission sources, 2007.

Source: EPA, National Emissions Inventory, Air Pollution Emissions Trends Data.

Figure 6.15
The smokestack in Sudbury, Ontario, is one of the world's tallest at 1250 ft (381 m).

Your Turn 6.14   **Coal Calculations**

a. Assume that coal has the chemical formula of $C_{135}H_{96}O_9NS$. Calculate the fraction and percent (by mass) of sulfur in the coal.

b. A certain power plant burns 1.00×10^6 tons of coal per year. Assuming the sulfur content calculated in part **a,** calculate the tons of sulfur released per year.

c. Calculate the tons of SO_2 formed from this amount of sulfur.

d. Once released into the atmosphere, the SO_2 is likely to react with oxygen to form SO_3. What can happen next if SO_3 encounters water droplets?

Answers

a. 0.0168, or 1.68%   **b.** 1.68×10^4 tons S

Having examined the chemistry of SO_2, we now return to that of NO and NO_2. In the next section, we describe NO emissions and the connection to automobiles and power plants. In two sections after that, we relate the bigger stories of nitrogen chemistry.

6.8 | Nitrogen Oxides and the Combustion of Gasoline

Those who live in southern California have experienced high levels of acid deposition. For example, in January 1982, the fog near the Rose Bowl in Pasadena was found to have a pH of 2.5. Breathing it must have been like inhaling a fine mist of vinegar! The acidity exceeded that of normal precipitation by at least 500 times. In this same year, the fog at Corona del Mar on the coast south of Los Angeles was 10 times more acidic than near the Rose Bowl, registering a pH of 1.5. However, the concentration of SO_2 was relatively low in both areas. Clearly, something else was the cause.

To find out what, we need to turn to the cars and trucks that jam the freeways of Los Angeles. At first glance, it may not be obvious how these thousands of vehicles contribute to acid precipitation. Gasoline burns to form CO_2 and H_2O, together with small amounts of CO, unburned hydrocarbons, and soot. But gasoline contains very little sulfur, so SO_2 is not the culprit. So what might be the source of the acidity?

Nitrogen oxides have already been identified as contributors to acid rain, but gasoline does not contain nitrogen. Therefore, logic and chemistry assert that nitrogen

Gasoline is a mixture of hydrocarbons. See Section 4.4.

You may have seen advertisements for "nitrogen-enriched" gasoline, referred to by one oil company as "a nitrogen-enriched cleaning system." The amount of nitrogen in this gasoline additive is quite low.

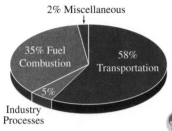

Figure 6.16

U.S. nitrogen oxides (NO_x) emission sources, 2007.

Source: EPA, National Emissions Inventory, Air Pollution Emissions Trends Data.

See Section 1.11 for more about equation 6.20.

oxides cannot be formed from burning gasoline. Literally, this is correct. Remember, however, that about 80% of air consists of N_2. This element is remarkably stable and for the most part unreactive. Nevertheless, if the temperature is high enough, nitrogen is able to react with a few elements. One of these is oxygen. Recall from Chapter 1 that with sufficient energy, nitrogen and oxygen combine to form nitrogen monoxide (nitric oxide).

$$N_2(g) + O_2(g) \xrightarrow{\text{high temperature}} 2\,NO(g) \qquad \textbf{[6.19]}$$

In an automobile, gasoline and air are drawn into the cylinders and compressed, bringing the N_2 and O_2 molecules closer together. The gasoline, once ignited, burns rapidly. The energy released powers the vehicle. But the unfortunate truth is that the energy also powers chemical equation 6.19.

The reaction of N_2 with O_2 is not limited to automobile engines. The same reaction occurs when air is heated in the furnace of a coal-fired electric power plant such that collectively these plants release huge amounts of NO_x. In the United States, the combustion of fossil fuels (particularly coal) in electric power plants accounts for over a third of the nitrogen oxide emissions (Figure 6.16). Transportation sources such as motor vehicles, aircraft, and trains account for over half. In an urban environment, an even greater proportion of NO_x arises from motor vehicles.

In the early 1990s, a green chemistry solution to reducing NO emissions and energy consumption was introduced by Praxair Inc. of Tarrytown, NY, a glass manufacturing company. The technology substitutes 100% oxygen for air in the large furnaces used to melt and reheat glass. Switching from air (78% nitrogen) to pure oxygen requires lower temperatures, thereby reducing NO production by 90% and cutting energy consumption by up to 50%. Glass manufacturers using the Praxair Oxy-Fuel technology save enough energy annually to meet the daily needs of 1 million Americans.

Once formed, nitrogen monoxide is highly reactive. As we noted in Chapter 1, through a series of steps, NO reacts with oxygen, the hydroxyl radical, and volatile organic compounds (VOCs) to form NO_2.

$$VOC + \cdot OH \longrightarrow A + O_2 \longrightarrow A' + NO \longrightarrow A'' + NO_2 \qquad \textbf{[6.20]}$$

The reactive intermediate species A, A′, and A″, present in trace amounts, are synthesized from the VOC molecules. The production of acid rain from NO_2 requires a trace amount of VOCs in the atmosphere.

Nitrogen dioxide is a highly reactive, poisonous, red-brown gas with a nasty odor. For our purposes, the most significant reaction of NO_2 is the one that converts it to nitric acid, HNO_3. Earlier, equation 6.14 was a simplification of this conversion. Actually a series of reactions occurs in the presence of sunlight. These take place in the air surrounding Los Angeles, Phoenix, Dallas, and other sunny metropolitan areas. A key player is the hydroxyl radical. Once formed in the atmosphere, the hydroxyl radical can rapidly react with nitrogen dioxide to yield nitric acid.

$$NO_2(g) + \cdot OH(g) \longrightarrow HNO_3(l) \qquad \textbf{[6.21]}$$

As you have already seen in equation 6.15, HNO_3 is a strong acid that dissociates completely in water to release aqueous H^+ and NO_3^-. The result is the alarmingly low pH values occasionally found in the rain and fog of cities like Los Angeles.

The bit of NO_x chemistry that we have described in this section fits into a larger picture involving agriculture, food, and actually all of the ecosystems of our planet. In the next section, we provide the details.

6.9 | The Nitrogen Cycle

Rarely a day goes by that you don't ingest food in one form or another. Clearly, you need to eat in order to stay alive. As you are reading this, men and women across the globe are producing food by planting fields of grain, harvesting fruits and vegetables by the truckload, and perhaps even growing oregano or chives on a sunny windowsill. To their credit, humans have become quite expert in raising both plants and animals.

However, producing food such as a sausage pizza releases NO_x and adds to the acidity of the environment. Human activity is affecting both where nitrogen compounds are found and how they move through the air, water, and land of our planet.

The connection between food production and NO_x emissions arises because nitrates act as fertilizers and promote plant growth. Plants depend on other elements as well, including carbon, hydrogen, phosphorus, sulfur, and potassium. Except for nitrogen, however, these other elements tend to be readily available in the environment for uptake by plants. Since usable forms of nitrogen are scarce, we need to supply them in the form of fertilizers.

You might be wondering how nitrogen levels can be so low in soils when N_2 makes up so much of our atmosphere. Although abundant, the nitrogen molecule is *not* in a chemical form that most plants can use. N_2 is far less reactive than O_2.

> Just as there is a "carbon cycle" (Chapter 3), there also is a "nitrogen cycle." Stay tuned!

> All living things, not just plants, require nitrogen.

Your Turn 6.15   Unreactive Nitrogen

How does the bond energy of the triple bond in N_2 compare with other bond energies?
Hint: See Table 4.4.

In order to grow, plants need access to a form of nitrogen that reacts more easily, such as the ammonium ion, ammonia, or the nitrate ion. These and other reactive forms are listed in Table 6.1. As you might imagine, the air pollutants NO and NO_2 are included in the list. These forms of nitrogen all occur naturally and are present on our planet in relatively small amounts. Other forms of reactive nitrogen also exist, such as a group of compounds known as "amines" that contain the $-NH_2$ group. We introduce amines when we need them for our study of polymers, proteins, and DNA in Chapter 12.

Your Turn 6.16   Reactive Nitrogen

Select one of the compounds in Table 6.1. Give evidence, either in the form of an observation or a chemical equation, that this compound is reactive.
Hint: Revisit Chapters 1 and 2. Also draw on your personal knowledge.

Although we categorized N_2 as generally unreactive, one reaction involving the nitrogen molecule is of utmost importance: biological nitrogen fixation. Plants such as alfalfa, beans, and peas "fix" (remove) N_2 from the atmosphere (Figure 6.17). To be more accurate, it is not the plants themselves, but rather the bacteria living on or near the roots of these plants that fix the nitrogen. As part of their metabolism, **nitrogen-fixing bacteria** remove nitrogen gas from the air and convert it to ammonia. When the

Table 6.1	Some Reactive Forms of Nitrogen
Name	**Chemical Formula**
nitrogen monoxide	NO
nitrogen dioxide	NO_2
nitrous oxide	N_2O
nitrate ion	NO_3^-
nitrite ion	NO_2^-
nitric acid	HNO_3
ammonia	NH_3
ammonium ion	NH_4^+

Note: These forms of nitrogen all are naturally occurring.

Figure 6.17

Nodules on the root of a soya plant that contain nitrogen-fixing bacteria.

ammonia dissolves in water, it produces the ammonium ion, NH_4^+. The reaction is shown in equations 6.5a and 6.5b. This ion is one of two forms of reactive nitrogen that most plants can absorb. Recall from Chapter 5 that compounds containing the ammonium ion tend to be water soluble. Here is the pathway:

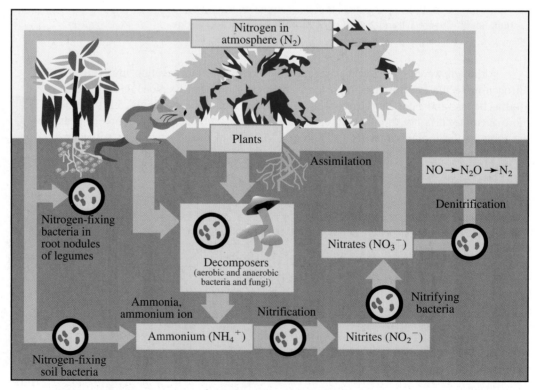

$$N_2 \xrightarrow[\text{nitrogen fixation}]{} NH_3 \xrightarrow{H_2O} NH_4^+ \qquad [6.22]$$

The other form of reactive nitrogen that plants can absorb is the nitrate ion. Again, compounds of the nitrate ion tend to be soluble. **Nitrification** is the process of converting ammonia in the soil to the nitrate ion. Two types of bacteria are involved along this pathway.

$$NH_4^+ \xrightarrow[\substack{\text{bacteria in} \\ \text{the soil}}]{} NO_2^- \xrightarrow[\substack{\text{bacteria in} \\ \text{the soil}}]{} NO_3^- \qquad [6.23]$$

Finally, to come full circle, bacteria again help with **denitrification,** the process of converting nitrates back to nitrogen gas. In so doing, these bacteria harness the energy released when the stable N_2 molecule forms. Recall from Chapter 4 the large amount of energy released in forming the triple bond found in the nitrogen molecule, a very stable molecule indeed!

Depending on the soil conditions, the pathway may occur in steps that include NO and N_2O. Thus, these reactive forms of nitrogen also can be converted to N_2 and released from the soil.

$$NO_3^- \xrightarrow[\substack{\text{bacteria in} \\ \text{the soil}}]{} NO \xrightarrow[\substack{\text{bacteria in} \\ \text{the soil}}]{} N_2O \xrightarrow[\substack{\text{bacteria in} \\ \text{the soil}}]{} N_2 \qquad [6.24]$$

All of these pathways are part of the **nitrogen cycle,** a set of chemical pathways whereby nitrogen moves through the biosphere. Figure 6.18 assembles pathways 6.22, 6.23, and 6.24 into a simplified version of the nitrogen cycle. In this cycle, all species are forms of reactive nitrogen except for N_2.

represents the bacteria responsible for chemical change. Bacteria of the genus *Nitrosomonas* convert NH_3 to NO_2^-. Bacteria of the genus *Nitrobacter* convert NO_2^- to NO_3^-.

Of the oxides of nitrogen, N_2O is emitted naturally in the greatest amount. It is a potent greenhouse gas. See Section 3.8.

Figure 6.18

The nitrogen cycle (simplified).

Remember that reactive forms of nitrogen are needed for plant growth. Because bacteria in the soil cannot supply ammonia, ammonium ion, or nitrate ion in the amounts needed for optimal plant growth, farmers use fertilizers. A few centuries ago, fertilizers were obtained by mining deposits of saltpeter (ammonium nitrate) from the deserts of Chile or by collecting guano, a nitrogen-rich deposit from bird and bat droppings in Peru. Neither source, however, was sufficient to meet the demand of the growing world population. An additional drain on the supply of nitrates was their use in the production of gunpowder and other explosives such as TNT. By the early 1900s, the search was on for a way to synthesize reactive nitrogen compounds from the abundant N_2 in the air.

How are fertilizers obtained in the large quantities needed for present-day agriculture? The answer lies in a second important reaction of N_2, one that literally captures it out of the air to synthesize ammonia:

$$N_2(g) + 3\,H_2(g) \longrightarrow 2\,NH_3(g) \qquad \text{[6.25]}$$

This famous chemical reaction is known as the Haber–Bosch process. It allows the economical production of ammonia, which in turn enables the large-scale production of fertilizers and nitrogen-based explosives. As a fertilizer, ammonia can be directly applied to the soil or can be applied as ammonium nitrate or ammonium phosphate. The green line that starts around 1910 in Figure 6.19 represents the large increase in global reactive nitrogen from the Haber–Bosch process.

Also notice the orange line on this graph that indicates the amount of reactive nitrogen formed by the burning of fossil fuels. Recall that at the high temperatures of combustion, N_2 and O_2 react to form NO. The purple line on the graph represents reactive nitrogen from all sources, or the sum of the orange, blue, and green lines. The overlaid red line for population, of course, comes as no surprise. The increases in reactive nitrogen from burning fossil fuels (energy production) and fertilization (food production) parallel the growth in world population (people production).

The reactive forms of nitrogen in this cycle continuously change chemical forms. Thus, the ammonia that starts out as a fertilizer may end up as NO, in turn increasing the acidity of the atmosphere. Or the NO may end up as N_2O, a greenhouse gas that is currently rising in atmospheric concentration. Or the ammonium ion, instead of being

In 1918, Fritz Haber received the Nobel Prize in Chemistry for synthesizing NH_3 from N_2 and H_2. In 1931, Carl Bosch received the prize for using this synthesis commercially.

Increased access to fertilizers and large-scale farming equipment enabled an unprecedented expansion of agriculture called the Green Revolution that began worldwide around 1945. Look for more on the Green Revolution in Section 11.12.

Greenhouse gases were introduced in Section 3.1 and further described in Section 3.8.

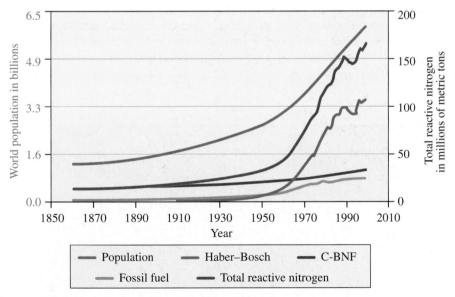

Figure 6.19

Global changes in reactive nitrogen produced by various sources (million metric tons, scale on the right). The top line is the world's population (billions, scale on the left).

Note: C-BNF is the reactive nitrogen created from the cultivation of legumes, rice, and sugarcane.

tightly bound to the soil, may end up being converted and leached out as the nitrite or nitrate ion, in turn contaminating a water supply.

Now we can begin to understand the full effects that NO emissions have on our environment. First, the oxides of nitrogen form ground-level ozone in the presence of sunlight, contributing to photochemical smog, as we saw in Chapter 1. Second, NO_x emissions are a form of reactive nitrogen, just like the fertilizers used for food production. NO is formed from unreactive N_2 in the air when fuels are burned. The more fuels are burned, the more unreactive N_2 is changed into a reactive form. Both NO_x and fertilizer use are affecting the balances within the nitrogen cycle on our planet. Finally, NO_x emissions increase the acidity of the precipitation falling from the sky.

6.10 | SO₂ and NOₓ—How Do They Stack Up?

Having identified SO_2 and NO_x as the two major contributors to acid precipitation, we now examine their production over time and strategies for controlling anthropogenic emissions. In addition to anthropogenic sources, these oxides are naturally produced. Some volcanoes steadily emit SO_2, and under certain conditions, they produce beautiful crystals of elemental sulfur (Figure 6.20). When they erupt, they can release large amounts of SO_2. The June 1991 eruption of Mount Pinatubo in the Philippines was 10 times as large as the 1980 eruption of Mount St. Helens in the United States, emitting between 15 and 30 million tons of sulfur dioxide into the stratosphere.

Oceans are a second natural source of sulfur emissions. Marine organisms produce dimethyl sulfide gas as a by-product. The dimethyl sulfide enters the troposphere and reacts with $OH\cdot$ to form SO_2. NO is formed any time that high heat causes nitrogen and oxygen gases to react; in nature, this occurs during lightning strikes and forest fires. Bacteria in soil convert nitrogen gas in the air into nitrogen oxides that plants can use to build mass.

Anthropogenic SO_2 and NO_x emissions outpace natural emissions. The amount of sulfur added to the atmosphere by humans is twice that from volcanoes, oceans, and other natural sources. The amount of nitrogen added as NO_x by humans is roughly four times that of natural sources such as lightning and the bacteria found in soils. In the United States, the annual anthropogenic emissions are on the order of 15 and 20 million tons for SO_2 and NO_x, respectively. Consult Figures 6.14 and 6.16 to review the sources of these emissions.

1 ton (short ton) = 2000 lb
= 0.9072 metric tons
(tonnes)

Figure 6.20

Elemental sulfur (*yellow*) occurring naturally in Hawaii Volcanoes National Park.

The levels of these two pollutants have changed dramatically over time. Before 1950, relatively small amounts of NO_x were present in rain, fog, and snow. Figure 6.21a shows that NO_x levels in the United States increased in the past, but now are leveling off. In contrast, SO_2 emissions in the United States have decreased substantially since their peak in 1974, a tribute to many efforts, including the 1990 Clean Air Act Amendments. In the final section of this chapter, we examine how costs, control strategies, and politics have influenced SO_2 and NO_x emissions in the United States.

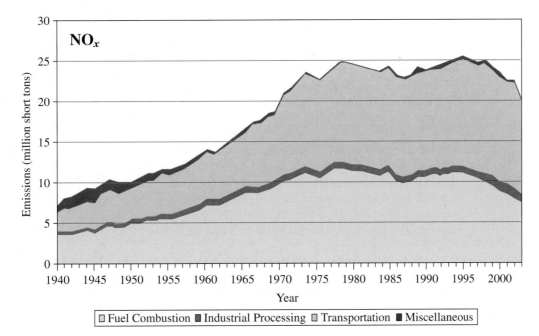

(a)

(b)

Figure 6.21

(a) U.S. nitrogen oxide emissions 1940–2003. (b) U.S. sulfur dioxide emissions 1940–2003.

Note: Fuel combustion refers to fossil fuel combustion, such as coal.

Source: EPA/OAR, *National Air Pollutant Emission Trends, 1900–1998,* with recent data added.

Your Turn 6.17   SO$_2$ and NO$_x$ Emissions

Figure 6.21a depicts NO$_x$ emissions in the United States from four sources. Which two did not change much between 1940 and 1995? In contrast, which two did? For SO$_2$ in this same period, which sources led to the large increase in emissions in the 1970s?

Globally, the levels of SO$_2$ and NO$_x$ also are changing over time. Because the NO$_x$ emissions originate from millions of small, unregulated, and mobile sources, they are difficult to track. In contrast, emissions of SO$_2$ can be estimated with a reasonable degree of accuracy. National data on fossil fuel consumption and the refining of metal ores that contain sulfur make this possible. To get an estimate, researchers start with the amount of fossil fuels (together with their sulfur content) produced in a country, then add in imports of fossil fuels, and finally subtract out exports. Metal refining is a bit trickier to estimate, as the amount of sulfur released depends on the technologies used (which are not always known). Nonetheless, it is possible to reach conclusions using these types of data.

One such estimate published in 2004 shows good news—a decline in world SO$_2$ emissions over the past decade. In the 1970s, Western Europe and North America shared the title of the world's largest emitters. As we saw earlier, the U.S. emissions levels then declined rapidly and those of Western Europe followed suit. Eastern Europe became the emissions leader, reaching its peak in 1989, and now likewise its emissions levels are dropping. In Western Europe, SO$_2$ emissions decreased as a result of environmental regulations, whereas in Eastern Europe economic depression was responsible for the emissions decline.

Today, the continent of Asia leads in SO$_2$ emissions. In 1970, the United States emitted about 30 million tons of sulfur dioxide and China about 10 million tons. In 1990, both countries released about 22 million tons. With the start of the year 2000, China emerged as the clear global leader in SO$_2$ emissions. However, China is closing some older inefficient coal plants, so the emissions from China have not risen as quickly as they could have.

China's economy continues to grow at a record pace, in part driven by developed countries' appetite for inexpensive manufactured goods. To meet the massive demand for electricity, China's coal output more than doubled in the first decade of this century. India appears to be headed down the same path as energy needs hamper its economic growth. China's coal consumption in 2008 was 43% of the world total, with the United States at 17% and India at 7%. Both China and India, with a combined one third of the world's population, have seen a massive growth in the numbers of motor vehicles. If China alone had the same number of cars per person as the United States it would have more than the world's total of 850 million cars currently on the road. The emissions worldwide of SO$_2$ and NO$_x$ are expected to grow to unprecedented levels. Clearly, a pattern of unchecked growth is not sustainable. As a result, many nations are pursuing sources of electricity powered by wind, geothermal, and the Sun. Such technologies that produce far less SO$_2$ and NO$_x$, coupled with conservation, are essential for our planet both now and in the future.

Even "green" technologies produce some SO$_2$ and NO$_x$. These gases are emitted during the manufacture, installation, and service of the wind turbines, solar cells, and electric transmission lines.

6.11 | Acid Deposition and Its Effects on Materials

As we have seen, much of the rain, mist, fog, and snow in the United States is more acidic than would be expected from carbon dioxide dissolution alone. In the worst cases, rain, fog, and dew can have a pH of 3.0 or lower. Does this really matter? To answer this question, we need to know something about the effects of acid deposition and how serious they really are.

National studies can help us. During the 1980s the U.S. Congress funded a national research effort called the National Acid Precipitation Assessment Program

(NAPAP). Over 2000 scientists were involved, with a total expenditure of $500 million. The project was completed in 1990, and the participating scientists prepared a 28-volume set of technical reports (NAPAP, *State of the Science and Technology*, 1991). Some of the material in the remainder of this chapter is drawn from the NAPAP Report. Other information is drawn from the proceedings of a conference in 2001 entitled "Acid Rain: Are the Problems Solved?" sponsored by the Center for Environmental Information. The purpose of that conference was "to put the acid rain problem squarely back on the forefront of the public agenda."

And we agree—acid rain should remain on the public agenda. One reason is the damage that acid rain causes (Table 6.2). In this section, we describe the effects of acid rain on metals, statues, and buildings. The effects of acid deposition on living things are explored in the section that follows.

Recall from Chapter 1 that about 80% of the elements on the periodic table are metals. Many of these metals are vulnerable to damage from acid precipitation. Metals typically are shiny and silvery; well, before they become tarnished or rusted.

Your Turn 6.18   Metals and Nonmetals

With the help of a periodic table, classify these elements as metals or nonmetals. Also give the chemical symbol for each.

a. iron   b. aluminum   c. fluorine
d. calcium   e. zinc   f. oxygen

Metals and nonmetals were defined in Section 1.6.

Although acid rain (pH 3–5) does not affect all metals, unfortunately iron is one that it does. Bridges, railroads, and vehicles of all kinds depend on iron and the steel that is made from it. Rods of steel are used to strengthen concrete buildings and roadways. In many parts of the country, decorative iron fences and latticework both ornament and protect city and rural homes.

Metals used in jewelry (gold, silver, and platinum) do not react with acidic depositions.

Table 6.2	Effects of Acid Rain and Recovery Benefits
Effects	**Recovery Benefits**
Materials Acid deposition contributes to the corrosion and deterioration of buildings, cultural objects, and cars. This decreases their value and increases the cost of correcting and repairing damage.	Less damage to buildings, cultural objects, and cars. Lowering future costs to correct and repair such damage. See Section 6.10.
Human Health Sulfur dioxide and nitrogen oxides in the air increase deaths from asthma and bronchitis.	Fewer visits to the emergency room, fewer hospital admissions, and fewer deaths. See Section 6.11.
Visibility In the atmosphere, sulfur dioxide and nitrogen oxides form sulfate and nitrate aerosols that impair visibility and affect enjoyment of national parks and other scenic views.	Reduced haze, therefore the ability to view scenery at a greater distance and with greater clarity. See Section 6.11.
Surface Waters Acidic surface waters injure animal life in lakes and streams. In more severe instances, some or all types of fish and other marine organisms die.	Lower levels of acidity in the surface waters and a restoration of animal and plant life in the more severely damaged lakes and streams. See Section 6.13.
Forests Acid deposition contributes to forest degradation by impairing the growth of trees and increasing their susceptibility to winter injury, insect infestation, and drought. It also causes leaching and depletion of natural nutrients in forest soil.	Less stress on trees, thereby reducing the effects of winter injury, insect infestation, and drought. Less leaching of nutrients from soil, thereby improving the overall forest health.

Source: Adapted from Emission Trends and Effects in the Eastern U.S., United States General Accounting Office, Report to Congressional Requesters, March, 2000.

This is an oxidation–reduction reaction. Look for more about this type of reaction in Chapter 8.

The problem with iron is that it rusts, as represented by this chemical equation.

$$4\,Fe(s) + 3\,O_2(g) \longrightarrow 2\,Fe_2O_3(s) \qquad [6.26]$$

Rusting is a slow process. Iron combines rapidly with oxygen only if you heat or ignite it, such as with a sparkler or fireworks. But at room temperature, the rusting of iron is a two-step process that requires the presence of hydrogen ions. Equation 6.26 is the overall equation for the process. The role of H^+ is evident in equation 6.27, the first step of the process. In this step, iron metal dissolves.

$$4\,Fe(s) + 2\,O_2(g) + 8\,H^+(aq) \longrightarrow 4\,Fe^{2+}(aq) + 4\,H_2O(l) \qquad [6.27]$$

Even pure water (pH = 7) has a sufficient concentration of H^+ to promote slow rusting. In the presence of acid, the rusting process is greatly accelerated. In the second step, the aqueous Fe^{2+} further reacts with oxygen.

$$4\,Fe^{2+}(aq) + O_2(g) + 4\,H_2O(l) \longrightarrow 2\,Fe_2O_3(s) + 8\,H^+(aq) \qquad [6.28]$$

Adding the two steps together gives equation 6.26. The solid product, Fe_2O_3, is the familiar reddish brown material that we call rust.

Your Turn 6.19   Rust Adds Up

Add equations 6.27 and 6.28 together to show the overall reaction for rust formation (equation 6.26).

Your Turn 6.20   Careful with the Charges

On our planet, the element iron is found in several different chemical forms. This section just mentioned three: Fe, Fe^{2+}, and Fe^{3+}.

a. Which one of these is the familiar silvery iron metal?
b. With respect to iron metal, have any of these gained outer (valence) electrons? If so, which one(s) and how many electrons?
c. Have any lost valence electrons? If so, which one(s) and how many?

Answer
b. With respect to Fe, neither Fe^{2+} nor Fe^{3+} has gained outer electrons.

Because iron is inherently unstable when exposed to the natural environment, billions of dollars are spent annually to protect exposed iron and steel in bridges, cars, buildings, and ships. Paint is the most common means of protection, but even paint degrades, especially when exposed to acidic rain and gases. Coating iron with a thin layer of a second metal such as chromium (Cr) or zinc (Zn) is another means of protection.

Automobile paint can be spotted or pitted by acid deposition. To prevent this damage, automobile manufacturers now use acid-resistant paints. It is an irony that automobiles emit the very chemical that mars their paint. Follow the NO from your car's tailpipe and you may find that its reaction product (nitric acid) eventually ends up in droplets hitting the hood of your car.

Acidic rain also damages statues and monuments made of marble. For example, those in the Gettysburg National Battlefield have suffered irreparable damage. Figure 6.22 shows a recognizable, but much deteriorated statue of George Washington in New York City. Marble limestone, composed mainly of calcium carbonate, $CaCO_3$, slowly dissolves in the presence of hydrogen ion.

$$CaCO_3(s) + 2\,H^+(aq) \longrightarrow Ca^{2+}(aq) + CO_2(g) + H_2O(l) \qquad [6.29]$$

In 1944

At present

Figure 6.22

Acid rain damaged this limestone statue of George Washington. It was erected in New York City in 1944.

Your Turn 6.21   **Damage to Marble**

Marble can contain both magnesium carbonate and calcium carbonate.

a. Analogous to equation 6.29, write the chemical equation for the reaction of acidic rain with magnesium carbonate.

b. Marble never contains sodium bicarbonate. Explain why.
Hint: See Section 5.8 on the solubilities of ionic compounds.

Your Turn 6.22   **Damage from SO_2**

Suppose that the acid represented in equation 6.29 by $H^+(aq)$ is sulfuric acid. Write the balanced chemical equation for the reaction of sulfuric acid with marble.

Visitors to the Lincoln Memorial in Washington, D.C., learn that the huge stalactites growing in chambers beneath the memorial are the result of acid rain eroding the marble, again a material containing either calcium carbonate or magnesium carbonate (or both). Other monuments in the eastern United States are suffering similar fates. Some limestone tombstones are no longer legible. Worldwide, many priceless and irreplaceable marble statues and buildings are being attacked by airborne acids (Figure 6.23). The Parthenon in Greece, the Taj Mahal in India, and the Mayan ruins at Chichén Itzá all show signs of acid erosion. Ironically, some of the acid deposition at these sites is due to the NO_x produced by the tour buses and vehicles with minimal emissions controls.

Consider This 6.23   **Deterioration and Damage**

Reexamine Figure 6.22. Although it may be tempting to blame acid rain for the damage, other agents may be at work. View possible other culprits for yourself by taking a photo tour of our nation's capitol, courtesy of a website on acid rain provided by the United States Geological Survey. A link is provided at the textbook's website. What kinds of damage do the photos show? What promotes damage by acid rain? What else has caused the deterioration?

Figure 6.23
Acid rain knows no geographic or political boundaries. Acid rain has eroded Mayan ruins at Chichén Itzá, Mexico.

Consider This 6.24   Acid Rain Across the Globe

The concerns of acid rain vary across the globe. Many countries in North America and Europe have websites dealing with acid rain. Search to locate one or use the links provided at the textbook's website. What are the issues in the country you selected? Does the acid deposition originate outside the borders of the country?

6.12 | Acid Deposition, Haze, and Human Health

Haze! This phenomenon, often an indication of acid deposition, results from tiny liquid droplets or solid particles suspended in air. If you live in a congested urban area, you often can see haze simply by looking at the buildings down a long street. If you live in the country, you may be familiar with the summer haze that sometimes settles over the fields. Airline passengers, as they peer down from a cruising altitude, often notice that the features and colors of the landscape are blurred. Ironically, you become more aware of the haze on the occasional clear day when it really does seem that you can see forever.

The causes of haze are well understood, but they differ from region to region. In large cities such as Beijing, China (Figure 6.24), coal-burning power plants produce the smoke and particulate matter that in turn create the haze. In rural regions, a different set of particulates, including soil dust and the soot of wood-burning stoves, add to the haze.

East or west, power plants emit NO_x and SO_2. Although both contribute to haze, for the purposes of illustrating acid deposition we focus on the latter. As we mentioned earlier, coal contains a few percent sulfur, and when the coal is burned, a steady stream of sulfur dioxide is released. Sulfur dioxide is colorless, so this gas is not what we are peering through as "haze." Rather, SO_2 is the precursor to the formation of this haze.

Let's focus on a molecule of SO_2 as it exits the tall smokestack of a coal-fired power plant. As it moves downwind, via a series of steps it forms an aerosol of sulfuric acid. The first is the reaction of SO_2 with oxygen to form SO_3, as we saw earlier in equation 6.18. Sulfur trioxide is also a colorless gas, but it has the property of being **hygroscopic,** that is, it readily absorbs water from the atmosphere and retains it. Next, a molecule of SO_3 can react rapidly with a water molecule to form sulfuric acid (see equation 6.12).

Similarly, acidic aerosols of nitric acid form from NO_x.

Figure 6.24

The Forbidden City in Bejing, China, as seen on two different days.

Your Turn 6.25   **Droplets of Acid**

As a review of the sulfur chemistry just described, write a set of chemical equations that start with elemental sulfur in coal and eventually produce sulfuric acid.

Many molecules of sulfuric acid form a tiny droplet, and many tiny, tiny droplets of sulfuric acid then coalesce to produce larger droplets. These droplets form an aerosol with droplets about a micrometer ($1 \ \mu m = 10^{-6}$ m) in diameter. These droplets of sulfuric acid do not absorb sunlight. Rather, they scatter (reflect) sunlight, reducing visibility. The aerosols of sulfuric acid, which can persist for several days, can travel hundreds of miles downwind, which is why the haze can become so widespread. In addition, these fine droplets of acid are stable enough that they enter our buildings and become part of the air that we breathe indoors.

You also may have heard of sulfate aerosols. Recall that sulfuric acid, H_2SO_4, ionizes to produce H^+, HSO_4^-, and SO_4^{2-}. The concentrations of each can be measured in an aerosol. But these acidic aerosols may react with bases to produce salts that contain the sulfate ion. Typically this base is ammonia or in aqueous form, ammonium hydroxide. Thus, the particles or droplets in an aerosol may be a mixture of sulfuric acid, ammonium sulfate, $(NH_4)_2SO_4$, and ammonium hydrogen sulfate, NH_4HSO_4. Reporting the concentration of sulfate and hydrogen sulfate ions (rather than simply the pH) gives a better indication of how much sulfuric acid was initially present.

Your Turn 6.26   Sulfate Aerosols

As a review of the acid–base chemistry just described, write balanced chemical equations that show how sulfate salts form in aerosols. Write reactants and products in their aqueous forms rather than dissociated into ions. For example, represent sulfuric acid as $H_2SO_4(aq)$.

a. The reaction of sulfuric acid with ammonium hydroxide to form ammonium hydrogen sulfate and water.
b. The reaction of sulfuric acid with ammonium hydroxide to form ammonium sulfate and water.

Hint: This requires 2 moles of base.

The summer sunshine plays a role in generating the hydroxyl radical in the atmosphere, which in turn catalyzes the SO_2-to-SO_3 conversion.

During hot and dry seasons, wild fires also contribute to haze.

Haze is most pronounced in summer when there is more sunlight to accelerate the photochemical reactions leading to the production of sulfuric acid. As a result, the average visibility in the eastern United States is now about 20 miles and occasionally as low as 1 mile. By contrast, visibility in the western states is now lessened from the natural visual range of about 200 miles to 100 miles or less. Where you formerly might have been able to see the mountains 100 miles away, these mountains may now have disappeared into the haze. The visibility in many national parks, including Yellowstone, the Grand Canyon, and the Great Smoky Mountains, has been affected.

In the United States, the Clean Air Act of 1970 and its subsequent amendments included provisions to improve the visibility in our national parks. Although the standards were set by federal law, the states were charged with their implementation. Visibility continued to drop in the national parks. In the last few days of his presidency, Bill Clinton signed a bill authorizing the EPA to issue regulations to help clear the skies in national parks and wilderness areas. These regulations, called the Regional Haze Rule (1999), required the hundreds of older power plants that emitted vast quantities of SO_2, NO_x, and particulates to retrofit their operations with pollution controls. A final set of amendments, the Clean Air Visibility Rule, was issued on June 15, 2005 by then President George W. Bush.

Consider This 6.27   Hazy at Mount Rainier?

Web cams! Live on the web, see for yourself the haze (or lack thereof) at Mount Rainier. Dozens of other places have web cams as well, and the EPA posts a list of these. A link is provided at the textbook's website.

a. During daylight hours, look up several web cams to see what's out there.
b. Find the current air quality for a location of your choice. Some sites provide this information together with the web cam photographs. For others, you can obtain the data from the EPA AIRNOW site.
c. How well does the air quality correlate with the visibility?

When you can see haze on the horizon, you are most likely also breathing it. Once inhaled, the acidic droplets attack the sensitive tissue of your lungs. People with asthma, emphysema, and cardiovascular disease are the most sensitive. Mortality rates may increase, especially for those with bronchitis and pneumonia. But even people in good health feel the irritating effects of the acidic aerosols. Thus breathing air contaminated with aerosols of sulfate and sulfuric acid clearly comes with a medical price tag! The low cost of electricity from burning coal fails to take into account the costs to the community of the health effects resulting from NO_x and SO_2 emissions.

A decrease in acidic aerosol levels would benefit many, both in real dollars and in quality of life. The problem is that the costs and savings are not directly borne by the same groups. Industry must pay to clean up; people must pay medical bills. Government, of course, is involved in paying both.

The EPA's Clean Air Visibility Rule of 2005 is expected to provide "substantial health benefits in the range of $8.4 to $9.8 billion each year—preventing an estimated 1600 premature deaths, 2200 nonfatal heart attacks, 960 hospital admissions, and more than 1 million lost school and work days." The cost-to-benefit ratio is thus exceedingly favorable. For some, the question is who is paying the cost, and who is reaping the benefit. For companies striving to reduce emissions, upgrading older equipment definitely comes with a cost. But there are savings as well. Fewer emissions translate to healthier communities (including employees) and lower health care costs and lost days at work. In terms of the Triple Bottom Line, reducing emissions translates to healthier ecosystems, healthier communities, and healthier economies.

Historically, polluted air has exacted a huge price. One of the worst recorded instances of pollution-related respiratory illness occurred in London in 1952. Periods of foggy bad air were nothing unusual to the British Isles, as factory chimneys had belched smoke into the air for several hundred years. But in December, 1952, the weather was colder than usual and people were burning large quantities of sulfur-rich coal in their home fireplaces. Due to unusual weather conditions, a deep layer of fog developed that trapped all the smoke and pollutants for five days, dropping visibility to practically zero. The deadly aerosol caused more than 4000 deaths, during its peak claiming 900 lives daily.

In 1948, a similar incident occurred in Donora, Pennsylvania, a steel mill town south of Pittsburgh. Again a layer of fog trapped industrial pollutants close to the ground. By noon, the skies had darkened with a choking aerosol of fog and smoke (Figure 6.25). An 81-year-old fireman who took oxygen door-to-door to the victims reported, "It may sound dramatic or exaggerated, but you could barely see." High concentrations of sulfuric acid and other pollutants soon caused widespread illness. During the fog, 17 people died, to be followed by 4 more later. Although Donora and London were extreme and unusual incidents from the past, people still breathe highly polluted air today. The U.S. EPA and the World Health Organization currently estimate that 625 million people are still exposed to unhealthy levels of SO_2 released by the burning of fossil fuels. Indeed, modern residents of Datong, China, the coal mining

Triple Bottom Line: healthy economies, Healthy ecosystems, and healthy communities. For more about the Triple Bottom Line, see Chapter 0.

(a)

(b)

Figure 6.25

(a) A 1948 news headline from Donora, Pennsylvania. **(b)** Donora at noon during the deadly smog of 1948.

center of the country, have reported winter air quality so poor that "even during the daytime, people drive with their lights on."

Although acidic fogs can be immediately hazardous to one's health, public concern is growing over the indirect effects of acid deposition. For example, the solubilities of certain toxic metal ions, including lead, cadmium, and mercury, are significantly increased in the presence of acids. These elements are naturally present on Earth, often tightly bound in minerals that make up soil and rock. Dissolved in acidified water and conveyed to the public water supply, these metals can pose serious health threats.

Clearly there is a connection between burning fossil fuels, acidic precipitation, and human health. An article written in the journal *Science* in 2001 by an international team of authors bluntly assessed the situation this way, "For every day that policies to reduce fossil-fuel combustion emissions are postponed, deaths and illness related to air pollution will increase." These words are as true today as they were in 2001. Health care costs will be reduced if the air we breathe is clean.

Similarly, studies by the EPA have estimated that the reductions in SO_2 and associated acid aerosol pollution called for by the Clean Air Act Amendments of 1990 should result in saving billions of dollars in health care costs over time. The savings would come principally from reduced costs to treat pulmonary diseases such as asthma and bronchitis and from a decrease in premature deaths.

6.13 | Damage to Lakes and Streams

Humans are not the only creatures bearing the costs of acidic precipitation. Organisms in the world's surface waters experience a change in environment when acidic precipitation fills lakes and streams. Healthy lakes have a pH of 6.5 or slightly above. If the pH is lowered below 6.0, fish and other aquatic life are affected (Figure 6.26). Only a few hardy species can survive below pH 5.0. At pH 4.0, lakes become essentially dead ecosystems.

Numerous studies have reported the progressive acidification of lakes and rivers in certain geographic regions, along with reductions in fish populations. In southern Norway and Sweden, where the problem was first observed, one fifth of the lakes no longer contain any fish, and half of the rivers have no brown trout. In southeastern Ontario, the average pH of lakes is now 5.0, well below the pH of 6.5 required for a healthy lake. In Virginia, more than a third of the trout streams are episodically acidic or at risk of becoming so.

Many areas of the midwestern United States have no problem with acidification of lakes or streams, even though the Midwest is a major source of acidic precipitation. This apparent paradox can be explained quite simply. When acidic precipitation falls on or runs off into a lake, the pH of the lake drops (become more acidic) unless the acid is neutralized or somehow used by the surrounding vegetation. In some regions, the surrounding soils contain bases that can neutralize the acid. The capacity of a lake or other body of water to resist a decrease in pH is called its **acid-neutralizing capacity.** The surface geology of much of the Midwest is limestone,

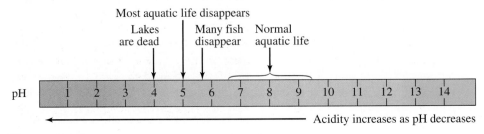

Figure 6.26

Aquatic life and pH.

$CaCO_3$. As a result, lakes in the Midwest have a high acid-neutralizing capacity because limestone slowly reacts with acid rain, as we saw earlier with marble statues and monuments (equation 6.29).

More importantly, the lakes and streams also have a relatively high concentration of calcium and hydrogen carbonate ions. This occurs as a result of the reaction of limestone with carbon dioxide and water.

$$CaCO_3(s) + CO_2(g) + H_2O(l) \longrightarrow \underset{\text{calcium ion}}{Ca^{2+}(aq)} + \underset{\text{hydrogen carbonate ion}}{2\ HCO_3^{-}(aq)} \qquad \textbf{[6.30]}$$

Because acid is consumed by the carbonate and hydrogen carbonate ions, the pH of the lake remains more or less constant.

Your Turn 6.28   The Bicarbonate Ion

The hydrogen carbonate ion produced in equation 6.30 also can accept a hydrogen ion.

a. Write the chemical equation.

b. Is the hydrogen carbonate ion functioning as an acid or a base?

Answer

a. $HCO_3^{-}(aq) + H^{+}(aq) \longrightarrow H_2CO_3(aq) \longrightarrow CO_2(g) + H_2O(l)$

In contrast to the Midwest, many lakes in New England and northern New York (as well as in Norway and Sweden) are surrounded by granite, a hard, impervious, and much less reactive rock. Unless other local processes are at work, these lakes have very little acid-neutralizing capacity. Consequently, many show a gradual acidification.

As it turns out, understanding the acidification of lakes is a good deal more complicated than simply measuring pH and acid-neutralizing capacities. One level of complexity is added by annual variations. Some years, for example, heavy winter snowfalls persist into the spring and then melt suddenly. As a result, the runoff may be more acidic than usual, because it contains all the acidic deposits locked away in the winter snows. A surge of acidity may enter the waterways at just the time when fish are spawning or hatching and are more vulnerable. In the Adirondack Mountains of northern New York, about 70% of the sensitive lakes are at risk for episodic acidification, in comparison with a far smaller percent that are chronically affected (19%). In the Appalachians, the number of episodically affected lakes (30%) is seven times those chronically affected.

When, if ever, will the lakes recover? The good news is that U.S. SO_2 emissions have been declining in recent years, and we have seen a corresponding decrease in the sulfate ion concentrations in the lakes of the Adirondacks. However, even though NO_x emissions have remained fairly constant, the amount of nitrates in the Adirondacks is increasing in more lakes than not. Thus, it appears that nitrogen saturation has occurred in the surrounding vegetation, with more of the acidity ending up in the lakes. The soil in the region of these lakes most likely has lost some of its acid-neutralizing capacity.

Recent findings are mixed. A March 2000 report to Congress puts it bluntly: "The lakes in the Adirondack Mountains are taking longer to recover than lakes located elsewhere and are likely to recover less or not recover, without further reductions of acid deposition." The Progress Report on Acid Rain issued by the EPA in 2004, however, reports some improvements. For example, in comparison with earlier years when over 10% of the lakes in the Adirondacks were acidic, today the value is closer to 8%. Similar improvements are documented in the Midwest, where now only about 1% of the lakes are acidic. In contrast, the lakes in New England and the Blue Ridge Mountains remain stubbornly acidic.

Conclusion

Emissions of acidic oxides—carbon dioxide, sulfur dioxide, and nitrogen oxides—are affecting the acidity of the world's oceans, rainfall, lakes, and rivers. In the United States, "acid rain" is not the dire plague once described by environmentalists and journalists. Nor is it a matter to be ignored. It is sufficiently serious that federal legislation, the Clean Air Act Amendments of 1990, were enacted to reduce SO_2 and NO_x emissions, precursors to acid deposition.

If you have learned anything from this chapter, we hope it has been the recognition that complex problems cannot be solved by simple or simplistic strategies. Any failure to acknowledge the intertwined relationships involving the combustion of coal and gasoline; the production of carbon, sulfur, and nitrogen oxides; and the reduced pH of seawater, fog, and precipitation is to deny some fundamental facts of chemistry. Knowledge of ecology and biological systems is needed as well, so that acid deposition can be understood in the context of entire ecosystems, a task that requires that experts from several disciplines collaborate.

Public health also is at issue. Economic analyses reveal that allocating funds to reduce sulfur and nitrogen emissions will have a huge payoff in terms of lower mortality rates, fewer illnesses, and higher quality of living.

One response that we as individuals and as a society might make to the problems of ocean acidification and acid precipitation has hardly been mentioned in this chapter, yet it is potentially one of the most powerful: to conserve energy and to transition to noncombustion sources of energy. Carbon dioxide, sulfur dioxide, and nitrogen oxides are by-products of our voracious demand for energy, especially for electricity and transportation. If our personal, national, and global appetite for fossil fuels continues to grow unchecked, our environment may well become a good deal warmer and a good deal more acidic. Moreover, the problem may be intensified as the supply of petroleum and low-sulfur coals diminishes and we become even more reliant on high-sulfur coal.

There are other sources of energy—nuclear fission, water and wind, renewable biomass, and the Sun itself. All currently are being utilized, and their use will no doubt increase. We explore nuclear fission in the next chapter. But we conclude this chapter with the modest suggestion that, for a multitude of reasons, the conservation of energy by industry and collectively by individuals could have profoundly beneficial effects on our environment.

Chapter Summary

Having studied this chapter, you should be able to:

- Define the terms *acid* and *base* and know how to use these definitions to distinguish acids from bases (6.1–6.3)

- Represent the dissociation (ionization) of acids and bases using chemical equations (6.1–6.2)

- Write neutralization reactions for acids and bases (6.3)

- Classify solutions as acidic, basic, or neutral based on their pH or concentrations of H^+ and OH^- (6.3–6.4)

- Calculate pH values given hydrogen or hydroxide ion in whole-number concentrations (6.4)

- Compare the pH of pure water, the pH of ordinary rain, the pH of acid rain, and the pH of seawater (6.4)

- Use chemical equations to relate increasing levels of carbonic acid in seawater to the dissolution of calcium carbonate shells (6.5)

- Locate on a map of the United States where the most acidic rain falls (6.6)

- Explain the role of sulfur oxides and nitrogen oxides in causing acid rain (6.7–6.8)

- Compare the causes of ocean acidification and acid precipitation (6.5–6.8)

- Explain why N_2 is a relatively inert element. Describe different forms of reactive nitrogen and how they are produced both naturally and by humans. Use the nitrogen cycle to explain the cascading effects of reactive nitrogen (6.9)

- Describe how the industrial production of ammonia and the acidic deposition of nitrates both contribute to the buildup of reactive nitrogen on our planet (6.9)

- List the different sources of NO_x and of SO_2 and explain the variations in the levels of these pollutants over the past 30 years (6.10)
- Describe the production of acidic aerosols and their effects on building materials and human health (6.12)

- Explain why acid rain control is a wise investment in terms of the benefits to human health (6.12)
- Describe nitrogen saturation and its consequences for lakes (6.13)

Questions

Emphasizing Essentials

1. This chapter opens with a discussion of ocean acidification.

   a. Seawater contains many salts, including sodium chloride. Write its chemical formula.

   b. Sodium chloride is soluble in water. What chemical process takes place when solid sodium chloride dissolves? *Hint:* See Section 5.8.

   c. Is sodium chloride an electrolyte? Explain.

2. Calcium carbonate is another salt. Write its chemical formula. Would you expect calcium carbonate to be soluble or insoluble in water? *Hint:* See Section 5.8.

3. Carbon dioxide is a gas found in our atmosphere.

   a. What is the approximate concentration?

   b. Why is its concentration in the atmosphere increasing?

   c. Draw the Lewis structure for the CO_2 molecule.

   d. Would you expect carbon dioxide to be highly soluble in seawater? Explain.

4. The term *anthropogenic emissions* is used in the opening section of this chapter. Explain its meaning.

5. a. Draw the Lewis structure for the water molecule.

   b. Draw Lewis structures for the hydrogen ion and the hydroxide ion.

   c. Write a chemical reaction that relates all three structures from parts **a** and **b.**

6. a. Give names and chemical formulas for five acids of your choice.

   b. Name three observable properties generally associated with acids.

7. Write a chemical equation that shows the release of one hydrogen ion from a molecule of each of these acids.

   a. HBr*(aq)*, hydrobromic acid

   b. H_2SO_3*(aq)*, sulfurous acid

   c. $HC_2H_3O_2$*(aq)*, acetic acid

8. a. Give names and chemical formulas for five bases of your choice.

   b. Name three observable properties generally associated with bases.

   c. Draw Lewis structures for each species in equation 6.3c.

9. Write a chemical equation that shows the release of hydroxide ions as each of these bases dissolves in water.

   a. KOH*(s)*, potassium hydroxide

   b. $Ba(OH)_2$*(s)*, barium hydroxide

10. Which gas dissolved in water to produce each of these acids?

    a. carbonic acid, H_2CO_3

    b. sulfurous acid, H_2SO_3

11. Consider these ions: nitrate, sulfate, carbonate, and ammonium.

    a. Give the chemical formula for each.

    b. Write a chemical equation in which the ion (in aqueous form) appears as a product.

12. Write a balanced chemical equation for each acid–base reaction.

    a. Potassium hydroxide is neutralized by nitric acid.

    b. Hydrochloric acid is neutralized by barium hydroxide.

    c. Sulfuric acid is neutralized by ammonium hydroxide.

13. In each pair, the $[H^+]$ is different. By what factor of 10?

    a. pH = 6 and pH = 8

    b. pH = 5.5 and pH = 6.5

    c. $[H^+] = 1 \times 10^{-8}$ M and $[H^+] = 1 \times 10^{-6}$ M

    d. $[OH^-] = 1 \times 10^{-2}$ M and $[OH^-] = 1 \times 10^{-3}$ M

14. Classify these aqueous solutions as acidic, neutral, or basic.

    a. HI*(aq)*

    b. NaCl*(aq)*

    c. NH_4OH*(aq)*

    d. $[H^+] = 1 \times 10^{-8}$ M

    e. $[OH^-] = 1 \times 10^{-2}$ M

    f. $[H^+] = 5 \times 10^{-7}$ M

    g. $[OH^-] = 1 \times 10^{-12}$ M

15. For parts **d** and **f** of the previous question, calculate the $[OH^-]$ that corresponds to the given $[H^+]$. Similarly, for parts **e** and **g,** calculate the $[H^+]$.

16. Which of these has the *lowest* concentration of hydrogen ions: 0.1 M HCl, 0.1 M NaOH, 0.1 M H_2SO_4, or pure water? Explain your answer.

17. Write a balanced chemical equation for the reaction of sulfur shown in Figure 6.13.

18. Assume that coal can be represented by the chemical formula $C_{135}H_{96}O_9NS$.

    a. What is the percent of nitrogen by mass in coal?

    b. If 3 tons of coal is burned, what mass of nitrogen in the form of NO is produced? Assume that all of the nitrogen in the coal is converted to NO.

    c. Actually more NO is produced than you just calculated. Explain.

19. In 2006, the United States burned about 1.1 billion tons of coal. Assuming that it was 2% sulfur by weight, calculate the tons of sulfur dioxide emitted.

20. Acid rain can damage marble statues and limestone building materials. Write a balanced chemical equation using an acid of your choice.

21. Calculate the number of tons of $CaCO_3$ needed to react completely with 1.00 ton of SO_2. *Hint:* See equation 6.30.

22. A garden product called dolomite lime is composed of tiny chips of limestone that contain both calcium carbonate and magnesium carbonate. This product is "intended to help the gardener correct the pH of acid soils," as it is "a valuable source of calcium and magnesium."

    a. Is the "calcium" in the form of calcium ion or calcium metal?

    b. Write a chemical equation that shows why limestone "corrects" the pH of acidic soils.

    c. Will the addition of dolomite lime to soils cause the pH to rise or fall?

    d. Plants such as rhododendrons, azaleas, and camellias should not be given dolomite lime. Explain.

Concentrating on Concepts

23. Professor James Galloway, an expert on acid rain, wrote "Human activity is not making the world acidic, rather it is making the world *more* acidic."

    a. Explain why the world is naturally acidic.

    b. Explain how humans are making the world more acidic.

    c. One large part of our planet is basic. Which one? *Hint:* Consult Figure 6.4.

24. A news reporter talked about how acidic the oceans currently are and that they were getting even more acidic. What email message would you send to the news station?

25. Assuming the ocean continues to acidify at the same rate that it has been over the past 200 years, how long will it take before the ocean actually becomes acidic?

26. Suppose you have a new mountain bike and accidentally spilled a can of carbonated cola on the metallic handle bars and paint.

    a. Soft drinks are more acidic than acid rain. About how many times more acidic? *Hint:* Consult Figure 6.4.

    b. In spite of the higher acidity, this spill is unlikely to damage your handle bars and paint (although the sugar probably isn't great on your gears). Why is damage unlikely?

27. a. On the label of a shampoo bottle, what does the phrase "pH-balanced" imply?

    b. Does the phrase "pH-balanced" influence your decision to buy a particular shampoo? Explain.

28. Judging by the taste, do you think there are more hydrogen ions in a glass of orange juice or in a glass of milk? Explain your reasoning.

29. The formula for acetic acid, the acid present in vinegar, is commonly written as $HC_2H_3O_2$. Many chemists write the formula as CH_3COOH.

    a. Draw the Lewis structure for acetic acid.

    b. Show that both formulas represent acetic acid.

    c. What are the advantages and disadvantages of each formula?

    d. How many hydrogen atoms can be released as hydrogen ions per acetic acid molecule? Explain.

30. Television and magazine advertisements tout the benefits of antacids. A friend suggests that a good way to get rich quickly would be to market "antibase" tablets. Explain to your friend the purpose of antacids and offer some advice about the potential success of "antibase" tablets.

31. In Your Turn 6.7, you listed the ions present in aqueous solutions of acids, bases, and common salts. Now add water, a molecular species, to this list.

    a. List all molecular and ionic species in order of decreasing concentration in a 1.0 M aqueous solution of NaOH.

    b. List all molecular and ionic species in order of decreasing concentration in a 1.0 M aqueous solution of HCl.

32. Many gases, including CO, CO_2, O_3, NO, NO_2, SO_2, and SO_3, are associated with exhaust from jet engines.

a. Which of these gases do jet engines emit *directly*?

b. Which ones form secondarily, that is, resulting from the emissions of part **a**?

33. Figure 6.14 offers information about SO_2 emissions from fuel combustion (mainly from electric power production) and from transportation. Figure 6.16 offers information about NO_x emissions from fuel combustion (again mainly from power production) and from transportation. Relative to fuel combustion and transportation, how do the emissions of SO_2 and NO_x differ? Explain this difference.

34. Almost equal *masses* of SO_2 and NO_x are produced by human activities in the United States.

    a. How does their production compare based on a *mole* basis? Assume that all the NO_x is produced as NO_2.

    b. Suggest reasons why the U.S. percentage of global emissions is greater for NO_x than for SO_2.

35. Reactive nitrogen compounds affect the biosphere both directly and indirectly through other chemicals they help form.

    a. Name a direct effect of reactive nitrogen compounds that is a benefit.

    b. Name two direct effects of reactive nitrogen compounds that are harmful to human health.

    c. Ozone formation is a harmful indirect effect. Explain the connection between reactive nitrogen compounds and the formation of ozone.

36. Explain why rain is naturally acidic, but not all rain is classified as "acid rain."

37. Some local newspapers give forecasts for pollen, UV Index, and air quality. Why do you suppose that no forecast for acid rain is provided?

38. Here are examples of what an individual might do to reduce acid rain. For each, explain the connection to producing acid rain.

    a. Hang your laundry to dry it.

    b. Walk, bike, or take public transportation to work.

    c. Avoid running dishwashers and washing machines with small loads.

    d. Add additional insulation on hot water heaters and pipes.

    e. Buy locally grown produce and locally produced food.

39. As mentioned in Section 6.6, rain samples now ar~ analyzed for acidity in the Central Analytic⁻¹ Laboratory in Illinois rather than out i~ ⁻⁻

    a. The pH values tend to be sli~⁻~ than in the field. Did the acidity ⁻⁻⁻⁻~ decrease?

    b. Speculate on the causes of the pH increase.

40. Speaking of field measurements, in January 2006 those living in the metropolitan St. Louis area experienced a severe hail storm. Hail stones were still visible a day later, as shown in the photo. A chemistry professor took samples of the hail, analyzed them in her laboratory and reported a pH of 4.8.

    a. How does this pH compare with the normal precipitation in the St. Louis area? To acidity levels further east and further west of St. Louis?

    b. What factors might lead to acid rain in St. Louis?

41. Mammoth Cave National Park in Kentucky is in close proximity to the coal-fired electric utility plants in the Ohio Valley. Noting this, the National Parks Conservation Association (NPCA) reported that this national park had the poorest visibility of any in the country.

    a. What is the connection between coal-fired plants and poor visibility?

    b. The NPCA reported "the average rainfall in Mammoth Cave National Park is 10 times more acidic than natural." From this information and that in your text, estimate the pH of rainfall in the park.

42. In the United States over the past few decades, emissions of ammonia have dramatically increased, although less so in the east than in the west.

    a. Show with a chemical equation that ammonia dissolves in rain to form a basic solution.

    b. Write the neutralization reaction for rain that contains both ammonia and nitric acid.

    c. Ammonium sulfate also is found in rain. Write a chemical equation that demonstrates how it could have formed.

43. Ozone in the troposphere is an undesirable pollutant, but stratospheric ozone is beneficial. Does nitric oxide, NO, have a similar "dual personality" in these two atmospheric regions? Explain. *Hint:* Consult Chapter 2.

44. The mass of CO_2 emitted during combustion reactions is much greater than the mass of NO_x or SO_2, but there is less concern about the contributions of CO_2 to acid rain than from the other two oxides. Suggest two reasons for this apparent inconsistency.

45. The average pH of precipitation in New Hampshire and Vermont is low, even though these states have relatively fewer cars and virtually no industry that emits large quantities of air pollutants. How do you account for this low pH?

46. Admittedly, global sulfur emissions are difficult to estimate, and you will find a range of values published in the literature. One set is shown in this figure for 1850–2000. These are estimates from a paper published in 2004 by David Stern at Rensselaer Polytechnic Institute. Gg stands for gigagrams, or 1×10^{12} grams.

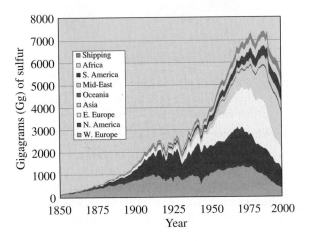

a. The figure shows total sulfur emissions. In what chemical form is this sulfur most likely to be?

b. According to the figure, in which years did the sulfur emissions peak?

c. List reasons for the decline in more recent years.

d. In 2000, which region of the world contributed the highest level of sulfur emissions?

e. Which regions were the largest contributors in the early 1970s?

47. The chemistry of NO in the atmosphere is complicated. NO can destroy ozone, as seen in Chapter 2. But remember from Chapter 1 that NO can react with O_2 to form NO_2. In turn, NO_2 can react in sunlight to produce ozone. Summarize these reactions, noting in which region of the atmosphere they each occur.

48. a. Efforts to control air pollution by limiting the emission of particulates and dust can sometimes contribute to an increase in the acidity of rain. Offer a possible explanation for this observation. *Hint:* These particulates may contain basic compounds of calcium, magnesium, sodium, and potassium.

b. In Chapter 2, stratospheric ice crystals in the Antarctic were involved in the cycle leading to the destruction of ozone. Is this effect related to the observations in part **a**? Explain.

Exploring Extensions

49. Discuss the validity of the statement, "Photochemical smog is a local issue, acid rain is a regional one, and the enhanced greenhouse effect is a global one." Describe the chemistry behind each issue. Do you agree that the magnitudes of the problems are really so different in scope?

50. In terms of taste, pH, and amount of dissolved gas, how does carbonated water differ from rain that contains dissolved carbon dioxide?

51. The compound $Al(OH)_3$ contains OH in its chemical formula. However, we do not write a reaction analogous to equation 6.3. Explain. *Hint:* Consult a solubility table.

52. In Your Turn 6.7, you listed the ions present in aqueous solutions of acids, bases, and common salts. In question 31, you added molecular substances to the list. To quantify this list,

a. calculate the molar concentration of all molecular and ionic species in a 1.0 M solution of NaOH.

b. calculate the molar concentration of all molecular and ionic species in a 1.0 M solution of HCl.

53. The chemical reaction in which NO reacts to form NO_2 in the atmosphere (see equation 6.20) involves intermediate species A′ and A″. Here are possible structures.

a. In each, what does the dot (·) represent? For the atom with the dot, redraw to show all valence electrons, both bonding and nonbonding pairs. *Hint:* This atom does not have an octet of electrons.

b. Name a chemical property that A′ and A″ have in common.

54. Equation 6.19 shows that energy (in the form of a hot engine or other source of heat) must be added to get N_2 and O_2 to react to form NO. A student wants to check this assertion and determine how much energy is required. Show the student how this can be done. *Hint:* Draw the Lewis structures for the reactants and products, noting that NO does not have an octet of electrons. The bond energy is 607 kJ/mol for the N=O double bond.

55. This chapter describes a green chemistry solution to reducing NO emissions for glass manufacturers.

    a. Identify the strategy.

    b. Which other industries similarly might make use of this green chemistry strategy?

56. One way to compare the acid-neutralizing capacity of different substances is to calculate the mass of the substance required to neutralize 1 mol of hydrogen ion, H^+.

    a. Write a balanced equation for the reaction of $NaHCO_3$ with H^+. Use it to calculate the acid-neutralizing capacity for $NaHCO_3$.

    b. If $NaHCO_3$ costs \$9.50/kg, determine the cost to neutralize one mole of H^+.

7 The Fires of Nuclear Fission

"For nearly 30 years, the inbox for new license applications at the U.S. Nuclear Regulatory Commission collected nothing but dust. These days, the inbox has no room for dust."

Mitch Jacoby, *Chemical & Engineering News*, August 24, 2009.

For decades, no new nuclear reactors have been built in the United States. But given the sudden surge in new license applications, construction could start within the decade. Electric utility companies project an increased demand for electric power over the next few decades. Those concerned with the operation of coal-fired power plants point out that nuclear reactors emit neither greenhouse gases nor air pollutants such as sulfur dioxide and nitrogen monoxide. And citizens recognize that nuclear power plants, unlike solar and wind installations, may offer a practical solution because they run in the dark and in the absence of wind.

People across the globe share the dream of clean and sustainable sources of energy for the future. Should we build more nuclear power plants as we move to achieve this? The answer depends on both whom you ask and when you ask them. Some long-time opponents of nuclear energy are now in favor of it. Similarly, some who once supported it are now questioning its societal costs, both to our current generation and those to come.

In part, the opposition to nuclear power is a result of the tremendous baggage that the word *nuclear* carries. The associations are disturbing: the bombing of Hiroshima, the radioactive fallout from atmospheric weapons testing, the tragedies of Chernobyl, the hazards of high-level radioactive waste, and the ultimate threat of nuclear annihilation. Probably no other topic in the physical sciences is more likely to provoke such an emotional response.

At the same time, people recognize the benefits of nuclear medicine. These days, you probably know someone who has undergone radiation therapy for cancer or who has had a diagnostic test with radioisotopes that bypassed both anesthesia and surgery. Perhaps even a classmate has had a thyroid scan using radioactive iodine.

Those who support or oppose nuclear power today have excellent tools with which to make compelling arguments. For example, a "cradle-to-cradle" analysis offers a more-inclusive picture of the economic, environmental, and societal costs of running a nuclear reactor, by taking into account what happens from the moment the uranium ore is mined to the ultimate fate of the spent nuclear fuel. A cradle-to-cradle analysis not only includes the high economic costs of construction but also the eventual decommissioning of the nuclear reactor.

Whether citizens (and politicians) support or oppose nuclear power, they still must deal with some real and pressing questions. If not nuclear, how are we going to produce electricity in the years to come? Do the benefits of nuclear power plants outweigh the costs and risks? How should we deal with the wastes that nuclear reactors produce? Can we prevent the diversion of nuclear materials to nuclear weapons? Is nuclear power sustainable? As was the case in earlier chapters, science and societal issues are tightly connected. In a moment, we will launch this chapter with an overview of nuclear power. But before we start, we ask you to consider your own position.

In 2009, the U.S. Energy Information Administration projected a 21% increase in the demand for electric power by 2030.

The term *cradle-to-cradle* was introduced in Chapter 0. In the case of nuclear power, cradle-to-grave may be more appropriate in that nuclear waste currently is stored rather than becoming the "cradle" for something else.

Decommissioning (shutting down) a nuclear plant is a complex operation. All parts must be analyzed and removed according to strict criteria.

Consider This 7.1 **Your Opinion of Nuclear Power**

a. Given a choice between purchasing electricity generated by a nuclear plant or by a coal-burning plant, would you choose one over the other? Explain.
b. What circumstances, if any, would change your position on the use of nuclear power for generating electricity?

Save your answers to these questions, because you will revisit them at the end of the chapter.

7.1 | Nuclear Power Worldwide

Most people flip a light switch without thinking of the energy source that caused the bulb to glow. But others, especially those who previously have lost electric power because of a storm or a power blackout, may not as easily take electricity for granted. All too well they know the sensation of hearing the switch click, but remaining in the dark.

Let's assume that the power is working and that you switch on your coffee pot. If you live in the United States, the odds are about 1 in 5 that a nuclear power plant

We have a high demand for electricity (and caffeine).

is providing the electricity to brew your coffee. But if you live in France, the odds are better than even. Nations of the world differ in the extent to which they employ nuclear power commercially. For example, in the United States, about 20% of the electric power is produced from just over 100 nuclear reactors, all licensed by the Nuclear Regulatory Commission (NRC). These reactors operate at 65 sites in 31 states. As you can see by Figure 7.1, the electricity generated by these nuclear plants has increased over the years, despite the drop in the number of operating reactors from its 1990 peak of 112. Nonetheless, in the United States, nuclear power remains a relatively steady 20% of the total production because the other sources have increased as well.

> The power increase over time in Figure 7.1 stems both from improved reactor efficiencies and upgrades to their components.

When you brew your coffee a decade from now, from where will the electricity come? No new nuclear plants have been built in the United States since 1978. Furthermore, nine nuclear plants ceased their operations, some of them before their licenses expired. These include what was once the nation's largest, the Zion nuclear power station on the shores of Lake Michigan. Reasons cited for plant closings included the competition of natural gas and the competitive pressures of energy deregulation.

Consider This 7.2   Nuclear Power State-by-State

The Nuclear Energy Institute provides a map showing the locations of all commercial nuclear power plants in the United States. Visit the textbook's website for links to help you to answer these questions.

a. Select any three states. For each, what percent of the state's electric energy is nuclear in origin?

b. As of 2008, four states generated more than half their electrical energy from nuclear power plants. Which ones?

c. Select a state with no nuclear power plants. How instead is the electricity generated?

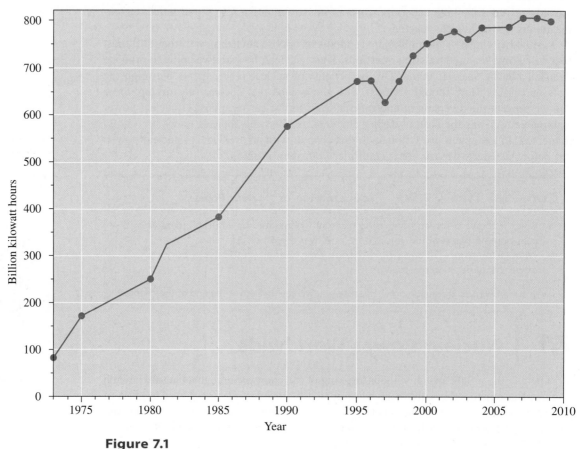

Figure 7.1

Nuclear power generation in the United States since 1973. *Source:* Energy Information Administration.

The construction and continued operation of nuclear plants is not only a matter of energy supply and demand, but also one of public acceptance. Depending on your age, you may have little recollection of the controversy that surrounded some nuclear power plants when they were proposed or being constructed. People have been lining up on one side or the other of the nuclear fence for quite some time.

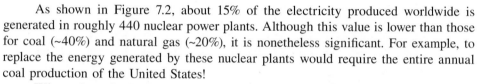

Consider This 7.3   **Take a Stand**

As you can see in this photo taken during the construction of the Seabrook nuclear power plant in New Hampshire (1977), signs are one way to convey your position. If a nuclear plant were being built near your community today, what would your sign say?

As shown in Figure 7.2, about 15% of the electricity produced worldwide is generated in roughly 440 nuclear power plants. Although this value is lower than those for coal (~40%) and natural gas (~20%), it is nonetheless significant. For example, to replace the energy generated by these nuclear plants would require the entire annual coal production of the United States!

Where does the United States rank in regard to nuclear power? Although the United States has more nuclear reactors than any other nation, these reactors are aging and, as of 2010, have an average age of just over 30 years. These reactors were designed with an expected operating life of 30–40 years, but with modifications can be expected to run an additional 20 years or so.

With only 20% of its electric power generated by nuclear power plants (Figure 7.3), the United States clearly does *not* lead the nuclear pack in terms of its energy portfolio. France does. As of early 2010, the French had 59 nuclear plants that generated over 75% of their electricity. With the exception of the Republic of South Korea, all of the countries that generate a third or more of their electricity via nuclear energy are in Europe. The Swiss generate about 40% of their electricity with only five reactors!

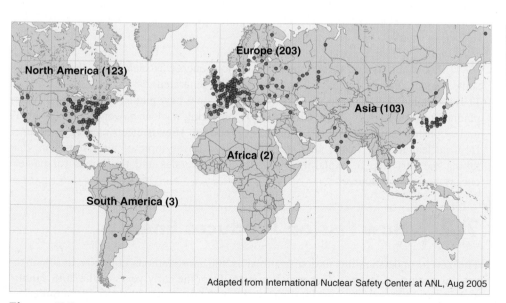

Some nuclear plants generate power using two or more nuclear reactors. For example, the Palo Verde plant pictured in Consider This 7.11 has three reactors.

Figure 7.2

Commercial nuclear reactors in operation worldwide as of December 2005. Some sites have more than one reactor.

Source: Argonne National Lab.

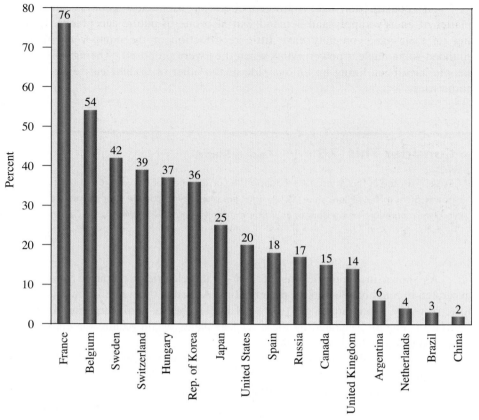

Figure 7.3

Percent of electric power generated by nuclear power reactors in selected countries, October 2009.

Source: World Nuclear Association.

Consider This 7.4   Nuclear Neighbors

a. Where are most of the nuclear reactors located? Use Figure 7.2 as your guide.
b. List three countries that currently do not have commercial nuclear reactors.
c. Suggest reasons for the different emphasis that countries place on nuclear energy.

Major commercial development of nuclear energy is clearly on the agenda of some nations. For example, although India only generated 2% of its electricity from 17 reactors in 2009, construction was underway for 6 new plants, with over 30 others planned or proposed. In 2009, China had 11 operating nuclear power reactors, with 17 under construction and over 110 planned or proposed. How is electric energy produced from nuclear fission? In the next section, we turn to the topic of nuclear fission, thus taking the first step in explaining both the controversies and the hopes for nuclear energy as a power source.

7.2 | How Fission Produces Energy

The key to understanding fission is probably the most famous equation in all of the natural sciences, $E = mc^2$. This equation dates from the early years of the 20th century and is one of the many contributions of Albert Einstein (1879–1955). It summarizes the equivalence of energy, E, and matter, or mass, m. The symbol c represents the speed of light, 3.0×10^8 m/s, so c^2 is equal to 9.0×10^{16} m²/s². The large value of c^2 means that it should be possible to obtain a tremendous amount of energy from a small amount of matter, whether in a power plant or in a weapon.

For over 30 years, Einstein's equation was a curiosity. Scientists believed that it described the source of the Sun's energy, but as far as anyone knew, no one on Earth had ever observed a transformation of a substantial fraction of matter into energy. But in 1938, two German scientists, Otto Hahn (1879–1968) and Fritz Strassmann (1902–1980), discovered otherwise. When they bombarded uranium with neutrons, they found what appeared to be the element barium (Ba) among the products. The observation was unexpected because barium has an atomic number of 56 and an atomic mass of about 137. Comparable values for uranium are 92 and 238, respectively. At first, the scientists were tempted to conclude that the element was radium (Ra, atomic number 88), a member of the same group in the periodic table as barium. But for Hahn and Strassmann, the chemical evidence for barium was too compelling to ignore.

The German scientists were unsure of how barium could have been formed from uranium, so they sent a copy of their results to their colleague, Lise Meitner (1878–1968), for her opinion (Figure 7.4). Dr. Meitner had collaborated with Hahn and Strassmann on related research, but was forced to flee Germany in March 1938 because of the Nazi government. When she received their letter, she was living in Sweden. She discussed the strange results with her physicist nephew, Otto Frisch (1904–1979), as the two of them went walking in the snow. In a flash of insight, she understood. Under the influence of the bombarding neutrons, the uranium atoms were splitting into smaller ones such as barium. The nuclei of the heavy atoms were dividing, like biological cells undergoing fission.

That word from biology is applied to a physical phenomenon in the letter that Meitner and Frisch published on February 11, 1939, in the British journal *Nature*. In the letter, entitled "Disintegration of Uranium by Neutrons: A New Type of Nuclear Reaction," the authors state the following:

> "Hahn and Strassmann were forced to conclude that isotopes of barium are formed as a consequence of the bombardment of uranium with neutrons. At first sight, this result seems very hard to understand. . . . On the basis, however, of present ideas about the behavior of heavy nuclei, an entirely different . . . picture of these new disintegration processes suggests itself. . . . It seems therefore possible that the uranium nucleus . . . may, after neutron capture, divide itself into two nuclei of roughly equal size. . . . The whole 'fission' process can thus be described in an essentially classical way."

Figure 7.4
Lise Meitner is pictured shortly after her arrival in New York in January, 1946.

Although just over a page long, this letter was immediately recognized for its significance. In fact, it would be difficult to think of a more important scientific communication. Niels Bohr (1885–1962), an eminent Danish physicist, learned of the news directly from Frisch and brought a copy of the letter to the United States on an ocean liner several days before its publication. Within a few weeks of Meitner and Frisch's letter in *Nature,* scientists in a dozen laboratories in various countries confirmed that the energy released by the splitting of uranium atoms was that predicted by Einstein's equation. Lise Meitner's contributions to the discovery of nuclear fission were honored by naming element 109 meitnerium.

Nuclear fission is the splitting of a large nucleus into smaller ones with the release of energy. Energy is released because the total mass of the products is slightly *less* than the total mass of the reactants. In spite of what you may have been taught, neither matter nor energy is individually conserved. Matter disappears and an equivalent quantity of energy appears. Alternatively, one can view matter as a very concentrated form of energy; nowhere is it more concentrated than in the atomic nucleus. Remember that an atom is mostly empty space. If a hydrogen nucleus were the size of a baseball, then its electron would be found within a sphere half a mile in diameter. Because almost all the mass of an atom is associated with its nucleus, the nucleus is incredibly dense. Indeed, a pocket-sized matchbox full of atomic nuclei would weigh over 2.5 billion tons! Given the energy–mass equivalence of Einstein's equation, the energy content of all nuclei is, relatively speaking, immense.

Only the nuclei of certain elements undergo fission and these only under certain conditions. Three factors determine whether a particular nucleus will split: its size, the numbers of protons and neutrons it contains, and the energy of the neutrons that bombard the nucleus to initiate the fission. For example, relatively light and stable atoms such as oxygen, chlorine, and iron do not split. Extremely heavy nuclei may fission spontaneously. And heavy nuclei, such as those of uranium and plutonium, can be made to split if hit hard enough with neutrons. Notably, one isotope of uranium fissions with neutrons of a more moderate speed, such as those employed in the reactor of a nuclear power plant.

Let's examine uranium more closely. *All* uranium atoms contain 92 protons. If these atoms are electrically neutral, these protons are accompanied by 92 electrons. In nature, uranium is found predominantly as two isotopes. The more abundant one (99.3%) contains 146 neutrons. The mass number of this isotope of uranium is 238, that is, 92 protons plus 146 neutrons. We represent this isotope as uranium-238, or more simply as U-238. The less abundant isotope (0.7%) contains 143 neutrons and 92 protons; namely, U-235.

The terms *mass number* and *isotope* were introduced in Section 2.2.

Your Turn 7.5   **Another Isotope of Uranium**

A trace amount of a third isotope, U-234, is also found in nature. How do U-238 and U-234 compare in terms of the number of protons and the number of neutrons?

More commonly, we specify an isotope with both its mass number and atomic number. The former is a superscript and the latter as a subscript, both written to the left of the chemical symbol. Using this convention, uranium-238 becomes:

$$\text{Mass number = number of protons + number of neutrons} \longrightarrow {}^{238}_{92}\text{U}$$
$$\text{Atomic number = number of protons} \longrightarrow$$

Similarly, U-235 is written as ${}^{235}_{92}\text{U}$. Although ${}^{235}_{92}\text{U}$ and ${}^{238}_{92}\text{U}$ differ by a mere three neutrons, this difference translates to one key difference in *nuclear* properties. Under the conditions present in a nuclear reactor, ${}^{238}_{92}\text{U}$ does *not* undergo fission, yet ${}^{235}_{92}\text{U}$ does.

The process of nuclear fission is initiated by neutrons and releases neutrons, as can be seen by this example.

$$\,_0^1n + \,_{92}^{235}U \longrightarrow [\,_{92}^{236}U] \longrightarrow \,_{56}^{141}Ba + \,_{36}^{92}Kr + 3\,_0^1n \qquad \textbf{[7.1]}$$

Let's examine the components from left to right. Initially, a neutron hits the nucleus of U-235. This neutron, $\,_0^1n$, has a subscript of 0, indicating no charge; the superscript is 1 because the mass number of a neutron is 1. The nucleus of $\,_{92}^{235}U$ captures the neutron, forming a heavier isotope of uranium, $\,_{92}^{236}U$. This isotope is written in square brackets indicating that it exists only momentarily. Uranium-236 immediately splits into two smaller atoms (Ba-141 and Kr-92) with the release of three more neutrons.

Nuclear equations are similar to, but not the same as "regular" chemical equations. To balance a nuclear equation, you count the protons and neutrons rather than counting atoms as you would do in a chemical equation. A nuclear equation is balanced if the sum of the subscripts (and of the superscripts) on the left is equal to the sum of those on the right. Coefficients in nuclear equations, such as the 3 preceding the $\,_0^1n$ in equation 7.1, are treated the same way as in chemical equations, multiplying the term that follows it. For example, examine the math to see why nuclear equation 7.1 is balanced.

Left
Superscripts: 1 + 235 = 236
Subscripts: 0 + 92 = 92

Right
141 + 92 + (3 × 1) = 236
56 + 36 + (3 × 0) = 92

When the nucleus of an atom of U-235 is struck with a neutron, many different fission products are formed. The activity that follows acquaints you with two other possibilities.

Your Turn 7.6   Other Examples of Fission

With the help of a periodic table, write these two nuclear equations. Both are initiated with a neutron.

a. U-235 fissions to form Ba-138, Kr-95, and neutrons.
b. U-235 fissions to form an element (atomic number 52, mass number 137), another element (atomic number 40, mass number 97), and neutrons.

Answer
a. $\,_0^1n + \,_{92}^{235}U \longrightarrow \,_{56}^{138}Ba + \,_{36}^{95}Kr + 3\,_0^1n$

Look again at nuclear equation 7.1. Both sides contain neutrons, which might lead you to think that we should cancel them. Although you might do this in a mathematical expression, don't do it here. The neutrons on both sides of the equation are important! The one on the left *initiates* the fission reaction; the ones on the right are *produced* by it. Each neutron produced can in turn strike another U-235 nucleus, cause it to split, and release a few more neutrons. This is an example of a **chain reaction,** a term that generally refers to any reaction in which one of the products becomes a reactant and thus makes it possible for the reaction to become self-sustaining. This particular rapidly branching nuclear chain reaction is self-sustaining and spreads in a fraction of a second (Figure 7.5). With exactly this chain reaction, the first controlled nuclear fission took place at the University of Chicago in 1942.

A **critical mass** is the amount of fissionable fuel required to sustain a chain reaction. For example, the critical mass of U-235 is about 15 kg, or 33 lb. Were this mass of pure U-235 to be brought together in one place, fission would spontaneously occur. And if the mass were held together, fission would continue. Nuclear weapons work on this principle, although the energy released quickly blows the critical mass apart, stopping the fission reaction. But as you will soon see, the uranium fuel in a

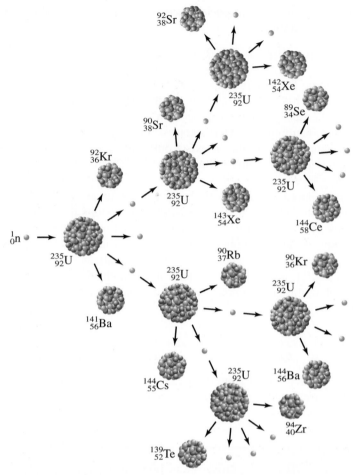

Figure 7.5

A neutron initiates the fission of uranium-235, starting a chain reaction.

Figures Alive! Visit the textbook's website to learn more about nuclear fission and chain reactions.

nuclear power plant is far from pure U-235 and is unable to explode like a nuclear bomb. There simply aren't enough neutrons around (and enough fissionable nuclei for these neutrons to hit) to produce an uncontrolled chain reaction characteristic of a nuclear explosion.

We mentioned earlier that energy is given off during fission because the mass of the products is slightly less than that of the reactants. However, from the nuclear equations we have just written, no mass loss is apparent because the sum of the mass numbers is the same on both sides. In fact, the actual mass does decrease slightly. To understand this, remember that the actual masses of the nuclei are not the mass numbers (the sum of the number of protons and neutrons); rather, they have measured values with many decimal places. For example, an atom of uranium-235 weighs 235.043924 atomic mass units. Were you to keep all six decimal places and compare the masses on both sides of the nuclear equation for the fission of U-235, you would find that the mass of the products is less by about 0.1%, or 1/1000th. As a consequence, the energy of the products is less than that of the reactants. This difference corresponds to the energy released.

How much energy would be released if all the nuclei in 1.0 kg (2.2 lb) of pure U-235 were to fission? We can calculate an answer by using an equation closely related to $E = mc^2$; namely, $\Delta E = \Delta mc^2$. Here the Greek letter delta (Δ) means "the change in," so now with a change in mass we can calculate a change in energy. Since 1/1000 of this mass is lost, the value for Δm, the change in mass, is 1/1000 of 1.0 kg, which is 1.0 g or 1×10^{-3} kg. Now substitute this value and $c = 3.0 \times 10^8$ m/s into Einstein's equation.

$$\Delta E = \Delta mc^2 = (1.0 \times 10^{-3} \text{ kg}) \times (3.0 \times 10^8 \text{ m/s})^2$$
$$\Delta E = (1.0 \times 10^{-3} \text{ kg}) \times (9.0 \times 10^{16} \text{ m}^2/\text{s}^2)$$

Completing the calculation gives an energy change in what may appear to be unusual units.

$$\Delta E = (9.0 \times 10^{13} \text{ kg·m}^2/\text{s}^2)$$

The unit kg·m^2/s^2 is identical to a joule (J). Therefore, the energy released from the fission of an entire kilogram of uranium-235 is a whopping 9.0×10^{13} J, or 9.0×10^{10} kJ.

As described in Section 4.2, the joule (J) is a unit of energy.
$1\ \text{J} = 1\ \text{kg·m}^2/\text{s}^2$

To put things into perspective, 9.0×10^{13} J is the amount of energy released by the explosion of about 22 metric kilotons of the explosive TNT. By way of comparison, this is roughly twice that of the atomic bombs dropped on Hiroshima and Nagasaki in 1945. This energy originates from the fission of a single kilogram of U-235, in which a mass of approximately 1 gram (0.1% mass change) was transformed into energy.

Your Turn 7.7   Coal Equivalence

Select a grade of coal from Table 4.1. What mass of coal would be needed to produce the same amount of energy as would the fission of 1 kg of U-235?

As it turns out, one cannot fission a kilogram or two of pure U-235 in one fell swoop. In an atomic weapon, for example, the energy that is released blasts the fissionable fuel apart in a fraction of a second, thus halting the chain reaction before all the nuclei can undergo fission. Nonetheless, the energy released is enormous—on the order of 10 kilotons of TNT for the atomic bomb dropped on the city of Hiroshima in 1945. Figure 7.6 shows an atomic explosion at the U.S. Nevada Test Site. Code-named

Figure 7.6
The nuclear test "Priscilla" was exploded on a dry lake bed northwest of Las Vegas, Nevada, on June 24, 1957.

Priscilla, this test in 1957 had more than twice the explosive power of the bombs at Hiroshima and Nagasaki in 1945.

Recognize, though, that the energy of nuclear fission can be harnessed. This is exactly the objective of a nuclear power plant. Here, the energy is slowly and *continually* released under controlled conditions, as we shall see in the next section.

7.3 | How Nuclear Reactors Produce Electricity

Chapter 4 described how a conventional power plant burns coal, oil, or some other fuel to produce heat. The heat is then used to boil water, converting it into high-pressure steam that turns the blades of a turbine. The shaft of the spinning turbine is connected to large wire coils that rotate within a magnetic field, thus generating electric energy. A nuclear power plant operates in much the same way, except that the water is heated not by combustion of a fuel, but by the energy released from the fission of nuclear "fuel" such as U-235. Like any power plant, a nuclear one is subject to the efficiency constraints imposed by the second law of thermodynamics. The theoretical efficiency for converting heat energy to work depends on the maximum and minimum temperatures between which the plant operates. This thermodynamic efficiency, typically 55–60%, is significantly reduced by other mechanical, thermal, and electric inefficiencies.

A nuclear power station has parts that are both nuclear and nonnuclear (Figure 7.7). The nuclear reactor is the hot heart of the power station. The reactor, together with one or more steam generators and the primary cooling system, is housed in a special steel vessel within a separate reinforced concrete dome-shaped containment building. The nonnuclear portion contains the turbines that run the electric generator. It also contains the secondary cooling system. In addition, the nonnuclear portion must be connected to some means of removing excess heat from the coolants. Accordingly,

The second law of thermodynamics has many versions. The one relevant here might be to say that it is impossible to convert heat completely into work in a process that is a cycle. See Section 4.2.

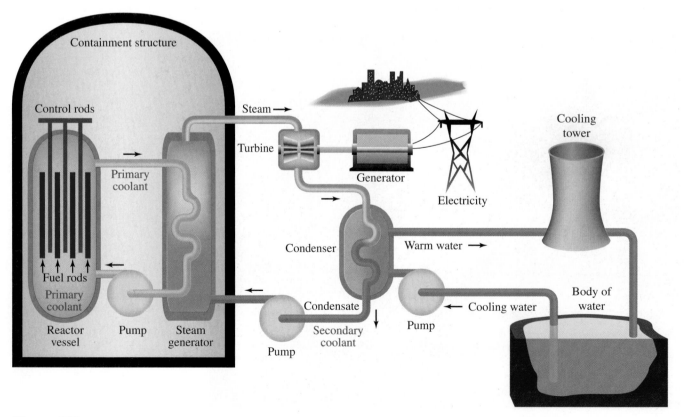

Figure 7.7

Diagram of a nuclear power plant.

a nuclear power station has one or more cooling towers or is located near a sizeable body of water (or both). Look back at Figure 4.2, which shows a diagram of a fossil fuel power plant. This plant also requires a means of removing heat, as shown by the stream of cooling water.

The uranium fuel in the reactor core is in the form of uranium dioxide (UO_2) pellets, each comparable in height to a dime, as shown in Figure 7.8. These pellets are placed end-to-end in tubes made of a special metal alloy, which in turn are grouped into stainless-steel-clad bundles (Figure 7.9). Each tube contains at least 200 pellets. Although a fission reaction, once started, can sustain itself by a chain reaction, neutrons are needed to induce the process (see equation 7.1 and Figure 7.5). One means of generating neutrons is to use a combination of beryllium-9 and a heavier element such as plutonium. The heavier element releases alpha particles, ^4_2He.

Figure 7.8

Nuclear fuel pellets and a U.S. dime.

$$^{238}_{94}\text{Pu} \longrightarrow \,^{234}_{92}\text{U} + \,^{4}_{2}\text{He} \qquad [7.2]$$
$$\text{alpha particle}$$

These alpha particles in turn strike the beryllium, releasing neutrons, carbon-12, and gamma rays, $^0_0\gamma$. Here is the nuclear equation.

Gamma rays were first introduced in Section 2.4.

$$^{4}_{2}\text{He} + \,^{9}_{4}\text{Be} \longrightarrow \,^{12}_{6}\text{C} + \,^{1}_{0}\text{n} + \,^{0}_{0}\gamma \qquad [7.3]$$
$$\text{gamma ray}$$

The neutrons produced in this way can initiate the nuclear fission of uranium-235 in the reactor core.

Your Turn 7.8   Poo-Bee and Am-Bee

A neutron source constructed with Pu and Be is called a PuBe, or "poo-bee," source. Similarly, the AmBe, or "am-bee," source is constructed from americium and beryllium. Analogous to the PuBe source, write the set of reactions that produce neutrons from an AmBe source. Start with Am-241.

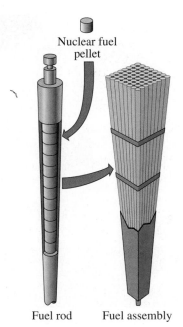

Nuclear fuel pellet

Fuel rod   Fuel assembly

Figure 7.9

Fuel pellet, fuel rod, and fuel assembly making up the core of a nuclear reactor (*left*).
The fuel assembly is submerged under water in an active reactor core (*right*).

Remember—one fission event produces two or three neutrons. The trick is to "sponge up" these extra neutrons, but still leave enough to sustain the fission reaction. A delicate balance must be maintained. With extra neutrons, the reactor runs at too high a temperature; with too few, the chain reaction halts and the reactor cools down. To achieve the needed balance, one neutron from each fission event should in turn cause another.

Metal rods interspersed among the fuel elements serve as the neutron "sponges." These **control rods,** composed primarily of an excellent neutron absorber such as cadmium or boron, can be positioned to absorb fewer or more neutrons. With the rods fully inserted, the fission reaction is not self-sustaining. But as the rods are gradually withdrawn, the reactor can "go critical" and become self-sustaining, running at different rates depending on the exact position of the control rods. Over time, fission products that absorb neutrons build up in the fuel pellets. To compensate, the control rods can be withdrawn. Eventually, the reactor fuel bundles must be replaced.

Your Turn 7.9   Earthquake!

Suppose a serious tremor were to occur, such as the one that rocked the Kashiwazaki-Kariwa facility in Japan in 2007, the largest nuclear power operation in the world. Reactors near the epicenter should automatically shut down. Should the software be programmed to fully insert the control rods into the reactor core, or should they be pulled out? Explain.

The fuel bundles and control rods are bathed in the **primary coolant,** a liquid that comes in direct contact with them and carries away heat. In the Byron nuclear reactor (Figure 7.10) and in many others, the primary coolant is an aqueous solution of boric acid, H_3BO_3. The boron atoms absorb neutrons and thus control the rate of fission and the temperature. The solution also serves as a **moderator** for the reactor, slowing the neutrons, thus making them more effective in producing fission. Another major function of the primary coolant is to absorb the heat generated by the nuclear reaction. Because the primary coolant solution is at a pressure more than 150 times normal atmospheric pressure, it does not boil. It is heated far above its normal boiling point and circulates in a closed loop from the reaction vessel to the steam generators, and back again. This closed primary coolant loop thus forms the link between the nuclear reactor and the rest of the power plant (see Figure 7.7).

Figure 7.10

The Byron nuclear power plant in Illinois. The two cooling towers (one with a cloud of condensed water vapor) are the most prominent features of this plant. The reactors, however, are located in the two cylindrical containment buildings with white roofs in the foreground.

The heat from the primary coolant is transferred to what is sometimes referred to as the **secondary coolant,** the water in the steam generators that does not come in contact with the reactor. At the Byron nuclear plant (see Figure 7.10), more than 30,000 gallons of water is converted to vapor each minute. The energy of this hot vapor turns the blades of turbines that are attached to an electric generator. To continue the heat transfer cycle, the water vapor is then cooled and condensed back to a liquid and returned to the steam generator. In many nuclear facilities the cooling is done using large cooling towers that commonly are mistaken for the reactors. The reactor buildings are not as large.

Cooling towers also are used in coal-fired plants.

Your Turn 7.10   Clouds (not mushroom-shaped)

Some days you can see a cloud coming out of the cooling tower of a nuclear power plant, as shown in Figure 7.10. What causes the cloud? Does it contain radioisotopes produced from the fission of U-235? Explain.

Nuclear power plants also use water from lakes, rivers, or the ocean to cool the condenser. For example, at the Seabrook nuclear power plant in New Hampshire, every minute about 400,000 gallons of ocean water flows through a huge tunnel (19 feet in diameter and 3 miles long) bored through rock 100 feet beneath the floor of the ocean. A similar tunnel from the plant carries the water, now 22 °C warmer, back to the ocean. Special nozzles distribute the hot water so that the observed temperature increase in the immediate area of the discharge is only about 2 °C. The ocean water is in a separate loop from the fission reaction and its products. The primary coolant (water with boric acid) circulates through the reactor core inside the containment building. However, this boric acid solution is kept isolated in a closed circulating system, which makes the transfer of radioactivity to the secondary coolant water in the steam generator highly unlikely. Similarly, the ocean water does not come in direct contact with the secondary system, so the ocean water is well protected from radioactive contamination. Clearly the electricity generated by a nuclear power plant is identical to the electricity generated by a fossil fuel plant; the electricity is not radioactive, nor can it be.

Consider This 7.11   The Palo Verde Reactors

One of the most powerful nuclear plants in operation in the United States is the Palo Verde complex in Arizona. At maximum capacity, just one of its three reactors generates 1243 million joules of electric energy every second. Calculate the total amount of electric energy produced per day and the loss of mass of U-235 each day.
Hint: Start by calculating the quantity of energy generated not per second, but per day. Then use the equation $\Delta E = \Delta mc^2$ and solve for the change in mass, Δm. Report the mass loss in grams.

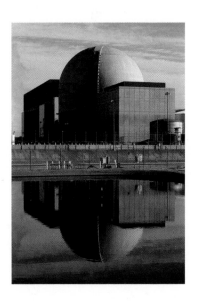

The topics we have been discussing—nuclear fission, uranium, nuclear fuel, nuclear weapons—all rest on an understanding of radioactivity. We now turn to this topic.

7.4 | What Is Radioactivity?

Our knowledge of radioactive substances is just over 100 years old. In 1896, the French physicist Antoine Henri Becquerel (1852–1908) discovered radioactivity. At the time, his research involved using photographic plates; film, of course, had not yet been invented. Prior to use, these plates were sealed in black paper to keep them from being exposed. By accident he left a mineral near one of these sealed plates and found that the plate's light-sensitive emulsion darkened. It was as though the plate had been

Figure 7.11

Marie Sklodowska Curie won two Nobel Prizes—one in chemistry, the other in physics—for her research on radioactive elements.

See Figure 2.7 for more information about the electromagnetic spectrum.

exposed to light! Becquerel immediately recognized that the mineral emitted powerful rays that penetrated the lightproof paper.

Further investigation by the Polish scientist Marie Sklodowska Curie (1867–1934) (Figure 7.11) revealed that the rays were coming from a constituent of the mineral—the element uranium. In 1899, Marie Curie applied the term **radioactivity** to the spontaneous emission of radiation by certain elements. Subsequent research by Ernest Rutherford (1871–1937) led to the identification of two major types of radiation. Rutherford named them after the first two letters of the Greek alphabet, alpha (α) and beta (β).

Alpha and beta radiation have strikingly different properties. A **beta particle (β)** is a high-speed electron emitted from the nucleus. It has a negative electric charge (1−) and only a tiny mass, about 1/2000 that of a proton or a neutron. If you are wondering how an electron (a beta particle) could possibly be emitted from a nucleus, stay tuned. We offer an explanation shortly.

In contrast, an **alpha particle (α)** is a positively charged particle emitted from the nucleus. It consists of two protons and two neutrons (the nucleus of a He atom) and has a 2+ charge since no electrons accompany the helium nucleus.

Gamma rays frequently accompany alpha or beta radiation. A **gamma ray (γ)** is emitted from the nucleus and has no charge or mass. It is a high-energy, short-wavelength photon. Just like infrared (IR), visible, and ultraviolet (UV) radiation, gamma rays are part of the electromagnetic spectrum and have energies similar to those of X-rays. Table 7.1 summarizes these three types of nuclear radiation.

The term *radiation* tends to be confusing, because people don't always specify whether they mean electromagnetic radiation or nuclear radiation. *Electromagnetic radiation* refers to all the different types of light: visible, infrared, ultraviolet, microwave, and, of course, gamma rays. For example, it is perfectly correct to say visible radiation instead of visible light. *Nuclear radiation,* however, refers to the radiation emitted by the nucleus, such as alpha, beta, or gamma radiation. Watch out for one more source of confusion. Gamma rays are *both* a type of electromagnetic radiation and of nuclear radiation. When emitted from the nucleus of a radioactive substance, we refer to gamma rays as nuclear radiation. In contrast, when emitted from a galaxy far away, we call these gamma rays electromagnetic radiation.

Your Turn 7.12   "Radiation"

For each sentence, use the context to decipher whether the speaker is referring to nuclear or electromagnetic radiation.

a. "Name a type of radiation that has a shorter wavelength than visible light."
b. "The gamma radiation from cobalt-60 can destroy a tumor."
c. "Watch out for UV rays! If you have lightly pigmented skin, this radiation may cause a sunburn."
d. "Rutherford detected the radiation emitted by uranium."

Answers
a. electromagnetic radiation   b. nuclear radiation

Table 7.1		Types of Nuclear Radiation		
Name	**Symbol**	**Composition**	**Charge**	**Change to the Nucleus That Emits It**
alpha	$^4_2$He or α	2 protons 2 neutrons	2+	mass number decreases by 4 atomic number decreases by 2
beta	$^0_{-1}$e or β	an electron	1−	mass number does not change atomic number increases by 1
gamma	$^0_0$γ or γ	a photon	0	no change in either the mass number or the atomic number

When either an alpha or beta particle is emitted, a remarkable transformation occurs—the atom that emitted the particle changes its identity. For example, earlier with the PuBe neutron source (see equation 7.2), you saw that alpha emission resulted in the nucleus of plutonium becoming that of uranium. Similarly, when uranium emits an alpha particle, it becomes the element thorium. This nuclear equation shows the process for uranium-238.

$$\underset{92}{\overset{238}{}}\text{U} \longrightarrow \underset{90}{\overset{234}{}}\text{Th} + \underset{2}{\overset{4}{}}\text{He} \qquad \textbf{[7.4]}$$

Notice that the sum of the mass numbers on both sides of the nuclear equation is equal: $238 = 234 + 4$. The same is true for the atomic numbers: $92 = 90 + 2$.

In some cases, the nucleus formed as the result of radioactive decay still is radioactive. For example, thorium-234, formed by the alpha decay of uranium-238, is radioactive. Thorium-234 undergoes subsequent beta decay to form protactinium (Pa).

$$\underset{90}{\overset{234}{}}\text{Th} \longrightarrow \underset{91}{\overset{234}{}}\text{Pa} + \underset{-1}{\overset{0}{}}\text{e} \qquad \textbf{[7.5]}$$

In contrast to alpha emission, with beta emission the atomic number *increases* by 1 and the mass number remains unchanged. Table 7.1 summarizes the changes that occur with both beta and alpha emission.

A model that can help you make sense of this seemingly unusual set of changes is to regard a neutron as a combination of a proton and an electron. Beta emission can be thought of as breaking a neutron apart. Equation 7.6 shows this process, giving us an explanation of how an electron can be emitted from the nucleus.

$$\underset{0}{\overset{1}{}}\text{n} \longrightarrow \underset{1}{\overset{1}{}}\text{p} + \underset{-1}{\overset{0}{}}\text{e} \qquad \textbf{[7.6]}$$

During beta emission, the mass number (neutrons plus protons) in the nucleus remains constant because the loss of the neutron is balanced by the formation of a proton. For example, a neutron in thorium "became" a proton in protactinium. Because of this proton, the atomic number increases by 1. Again, this model can help you to better visualize beta emission, but may not be exactly what is occurring.

Your Turn 7.13   Alpha and Beta Decay

a. Write a nuclear equation for the beta decay of rubidium-86 (Rb-86), a radioisotope produced by the fission of U-235.

b. Plutonium-239, a toxic isotope that causes lung cancer, is an alpha emitter. Write the nuclear equation.

Answer

a. $\underset{37}{\overset{86}{}}\text{Rb} \longrightarrow \underset{38}{\overset{86}{}}\text{Sr} + \underset{-1}{\overset{0}{}}\text{e}$

As we noted earlier, a nucleus may decay to produce another radioactive nucleus. In some cases we can predict this, because *all* isotopes of *all* elements with atomic number 84 (polonium) and higher are radioactive. Thus all the isotopes of uranium, plutonium, radium, and radon are radioactive because these elements all have atomic numbers greater than 83.

What about the lighter elements? Some of these are naturally radioactive, such as carbon-14, hydrogen-3 (tritium), and potassium-40. Whether an isotope is radioactive (a radioisotope) or stable depends on the ratio of neutrons to protons in its nucleus. With each emission of an alpha or a beta particle, this neutron-to-proton ratio changes. Eventually a stable ratio is achieved, and the nucleus is no longer radioactive. Most of the atoms that make up our planet are *not* radioactive. They are here today, and you can count on their being here tomorrow, although possibly not located in the same spot you last saw them (such as the atoms that make up your car keys).

In some cases, radioisotopes may decay many times before producing a stable isotope. For example, the radioactive decay of U-238 and Th-234 (see equations 7.4

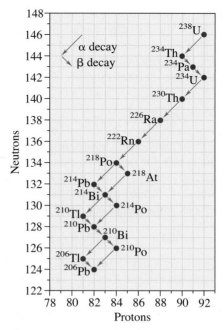

Figure 7.12

The naturally occurring radioactive decay series of uranium-238.

and 7.5) are the first two steps of a 14-step sequence! As shown in Figure 7.12, lead-206 is the end product in this sequence. Similarly, lead-207 is the end product in a different sequence of 11 steps that begins with U-235. Each of these sequences is called a **radioactive decay series,** that is, a characteristic pathway of radioactive decay that begins with a radioisotope and progresses through a series of steps to eventually produce a stable isotope. Radon, a radioactive gas, is produced midway in both the U-238 and U-235 decay series. Thus, wherever uranium is present, so is radon.

7.5 | Looking Backward to Go Forward

We need to examine the legacy of nuclear power—what has worked and what has not. All nuclear plants use the process of fission to produce energy; all produce radioactive fission products. Have these radioactive products posed a danger in the past? Are they likely to in the future? In this section, we consider a significant part of the legacy; that is, the scenario of an accidental release of radioisotopes into the environment.

In 1979, a film called *The China Syndrome* portrayed a near-disaster in a fictitious nuclear power plant. The heat-generating fission reaction went critical and a meltdown of the reactor was imminent. Supposedly, the heat was capable of melting the underlying rock all the way to China. But in the nick of time, the safety features of the system prevailed and no meltdown occurred.

Seven years later on April 26, 1986, the engineers of the very real Chernobyl power plant in Ukraine, then part of the Soviet Union, were less fortunate (Figure 7.13). This plant had four reactors—two built in the 1970s and two more in the 1980s—all near the town of Chernobyl (pop. 12,500). Water from the nearby Pripyat River was used to cool the reactors. Although the surrounding region was not heavily populated, nonetheless approximately 120,000 people lived within a 30-km radius.

Chernobyl stands as the world's worst nuclear power plant accident. What went wrong? During an electric power safety test at the Chernobyl Unit 4 reactor, operators deliberately interrupted the flow of cooling water to the core as part of the test. The temperature of the reactor rose rapidly. In addition, the operators had left an insufficient number of control rods in the reactor (that couldn't be reinserted quickly enough), and the steam pressure was too low to provide coolant (due to both operator error and faulty design). A chain of events quickly produced a disaster. An overwhelming power surge

Chernobyl is the transliteration of the Russian pronunciation. Чорнобиль (Chornobyl) is the Ukrainian word.

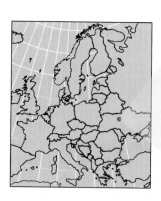

Figure 7.13

Chernobyl, Ukraine, in the former Soviet Union.

The red symbol on Figure 7.13 is one used to denote nuclear radiation hazards. Here is an example.

produced heat, rupturing the fuel elements, and releasing hot reactor fuel particles. These, in turn, exploded on contact with the coolant water, and the reactor core was destroyed in seconds. The graphite used to slow neutrons in the reactor caught fire in the heat. When water was sprayed on the burning graphite, the water and graphite reacted chemically to produce hydrogen gas, which exploded when it reacted with oxygen in the air.

$$2\ H_2O(l) + C(graphite) \longrightarrow 2\ H_2(g) + CO_2(g) \qquad \textbf{[7.7]}$$

$$2\ H_2(g) + O_2(g) \longrightarrow 2\ H_2O(g) \qquad \textbf{[7.8]}$$

The explosion at Chernobyl was produced by combustion. It was a chemical reaction, not a nuclear one.

The explosion blasted off the 4000-ton steel plate covering the reactor (Figure 7.14). Although a "nuclear" explosion never occurred, the fire and explosions of hydrogen blew vast quantities of radioactive material out of the reactor core and into the atmosphere.

Figure 7.14

An aerial view of the Chernobyl Unit 4 reactor taken shortly after the chemical explosion.

Fires started in what remained of the building. In a short time, the plant lay in ruins. The head of the crew on duty at the time of the accident wrote: "It seemed as if the world was coming to an end . . . I could not believe my eyes; I saw the reactor ruined by the explosion. I was the first man in the world to see this. As a nuclear engineer I realized the consequences of what had happened. It was a nuclear hell. I was gripped with fear." (*Scientific American,* April 1996, p. 44)

The disaster continued. As the reactor burned, for 10 days it continued to spew large quantities of radioactive fission products into the atmosphere. Nearly 150,000 people living within 60 km of the power plant were permanently evacuated after the meltdown. People in nearby regions reported an odd, bitter, and metallic taste as they inhaled the invisible particles. The release of radioactivity was estimated to be on the order of 100 of the atomic bombs dropped on Hiroshima and Nagasaki. The radioactive dust cut a swath across Ukraine, Belarus, and up into Scandinavia, affecting some who never had benefited from the power plant but nonetheless shared in its risks.

The human toll was immediate. Several people working at the plant were killed outright, and another 31 firefighters died in the cleanup process from acute radiation sickness, a topic we examine in a later section. An estimated 250 million people were exposed to levels of radiation that ultimately may shorten their lives. Included in this figure are 200,000 "liquidators," people who buried the most hazardous wastes and constructed a 10-story concrete structure ("the sarcophagus") to surround the failed reactor.

One of the particularly hazardous radioisotopes released was iodine-131. It is a beta emitter with an accompanying gamma ray.

$$^{131}_{53}\text{I} \longrightarrow \,^{131}_{54}\text{Xe} + \,^{0}_{-1}\text{e} + \,^{0}_{0}\gamma \qquad\qquad \textbf{[7.9]}$$

If ingested, I-131 can cause thyroid cancer. In the contaminated area near Chernobyl, the incidence of thyroid cancer increased sharply, especially for those younger than age 15 (Figure 7.15). As of 2001, more than 700 children in Belarus, a neighboring country, were treated for thyroid cancer. Fortunately, with treatment, the survival rate for thyroid cancer is high and most have survived. As Dr. Akira Sugenoya, a Japanese physician who volunteered his expertise in Belarus to treat the children suffering from thyroid cancer, remarked "The last chapter of the terrible accident is far from written."

The thyroid gland incorporates iodine in the form of iodide ion to manufacture thyroxin, a hormone essential for growth and metabolism.

Figure 7.15

The level of radioactivity in the thyroid gland of this child is being checked at a clinic north of Minsk, near the Chernobyl nuclear reactor.

Your Turn 7.14   "Iodine"

When people speak of iodine, they may be referring to an iodine atom, an iodine molecule, or an iodide ion, depending on the context.

a. Draw Lewis structures to distinguish among these chemical forms.
b. Which is the most chemically reactive and why?
c. Which chemical form of iodine-131 is implicated in thyroid cancer?

Answer
c. The element iodine (including radioactive I-131) is taken up by the thyroid gland in the chemical form of I⁻, the iodide ion.

Given the demonstrable problems with their design, the four reactors at Chernobyl have been shut down. On Friday, December 15, 2000, the control rods slid permanently into the core at Unit 3, the last remaining reactor operating at Chernobyl. Ukrainian President Leonid Kuchma reported, "This decision came from our experience of suffering. We understand that Chernobyl is a danger for all of humanity and we forsake a part of our national interests for the sake of global safety."

Today, most of the 1000 square miles of contaminated land in the vicinity of Chernobyl has not yet returned to farming. One small exception in Belarus is the land near Viduitsy, as shown in Figure 7.16. A 2005 report indicated that the summer crop of rye and barley harvested tested negatively for radioisotopes.

Figure 7.16

This area in Belarus is gradually being returned to agriculture.

Consider This 7.15   Chernobyl's Legacy

In 2006, twenty years after the accident, *Chernobyl's Legacy: Health, Environmental and Socio-economic Impacts* was issued. This report, an initiative of the International Atomic Energy Agency, was written in cooperation with many other international agencies. The textbook's website provides a link to the document.

a. What diseases resulted from the radiation exposure? Describe two of them and note which people were affected. When possible, cite the number of people affected.

b. Why is it not possible to reliably assess the number of fatal cancers caused by the accident?

Underlying the solemn facts of Chernobyl lies the inevitable question: "Could it happen here?" America's closest brush with nuclear disaster occurred in March 1979, when the Three Mile Island power plant near Harrisburg, Pennsylvania, lost coolant and a partial meltdown occurred. Although some radioactive gases were released during the incident, no fatalities resulted. A 20-year study concluded in 2002 that the total cancer deaths among the exposed population were not higher than those of the general population. In spite of the initial failure, the system held and the damage was contained. Since then, refinements in design and safety were made to existing reactors and those under construction. Nuclear engineers agree that no commercial nuclear reactors in the United States have the design defects that led to the Chernobyl catastrophe.

Consider, for example, the Seabrook nuclear power plant in New Hampshire that was hailed as an example of state-of-the-art engineering when it was built. The energetic heart of the station is a 400-ton reaction vessel with 44-foot high walls made of 8-inch thick carbon steel. Unlike the Chernobyl plant, a reinforced concrete and dome-shaped containment building must surround all reactors in the United States. As the name suggests, the containment structure is built to withstand accidents and prevent the release of radioactive material. The inner walls of the building are several feet thick and made of steel-reinforced concrete; the outer wall is 15 inches thick. The containment building is constructed to withstand hurricanes, earthquakes, and high winds.

Nonetheless, if you live near a nuclear power plant, you probably read about small reactor "incidents" in the daily news. For example, The *New York Times* regularly reports on the Indian Point reactor Units 2 and 3 just north of Manhattan on the Hudson River. Reactor 1 opened in 1962, was closed in 1974, but has not yet been decommissioned. Many of the incidents relate to the spent nuclear fuel, a topic that we take up in Section 7.9.

Could a nuclear meltdown happen in some other region of the world? That possibility does exist, because such disasters result from the complex interplay of faulty plant design, human error, and political instability. Each of these factors must be minimized to keep a nuclear power plant operating safely. Although the nuclear units in many parts of the world get high rankings on all three factors, this is not the case everywhere. For example, in July 2001 the German government urged the closing of a Czech nuclear power plant near the German border because of safety concerns. Several reactors in Russia have long histories of safety violations and raise similar concerns. A plume of radioactive dust easily crosses international boundaries, and so the concerns of neighboring nations are well placed.

A related and far scarier question relates to acts of terrorism. Are nuclear reactors being considered as targets? This question must be taken seriously. Fortunately, as this book went to press, no incidents have occurred. The impact of one or more fully fueled commercial jets on a containment dome, however, is another matter entirely. Were the dome to be breached, the results could truly be catastrophic, exceeding the radioactive releases of Chernobyl.

Today, about 20% of the world's nuclear reactors are located in regions of seismic activity, such as the Pacific Rim. So even before the threat of terrorism, nuclear reactors had to be constructed to withstand a shock. Reactors are fitted with seismic detectors that immediately shut the reactor down if a quake occurs.

Skeptical Chemist 7.16   More About the Pacific Rim

We just stated that about a fifth of the world's reactors are in regions of seismic activity, such as the Pacific Rim. Is this true? Use Figure 7.2, your knowledge of earthquake zones, and any other information you may need to look up on the web to check the accuracy of this statement. See also if you can find the details of how reactors are built to withstand seismic shocks.

Today, nuclear plants and their past operations continue to be under intense scrutiny, hence the title of this section, "Looking Backward to Go Forward." Indeed we must look to the past in order to gain the wisdom we need to move ahead. Undoubtedly, nuclear energy will be part of our future.

7.6 | Radioactivity and You

It would be a serious mistake to dismiss nuclear radiation as harmless. The evidence of the past makes this quite clear. Marie Curie, for example, died of a blood disorder that most likely was induced by her exposure to radiation. Unfortunately she and many of the scientists who first studied radioactive substances were not aware of their dangers.

Nuclear radiation can be hazardous because alpha particles, beta particles, and gamma rays have sufficient energy to ionize the molecules they strike. As you might expect, the same is true for X-rays because their energies are in the same range as those of gamma rays. For this reason, all of these types of "radiation" are referred to as *ionizing radiation*. For example, when a beta particle penetrates your tissue, it may hit a water molecule and knock out an electron.

X-rays and gamma rays are essentially the same thing but named by their source. X-rays are generated by a machine; gamma rays are emitted from a radioactive nucleus.

$$H_2O \xrightarrow{\text{ionizing radiation}} H_2O^+ + e^- \qquad [7.10]$$

The species formed, H_2O^+, has an unpaired electron and is highly reactive. In your body, H_2O^+ further reacts, often with another water molecule. The products in turn can react with still other molecules, including your DNA. This cascading set of radiation-induced molecular changes can range from being perfectly harmless to those causing the death of the cell.

Section 11.2 points out that your body is about 60% water, making it likely radiation penetrating your skin will hit a water molecule.

Consider This 7.17   Free Radicals

Species with unpaired electrons (free radicals) are highly reactive. Below we have rewritten equation 7.10, indicating the unpaired electron on H_2O^+ with a dot.

$$H_2O \xrightarrow{\text{ionizing radiation}} [H_2O\cdot]^+ + e^-$$

In turn, the product may react with another water molecule.

$$[H_2O\cdot]^+ + H_2O \longrightarrow H_3O^+ + HO\cdot$$

Draw Lewis structures for all reactants and products, labeling which ones are free radicals.

Look for more about free radicals in other contexts.

Chapter 1: $\cdot OH$, formation of NO_2 and then tropospheric ozone

Chapter 2: $Cl\cdot$, $ClO\cdot$, and $\cdot NO$, depletion of stratospheric ozone

Chapter 6: $\cdot OH$, formation of SO_3 in acid rain

Chapter 9: $R\cdot$, polymerization of ethylene

Rapidly dividing cells are particularly susceptible to damage by ionizing radiation. As a consequence, nuclear radiation can be used to *treat* certain kinds of cancer, such as prostate cancer and breast cancer. Radioactivity can treat other diseases as well. For example, in Graves' disease, the thyroid is hyperactive. Patients receive small amounts of radioactive I-131 orally in the form of potassium iodide. Because dietary iodine is incorporated into the thyroid gland, the same is true for radioactive iodine. The radiation it emits destroys the overactive tissue, in whole or in part (Figure 7.17). To restore normal thyroid function, most patients need a supplement of a synthetic form of thyroxin, the iodine-containing hormone normally secreted by the thyroid.

End-of-chapter question #51 relates to taking potassium iodide tablets to treat exposure to I-131.

But radiation also can damage healthy rapidly dividing cells, such as those in the bone marrow, the skin, hair follicles, stomach, and intestine. People who receive radiation treatments for cancer often experience a host of side effects that relate to the damage of *healthy* cells. Collectively, these side effects are termed **radiation sickness,** the illness characterized by early symptoms of anemia, nausea, malaise, and susceptibility to infection that are the result of a large exposure to radiation. Radiation sickness affected those near the Chernobyl accident, as well as those who survived the initial firestorm from the bombs dropped at Hiroshima and Nagasaki. Radiation-induced transformations of DNA also can produce genetic mutations, some that occasionally lead to cancer or birth defects.

Today, considerable care is taken to protect workers from nuclear radiation. This is accomplished in many ways, including by using shields made from a dense metal such as lead. Remember, though, that our world (and our bodies) naturally contains radioactive substances, so that radiation levels can never be reduced to zero. **Background radiation** is the level of radiation, on average, that is present at a particular location. It can arise from both natural and human-made sources. The level of background radiation depends primarily on the rocks and soils where you live. Although uranium averages about 4 ppm in the Earth's crust, some types of rocks have a higher concentration than others. It also depends on what type of dwelling you live in, because building materials taken from the Earth contain tiny and again varying amounts of uranium and its decay products. As you can see from Figure 7.18, about 80% of your radiation exposure is natural in origin. The largest natural source is radon, a radioactive gas released in the decay series of uranium in the soils and rocks (see Figure 7.12).

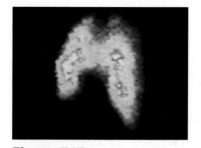

Figure 7.17

A thyroid image produced with I-131. Radioactive iodine has concentrated in the red and yellow areas.

Consider This 7.18   Radon and You

Radon-222 is an alpha emitter.

a. Write the nuclear equation for the radioactive decay of radon-222.
b. Do you expect the product to be radioactive?
c. Given your previous answer and that the product is a solid, explain why radon can cause lung cancer.

Answers

a. $^{222}_{86}Rn \longrightarrow \,^{218}_{84}Po + \,^4_2He$

b. Polonium-218 is radioactive, as are all elements of atomic number 84 or higher.

See Section 1.13 for more about radon as an indoor air pollutant.

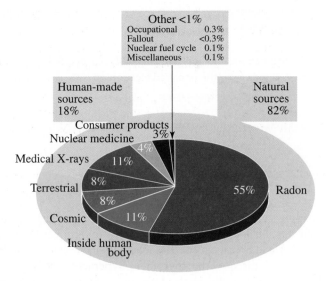

Figure 7.18

Sources of ionizing radiation exposure for people living in the United States.

Source: National Council on Radiation Protection and Measures (NCRP) Report No. 93, *Ionizing Radiation Exposure of the Population of the United States,* 1987.

We use different units to measure the radioactivity of a sample and the damage that exposure to a dose of radiation can cause. For the former, radioactivity is assessed by counting the number of disintegrations of a sample (alpha, beta, or gamma emissions) in a given time period. The **curie (Ci),** named in honor of Marie Curie, is a measure of the radioactivity of a sample, approximately equivalent to the activity of one gram of radium.

$$1 \text{ curie (Ci)} = 3.7 \times 10^{10} \text{ disintegrations/second}$$

Since radium is highly radioactive, one curie is a large amount! Accordingly, people typically refer to radioactivity levels in terms of *milli*curies (mCi), *micro*curies (μCi), *nano*curies (nCi), or even *pico*curies (pCi). For example, household radon measurements are quoted in picocuries, as you will see in Your Turn 7.20. Chemists working with radioisotopes in the lab typically use millicurie or microcurie amounts. If a laboratory worker spilled an amount as large as 100 mCi, serious cleanup procedures would be needed. In contrast, a spill of 100 μCi would be considered far less dangerous.

To put these values in perspective, the explosion at Chernobyl spewed 100–200 *million* curies into the atmosphere. In terms of the amount of radioactivity itself, this is the equivalent of dispersing 100–200 million grams of radium. At the time of the accident, the levels of radioactivity near Chernobyl were from 5 to over 40 Ci per square kilometer. The amount of radiation released by the atomic bombs that exploded on Nagasaki and Hiroshima was lower by two orders of magnitude.

pico
1×10^{-12}
1/1,000,000,000,000

nano
1×10^{-9}
1/1,000,000,000

micro
1×10^{-6}
1/1,000,000

milli
1×10^{-3}
1/1000

Consider This 7.19   Assessing Radioactive Releases

It is not sufficient to just report the amount of a radioactive release. Rather, the *identity* of the radioisotopes also should be reported. Explain why, using the nuclear fission products I-131, Sr-90, and Cs-137 as examples.

Answer

Radioactive Cs-137 is particularly dangerous because it is taken up in the food chain in the form of Cs$^+$, the cesium ion. The same is true for Sr-90 in the form of Sr^{2+}, which is a calcium mimic, and I-131, which accumulates in the thyroid gland. Cs-137 and Sr-90 also are dangerous because they have half-lives long enough to persist for decades. Look for more about half-life in Section 7.8 and more about Sr-90 in Consider This 7.28.

When measuring the damage that exposure to a dose of radiation can cause, we use a different set of units. We also switch our focus from the *sample* that is undergoing radioactive decay to the *tissue* that is absorbing the dose from the radioisotope. In part, the damage depends on the total amount of energy that the tissue absorbs. The **rad,** short for "radiation absorbed dose," is a measure of the energy deposited in tissue and is defined as the absorption of 0.01 joule of radiant energy per kilogram of tissue. Thus, if a 70-kg person were to absorb 0.70 J of energy, a dose of 1 rad would be received. Although this is not very much energy, the energy is localized right where the radiation hits.

The biological damage is more than simply a matter of the energy deposited. To estimate the physiological damage, another unit multiplies the number of rads by a factor Q determined by the type of radiation, the rate the radiation is delivered, and the type of tissue. Less damaging types, including beta, gamma, and X-rays, are arbitrarily assigned a Q of 1. Highly damaging radiation, such as alpha particles and high-energy neutrons, have a Q as high as 20. The **rem,** short for "roentgen equivalent man," is a measure of the dose of radiation received that takes into account the damage caused to human tissue. It is calculated by multiplying Q by the number of rads.

> Q is called the relative biological effectiveness, sometimes referred to as RBE.

$$\text{Number of rems} = Q \times (\text{number of rads})$$

Thus for beta or gamma radiation ($Q = 1$), a dose of 10 rads is 10 rems. In contrast, a 10-rad dose of alpha radiation ($Q = 20$) may be as high as 200 rems. Thus, it takes approximately 20 times as many beta particles to do the same damage as a given number of alpha particles. This makes sense, as alpha particles are larger and deposit a greater amount of energy in the target tissue.

The scientific community in the United States is moving toward using the sievert rather than the rem. The **sievert (Sv)** is a unit used internationally equal to 100 rem. Alternatively, 1 rem equals 0.0100 Sv. Because most doses of radiation are significantly less than a sievert (or a rem), smaller units such as microsieverts (μSv) and millirems (mrem) are employed.

> 1 microsievert = 0.1 millirem

$$1 \text{ microsievert } (\mu Sv) = 1/1,000,000 \text{ of a sievert} = 1 \times 10^{-6} \text{ Sv}$$
$$1 \text{ millirem (mrem)} = 1/1000 \text{ of a rem} = 1 \times 10^{-3} \text{ rem}$$

Table 7.2 shows a sample calculation for an annual radiation dose. The values (in microsieverts) were selected for an adult nonsmoker living in the Midwest. In the next activity, you can do your own calculation.

Your Turn 7.20   Your Personal Radiation Dose

a. Thanks to a web page posted by the EPA, you can calculate the radiation dose that you receive in a year. Are you above or below the national average?
Hint: Remember to convert from millirems (mrem) to microsieverts (μSv).

b. The Los Alamos National Laboratory also hosts a web page that allows you to estimate your annual radiation dose. Use it to perform a similar calculation. How do your results compare with the previous part? Explain any differences.

Links to both radiation dose calculators are provided at the textbook's website.

Nearly 3000 μSv, or about four fifths of the 3600 μSv absorbed per year by a typical (nonsmoking) U.S. resident, comes from natural background sources. As Table 7.2 shows, radon is the major source with additional natural contributions from cosmic rays, soil, and rock. The remainder of the annual dose comes from human sources, typically medical procedures such as diagnostic X-rays.

Table 7.2 also reveals that the radiation from a properly operating nuclear power plant is negligible. In fact, it is less than that of the naturally occurring radioisotopes in your own body. For example, about 0.01% of all the potassium ions (K^+) in your body are radioactive K-40. These K-40 ions give off about 200 μSv per year, approximately

Table 7.2	Annual Radiation Dose (Sample Calculation)*	
Sources of Radiation		**(µSv/yr)**
1. Cosmic radiation		
a. Sea level (U.S. average)		260
b. Additional dose if you are above sea level		
up to 1000 m (3300 ft) add 20 µSv		20
1000–2000 m (6600 ft) add 50 µSv		
2000–3000 m (9900 ft) add 90 µSv		
3000–4000 m (13,200 ft) add 15 µSv		
4000–5000 m (16,500 ft) add 21 µSv		
2. Building material(s) used in your dwelling		
Stone, brick, or concrete add 70 µSv		
Wood or other add 20 µSv		20
3. Rocks and soil		460
4. Food, water, and air (K and Rn)		2400
5. Fallout from nuclear weapons testing		10
6. Medical and dental X-rays		
a. Chest X-ray, add 100 µSv each		0
b. Gastrointestinal tract X-ray, add 5000 µSv each		0
c. Dental X-rays, add 100 µSv each		100
7. Airplane travel		
5-hour flight at 30,000 feet, add 30 µSv/flight		300
8. Other		
a. Live within 50 miles of a nuclear plant, add 0.09 µSv		0.09
b. Live within 50 miles of a coal-fired power plant, add 0.3 µSv		0.3
c. Use a computer terminal, add 1 µSv		1
d. Watch TV, add 10 µSv		10
Total Annual Radiation Dose		**3581**
U.S. average = 3600 µSv/yr		

* Sample calculation is for an adult nonsmoker living in the Midwest.
If you smoke one pack of cigarettes per day, add 10,000 µSv.

Sources: Adapted from information provided by the U.S. Environmental Protection Agency and the American Nuclear Society.

2000 times the exposure from living within 50 miles of a nuclear power plant. Although bananas are rich in K^+, a steady diet of them will not increase your personal radioactivity because the potassium ion continuously moves in and out of your body.

In addition to K-40, our food contains other naturally occurring radioisotopes. One is carbon-14. This radioisotope is produced in our upper atmosphere by the interaction of nitrogen atoms and molecules with cosmic rays. Carbon-14 gets incorporated into carbon dioxide molecules that diffuse down into the troposphere where we live and breathe. The late Isaac Asimov, a prolific science writer, pointed out that a human body contains approximately 3.0×10^{26} carbon atoms, of which 3.5×10^{14} are C-14. Each breath you inhale contains carbon dioxide, including about 3.5 million carbon dioxide molecules that contain C-14 atoms. This number of atoms is so insignificant that Table 7.2 contains no entry for radioactive carbon-14.

Your Turn 7.21   Radioactive Carbon and You

Assume that Isaac Asimov's figures are correct, and that 3.5×10^{14} of the 3.0×10^{26} carbon atoms in your body are radioactive. Calculate the percent that is C-14.

Table 7.3	Physiological Effects of a Single Dose of Radiation	
Dose (Sv)	**Dose (rem)**	**Likely Effect**
0–0.25	0–25	no observable effect
0.25–0.50	25–50	white blood cell count decreases slightly
0.50–1.00	50–100	significant drop in white blood cell count, lesions
1.00–2.00	100–200	nausea, vomiting, loss of hair
2.00–5.00	200–500	hemorrhaging, ulcers, possible death
>5.00	>500	death

The likely effects of a single dose of radiation are described in Table 7.3. Below a single dose of 0.25 Sv, or 250,000 μSv, no immediate physiological effects are observable. A dose of 0.25 Sv is nearly 70 times the average annual exposure!

The long-term effects of low doses of radiation, however, are still under debate. The issue is how to extrapolate from the known high-dose data to lower doses. Extrapolation is necessary because, of course, we cannot do experiments on humans to make reliable measurements. Additionally, the effects of low doses would be small and would only show up over a long time span.

Two radiation dose-response models are illustrated in Figure 7.19. The first, the more conservative of the two, is the **linear, nonthreshold model.** This model assumes a linear relationship between the adverse effects and the radiation dose, with radiation being harmful at all doses, even low ones. Thus, if the adverse effect is to get cancer, doubling the radiation dose doubles the incidence of cancer, and tripling it causes three times as much. No cellular repair of the damage caused by radiation is assumed to take place, even at low doses. Although not illustrated here, other models also exist. For example, **hormesis** is the concept that low doses of a harmful substance (such as radiation) may actually be beneficial.

The model represented by Figure 7.19(b) makes a different assumption about low doses. Here, no observable adverse effects occur until a certain threshold is reached. Presumably, cellular repair can take place so that the response curve initially remains flat. By this model, low doses of radiation are safe.

Currently, the linear, nonthreshold model is being used by the EPA and other federal agencies in setting exposure standards. An exhaustive report compiled in 2002 by the National Council on Radiation Protection and Measurement states that there is "no conclusive evidence" on which to reject the linear, nonthreshold model, also noting

End-of-chapter question #58 allows you to further investigate the hormesis phenomenon.

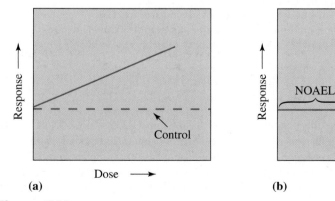

(a)   **(b)**

Figure 7.19

Two dose-response curves for radiation. **(a)** Linear, nonthreshold model. **(b)** Threshold model. *Note:* NOAEL stands for no observed adverse effect level.

Source: Environmental Science & Technology, "Redrawing the Dose-Response Curve," Volume 38, No. 5, March 1, 2004, page 90A.

that it may never be possible to "prove or disprove the validity." Steve Page, the director of EPA's Office of Air Quality Planning and Standards, adds that with the nonthreshold model "the risk from radiation is within the allowable range from toxic chemicals, 1-in-10,000 to 1-in-a-million chances of developing cancer." As a conservative model, it is more costly to implement and hence more controversial.

Consider This 7.22   **Standards and Your Job**

The ramifications of adopting a specific radiation dose-response model are both biological and economic. By using the nonthreshold model, are we being "better safe than sorry" or are we wasting a lot of money protecting ourselves needlessly? As a health professional who must operate under the stricter federal limits for radiation safety, draw up a list of points that you could use with the public to support the stricter standards.

We conclude this section by pointing your attention to a small entry in Figure 7.18: *the nuclear fuel cycle*, a term we define in the next section. Although the dose you receive from this source is small, nuclear fuel warrants our closer attention. In the next section, we explore the connection between nuclear fuel and weapons. In the one that follows, we begin a much longer discussion of nuclear half-lives that will enable us to tackle questions relating to nuclear waste.

7.7 | The Weapons Connection

Although both nuclear power plants and atomic bombs derive their energy from fission, each requires that energy be released at a different rate. A nuclear power plant needs a slow, controlled energy release; in contrast, a nuclear weapon requires one that is rapid and uncontrolled. In either case, the fission reaction is essentially the same. Both are fueled by **enriched uranium,** that is, uranium that has a higher percent of U-235 than its natural abundance of about 0.7%. The difference lies in the *extent* of the enrichment. Commercial nuclear power plants typically operate with 3–5% U-235, whereas atomic weapons use fuel that may be as high as 90% U-235. The latter is sometimes referred to as highly enriched or weapons-grade uranium.

> Although hydrogen bombs are initiated by nuclear fission, they derive their energy primarily from nuclear fusion, a topic not explored in this chapter.

> Most commercial reactors worldwide use enriched uranium as fuel. However, some British and Canadian reactors are designed to run on natural (unenriched) uranium.

Your Turn 7.23   **Enriched Uranium**

The fuel pellets in a nuclear power plant are enriched to 3–5% uranium-235.

a. Another isotope of uranium is present in the pellets. Which one?
b. Is this isotope fissionable under the conditions in a nuclear reactor?
c. After use in a reactor, spent fuel pellets contain radioisotopes of many different elements, including strontium, barium, krypton, and iodine. Explain the origin of these radioisotopes.

 Hint: See **Figures Alive!** for more about fuel pellets.

In a nuclear reactor, the concentration of fissionable U-235 is low. Most of the neutrons given off during fission of U-235 are absorbed by U-238 nuclei in the fuel pellets and by other elements such as cadmium and boron in the control rods. Consequently, the neutron stream cannot build up sufficiently to cause a nuclear explosion. In contrast, atomic weapons use highly enriched uranium in which neutrons are likely to encounter another U-235 nucleus. As we noted earlier, an explosive fission reaction (that is, the explosion of an atomic bomb) occurs only if a critical mass of U-235 (about 33 lb) is quickly assembled all in one place.

Enriching uranium is no easy task! U-235 and U-238 behave essentially the same in all chemical reactions, so they cannot be separated by using one. Rather, the trick to separating them lies in their tiny mass difference of three neutrons. How can this difference be exploited to achieve a separation? On average, lighter gas molecules move faster than heavier ones. Therefore gas molecules containing U-235 should travel slightly more rapidly than their analogs containing U-238. One way to separate molecules is by **gaseous diffusion,** a process in which gases with different molecular weights are forced through a series of permeable membranes. Lighter gas molecules diffuse more rapidly through the membranes than do heavier ones.

But uranium ore clearly is not a gas; rather, it is a mineral that contains UO_3 and UO_2. Most other uranium compounds are solids as well. However, the compound uranium hexafluoride (UF_6) has a notable property. Known as "hex," UF_6 is a solid at room temperature but readily vaporizes when heated to 56 °C (about 135 °F). To produce hex, the uranium ore is converted to UF_4, which in turn is reacted with more fluorine gas.

$$UF_4(g) + F_2(g) \longrightarrow UF_6(g) \qquad [7.11]$$

Equation 7.11 is a *chemical* equation, not a *nuclear* equation.

On average, a $^{235}UF_6$ molecule travels about 0.4% faster than a $^{238}UF_6$ molecule. If the gaseous diffusion process is allowed to occur repeatedly through a long series of permeable membranes, significant amounts of $^{235}UF_6$ and $^{238}UF_6$ can be separated. Prior to World War II and then during the Cold War, U.S. scientists separated uranium isotopes by gaseous diffusion at the Oak Ridge National Laboratory in Tennessee.

The process of gaseous diffusion is used to enrich uranium at commercial plants in a few nations, including France and the United States. However, enrichment now is more commonly carried out using large gas centrifuges. Like gaseous diffusion, the gas centrifuge process makes use of the small mass difference between $^{235}UF_6$ and $^{238}UF_6$. However, it has the advantage of requiring significantly less energy than gaseous diffusion for the same degree of enrichment. For this reason, newer commercial enrichment plants use centrifuges. For example, in 2006, the Nuclear Regulatory Commission issued a license to a consortium of U.S. and European energy companies to build a state-of-the-art gas centrifuge facility near Eunice, New Mexico (Figure 7.20). In 2010, officials held a ribbon-cutting ceremony at the plant; it is expected to reach full capacity in 2015.

Regardless of the enrichment method, once the U-235 has been separated, the U-238 that remains is now "depleted." Nicknamed DU, **depleted uranium** is composed almost entirely of U-238 (~99.8%) because much of the U-235 that it once naturally

Figure 7.20
The new state-of-the-art uranium enrichment plant to fuel commercial reactors in the United States. It is located near Eunice, New Mexico, and was built by URENCO, an international nuclear fuel company.

Source: URENCO.

contained has been removed. Estimates indicate that over 1 billion metric tons of DU is stored currently in the United States. In recent years, the military has deployed DU in armor-piercing munitions. The next activity gives you an opportunity to learn more about DU.

🕸 Consider This 7.24   Depleted Uranium

Depleted uranium is used to tip antitank shells. These first were used in the Gulf War in 1991 and later in other armed conflicts, including Kuwait, Bosnia, Afghanistan, and Iraq. Research the properties of DU in order to explain why it can pierce armor. Also summarize the controversies involved.

At enrichment levels of 3–5%, nuclear fuel rods cannot be incorporated into functional atomic bombs. However, the technology to transform the uranium ore into weapons-grade uranium (about 90% U-235) is essentially identical to that used to produce reactor-grade fuel. In recognition of this fact, only certain countries are authorized to produce enriched uranium according to the Nuclear Non-Proliferation Treaty of 1968. That agreement bestows on signatory sovereign nations the right to pursue nuclear power (and hence uranium enrichment) for peaceful purposes. Iran, a signer of the treaty, restarted its uranium enrichment program in the summer of 2005, despite protests from the United States and other countries.

A more likely scenario for clandestine weapon manufacturing would be to use the plutonium-239 formed from U-238 in a conventional reactor. Analogous to U-235 in equation 7.1, U-238 absorbs a neutron and forms the unstable species U-239. In this case, fission does *not* occur, but rather in a matter of hours U-239 undergoes beta decay.

$$\,^1_0n + \,^{238}_{92}U \longrightarrow [\,^{239}_{92}U] \longrightarrow \,^{239}_{93}Np + \,^0_{-1}e \qquad [7.12]$$

The new element formed, neptunium-239, also is a beta emitter and decays to form plutonium-239.

$$\,^{239}_{93}Np \longrightarrow \,^{239}_{94}Pu + \,^0_{-1}e \qquad [7.13]$$

This transformation was discovered early in 1940. The chemical and physical properties of plutonium were determined with an almost invisible sample of the element on the stage of a microscope. The chemical processes devised on such minute samples were scaled up a billion-fold and used to extract plutonium from the spent fuel pellets from a reactor built on the Columbia River at Hanford, Washington. The plutonium was chemically separated from the uranium and used in the first test explosion of a nuclear device on July 16, 1945, near Alamogordo, New Mexico. The bomb dropped on Nagasaki a little less than a month later also was fueled by plutonium.

Depleted uranium, enriched uranium, and plutonium all are components of the **nuclear fuel cycle,** a way of conceptualizing all the different processes that can happen when uranium ore is mined, processed, used to fuel a reactor, and then dealt with as waste. Examine Figure 7.21 to see the connection with plutonium. As equations 7.12 and 7.13 show, plutonium-239 is produced ("bred") in nuclear reactors and thus is a component of the spent fuel. Reactors can be designed to breed more (or less) plutonium. We return to the nuclear fuel cycle later in the chapter.

Look for more about breeder reactors in Section 7.9.

Plutonium-239 poses an international security problem because the plutonium produced in nuclear power reactors could possibly be incorporated into nuclear bombs. Given the risks associated with Pu-239 and U-235, it is essential that both national and international organizations carefully monitor the supplies and distribution of these two isotopes throughout the world. Safeguarding existing nuclear materials has taken on a new meaning since the end of the Cold War and the demise of the former Soviet Union. One part of the problem is the plutonium and highly enriched uranium in Russia's nuclear arsenal (about 20,000 warheads). Another problem stems from Russia's legacy from the Cold War, a stockpile of highly enriched uranium and plutonium

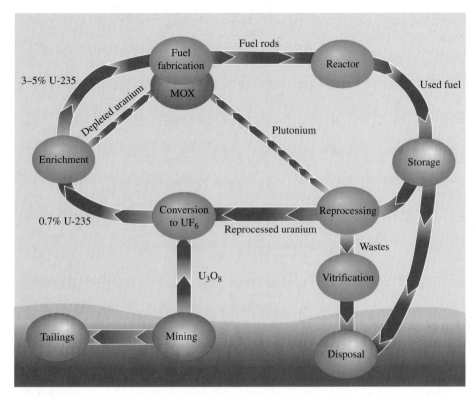

Figure 7.21

One representation of the nuclear fuel cycle. Currently, some of the cycle parts are not in operation. *Note:* MOX stands for mixed oxide fuel. U_3O_8 is uranium oxide, or "yellow cake," a uranium compound produced from the refining of uranium ore.

Source: World Nuclear Association, 2009.

Figure 7.22

A smuggled canister of military grade Pu-239 captured in Germany.

Look for more about yellow cake and parts of the nuclear fuel cycle in Section 7.10.

(about 600 metric tons). Both of these pose a threat to world security (Figure 7.22). The fissionable materials stored in labs, research centers, and shipyards across the former Soviet Union are vulnerable to theft. These 600 tons of fissionable material translate into the capacity to construct approximately 40,000 new nuclear weapons.

The world community clearly recognizes the dangers of nuclear trafficking and the need for effective safeguards. In recognition of the global threat, the Nobel Peace Prize for 2005 was shared equally between the International Atomic Energy Agency (IAEA) and its Director General, Mohamed ElBaradei. The Nobel committee cited ". . . their efforts to prevent nuclear energy from being used for military purposes and to ensure that nuclear energy for peaceful purposes is used in the safest possible way." The future safety of nations, if not of our planet, may depend on our ability to safeguard and ultimately recycle plutonium and highly enriched uranium.

7.8 | Nuclear Time: The Half-Life

How long does a radioactive sample "last"? The answer depends on the radioisotope. Some radioisotopes decay quickly over a short period of time; others undergo radioactive decay much more slowly. Each radioisotope has its own **half-life ($t_{1/2}$),** the time required for the level of radioactivity to fall to one half of its initial value. For example, plutonium-239, an alpha emitter formed in nuclear reactors fueled with uranium, has a half-life of about 24,110 years. Accordingly, it will take 24,110 years for the radioactivity of a sample of Pu-239 to halve. After a second half-life (another 24,110 years), the level of radioactivity will be one fourth of the original amount. And in three half-lives (72,330 years), the level will be one eighth (Figure 7.23). From these times, you can see that it takes a very long time for the amount of plutonium to decrease!

Revisit nuclear equations 7.12 and 7.13 to see how Pu-239 can be produced from U-238 in a nuclear reactor.

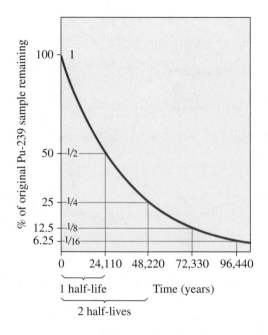

Figure 7.23

Decay of a sample of Pu-239 over time. **Figures Alive!** has an interactive version of this decay curve.

Other radioisotopes decay even more slowly. For example, the half-life of U-238 is 4.5 billion years. Coincidentally, this is approximately the age of the oldest rocks on Earth, a determination made by measuring their uranium content. The half-life for each particular isotope is a constant and is independent of the physical or chemical form in which the element is found. Moreover, the rate of radioactive decay is essentially unaltered by changes in temperature and pressure. From Table 7.4 you can see that half-lives range from milliseconds to millennia.

From Table 7.4, you also can see that Pu-239 and Pu-231 have different half-lives. Other isotopes of plutonium have different half-lives as well. For example, in 1999 Carola Laue, Darleane Hoffman, and a team at Lawrence Berkeley National Laboratory characterized plutonium-231. These researchers had to work fast

Table 7.4	Half-life of Selected Radioisotopes	
Radioisotope	Half-life ($t_{1/2}$)	Found in the spent fuel rods of nuclear reactors?
uranium-238	4.5×10^9 years	Yes. Present originally in fuel pellet.
potassium-40	1.3×10^9 years	No.
uranium-235	7.0×10^8 years	Yes. Present originally in fuel pellet.
plutonium-239	24,110 years	Yes. See equation 7.13.
carbon-14	5715 years	No.
cesium-137	30.2 years	Yes. Fission product.
strontium-90	29.1 years	Yes. Fission product.
thorium-234	24.1 days	Yes. Small amount generated in natural decay series of U-238.
iodine-131	8.04 days	Yes. Fission product.
radon-222	3.82 days	Yes. Small amount generated in natural decay series of U-238.
plutonium-231	8.5 minutes	No. Half-life is too short.
polonium-214	0.00016 seconds	No. Half-life is too short.

because the half-life of Pu-231 is a matter of mere minutes! In general, each radioisotope has its own unique half-life, including when the isotopes are for the same element.

We can use the half-life ($t_{1/2}$) of a radioisotope to determine the percent of a sample that remains at some later point in time. For example, once Pu-231 is generated in a laboratory, what percent of the original sample *remains* after 25 minutes? To answer this, first recognize that 25 minutes is roughly three half-lives or 3×8.5 minutes. After one half-life, 50% of the sample has decayed and 50% remains. After two half-lives, 75% of the sample has decayed and 25% remains. And after three half-lives, 87.5% has decayed and 12.5% remains. These values are not exact, because 25 minutes is not exactly three half-lives. Nonetheless, quick back-of-the-envelope calculations can be useful.

This question also could have been phrased in this way: "After 25.5 minutes what percent of a sample of Pu-231 has *decayed*?" This question requires one more step. To find the amount decayed, simply subtract the percent that remains from 100%. If 12.5% remains, then $100\% - 12.5\% = 87.5\%$ has decayed. Table 7.5 summarizes these changes for any radioisotope.

Your Turn 7.25   Here today . . .

. . . and gone tomorrow? People sometimes use the value of 10 half-lives to indicate when a radioisotope will be gone, that is, when only a negligible amount of it will be present. What percent of the original sample remains after 10 half-lives? Add rows to Table 7.5 so that it shows the mathematics of decay to 10 half-lives.

Let's do another back-of-the-envelope calculation with a different radioisotope. For example, if you had a sample of U-238 ($t_{1/2} = 4.5 \times 10^9$ years), what percent of it would remain after 25 minutes? To answer this, recognize that minutes, days, or even months would be a mere instant in the span of a 4.5-billion-year half-life. Thus, essentially all of the uranium-238 would remain. The next two activities offer you more practice with half-life calculations.

Your Turn 7.26   Tritium Calculation

Hydrogen-3 (tritium, H-3) sometimes is formed in the primary coolant water of a nuclear reactor. Tritium is a beta emitter with $t_{1/2} = 12.3$ years. For a given sample containing tritium, after how many years will about 12% of the radioactivity remain?

Table 7.5	Half-life Calculations	
# of Half-lives	% Decayed	% Remaining
0	0	100
1	50	50
2	75	25
3	87.5	12.5
4	93.75	6.25
5	97.88	3.12
6	98.44	1.56

Your Turn 7.27   Radon Calculation

Radon-222 is a radioactive gas produced from the decay of radium, a radioisotope naturally present in many rocks.

   a. What is the most likely origin for the radium present in rocks?
      Hint: See Figure 7.12.
   b. Radon activity is usually measured in picocuries (pCi). Suppose that the radioactivity from Rn-222 in your basement were measured at 16 pCi, a high value. If no additional radon entered the basement, how much time would pass before the level dropped to 0.50 pCi?
      Hint: In dropping from 16 to 1 pCi, the radioactivity level halves four times: 16 to 8 to 4 to 2 to 1.
   c. Why is it incorrect to assume that no more radon will enter your basement?

One final difficulty with reactor waste is that the fission products, if released, may enter and accumulate in your body, with potentially fatal consequences. One culprit is strontium-90, a radioactive fission product that entered the biosphere in the 1950s from the atmospheric testing of nuclear weapons. Strontium ions are chemically similar to calcium ions; both elements are in Group 2A of the periodic table. Hence, like Ca^{2+}, Sr^{2+} accumulates in milk and in bones. Thus, once ingested, radioactive strontium with its half-life of 29 years poses a lifelong threat. Like I-131, Sr-90 was among the harmful fission products released in the vicinity of the Chernobyl reactor.

Your Turn 7.28   Strontium-90

Sr-90 is one of the fission products of U-235 listed in Table 7.4. It forms in a reaction that produces three neutrons and another element. Write the nuclear equation.
Hint: Remember to include the neutron that induces the fission of U-235.

On a cheerier note, we end this section with carbon-14, a radioisotope mentioned in the previous section. Carbon-14 has a half-life of 5715 years and decays to nitrogen-14 through the process of beta decay. Our atmospheric carbon dioxide contains a constant steady-state ratio of one radioactive C-14 atom for every 10^{12} atoms of nonradioactive C-12. Living plants and animals incorporate the isotopes in that same ratio. However, when the organism dies, exchange of CO_2 with the environment ceases. Thus, no new carbon is introduced to replace the C-14. As a consequence, the concentration of C-14 in any material that once was alive decreases with time, halving every 5715 years.

In the 1950s, W. Frank Libby (1908–1980) first recognized this decrease by experimentally measuring the C-14/C-12 ratio in a sample. The ratio provided an estimate of when the organism died. Human remains and many human artifacts contain carbon, and fortunately, the rate of decay of C-14 is a convenient one for measuring human activities. Charcoal from prehistoric caves, ancient papyri, mummified human remains, and suspected art forgeries have all revealed their ages by this technique. The C-14 technique provides ages that agree to within 10% of those obtained from historical records, thus validating the legitimacy of the radiocarbon-dating technique.

With high confidence, carbon-14 dating was used to establish the age of the famous Shroud of Turin as approximately 1300 CE.

Skeptical Chemist 7.29   Ancient Shroud

Using carbon-14 dating, a burial cloth from a tomb was estimated to have an age of 100,000 years. Does this determination seem reasonable to you, given that the half-life of C-14 is 5715 years?
Hint: After more than 10 half-lives have passed, consider the amount of radioisotope that remains. Revisit Your Turn 7.25 to see this.

The nuclear decay process cannot be hastened; we can neither make the nuclear clocks run fast nor slow. Radiocarbon dating depends on the unerring "ticking" of carbon-14. The same principles apply to nuclear waste. As we see in the next section, we can do nothing to make any particular radioisotope decay more quickly. We have to deal with what we have—perhaps for millennia.

7.9 | Nuclear Waste: Here Today, Here Tomorrow

Of the issues surrounding nuclear power, safely dealing with nuclear waste is the most pressing. There is no apparent "silver bullet" (or silver waste canister). In a June 1997 *Physics Today* article, John Ahearne, past chair of the U.S. Nuclear Regulatory Commission, reminds us that, ". . . Like death and taxes, radioactive waste is with us—it cannot be wished away. . . ."

Before we discuss the options, it makes sense first to define the types of nuclear waste materials. **High-level radioactive waste (HLW),** as the name implies, has high levels of radioactivity and, because of the long half-lives of the radioisotopes involved, requires essentially permanent isolation from the biosphere. HLW comes in a variety of chemical forms, including ones that are highly acidic or basic. It also can contain toxic metals. Thus, HLW is sometimes labeled as a "mixed waste" in that it is hazardous *both* because of the chemicals *and* their radioactivity. Furthermore, as we noted in Section 7.7, this waste also poses a national security risk because it contains plutonium that could be extracted and used to construct a nuclear weapon. Huge quantities of HLW also were created during the Cold War because reactor fuel was reprocessed to produce plutonium for nuclear warheads. This military waste tends to be in the inconvenient form of solutions, suspensions, slurries, and salt cake stored in barrels, bins, and underground tanks.

In contrast, **low-level radioactive waste (LLW)** contains smaller quantities of radioactive materials than HLW and specifically excludes spent nuclear fuel. LLW includes a wide range of materials, including contaminated laboratory clothing, gloves, and cleaning tools from medical procedures using radioisotopes, and even discarded smoke detectors. As you might guess, the hazards associated with LLW are significantly less than those from HLW. Nearly 90% of the volume of all nuclear waste is low level.

Consider This 7.30   **Compact It! Incinerate It!**

An elected state representative visited a chemistry class to address questions relating to the radioactive waste that was being sent to local landfills. She proposed compacting the waste to reduce its radioactivity and then incinerating it. In her view, this was preferable to filling the landfills with radioactive waste. Assume that you were one of her staff members. Draft a tactful memo to set her straight.

Commercial and military nuclear power plants are the primary source of HLW. For example, each of the 100+ commercial nuclear reactors in the United States produces about 20 tons of spent fuel annually. **Spent nuclear fuel (SNF)** is the radioactive material remaining in fuel rods after they have been used to generate power in a nuclear reactor. After removal from the reactor, the spent fuel rods are still "hot," both in temperature and their radioactivity. These rods contain primarily U-238 with about 1% of the U-235 that did not fission. They also contain fission products, that is, many highly radioactive isotopes including iodine-131, cesium-137, and strontium-90. In addition, the spent fuel rods contain plutonium. Pu-239 is formed from U-238, as we saw earlier in equation 7.13. Refer back to Table 7.4 for the values of half-lives for some radioisotopes present in spent nuclear fuel.

Figure 7.24

A cask containing spent fuel is lowered into a deep underwater storage pool at the Savannah River Site in Aiken, South Carolina. This site is for interim storage.

Source: U.S. Department of Energy, Office of Civilian Radioactive Waste Management.

At each nuclear reactor in the United States, approximately 30% of the fuel rods are replaced annually on a rotating schedule. After the spent rods are removed from the reactor, they are transferred to deep basins of water for temporary storage (Figure 7.24). The water serves both to cool the fuel and to absorb alpha and beta radiation, thus shielding any nearby workers.

These pools are not intended for permanent storage of spent nuclear fuel. They are expensive to operate, and some corrosion of the metal rods occurs under water. Furthermore, most of the pools in the United States have reached their capacity, so the fuel rods need to be removed to make room for new ones. So for many reasons, after a year or so of "wet" storage in pools, the spent nuclear fuel is removed, dried, and transferred to casks.

Dry cask storage typically involves putting the spent reactor fuel into leak-tight steel cylinders that are enclosed by additional layers of steel or concrete to provide additional shielding. These casks are then stored in a concrete vault. Like the deep pools, this option for storage is temporary and requires continual maintenance (Figure 7.25).

Today, almost all reactor waste is being stored on site where it was generated. The storage facilities, not built for the long term, are hardly ideal. In the 1950s and early 1960s, the plan had been to reprocess the spent fuel to extract plutonium and uranium from it and to recycle these elements as nuclear fuel. On-site storage capacity for spent fuel rods was designed with such reprocessing in mind. However, only one of several planned reprocessing plants ever went into operation and then only briefly (1967–1975). Thus, reprocessing never was capable of keeping up with the rate of spent fuel production, about 2000 tons every year. In 1977, then President Jimmy Carter, a nuclear engineer, declared a moratorium on commercial nuclear fuel reprocessing that continues to this day.

The restrictions vary by nation. France, the United Kingdom, Germany, and Japan all reprocess some of their SNF. An option that is becoming more and more attractive is the use of a **breeder reactor,** a nuclear reactor that can produce more fissionable fuel (usually Pu-239) than it consumes (usually U-235). This seems like a dream come true to an energy-hungry planet. Imagine if your car synthesized gasoline as you drove! Scientists in the United States and elsewhere have figured out how to recover plutonium from the spent fuel of breeder reactors. In the 1970s, several factors led the United States to stop pursuing breeder reactor technology, including the objection

Figure 7.25

The dry cask system for storage of nuclear waste at a reactor site.

Source: U.S. Nuclear Regulatory Commission.

to plutonium reprocessing and the more complicated reactor design required. We take a more detailed look at the possible future of reprocessing in the final section of this chapter.

In the absence of reprocessing, two options exist for the storage of HLW: monitored storage on or near the surface and storage in geological repositories deep underground. These differ in a key variable: *active management* (Figure 7.26). In surface storage, human societies over thousands of years must commit resources to maintaining the integrity of the wastes. In geological repository storage, the wastes may be accessible and retrievable (although less easily) or sealed "forever," requiring minimal human vigilance. In a report published by the National Academies in 2000, the option of deep-underground storage was favored, noting that it was not prudent to assume that future societies on Earth would be able to maintain surface storage facilities.

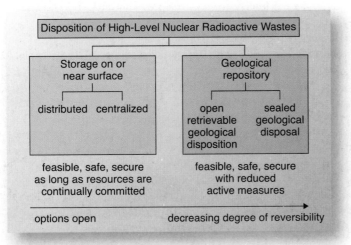

Figure 7.26

Strategies for high-level nuclear waste deposition.

Source: Disposition of High-Level Wastes and Spent Nuclear Fuel, National Academy Press, 2000.

Figure 7.27

Encapsulating reprocessed HLW in glass canisters (vitrification).

With either strategy for storage, HLW must remain isolated from the groundwater for at least 10,000 years to allow the high levels of radioactivity to decrease significantly. Some plans employ a method known as **vitrification,** in which the spent fuel elements or other mixed waste are encased in ceramic or glass. First, the waste is dried, pulverized, and then mixed with finely ground glass and melted at about 1150 °C. Then the molten glass and wastes are poured into stainless-steel canisters, cooled, and capped for on-site storage. More than 1 million pounds of waste already has been treated in this way and awaits the development of a long-term underground repository (Figure 7.27). The radioactivity remains, but the nuclear materials are trapped in solid glass.

Whatever physical form it may take, at present nuclear waste has no final resting place. In 1997, the Nuclear Waste Policy Amendments Act designated Yucca Mountain (Figure 7.28) in Nevada as the sole site to be studied as an underground long-term, high-level nuclear waste repository. In the years that followed, billions were spent to fund the development of such a repository. As of 2010, however, it appeared that the Yucca Mountain depository would not become operational.

Congress (or its successor) may still be debating the issue 24,110 years from now, when the plutonium-239 we create today completes its *first* half-life. Other disposal methods seem even less promising. Disposal in deep-sea clay sediments beneath 3000–5000 meters of water was investigated. Proposals to bury the radioactive waste under the Antarctic ice sheet or to rocket it into space have largely been discredited. But one thing is sure. Whatever disposal methods are ultimately adopted, they must be effective over an extremely long time. Nuclear engineers William Kastenberg and Luca Gratton conclude their 1997 article in *Physics Today* with a sobering thought that still rings true today:

> "For a high-level waste depository of the type proposed for Yucca Mountain, it is clear that natural processes will eventually redistribute the waste materials. Present design efforts are directed toward ensuring that, at worst, the degraded waste configurations will eventually resemble stable, natural ore deposits, preferably for periods exceeding the lifetimes of the more hazardous radionuclides. Perhaps that's the best we can hope for."

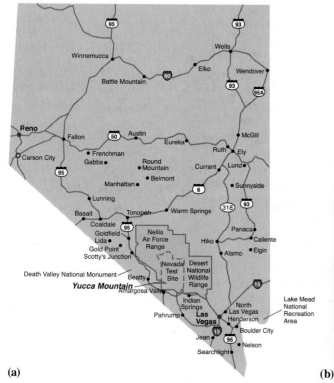

(a)

(b)

Figure 7.28

(a) Map of Yucca Mountain and state of Nevada. **(b)** Yucca Mountain, looking south into the desert.

Source: **(a)** U.S. Department of Energy.

Consider This 7.31   Nuclear Waste Warning Markers

On February 15, 2007, the International Atomic Energy Agency (IAEA) unveiled the new symbol pictured to the right to warn the public about the dangers of radiation. You can read the details at the IAEA website.

a. Describe what this new symbol conveys to you.
b. Suppose you were asked to design markers to be installed near an underground nuclear waste repository. These markers must warn future generations of the existence of nuclear waste and must last for at least 10,000 years (more than four times the age of the pyramids of Egypt). The message must be intelligible to Earthlings of the future. Try your hand at designing these warning markers, keeping in mind the changes that have occurred in *Homo sapiens* during the past 10,000 years and those that might occur in the next 10 millennia.

7.10 | Risks and Benefits of Nuclear Power

In earlier sections, we noted the markedly different degree to which countries around the world employ nuclear power to generate electricity. Regardless of whether a country contains many nuclear-powered electric generators or only a few, the associated risks and benefits must be weighed by all involved.

Although a risk–benefit analysis may seem daunting, we actually make these kinds of assessments every day. Many types of risk are possible. Risks can be voluntary, such as those associated with wind surfing or bungee jumping; or they can be involuntary, such as inhaling someone else's cigarette smoke. When we drive a car, we can control some of the risks by using defensive driving techniques. But aboard a flight to Toledo or Tokyo, we have no control over the increased risk of radiation exposure at cruising altitude. Counterbalancing risks are many types of benefits. These include the improvement of health, increased personal comfort or quality of life, cost savings, or a lower ecological footprint.

> The 19th century poet William Wordsworth spoke of technological risks and benefits as "... Weighing the mischief with the promised gain ..." He was speaking, in this case, about the railroad, a technology new in his time.

Consider This 7.32   Informed Citizens

"Nuclear plants are no problem! I get a higher dose of radiation from flying for 5 hours than I do from working in a nuclear power plant. No worker has ever lost his or her life on the job."

a. In what sense is this assessment valid?
b. In what sense is this nuclear worker missing the point?

Hint: Use Table 7.2 to help formulate your answer. When it comes to radiation, also remember that the identity of the radioisotope (and whether it is inside or outside of your body) is a factor to consider.

There is no such thing as zero risk! Everyday living inevitably involves risks and their related benefits: crossing a street, riding a motorcycle or in a car, cooking a meal or eating one, and even the simple act of getting up in the morning. Because there is some element of risk in everything we do, automatically we make judgments about what level of risk we consider acceptable. Most people do not intentionally put themselves at high risk. On the other hand, expecting "zero risk" in whatever we do is impossible to achieve.

Clearly, one of the desired benefits of nuclear power is electricity. Along with this, we desire minimal risks, including those to local, regional, and global economies, those to workers in all parts of the nuclear fuel cycle (see Figure 7.21), and those to the environment. This should sound reminiscent of the Triple Bottom Line: the nuclear

Figure 7.29

The "green zone" in which we have healthy ecosystems, healthy economies, and healthy communities.

During the useful lifetime of the nuclear fuel rods, a conventional power plant would need 200 train cars each carrying 15 tons of coal **every day** to produce the same amount of electricity.

power industry should promote health for our economy, for our communities, and for our ecosystems. In essence, we want to operate in the "green zone" where all three overlap (Figure 7.29, reproduced from Chapter 0 of this book).

How does nuclear power stack up against the alternatives? This is no easy question! Even so, we explore the answers, at least in part. For example, one alternative to "burning" uranium in a nuclear reactor is to burn coal in a conventional power plant. Here are some of the risks associated with coal-fired power plants, including some we described earlier in Chapter 4.

- **Mine worker safety**
  Over 100,000 workers have been killed in American coal mines since 1900, most prior to the 1950s when higher safety standards were instituted. Once mined, however, the coal does not require further refining. Mine workers also die from black lung disease. Although the overall rate in the United States is dropping, it still claims the lives of several hundred per year.

- **Greenhouse gas generation**
  Coal-fired power plants produce carbon dioxide, a waste product of combustion. Annually, a 1000-MW coal-fired electric power plant releases about 4.5 million tons of CO_2. The total release in the United States from burning coal is on the order of 2 billion tons yearly.

- **Air pollutant generation**
  A typical 1000-MW coal-fired power plant burns over 10,000 tons of coal and could easily release 300 tons of SO_2 and perhaps 100 tons of NO_x daily. Deaths attributed to poor air quality are numbered annually in the tens of thousands.

- **Ash generation**
  In a year, a 1000-MW coal-fired plant generates about 3.5 million cubic feet of waste ash, a substantial volume. Revisit Figure 4.8 (reproduced in margin) to see the devastation caused when millions of gallons of fly ash sludge spilled down a valley in Tennessee.

- **Mercury release**
  Coal contains trace quantities of mercury. When coal is burned, mercury is released into the air in the form of elemental mercury. Although Hg emissions are slowly dropping, the amount emitted yearly worldwide is still on the order of hundreds of tons. Each year, about 50 tons of mercury is released in the United States.

- **Uranium and thorium release**
  Trace quantities of uranium in coal can be as high as 10 ppm, and the amount of thorium is usually higher. In the United States, at an annual coal consumption of over 1100 million tons, over 1300 tons of uranium and 2600 tons of thorium is being released into the environment yearly, exceeding the amount of uranium consumed in nuclear plants. Although much of the radioactive metals are collected in the fly ash, this too must be disposed of.

By way of contrast, here are some of the risks associated with nuclear power plants. Note that there is some overlap.

- **Mine worker safety**
  Uranium ore is mined and then chemically processed at a uranium mill to produce "yellow cake" (Figure 7.30). Mine and mill workers are at risk for cancer (especially lung cancer) both from the uranium dust and from the radon that it emits. After World War II in the United States, most uranium mining took place on the Colorado Plateau. Hundreds of workers later died of lung cancer. Family members of the workers sometimes were affected as well, as the workers carried home the uranium dust. Today, a stricter set of safety regulations for ventilation and radiation exposure is in place.

- **Greenhouse gas generation**
  Nuclear power plants produce no carbon dioxide, although CO_2 emissions are associated with the mining, milling, enriching, and transporting of uranium

Figure 7.30

A sample of yellow cake, U_3O_8. This product is then refined to produce uranium metal, which in turn is enriched in U-235 (see Section 7.7).

and handling of spent reactor fuel. CO_2 emissions also accompany cement manufacture, used in power plant construction.

- **High-level nuclear waste generation**
  A 1000-MW nuclear power reactor produces about 70 cubic feet of high-level waste (HLW) per year. See previous section. In the United States, the total is on the order of 2000 tons annually.

- **Releases of fission products**
  In the past decades, any releases from nuclear power plants have been relatively tiny. Revisit Table 7.2 to see that living within 50 miles of a nuclear reactor contributes essentially nothing to your annual radiation dose.

- **Mine tailings and mill waste**
  Uranium mining and milling operations produce radioactive tailings and waste. These rock tailings, because they contain uranium, also emit radon, so must be capped. Mining spills have occurred in the United States, most notably in 1979 at Church Rock, New Mexico. Today this is a Superfund cleanup site.

About 1 pound of CO_2 is emitted for each pound of cement manufactured.

Recall from Section 7.9 that high-level radioactive waste (HLW) requires essentially permanent isolation from the biosphere.

As we noted early in this chapter, nuclear energy carries tremendous emotional overtones. In part, these stem from mystery, misunderstandings, and the powerful image of the mushroom-shaped cloud. The possibility of a major disaster, however remote, looms large in human consciousness. The accidents at Three Mile Island and Chernobyl, though hardly equivalent, have made the public wary. We have limited trust in technology and perhaps even less in people. We are apprehensive about human error in the design, construction, and management of nuclear power plants. After all, human errors and technicians' responses to them were the weak points in the prescribed safety procedures that caused the accidents at Three Mile Island and Chernobyl.

What about the risks in comparison with other sources of energy, such as wind, solar, and geothermal? The next activity offers an opportunity to explore some of the risks and benefits of wind power, a topic otherwise not discussed in *Chemistry in Context*.

Look for the details of solar energy in Chapter 8.

Skeptical Chemist 7.33   Wind Power Safety

The author of a 2009 article on wind versus nuclear energy wrote: "The wind turbine industry, on the other hand, has quite a treacherous track record."

a. According to 2009 data from the World Nuclear Association, nuclear plants account for about 15% of the generation of electricity worldwide. How does wind compare?
b. Prepare a bullet point list of the risks associated with constructing wind turbines.
c. Now list the benefits of wind power. What do you conclude?

The risks associated with energy produced by nuclear plants clearly are different from those associated with other types of power generation. Remember, though, that "zero risk" is impossible for any energy source. Clearly, conservation and the efficient use of natural resources and energy are the best options of all. With this thought in mind, we now address the more general question of whether we can design nuclear power plants that meet the criteria for sustainability in the future.

7.11 | A Future for Nuclear Power

"People across the globe share the dream of clean and sustainable sources of energy for the future. Should we build more nuclear power plants to achieve this? The answer depends on both whom you ask and when you ask them." These words came from the opening section of the chapter. Does this dream include nuclear energy?

If you had asked this question in the United States back in the early 1960s, the answer would have been yes. At this time, the United States experienced a dramatic growth in the nuclear power industry, one that lasted until 1979 when the malfunction at Three Mile Island occurred. The fear that accompanied this incident certainly contributed to the end of the growth phase. More important at that time, however, were the economics of nuclear energy. With the retreat of fossil fuel prices and the added costs of nuclear safety and oversight imposed in the 1980s, it simply was not economically feasible for utilities to construct new nuclear plants.

What are the economic realities today? Again, the answer depends on whom you ask and when you ask them. Two things, however, are clear. The first is that any new reactors will be built with new and improved designs. And the second is that these designs will have a higher price tag.

New designs most likely will involve higher operating temperatures. As we noted earlier, current nuclear power stations, like any power plant, are designed to produce heat. This heat is used to produce high-pressure steam that in turn drives the blades of a turbine to produce electricity. As we pointed out in an earlier section, the theoretical efficiency depends on the maximum and minimum temperatures between which the plant operates. Clearly, then, one way to increase the efficiency is to increase the maximum temperature of the steam.

Most current reactors that cool the reactor core with water as a primary coolant operate with a temperature in the range of 300 °C and at a pressure of over 100 atmospheres. The primary coolant transfers heat to the secondary coolant to boil water to produce steam. What if the primary coolant could operate at an even higher temperature? Here is where chemists who design materials are called to action. Running reactors at higher temperatures brings two challenges: faster corrosion of parts and higher levels of radiation. Both of these damage the materials in the core of the reactor.

Your Turn 7.34   The Hot Hub of a Nuclear Reactor

Revisit Figure 7.7, the schematic diagram of a pressurized water reactor.

a. The primary coolant has been termed "a heat transfer medium." What is the source of this heat?

b. Which parts of this reactor are exposed to the highest temperatures?

As reported in the August 24, 2009, issue of *Chemical & Engineering News* (Figure 7.31), the weekly magazine of the American Chemical Society:

> "Corrosion is a fact of life," says Paul A. Sherburne, a technical consultant at the Lynchburg, Virginia, facility of French power plant manufacturer Areva. And life is expected to be even tougher for components in future reactors, such as ones incorporating lead or lead-bismuth coolant and running at 800 °C and the so-called Very High Temperature Reactor that is designed to run at 1000 °C with helium coolant."

Clearly, chemists will continue to be at the table as the designs for the new reactors progress. Again, any new reactors will be built with new and improved designs. With these will come technological challenges!

Price tag is the other consideration. Not only is there the cost of the reactor itself to consider, but also that of the fissionable fuel, nuclear waste, and ultimately decommissioning the nuclear plant itself. One who reports that the economics of nuclear power are flawed is Lester R. Brown, author of *Plan B, Mobilizing to Save Civilization*.

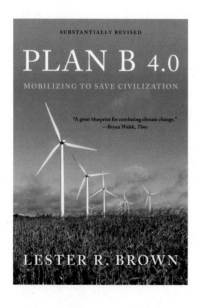

> In an excellent recent analysis, "The Nuclear Illusion," Amory B. Lovins and Imran Sheikh put the cost of electricity from a new nuclear power plant at 14¢ per kilowatt hour and that from a wind farm at 7¢ per kilowatt hour. This comparison includes the costs of fuel, capital, operations and maintenance, and transmission and distribution. It does not include the additional costs for disposing of nuclear waste, insuring plants against an accident, and decommissioning the plants when they wear out. Given this huge gap, the so-called nuclear revival can succeed only by unloading these costs onto taxpayers. If all the costs of generating nuclear electricity are included in the price to consumers, nuclear power is dead in the water.

Figure 7.31

This cover shows the bundles of fuel rods (containing uranium dioxide fuel pellets) that generate the intense heat that must be transferred to the primary coolant.

Consider This 7.35   Simple Logic?

"Ultimately, the replacement of old, highly polluting coal-fired power plants by nuclear reactors is essentially no different from deciding, after putting sentimental considerations aside, to replace your inexpensive and reliable—but obsolete—1983 Olds Omega with a 2007 Toyota Camry or BMW 3 Series sedan." *Discover Magazine*, August 2007, page 61.

a. Explain the logic of this science writer's point of view.
b. In light of the differences of the new reactors from the old, point out the flaws in the logic. Also feel free to comment on the author's choice of replacement vehicle.

What about the social and environmental realities today? These are every bit as important as the economic ones, if nuclear power is to be sustainable. For example, the economic analysis of Lester Brown that we just cited connects to the environmental issue of nuclear waste from spent fuel rods. Earlier we described the nuclear fuel cycle (see Figure 7.21). As the caption to this figure notes, today some parts of the cycle are not in operation. In particular, we are not reprocessing the nuclear waste of commercial power plants. In fact, we are not even storing it in a way that guarantees long-term separation from ecosystems.

Communities today have informed citizens who can speak to these environmental issues. Their voices are heard, especially those who live near proposed reactor sites. One of the social barriers to nuclear energy is the NIMBY problem—not in my backyard. Politicians are equally sensitive to it—not in my term of office. Citizens and politicians alike will certainly continue to debate the future of nuclear power. Actually, so will students and their teachers.

So where does that leave us? Our need for energy expands daily, as does the mass of radioactive waste with which we must cope. The era of cheap oil is over and the era of climate change has dawned. Yet the both real and perceived hazards associated with radioactivity, with mining and enriching uranium, and with nuclear weapons still remain. This presents a classic risk–benefit situation, and the final compromise has yet to be reached. For now, it is clear that nuclear power is not the cure-all for the world's energy woes. It may even be the cause of some environmental and societal woes. Even so, it is likely to remain a piece of the energy pie in the years to come.

Consider This 7.36   Second Opinion Survey

Now that you are near the end of your study of nuclear power, return to the personal opinion survey of Consider This 7.1 and answer these questions again. Then compare your new answers with your earlier ones. Are there any striking differences in your opinions? If so, what changed and how do you account for the difference(s)?

Conclusion

Over 50 years have passed since the first commercial nuclear power plant began producing electricity in the United States. The glittering promise of boundless, unmetered electricity, drawn from the nuclei of uranium atoms, has proved illusory. But the needs of our nation and our world for safe, abundant, and inexpensive energy are far greater today than they were in 1957. Therefore, scientists and engineers continue their atomic quest.

Where the search will lead is uncertain, but it is clear that people and politics will have a major say in ultimately making the decision. Reason, together with a regard for those who will inhabit our planet in both the near and far future, must govern our actions. Maybe Homer Simpson was right when he proclaimed, ". . . Lord, we are especially thankful for nuclear power, the cleanest, safest energy source there is. Except for solar, which is just a pipe dream." As it just so happens, we explore a little of the rationality behind that pipe dream—and other alternative energy sources as well—in the next chapter.

Chapter Summary

Having studied this chapter, you should be able to:

- Give an overview of the past and current use of nuclear power in the United States or another country of your choice (7.1)

- Report on the use of nuclear power for electricity generation worldwide (7.1)

- Explain the process of nuclear fission, the role of neutrons in sustaining a chain reaction, and the source of the energy it produces (7.2)

- Compare and contrast how electricity is produced in a conventional power plant and in a nuclear power plant (7.3)

- Compare the processes of alpha, beta, and gamma decay in terms of the changes that occur in the nucleus of the radioactive atom (7.4)

- Interpret the meaning of the word *radiation*, depending on the context (7.4)

- Explain how the radioactive decay of uranium-238 leads to the production of a series of radioisotopes. Also explain why naturally occurring radioisotopes such as carbon-14 and hydrogen-3 are *not* part of this series (7.4)

- Describe the accident at Chernobyl and explain why radioactive iodine was released and was hazardous to people (7.5)

- Rank the sources that contribute to your annual dose of radiation, both natural and human-made (7.6)

- Explain why nuclear radiation is also termed *ionizing radiation*. In your body, explain the connection between ionizing radiation and the production of free radicals (7.6)

- Some units describe the radioactive sample; others describe the damage it does to tissue. Use the curie, the rad, and the rem to illustrate this (7.6)

- The terms *enriched uranium* and *depleted uranium* are confusing to people. Describe this pair of terms in such a way that the general public could more easily grasp the similarities and differences (7.7)

- Do "back-of-the-envelope" half-life calculations for radioisotopes, being able to quickly determine how much radioactivity is left after time has passed (7.8)

- Apply the concept of half-life to the storage of nuclear waste (7.8)

- Evaluate radioisotopes in terms of their health hazards, discussing factors such as half-life, type of radioactive

decay, effect once in the body, and route of entry into the body. For example, compare radon-222, iodine-131, and strontium-90 (7.8)

- Describe the issues associated with the production and storage of high-level radioactive waste, including spent nuclear fuel (7.9)

- Take an informed stand on how high-level radioactive wastes should be handled and stored (7.9)

- Evaluate news articles on nuclear power and nuclear waste with confidence in your ability to understand the scientific principles involved (7.9–7.11)

- Describe the connections between nuclear power and nuclear weapons proliferation (7.9)

- Assess the risks and benefits in regard to the use of nuclear power (7.10)

- Take an informed stand on the use of nuclear power for electricity production (7.11)

- Outline the factors that favor or oppose the growth of nuclear energy in the next decade (7.11)

Questions

Emphasizing Essentials

1. Name two ways in which one carbon atom can differ from another. Then name three ways in which *all* carbon atoms differ from *all* uranium atoms.

2. The representations ^{14}N or ^{15}N give more information than simply the chemical symbol N. Explain.

3. **a.** How many protons are in the nucleus of this isotope of plutonium: $^{239}_{94}Pu$?

   b. The nuclei of all atoms of uranium contain 92 protons. Which elements have nuclei with 93 and 94 protons, respectively?

   c. How many protons do the nuclei of radon-222 contain?

4. Determine the number of protons and neutrons in each of these nuclei.

   a. ^{14}C, a naturally occurring radioisotope of carbon

   b. ^{12}C, a naturally occurring stable isotope of carbon

   c. ^{3}H, tritium, a naturally occurring radioisotope of hydrogen

   d. Tc-99, a radioisotope used in medicine

5. $E = mc^2$ is one of the most famous equations of the 20th century. Explain the meaning of each symbol in it.

6. Give an example of a nuclear equation and of a chemical equation. In what ways are the two equations alike? Different?

7. This nuclear equation represents a plutonium target being hit by an alpha particle. Show that the sum of the subscripts on the left is equal to the sum of the subscripts on the right. Then do the same for the superscripts.

$$^{239}_{94}Pu + {}^{4}_{2}He \longrightarrow [{}^{243}_{96}Cm] \longrightarrow {}^{242}_{96}Cm + {}^{1}_{0}n$$

8. For the nuclear equation shown in the previous question,

   a. suggest the origin of the $^{4}_{2}He$ particle.

   b. $^{1}_{0}n$ is a product. What does this symbol represent?

   c. Curium-243 is written in square brackets. What does this notation convey? *Hint:* See equation 7.1.

9. Californium, element number 98, was first synthesized by bombarding a target with alpha particles. The products were californium-245 and a neutron. What was the target isotope used in this nuclear synthesis?

10. Explain the significance of neutrons in initiating and sustaining the process of nuclear fission. In your answer, define and use the term *chain reaction*.

11. Nuclear fission occurs through many different pathways. For the fission of U-235 induced by a neutron, write a nuclear equation to form:

   a. bromine-87, lanthanum-146, and more neutrons.

   b. a nucleus with 56 protons, a second with a total of 94 neutrons and protons, and 2 additional neutrons.

12. This schematic diagram represents the reactor core of a nuclear power plant.

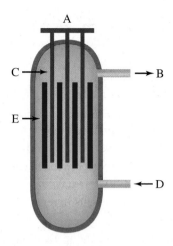

Match each letter in the figure with one of these terms.

   fuel rods

   cooling water into the core

   cooling water out of the core

   control rod assembly

   control rods

13. Identify the segments of the nuclear power plant diagrammed in Figure 7.7 that contain radioactive materials and those that do not.

14. Explain the difference between the primary coolant and the secondary coolant. The secondary coolant is not housed in the containment dome. Why not?

15. Boron can absorb neutrons.

  a. Write the nuclear equation in which boron-10 absorbs a neutron to produce lithium-7 and an alpha particle.

  b. Boron, like cadmium, can be used in control rods. Explain.

16. What is an alpha particle and how is it represented? Answer these same questions for a beta particle and a gamma ray.

17. Plutonium-239 decays by alpha emission (no gamma ray), and iodine-131 decays by beta emission (an accompanying gamma ray).

  a. Write the nuclear equation for each.

  b. Plutonium is most hazardous when inhaled in particulate form. Explain.

  c. Iodine-131 can be hazardous if ingested. Where do all isotopes of iodine accumulate in the body?

  d. Would you expect the radioactivity of a sample of each isotope to decrease to background level on a timescale of hours, days, years, or thousands of years? Explain. *Hint:* See Table 7.4.

18. Radioactive decay is accompanied by a change in the mass number, a change in the atomic number, a change in both, or a change in neither. For the following types of radioactive decay, which change(s) do you expect?

  a. alpha emission

  b. beta emission

  c. gamma emission

19. Figure 7.12 shows the radioactive decay series for U-238. Analogously, U-235 decays through a series of steps (α, β, α, β, α, α, α, β, α, β, α) to reach a stable isotope of lead. For practice, write nuclear reactions for the first six. Although some steps are accompanied by a gamma ray, you may omit this. *Hint:* The result is an isotope of radon.

20. Given that the average U.S. citizen receives 3600 μSv of radiation exposure per year, use the data in Table 7.2 to calculate the percentage of radiation exposure the average U.S. citizen receives from each of these sources.

  a. food, water, and air

  b. a dental X-ray twice a year

  c. the nuclear power industry

21. What percent of a radioactive isotope would remain after two half-lives, four half-lives, and six half-lives? What percent would have decayed after each period?

22. Estimate the half-life of radioisotope X from this graph.

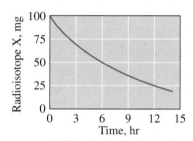

23. a. Is depleted uranium (DU) radioactive? Explain.

  b. Is spent nuclear fuel (SNF) radioactive? Explain.

Concentrating on Concepts

24. Alchemists in the Middle Ages dreamed of converting base metals, such as lead, into precious metals—gold and silver. Why could they never succeed? Today could we convert lead into gold? Explain.

25. Make a time line of nuclear history, putting at least a dozen dates on your line. For example, start with Becquerel's discovery of radioactivity in 1896. Other candidates for inclusion are Chernobyl, Hiroshima, the opening of the first commercial reactor, the discovery of various medical isotopes, the use of uranium glazes in Fiestaware, and the Nuclear Test Ban Treaty.

26. The isotopes U-235 and U-238 are alike in that they are both radioactive. However, these two isotopes have very different abundances in nature. List their natural abundances and explain the significance of this difference.

27. Consider the uranium fuel pellets used in commercial nuclear power plants.

    a. Describe one way in which U-235 and U-238 can be separated.

    b. Why is it necessary to enrich the uranium for use in the fuel pellets?

    c. Fuel pellets are enriched only to a few percent, rather than to 80–90%. Name three reasons why.

    d. Explain why it is not possible to separate U-235 and U-238 by chemical means.

28. a. Why must the fuel rods in a reactor be replaced every few years?

    b. What happens to the fuel rods after they are taken out of the reactor?

29. At full capacity, each reactor in the Palo Verde power plant uses only a few pounds of uranium to generate 1243 megawatts of power. To produce the same amount of energy would require about 2 million gallons of oil or about 10,000 tons of coal in a conventional power plant. How is energy produced in the Palo Verde plant, compared with conventional power plants?

30. One important distinction between the Chernobyl reactors and those in the United States is that those in Chernobyl used graphite as a moderator to slow neutrons, whereas U.S. reactors use water. In terms of safety, give two reasons why water is a better choice.

31. If you look at nuclear equations in sources other than this textbook, you may find that the subscripts have been omitted. For example, you may see an equation for a fission reaction written this way.

$$^{235}U + \ ^{1}n \longrightarrow [^{236}U] \longrightarrow \ ^{87}Br + \ ^{146}La + 3 \ ^{1}n$$

    a. How do you know what the subscripts should be? Why can they be omitted?

    b. Why are the superscripts *not* omitted?

32. Using the model of a neutron presented in equation 7.6, explain how a high-speed electron can be ejected from the nucleus in beta decay.

33. Coal can contain trace amounts of uranium. Explain why thorium must be found in coal as well.

34. Suppose somebody tells you that a radioisotope is gone after 10 half-lives. Critique this statement, explaining why it could be a reasonable assumption for a small sample, but might not be for a large one.

35. "Bananas are radioactive!" A vice president of nuclear services made this comment in a public lecture in the context of comparing the different sources of radiation to which people are exposed.

    a. Why might he have made such an assertion?

    b. Suggest a better way to have phrased this.

    c. Should you stop eating bananas because they are radioactive? Explain.

36. a. A website describing an X-ray procedure reports, "Despite its negative connotations, people are exposed to more radiation on a daily basis than they may realize. For example, infrared radiation is released whenever there is extreme heat. The Sun generates ultraviolet radiation, and a little exposure to it will tan a lighter skinned person. In addition, the body contains naturally radioactive elements." Examine the three examples given in this explanation. Do they refer to nuclear or electromagnetic radiation?

    b. What type of radiation do you avoid in attending class at MATC, Madison Area Technical College?

37. Consider this representation of a Geiger–Müller counter (also called a Geiger counter), a device commonly used to detect ionizing radiation. The probe contains a gas under low pressure.

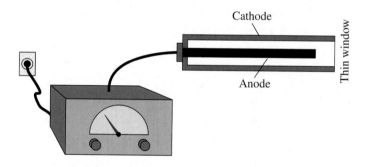

    a. How does radiation enter the Geiger–Müller counter?

    b. Why does this device only detect radiation that is capable of ionizing the gas contained in the probe?

    c. What other methods are used to detect the presence of ionizing radiation?

38. Rapidly dividing cells are present in several places in the adult body. These include the skin, the hair follicles, the stomach and intestines, the lining of your mouth, and your bone marrow. Match the symptoms listed in Table 7.3 with the type of cell affected by the radiation.

39. Exposure to ionizing radiation can cause cancer. A beam of ionizing radiation also can be used to cure certain types of cancer. Explain.

40. Fluorine has only one naturally occurring radioisotope, F-19. If fluorine also occurred in nature as F-18, would this necessarily complicate the separation of $^{238}UF_6$ and $^{235}UF_6$? Explain.

41. It is generally believed that terrorists would be more likely to construct a nuclear bomb using Pu-239 reclaimed from breeder reactors than using U-235. Use your knowledge of nuclear chemistry to explain why.

42. Weapons-grade plutonium is almost completely Pu-239. In contrast, the plutonium produced in the normal operation of a water-cooled power reactor (reactor-grade plutonium) generally has a higher concentration of heavier isotopes such as Pu-240 and Pu-241. Propose an explanation for this observation.

43. a. What are the characteristics of high-level radioactive waste (HLW)?

    b. Explain how low-level waste (LLW) differs from HLW.

Exploring Extensions

44. In Consider This 7.1, you were asked to answer several questions about nuclear power. Ask the same questions of someone at least one generation older than you and someone younger. In comparison with your answer, what similarities and differences do you find?

45. The film King Corn (see Consider This 12.2) opens with a scene from Professor Stephen Macko at the University of Virginia, a forensic chemist. He analyzed hair samples from two college students, reporting that much of the carbon in their body originated from corn. His analysis was based on carbon-13.

    a. Is this a stable or a radioactive isotope of carbon?

    b. What can be learned about your diet by analyzing samples of hair?

46. Explain the term *decommission*, as in "decommissioning a nuclear power plant." What technical challenges are involved? The resources of the web can help you.

47. Einstein's equation, $\Delta E = \Delta mc^2$ applies to chemical reactions as well as nuclear ones. An important chemical change studied in Chapter 4 was the combustion of methane, which releases 50.1 kJ of energy for each gram of methane burned.

    a. What mass loss corresponds to the release of 50.1 kJ of energy?

    b. To produce the same amount of energy, what is the ratio of the mass of methane burned in a chemical reaction to the mass loss converted into energy according to the equation $\Delta E = \Delta mc^2$?

    c. Use your results in parts **a** and **b** to comment on why Einstein's equation, although correct for both chemical and nuclear changes, usually is only applied to nuclear changes.

48. When 4.00 g of hydrogen nuclei undergoes fusion to form helium in the Sun, the change in mass is 0.0265 g and energy is released. Use Einstein's equation, $\Delta E = \Delta mc^2$, to calculate the energy equivalent of this mass change.

49. Under conditions like those on the Sun, hydrogen can fuse with helium to form lithium, which in turn can form different isotopes of helium and of hydrogen. The mass of one mole of each isotope is given.

$$^2_1H + \,^3_2He \longrightarrow [^5_3Li] \longrightarrow \,^4_2He + \,^1_1H$$

2.01345 g   3.01493 g   4.00150 g   1.00728 g

    a. In grams, what is the mass difference between the reactants and the products?

    b. For one mole of reactants, how much energy (in joules) is released?

50. Lise Meitner and Marie Curie were both pioneers in developing an understanding of radioactive substances. You likely have heard of Marie Curie and her work, but may not have heard of Lise Meitner. How are these two women related in time and in their scientific work?

51. Taking potassium iodide tablets can protect your thyroid from exposure to radioactive iodine, thus reducing your risk of thyroid cancer.

    a. Give the chemical formula for potassium iodide.

    b. By what mechanism does potassium iodide protect you?

    c. How long does the protection last?

    d. Are the tablets expensive? *Hint:* The FDA website is a good source of information for parts **b** and **c.**

52. A stockpile of approximately 50 metric tons of plutonium exists in the United States as a result of disassembling warheads from the nuclear arms race. What is the likely fate of this plutonium? *Hint:* Search for *plutonium disposal*. Try also including *United States* and *Department of Energy* in your search string.

    a. Some propose that the plutonium be sent to local nuclear power plants to "burn" as fissionable fuel. What are the advantages and disadvantages of such a course of action?

    b. Others propose that it be stored permanently in a repository. Again, list the advantages and disadvantages.

53. Advertisements for Swiss Army watches stress their use of tritium. One ad states that the "... hands and numerals are illuminated by self-powered tritium gas, 10 times brighter than ordinary luminous dials. . . ." Another advertisement boasts that the "... tritium hands and markers glow brightly making checking your time a breeze, even at night. . . ." Evaluate these statements and, after doing some web research, discuss the chemical form of tritium in these watches, and what its role is.

54. Nuclear weapons are not the only threat. Consider also "the dirty bomb," a device that employs a conventional explosive to disperse a radioactive substance. No fission is involved with a dirty bomb; only a conventional explosive.

    a. What radioisotopes might be used in a dirty bomb?

b. A brochure on nuclear terrorism makes this assertion: "A nuclear weapon, if exploded, would create more radioactive substances than originally present in the weapon. In contrast, if a dirty bomb were to be exploded, the amount of radioactivity would be the same before, during, and right after the explosion." This statement is accurate. Explain why.

55. According to Table 7.2, smoking 1.5 packs of cigarettes a day adds 15,000 μSv to your annual radiation dose.

    a. Polonium-210 is the radioactive element largely responsible. What is its mode of radioactive decay and its half-life?

    b. Why is polonium-210 found (in tiny amounts) in tobacco?

56. MRI, or magnetic resonance imaging, is an important tool for some types of medical diagnoses.

    a. The science behind MRI is complex. You should, however, be able to pin down whether or not MRI uses ionizing radiation to produce an image. Does it?

    b. How does an MRI compare with a CT (computed tomography) scan, in terms of the image produced and the radiation used?

    c. MRI is based on NMR, nuclear magnetic resonance. Speculate why the abbreviation MRI is used to denote the medical tool rather than NMR.

57. Deciding where to locate a nuclear power plant requires analysis of both risks and benefits associated with the plant. If you were to play the role of a CEO of a major electric utility considering whether to pursue permits for the construction of a nuclear power plant in your area, what risks and benefits would you cite?

58. The hormesis phenomenon, defined in Section 7.6, is that toxic substances in small amounts can increase one's resistance to the same substance in large amounts. Analogous to the dose-response curves of Figure 7.19, the figure here indicates the zone of hormesis where the curve dips *below* the control line into a therapeutic region. Use the resources of the web to investigate hormesis. Prepare a summary of your findings. One starting point is an article, "Is Radiation Good for You?" at the science website *The Why? Files*.
Hint: NOAEL means no observed adverse effect level.

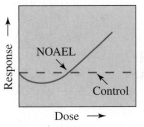

8

Energy from Electron Transfer

"In going on with these Experiments, how many pretty systems do we build, which we soon find ourselves oblig'd to destroy! If there is no other Use discover'd of Electricity, this, however, is something considerable, that it may help to make a vain Man humble."

Benjamin Franklin, Statesman, Scientist, Inventor, Diplomat (1706–1790)

We hope you never tempt fate by flying a kite in a thunderstorm with a key attached to the string. Ben Franklin probably never sent a key aloft either, although he did prove that lightning was electric in nature. As Franklin himself commented, the power of electricity is something *"to make a vain Man humble."* Indeed, a bolt of lightning is one of nature's grandest displays of electrical energy. You may have jumped at the crack of a nearby lightning strike. Inside your body, a cascade of electrons as part of cellular metabolism powered your startled response. Thus both a lightning strike and your physiological response to it involve processes driven by the flow of electrons. Clearly, our world is naturally electric!

People also have built additional electric systems. We rely on a flow of electrons—better known as electricity—to heat or cool our living and workspaces, to provide light to read by, and to power our TV sets. For most of us, the electricity we use is generated at centralized power plants, such as those fueled by fossil fuels (see Chapter 4) or fissionable isotopes (see Chapter 7). To a lesser extent, we also rely on wind, sun, and geothermal, as well as the potential energy of water trapped by dams as sources to generate electric power.

Additionally, we have created convenient-sized portable sources of electricity, better known as batteries. These long-lasting and reliable devices fill a special energy niche. They power our cell phones, MP3 players, laptops, and perhaps even our hearing aids and motorized wheelchairs. In order to start thinking about your own personal battery use, complete this activity.

In 1752, Benjamin Franklin demonstrated that lightning is electricity. He also designed the lightning rod.

Look for more about metabolism in Chapter 11, the food chapter.

Consider This 8.1   Personal Battery Use

Many devices, both large and small, contain electrochemical cells that people usually call "batteries." Create a table with four columns. Head these with the labels: Device, Battery Use, Rechargeable, and Recyclable.

a. Fill in the "Device" column with at least four items powered by batteries.
b. Some devices use a battery as the main source of power; others use it as a backup. Categorize the items in your table.
c. Some batteries are rechargeable; others are not. Again categorize the items.
d. When the battery runs down, do you throw it out, recycle it, or give it to a dealer to recycle? Fill out the last column of your table. We explore the challenges of battery recycling later in this chapter.

If you are like many students we know, you may carry a cellular phone (lithium-ion battery, rechargeable), own a wristwatch (mercury battery), snap photos with a digital camera (nickel–cadmium, or Ni-Cd, battery, rechargeable), and punch numbers into a calculator (alkaline battery). You may even own a laptop computer with a lithium-ion rechargeable battery and perhaps drive a car with a lead–acid storage battery.

Not everyone in the world has access to these consumer items. In fact, the International Energy Agency estimated in 2005 that 1.6 billion people, approximately a quarter of the world's population, lacked access to electricity. Increasingly, people worldwide want electric lights and appliances. Indeed, in 2009 the U.S. Energy Information Administration projected that electricity will supply an increasing share of the world's total energy demand and is the fastest growing use of energy worldwide.

This growth will place an ever-increasing demand on our natural resources. There are practical limits to the long-term availability of both fossil fuels and the metals used to power batteries. Fissionable isotopes, though available, are a difficult fuel to obtain. Moreover, all fuels come with environmental and societal price tags, sometimes referred to as their "external costs." The combustion of coal, petroleum products, and natural gas releases vast quantities of carbon dioxide, a significant contributor to global warming. The combustion of fossil fuels also releases sulfur dioxide and nitrogen oxides, leading to decreases in air quality and increases in health costs. Processing uranium or breeding plutonium create both low-level and high-level nuclear waste. Additional high-level nuclear waste from spent nuclear fuel must be safely stored for generations to come (see Chapter 7).

The conclusion seems obvious. If we are to continue to inhabit this planet and not compromise the ability of future generations to meet their needs, we must develop and depend on other sources of energy. We also must better match our current batteries (and other sources of electricity) with their end uses. The principles of green chemistry and cradle-to-cradle stewardship can help inform our national approaches as well as our day-to-day activities.

Electron transfer! In this chapter we look at several power sources that derive energy through electron transfer technology. These include batteries for portable devices, automobiles, fuel cells, and solar photovoltaic power. We begin with the basics of batteries.

8.1 | Batteries, Galvanic Cells, and Electrons

Batteries are a big and growing business worldwide due to consumer demand for products that require them (Figure 8.1). Many consumer products require batteries, spurring continued growth in the battery industry. Although we commonly use the word *battery*, a standard flashlight "battery" is more correctly called a **galvanic cell.** This is a type of electrochemical cell that converts the energy released in a spontaneous chemical reaction into electrical energy. A collection of several galvanic cells wired together constitutes a true battery.

All galvanic cells produce useful energy through the transfer of electrons from one substance to another. For this transfer process, you can write an overall chemical equation. In turn, this can be divided—split if you like—into two parts. One is for **oxidation,** a process in which a chemical species loses electrons. The other is for **reduction**, a process in which a chemical species gains electrons. We refer to these two parts as "half-reactions" in the sense that each represents half of the overall process occurring in the galvanic cell. More formally, a **half-reaction** is a type of chemical equation that shows the electrons either lost or gained by the reactants.

Half-reactions are a bit different from the chemical equations that we used earlier in this text. First, they always occur in pairs. Secondly, they include electrons! Even though electrons cannot really be poured from a bottle into a flask, it still is helpful to show them in half-reactions so that you can better understand what is taking place. Note that the electrons show up either on the right or the left side of the half-reaction, but not on both. If on the right side, then the reactant has lost electrons; this is an oxidation half-reaction. In contrast, if the electrons are on the left side of the half-reaction, then the reactant is gaining electrons, and this is a reduction half-reaction.

Think of a battery as a collection of related things, such as a battery of tests or a battery of artillery cannon.

oxidation = loss of electrons
reduction = gain of electrons
A mnemonic device, LEO the lion says GER, can help you to remember that Loss of Electrons is Oxidation and Gain of Electrons is Reduction.

Frank and Ernest

© Tom Thaves. Reprinted with permission.

Figure 8.1
A humorous although realistic view of how batteries link to consumer products.

As an example, consider a simplified version of the reaction that takes place in a nickel–cadmium (Ni-Cd, or "nicad") battery:

oxidation half-reaction: $Cd \longrightarrow Cd^{2+} + 2\,e^-$ **[8.1]**

reduction half-reaction: $2\,Ni^{3+} + 2\,e^- \longrightarrow 2\,Ni^{2+}$ **[8.2]**

Ni-Cd is an abbreviation, not a chemical formula for the nickel–cadmium battery. It is pronounced "NYE-cad."

In this case, two electrons are given off, or "lost," in the oxidation half-reaction. Where do they go? These electrons were transferred to the ion being reduced. The number of electrons given off during oxidation must equal the number of electrons gained through reduction for the overall equation to balance. For this reason, the coefficient "2" appears in the reduction half-reaction (see equation 8.2).

We now can add the two half-cell reactions to obtain the overall equation:

$$2\,Ni^{3+} + Cd + 2\,e^- \longrightarrow 2\,Ni^{2+} + Cd^{2+} + 2\,e^- \qquad \textbf{[8.3]}$$

The electrons that appear on both sides of equation 8.3 cancel, as the electrons "lost" by the cadmium metal are gained by the nickel ions. So we can rewrite the overall cell equation as:

overall cell equation: $2\,Ni^{3+} + Cd \longrightarrow 2\,Ni^{2+} + Cd^{2+}$ **[8.4]**

Your Turn 8.2   **Electrons in Half-Reactions**

Categorize each as an oxidation half-reaction or a reduction half-reaction. Explain your reasoning.

a. $Al^{3+} + 3\,e^- \longrightarrow Al$

b. $Zn \longrightarrow Zn^{2+} + 2\,e^-$

c. $Mn^{7+} + 3\,e^- \longrightarrow Mn^{4+}$

d. $2\,H_2O \longrightarrow 4\,H^+ + O_2 + 4\,e^-$

e. $2\,H^+ + 2\,e^- \longrightarrow H_2$

Answer

a. Reduction. The aluminum ion gained three electrons to become aluminum in its elemental form; that is, aluminum metal (no charge).

Figure 8.2

This 7.2-V Ryobi portable power drill comes with two Ni-Cd battery packs and a recharging unit.

Potential energy was first introduced in Section 4.1.

The movement of electrons through an external circuit produces **electricity,** the flow of electrons from one region to another that is driven by a difference in potential energy. The electrochemical reaction provides the energy needed to drive a cordless razor, a power tool, or countless other battery-operated devices. The chemical species oxidized and reduced in the cell must be connected in such a way to allow electrons released during the oxidation to transfer to the reactant being reduced while following an appropriate electrical path for the desired application.

Electrodes, electrical conductors within a cell that serve as sites for chemical reactions, facilitate this electron transfer. At the **anode,** oxidation takes place and is the source of electrons in the current flow. At the **cathode,** reduction takes place. The cathode receives the electrons sent from the anode through the external circuit to complete the reduction. Once the electrical circuit is completed, then a **voltage** can be measured across the cell, that is, the difference in electrochemical potential between the two electrodes. Voltage is measured in units called volts (V). The greater the difference in potential between the two electrodes, the higher the voltage and the greater the energy associated with the electron transfer. For example, with a Ni-Cd cell, the maximum difference in electrochemical potential under the conditions specified is measured as 1.2 V. In contrast, alkaline cells deliver 1.5 V, mercury cells 1.35 V, and lithium ion cells are capable of potentials in excess of 4 V! In order to produce the higher voltages necessary to power larger devices (for example, power tools or automobile starter motors) several cells must be connected (Figure 8.2).

The chemical reaction that takes place in a Ni-Cd cell is more complicated than represented in equations 8.1–8.4. Cadmium metal, at the anode, contains atoms of cadmium that are oxidized to Cd^{2+}. These in turn combine with OH^- to form $Cd(OH)_2$.

anode = oxidation
cathode = reduction
Note that one pair of terms begins with a vowel and the other set begins with a consonant.

The unit "volt" honors the Italian physicist Alessandro Volta (1745–1827). He is credited with inventing the first electrochemical cell in 1800.

Electrolytes were first introduced in Section 5.6.

Simultaneously, Ni^{3+}, present in the hydrated form of $NiO(OH)$ on the nickel cathode, is reduced to Ni^{2+} in the chemical form of $Ni(OH)_2$. A water-based electrolyte paste containing a highly concentrated solution of the strong base $NaOH$ or KOH separates the electrodes and allows the flow of charge.

oxidation half-reaction (anode):

$$Cd(s) + 2\,OH^-(aq) \longrightarrow Cd(OH)_2(s) + 2\,e^- \qquad \textbf{[8.5]}$$

reduction half-reaction (cathode):

$$2\,NiO(OH)(s) + 2\,H_2O(l) + 2\,e^- \longrightarrow 2\,Ni(OH)_2(s) + 2\,OH^-(aq) \qquad \textbf{[8.6]}$$

overall cell equation (sum of the two half-reaction):

$$Cd(s) + 2\,NiO(OH)(s) + 2\,H_2O(l) \longrightarrow 2\,Ni(OH)_2(s) + Cd(OH)_2(s) \qquad \textbf{[8.7]}$$

These three equations show exactly the same transfer of electrons represented in equations 8.1–8.4, but now different states and chemical forms are indicated. Figure 8.3 illustrates an inner view of a Ni-Cd galvanic cell (a "battery").

Your Turn 8.3   Checking Balance and Charge

Consider equations 8.1–8.7, which include both chemical equations and half-reactions.

   a. Is each equation balanced in terms of the number of atoms; that is, by the law of conservation of matter and mass? Explain.
   b. Does each equation have the same amount of electrical charge on both sides? Explain.
   c. Name a quick way to distinguish an overall cell equation from a half-reaction.

A Ni-Cd "battery" is really a single galvanic cell, rather than several (= a battery). But given the common use of the word battery, from now on we will simply refer to it as a battery.

A Ni-Cd "battery" is rechargeable, an added advantage for many applications. A rechargeable battery employs electrochemical reactions that can run in both directions. The transfer of electrons takes place both during the forward (discharging) and the reverse (recharging) processes.

$$Cd(s) + 2\,NiO(OH)(s) + 2\,H_2O(l) \underset{\text{recharging}}{\overset{\text{discharging}}{\rightleftharpoons}} 2\,Ni(OH)_2(s) + Cd(OH)_2(s) \quad \textbf{[8.8]}$$

What feature makes a battery rechargeable? The key is that both the reactants and products are solids. Furthermore, the solid products cling to a stainless-steel grid within the battery rather than dispersing. If a voltage is applied to this grid, these products can be converted back to reactants, thus recharging the battery. Although a rechargeable battery can be discharged and recharged many times, eventually the accumulation of impurities, breakdown of the separators, or generation of unwanted side-reaction by-products ends its useful life.

Batteries come in many shapes and sizes, each one uniquely matched to its use. For example, in an application like a hearing aid, the size and weight of the cell is of paramount importance. In contrast, an automobile battery must last for years and perform over a range of temperatures. To be successful in the eyes of today's consumers, batteries must be affordable, last a reasonable length of time, and be safe to use and recharge. Ultimately, to be successful in the years to come, batteries also must be designed so that their materials can be recycled in a sustainable way.

Most electrochemical cells convert chemical energy into electric energy with an efficiency of about 90%. Compare this with the much lower efficiencies of 30–40% that characterize coal-fired power plants that generate electricity. Recognize, though, that electricity from these plants is used to recharge batteries. This is but one of many incentives to explore renewable energy sources.

With the exception of lead–acid batteries used for automobile starter motors, aqueous solutions usually are too hazardous to use in batteries because sooner or later they leak from the battery casing. For example, you may have seen the corrosive mess inside of a flashlight or child's toy from a leaking battery. However, in the chemistry

Cathode, $NiO(OH)(s)$ | Anode, $Cd(s)$

Separator, $KOH(aq)$ paste

Figure 8.3

Representation of a Ni-Cd galvanic cell showing how the components are layered to increase the surface area of the electrodes.

laboratory, you can safely use aqueous solutions to build a galvanic cell. Many different combinations are possible! For example, ask your instructor if you can set up the copper–zinc cell shown in the next activity.

Your Turn 8.4   Galvanic Cells in the Lab

As this cell operates, a reddish coating of copper metal begins to appear on the surface of the copper cathode. The overall equation for the chemical reaction is given below the diagram.

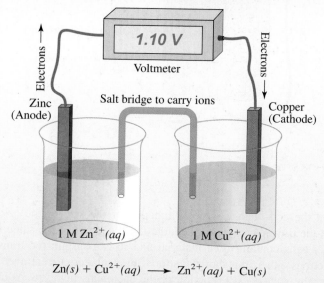

$$Zn(s) + Cu^{2+}(aq) \longrightarrow Zn^{2+}(aq) + Cu(s)$$

a. Write the oxidation half-reaction taking place at the anode.
b. Write the reduction half-reaction taking place at the cathode.
c. This laboratory galvanic cell is not rechargeable. Explain why.

The overview of galvanic cells provided in this section is the start of a much longer story. We continue the tale in the next section.

8.2 | Other Common Galvanic Cells

Almost everyone has inserted an alkaline battery into a flashlight, calculator, or digital camera. You may recognize the ones shown in Figure 8.4. One end of each of these batteries is marked with a + sign; the other end with a − sign. These markings point to the fact that electron transfer is at work. Alkaline cells each produce 1.5 V, but the larger ones can sustain a current through the external circuit for a longer time. The **current**, or rate of electron flow, is measured in amperes (amps, A) or likely in milliamps (mA) for smaller cells.

The voltage of a battery is primarily determined by its chemical composition. The alkaline cell is based on chemical reactions involving zinc and manganese (Figure 8.5). The cell is called "alkaline" because it operates in a basic, rather than acidic medium. The half-reactions for this cell are:

oxidation half-reaction (anode):

$$Zn(s) + 2\,OH^-(aq) \longrightarrow Zn(OH)_2(s) + 2\,e^- \qquad \textbf{[8.9]}$$

reduction half-reaction (cathode):

$$2\,MnO_2(s) + H_2O(l) + 2\,e^- \longrightarrow Mn_2O_3(s) + 2\,OH^-(aq) \qquad \textbf{[8.10]}$$

overall cell equation (sum of the two half-reactions):

$$Zn(s) + 2\,MnO_2(s) + H_2O(l) \longrightarrow Zn(OH)_2(s) + Mn_2O_3(s) \qquad \textbf{[8.11]}$$

You can review alkaline (basic) solutions in Section 6.4.

Current is measured in amperes (amps, A) to honor André Ampère (1775–1836). He was a self-taught French mathematician who worked on electricity and magnetism.

Figure 8.4

These alkaline cells of size AAA to D all produce 1.5 V.

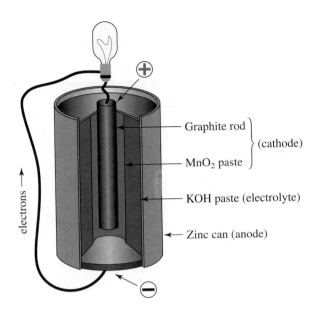

Figure 8.5

Representation of an alkaline cell.

Note that the cell voltage is not a function of the size of the cell. All alkaline batteries, from the tiny AAA size to the large D cells, produce the same voltage, 1.5 V. However, larger cells contain more material and so can sustain the transfer of a large number of electrons either in a short burst or with a smaller current over a longer time.

However, the voltage *is* a function of the chemicals involved. You can see from the examples listed in Table 8.1 that different voltages are produced using different chemical systems. Only a few volts are possible with a single galvanic cell. But as we noted in the previous section, higher voltages are possible by connecting cells. For example, in order to run a 14.4 or 19.2 V power drill, manufacturers sell a "battery pack" that contains multiple cells.

Compact, long-lasting cells may find their way into your body. For example, the widespread use of cardiac pacemakers is largely due to the improvements made in the electrochemical cells rather than in the pacemakers themselves. Lithium–iodine cells are so reliable and long-lived that they are often the battery of choice for this application, lasting as long as 10 years before needing to be replaced.

Lithium batteries take advantage of the low density and large oxidation potential of lithium metal to make a lightweight battery with large energy output.

Table 8.1	Some Common Galvanic Cells		
Type	**Voltage (V)**	**Rechargeable**	**Examples of Uses**
alkaline	1.5	no	flashlights, small appliances, some calculators
lithium–iodine	2.8	no	pacemakers
lithium–ion	3.7	yes	laptop computers, cell phones, digital music players, power tools
lead–acid	2.0	yes	automobile batteries
nickel–cadmium (Ni-Cd)	1.3	yes	toys and portable electronic devices including digital cameras, power tools
nickel–metal hydride (NiMH)	1.3	yes	replacing Ni-Cd for many uses; hybrid vehicles batteries
mercury	1.3	no	once used widely in cameras, watches, and hearing aids, but now are banned or being phased out

Consider This 8.5   Can You Swing a Hammer?

Shifting baselines, introduced in Chapter 0, is the idea that what people perceive as normal has changed over time, especially with regard to the ecosystems on our planet. The idea also can be applied more generally.

a. "Carpenters today no longer know how to swing a hammer!" Of course a good carpenter can still drive a nail with a single blow, but this lament indeed has some truth to it. Power tools now provide much of the muscle. With an eye to shifting baselines, interview someone old enough to be able to tell you stories of carpentry before power tools. Write a brief summary.

b. Power tools are but one of the many uses for batteries listed in Table 8.1. Select a different use and propose at least three ways in which what people consider as normal has shifted because batteries are used.

Parking meters were unknown before 1935.

Found under the front hood of a car, the lead–acid battery is the workhorse of today's rechargeable batteries. It is an excellent example of a battery that has changed what is "normal" for people now as compared with a century ago, as it was one of several technological advances that allowed the rise of the automobile. Cars indeed have changed our world! For example, parking ramps and parking meters were unknown a hundred years ago. You could not have bought an infant car seat or windshield washer fluid, because these items did not yet exist. Gas stations did not dot the landscape. And the waste products of gasoline combustion were not dirtying the air we breathe or warming our planet.

Today, however, we take the lead–acid battery for granted. In an automobile, it powers an electric motor that replaced the hand crank that people once used to start their cars. It is a true battery since it consists of six electrochemical cells, each generating 2.0 V for a total of 12.0 V (Figure 8.6). Here is the overall chemical equation, the sum of the two half-reactions.

$$\underset{\text{lead}}{Pb(s)} + \underset{\text{lead dioxide}}{PbO_2(s)} + \underset{\text{sulfuric acid}}{2\,H_2SO_4(aq)} \underset{\text{recharging}}{\overset{\text{discharging}}{\rightleftharpoons}} \underset{\text{lead(II) sulfate}}{2\,PbSO_4(s)} + \underset{\text{water}}{2\,H_2O(l)} \qquad \textbf{[8.12]}$$

As the arrows in equation 8.12 indicate, when the chemical reaction proceeds to the right, the battery is discharging. For example, using the battery to start the car

Lead oxide is the common name for lead (IV) oxide. We will use the common name in this chapter.

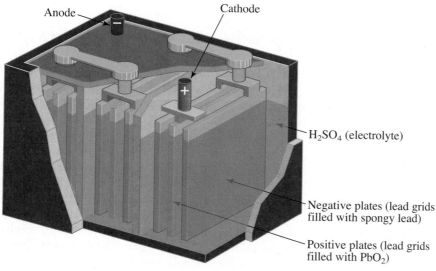

Anode —
Cathode —
H_2SO_4 (electrolyte)

Negative plates (lead grids filled with spongy lead)

Positive plates (lead grids filled with PbO_2)

Figure 8.6
Cutaway view of a lead–acid storage battery.

Older lead–acid batteries required addition of water and therefore were designed with removable screw caps.

discharges the battery. Running the lights or radio without the engine on also discharges it. But once the engine is running, an alternator, turned by the engine, provides the current needed to reverse the chemical reaction and recharge the battery. Fortunately, the battery can discharge and be recharged many times before it needs to be replaced. A high-quality battery can perform for five years or more!

Your Turn 8.6   The Battery in Your Car

Let's take a closer look at the lead–acid storage battery, the one found in most cars (see equation 8.12).

a. Lead occurs in this equation as Pb, PbO_2, and $PbSO_4$, all solids. In which of these is lead in an ionic form? What is the ion? Which one of these represents lead in its metallic form?

b. When lead is converted from its metallic form to an ionic form, are electrons lost or gained? Is this oxidation or reduction?

c. When the battery is discharging, is lead oxidized or reduced?

Answer

a. Pb(s) is lead in its elemental form (a metal), whereas both $PbO_2(s)$ and $PbSO_4(s)$ are compounds that contain the lead ion, Pb^{2+}.

Because lead–acid batteries have the advantage of being rechargeable and low in cost, they may be used together with wind turbine electric generators. The generator recharges the batteries during favorable winds; the batteries discharge during unfavorable wind conditions. You can also find lead–acid batteries in environments where the emissions from internal combustion engines cannot be tolerated. The forklifts in warehouses, the passenger carts in airports, and the electric wheelchairs in supermarkets typically are powered by lead–acid storage batteries. Their weight may even be an advantage in stabilizing these vehicles.

In an automobile, however, the weight of the lead–acid battery is a disadvantage. Another disadvantage is the chemical components of the battery. The anode (lead metal), cathode (lead dioxide), and the electrolyte (sulfuric acid solution) pose disposal challenges as toxic or corrosive chemicals. If batteries are to be used sustainably, we must meet these challenges. The next section speaks more directly to the components of batteries and where they do (and should) end up when the battery is disposed of.

8.3 | Battery Ingredients: Cradle-to-Cradle

Can you recall when you last went an entire day without using a device powered by a battery? Batteries have made cell phones, MP3 players, laptops, and hand-held calculators so commonplace that we tend to take them for granted. Developing technologies also rely heavily on batteries. For example, batteries are a key component of hybrid vehicles. Off-grid scale solar energy installations also require the use of batteries to deliver power at night.

Yet the battery in your cell phone, car, or even in a solar installation costs more than what you pay for it in the store. There also is an environmental price tag; that is, an "external cost" that is borne by all. Part of this stems from the "ingredients" found in just about any battery, namely, one or more metals. These metals must be mined from the Earth and refined from the ores in which they occur. The process of mining is energy-intensive and produces mine tailings and other waste. The refining process also requires energy and produces pollutants. For example, metal refining often results in the release of sulfur dioxide because so many metals occur naturally as sulfides. The next activity gives you the opportunity to examine the details of a metal-refining process.

Smelting is the process of heating and chemically processing an ore. The smelting of sulfide ores was mentioned in the context of air quality (see Section 1.11) and acid rain (see Section 6.7).

Your Turn 8.7   Metal Refining (Smelting)

The Ni-Cd battery in your digital camera requires two metals: nickel and cadmium. These metals are smelted from sulfur-containing ores such as NiS and CdS.

a. Name three attributes that distinguish a metal from a nonmetal.
Hint: Review Sections 1.6 and 5.6.

b. To produce elemental nickel, oxygen gas is reacted with an ore of nickel and sulfur, represented as NiS.

$$NiS(s) + O_2(g) \longrightarrow Ni(s) + SO_2(g)$$

Is Ni in the ore oxidized or reduced?

c. Write the analogous chemical equation for cadmium, and again identify the species oxidized and reduced.

d. Why is the release of SO_2 a serious problem?
Hint: Revisit both Chapter 1 and Chapter 6.

Answer

b. NiS contains Ni^{2+}, which is reduced. This ion gains two electrons to form Ni metal.

The environmental price tag also includes the disposal of "dead" batteries. Even rechargeable batteries eventually have to be replaced because at some point the voltage drops below usable levels, and electrons no longer flow. Although the battery may be dead, the chemicals still can be hazardous. Thus communities eventually either must pay the cost of cleaning up the landfills where batteries are improperly disposed of, or they must pay to properly recycle the batteries.

Remember in Chapter 1 that we commented on the logic of prevention. It makes more sense not to dirty the air we breathe than it does to clean it up after the fact. In essence, take off your muddy boots rather than laundering the carpet! Thus we urge you to use this same logic when it comes to batteries. Companies should take responsibility—as should you—for items from the moment the natural resources used to make them were taken out of the ground to their point of ultimate "disposal." Throwing batteries that contain mercury, lead, or cadmium into the trash, ultimately into the landfill, is a poorly planned scenario. An easy way to reduce the number of batteries that wind up as trash is to switch to using rechargeable ones. Investing in a battery charger should easily pay for itself if used properly.

Another way to reduce battery waste is to think "cradle-to-cradle." The end of the life cycle of one item should dovetail with the beginning of the life cycle of another, so that everything is reused rather than added to the waste stream. If each battery served as the starting material for a new product, then the metals these batteries contain would not be lost to the landfill. This also is called "closed-loop recycling."

The economics make sense, especially when it is cheaper to extract and reuse a metal (such as from a discarded battery) than to mine new ore and refine it. With a cradle-to-cradle approach, the item to be recycled is sent to a company that pulls out the desired metal and then returns it to another manufacturer. Unfortunately, far too few batteries are recycled in this manner worldwide.

Keeping toxic materials out of the environment also makes sense. For example, the components of an automobile battery—metallic lead, lead dioxide (PbO_2), and sulfuric acid—are toxic or corrosive. Other metals commonly used in batteries, including cadmium and mercury, are equally if not more toxic. Disposing of these batteries in landfills eventually contaminates the land, the surface water, and ultimately the groundwater with these metals. In addition, the metals become lost to the manufacturing supply chain as they are too widely dispersed to effectively mine. The next activity explores a possible future scenario if we continue along this path.

The term *cradle-to-cradle* was introduced in Section 0.4.

Lead toxicity was mentioned earlier both in the context of paint (see Section 1.13) and water quality (see Section 5.10).

Consider This 8.8   Could Metals Become Extinct?

In 2009, this very question was posed in an article called "Future of Metals." It was published in *Chemical and Engineering News*, the weekly magazine of the American Chemical Society.

a. Why is it not possible for metals to become extinct, at least as such?
b. Nonetheless, the author has a point. Explain.
c. The authors pinpointed copper, zinc, and platinum as the currently "endangered species." Which, if any, of these metals currently are in demand for batteries?

The lead–acid battery represents one success story. Today most state laws require retailers that sell lead–acid batteries to collect them for recycling. The EPA reports that since 1988, greater than 90% of lead–acid batteries have been recycled in the United States. The next activity enables you to learn more about battery recycling.

Consider This 8.9   Battery Recycling

What can you do to keep the metals used in batteries from being lost to a landfill? The answer depends on the battery type. The textbook's website provides some useful links to get you started.

a. Which types of batteries currently are more commonly recycled: rechargeable ones or the nonrechargeable (single-use) ones?
b. Why is recycling a Ni-Cd battery more critical than recycling an alkaline one?
c. List some reasons why household battery-recycling programs have not been as effective as those for recycling car batteries.

Lithium (stored in oil)

Sodium (removed from oil, being cut)

Potassium (in sealed glass tube)

Rubidium (in sealed glass tube)

Figure 8.7

Selected Group 1A elements.

In this same article on the future of metals, Thomas Graedel, an industrial ecologist at Yale, points out that "Metals have limits in the same way that crude oil and clean water do." His point is well taken. If metals are to remain available for use in the future, we all need smarter battery designs that allow for efficient recycling of the metals they contain. It makes no sense for cadmium, mercury, nickel, and lead to end up in landfills. Rather, they should end up in new batteries.

There is precedent and good reason for recycling metals. Recall from Chapter 1 that catalytic converters typically contain platinum. Chemical and petroleum industries already have set protocols in place for recycling platinum catalysts. The item to be recycled is sent to a company that extracts the platinum and then returns the metal for reuse by a manufacturer. Although rechargeable batteries can and are being recycled, such is not yet the norm for single-use batteries.

Lithium is an interesting case in point. Recall from Chapter 2 that lithium is an alkali metal in Group 1A of the periodic table, just like sodium and potassium (Figure 8.7), and that Li, Na, and K all are highly reactive metals, with one outer electron. Finally, recall from Chapter 5 that these metals occur in nature as ions: Li^+, Na^+, and K^+.

However, lithium is different from other Group 1A elements in several important ways. In comparison with the atoms of sodium and potassium, those of lithium are smaller and lighter. Having less mass is an advantage when it comes to building portable batteries. And being smaller also is an advantage in that lithium ions are small enough to fit within certain types of electrode materials, in contrast to Na^+ and K^+, which may be too large. Lithium also is far less abundant in the Earth's crust than sodium and potassium, both of which are readily available. Furthermore, lithium deposits tend to be found in remote locations, such as the one shown in Figure 8.8. At present, lithium is mined primarily from salt lakes (brine lakes) that once were ancient sea beds.

The future availability of lithium is a key point for discussion. As you will see in the next section, the batteries under development for the next generation of hybrid electric cars are slated to contain lithium. The use of these batteries in millions of

Figure 8.8

One of the world's largest lithium deposits, a salt lake in a desert area of Chile. Lithium is in the form of soluble chloride and carbonate salts, $LiCl$ and Li_2CO_3.

cars—each with about 4.5 kg (10 lb) of lithium per battery pack—may severely test our ability to supply lithium to battery manufacturers. At the moment, people are arguing over whether we will be able to do this. On one hand, the lithium on our planet appears to be present in sufficient quantity to meet our needs. On the other, only certain of the lithium deposits are of high enough quality and in accessible enough regions to be extracted economically.

Clearly, it is in everybody's best interest to follow the principles of green chemistry as we manufacture batteries now and in the future. This means that we must minimize or avoid the use of toxic metals in batteries *and* use smart battery designs that enable metals in short supply to be efficiently recycled. With an eye to how batteries can be one component of a more sustainable energy picture, we now turn to the topic of hybrid vehicles.

Check green chemistry principles #3 and #5 printed on the inside front cover.

8.4 | Hybrid Vehicles

As concerns grow about the cost and availability of gasoline and about the pollutants that gasoline-powered vehicles emit, more car owners are considering **hybrid electric vehicles (HEVs),** better known as "hybrids." These vehicles are propelled by a combination of a conventional gasoline engine and an electric motor run by batteries. Honda and Toyota have led the way in developing hybrids (Table 8.2). In 1999, the Honda Insight, a small two-seater, was the first sold in the United States. The Toyota Prius (Figure 8.9) became available in Japan in 1997 and then three years later in the United States. Today, other manufacturers now produce hybrid cars, SUVs, trucks, and even luxury models.

In Latin, *prius* means "to go before."

Consider This 8.10   The NiMH Battery

As of 2010, the Prius gasoline–electric hybrid car employed two batteries, one a conventional lead–acid storage battery and the other a nickel–metal hydride (NiMH).

a. Which features of NiMH batteries make them superior to lead–acid storage batteries?

b. The next generation of batteries in hybrid cars may use lithium. Find out which improvements are being sought.

(a)

(b)

Figure 8.9

a. A 2010 Toyota Prius, a gasoline–electric hybrid. **b.** The NiMH batteries under the back seat and in the trunk area.

In delivering about 50 miles to the gallon, the Prius burns about half the gasoline and thus emits about half of the carbon dioxide of a conventional car. It has a 1.8-L gasoline engine working together with nickel–metal hydride batteries, an electric motor, and an electric generator. The electric motor draws power from the batteries to start the car moving and to power it at low speeds. Using a process called regenerative braking, the energy of the car's motion is transferred to the alternator, which in turn charges the batteries during deceleration and braking. The gasoline engine assists the electric motor during normal driving, with the batteries boosting power when extra acceleration is needed.

Given that each gallon of gasoline burned releases about 18 pounds of CO_2 into the atmosphere, the average vehicle emits from 6 to 9 tons of CO_2 each year. Unlike other vehicle emissions, such as NO and CO, pollution control technologies currently do not reduce CO_2 emissions. Rather, we must reduce the amount of carbon dioxide either by burning less fuel *or* by burning a fuel such as H_2 that does not contain carbon. Doing the math, each increase of 5 miles per gallon a year (for example, improving from 20 to 25 miles per gallon) can reduce CO_2 emissions by about 18 tons over a vehicle's lifetime. This calculation assumes a vehicle lifetime of 200,000 miles, which results in burning about 2000 fewer gallons of gasoline by increasing the fuel efficiency by 5 miles per gallon.

Kinetic energy was first introduced in Section 4.1.

Many hybrids, unlike conventional gasoline-powered cars, deliver better mileage in city driving than at highway speeds.

Table 8.2	Fuel Economy Leaders for the 2010 Model Year	
Rank	Manufacturer/Model	Miles per Gallon (city/highway)
1	Toyota Prius (hybrid)	51/48
2	Ford Fusion Hybrid FWD Mercury Milan Hybrid FWD	41/36
3	Honda Civic Hybrid	40/45
4	Honda Insight (hybrid)	40/43
5	Lexus HS250h (hybrid)	35/34

Source: U.S. EPA.

Skeptical Chemist 8.11 — Yes, Tons of CO$_2$!

Could an automobile really emit 7 tons of carbon dioxide in a year? Do a calculation to prove or disprove this. State all the assumptions that you make.

Automobile manufacturers now provide more than one option for fuel-efficient, low-emission vehicles, allowing customers to choose the option that best meets their transportation needs. One such option is a **plug-in hybrid electric vehicle (PHEV).** These vehicles use rechargeable batteries for short daily commutes to run an electric motor and switch to a combustion engine to travel longer distances. The electric energy provided by the battery decreases the direct emissions out the tailpipe, a major selling point. In addition, proponents argue that PHEVs would cost 2–4¢/mile to operate, a bargain compared with the 8–20¢/mile cost of a regular car. However, according to a 2009 study by the U.S. National Research Council, PHEVs are still a few decades away from mass introduction into the U.S. automobile market.

In 2010, the cost to manufacture a plug-in hybrid was estimated at about $18,000 more than that of an equivalent gasoline-powered vehicle. The larger lithium-ion batteries were one factor in driving up the price tag. To recoup the upfront cost of a PHEV, given the current price of gasoline it would take several decades before the saving in fuel cost, a mile driven on electricity versus a mile driven on gasoline, could be seen. As a result, PHEVs are not expected to be players in reducing gasoline consumption or carbon emissions until the energy picture changes. However, nearly every major automotive manufacturer currently has major research and development teams working on PHEVs. Most of these companies agree that in the future, plug-in electric cars could easily number in the millions on U.S. city streets, provided that battery technologies continue to improve and suitable economic incentives are developed. Even so, the numbers are daunting. Assuming that the number of vehicles continues to increase, these PHEVs would still be outnumbered by the 300 million cars that we expect to find on U.S. highways a few decades from now.

With such advantages offered by both HEVs and PHEVs, has the United States turned into a hybrid nation? The answer most certainly is no. Given the recent world recession, a shake-up in the automotive industry, and current gasoline prices, hybrids remain an interesting choice but not a primary one. At the end of 2009, U.S. sales of light vehicles were down 24% from the previous year, and the sales of hybrid vehicles (with a market share of just under 2%) dropped by 11%. It is also interesting to note that in 2009, China overtook the United States as the world's biggest market for automobiles. With a higher demand for cars by Chinese consumers, it remains to be seen which types of cars and trucks will be produced.

Consider This 8.12 — Electric Vehicles

A car totally powered by electricity has been in the human imagination, on the drawing boards, and even in a few automobile showrooms during the past decade.

a. List three advantages that an electric vehicle (EV) would have over an HEV or a PHEV.
b. Give three reasons why EVs are not yet the norm today. The textbook's website offers some helpful links to get you started.

At this point, a major limitation to EV, HEV, and PHEV development is in the battery technology and in the economics of battery development and cost of the vehicle. It remains to be seen how hybrid vehicles will affect your mode of transportation. In the next section, we examine another way in which we might power our vehicles, hydrogen fuel cells.

8.5 | Fuel Cells: The Basics

With fuel cells, we take another step on our journey to find fuels that release high amounts of energy and low amounts of pollutants. In Chapter 4, we compared the energy released, gram for gram, in the combustion of coal, hydrocarbons, and other combustible fuels. As we saw, methane was clearly the winner. Assuming the combustion products to be $CO_2(g)$ and $H_2O(g)$, the heats of combustion of coal (anthracite or bituminous) and *n*-octane, $C_8H_{18}(l)$, (a major component in gasoline) are 30 and 45 kJ per gram of fuel, respectively. In comparison, the heat of combustion of methane is 50 kJ/g.

However, when paired against methane, hydrogen easily wins the competition, as you can see from this equation.

$$H_2(g) + \tfrac{1}{2}\,O_2(g) \longrightarrow H_2O(g) + 249 \text{ kJ} \qquad \textbf{[8.13]}$$

In equation 8.13, the heat of combustion, 249 kJ, is per mole of H_2. This is equivalent to 124.5 kJ per gram of H_2.

Therefore, hydrogen releases almost three times as much energy as methane per gram when burned! In addition to its superior energy production, using hydrogen raises another tantalizing prospect—the powering of motor vehicles with a fuel that would produce only water vapor as a product. Neither CO nor CO_2 would be produced, although depending on the engine conditions and temperatures, some NO conceivably could form.

Skeptical Chemist 8.13   Hydrogen Versus Methane

Is hydrogen really that good of a fuel? Use the bond energy values from Table 4.4 to find out. Clearly show how you performed your calculation, noting any assumptions you needed to make. Does the value you calculated match that of equation 8.13?

Hint: You'll find most of the work done for you in Section 4.6, except that the calculation is done on a mole basis.

As with other flammable fuel sources, like methane or gasoline, when hydrogen is directly mixed with oxygen, a mere spark can set off an explosion. With its 7 million cubic feet of hydrogen gas, the Hindenburg was to airspace as the Titanic was to the high seas. When the air ship caught fire in 1937 and plunged its passengers and crew to their deaths, hydrogen was indelibly stamped in our consciousness as an explosive fuel.

But suppose someone were to suggest a way to combine H_2 and O_2 to form H_2O without the hazards of combustion. Furthermore, let's suppose that this person also claimed that the reaction could be carried out with no direct contact between the hydrogen and the oxygen. The Skeptical Chemist might well dismiss such assertions as sheer nonsense—an outright impossibility. And yet, the operation of a fuel cell is a case in point. A **fuel cell** is an electrochemical cell that produces electricity by converting the chemical energy of a fuel directly into electricity without burning the fuel. William Grove, an English physicist, invented fuel cells in 1839. However, these cells remained a mere curiosity until the dawn of the Space Age. It was only in the 1980s, when the U.S. Space Shuttle carried three sets of 32 cells fueled with hydrogen, that fuel cells came into public view. The electricity generated by these cells powered the lights, motors, and computers onboard the shuttle.

Unlike conventional batteries such as those in flashlights, under the hood of a car, or powering your laptop computers, fuel cells operate on an external supply of fuel that is electrochemically oxidized inside the fuel cell. They also require an external supply of oxygen gas or other "oxidant" material to accept the electrons that are lost by the fuel. With the supply of fuel and oxidant continually being replenished, these "flow batteries" produce electricity. They do not run down or need to be recharged in the same manner as conventional batteries do. Check out the location of the hydrogen fuel supply in Figure 8.10, a schematic drawing of a fuel cell vehicle (FCV). The hydrogen tanks are pressurized up to 5000 pounds/inch^2 (psi). FCVs can travel up to 200 miles before needing to be refueled.

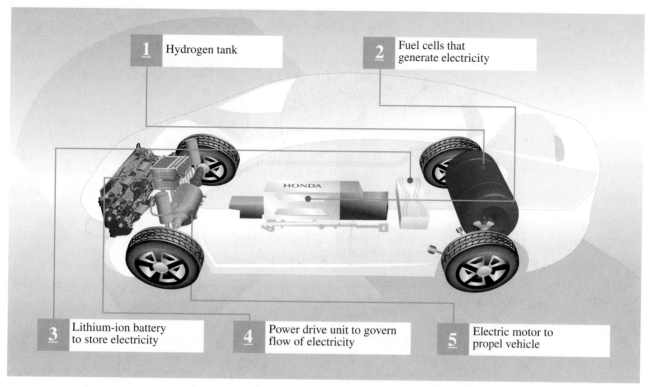

Figure 8.10

Schematic of Honda FCX, powered by fuel cells, that shows the location of the fuel cell, hydrogen storage tank, and lithium-ion storage battery.

Source: American Honda Motor Company.

What may surprise you about fuel cells is that the chemicals being oxidized and reduced are physically separated; that is, they do not come in direct contact with each other. Oxidation still occurs at the anode and reduction at the cathode. However, instead of the anode *itself* being the source from which electrons are released, the anode is merely an electric conductor that provides a physical location in the cell at which the oxidation of the fuel takes place. Similarly, the cathode is an electric conductor where reduction of the oxygen takes place and does not enter into the reaction itself.

The electrolyte that separates the anode from the cathode serves the same purpose as in a traditional electrochemical cell; that is, to allow the flow of ions and hence the flow of charge. The earliest commercially available fuel cells used a strong corrosive acid, H_3PO_4, as an electrolyte. As a result, these fuel cells were closed systems that fully contained the liquid, not unlike the closed system of a conventional alkaline battery. Current designs of fuel cells are open systems that require a continued flow of fuel and oxidant, adding complexity and cost.

Today, fuel cells based on different electrolyte materials have been developed for a variety of applications. One type incorporates a solid polymer electrolyte separating the reactants. We use this type to explain the general operation of fuel cells. The polymer electrolyte membrane, also called a proton exchange membrane (PEM), is permeable to H^+ ions and is coated on both sides with a platinum-based catalyst. These electrolytes operate at reasonably low temperatures, typically from 70 °C to 90 °C, and transfer electrons to rapidly provide electric power. As a result, PEM fuel cells currently are popular with automakers for new fuel cell prototype vehicles and for personal consumer applications. A typical design is shown in Figure 8.11.

In fuel cells hydrogen is used as the fuel in conjunction with oxygen; the oxidation and reduction half-reactions are represented by equations 8.14 and 8.15. As a molecule of hydrogen (H_2) passes through the membrane, it is oxidized and loses two electrons to form two hydrogen ions.

Look for more about polymers in Chapter 9.

The H^+ ion is a proton, the simplest cation (see Section 6.1).

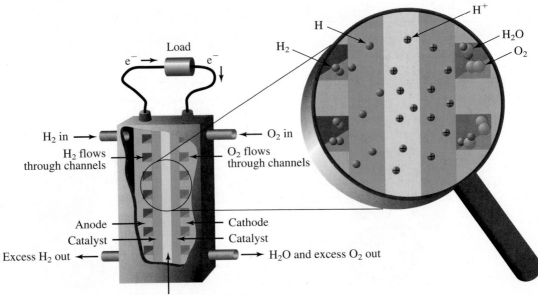

Figure 8.11

A PEM fuel cell in which H_2 and O_2 combine to form water without combustion.

 Figures Alive! Visit the textbook's website to learn more about the chemistry of a PEM fuel cell.

oxidation half-reaction (anode):

$$H_2(g) \longrightarrow 2\,H^+(aq) + 2\,e^- \qquad [8.14]$$

The hydrogen ions, H^+, flow through the proton exchange membrane and combine with oxygen (O_2). At the same time, they combine with two electrons to form water.

reduction half-reaction (cathode):

$$1/2\,O_2(g) + 2\,H^+(aq) + 2\,e^- \longrightarrow H_2O(g) \qquad [8.15]$$

As with galvanic cells, the overall cell equation is the sum of the two half-reactions.

$$H_2(g) + 1/2\,O_2(g) + 2\,H^+(aq) + 2\,e^- \longrightarrow 2\,H^+(aq) + H_2O(g) + 2\,e^- \qquad [8.16]$$

The $2\,e^-$ and $2\,H^+$ appearing on both sides of the arrow can be cancelled.

overall cell equation (sum of the two half-reactions):

$$H_2(g) + 1/2\,O_2(g) \longrightarrow H_2O(g) \qquad [8.17]$$

In Chapters 4 and 7, we showed this chemical equation with whole-number coefficients as $2\,H_2(g) + O_2(g) \rightarrow 2\,H_2O(g)$.

The electrons flowing from the anode to the cathode of a fuel cell move through an external circuit to do work, which is the whole point of the device. Thus, in a fuel cell, a transfer of electrons occurs from H_2 to O_2. This occurs with no flame, with relatively little heat, and without producing any light. Because of these characteristics, the reaction is not classed as combustion. If only the power-producing step is considered (admittedly omitting other parts of the energy picture), hydrogen fuel cells are judged to be a more environmentally friendly way to produce electricity than are coal-fired or nuclear power plants. No carbon-containing greenhouse gases are produced, no air pollutants are emitted, and no spent nuclear fuel needs to be disposed of. Water is the only chemical product if hydrogen is the fuel, an added benefit for the astronauts on the Space Shuttle, who relied on it as their source of water while in space.

Note that it *requires* energy to produce H_2 from compounds containing hydrogen.

The overall cell equation (see equation 8.17) releases 249 kJ of energy per mole of water formed. But instead of liberating most of this energy in the form of heat, the fuel cell converts 45–55% of it to electric energy. This direct production of electricity eliminates the inefficiencies associated with using heat to do work to produce electricity. Internal combustion engines are only 20–30% efficient in deriving energy from fossil fuels. Table 8.3 shows a comparison of fuel combustion with fuel cell technology.

Table 8.3	Combustion Versus Hydrogen Fuel Cell Technology			
Process	**Fuel**	**Oxidant**	**Products**	**Other Considerations**
combustion	hydrocarbons, alcohols, H_2, wood, etc.	O_2 from air	H_2O, CO/CO_2, heat, light, and possibly even sound	rapid process, flame present, lower efficiency, most useful for producing heat
hydrogen fuel cell	H_2	O_2 from air	H_2O, electricity, and some heat	slower process, no flame, quiet, higher efficiency, most useful for generating electricity

Consider This 8.14   Revisiting the PEM Fuel Cell

Spend time exploring the animations of a PEM fuel cell at Figures Alive! on the textbook's website. Then answer these questions.

a. How is a fuel cell different from other batteries described earlier in this chapter?
b. Can a PEM fuel cell be recharged? Explain.
c. Why is the combination of H_2 and O_2 in a fuel cell not classified as combustion? Explain.

Just as batteries, motors, and electric generators come in different sizes and types, so do fuel cells. Though the fuels and principles of operation are essentially the same, different electrolytes give each type of fuel cell unique characteristics that are appropriate for a given application. Many companies are experimenting with fuel cell vehicles. Like EVs, fuel cell vehicles (FCVs) are powered by electric motors. But they differ in that FCVs create their own electricity, whereas EVs draw electricity from an external source, storing that energy in an onboard battery.

An alternative to hydrogen gas is to use a hydrogen-rich fuel such as methanol or natural gas. These fuels must be converted into hydrogen gas by a reformer. This device uses heat, pressure, and catalysts to run a chemical reaction that yields hydrogen as one of the products (Figure 8.12). As liquid fuels, methanol and ethanol could be pumped at conventional gas stations. However, the onboard reformers add cost and maintenance demands to the vehicle. They also emit greenhouse gases and other air pollutants generated in the reforming process.

Section 4.7 discussed reforming in a different context, that of reforming *n*-octane to produce iso-octane, an isomer that burns more smoothly.

As a source of electricity, fuel cells have a broad range of applications. Hospitals, airports, banks, police stations, and military installations all now make use of them for standby and backup power applications. Fuel cells are a form of **distributed generation**, that is, they generate electricity on-site right where it is used, avoiding the losses of energy that occur over long electric transmission lines. As such, they serve as an alternative to central electric utility power plants. Also under development is the powering of portable electronic devices such as cell phones and laptop computers with

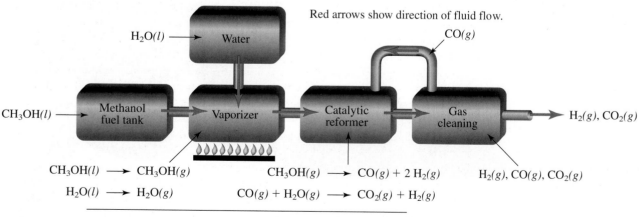

Figure 8.12

Hydrogen obtained from methanol via a reforming process.

miniature fuel cells. Such devices offer an advantage over batteries because they would not require time-consuming electric recharging but rather could be refueled by simply swapping out or refilling a fuel cartridge.

Before our societies can fully benefit from hydrogen fuel cell technologies, scientists and engineers need to meet several technological challenges. The first is to store, transport, and eventually distribute hydrogen to the consumer. A second challenge is to produce enough hydrogen to meet the projected demand. The next section examines both of these challenges in more depth.

8.6 | Hydrogen for Fuel Cell Vehicles

Imagine needing to refuel your fuel cell vehicle with hydrogen at a "gas station." Currently, refueling stations are few and far between. In early 2010, the National Hydrogen Association reported only 78 fueling stations were operational in the United States and Canada. Looking ahead, about 40 more were on the drawing boards. An FCV can travel about 300 miles before refueling, which is certainly competitive with mileage achieved with a conventional gasoline-fueled engine.

Since hydrogen is a gas, it requires a different system for storage and transfer from that used for gasoline. As a gas, hydrogen also takes up a lot of space. For example, at sea level and room temperature, H_2 occupies a volume of about 11 L (almost 4 gal) per gram! In order to avoid having an enormous fuel tank, your vehicle must store hydrogen in a gas cylinder under pressure. To replenish the hydrogen in this cylinder, you must refuel with an airtight connection through a hose that can withstand higher pressures, as shown in Figure 8.13. Although the refueling process does require a different system than a gasoline pump used for a gasoline-powered vehicle, the process is similar in that there is a nozzle and you squeeze a trigger to start the flow of hydrogen.

Instead of compressing H_2 into metal cylinders, which in the past have been heavy and somewhat unwieldy, chemical engineers are investigating other methods for storing and transporting H_2 that could reduce space and the need for high-pressure gas compression. One promising technology is that some compounds, if subjected to hydrogen under high pressure, can absorb the hydrogen molecules like a sponge absorbs water. Then, by either reducing the hydrogen pressure or increasing the temperature, the H_2 can be re-released on demand (Figure 8.14). For example, metal hydrides can perform

By way of comparison, 12 L of gasoline has a mass of 9 kg.

Figure 8.13
Refueling a Honda FCX Clarity, a hydrogen-powered vehicle.

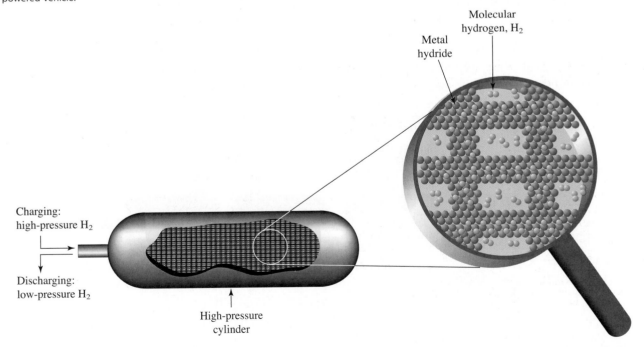

Figure 8.14
Absorption and release of hydrogen from a metal hydride.

in this way. Lithium hydride, LiH, is one example. A chemical formula of LiH may appear strange to you and for good reason. The problem is not the lithium ion (Li^+), as this should be an old friend by now. Rather, it is the hydride ion, H^-, a chemical species that differs markedly from the hydrogen ion, H^+. The hydride ion, with two electrons instead of one for neutral hydrogen, plays an important role in battery chemistry. Unlike H^+, the hydride ion is not stable in aqueous solution and hence we had no need to mention it earlier in our chapter on water chemistry.

Metal hydride storage systems are ideally suited for PEM fuel cells that require high-purity hydrogen. Because metal hydrides are selective and absorb only hydrogen and not larger gas molecules such as CO, CO_2, or O_2, they act simultaneously as a storage material and a way of filtering out other gases. New storage technologies must meet the challenge of taking up less vehicle space needed for people and cargo while allowing designers to put more fuel on board for longer-range travel.

A second challenge is the projected demand for hydrogen as a fuel. Where is all the hydrogen going to come from? On one hand, things look promising because hydrogen is the most plentiful element in the universe. Over 93% of all atoms are hydrogen atoms! Although hydrogen is not nearly this abundant on Earth, still there is an immense supply of the element. On the other hand, essentially all of the hydrogen on our planet is in some form other than H_2. Hydrogen gas is too reactive to exist for long in this form and so primarily is found in its oxidized form of H_2O, better known as water. Therefore, to obtain hydrogen for use as a fuel, we must form it from water or other hydrogen-containing compounds, a process that requires energy.

Fossil fuels, including natural gas and coal, as hydrocarbons, are one possible source of hydrogen. In particular, methane, the major component of natural gas, currently is the chief source of hydrogen. Hydrogen can be produced from CH_4 via an endothermic reaction with steam.

$$165 \text{ kJ} + CH_4(g) + 2 H_2O(g) \longrightarrow 4 H_2(g) + CO_2(g) \qquad \textbf{[8.18]}$$

Another possible way of producing hydrogen from methane is via a reaction with carbon dioxide.

$$247 \text{ kJ} + CO_2(g) + CH_4(g) \longrightarrow 2 H_2(g) + 2 CO(g) \qquad \textbf{[8.19]}$$

You can see the downside of this reaction—it requires a significant energy input. But the Hydrogen Energy Corporation now uses a solar mirror array that can focus sunlight to heat the reactants, CO_2 and CH_4. Not only can this technology produce hydrogen, but it can also do so from a waste gas generated by a landfill.

Your Turn 8.15   Back to Bond Energies

a. Use the average bond energy values in Table 4.4 to check the energy required by the reaction in equations 8.18 and 8.19. Show your work.
b. Are the chemical reactions endothermic or exothermic?
c. Did your calculated value match the values given in the equation? Explain.
   Hint: Revisit Section 4.6.

Answer

a. In equation 8.18, 4 mol of C–H bonds and 4 mol of O–H bonds are broken

   = 4 mol (416 kJ/mol) + 4 mol (467 kJ/mol)
   = 1664 kJ + 1868 kJ
   = 3532 kJ

In equation 8.18, 4 mol of H–H bonds and 2 mol of C=O bonds are formed

   = 4 mol (436 kJ/mol) + 2 mol (803 kJ/mol)
   = 1744 kJ + 1606 kJ
   = 3350 kJ

For the overall reaction, (+3532 kJ) + (−3350 kJ) = 182 kJ. A similar calculation for equation 8.19 gives 252 kJ.

Still, each of the reactions just described contains a major flaw; either carbon dioxide or carbon monoxide is produced. Is there another source of hydrogen around? In Jules Verne's 1874 novel, *Mysterious Island,* a shipwrecked engineer speculates about the energy resource that will be used when the world's coal supply has been used up. "Water," the engineer declares, "I believe that water will one day be employed as fuel, that hydrogen and oxygen which constitute it, used singly or together, will furnish an inexhaustible source of heat and light."

Is this simply science fiction, or is it energetically and economically feasible to break water into its elemental components? To assess the credibility of the claim by Verne's engineer, we need to examine the energy requirements of this chemical reaction. In Section 8.4, we noted that the formation of 1 mol of water from hydrogen and oxygen releases 249 kJ of energy (see equation 8.13). An identical quantity of energy must be absorbed to reverse the reaction to produce hydrogen.

$$249 \text{ kJ} + H_2O(g) \longrightarrow H_2(g) + 1/2\, O_2(g) \qquad [8.20]$$

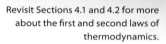

To decompose liquid water into H_2 gas and O_2 gas, the additional energy required to vaporize the water also must be supplied. See Section 5.1.

The most convenient method of decomposing water into hydrogen and oxygen is by **electrolysis,** the process of passing a direct current of electricity of sufficient voltage through water to decompose it into H_2 and O_2 (Figure 8.15). This process takes place in an **electrolytic cell,** a type of electrochemical cell in which electric energy is converted to chemical energy. An electrolytic cell is the opposite of a galvanic cell, where chemical energy is converted to electric energy. When water is electrolyzed in an electrolytic cell, the volume of hydrogen generated is twice that of the oxygen as shown in equation 8.20. This suggests that a water molecule contains twice as many hydrogen atoms as oxygen atoms, testimony to the formula H_2O.

Water electrolysis requires about half the energy input per mole of H_2 than does using methane to produce hydrogen and produces no CO_2 (see equation 8.19). From a thermodynamic point of view, it takes energy to split water into oxygen and hydrogen. Figure 8.16 shows the energy differences involved.

Revisit Sections 4.1 and 4.2 for more about the first and second laws of thermodynamics.

Of course, the question remains: How will the electricity be generated for large-scale electrolysis? Most electricity in the United States is produced by burning fossil fuels in conventional power plants. If we only had to contend with the first law of thermodynamics, the best we could possibly achieve would be to burn an amount of fossil fuel equal in energy content to the hydrogen produced in electrolysis. But we must also deal with the consequences of the second law of thermodynamics. Because

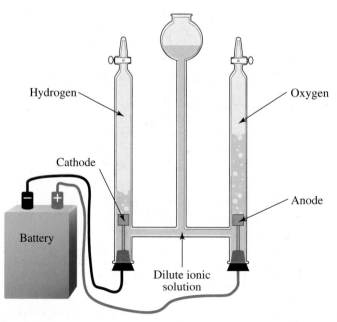

Figure 8.15

Electrolysis of water.

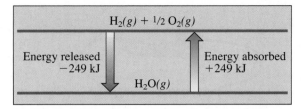

$$H_2(g) + \tfrac{1}{2}\,O_2(g)$$

Energy released
−249 kJ

Energy absorbed
+249 kJ

$$H_2O(g)$$

Figure 8.16

Energy differences in the hydrogen–oxygen–water system.

of the inherent and inescapable inefficiency associated with transforming heat into work, the maximum possible efficiency of an electric power plant is 63%. When we add the additional energy losses caused by friction, incomplete heat transfer, and transmission over power lines, it would require at least twice as much energy to produce the hydrogen than we could obtain from its combustion. This is comparable to buying eggs for 10¢ each and selling them for 5¢, which is no way to do business.

Another way to produce hydrogen is to use heat energy to decompose water. Simply heating water to decompose it thermally into H_2 and O_2 is not commercially promising. To obtain reasonable yields of hydrogen and oxygen, temperatures of over 5000 °C are required. To attain such temperatures is not only extremely difficult, but also requires enormous amounts of energy—at least as much as is released when the hydrogen burns. Thus, we have again reached a point where we are investing a great deal of time, effort, money, and energy to generate a quantity of hydrogen that, at best, returns only as much energy as we invested. In practice, a good deal less energy results.

Instead of burning fossil fuel to generate the enormous amount of heat needed to split water, another option is to use a sustainable source of energy, the radiant energy of Sun. Photons of visible light have enough energy to split water. Unfortunately, water doesn't absorb light at these wavelengths (which is why water is colorless). New materials are being designed to use the power of the Sun to help drive the splitting of water. One photoelectrochemical cell, or a galvanic cell, is constructed containing a Pt anode and a cathode covered with nanoparticles of TiO_2 and coated with dye molecules. The dye molecules are tuned to absorb light in the most intense part of the solar spectrum. When submerged in an aqueous electrolyte solution and exposed to light, some of the electrons in the dye are promoted to higher energy states, high enough that they are transferred quickly to the TiO_2. Once there, the electrons can leave the electrode and move through an electric circuit (Figure 8.17).

The loss of electrons, as you have learned, corresponds to oxidation, and in this case the oxygen in water can be oxidized to O_2. After passing through the external circuit, the electrons arrive at the platinum anode where they reduce hydrogen ions to H_2. Efficiencies of modern devices are less than 10% but are expected to increase.

Figure 2.9 shows UV radiation breaking bonds within molecules.

Consider This 8.16   Light That Splits Water

The energy required for equation 8.20 corresponds to a wavelength of 420 nm.

  a. Which region of the electromagnetic spectrum does this fall in?
    Hint: Refer to Figure 2.7.
  b. It is advantageous to use light energy directly, as opposed to the heat energy of the Sun to split water. Explain.

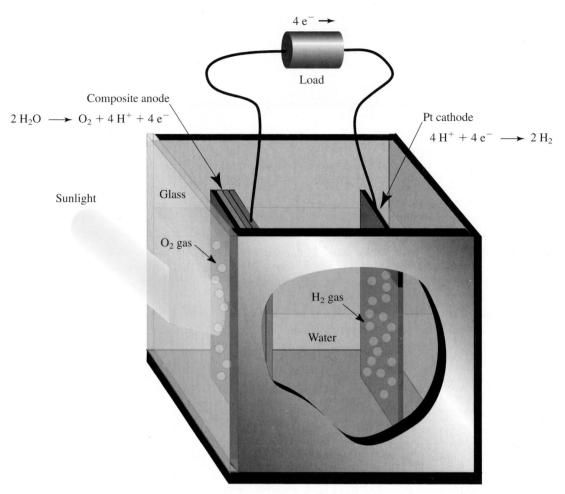

$$4\,e^{-} \rightarrow$$

Load

Composite anode

$$2\,H_2O \longrightarrow O_2 + 4\,H^+ + 4\,e^-$$

Pt cathode

$$4\,H^+ + 4\,e^- \longrightarrow 2\,H_2$$

Sunlight

Glass

O_2 gas

H_2 gas

Water

Figure 8.17

A schematic diagram of a photoelectrochemical cell for splitting water.

In a vividly green example of green chemistry, scientists are looking to biological organisms to produce hydrogen. Certain species of unicellular green algae produce hydrogen gas during photosynthesis (Figure 8.18). The advantage here is that sunlight provides the energy, rather than fossil fuel combustion. At present, the efficiency of the process is far too low to be commercially viable. However, having new strains of algae that are more efficient in utilizing sunlight could tip the economic balance. A current area of research is to genetically engineer such types of algae—both a promising and a controversial area of inquiry.

Green plants, including algae, harness the energy of the Sun in order to grow and reproduce. However, we humans are not nearly as efficient in doing so as plants. In the next section, we turn to the successful technology we have developed to generate electricity from sunlight—photovoltaics—perhaps better known as "solar cells."

8.7 | Photovoltaic Cells: The Basics

It surely would make sense to take advantage of sunlight, a renewable energy source. The rays of our Sun hit the Earth every hour with enough energy to meet the world's energy demand for an entire year! Currently, however, less than 1% of the electric power generated in the United States comes directly from solar energy. Why does solar energy currently account for such a small part of the larger energy picture?

Although remarkable amounts of sunshine hit the Earth daily, the rays do not strike any one site on the planet for 24 hours a day, 365 days a year. Furthermore, some parts of the planet receive too low an intensity of light to be practical for solar

Look for more about genetic engineering, including its controversies, in Chapter 12.

Figure 8.18

Some kinds of algae can produce hydrogen via photosynthesis.

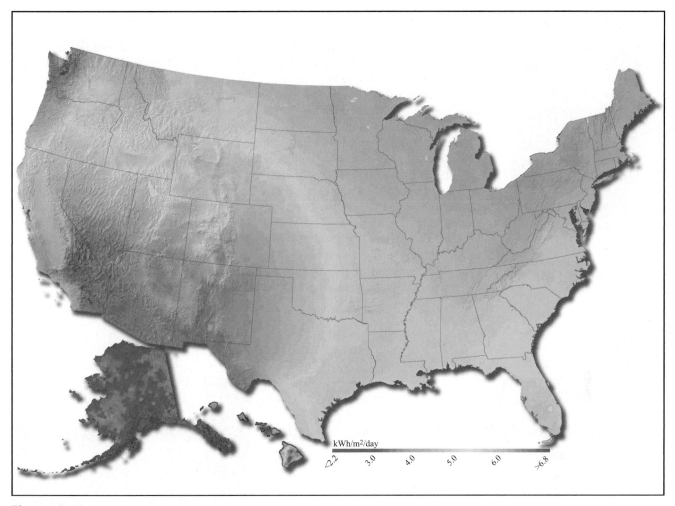

Figure 8.19

The average amount of daily solar energy received by a fixed photovoltaic panel oriented due south.

Source: Billy Roberts, National Renewable Energy Laboratory (NREL) for the U.S. Department of Energy, 2008.

collecting. The differences arise because of the geographical location and because of local factors such as cloud cover, aerosols, smog, and haze. For example, examine the map shown in Figure 8.19. The data in the figure are reported in kilowatt-hours (kWh, a unit of energy) per square meter per day for a flat-panel solar collector that is stationary. The next activity helps you explore the differences in daily solar energy over a calendar year.

1 kWh = 3,600,000 joules

The energy values in Figure 8.19 would be higher if the panels tracked the path of the Sun, rather than being stationary.

Consider This 8.17   Solar Maps

Thanks to a website provided by the U.S. National Renewable Energy Laboratory, you can view solar maps for different parts of the United States.

a. Select a state of your choice and view the data for each month of the year. What do you notice about how energy varies throughout the year?

b. It should come as no surprise that California, Arizona, New Mexico, and Texas lead the United States in average annual solar radiation. Why do some parts of these states have higher values than others?

The challenge, then, is to locate the areas in which the incident average solar energy is high and to collect this energy in sufficient quantities to produce electricity. One possibility is to convert sunlight *directly* into electricity, the topic of this section.

Figure 8.20

Photovoltaic (solar) cells are used to improve security, enhance safety, and direct pedestrians and vehicles.

Another is to trap the heat generated from solar radiation, a topic to be addressed in the next section.

One way to tap into the Sun's energy is to use a **photovoltaic cell (PV),** a device that converts light energy directly to electric energy, sometimes called a solar cell. It takes only a few PV cells to produce enough electricity to power your calculator or digital watch. Other common uses of photovoltaic cells include communication satellites, highway signs, security and safety lighting (Figure 8.20), automobile recharging stations, and navigational buoys. Cost savings can be substantial. For example, using solar cells rather than batteries in navigational buoys saves the U.S. Coast Guard several million dollars annually through reduced maintenance and repair.

If more power is required, PV cells can be combined into modules or arrays to make up solar panels, as shown in Figure 8.21. Many people today power their homes and businesses with solar PV systems. Depending on the size of a home, it may use a dozen or more solar panels for power. These panels are usually mounted facing due south. Installing them on a system that rotates to track the Sun's path, thus maximizing their exposure to sunlight, optimizes the efficiency, but has a higher initial cost. For electric utility or industrial applications, hundreds of solar arrays are interconnected to form a large-scale PV system, such as the one shown in a field in Bavaria, Germany (see Figure 8.21).

Figure 8.21

a. Arrangement of photovoltaic cells used to make a module and an array. **b.** A silicon solar array installed on a roof. **c.** An aerial view of the Solarpark Gut Erlasee in Bavaria, Germany. At peak capacity, it can generate 12 MW. A typical nuclear power plant generates 1000 MW of electricity.

Source: NREL.

How does a photovoltaic cell generate electricity? The answer lies in the behavior of the electrons in the cell material. When light shines onto a PV cell, it may pass right through the cell, be reflected, or be absorbed. If absorbed, the energy may cause an excitation of the electrons in the atoms of the cell. These excited electrons escape from their normal positions in the cell material and become part of an electric current.

Only certain materials behave this way in the presence of light. Photovoltaic cells are made from a class of materials called **semiconductors,** materials that have a limited capacity of conducting an electric current. Most semiconductors are made from a crystalline form of silicon, a metalloid. To induce a voltage in a PV cell, two layers of semiconducting materials are placed in direct contact. An ***n*-type semiconductor** is a layer with an abundance of electrons. The ***p*-type semiconductor** is the other layer with a deficit of electrons, sometimes referred to as "holes". In order to generate an electric current, the light hitting the PV cell must have enough energy to set the electrons in motion from the *n*-type side to the *p*-type side through the electric circuit. The transfer of electrons generates a current of electricity that can be intercepted to do all the things electricity does, including being stored in batteries for later use. As long as the cell is exposed to light, the current continues to flow, powered only by solar energy.

The element silicon was one of the first semiconducting materials developed for use in computers and in PV cells. In fact, many of the high-tech businesses that developed semiconductors were clustered in California's "Silicon Valley." A crystal of silicon consists of an array of silicon atoms, each bonded to four other atoms by means of shared pairs of electrons (Figure 8.22a). These shared electrons are normally fixed in the bonds and unable to move through the crystal. Consequently, silicon is not a very good electric conductor under ordinary circumstances. However, if a valence electron absorbs sufficient energy, it can be excited and released from its bonding position (Figure 8.22b). Once freed, the electron can move throughout the crystal lattice, making the silicon an electric conductor.

In actuality, pure silicon semiconductors do not allow an electric current to flow unless they are doped. **Doping** is a process of intentionally adding small amounts of other elements, "dopants" (or sometimes called impurities), to pure silicon. These dopants are chosen for their ability to facilitate the transfer of electrons. For example, about 1 ppm of gallium (Ga) or arsenic (As) is often introduced into the silicon. These two elements and others from the same groups on the periodic table are used because their atoms differ from silicon by a single outer electron. Silicon has four electrons in its outer energy level, gallium has three, and arsenic has five. Thus, when an atom of As is introduced in place of Si in the silicon lattice, an extra electron is added. The replacement of a Si atom with a Ga atom means that the crystal is now one electron "short." Figure 8.23 illustrates doped *n*- and *p*-type semiconductors. Both types of doping

In 1839, A. E. Becquerel, a French physicist, discovered the process of using sunlight to produce electricity in a solid material.

Conductivity in aqueous ionic solution was discussed in Section 5.6. Metalloids (semimetals) were introduced in Section 1.6.

At most, an individual solar cell produces 0.5 V.

A crystalline structure has a regular repeating array of atoms or ions, as shown in Figure 5.18 for NaCl.

Ga is in Group 3A.
Si is in Group 4A.
As is in Group 5A.

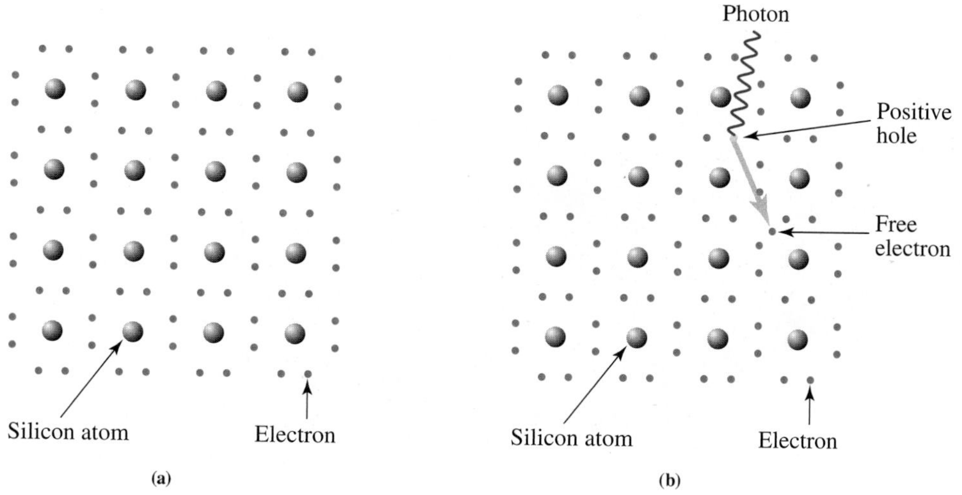

Silicon atom   Electron   Silicon atom   Electron

(a)   (b)

Figure 8.22

a. Schematic of bonding in silicon. **b.** Photon-induced release of a bonding electron in a silicon semiconductor.

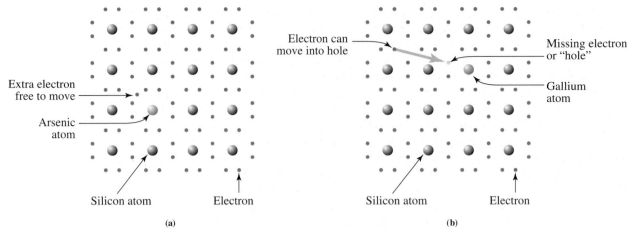

Figure 8.23

a. An arsenic-doped *n*-type silicon semiconductor. **b.** A gallium-doped *p*-type silicon semiconductor.

increase the electric conductivity of silicon because electrons can now move from an electron-rich to an electron-deficient environment.

Your Turn 8.18   Other Dopants

Some solar cell designs use phosphorus and boron to dope silicon crystals.

a. Which will form an *n*-type semiconductor? Explain your reasoning.
b. Which will form a *p*-type semiconductor? Explain your reasoning.

A photovoltaic cell typically includes multiple layers of doped *n*- and *p*-type semiconductors in close contact (Figure 8.24). The *p*–*n* junctions not only make possible the conduction of electricity but also ensure that the current flows in a specific direction through the cell. Only photons with enough energy can knock electrons free from the dopants. These electrons then become part of the electric circuit. For a PV cell to convert as much sunlight as possible into electricity, the semiconductors must be constructed in such a way to make the best use of the photon's energy. If not, the energy of the Sun is lost as heat or not trapped at all.

The fabrication of photovoltaic cells poses some significant challenges. The first is that although silicon is the second most abundant element in Earth's crust, it is most

All photovoltaic cells made from silicon are "doped."

"Sandwiches" of n- and p-type semiconductors are used in transistors and other miniaturized electronic devices that have revolutionized communication and computing.

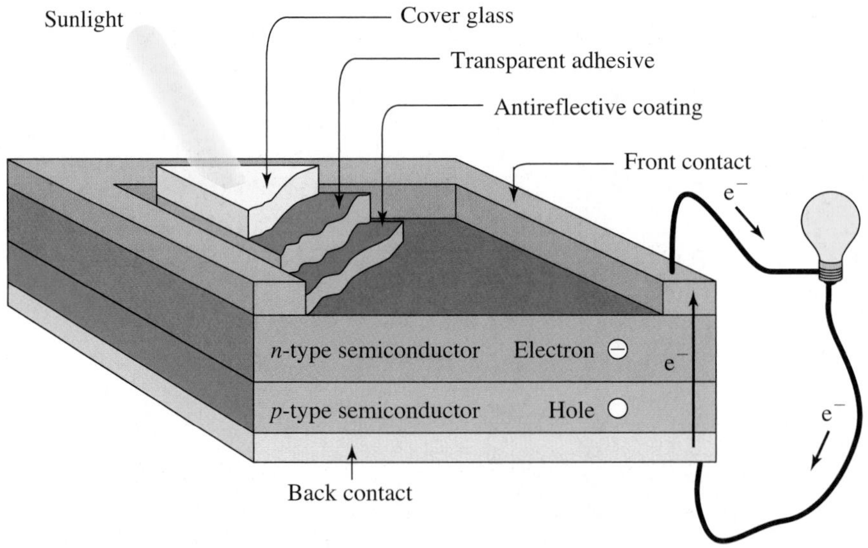

Sunlight — Cover glass

Transparent adhesive

Antireflective coating

Front contact

e⁻

n-type semiconductor   Electron ⊖

p-type semiconductor   Hole ○

e⁻

e⁻

Back contact

Figure 8.24

Schematic diagram of one layer of a solar cell showing the *n*-type and *p*-type semiconductors.

frequently found combined with oxygen as silicon dioxide, SiO_2. You know this material by its common name, sand, or more correctly as quartz sand. The good news is that the starting material from which silicon is extracted is cheap and abundant. The not-so-good news is that processes to extract and purify silicon are expensive. Many of the early PV cell designs required ultrapure 99.999% silicon.

A second challenge is that the direct conversion of sunlight into electricity is not very efficient. A photovoltaic cell could, in principle, transform up to 31% of the radiant energy to which it is sensitive into electricity. However, some of the radiant energy is reflected by the cell or absorbed to produce heat instead of an electric current. Typically, a commercial solar cell now has an efficiency of only 15%, but even this is a significant increase over the first solar cells built in the 1950s, which had efficiencies of less than 4%. In Chapter 4, we lamented the 35–50% efficiency of converting heat to work in a conventional power plant. It might seem that we should be even more distressed at the lower limits that can be achieved by photovoltaics. Remember, however, that the first use of solar cells was to provide electricity in NASA spacecraft. For that application, the intensity of radiation was so high that low efficiency was not a serious limitation and costs were not of paramount concern. For commercial use on Earth, costs and efficiency are issues. Our Sun is an essentially unlimited energy source, and converting its energy to electricity, even inefficiently, is free from many of the environmental problems associated with burning fossil fuels or storing spent fuel from nuclear fission. These considerations add impetus to research and development of solar cells.

One approach to increasing commercial viability is to replace crystalline silicon with the noncrystalline form of the element. Photons are more efficiently absorbed by less highly ordered Si atoms, a phenomenon that permits reducing the thickness of the silicon semiconductor to 1/60th or less of its former value. The cost of materials is thus significantly reduced.

Other researchers are developing multilayer solar cells. By alternating thin layers of *p*-type and *n*-type doped silicon, each electron has only a short distance to travel to reach the next *p–n* junction. This lowers the internal resistance within the cell and raises the efficiency of the cells. Maximum theoretically predicted efficiencies could improve to 50% for 2 junctions, to 56% for 3 junctions, and to 72% for 36 junctions. As of 2007, the maximum efficiency actually demonstrated with a multijunction solar cell was 40.7%. Figure 8.25 gives a sense of just how thin these layers actually are.

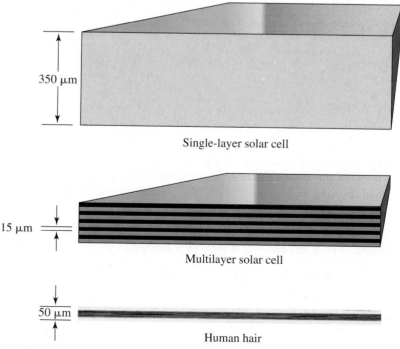

Figure 8.25

A comparison of the relative thickness of a solar cell layer, either in a single-layer or multilayer cell, to the diameter of an average human hair. *Note:* 1 μm = 10⁻⁶ m.

Figure 8.26

Thin-film solar tiles on a roof.

Source: NREL.

Multilayer technology, compared with single-cell technology, uses smaller quantities of silicon, and the production process can become highly automated.

Thin-film solar cells are made from amorphous silicon or nonsilicon materials such as cadmium telluride (CdTe). These thin-films use layers of semiconductor materials only a few micrometers thick. For comparison, a typical human hair is about 50 μm! Thin-film solar cells can even be incorporated into rooftop shingles and tiles, building facades, or the glazing for skylights due to their flexibility compared with more rigid traditional cells (Figure 8.26). Other solar cells are being made using various materials, such as solar inks using conventional printing press technologies, solar dyes, and conductive plastics. Solar modular units use plastic lenses or mirrors to concentrate sunlight onto small but very highly efficient PV materials. Utilities and industries experimenting with these solar lens materials find that despite their higher cost, using a small amount of these more efficient materials is becoming more cost-effective.

Your Turn 8.19   Solar PV Use

How are people today using solar photovoltaics? Answer this question for each group listed below. The links provided at the textbook's website can help you get started.

a. farmers and ranchers   **b.** small-business owners   **c.** homeowners

Answer
a. Uses include pumping water for livestock and lighting in areas without electricity. Even on farms and ranches with electricity, solar PVs can reduce electric utility bills.

Long-range prospects for photovoltaic solar energy are encouraging. Its cost is decreasing while the cost of electricity generated from fossil fuels is increasing. But there is still the question of land use. At currently attainable levels of operating efficiency, the electricity needs of the United States have been estimated to require a photovoltaic generating station covering an area of 85 miles by 85 miles, roughly the size of New Jersey.

Consider This 8.20   If Not New Jersey, . . .

Last we heard, New Jersey was not volunteering to be converted wholesale into a solar farm to power the rest of the United States.

a. Would New Jersey be a reasonable geographic location?
    Hint: Revisit Figure 8.19.
b. Which locations in the United States show the most promise for solar energy collection?
c. Location isn't everything. Name two other factors that come into play in dedicating land to solar energy collection.

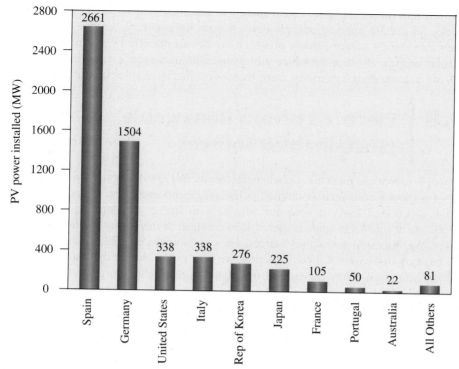

Figure 8.27

Installed PV power around the world, 2008.

Source: International Energy Agency.

Consider This 8.21   A Million Solar Roofs

Distributed generation! Use the web to learn more about how people are using solar energy locally, such as the Million Solar Roofs project. Then propose a solar energy project in a community of your choice. List at least five factors to consider before proceeding with the project.

Although photovoltaic power is steadily growing, it still represents a minute fraction of global power supplies. As 2008 drew to a close, over 5500 MW of installed PV capacity was installed for those countries reporting to the International Energy Agency (Figure 8.27).

Because of the diffuse nature of sunlight, photovoltaic technology is well suited to distributed generation, just as was described earlier for fuel cells (see Section 8.5). More than a third of Earth's population is not hooked into an electric network because of the costs associated with constructing and maintaining equipment and supplying the fuel to generate the electricity. Because PV installations are relatively maintenance-free, they are particularly attractive for electric generation in remote regions. For example, the highway traffic lights in certain parts of Alaska, far from power lines, operate on solar energy. A similar but more significant application of photovoltaic cells may be to bring electricity to isolated villages in developing countries. In recent years, more than 200,000 solar lighting units have been installed in residential units in Colombia, the Dominican Republic, Mexico, Sri Lanka, South Africa, China, and India. Photovoltaic cells currently are affecting the lives of millions of people across our planet (Figure 8.28).

Electricity generated by photovoltaic cells during the day must be stored using batteries for use at night. Nevertheless, the direct conversion of sunlight to electricity has many advantages. In addition to relieving some of our dependence on fossil fuels, an economy based on solar electricity would reduce the environmental damage of extracting and transporting these fuels. Furthermore, it would help to lower the levels of air pollutants such as sulfur oxides and nitrogen oxides. It would also help avert the dangers of global warming by decreasing the amount of carbon dioxide

Figure 8.28

Photovoltaics can power water pumps in remote areas of the world where there is no access to electricity.

Source: NREL.

released into the atmosphere. Fossil fuels will certainly remain the preferred form of energy for certain applications. However, for the longer term, we can turn to many renewable energy sources, many of which are driven directly by the Sun or as result of solar heating of our atmosphere and water. The next section takes a brief look at how we can generate electricity from these sustainable renewable resources.

8.8 | Electricity from Renewable (Sustainable) Sources

No single source can meet our global energy needs. We also know that no energy source comes without a cost, such as mining, pollution, greenhouse gases, or setting up distribution networks. Clearly it is to our advantage to further develop and add a greater percentage of renewable sources than it is to continue to rely on fossil fuels and nuclear power. We discussed renewable sources, such as biofuels, ethanol, biodiesel, garbage, and biomass in Chapter 4. In this section, we turn to the heat of the Sun, wind, water, and the heat given off by the core of our planet as renewable energy sources.

Solar Thermal

In addition to emitting light, the Sun gives off heat. Concentrating the Sun's energy to heat water is called a solar-thermal process and is also known as concentrating solar power (CSP). Unlike solar voltaic technologies that rely on radiant energy to knock electrons loose from a semiconductor, CSP depends on solar collector devices such as the ones shown in Figure 8.29. These mirrored arrays concentrate sunlight in much the same way that a magnifying glass can focus light to burn a hole in a piece of paper.

Your Turn 8.22   Solar-Thermal Collectors I

Revisit Figure 4.2, reproduced here in miniature. It is a diagram of an electric power plant illustrating the conversion of energy from the combustion of fuels into electricity.

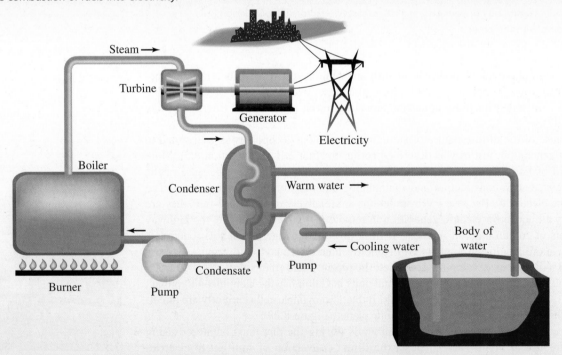

a. What part of this figure would change if solar energy were to be used?
b. Name two advantages that collecting solar energy locally (distributed generation) offers over a centralized system. Then answer this question in the reverse.

(a)

(b)

Figure 8.29

a. An aerial view of the Solar Millennium Andasol project in Spain, which will have an approximate capacity of 150 MW of solar-thermal power. **b.** A close-up of a portion of the mirrored array.

Source: Chemical and Engineering News, February 1, 2010.

 Consider This 8.23   **Solar-Thermal Collectors II**

All solar collectors focus and concentrate the Sun's rays for the purpose of producing heat. However, they do so in different ways.

  a. Describe the designs for three different types of collectors.
  Hint: Check the textbook's website for helpful links.
  b. How is each design matched to its end use? As part of your answer, include the scale of use—that is, for a single home, for a community, or for a business.
  c. Name at least one limitation for each.

Wind

The Sun's heat ultimately drives the large-scale movements of the air on our planet that we know better as "wind." For centuries, humans relied on various forms of windmills that, in turn, spun wheels to grind grain or pump water. Wind turbines today make use of

Figure 8.30

Pakini Nui Wind Farm, completed in 2007, now supplying 20.5 MW of power.

large blades, sometimes nicknamed by the locals as the "pinwheels" that dot the landscape. These spin a shaft that turns a generator to produce electricity. Wind farms are located around the world in order to take advantage of prevailing winds. Such a farm is shown at Ka Lae (South Point) on the Big Island of Hawaii (Figure 8.30).

Water

For centuries, humans have harnessed the movement of water with devices such as water wheels. When water flows over a wheel, this wheel can turn other wheels, including stones that can grind grain into flour. Similarly, small and larger scale hydroelectric dams harness the movement of water. When water falls across turbine blades, the potential energy of water trapped in reservoirs is converted into kinetic energy which in turn is converted to electricity. Although worldwide a few dams are still being constructed, most large reservoirs of water are already in the service of hydroelectric projects.

The movement of ocean water in tides, currents, and waves can be harvested by a variety of principles to generate electricity. Some involve turning blades of a turbine, others involve forcing compressed air through a turbine. All involve the kinetic energy of motion to turn a generator to produce electricity.

Geothermal

Another renewable energy source is the heat given off from the core of our planet. Literally "earth heat," geothermal energy relies on drilling into underground reservoirs containing hot water or steam and thus drawing heat from the Earth. These heated sources of water can then be used to drive generators to produce electricity or the hot water may be used directly to heat a home. Geothermal works well in locations known for volcanic activity that have "hot rock," such as Hawaii, which generates 25% of its energy from geothermal sources.

 Consider This 8.24   **Our Energetic Future**

We can obtain renewable energy from the wind, the oceans, or geothermal sources, not only from the Sun or biomass. Pick one of these renewable energy sources and learn more about the technologies available to harness it.

a. Name the geographic restrictions (if any) to its use.
b. Prepare a list of the reasons to support this technology. Prepare a similar list for the nay-sayers.
c. Predict how this technology will affect the energy production capacity where you live.

This section merely offered a renewable energy sampler, and no world view of energy resources would be complete without considering these and other sustainable sources of energy. To increase the share that renewable sources occupy on the world's energy scene, their economics, availability, and ease of use must be improved.

Conclusion

We look to many different forms of electron transfer to meet our energy needs. Batteries can store chemical energy and convert it to a flow of electrons useful for many applications. Hybrid vehicles use new battery technologies combined with internal combustion engines to improve fuel efficiency. Fuel cells are one of the most efficient ways to produce electricity and may become a major energy source for future personal power use, transportation, and perhaps even large-scale electricity production.

Photovoltaic cells can tap the energy of the Sun. Advances in research together with changes in global economies may make it both fiscally and energetically feasible to use solar energy to extract hydrogen from water or another hydrogen-containing compound. Thus in the years to come, we look forward to new developments that will improve each of these energy options, making their use suitable not only for our generation but for generations to come.

We hope that our discussions of energy in this book have provided you with enough background that you have gained a perspective on the complexity of energy issues that we face. We also hope you are in a position to take stock of the situation and look ahead to the future. A few facts seem beyond debate. The world's thirst for energy will not abate; it most assuredly will continue to grow. Moreover, the ways in which we currently generate energy are not sustainable. The coal, petroleum, and natural gas from which we now derive the vast majority of our power are destined to become scarce or difficult to extract in the not-too-distant future. In addition, their use is not without an environmental cost. Nuclear power, though not directly responsible for the emission of greenhouse and acid-rain-causing gases, carries its own risks and challenges as a long-term energy solution. A transformation is required.

But the laws of thermodynamics and human nature are such that these transformations will not occur spontaneously. A transition to sustainable energy alternatives and their applied uses cannot be developed without hard work and the investment of intellect, time, and money. The sacrifices and compromises that "kicking our fossil fuel habit" require depend heavily on the choices we make today. We need to establish global, national, and personal priorities and to muster and sustain the will to act on them.

We have been the beneficiaries of a bountiful resource base from the Earth. In turn, we have an obligation to ensure energy sources for generations yet to come.

Chapter Summary

Having studied this chapter, you should be able to:

- Discuss the principles governing the transfer of electrons in galvanic cells, including the processes of oxidation and reduction (8.1)

- Identify oxidation and reduction half-reactions and be able to distinguish which chemical species is oxidized and which is reduced (8.1)

- Describe the design, operation, applications, and advantages of several different types of batteries (8.1–8.3)

- Compare and contrast the principles, advantages, and challenges of producing and using hybrid vehicles (8.4)

- Describe the design, operation, applications, and advantages of typical fuel cells (8.5)

- Explain the energy costs and gains of producing hydrogen and using it as a fuel (8.6)

- Describe the principles governing the operation of photovoltaic (solar) cells and their current and future uses (8.7)

- Describe advantages of renewable energy sources over traditional energy sources and how they are tied to energy through electron transfer (8.8)

Questions

Emphasizing Essentials

1. a. Define the terms *oxidation* and *reduction*.

   b. Why must these processes take place together?

2. Which of these half-reactions represent oxidation and which reduction? Explain your reasoning.

   a. $Fe \longrightarrow Fe^{2+} + 2\,e^-$

   b. $Ni^{4+} + 2\,e^- \longrightarrow Ni^{2+}$

   c. $2\,Cl^- \longrightarrow Cl_2 + 2\,e^-$

3. Which chemical species gets oxidized and which gets reduced in the following overall chemical equation:

   $$2\,Zn(s) + O_2(g) \longrightarrow 2\,ZnO(s)$$

4. What is the difference between a galvanic cell and a true battery? Give an example for each.

5. Two common units associated with electricity are the volt and the amp. What does each unit measure?

6. Consider this galvanic cell. A coating of impure silver metal begins to appear on the surface of the silver electrode as the cell discharges.

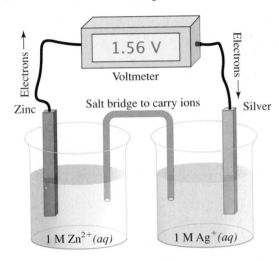

1.56 V

Voltmeter

Electrons →

Electrons ↓

Zinc

Salt bridge to carry ions

Silver

$1\,M\,Zn^{2+}(aq)$

$1\,M\,Ag^+(aq)$

   a. Identify the anode and write the oxidation half-reaction.

   b. Identify the cathode and write the reduction half-reaction.

7. In the lithium–iodine cell, Li is oxidized to Li^+; I_2 is reduced to $2\,I^-$.

   a. Write the oxidation half-reaction and the reduction half-reaction that take place in this cell.

   b. Write the overall cell equation.

   c. Identify the half-reactions that occur at the anode and at the cathode.

8. a. How does the voltage from a tiny AAA alkaline cell compare with that from a large D alkaline cell? Explain.

   b. Can both batteries sustain the flow of electrons for the same amount of time? Explain.

9. Identify the type of galvanic cell commonly used in each of these consumer electronic products. Assume none uses solar cells.

   a. laptop computer

   b. MP3 player

   c. digital camera

   d. calculator

10. The mercury battery has been used extensively in medicine and industry. Its overall cell reaction can be represented by this equation.

    $$HgO(l) + Zn(s) \longrightarrow ZnO(s) + Hg(l)$$

    a. Write the oxidation half-reaction.

    b. Write the reduction half-reaction.

    c. Why is the mercury battery no longer in common use?

11. a. What is the function of the electrolyte in a galvanic cell?

    b. What is the electrolyte in an alkaline cell?

    c. What is the electrolyte in a lead–acid storage battery?

12. These two *incomplete* half-reactions in a lead–acid storage battery do not show the electrons lost or gained. The reactions are more complicated, but it is still possible to analyze the reactions that take place.

    $$Pb(s) + SO_4^{2-}(aq) \longrightarrow PbSO_4(s)$$
    $$PbO_2(s) + 4\,H^+(aq) + SO_4^{2-}(aq) \longrightarrow$$
    $$PbSO_4(s) + 2\,H_2O(l)$$

    a. Balance both equations with respect to charge by adding electrons as needed.

    b. Which half-reaction represents oxidation and which reduction?

    c. One of the electrodes is made of lead; the other is lead dioxide. Which is the anode and which is the cathode?

13. During the conversion of $O_2(g)$ to $H_2O(l)$ in a fuel cell (see equation 8.13), the following half-reaction takes place.

    $$1/2\,O_2(g) + 2\,H^+(aq) + 2\,e^- \longrightarrow H_2O(l)$$

    Does this half-reaction represent an example of oxidation or reduction? Explain.

14. How does the reaction between hydrogen and oxygen in a fuel cell differ from the combustion of hydrogen and oxygen?

15. This diagram represents the hydrogen fuel cell that was used in some of the earlier space missions.

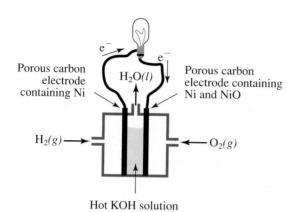

Hot KOH solution

The chemistry in the hydrogen-oxygen fuel cell can be represented by these half-reactions:

$$H_2(g) \longrightarrow 2\,H^+(aq) + 2\,e^-$$
$$\tfrac{1}{2}\,O_2(g) + 2\,H^+(aq) + 2\,e^- \longrightarrow H_2O(l)$$

Which half-reaction takes place at the anode and which at the cathode? Explain.

16. What is a PEM fuel cell? How does it differ from the fuel cell represented in the previous question?

17. In addition to hydrogen, methane also has been studied for use in PEM fuel cells. Balance the given oxidation and reduction half-reactions, and write the overall equation for a methane-based fuel cell.

Oxidation half-reaction:

$$\_\_\,CH_4(g) + \_\_\,OH^-(aq) \longrightarrow$$
$$\_\_\,CO_2(g) + \_\_\,H_2O(l) + \_\_\,e^-$$

Reduction half-reaction:

$$\_\_\,O_2(g) + \_\_\,H_2O(l) + \_\_\,e^- \longrightarrow \_\_\,OH^-(aq)$$

18. Relative to a vehicle with an internal combustion engine, list two advantages offered by hydrogen FCVs.

19. Potassium and lithium both are reactive Group 1A metals. Both form hydrides, highly reactive compounds.

    a. Potassium reacts with H_2 to form potassium hydride, KH. Write the chemical equation.

    b. KH reacts with water to produce H_2 and potassium hydroxide. Write the chemical equation.

    c. Offer a reason why LiH (rather than KH) has been proposed as a means of storing H_2 for use in fuel cells.

20. What challenges keep hydrogen fuel cells from being a primary energy source for vehicles?

21. Every year, 5.6×10^{21} kJ of energy comes to Earth from the Sun. Why can't this energy be used to meet all of our energy needs?

22. This *unbalanced* equation represents the last step in the production of pure silicon for use in solar cells.

$$\_\_\,Mg(s) + \_\_\,SiCl_4(l) \longrightarrow \_\_\,MgCl_2(l) + \_\_\,Si(s)$$

    a. How many electrons are transferred per atom of pure silicon formed?

    b. Does the $SiCl_4(l)$ get oxidized or reduced in the formation of $Si(s)$? Explain your reasoning.

23. The symbol • represents an electron and the symbol ⚫ represents a silicon atom. The darker purple sphere in the center of the diagram represents either a gallium or an arsenic atom. Does this diagram represent a gallium-doped *p*-type silicon semiconductor or an arsenic-doped *n*-type silicon semiconductor? Explain your answer.

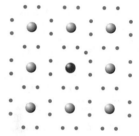

24. Describe the main reasons why solar cells have solar energy conversion efficiencies significantly less than the theoretical value of 31%.

Concentrating on Concepts

25. Explain the significance of the title of this chapter, "Energy from Electron Transfer."

26. Consider these three sources of light: a candle, a battery-powered flashlight, and an electric light bulb. For each source, provide

    a. the origin of the light.

    b. the immediate source of the energy that appears as light.

    c. the original source of the energy that appears as light. *Hint:* Trace this back stepwise as far as possible.

    d. the end-products and by-products produced from using each.

    e. the environmental costs associated with each.

    f. the advantages and disadvantages of each light source.

27. Explain the difference between a rechargeable battery and one that must be discarded. Use a Ni-Cd battery and an alkaline battery as examples.

28. What is the difference between an electrolytic cell and a fuel cell? Explain, giving examples to support your answer.

29. List some differences between a lead–acid storage battery and a fuel cell.

30. "The Earth is a metal-rich rock. I can't see the human race running out of metals when it will be possible to mine in new places or recycle or simply reduce consumption. We probably won't be able to live on the planet due to global warming or other environmental

problems before we run into a metal supply problem."
These comments were offered by geologist Maurice A.
Tivey of Woods Hole Oceanographic Institution in an
article in *Chemical & Engineering News* published in
June 2009.

   a. Do you agree with the writer's sentiment about not
running out of metals? Explain.

   b. Name two challenges connected with increasing the
recycling of batteries.

31. The company ZPower is promoting its silver–zinc
batteries as replacements for lithium–ion batteries in
laptops and cell phones.

   a. What advantages do silver–zinc batteries have over
the current lithium–ion batteries?

   b. Write the oxidation and reduction half-reactions
using this overall cell equation as a guide. Indicate
which reactant gets oxidized and which gets reduced.

$$Zn + Ag_2O \longrightarrow ZnO + 2\,Ag$$

32. The battery of a cell phone discharges when the phone
is in use. A manufacturer, while testing a new "power
boost" system, reported these data.

Time, min.sec	Voltage, V
0.00	6.56
1.00	6.31
2.00	6.24
3.00	6.18
4.00	6.12
5.00	6.07
6.35	6.03
8.35	6.00
11.05	5.90
13.50	5.80
16.00	5.70
16.50	5.60

   a. Prepare a graph of these data.

   b. The manufacturer's goal was to retain 90% of its
initial voltage after 15 minutes of continuous use.
Has that goal been achieved? Justify your answer
using your graph.

33. Assuming that HEVs are available in your area, draw
up a list of at least three questions you would ask the
auto dealer before deciding to buy or lease one. Offer
reasons for your choices.

34. You never need to plug in Toyota's gasoline–battery
hybrid car to recharge the batteries. Explain.

35. Prepare a list of the environmental costs and benefits
associated with HEVs, PHEVs, and EVs. Compare that
list with the environmental costs and benefits of
vehicles powered by gasoline. On balance, which
energy source do you favor, and why?

36. What is *the tragedy of the commons*? How does this
concept apply to our practice of using metals such as
mercury and cadmium in batteries?

37. Hydrogen is considered an environmentally friendly
fuel, producing only water when burned in oxygen.
Name two positive effects that the widespread use of
hydrogen would have on urban air quality.

38. Fuel cells were invented in 1839 but never developed into
practical devices for producing electric energy until the
U.S. space program in the 1960s. What advantages did
fuel cells have over previous power sources?

39. Hydrogen and methane both can react with oxygen in a
fuel cell. They also can be burned directly. Which has
greater heat content when burned, 1.00 g of H_2 or
1.00 g of CH_4? *Hint:* Write the balanced chemical
equation for each reaction and use the bond energies in
Table 4.4 to help answer this question.

40. Engineers have developed a prototype fuel cell that
converts gasoline to hydrogen and carbon monoxide.
The carbon monoxide, in contact with a catalyst, then
reacts with steam to produce carbon dioxide and more
hydrogen.

   a. Write a set of reactions that describes this prototype
fuel cell, using octane (C_8H_{18}) to represent the
hydrocarbons in gasoline.

   b. Speculate as to the future economic success of this
prototype fuel cell.

41. How can the principles of green chemistry be applied
during the development of new technologies for batteries,
photovoltaic cells, and fuel cells? Give three specific
examples.

42. Consider this representation of two water molecules in
the liquid state.

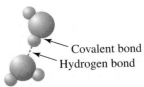

Covalent bond
Hydrogen bond

   a. What happens when water boils? Does boiling break
covalent bonds within molecules or does it disrupt
hydrogen bonds between molecules? *Hint:* Revisit
Chapter 5.

   b. What happens when water undergoes electrolysis?
Does this break covalent bonds within molecules or
does it disrupt hydrogen bonds between molecules?

43. Why isn't the electrolysis of water the best method to
produce hydrogen gas?

44. Small quantities of hydrogen gas can be prepared in the
lab by reacting metallic sodium with water, as shown in
this equation.

$$2\,Na(s) + 2\,H_2O(l) \longrightarrow H_2(g) + 2\,NaOH(aq)$$

   a. Calculate the grams of sodium needed to produce
1.0 mol of hydrogen gas.

b. Calculate the grams of sodium needed to produce sufficient hydrogen to meet an American's daily energy requirement of 1.1×10^6 kJ.

c. If the price of sodium were $165/kg, what would be the cost of producing 1.0 mol of hydrogen? Assume the cost of water is negligible.

45. a. As a fuel, hydrogen has both advantages and disadvantages. Set up parallel lists for the advantages and disadvantages of using hydrogen as the fuel for transportation and for producing electricity.

b. Do you advocate the use of hydrogen as a fuel for transportation or for the production of electricity? Explain your position in a short article for your student newspaper.

46. Fossil fuels have been called ". . . Sun's ancient investment on Earth." Explain this statement to a friend who is not enrolled in your course.

47. The cost of electricity generated by solar thermal power plants currently is greater than that of electricity produced by burning fossil fuels. Given this economic fact, suggest two strategies that might be used to promote the use of environmentally cleaner electricity from photovoltaics.

48. Name two current applications of photovoltaic cells *other* than the production of electricity in remote areas.

Exploring Extensions

49. Although Alessandro Volta is credited with the invention of the first electric battery in 1800, some feel this is a reinvention. Research the "Baghdad battery" to evaluate the merit of this claim.

50. Oxidation and reduction also take place during combustion, the process of burning a fuel in oxygen. Because no metal electrodes are present during combustion, the electron transfer is harder to track. In this case, oxidation occurs when a chemical species loses H atoms or gains O atoms. Similarly, reduction occurs when a chemical either gains H atoms or loses O atoms.

a. Use these new definitions to determine which species is oxidized and which is reduced in equation 8.17, the combustion of hydrogen. Explain.

$$H_2(g) + 1/2 \, O_2(g) \longrightarrow H_2O(g)$$

b. Determine which species is oxidized and which is reduced in each of the following combustion reactions. Explain.

$$C + O_2 \longrightarrow CO_2$$
$$2 \, C_8H_{18} + 17 \, O_2 \longrightarrow 16 \, CO + 18 \, H_2O$$

51. If all of today's technology presently based on fossil fuel combustion were replaced by hydrogen fuel cells, significantly more H_2O would be released into the environment. Is this of concern? Research other consequences that might be anticipated from switching to an economy powered by hydrogen, a so-called "hydrogen economy."

52. Iceland is taking bold steps to cut its ties to fossil fuels. Part of the plan is to demonstrate that the country can produce, store, and distribute hydrogen to power both public and private transportation.

a. Name three factors that motivate Iceland to cut its ties to fossil fuels.

b. What tangible outcomes have resulted to date?

c. Can lessons learned in Iceland be relevant where you live? Explain.

53. At the cutting edge of technology the line between science and science fiction often blurs. Investigate the "futuristic" idea of putting mirrors in orbit around Earth to focus and concentrate solar energy for use in generating electricity.

54. Although silicon, used to make solar cells, is one of the most abundant elements in Earth's crust, extracting it from minerals is costly. The increased demand for solar cells has some companies worried about a "silicon shortage." Find out how silicon is purified and how the PV industry is coping with the rising prices.

55. Figure 8.21 shows an array of photovoltaic cells installed at the Solarpark Gut Erlasee in Bavaria, Germany.

a. At present, where is the largest photovoltaic power plant located in your country?

b. Name two other locations of large-scale photovoltaic cell installations.

c. Name two factors that promote a centralized array rather than individual rooftop solar units.

9 The World of Polymers and Plastics

On the trail.

"Hey, I'm out of here this weekend. I sure need a break from studying. I've got a date with my mountain bike before the cold weather really sets in. It doesn't matter to me if it rains. Check out the new biking gear I just bought. I have my Thinsulate-lined Gore-Tex jacket that will keep me dry. And if it gets cold? No problem. I have a C-Tech polyester microfiber jersey and new Porelle Drys socks to keep me warm. My Spandura tights are 18 times tougher than my older nylon/Lycra ones. And I can stash the rest of my gear in my lightweight nylon backpack with polyurethane padded straps for comfort. Like I said, I'm outta here. It will feel great to get away from the synthetic world and commune with Ma Nature for a change. Catch ya later."

Our biker is right. When she hits the road she can get away from school, exams, and the stress that may accompany these. But she cannot get away from the world of synthetics—synthetic polymers, that is. Polymers are everywhere. They are present in synthetic forms, such as Gore-Tex and nylon. They also are present in natural forms, such as proteins and cellulose, the structural material in plants. While both are important, synthetic polymers are the focus of this chapter.

Consider This 9.1   Polymers for Fun

Choose a favorite activity—on the land, water, or in the air. Search the web to find a company that manufactures or sells equipment for this activity. Which polymers are mentioned on the website? Make a table of their names, the equipment in which they are used, and any desirable properties cited as advertising points.
Hint: Some polymer names start with *poly*, such as polyester or polypropylene. Others have trade names, such as Gore-Tex, Orlon, or Styrofoam. Still other polymers are coatings and resins and may be mentioned as epoxides or acrylics.

Recreation has been revolutionized by the introduction of synthetic polymers and plastics. Football is played on artificial turf by players wearing plastic helmets. Tennis balls are made from synthetic polymers; so are parts of tennis racquets. Carbon fibers embedded in plastic resins provide the strength and flexibility required in trail bikes, fishing rods, golf-club shafts, and sailboat hulls. Hockey players can skate on rinks of Teflon or high-density polyethylene. Most canoes now are made of Kevlar, Royalex ABS sandwiched between vinyl, or polyethylene synthetic polymers, rather than birch bark, wood, or aluminum.

Consider This 9.2   Tennis Analysis

Examine this photo of a tennis player. The clothing, racket, ball, and net may all contain synthetic materials. Describe the properties of these materials that make them well suited for their intended use.

At this moment, you probably are wearing or carrying at least a dozen materials that did not exist 75 years ago, with some new in the last decade. Your running shoes alone contain a variety of polymers: the sole, the trim, the foam padding, the upper, the laces, and even the lace tips. These polymers add cushioning, support, and shock absorption as you walk, jog, or run. Your shirt and pants may contain synthetic fibers as well. These can add a bit more stretch, minimize wrinkles, repel rain or stains, or provide additional strength. Your clothing also may be made of microfibers, that is, polymers made into small-diameter threads. These fine threads can add desirable properties such as the ability to insulate from the cold and breathability.

Also look for polymers in the world around you. For example, think about your pen and cell phone. Most likely both have plastic cases. The same is true for calculators and computers. And of course the CDs and DVDs you play are made of polymers.

Polymers such as these come primarily from a single raw material: petroleum. As you learned earlier, our planet's supply of petroleum is limited and most of it is refined to produce fuels. Only a small percent is used to manufacture polymers and other important chemicals. In principle, polymers can be made from any carbon-containing starting material. Crude oil, however, remains the most convenient and economical.

Chemical companies now can make "eco-friendly" polymers from renewable materials such as wood, cotton fibers, straw, starch, and sugar. This work has implications for land use, crop productivity, and no doubt much more. For example, the NatureWorks unit of the company Cargill has introduced a plastic made from corn glucose that can be used

Dmitri Mendeleev, the great Russian chemist who proposed the periodic table, remarked that burning petroleum as a fuel "would be akin to firing up a kitchen stove with bank notes."

for clothing, food packaging, and even the plastic parts of automobiles. In 2006, spokes-woman Ann Tucker reported "triple-digit sales growth for the past two years." Another area of growth is "eco-dinnerware" that is designed with an eye toward minimizing the use of petroleum and maximizing disposal options. For example, the company EarthShell produces plates and bowls from renewable materials such as potato starch that are more readily composted. These now are on the shelves at some large food stores. In August 2006, EarthShell CEO Vincent Truant was quoted in *USA Today* as saying, "People prefer a product from a Midwest cornfield than a Middle East oil field."

Consider This 9.3   Eco-Friendly Plates

Do you want to eat your tofu off something other than a Styrofoam plate? Search on the web for NatureWorks from Cargill, Sorona from DuPont, eco-dinnerware from EarthShell, or for other bio-based polymers. What do these polymers promise? Which products are now on the market? Are they biodegradable?

To understand the complexities surrounding the sources and uses of polymers, we first must understand their structure and how they are synthesized. We will focus on these in the next few sections and then will return to the issues of polymer use in our society.

9.1 | Polymers: Long, Long Chains

Rayon, nylon, Lycra, polyurethane, Teflon, Saran, Styrofoam, Formica! These seemingly different materials all are synthetic polymers. What they have in common is evident at the molecular level. **Polymers** are large molecules consisting of a long chain or chains of atoms covalently bonded together. A polymer molecule can contain thousands of atoms and have a molar mass of over a million grams. Given their size, polymers are referred to as **macromolecules,** that is, molecules of high molecular mass that have characteristic properties because of their large size.

Monomers (*mono* meaning "one"; *meros* meaning "unit") are the small molecules used to synthesize the larger polymeric chain. Each monomer is analogous to a link of the chain. The polymers (*poly* means "many") can be formed from the same type of monomer or from a combination of monomers. The long chain shown in Figure 9.1 may help you to imagine a polymer made from identical monomers, that is, identical links in a chain.

Keep in mind that chemists did not invent polymers. Natural polymers are found both in plants and animals. For example, wood, wool, cotton, starch, natural rubber,

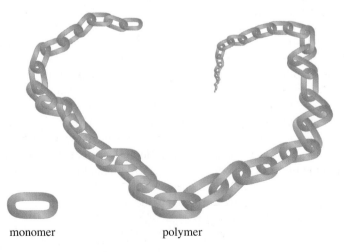

monomer   polymer

Figure 9.1

Representations of a monomer (*single link*) and a polymer (*long chain*) made from one type of monomer.

Figure 9.2

Oak logs and grass have in common the natural polymer cellulose. The monomer is glucose.

skin, and hair are all natural polymers. Like synthetic polymers, natural ones exhibit a stunning variety of properties. They give strength to an oak tree, delicacy to a spider's web, softness to goose down, and flexibility to a blade of grass (Figure 9.2).

When chemists first created polymers, they used natural ones such as cotton and rubber as models. Indeed, many synthetic polymers originally were substitutes for expensive or rare materials that occurred naturally. As the number of new polymers grew, their uses expanded dramatically. Many applications make use of the fact that plastics deliver comparable strength at a lower weight. For example, polymers generally have a density of 1–2 g/cm^3 compared with that of approximately 8 g/cm^3 for steel. Thus an automobile body built from polymers would weigh far less than its steel counterpart and would require less energy (and therefore less fuel) to move. Similarly, plastic packaging reduces weight, eliminates breakage, and helps save fuel during shipping.

> Density was discussed in Section 5.2.

Synthetic polymers sometimes are called plastics, a term that applies to materials with a broad range of properties and applications. The word *plastic* is both an adjective, "capable of being molded," and a noun, "something capable of being molded." More specifically, the *Merriam-Webster Collegiate Dictionary,* 11th edition, refers to plastics as "any of numerous organic synthetic or processed materials that are mostly . . . polymers of high molecular weight and that can be molded, cast, extruded, drawn, or laminated into objects, films, or filaments." As it turns out, some metals have plastic-like properties because they can be "cast, extruded, and drawn." Therefore, the word *plastic* has many applications beyond that of describing synthetic polymers. We will generally use the word *polymer* in this chapter. Either way, plastic or polymer, we are talking about a large molecule that has been synthesized from smaller ones.

9.2 | Adding up the Monomers

How do monomers combine to make a polymer? In the previous section, we used a chain to represent a polymer, but made no mention of how the chain was formed. In this section, we will provide the details. As you may surmise, a polymer has no chain links as such. Rather, covalent chemical bonds connect the monomers.

Polyethylene will serve as our first example. As the name indicates, polyethylene is synthesized from the monomer ethylene, $H_2C=CH_2$. Ethylene is the common name for ethene, the smallest member in the family of hydrocarbons containing a $C=C$ double bond. In the polymerization reaction, n molecules of ethylene combine to form polyethylene.

$$n \begin{array}{c} H \\ \diagdown \\ C \end{array}\!\!=\!\!\begin{array}{c} H \\ \diagup \\ C \\ \diagdown \\ H \end{array} \quad \xrightarrow{\text{R}\cdot} \quad \left[\begin{array}{cc} H & H \\ | & | \\ C & C \\ | & | \\ H & H \end{array} \right]_n$$

[9.1]

Closely examine each part of equation 9.1. On the left is the ethylene monomer. The coefficient *n* in front of it specifies the number of molecules that react. In turn, this specifies the length and the molecular mass of the polymer. Molecular masses of polyethylene generally fall between 10,000 and 100,000 g/mol, but can run into the millions. On the right, the *n* also appears as a subscript in the product, indicating that each monomer becomes part of the long chain. The large square brackets enclose the repeating unit of the polymer.

Polyethylene is the sole product. It contains exactly the same number and kinds of atoms as did the monomers. Literally, the monomers add to one another to form a long chain of *n* units. As a result, we call this **addition polymerization:** the monomers add to the growing polymer chain in such a way that the product contains all the atoms of the starting material. No other products are formed.

In equation 9.1, also notice the R• over the arrow. To appreciate its significance, you need to know a bit more about the ethylene monomer. Often produced at oil refineries, this compound is a colorless gas with a faint gasoline-like odor. It is distinctly different from the odorless solid polyethylene (Figure 9.3a). As hydrocarbons, however, ethylene and polyethylene share the property of being flammable, with ethylene being much more highly so. Although not classified as an air pollutant, ethylene still is a VOC, that is, a volatile organic compound. As you learned in Chapter 1, any VOC released into the atmosphere can contribute to the buildup of photochemical smog. For several reasons, then, safety precautions are needed when transporting ethylene from the refineries to the sites where polyethylene is manufactured. Since it is inconvenient to transport ethylene as a gas (think how large the container would have to be), it first is subjected to pressure and low temperature. The resulting liquid then can be transported in tank cars that bear a label like the one shown in Figure 9.3b. But under these conditions will the ethylene polymerize in the tank car? Fortunately, no. Clearly the end user of the ethylene would be distressed to receive a tank car full of solid polyethylene! Perhaps you now can better appreciate the R• that appears over the arrow in equation 9.1. In the absence of a catalyst such as R•, ethylene does not polymerize.

The catalyst R• warrants further examination. As a free radical, it has an unpaired electron and is highly reactive. As shown in Figure 9.4, R• attaches to a $H_2C=CH_2$ molecule to initiate the chain-forming process. Also recall that the double bond in ethylene contains *four* electrons. After an ethylene molecule reacts with R•, *two* of these electrons remain, leaving a single bond. The other two electrons move (as the red arrows indicate) to form two new single bonds, one to R and the other where the next ethylene molecule adds to the chain. Thus the chain grows.

The reactive hydroxyl radical, •OH, was discussed earlier in Sections 1.11, 2.3, and 6.7.

(a)   (b)

Figure 9.3

(a) Bottles made from polyethylene. (b) A sign posted on a railway tank car that transports liquefied ethylene. The 1038 identifies it as ethylene, the red diamond indicates high flammability, and the 2 indicates moderate reactivity.

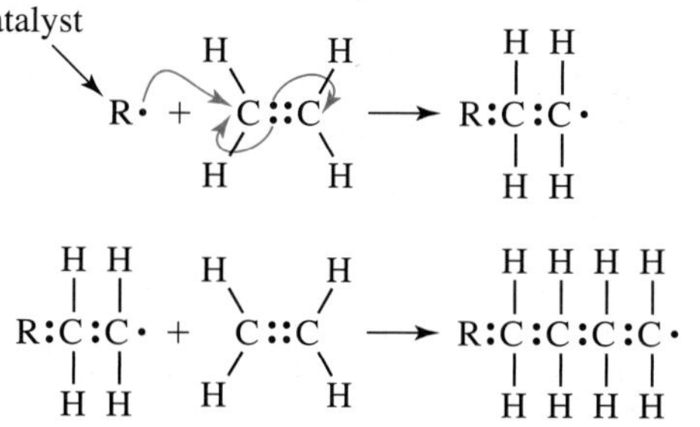

Initiating
free-radical
catalyst

$$R\cdot + \quad \overset{H}{\underset{H}{C}}{::}\overset{H}{\underset{H}{C}} \longrightarrow R{:}\overset{H}{\underset{H}{C}}{:}\overset{H}{\underset{H}{C}}\cdot$$

$$R{:}\overset{H}{\underset{H}{C}}{:}\overset{H}{\underset{H}{C}}\cdot + \quad \overset{H}{\underset{H}{C}}{::}\overset{H}{\underset{H}{C}} \longrightarrow R{:}\overset{H}{\underset{H}{C}}{:}\overset{H}{\underset{H}{C}}{:}\overset{H}{\underset{H}{C}}{:}\overset{H}{\underset{H}{C}}\cdot$$

Figure 9.4

The polymerization of ethylene.

 Figures Alive! Visit the textbook's website to see an animated version of this figure. Look for the Figures Alive! icon in this chapter as a guide to related activities.

As each monomer of ethylene adds, two things happen: the double bond between the carbons converts to a single bond, and a new bond forms between the monomer and the growing chain. This process is repeated many times over in many chains simultaneously. Occasionally, the ends of two polymer chains join and stop the chain growth. The process also can be stopped by adding specific compounds to the reaction that "cap" the reactive chain end. And of course the process stops if you run out of monomers. The result of all this chemistry is that gaseous ethylene is converted to solid polyethylene.

Although we placed R• over the arrow in equation 9.1, in truth, we also could have represented the reaction like this.

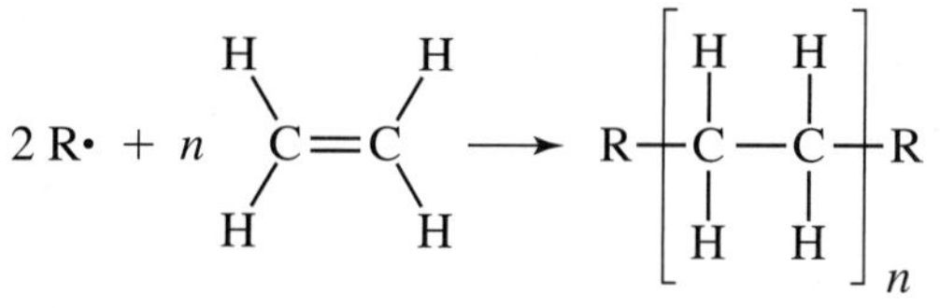

Industrial chemists use several synthetic routes to produce polyethylene. The most common routes employ a metal catalyst and mild temperatures.

$$2\,R\cdot + n \quad \overset{H}{\underset{H}{C}}{=}\overset{H}{\underset{H}{C}} \longrightarrow R{-}\left[\overset{H}{\underset{H}{\overset{|}{C}}}{-}\overset{H}{\underset{H}{\overset{|}{C}}}\right]_n{-}R \qquad [9.2]$$

However, we will continue with our practice of omitting R• as a reactant as we did in equation 9.1. The R group that "caps" the end of the molecule can be considered chemically insignificant with respect to the long chain.

The numerical value of n and hence the length of the chain can vary. During the manufacturing process, n will be adjusted in order to create specific properties for the polymer. Moreover, within a single sample the individual polymer molecules can have varying lengths. In every case, however, the molecules contain a chain of carbon atoms that are attached to one another by single bonds. In essence, any given polyethylene molecule is like an octane molecule, but much, much longer.

Your Turn 9.4   Polymerization of Ethylene

In equation 9.1 for the polymerization of ethylene, suppose that $n = 4$.

a. Rewrite equation 9.1 to indicate this.
b. Draw the structural formula of the product without using brackets. Remember to put R groups at the ends of the chain.
c. In terms of its molecular structure, how does the product differ from octane?

Answer

c. Octane is C_8H_{18}. Although the product molecule similarly has eight carbon atoms, it has two fewer hydrogen atoms and two R groups at the ends of the molecule.

9.3 | Polyethylene: A Closer Look

Polyethylene has a wide variety of uses. For example, it is found in plastic milk jugs, detergent containers, baggies, and packing materials (Figure 9.5). Yet, as we have seen in the previous section, all polyethylene is made from the same starting material, $H_2C=CH_2$. How can this one monomer form polymers that can be used in so many different ways?

Your Turn 9.5   Polyethylene Hunt

Take time to explore the properties of polyethylene. As we will see in the next section, polyethylene containers, baggies, and packaging materials are marked either as low density (LDPE) or high density (HDPE). Using this code as your guide, locate some items made of polyethylene and note their properties. Do LDPE and HDPE differ in flexibility? Is one more translucent? Is one more often colored with a pigment than the other?

$\boxed{4}$ LDPE $\boxed{2}$ HDPE

The different properties of polyethylene largely stem from differences in the long molecules that compose it. Relatively speaking, these molecules are very long indeed. Imagine a polyethylene molecule to be as wide as a piece of spaghetti. If this were the case, the molecule would be almost a half mile long! To continue the analogy, the polyethylene used to make plastic bags contains molecular chains arranged somewhat like cooked spaghetti on a plate. The strands are jumbled up and not very well aligned, although in some regions the molecular chains run in parallel. Moreover, the polyethylene chains, like the spaghetti strands, are not bonded to one another.

Recall that we used hydrogen bonding to describe the attractive force *between* water molecules in the liquid phase. The hydrogen bond is not a covalent bond. Rather, it is an **intermolecular force,** that is, an attraction between two molecules resulting from the interactions of the electron clouds and nuclei. These attractive forces are different from the covalent bonds that exist *within* each molecule in which electrons are truly shared between two atoms. Intermolecular attractive forces are much weaker than covalent bonds, but they still have an effect on the behavior of a group of atoms. In water, these forces help to keep the molecules very close together, sliding past and around one another as the liquid flows, unlike in the gas phase where the molecules are far apart and bump into one another less frequently.

Polymers that contain only hydrogen and carbon, such as HDPE, LDPE, and polypropylene (PP), do not have hydrogen bonds. Rather, there is another type of intermolecular attractive force that keeps the molecules close to one another. Each atom

Hydrogen bonding was explained in Section 5.2. The unusual properties of water also were mentioned.

Figure 9.5

Packaging made from polyethylene.

in the long polymeric chain contains its own electrons. But these electrons can be attracted to the atoms on neighboring molecular chains, and the magnitude of the attraction between strands of polyethylene is a direct result of the large number of atoms involved. The attraction is a bit like that between the two halves of Velcro. The bigger the surface area of one Velcro strip, the better it will hold to the other. The intermolecular forces holding polyethylene together are called **dispersion forces,** and they are attractions between molecules that result from a distortion of the electron cloud that causes an uneven distribution of the negative charge. Dispersion forces can be quite significant in very large molecules such as polymers.

Evidence of the molecular arrangement of polyethylene can be obtained by doing a short experiment. Cut a strip from a heavy-duty transparent polyethylene bag, grab the two ends of the strip, and pull. A fairly strong pull is required to start the plastic stretching, but once it begins, less force is needed to keep it going. The length of the plastic strip increases dramatically as the width and thickness decrease (Figure 9.6a). A little shoulder forms on the wider part of the strip and a narrow neck almost seems to flow from it in a process called "necking." Unlike the stretching of a rubber band, the necking effect is not reversible, and eventually the plastic thins to the point where it tears.

(a)

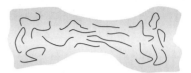

(b)

Figure 9.6

(a) A plastic bag stretched until it "necks." **(b)** A representation of "necking" at the molecular level.

Consider This 9.6   "Necking" Polyethylene

Necking permanently changes the properties of a piece of polyethylene.

a. Does necking affect the number of monomer units, n, in the average polymer?
b. Does necking affect the bonding between the monomer units within the polymer chain?

Figure 9.6b represents the necking of polyethylene from a molecular point of view. As the strip narrows down, the previously mixed-up molecular chains shift, slide, and align parallel to one another in the direction of pull. In some plastics, such stretching, or "cold drawing," is carried out as part of the manufacturing process to alter the three-dimensional arrangement of the chains in the solid. Of course, as the force and stretching continue, the polymer eventually reaches a point at which the strands can no longer realign, and the plastic breaks. Paper, a natural polymeric material, tears when pulled because the strands (fibers) in paper are rigidly held in place and are not free to slip like the long molecules in polyethylene.

A strategy to control the molecular structure and physical properties of polymers is to regulate the branching of the polymer chain. This approach is used to produce high-density polyethylene (HDPE) and low-density polyethylene (LDPE) (Figure 9.7).

As you may have discovered in Your Turn 9.5, the plastic bags dispensed in the produce aisles of supermarkets are often LDPE. These bags are stretchy, transparent, and not very strong. This low-density form was the first type of polyethylene to be manufactured. A study of its structure reveals that the molecules consist of about 500 monomeric units and that the central polymer chain has many side branches, like the limbs radiating from a central tree trunk.

About 20 years after the discovery of LDPE, chemists were able to adjust reaction conditions to prevent branching. Thus HDPE was born. In their Nobel Prize-winning research, Karl Ziegler (1898–1973) and Giulio Natta (1903–1979) developed new catalysts that enabled them to make linear (unbranched) polyethylene chains consisting of about 10,000 monomer units. Having no side branches, these long chains can arrange parallel to one another (see Figure 9.7a). The structure of HDPE is thus more crystalline than the irregular tangle of the polymer chains in LDPE. The more highly ordered structure of HDPE gives it greater density, rigidity, strength, and a higher melting point than LDPE. Furthermore, the high-density form is opaque; the low-density form tends to be transparent.

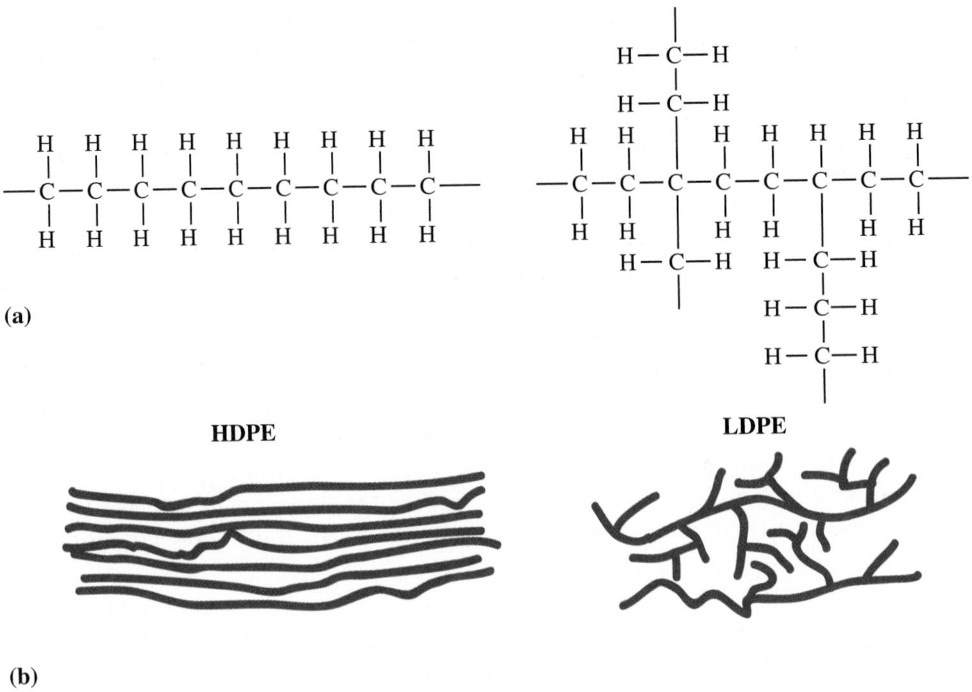

(a)

HDPE   **LDPE**

(b)

Figure 9.7

High-density (linear) polyethylene and low-density (branched) polyethylene.
(a) Details of bonding. **(b)** Schematic representations.

Consider This 9.7   HDPE and LDPE

The densities of HDPE and LDPE are actually quite close: 0.96 g/cm^3 and 0.93 g/cm^3, respectively. Nonetheless, this density difference results in different properties. Use Figure 9.7 to rationalize why the density of HDPE is slightly greater than that of LDPE.

Given these differences in properties, HDPE and LDPE have different applications. High-density polyethylene is used to make toys, containers, stiff or "crinkly" plastic bags, and heavy-duty pipes. A newer use of HDPE was spurred by surgery patients who have HIV/AIDS. Without suitable protection, surgeons would run the risk of acquiring the disease through contact with a patient's blood. Allied-Signal Corporation has produced a linear polyethylene fiber called Spectra that can be fabricated into liners for surgical gloves. Spectra gloves are said to have 15 times more resistance to cuts than medium-weight leather work gloves, but are so thin that a surgeon can retain a keen sense of touch. A sharp scalpel can be drawn across the glove with no damage to either the fabric or the hand inside. Such strength is in marked contrast to the properties of the common polyethylene plastic grocery bag.

Consider This 9.8   Shopping for Polymers

The *Macrogalleria* is an intriguing website that contains a virtual shopping mall with stores selling items made from polymers. Visit the stores to find at least six items made of LDPE or HDPE. Make your selections from several different shops. A link to the site is provided at the textbook's website.

It would be a mistake to conclude that polyethylene is restricted to the extremes represented by highly branched or strictly linear forms. By modifying the extent and location of branching in LDPE, its properties can be varied from the soft and wax-like coatings on paper milk cartons to stretchy plastic food wrap. Because of its linearity and close packing, HDPE is sufficiently rigid to be used for plastic milk bottles. Unfortunately, consumers sometimes are unaware of the consequences of such structural tinkering. For example, the higher melting point of HDPE (130 °C) means that it can go through the dishwasher safely. In contrast, objects made of LDPE (melting point of 120 °C) may melt if near the heating element. The hot water of a dishwasher will melt neither HDPE nor LDPE.

Finally, one of the first and most important uses of polyethylene was a consequence of its being a good electrical insulator. During World War II, polyethylene was used by Allied Forces to coat electrical cables in aircraft radar installations. Sir Robert Watt, who discovered radar, described polyethylene's critical importance.

"The availability of polythene [polyethylene] transformed the design, production, installation, and maintenance problems of airborne radar from the almost insoluble to the comfortably manageable . . . A whole range of aerial and feeder designs otherwise unattainable was made possible, a whole crop of intolerable air maintenance problems was removed. And so polythene played an indispensable part in the long series of victories in the air, on the sea, and on land, which were made possible by radar."

Quoted by J. C. Swallow in "The History of Polythene" from
Polythene—The Technology and Uses of Ethylene Polymers (2nd ed.)
A. Renfrew, Editor. London: Iliffe and Sons, 1960.

Polyethylene is called polythene in the United Kingdom.

Consider This 9.9   Other Types of Polyethylene

In addition to LDPE and HDPE, polyethylene is manufactured as MDPE and LLDPE. Use the web to find out about these and other types. How do the properties of these differ?

9.4 | The "Big Six": Theme and Variations

Today, more than 60,000 synthetic polymers are known. Since 1976, the United States has produced a larger volume of synthetic polymers than the volumes of steel, copper, and aluminum combined. Although many polymers were developed for specialized uses, six account for roughly 75% of those used in the United States. We refer to these everyday polymers as the "Big Six." These six polymers, listed in Table 9.1, are polyethylene (low-density and high-density), polypropylene, polystyrene, polyvinyl chloride, and polyethylene terephthalate. Pronounce this last word as "ter-eh-THAL-ate" (the "ph" is silent).

Table 9.1 also lists some of the important properties of these six polymers. All are solids that can be colored with pigments. Although all are insoluble in water, some dissolve or soften in the presence of hydrocarbons, fats, and oils. These six polymers are **thermoplastic,** meaning that with heat they can be melted and reshaped over and over again. However, they exhibit a range of melting points depending on the route by which they were manufactured. In general, polyethylene has a relatively low melting point, with LDPE and HDPE melting at about 120 °C and 130 °C, respectively. Polypropylene (PP) has a higher melting point of 160–170 °C.

Depending on the arrangement of their molecules, polymers have varying degrees of toughness and strength. At the microscopic level, in some parts of a polymer the molecules may have a very orderly and repeating pattern, such as one would find in a crystalline solid. In these **crystalline regions,** the long polymer molecules are arranged neatly and tightly in a regular pattern. In other parts of the

Table 9.1	The Big Six		
Polymer	**Monomer**	**Properties of Polymer**	**Uses of Polymer**
Polyethylene (LDPE) 4 LDPE	Ethylene $H_2C{=}CH_2$	Translucent if not pigmented. Soft and flexible. Unreactive to acids and bases. Strong and tough.	Bags, films, sheets, bubble wrap, toys, wire insulation.
Polyethylene (HDPE) 2 HDPE	Ethylene $H_2C{=}CH_2$	Similar to LDPE. More rigid, tougher, slightly more dense.	Opaque milk, juice, detergent, and shampoo bottles. Buckets, crates, and fencing.
Polyvinyl chloride 3 PVC, or V	Vinyl chloride $CHCl{=}CH_2$	Variable. Rigid if not softened with a plasticizer. Clear and shiny, but often pigmented. Resistant to most chemicals, including oils, acids, and bases.	Rigid: Plumbing pipe, house siding, charge cards, hotel room keys. Softened: Garden hoses, waterproof boots, shower curtains, IV tubing.
Polystyrene 6 PS	Styrene	Variable. "Crystal" form transparent, sparkling, somewhat brittle. "Expandable" form lightweight foam. Both forms rigid and dissolve in many organic solvents.	"Crystal" form: Food wrap, CD cases, transparent cups. "Expandable" form: Foam cups, insulated containers, food packaging trays, egg cartons, packaging peanuts.
Polypropylene 5 PP	Propylene	Opaque, very tough, good weatherability. High melting point. Resistant to oils.	Bottle caps. Yogurt, cream, and margarine containers. Carpeting, casual furniture, luggage.
Polyethylene terephthalate 1 PETE, or PET	Ethylene glycol $HO-CH_2CH_2-OH$ Terephthalic acid	Transparent, strong, shatter-resistant. Impervious to acids and atmospheric gases. Most costly of the six.	Soft-drink bottles, clear food containers, beverage glasses, fleece fabrics, carpet yarns, fiber-fill insulation.

Note: The structures of the first five monomers differ only by the atoms shown in blue.

same polymer, you can find **amorphous regions.** Here, the long polymer molecules are found in a random, disordered arrangement and their packing is much looser. Because of their structural regularity, the crystalline regions impart toughness and resistance to abrasion. This accounts for the strength of HDPE and PP. Although some polymers are highly crystalline, most still include amorphous regions as well. These regions impart flexibility. For example, the amorphous regions in PP give it the ability to be bent without breaking. The range of properties among polymers means that they are differently suited for specific applications. The next exercise provides an opportunity to match polymers with their uses.

Consider This 9.10   Uses of the "Big Six"

Use the information in Table 9.1 about the Big Six polymers to determine:

a. Which polymer would not be suitable for margarine tubs because it softens with oil.

b. Which ones are transparent and which one is used in soft-drink bottles.

c. Which one is tough and used for bottle caps. Name another application in which toughness counts.

d. Which ones are listed as unreactive/impervious to acids and thus could serve as orange juice containers.

e. Which one has the lowest melting point. What are the implications for leaving items made from this polymer in a car on a sunny day?

Table 9.1 also shows that six monomers are used to make six different polymers, but perhaps not in the way you would expect. Two of the polymers use the same monomer, ethylene. Three others each use a monomer closely related to ethylene: styrene, vinyl chloride, and propylene. The sixth polymer, polyethylene terephthalate, is the odd one of the six being made using two different monomers. We will return to polyethylene terephthalate, but after first spending time with the other five members of this sextet.

In this section, we will more closely examine the three monomers that structurally are related to ethylene. Here they are together with ethylene.

$$
\underset{\text{ethylene}}{\overset{H}{\underset{H}{>}}C=C\overset{H}{\underset{H}{<}}} \qquad
\underset{\text{vinyl chloride}}{\overset{H}{\underset{H}{>}}C=C\overset{H}{\underset{Cl}{<}}} \qquad
\underset{\text{propylene}}{\overset{H}{\underset{H}{>}}C=C\overset{H}{\underset{CH_3}{<}}} \qquad
\underset{\text{styrene}}{\overset{H}{\underset{H}{>}}C=C\overset{H}{\underset{C_6H_5}{<}}}
$$

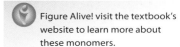

Figure Alive! visit the textbook's website to learn more about these monomers.

In the case of vinyl chloride, one of the hydrogen atoms in ethylene is replaced by a chlorine atom. In propylene, a methyl group ($-CH_3$) substitutes for a hydrogen atom. And in styrene, the substituent is $-C_6H_5$, known as a phenyl group.

As you might suspect, all three undergo addition polymerization just like ethylene. But the substituent introduces variety. To see this, let us examine the polymerization of n molecules of vinyl chloride to form polyvinyl chloride (PVC).

$$
n\ \overset{H}{\underset{H}{>}}C=C\overset{H}{\underset{Cl}{<}} \xrightarrow{R\cdot}
\left[\begin{array}{cc} H & H \\ | & | \\ -C & -C- \\ | & | \\ H & Cl \end{array} \right]_n \qquad \textbf{[9.3]}
$$

The chlorine atom creates a type of asymmetry in the monomer. Visualize this by arbitrarily thinking of the carbon atom bearing two hydrogen atoms as the "head" and the carbon with the chlorine atom as the "tail."

When vinyl chloride molecules add to form polyvinyl chloride, the molecules can orient in one of three ways: head-to-tail, alternating head-to-head/tail-to-tail, and randomly (Figure 9.8). In the repeating head-to-tail arrangement, chlorine atoms are on alternate carbons. In a head-to-head/tail-to-tail arrangement, chlorine atoms are on adjacent carbons. In the random arrangement, an irregular mixture of the previous two occurs. The head-to-tail arrangement is the usual product for polyvinyl chloride.

The arrangement of monomers in the chain is one factor that affects the flexibility of the polymer. Thus each arrangement of PVC has somewhat different properties, with the most regular one, repeating head-to-tail, being the stiffest because the molecules pack more easily together to form crystalline regions. Stiff PVC finds use in drain and sewer pipes, credit cards, house siding, furniture, and various automobile parts. The random arrangement is still stiff, but somewhat less so.

$$
\underset{\substack{\text{tail} \qquad \text{head}}}{\overset{H}{\underset{Cl}{>}}C=C\overset{H}{\underset{H}{<}}}
$$

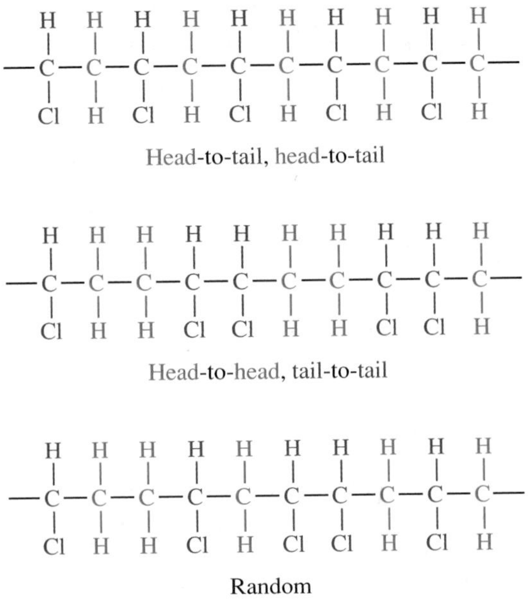

Head-to-tail, head-to-tail

Head-to-head, tail-to-tail

Random

Figure 9.8

Three possible arrangements of the monomers in PVC.

PVC can be further softened with **plasticizers,** compounds that are added in small amounts to polymers to make them softer and more pliable. Plasticizers work by fitting in between the large polymer molecules, thus disrupting the regular packing of the molecules. Flexible PVC that contains plasticizers is familiar in shower curtains, "rubber" boots, garden hoses, clear IV bags for blood transfusions, artificial leather ("patent leather"), and flexible insulation coatings on electrical wires.

Next, let us examine the polymerization of *n* molecules of propylene to form polypropylene. Again, we have addition polymerization, and again several arrangements of the monomer are possible. A particularly useful form of polypropylene is the repeating head-to-tail, head-to-tail arrangement. This regularity imparts a high degree of crystallinity and makes the polymer strong, tough, and able to withstand high temperatures. These properties are reflected in the uses. For example, polypropylene is found as the strong fibers of indoor–outdoor carpeting as well as in tough videocassette cases. Strength and chemical resistance make polypropylene a good choice for applications in which structural ruggedness is required.

Just as ethylene also is called ethene, propylene also is called propene.

Your Turn 9.11   From Propylene to Polypropylene

a. Analogous to equation 9.3, write the polymerization reaction of *n* monomers of propylene to form polypropylene.
   Note: To get the repeating head-to-tail, head-to-tail arrangement, a special catalyst is used, so just put "catalyst" over the arrow.
b. Analogous to Figure 9.8, show a random arrangement of the monomers in a segment of polypropylene.

Consider This 9.12   Polypropylene, "The Tough One"

Polypropylene, one of the Big Six, is used to construct items for which toughness counts. This polymer may not be as familiar to you as polyethylene and PET, because polypropylene items sometimes do not carry a recycling symbol and are not collected curbside. Search the web to find five items manufactured from polypropylene.

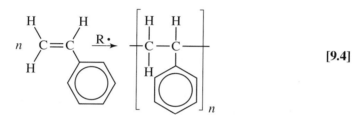

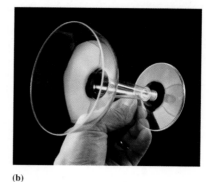

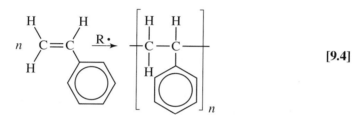

Figure 9.9

(a) The random arrangement of polystyrene.
(b) Partyware made from "crystal" (general purpose) polystyrene.

Finally, let us examine the polymerization of *n* molecules of styrene to form polystyrene. Styrene usually polymerizes to polystyrene with the head-to-tail arrangement. Here is the addition polymerization, and in this case *n* equals about 5000.

[9.4]

Figure Alive! Visit the textbook's website to learn more about other addition polymers.

Polystyrene is a low-cost plastic with several familiar uses. Transparent cases for CDs and clear rigid plastic party glasses are both made from polystyrene. In this form, sometimes referred to as general purpose or "crystal" polystyrene, the polymer is hard and brittle. Have you ever squeezed too hard on your clear plastic party glass causing it to split down the side? It probably was polystyrene. "Crystal" polystyrene is in the random arrangement shown in Figure 9.9.

The familiar foam hot beverage cup, foam egg carton, and foam meat packaging tray are examples of another type of polystyrene (PS), sometimes referred to as "expandable." Such items are made from "expandable" polystyrene beads. These beads contain 4–7% of a **blowing agent,** that is, either a gas or a substance capable of producing a gas to manufacture a foamed plastic. For PS, the blowing agent is typically a low-boiling liquid such as pentane. If the beads are placed in a mold and heated with steam or hot air, the pentane vaporizes. In turn, the expanding gas expands the polymer. The expanded particles are fused together into the shape determined by the mold. Because it contains so many bubbles, this plastic foam not only is light, but also is an excellent thermal insulator. The hard, transparent version of polystyrene is made by molding the melted polymer *without* the blowing agent. It is used to fabricate the cases of computers and TVs, in addition to CD cases and partyware.

At one time, chlorofluorocarbons were included in the list of compounds used as blowing agents. Concern over the involvement of CFCs in the destruction of stratospheric ozone led to their phaseout by 1990. Gaseous pentane (C_5H_{12}) and carbon dioxide are now frequently used for this purpose. The Dow Chemical Company developed a new process that uses pure carbon dioxide as a blowing agent to produce Styrofoam for packaging material. Using the Dow 100% CO_2 technology eliminates the use of 3.5 million pounds of CFC-12 or HCFC-22 as blowing agents. The CO_2 used in the process is a by-product from existing commercial and natural sources, such as ammonia plants and natural gas wells. Thus, it does not contribute additional CO_2, a greenhouse gas. Because it is nonflammable, carbon dioxide is preferred as a blowing agent over pentane, which is flammable.

Styrofoam is a brand name of polystyrene foam insulation, produced by the Dow Chemical Company.

9.5 | Condensing the Monomers

Monomers make the polymer! As we just saw, structural changes in the monomer lead to changes in the properties of the polymer. In this section, we will expand on this theme and offer another variation.

To understand different monomers, it is helpful if you have a way to categorize the common groups of atoms that they may contain. These common groups are known as **functional groups.** They are distinctive arrangements of groups of atoms that impart characteristic chemical properties to the molecules that contain them (Table 9.2).

Earlier you met the phenyl group, $-C_6H_5$, in the styrene monomer. This group consists of six carbon atoms arranged to form a hexagon. Typically, in representations, the C and H atoms in the phenyl group are omitted, and the entire structure is represented by a hexagon with a circle inside. Here are three representations of the phenyl group.

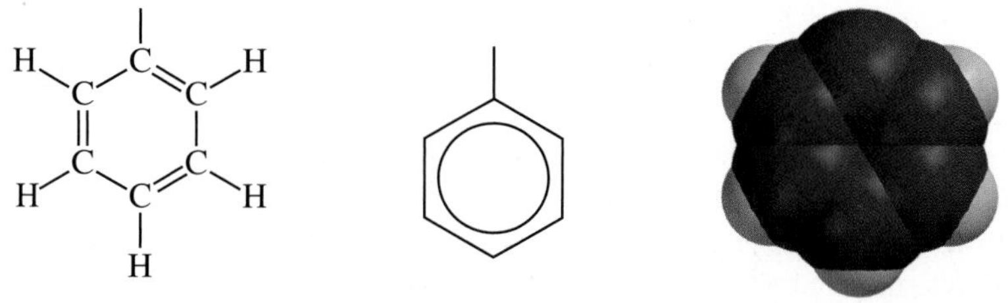

Chapter 10 discusses functional groups in more depth.

Another example of a functional group is the hydroxyl group, $-OH$. This group is found in all alcohol molecules. The carboxylic acid group, $-COOH$, also has this $-OH$ group within it:

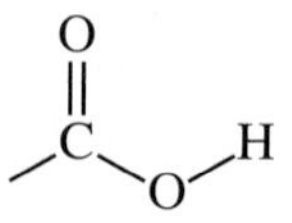

Table 9.2	Selected Functional Groups	
Name	**Chemical Formula**	**Structural Formula**
hydroxyl	–OH	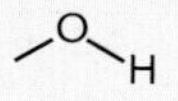
carboxylic acid	–COOH	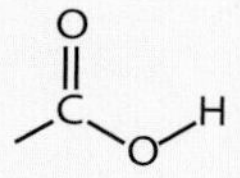
ester	–COOC–	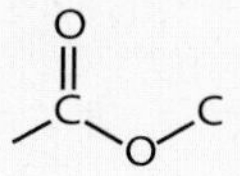
amine	–NH₂	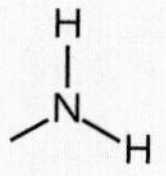
amide	–CONH₂	
phenyl	–C₆H₅	

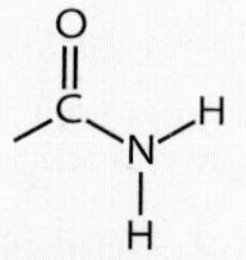

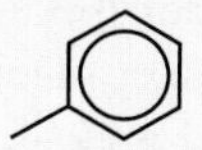

However, the –OH within a –COOH group is *not* an alcohol. Think of the entire carboxylic acid group, –COOH, as a unit.

The ester functional group is closely related to the carboxylic acid group. The general structure is:

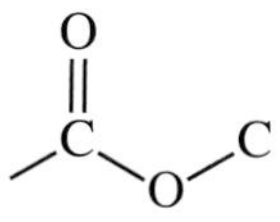

These functional groups and others will be discussed in greater detail later. This set will be sufficient for you to understand the basics of condensation polymers, the topic of this section.

Polyethylene terephthalate (PET or PETE) is formed by **condensation polymerization,** a process in which the monomers join by eliminating (splitting out) a small molecule such as water. Thus, a condensation polymerization has two products: the polymer itself plus molecules released during the polymer's formation. Many polymers are formed by condensation reactions. Natural ones include cellulose, starch, wool, silk, and proteins; synthetics include nylon, Dacron, Kevlar, ABS, and Lexan.

Polyethylene terephthalate is a **copolymer,** a combination of two or more different monomers—ethylene glycol and terephthalic acid (see Table 9.1). With two monomers, we have double the trouble or double the fun, depending on how you wish to approach this. One monomer, ethylene glycol, $HOCH_2CH_2OH$, contains two hydroxyl groups, one on each carbon atom. The other monomer, terephthalic acid, contains two carboxylic acid groups, one on each side of the benzene ring. Thus each monomer is armed with two functional groups.

To understand how copolymers can form, let us first consider just one molecule of each monomer. In the equation that follows, keep your eye on the OH in the carboxylic acid and the H in the alcohol. These are released to form a water molecule.

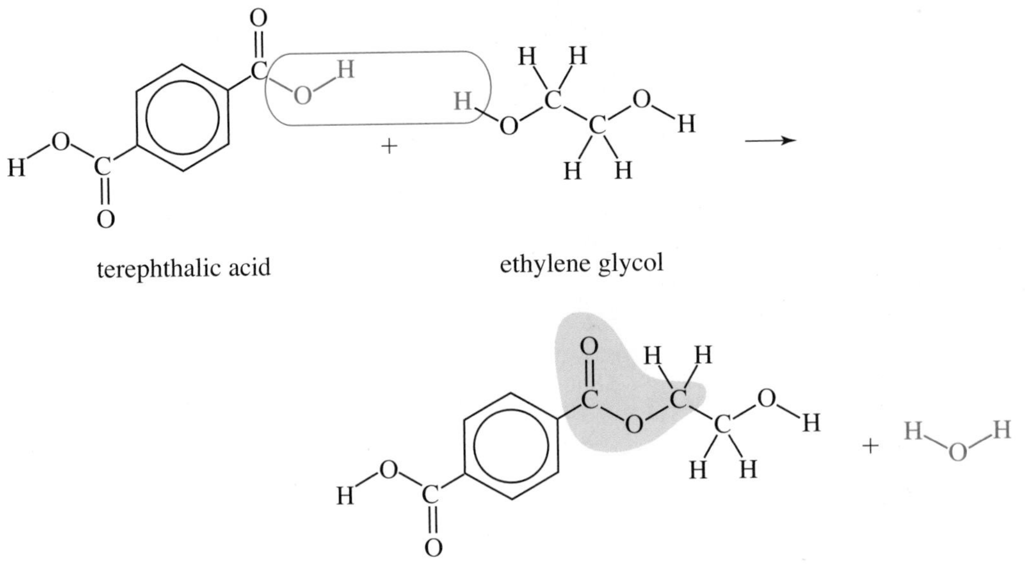

terephthalic acid   ethylene glycol   [9.5]

The remaining portions of the alcohol and the acid connect through the ester functional group. This ester linkage is highlighted in blue.

The product of equation 9.5 still has groups remaining that can react: –COOH on the left end and –OH on the right. The carboxylic acid group can react with the hydroxyl group of another ethylene glycol molecule; likewise, the hydroxyl group can react with a carboxylic acid group of another terephthalic acid molecule. Each time, a molecule of water is released and an ester is formed. This process, represented in Figure 9.10, occurs many times over to yield polyethylene terephthalate. This polymer is a type of polyester, as all the monomers are joined with ester functional groups.

Consider This 9.15   Esters and Polyesters

You have seen that terephthalic acid and ethylene glycol can react. Now consider acetic acid and ethyl alcohol (ethanol):

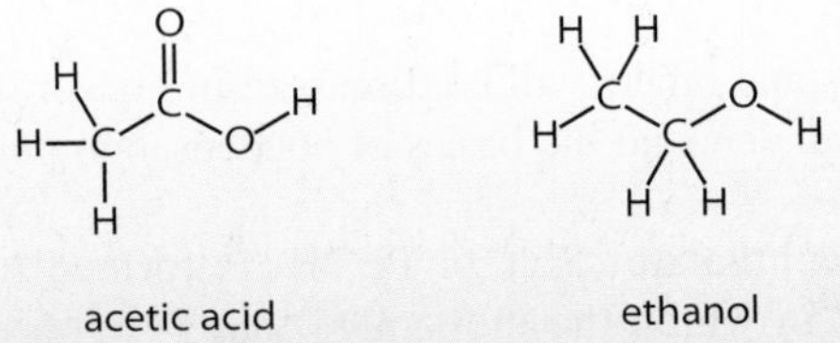

acetic acid   ethanol

a. Show how this carboxylic acid and alcohol can react to form an ester.
   Hint: Remember the water molecule is formed as a product.
b. Could acetic acid and ethanol react to form a polyester? Explain your reasoning.

The most common use for PET is in beverage bottles because PET is semi-rigid, colorless, and gas-tight (Figure 9.11). Narrow, thin ribbons of it (under the trade name Mylar) are coated with metal oxides and magnetized to make audiotapes and video-tapes. Artificial hearts contain parts made of PET. Photographic and X-ray film are made from PET, and containers are made from it for medical supplies to be sterilized by irradiation.

PET is not the only polyester in town. By varying the number and type of carbon atoms in the monomers, chemists have synthesized different polyester polymers. Trade names include Dacron, Polartec, Fortrel, and Polarguard. Polyester can easily be spun into fibers that are easy to wash and dry quickly. Since their introduction, polyester fibers have found many uses in fabrics. Dacron and other polyesters are frequently

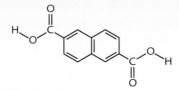

Figure 9.10

The PET polymer grows as molecule of ethylene glycol (right) and a molecule of terephthalic acid (left) are added. The ester functional group is highlighted in blue.

mixed with cotton or wool to make fabric blends. Polartec and Polartec fleece are commonly used in outer vests, jackets, hats, blankets, scarves, and gloves. The next activity shows the two monomers for polyethylene naphthalate (PEN), a polymer similar to PET, but with better temperature resistance.

Consider This 9.16 **From PET to PEN**

In both PET and PEN, the alcohol monomer is ethylene glycol, but the organic acid monomers differ slightly. Here is the organic acid monomer in PEN, naphthalic acid.

Use structural formulas to show the reaction of two molecules of naphthalic acid with two molecules of ethylene glycol.

Figure 9.11

Two-liter soft-drink bottles made from PET.

9.6 | Polyamides: Natural and Nylon

No discussion of condensation polymerization can be complete without examining two other classes of polymers. One is proteins, which are natural polymers; the other is nylon, a synthetic substitute that brilliantly duplicates some of the properties of silk, a naturally occurring protein. Other natural proteins can be found in our skin, fingernails, hair, and muscle tissue.

Amino acids are the monomers from which our body builds proteins. Each amino acid molecule contains two functional groups: an amine group ($-NH_2$) and a carboxylic acid group ($-COOH$). Twenty different amino acids occur naturally, each differing in one of the groups attached to the central carbon atom. This side chain is represented with an R, as shown in this general structural formula for an amino acid.

See Sections 11.7, 12.1, and 12.4 for more about amino acids and proteins.

In some amino acids, R consists of only carbon and hydrogen atoms; in others, R may include additional atoms, such as oxygen, nitrogen, and even sulfur. Some R groups have acidic properties; others are basic.

As monomers, amino acids join to form a long chain via condensation polymerization. However, there are three key differences between the condensation polymer PET and any given protein: (1) PET monomers are linked by the ester functional group. In contrast, both proteins and nylon are **polyamides;** that is, condensation polymers that contain the amide functional group. (2) PET contains just two monomers, ethylene glycol and terephthalic acid, that must be in a 1:1 ratio. In contrast, proteins can contain up to 20 different amino acid monomers that can be in any ratio. And (3) as monomers, each amino acid has two *different* functional groups, an amine and a carboxylic acid. In PET, each monomer contained two of the *same* functional group: an alcohol or a carboxylic acid.

Let's see how these three differences play out. Here is a representation of the reaction between two amino acids, one with the side chain R and the other with R′.

[9.6]

peptide bond

In this reaction, an amide is formed and a molecule of water is eliminated. This amide contains a C–N bond known as a **peptide bond,** the covalent bond that forms when the –COOH group of one amino acid reacts with the –NH$_2$ group of another, thus joining the two amino acids. In the sophisticated chemical factories called biological cells, this condensation reaction is repeated many times over to form the long polymeric chains called proteins. Given that there are 20 different naturally occurring amino acids as building blocks, a great variety of proteins can be synthesized. Some contain hundreds of these 20 amino acids, whereas others contain only a few.

Chemists often attempt to replicate the chemistry of nature. For example, a brilliant chemist working for the DuPont Company, Wallace Carothers (1896–1937) (Figure 9.12) was studying many polymerization reactions, including the formation of peptide bonds (see equation 9.6). Instead of using amino acids, Carothers tried combining adipic acid and hexamethylenediamine, also known as 1,6-diaminohexane.

adipic acid   hexamethylenediamine

Figure 9.12

Wallace Carothers, the inventor of nylon.

Note that the molecule of adipic acid has two carboxylic acid groups, one on each end; similarly, hexamethylenediamine has two amine functional groups. As in the case of protein synthesis, the acid and amine groups react to eliminate water and form an amide. But in this instance, the polymer consisted of two alternating monomers. This polymer is known as nylon.

adipic acid   hexamethylenediamine

$+ \ H_2O$ **[9.7]**

DuPont executives decided the new polymer had promise, especially after company scientists learned to draw it into thin filaments. These filaments were strong and smooth and very much like the protein spun by silkworms. Therefore, nylon was first introduced to the world as a substitute for silk. The world greeted it with bare legs and open pocketbooks. Four million pairs of nylon stockings were sold in New York City on May 15, 1940, the first day they became available (Figure 9.13). But, in spite of consumer passion for "nylons," the civilian supply soon dried up, as the polymer was diverted from hosiery to parachutes, ropes, clothing, and hundreds of other wartime uses. By the time World War II ended in 1945, nylon had repeatedly demonstrated that it was superior to silk in strength, stability, and rot resistance. Today this polymer, in its many modifications, continues to find wide applications in clothing, sportswear, camping equipment, the workroom, the kitchen, and the laboratory.

Your Turn 9.17   Kevlar

Kevlar is a condensation polymer used to make bulletproof vests. Like PET, one of the monomers is terephthalic acid. The other monomer, phenylenediamine, contains two amine functional groups.

terephthalic acid   phenylenediamine

Draw a segment of a Kevlar molecule built from two of each of these monomers.

Figure 9.13
Customers eagerly lined up to buy nylon stockings in 1940, when they were first available commercially.

9.7 | Recycling: The Big Picture

This chapter opened noting that most polymers are synthesized from petroleum feedstocks. But on a day-to-day basis, people seem to be a good deal more concerned about where plastics go than where they came from. The next activity will enable you to do your own tally.

Consider This 9.18   **Plastic You Toss**

Keep a journal of all the plastic you throw away (not recycle) in one week. Include plastic packaging from food and other products that you purchase. Estimate the weight of this plastic—is it a few ounces, a pound, or more? Keep the journal handy because you will be asked to review it later.

Let's put the plastic you discard in perspective. Fortunately, the EPA has been keeping statistics about municipal solid waste—better known as garbage—for over 30 years. **Municipal solid waste** (MSW) includes everything you discard or throw into your trash, including food scraps, grass clippings, and old appliances. MSW does not include industrial waste or waste from construction sites. For the past few decades, the amount of MSW has averaged just under 4.5 pounds per person per day in the United States. The largest single item is paper, as you can see from Figure 9.14. In fact, materials of biological origin such as paper, wood, food scraps, and yard trimmings make up the majority of the materials we toss. What about plastic? In this section we will address the big picture: How much plastic do we recycle and how much must we dispose of? In the next section, we will delve into the details of recycling.

First the good news. The EPA reports that, on average, we each recycle a pound of stuff a day. Furthermore, the percentage of MSW being recycled is increasing. Items that people recycle in bins or at curbside include aluminum cans, office paper, cardboard, glass, and plastic containers. In addition, waste such as grass clippings and food scraps gets composted. This amounts to about a third of a pound. Given these reductions in waste, the amount that goes to the landfill is now closer to 3 pounds per person per day.

However, a 2005 report issued by the EPA lessens any tendency for us to rest on our laurels. It states: "Despite sustained improvements in waste reduction, household waste remains a constant concern because trends indicate that the overall tonnage we create continues to increase." The tonnage is indeed large. Our current MSW per year is just short of 240 million tons and growing.

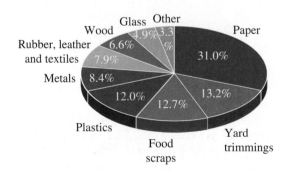

Figure 9.14

What's in your garbage? Prior to recycling, composition by weight of municipal solid waste.

Source: U.S. Environmental Protection Agency, EPA 530 F-08-018, November 2008.

How much of this is plastic? Consult Figure 9.14 to see that roughly 12% of what we discard is plastic. The EPA tallies the data for three kinds of plastics. The first is durable items, such as plastic furniture and garden hoses. The second is nondurable items such as textiles, plastic pens, and safety razors. The third is containers and packaging, such as beverage bottles and food containers. Doing the math, about 12% of our nation's 240 million tons of waste per year is plastic, or about 29 million tons. Depending on the type of plastic, our efficiency in recycling may be low or high, as the next activity will reveal.

Consider This 9.19   Recycling Scorecard

According to the EPA, here is our recycling scorecard for 2008.

Type of Plastic	Weight Generated (millions of tons)	Weight Recovered (millions of tons)
Durable goods	10.52	0.39
Nondurable goods	6.52	Negligible
Containers/Packaging	13.01	1.73

a. For each type of plastic, calculate the plastic recovered as a percent of the waste generated.

b. Nondurable goods have very low recycling rates. Name some nondurable items and suggest reasons why.

If you did the calculations, you saw that we recycle plastics at a surprisingly low rate. This may seem at odds with all the milk jugs and plastic bottles you see being recycled in your own community. Indeed, you are correct. Certain plastics are recycled much more consistently than others. For example, polyethylene milk containers are recycled at a rate of 32% and clear PET soft drink bottles at 25%. Pitching these plastics into recycling bins adds up!

Skeptical Chemist 9.20   Your Personal Tally

Earlier, in Consider This 9.18, you kept a journal of all the plastic you discarded in a week. Do the figures just cited by the EPA ring true? That is, do you throw out far more plastic than you recycle? In your deliberations, remember that you may not have discarded any durable goods (garden hoses and the like) on a particular day. Factor in your use of polymers such as nylon, polyester, and polyethylene as part of the bigger picture.

The plastic that we do not recycle or reuse eventually ends up in one of two places: a landfill (the usual "out of sight, out of mind" approach) or an incinerator. Both are problematic. Although landfill space is still available and in some areas even plentiful, landfills nonetheless have drawbacks. They take up space in congested areas, they have costs associated with their construction and upkeep, they may leak, and they emit methane, a greenhouse gas.

Another complication is that the majority of plastics do not readily biodegrade in the landfill (or anywhere else). Most bacteria and fungi lack the enzymes necessary to break down synthetic polymers. Some microbes, however, possess the enzymes to break down naturally occurring polymers such as cellulose. For example, in Chapter 3 you read about the release of methane by cattle. Actually, the methane is produced when bacteria obtain energy by decomposing cellulose in the cow's rumen. In the same chapter, you also saw that methane is generated by natural decomposition of organic materials in landfills, another result of bacterial activity.

Chemists can engineer biodegradability into some polymers by introducing certain bonds or groups into the chain. For example, research scientists at DuPont recently developed a polymer called Biomax that decomposes in about eight weeks in a landfill. This new polymer is a chemical relative of PET, but differs in that smaller amounts of other monomers are included with the usual ones, terephthalic acid and ethylene glycol. These additional monomers create weak spots in the polymers that are susceptible to degradation by moisture. Once water does its job of breaking the polymer into smaller chains, microorganisms feed off the smaller chains to produce CO_2 and water. Biomax, not yet in wide use, nonetheless has many potential applications: fast-food packaging, lawn bags, and the liners of disposable diapers. In 2003, Biomax was awarded the Compostable Logo for products "designed to compost quickly, completely, and safely" (Figure 9.15).

Achieving biodegradability in polymers raises some concerns, as an EPA report cautions.

> *"Before the application of these technologies can be promoted, the uncertainties surrounding degradable plastics must be addressed. First, the effect of different environmental settings on the performance (e.g., degradation rate) of degradables is not well understood. Second, the environmental products or residues of degrading plastics and the environmental impacts of degradables on plastic recycling are unclear."*

Part of the difficulty is that even natural polymers do not decompose completely in landfills. Modern waste disposal facilities are covered and lined to prevent leaching of waste and waste by-products into the surrounding ground. Landfill linings and coverings also create anaerobic (oxygen-free) conditions that impede bacterial and fungal action. As a result, many supposedly biodegradable substances decompose slowly or not at all. Excavation of old landfills has unearthed old newspapers that are still readable (Figure 9.16) and five-year-old hot dogs that, while hardly edible, are at least recognizable.

Figure 9.15

Logo awarded by the Biodegradable Products Institute.

Figure 9.16

Some buried wastes can remain intact for a long time. This newspaper from 1952 was excavated 37 years later.

Consider This 9.21   **Landfill Liners**

Landfill liners include natural clay and human-made plastics. For example, thick sheets of high-density polyethylene may be employed. Even the best HDPE liners, however, can crack and degrade.

  a. From Table 9.1, which types of chemicals soften HDPE?
  b. Name five substances sent to the landfill that over time could degrade a HDPE liner.

Answer
  b. Cooking oil, shoe polish, and alcohol, to name a few.

What about incineration? Because the Big Six and most other polymers are primarily hydrocarbons, incineration is an excellent way to dispose of them. The chief products of combustion are carbon dioxide, water, and a good deal of energy. In fact, pound for pound, plastics have a higher energy content than does coal. Although plastics account for only 12% of the weight of municipal solid waste, they have approximately 30% of its energy content.

But incineration of plastics is not without its drawbacks. The repeated message of Chapters 1–4, that burning does not destroy matter, applies here as well. The gases produced by combustion may be "out of sight," but they had best not be "out of mind." Burning plastics produces CO_2, a greenhouse gas. Of special concern in incineration are chlorine-containing polymers such as polyvinyl chloride that release hydrogen chloride during combustion. Because HCl dissolves in water to form hydrochloric acid, such smokestack exhaust could make a serious contribution to acid rain. Chlorine-containing plastics sometimes burn to produce toxic gases. Moreover, some plastic products have inks containing lead and cadmium. These toxic metals concentrate in the ash left after incineration and thus contribute to a secondary disposal problem. However, if carefully monitored and controlled, incineration can lead to a large reduction in plastic waste, generate much-needed energy, and have only a small negative effect on the environment.

Refer back to Section 4.10 to see how garbage safely can be burned to produce electricity.

Your Turn 9.22   **Burning a Plastic**

Polypropylene burns completely to produce carbon dioxide and water. Write a balanced chemical equation. Assume an average chain length of 2500 monomers.

Given the problems associated with landfill disposal and incineration of natural and synthetic polymers, attention has logically turned to *recycling* both. Although recycling polymers does not literally dispose of them as does incineration or biodegradation, it helps to reduce the amount of new plastic entering the waste stream. However, in contrast to incineration, recycling polymers requires an input of energy. Furthermore, if the waste plastic is dirty or of low quality, more energy may be needed to recycle it than to produce a comparable quantity of new virgin plastic. Nonetheless, recycling is one of several ways to divert plastic from landfills and incinerators. In the next section, we examine the details.

9.8 | Recycling Plastics: The Details

In the United States, each year we generate more and more tons of municipal solid waste. The overall rate of recycling in recent years, however, has been just over 30%. The graph in Figure 9.17 shows the values over the years for *total* recycling, not just for plastics. To see how the recycling of *plastics* fared over time, we must examine not

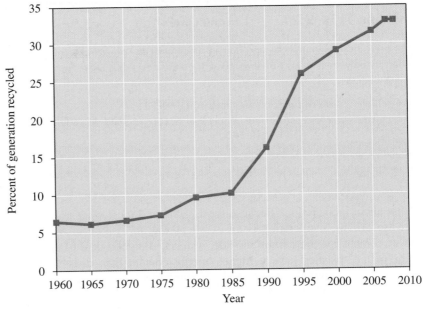

Figure 9.17

Municipal solid waste (MSW) recycling from 1960 to 2008.

Source: U.S. Environmental Protection Agency, EPA 530 F-009-021, November 2009.

just one plastic, but a whole host of them. Some plastics can be easily recycled; others present nothing short of a logistical nightmare. The devil is in the details!

According to the American Plastics Council, the amount of plastic we recycle is steadily increasing. For example, examine Table 9.3 to see the success we are achieving with PET and HDPE. It shows that we are adept in dropping PET pop bottles into recycling bins and in setting out HDPE milk jugs at the curb. The American Plastics Council attributes the success with PET and HDPE to several factors, including increased curbside pick-ups, the rising recycled value of plastic, and successful recycling legislation in California and New York City. In contrast, LDPE and polyvinyl chloride are barely on the radar screen, as these plastics are far more difficult to reuse. Many factors must work synergistically to reduce the plastic waste stream.

Skeptical Chemist 9.23   Pounds or Tons Recycled

Examine the values in Table 9.3 from the American Plastics Council (APC).

a. Are these values comparable to those quoted by the EPA in Consider This 9.19? Assume that the tons quoted were short tons, that is, 2000 lb per ton.

b. Why do you think the EPA reported in one unit, and the APC reported in another?

Answer

a. 1451 + 937 + 0.8 + 21.2 = 2410 million pounds recycled, or 1.20 million tons. This is in the ballpark of the 1.73 million tons quoted by the EPA, especially considering that one set of data is for 2008, one year later. Also, data for polystyrene is missing in the APC set.

But counting individual pop bottles begs the bigger question relating to the overall recycling rate: Are *proportionately* more (or fewer) pop bottles landing in the recycling bin? To answer this, we need data that compare the amount recycled with the total plastic waste stream. According to the Container Recycling Institute,

Table 9.3	Recycled Plastic Bottles in 2008	
Plastic	Amount Recycled in 2008 (million pounds)	Change from 2003 (million pounds)
PET	1451	+448
HDPE	937	+33
PVC	0.4	−0.5
PP	21.2	+15.2
LDPE	0.4	+0.1

Source: American Chemistry Council, 2008, U.S. National Post-Consumer Plastic Bottle Recycling Report.

the news is not good. In 1995, one out of every three containers was recycled; today the ratio stands closer to one in seven. Thus while *more* containers are being recycled, this is offset by the fact that *more* containers are being manufactured. Figure 9.18 shows the bad news, at least for the United States. The values shown on these graphs translate to billions of plastic soda bottles in the landfills, in the incinerators, or more likely, in both. Not just Sweden, but many European countries are more successful.

Can bottle bills help? Certainly in some regards, such as in reducing the amount of litter and municipal solid waste. These bills require consumers to pay a deposit for their beverages at the time of purchase that is later refunded on return of the container. In essence, a bottle bill is a "container deposit law." Such refunds for beer and soft-drink bottles are nothing new. Older people may remember as children the incentive to collect glass bottles in order to claim 1- or 2-cent refunds at the local grocery store. Back in the 1950s and 1960s, of course, there were no plastic beverage bottles. The aluminum can was introduced gradually during this time. According to the Container Recycling Institute, "By 1970, cans and one-way bottles had increased to 60% of beer market share, and one-way containers had grown from just 5% in 1960 to 47% of the soft-drink market. British Columbia enacted the first beverage container recovery system in North America in 1970." So dealing with bottle waste, plastic or otherwise, is nothing new.

Today, several states have bottle bills that apply to some or all beverage containers. If you live in Oregon, your state was the first to enact such a bill in 1970. As of 2011, eleven states have bottle bills, and others have legislation pending (Figure 9.19). In addition, several provinces in Canada have placed deposits on beverage containers. Bottle bills are not without controversy, and you can explore the origins in the next activity.

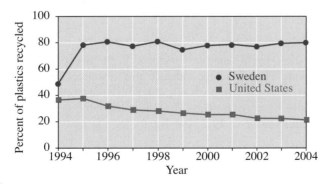

Figure 9.18

Recycling of all plastics in United States and Sweden, 1994–2004.

Source: Container Recycling Institute, including data derived from the American Plastics Council, the National Association for PET Container Resources, and AB Svenska Returpack.

Figure 9.19

States with a bottle bill or a current bottle bill campaign.

Source: Container Recycling Institute, 2007.

Consider This 9.24   Bottle Bill Controversies

Cartoons by Mark Wilson appear regularly in newspapers in upstate New York. In this one from June 2, 2005, what position is he taking? Explain why some grocers and bottle associations stand strongly against bottle bills, yet some consumer groups argue strongly for them. To get you started, links are provided at the textbook's website.

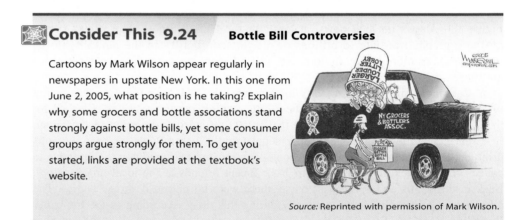

Source: Reprinted with permission of Mark Wilson.

Figure 9.20

Solid waste management hierarchy.

Source: Adapted from www.epa.gov/ epawaste/nonhaz/municipal/index.htm

What other than legislation can help reduce the plastic waste stream? Source reduction and reuse also are of key importance. In fact, these are the preferred ways to manage solid waste, as indicated by the hierarchy from the EPA (Figure 9.20).

Your Turn 9.25   Management Hierarchy

As just mentioned, some states have deposit-refund systems. Where do these fall on the solid waste management hierarchy?

Hint: Your answer may vary, depending on whether the deposits are on glass or plastic bottles.

Source reduction means using less material in the first place and thus generating less waste in the end. The advantages include that resources are conserved, pollution is reduced, and any toxic materials are minimized. As an example, consider the reductions now possible in beverage bottles. The 2-L soda bottle now uses 25% less plastic than when it was introduced in 1975; the 1-gal milk jug now weighs 30% less than a decade ago.

Another reduction has occurred with polystyrene (PS). This plastic is used extensively in food packaging and beverage cups. As a result, it tends to be contaminated by food residues, making it difficult to clean and recycle. Enter source reduction. The Polystyrene Packaging Council reports that since 1974, 9% less of the plastic is now needed to manufacture the products. All told, according to the Council this reduction has "eliminated the need for more than 2900 billion pounds of polystyrene."

Polystyrene also can be reused and with good reason. Packing "peanuts," while a tiny part of the waste stream, are a huge nuisance. These peanuts seem to end up just about everywhere, including storm drains, sewers, and waterways. Reuse is the option of choice, and in some parts of the country, the rate is as high as 30%. Similarly, different types of insulated polystyrene foam containers are reused multiple times.

The reuse option also is an important factor in the paper versus plastic debate. As we pointed out in the previous section, only just over a tenth of municipal solid waste is plastic. In contrast, paper and cardboard make up the largest percentage of municipal solid waste (35%). This raises a question that is often discussed: Which constitutes the lesser environmental burden, paper or plastic? One issue to consider is that 1000 plastic bags weigh 17 lb and have a volume of 1219 in^3 (about 2/3 of a cubic foot). The same number of paper grocery bags weighs 122 lb and takes up 8085 in^3, or about 4.5 ft^3. Figure 9.21 shows the volume relationship to scale.

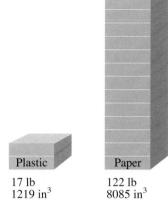

Plastic
17 lb
1219 in^3

Paper
122 lb
8085 in^3

Figure 9.21

Relative volumes occupied by 1000 plastic bags and by 1000 paper bags.

Consider This 9.26   **Neither Paper nor Plastic?**

When you are in a supermarket, which do you usually request, plastic or paper grocery bags? List the advantages and disadvantages of each. How often do you bring your own bags, thereby using neither?

Clearly when it comes to paper or plastic, the answer is neither. Rather, the best method is source reduction by bringing your own sacks to the grocery store.

Returning to our discussion of the solid waste management hierarchy (see Figure 9.20), we come to recycling. Observe that recycling is not the preferred option; rather, it occupies a middle position. For recycling to be successful and self-sustaining, a number of factors must be coordinated. These involve not only science and technology, but also economics and sometimes politics. True recycling involves a closed loop (Figure 9.22) in which plastics are collected, sorted, and then converted into products that consumers buy, use, and later recycle.

In order to recycle, it is necessary to collect the plastic. We already have mentioned several options: collecting at curbside, at local drop-off centers, and through bottle bill programs involving a deposit and refund. For recycling to be successful, a dependable supply of used plastic must be consistently available at designated locations.

Once collected, the plastic needs to be transported to a facility at which it can be sorted and prepared for some marketable commodity. The codes that appear on plastic objects (see Table 9.1) help facilitate the sorting process. Because of the large volume of material, automated sorting methods have been developed. Once sorted, almost any polymer that is not extensively cross-linked can be melted. The molten polymer can be used directly in the manufacturing of new products. Alternatively, it can be solidified, pelletized, and stored for future use.

If a mixture of various polymers is melted, the product tends to be darkly colored and have different properties depending on the nature of the mixture. This type of reprocessed material is generally good enough for lower grade uses such as parking lot bumpers, disposable plastic flower pots, and cheap plastic lumber. Such mixed material is obviously not as valuable as the pure, homogeneous recycled polymer. This underscores the importance of sorting plastics. For similar reasons, manufacturers prefer to use only a single polymer in a product to avoid the need to separate.

Successful recycling requires visionaries. The chemist Nathaniel Wyeth (Figure 9.23) used his creativity to develop the plastic soda bottle in contrast with his artist brother Andrew Wyeth, who expressed his creativity on canvas. Nathaniel Wyeth was quoted as saying: "One of my dreams is that we're going to be able to melt the returned bottles down, mix them with reinforcing fibers, and make car bodies out of them. Then, once the car has served its purpose, rather than put it in the junk pile, melt the car down and make bottles out of it."

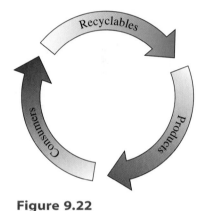

Figure 9.22

True recycling involves a never-ending loop.

Source: Reprinted by permission of the National Association for PET Container Resources.

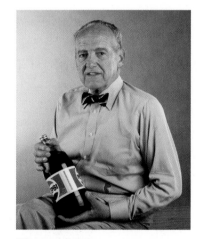

Figure 9.23

Nathaniel Wyeth (1911–1990), with his invention of the PET soda bottle still in use today.

A variant on this method of reprocessing plastics is to decompose the polymers into simpler molecules, in some cases back to the actual original monomers. Fanciful as Nathaniel Wyeth's bottles-to-cars-to-bottles idea might seem, it is nonetheless a reality. DuPont chemists were nominated for a Presidential Green Chemistry Challenge Award for improving on Wyeth's dream. They developed a proprietary process for treating postconsumer PET that, like pulling beads apart from a necklace, unlocks (depolymerizes) the polymer back to its original monomers. These can then be reused to make new PET for other products. Mary Johnson, a DuPont employee, says "Because these monomers retain their original properties, they can be reused over and over again in any first-quality application. A popcorn bag can become an overhead transparency, then a polyester peanut butter jar, then a snack food wrapper, then a roll of polyester film, then a popcorn bag again."

Given a vision and a supply of plastic (ideally clean and sorted), the manufacturers can get to work. The items produced contain varying percentages and types of recycled materials. The terminology is confusing. **Recycled-content products** are those produced from materials that otherwise would have been in the waste stream. These include items manufactured from discarded plastic as well as rebuilt items, such as plastic toner cartridges that are refilled. Trash bags, laundry detergent bottles, and carpeting are common plastic items that may qualify as recycled-content products. Some playground equipment and park benches also are made from discarded plastic.

"Recycled" products are now beginning to provide the origin of the recycled material. **Postconsumer content** is material that previously was used individually or industrially that otherwise would have been discarded as waste. **Preconsumer content** is waste left over from the manufacturing process itself, such as scraps and clippings. Preconsumer plastic, for example, could have been sent to the landfill instead of being "recycled." Finally, **recyclable products** are simply ones that you can recycle. They do not necessarily contain any recycled materials.

Your Turn 9.27   Recycled and Recyclable

Give three examples of recyclable items that you might purchase. Also give three examples of recycled-content products. Can an item fall into both categories?

To complete the cycle shown in Figure 9.22, the recycled items are marketed and (ideally) purchased by consumers. A company or a city would be well advised, however, to determine or create the demand for recycled polymers before completing all the other steps. Without a product and buyers, recycling programs are doomed to fail. In fact, recycling laws in a number of cities have not been implemented and enforced because one of the links in this polymeric chain of supply, collecting, sorting, processing, manufacturing, and marketing was missing. Without all the activities needed for recycling, the system will not work unless it is heavily subsidized. More municipalities are becoming willing to provide the necessary funds as the market for post-consumer plastics grows stronger and prices increase enough to justify significant recycling activity.

Since PET is more successfully recycled than most plastics, we should take a closer look. Approximately 1 billion pounds of PET is recycled annually in the United States alone! But PET soft-drink bottles need special handling before they can be melted and reused. The bottles usually are sorted to remove tinted PET containers and PVC containers. The latter, if left in the batch, can weaken the final product. Any labels, bottle caps, or food that adhered to the plastic also must be separated or scrubbed off. Bottle caps, for example, are usually made out of the tougher polypropylene. The next exercise shows how PET can be separated from other polymers by density. This is helpful in the case of PET mixed with PVC, as these both can be transparent and identical to the eye.

Consider This 9.28   Sink or Float?

When placed in a liquid, a plastic will float if its density is less than that of the liquid and sink if it is greater. Here is the density for PET and three other plastics that are likely to be found with it in a recycling bin.

Plastic	Density (g/cm³)
PET	1.38–1.39
HDPE	0.95–0.97
PP	0.90–0.91
PVC	1.18–1.65

The densities of six liquids at the same temperature are:

Liquid	Density (g/mL)
methanol	0.79
an ethanol/water mixture	0.92
a different ethanol/water mixture	0.94
water	1.00
saturated solution of $MgCl_2$	1.34
saturated solution of $ZnCl_2$	2.01

Given a mixed sample containing PET, propose a way to separate it from the other three plastics.

Figure 9.24
Activewear made from recycled PET.

Much of recycled PET is converted into polyester fabrics, including carpeting, T-shirts, the popular "fleece" used for jackets and pullovers (Figure 9.24), and the fabric uppers in jogging shoes. Five recycled 2-L bottles can be converted into a T-shirt or the insulation for a ski jacket; it takes just about 450 such bottles to make polyester carpeting for a 9 × 12 foot room. Currently only a small percentage of beverage bottles are made from recycled PET, but the amount is expected to increase in the next decade.

In this section we have explored the complexities of recycling. The technologies are new and rapidly developing as the markets shift. And recycling is not the only game in town. There is no single, best solution to the problems posed by plastic waste, or more generally by *all* solid waste. Incineration, biodegradation, reuse, recycling, and source reduction all provide benefits, and all have associated costs. Therefore, it is likely that the most effective response will be the development of an integrated waste management system that will employ all of these strategies. The goal of such an integrated system would be to optimize efficiency, conserve energy and material, and minimize cost and environmental damage.

Consider This 9.29   Responsible Consumerism

Unless consumers buy products made from recycled plastics, manufacturers will have little financial incentive to produce such products, and this may threaten the viability of recycling. Use the web to find five or more products from recycled plastics, other than those mentioned in this chapter. Identify the polymer(s) used in each product.

Conclusion

Synthetic polymers are at the very center of modern living, yet their existence depends on a precious resource that rapidly is being consumed—crude oil. We have come to not only depend on synthetic polymers, but in many cases to take them for granted to the point of being wasteful. Once more we revisit the issue of lifestyle. Over time, chemists have created an amazing array of polymers and plastics—new materials that have made our lives more comfortable and more convenient. Many of these plastics represent a significant improvement over the natural polymers they replace. Furthermore, many products we use today would be impossible without synthetic polymers and plastics. There would be no DVDs, no cell phones, no breathable contact lenses, no Polartec clothing, no kidney dialysis apparatus, and no artificial hearts. We have become dependent on polymers, and it would be difficult if not impossible to abandon their use. The chemical industry has given consumers what they want. But there now appears to be more of it than we would like or perhaps than we can deal with responsibly. We must learn to cope with this glut of plastic waste while saving matter and energy for tomorrow. To create a new world of plastics and polymers will require the intelligence and efforts of policy planners, legislators, economists, manufacturers, consumers, and of course, chemists.

Chapter Summary

Having studied this chapter, you should be able to:

- Give examples of both natural and synthetic polymers (9.1)

- Understand at the molecular level the relationship between polymers and the monomers from which they are synthesized (9.1)

- Understand the molecular mechanism of addition polymerization (9.2)

- Compare and contrast low-density polyethylene and high-density polyethylene, both at the molecular level and in terms of their properties (9.3)

- Recognize the molecular structures for each of the Big Six polymers and be able to draw the structures of the monomers from which they were made (9.4)

- Match the properties of the Big Six polymers with their uses (9.4):

    low-density polyethylene (LDPE) and high-density polyethylene (HDPE) (9.3)

    polyvinyl chloride (PVC) (9.4)

    polystyrene (PS) (9.4)

    polypropylene (PP) (9.4)

    polyethylene terephthalate (PET) (9.5)

- Compare and contrast condensation polymerization with addition polymerization (9.5)

- Be able to name and draw structural formulas for different functional groups (9.5)

- Explain the relationship between amino acids and proteins (9.6)

- Use structural formulas to write the chemical equation for the synthesis of nylon (9.6)

- Explain and interpret trends in plastics recycling over the past decade (9.7)

- Relate the technical, economic, and political issues in methods for disposing of waste plastic: incineration, biodegradation, reuse, recycling, and source reduction (9.7–9.8)

- Discuss the different activities involved in recycling and their inherent complexities (9.8)

- Explain the waste reduction hierarchy, and why source reduction and reuse are preferred (9.8)

Questions

Emphasizing Essentials

1. Give two examples each of natural and of synthetic polymers.

2. Polymers sometimes are referred to as macromolecules. Explain.

3. Equation 9.1 contains an n on both sides of the equation. The one on the left is a coefficient; the one on the right is a subscript. Explain.

4. In equation 9.1, explain the function of the R• over the arrow.

5. Describe how each of these strategies would be expected to affect the properties of polyethylene. Also provide an explanation at the molecular level for each effect.

    a. increasing the length of the polymer chain

    b. aligning the polymer chains with one another

    c. increasing the degree of branching in the polymer chain

6. Figure 9.3a shows two bottles made from polyethylene. How do the two bottles differ at the molecular level?

7. Ethylene (ethene) is a hydrocarbon. Give the names and structural formulas of two other hydrocarbons that, like ethylene, can serve as monomers.

8. Why is a repeating head-to-tail arrangement not possible for ethylene?

9. Determine the number of $H_2C=CH_2$ monomeric units, *n*, in one molecule of polyethylene with a molar mass of 40,000 g. How many carbon atoms are in this molecule?

10. A structural formula for styrene is given in Table 9.1.

   a. Redraw it to show all of the atoms present.

   b. Give the chemical formula for styrene.

   c. Calculate the molar mass of a polystyrene molecule consisting of 5000 monomers.

11. Vinyl chloride polymerizes to form PVC in several different arrangements, as shown in Figure 9.8. Which is shown here?

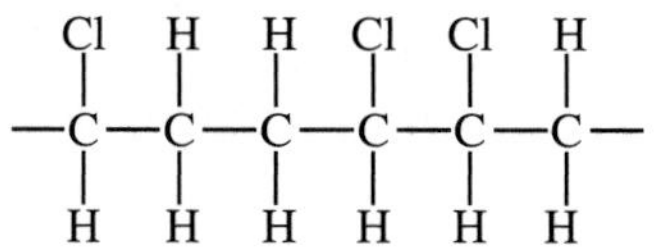

12. Here are two segments of a larger PVC molecule. Do these two structures represent the same arrangement? Explain your answer by identifying the orientation in each arrangement. *Hint:* See Figure 9.8.

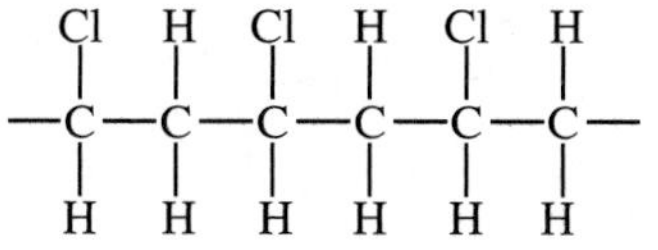

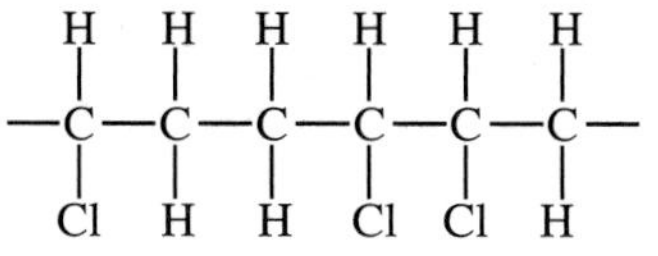

13. Butadiene, $H_2C=CH-HC=CH_2$, can be polymerized to make Buna rubber. Would this be by addition or condensation polymerization?

14. Which of the "Big Six" most likely would be used for these applications?

   a. clear soda bottles

   b. opaque laundry detergent bottles

   c. clear, shiny shower curtains

   d. tough indoor–outdoor carpet

   e. plastic baggies for food

   f. packaging "peanuts"

   g. containers for milk

15. Polyethylene is the most widely used synthetic polymer, but it is not the plastic of choice for margarine containers. Similarly it is not used for soft-drink bottles. Explain.

16. Plastics are widely used as containers. Check the recycling code on 10 containers of your choice (see Table 9.1). In your sample, which polymer was the most widely used?

17. Name the functional group(s) in each of these monomers.

   a. styrene

   b. ethylene glycol

   c. terephthalic acid

   d. the amino acid where R=H

   e. hexamethylenediamine

   f. adipic acid

18. Circle and identify all the functional groups in this molecule:

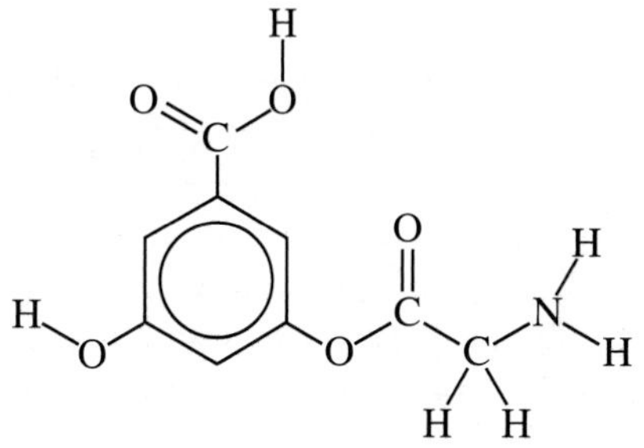

19. Kevlar is a type of nylon called an *aramid*. It contains rings similar to that of benzene. Because of its great mechanical strength, Kevlar is used in radial tires and in bulletproof vests. Your Turn 9.17 gives the structures for the two monomers, terephthalic acid and phenylenediamine. Name the functional groups in both the monomers and in the polymer.

20. Explain how a copolymer is a subset of the more general term *polymer*. Give an example of a polymer that is a copolymer, and one that is not.

21. Silk is an example of a natural polymer. Name three properties that make silk desirable. Which synthetic polymer has a chemical structure modeled after silk?

22. The Dow Chemical Company has developed a process that uses CO_2 as the blowing agent to produce Styrofoam packaging material.

   a. What is a blowing agent?

   b. What compound does CO_2 likely replace in the process, and why is this substitution environmentally beneficial?

23. Consider these data:

Year	U.S. Population (millions)	Plastics Produced in the United States (billions of pounds)
1997	269	89
2003	290	107

In the United States,

a. How many pounds of plastic were produced in 1997? In 2003?

b. How many pounds of plastic were produced per person in these same years?

c. Between 1997 and 2003, what is the percent change in the number of pounds of plastic produced per person?

Concentrating on Concepts

24. You were asked in Consider This 9.18 to keep a journal of all the plastic products you *throw away* in one week. Now consider all of the plastic items that you *recycle* in one week. Are there any from your first list of items thrown away that could be on your second list? Explain.

25. Draw a diagram to show the relationships among these terms: *natural, synthetic, polymer, nylon, protein.* Add other terms as needed.

26. Celluloid was the first commercial plastic, developed in response to the need to replace ivory for billiard balls and piano keys. Speculate on the properties of celluloid that made it a successful substitute for ivory in these products.

27. Glucose from corn is the source of some new bio-based polymer materials. Glucose also is the monomer in cellulose. Earlier in this text you encountered glucose in the chemical reaction of photosynthesis. What is photosynthesis and from what is glucose produced?

28. The properties of a plastic are a consequence of more than just its chemical composition. Name two other features of a polymer chain that can influence its properties.

29. Many monomers contain a C=C double bond. Select one and draw its structural formula together with the corresponding polymer. Describe the similarities and differences between the monomer and the polymer.

30. What structural features must a monomer possess to undergo addition polymerization? Explain, giving an example. Do the same for condensation polymerization.

31. This equation represents the polymerization of vinyl chloride. At the molecular level as the reaction takes place, how does the Cl–C–H bond angle change?

$$n \quad \underset{H}{\overset{H}{C}} = \underset{Cl}{\overset{H}{C}} \quad \xrightarrow{R\cdot} \quad \left[\underset{H}{\overset{H}{C}} - \underset{Cl}{\overset{H}{C}} \right]_n$$

32. Polyacrylonitrile is a polymer made from the monomer acrylonitrile, CH_2CHCN.

a. Draw a Lewis structure of this monomer.

b. Polyacrylonitrile is used in making Acrilan fibers used widely in rugs and upholstery fabric. What danger do rugs or upholstery made of this polymer create in the case of house fires?

33. Roy Plunkett, a DuPont chemist, discovered Teflon while experimenting with gaseous tetrafluoroethylene. Here is the monomer.

$$\underset{F}{\overset{F}{\diagdown}} C = C \underset{F}{\overset{F}{\diagup}}$$

a. Analogous to equation 9.1, write the chemical reaction for the polymerization of *n* molecules of tetrafluoroethylene to form Teflon.

b. Why is a repeating head-to-tail arrangement not possible for this polymer?

34. Equation 9.1 shows the polymerization of ethylene. From the bond energies of Table 4.2, is this reaction endothermic or exothermic?

35. Would your answer from the previous question differ if tetrafluoroethylene were used as the monomer? See Question 33 for the monomer.

36. Do you expect the heat of combustion of polyethylene, as reported in kilojoules per gram (kJ/g), to be more similar to that of hydrogen, coal, or octane, C_8H_{18}? Explain your prediction.

37. Recycling is not the same as waste prevention. Explain.

38. This graph shows U.S. production of plastics from 1977 through 2003.

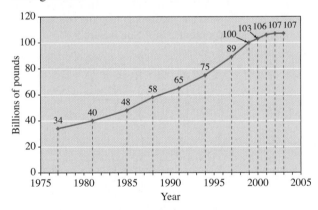

a. What is the approximate increase in plastic production for any five-year period before 2003?

b. How many years were required for the production in 1977 to double?

c. Redraw this as a bar graph that shows the relationship between the year and the pounds of plastics produced. Discuss whether the bar graph is easier than the line graph to establish the doubling time.

d. Suggest factors that may have contributed to changes in production from 2001 to 2003.

39. Consider the polymerization of 1000 ethylene molecules to form a large segment of polyethylene.

$$1000 \ CH_2 = CH_2 \xrightarrow{R\cdot} \ (CH_2CH_2)_{1000}$$

a. Calculate the energy change for this reaction.
   Hint: Use Table 4.2 of bond energies.

b. To carry out this reaction, must heat be supplied or removed from the polymerization vessel? Explain.

40. Here is the structural formula for Dacron, a condensation polyester:

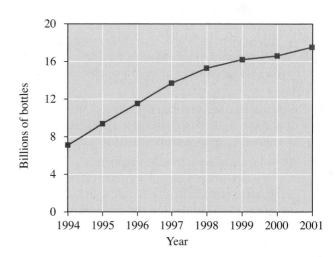

Dacron is formed from two monomers, one with two hydroxyl groups (−OH) and the other with two carboxylic acids (−COOH). Draw the structural formulas for the alcohol and the acid monomers used to produce Dacron.

41. Catalysts are used to help control the average molar mass of polyethylene, an important strategy to control polymer chain length. During World War II, low-pressure polyethylene production used varying mixtures of triethylaluminum, $Al(C_2H_5)_3$, and titanium tetrachloride, $TiCl_4$, as a catalyst. Here are some data showing how the molar ratio of the two components of the catalyst affects the average molar mass of the polymer produced.

Moles $Al(C_2H_5)_3$	Moles $TiCl_4$	Average Molar Mass of Polymer, g
12	1	272,000
6	1	292,000
3	1	298,000
1	1	284,000
0.63	1	160,000
0.53	1	40,000
0.50	1	21,000
0.20	1	31,000

a. Prepare a graph to show how the molar mass of the polymer varies with the mole ratio of $Al(C_2H_5)_3/TiCl_4$.

b. What conclusion can be drawn about the relationship between the molar mass of the polymer and the mole ratio of $Al(C_2H_5)_3/TiCl_4$?

c. Use the graph to predict the molar mass of the polymer if an 8:1 ratio of $Al(C_2H_5)_3$ to $TiCl_4$ were used.

d. What ratio of $Al(C_2H_5)_3$ to $TiCl_4$ would be used to produce a polymer with a molar mass of 200,000?

e. Can this graph be used to predict the molar mass of a polymer if either pure $Al(C_2H_5)_3$ or pure $TiCl_4$ were used as the catalyst? Explain.

42. When you try to stretch a piece of plastic bag, the length of the piece of plastic being pulled increases dramatically and the thickness decreases. Does the

same thing happen when you pull on a piece of paper? Why or why not? Explain on a molecular level.

43. Consider Spectra, Allied-Signal Corporation's HDPE fiber, used as liners for surgical gloves. Although the Spectra liner has a very high resistance to being cut, the polymer allows a surgeon to maintain a delicate sense of touch. The interesting thing is that Spectra is *linear* HDPE, which is usually associated with being rigid and not very flexible.

a. Suggest a reason why branched LDPE cannot be used in this application.

b. Offer a molecular level reason for why linear HDPE is successful in this application.

44. One limitation of the Big Six is the relatively low temperatures, 90–170 °C, at which they melt (see Table 9.1). Suggest ways to raise the upper temperature limits while maintaining the other desirable properties of these substances.

45. All the Big Six polymers are insoluble in water, but some dissolve or at least soften in hydrocarbons (see Table 9.1). Use your knowledge of molecular structure and solubility to explain this behavior.

46. When Styrofoam packing peanuts are immersed in acetone (the primary component in some nail-polish removers), they dissolve. If the acetone is allowed to evaporate, a solid remains. The solid still consists of Styrofoam, but now it is solid and much denser. Explain. *Hint:* Remember that Styrofoam is made with foaming agents.

47. This figure, entitled "Plastic Soda Bottles Wasted," was adapted from one shown in *Beverage World* magazine. The *y*-axis is in units of billions of beverage bottles.

a. Which polymer is used in clear plastic beverage bottles?

b. If recycled, to what uses can this polymer be put?

c. More bottles are recycled each year, yet this graph still shows an increase in bottles wasted. Explain.

Exploring Extensions

48. a. Name two functional groups not discussed in this chapter. Give an example of a molecule containing each one. *Hint:* You might want to look ahead to Chapter 10.

   b. Find the structural formula for the acetone molecule. What functional group does it contain?

49. Cotton, rubber, silk, and wool are natural polymers. Consult other sources to identify the monomer in each of these polymers. Which are addition polymers and which are condensation polymers?

50. Until recently, most PET beverage bottles were colorless. Now, however, Dasani water is sold in blue PET, as are other beverages. Similarly, Sprite is sold in green PET. Find out what effect, if any, the blue and green PET is having on recycling.

51. A Teflon ear bone, fallopian tube, or heart valve? A Gore-Tex implant for the face or to repair a hernia? Some polymers are biocompatible and now used to replace or repair body parts.

   a. List four properties that would be desirable for polymers used *within* the human body

   b. Other polymers may be used outside your body, but in close contact with it. For example, no surgeon is needed for you to use your contact lenses—you insert, remove, clean, and store them yourself. From which polymers are contact lenses made? What properties are desirable in these materials? Either a call to an optometrist or a search on the web may provide some answers.

   c. What is the difference in the material used in "hard" and "soft" contact lenses? How do the differences in properties affect the ease of wearing the contact lenses?

52. PVC, also known as "vinyl," is a controversial plastic. Do a risk–benefit analysis of using PVC from either the standpoint of a consumer or from a worker in the vinyl industry.

53. The plasticizers used to soften PVC are controversial as well. A common class of plasticizers is phthalates (THAL-ates), esters of phthalic (THAL-ic) acid.

   a. Phthalic acid is an isomer of terephthalic acid, one of the two monomers used to synthesize PETE. The structure of phthalic acid is similar to that of terephthalic acid except that the acid groups on the benzene ring are adjacent to each other. Draw the structural formula for phthalic acid.

   b. Write the chemical equation in which phthalic acid reacts with two molecules of ethanol to form a double ester.

   c. The plasticizers DINP and DEHP (here the P stands for phthalate) use a longer chain alcohol than ethanol, resulting in an ester with longer side-chains. Given

the role that plasticizers play, why do you think a longer chain alcohol is needed?

   d. Use the resources of the web to research why plasticizers such as DINP and DEHP are controversial.

54. Who first synthesized Kevlar? What was the background and academic training of these scientists? Was the potential for using this polymer in radial tires immediately understood? What are other applications of Kevlar? Write a short report on the results of your findings. Be sure to cite your sources.

55. Isoprene is the monomer that forms natural rubber. Here is its structural formula, with the carbons numbered.

$$CH_2 \overset{2}{\underset{1}{=}} \overset{CH}{\underset{3}{C}} \overset{}{=} \underset{4}{CH_2}$$
$$\underset{CH_3}{|}$$

When isoprene monomers add to form polyisoprene (natural rubber), the polymer has a C=C double bond between carbon atoms 2 and 3. How does this double bond form? *Hint:* Each double bond has four electrons in it. When a new single bond is formed between two monomers, that single bond only needs two electrons in it, one from each of the monomers joined by that new bond.

56. a. What are the structures of the monomers used in SBR synthetic rubber?

   b. How are natural and synthetic rubber alike, and how do they differ?

57. Synthetic rubber is usually formed through addition polymerization. An important exception is silicone rubber, which is made by the condensation polymerization of dimethylsilanediol. This is a representation of the reaction.

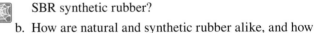

$$n\ \text{HO}-\underset{\underset{CH_3}{|}}{\overset{\overset{CH_3}{|}}{Si}}-\text{OH} \longrightarrow \left[\text{O}-\underset{\underset{CH_3}{|}}{\overset{\overset{CH_3}{|}}{Si}}-\text{O} \right]_n + n\ H_2O$$

   a. Predict some of the properties of this polymer. Explain the basis for your predictions.

   b. Silly Putty is a popular form of silicone rubber. What are some of the properties of Silly Putty?

58. A few decades ago, recycling personal computers was not a concern because not enough of them were around to matter. Today, however, there is good

reason to keep keyboards, monitors, and "mice" out of the landfill.

a. Which polymers do your computer and its accessories contain?

b. What are the options for recycling the plastics in computers?

59. Free-radical peroxides promote the polymerization of ethylene into polyethylene. They also play a key role in tropospheric smog formation. Use the web to learn more about how the peroxides promote ethylene polymerization and how peroxides are involved with photochemical smog formation in the troposphere. Write a brief report comparing the types of peroxides important with each of these cases. Cite all sources.

60. In 2007, Cargill won a green chemistry award for using soybeans instead of petroleum to produce polyols. What is a polyol, and how are polyols used to produce "soybean plastics"?

10 Manipulating Molecules and Designing Drugs

An ancient medicinal plant, *Ephedra sinica*, also called *ma huang* (above). Shown as a tincture, twigs, powdered twigs, and dried roots (below).

Drugs. This word elicits hope, relief, fear, intrigue, outrage, or maybe simply disdain. Pharmaceuticals (drugs) are substances intended to prevent, moderate, or cure illnesses. Medicinal chemistry is the science that deals with the discovery or design of new therapeutic chemicals and their development into useful medicines.

Modern pharmacology has its origins in folklore, and the history of medicine is full of herbal and folk remedies. The use of herbs, roots, berries, and barks for relief from illness can be traced to antiquity as illustrated in documents recorded by ancient Chinese, Indian, and Near East civilizations. The Rig-Veda (compiled in India between 4500 and 1600 BCE), one of the oldest repositories of human learning, refers to the use of medicinal plants. The Chinese Emperor Shen Nung prepared a book of herbs over 5000 years ago. In it, he described a plant called *ma huang* (now called *Ephedra sinica*), used as a heart stimulant. This plant contains ephedrine, a drug we will consider later in this chapter.

More recently, chemists have designed, synthesized, and characterized a vast array of prescription and over-the-counter drugs. Today, drugs help patients regulate their blood sugar, blood pressure, cholesterol, and allergies. They help AIDS patients stay alive while scientists search for a cure. Effective anticancer drugs and powerful analgesics now exist. Other drugs can even manage mental disorders that once were thought to be untreatable.

People have long taken drugs for the purpose of altering their perceptions and moods. The famed philosopher Nietzsche said that no art could exist without intoxication. Many writers and artists have found that drugs act as a creative and destructive force in their lives and work. People abuse drugs primarily because of the promise of instant relief or pleasure and the possibility of heightened awareness. It is a common misconception that today's problem with drug abuse is a recent phenomenon. The reality is that human history has been marked with drug use and abuse.

In discussing drugs, we will consider these questions: Where do the ideas and resources to develop new drugs come from? What is the process by which pharmaceuticals make it to market? Why does a drug have a certain effect, and which features of its molecular structure contribute to the biological activity? How do drugs move from being available by prescription only to being sold over the counter? What are the merits and pitfalls of herbal medicines, and are naturally occurring drugs "safer" than synthetic ones? Which drugs are most commonly abused? Chemistry concepts key to answering these questions will be presented in this chapter. Understanding some basic chemistry can go a long way toward staying healthy in today's complicated world.

Consider This 10.1   Today's Drugs

a. Consider the modern pharmaceuticals prescribed in the United States today. List what you think are the top five most frequently prescribed drugs.
b. List what you think are the top five most frequently abused drugs.
c. Share your lists with a small group of students. Do some of the drugs that you or others listed appear in both parts **a** and **b**?

10.1 | A Classic Wonder Drug

In the fourth century BCE, Hippocrates, perhaps the most famous physician of all time, described a "tea" made by boiling willow bark in water. The concoction was said to be effective against fevers. Over the centuries, that folk remedy, common to many different cultures, ultimately led to the synthesis of a true "wonder drug," one that has aided millions of people.

One of the first systematic investigators of willow bark (Figure 10.1) was Edmund Stone, an English clergyman. His report to the Royal Society (1763) set the stage for a series of further chemical and medical investigations. Chemists were subsequently

Figure 10.1

The white willow tree, *Salix alba*, source of a miracle drug.

able to isolate small amounts of yellow, needle-shaped crystals of a substance from the willow bark extract. Because the tree species was *Salix alba,* this new substance was named salicin. Experiments showed that salicin could be chemically separated into two compounds. Clinical tests provided evidence that only one of these components reduced fevers and inflammation. It also was demonstrated that the active component was converted to an acid in the body. Unfortunately, the clinical testing revealed some troubling side effects. The active component not only had a very unpleasant taste, but also its acidity led to acute stomach irritation in some individuals.

Acid–base neutralization reactions were discussed in Section 6.3.

The active acidic compound was used to treat pain, fever, and inflammation. But recognizing its serious side effects, chemists set out to find a derivative that still would be effective but not cause stomach distress and lack the undesirable taste. The first attempt took a very simple approach. The acid was neutralized with a base, either sodium hydroxide or calcium hydroxide, to form a salt of the acid. It turned out that the resulting salts had fewer side effects than the parent compound. Based on this finding, chemists correctly concluded that the acidic part of the molecule was responsible for the undesirable properties. Consequently, the next step was to seek a structural modification that would lessen the acidity of the compound without destroying its medicinal effectiveness.

One of the chemists working on the problem was Felix Hoffmann, an employee of a major German chemical firm. Hoffmann's motivation was more than just scientific curiosity or assigned task. His father regularly took the acidic compound as treatment for arthritis. It worked, but he suffered nausea. The younger Hoffmann succeeded in converting the original compound into a different substance, a solid that reverted back to the active acid once it was in the bloodstream. This molecular modification greatly reduced nausea and other adverse reactions; a new drug had been discovered (1898).

Extensive hospital testing of Hoffmann's compound began along with simultaneous preparation for its large-scale manufacture by a well-known pharmaceutical company. The new drug itself could not be patented because it was already described in the chemical literature. However, the company hoped to recoup its investment by patenting the manufacturing process. Clinical trials showed the drug to be nonaddicting. Its toxicity is classified as low, but 20–30 g ingested at one time may be lethal. At the suggested dose of 325–650 mg every 4 hours, it is a remarkably effective antipyretic (fever-reducing), analgesic (pain-relieving), and anti-inflammatory agent. Data from clinical tests uncovered the side effects noted in Table 10.1. The drug was also found to inhibit blood clotting and to cause at least some small, almost always medically insignificant, amounts of stomach bleeding in about 70% of users.

Table 10.1	Side Effects of the "Wonder Drug"	
Symptoms	Frequency	Severity*
drowsiness	rare	4
rash, hives, itch	rare	3
diminished vision	rare	3
ringing in the ears	common	5
nausea, vomiting, abdominal pain	common	2
heartburn	common	4
black or bloody vomit	rare	1
blood in the urine	rare	1
jaundice	rare	3
shortness of breath	rare	3

*The severity scale ranges from 1, life-threatening, seek emergency treatment immediately to 5, continue the medication and tell the physician at the next visit.

Source: H. W. Griffith, *The Complete Guide to Prescription and Non-Prescription Drugs,* 1983, HP Books, Tucson, Arizona.

Consider This 10.2   **Miracle Drug**

In the United States, the final step for approval of a drug is the submission of all of its clinical test results to the Food and Drug Administration (FDA) for a license to market the product.

   a. If you were an FDA panel member presented with the information in Table 10.1, would you vote to approve this drug that treats pain, fever, and inflammation?
   b. If approved, should this drug be released as an over-the-counter drug or should its availability be restricted as a prescription drug? Write a one-page position paper.

Perhaps you have already guessed the identity of the miracle drug related to willow bark tea. Its chemical name, 2-acetyloxybenzoic acid or (more commonly) acetylsalicylic acid, may not help much. But the power of advertising is such that, had we revealed that the firm that originally marketed the drug was the Bayer division of I. G. Farben, we would have let the tablet out of the bottle. The compound in question is the world's most widely used drug, even a century after its discovery. People in the United States annually consume nearly 80 billion tablets of this miracle medicine. You know it as aspirin.

Admittedly, we have only given the highlights in the history of aspirin. Most of the development, testing, and design of aspirin occurred in the 18th and 19th centuries. Stone's letter to the Royal Society was written in 1763, and Felix Hoffmann's modification of salicylic acid to yield aspirin was done in 1898. Furthermore, the clinical testing of aspirin was somewhat less systematic than our account implies. But the basic facts and the steps that led to aspirin's full development are essentially correct. We must also add one more very important fact. Aspirin did not have to receive drug approval before being put on the market; no such certifying process was in place at the time. Had approval based on clinical test results been necessary, it is quite likely that aspirin would have been available only on a prescription basis.

Consider This 10.3   **What Should a Drug Be Like?**

Make a list of the properties you think a drug should have. Then compare your list with those of your classmates. Note similarities and differences.

   a. Are any items missing from your list that you now think you should include?
   b. Are any items present on your list that you now think you should delete?

10.2 | The Study of Carbon-Containing Molecules

One of carbon's interesting properties is its ability to form a wide variety of molecules. This element is so ubiquitous in nature that one of the largest subdisciplines of chemistry, **organic chemistry,** is devoted to the study of carbon compounds. The name *organic* is historical and suggests a biological origin for the substances under investigation, but this is not necessarily true. In practice, most organic chemists confine themselves to compounds in which carbon is combined with a relatively small number of other elements: hydrogen, oxygen, nitrogen, sulfur, chlorine, phosphorus, and bromine. Even with this restriction, over 12 million of the 27 million total known compounds are considered organic. The chemical behavior (i.e., properties and reactivity) of organic compounds enables us to organize them into a relatively small number of categories. As a result, in this chapter we concentrate on only a few and stress their important roles within living things.

To specify an organic compound from among the myriad of possibilities, you must be able to name it correctly. An international committee called the International Union of Pure and Applied Chemists (IUPAC) established and periodically updates a formal set of nomenclature rules so each of the known compounds can be uniquely named. However, many of these compounds have been known for a long time by common names such as alcohol, sugar, or morphine. When a headache strikes, even chemists do not call out for 2-acetyloxybenzoic acid; they simply say "Give me some aspirin!" Likewise, prescriptions specify penicillin-N rather than 6[(5-amino-5-carboxy-1-oxopentyl)amino] 3,3-dimethyl-7-oxopentyl-4-thia-1-azabicyclo[3.2.0]heptane-2-carboxylic acid. A mouthful like this is the cause of great merriment to those who like to satirize chemists. Nonetheless, chemical names are important and unambiguous to those who know the system. You can rest easy because in this chapter, we will use common names in almost all cases.

An incredible variety of organic compounds exists because of the remarkable ability of carbon atoms to bond in multiple ways both to other carbon atoms and to atoms of other elements. To better understand such possibilities, we need a few basic rules for bonding in organic molecules. You used one of these in Chapter 2, the octet rule. When bonded, each carbon atom has a share in eight electrons, an octet. Eight electrons can be arranged to form four bonds, with a pair of shared electrons in each covalent bond. The most common bonding arrangements for these four bonds around a carbon atom are (a) four single bonds, (b) two single bonds and one double bond, and (c) one single bond and one triple bond. These arrangements are illustrated in Figure 10.2.

> The octet rule was discussed in Section 2.3.

Other elements exhibit different bonding behavior in organic compounds. A hydrogen atom is always attached to another atom by a single covalent bond. An oxygen atom typically attaches either with two single bonds (to two different atoms) or one double bond (to a single atom). A nitrogen atom commonly forms three single bonds (to three different atoms), but also can form either a triple bond (to one other atom), or a single and a double bond.

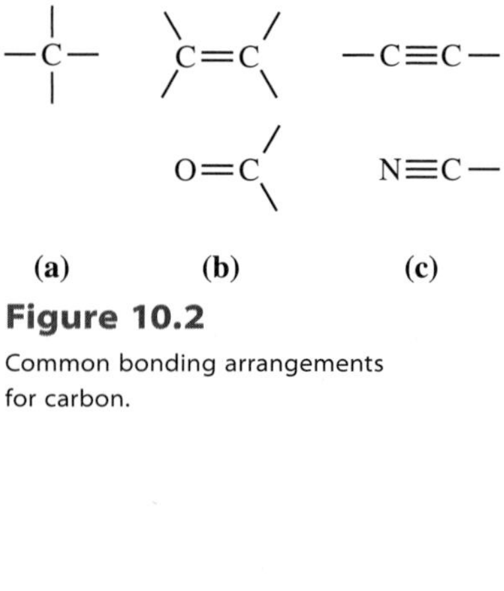

(a)   (b)   (c)

Figure 10.2

Common bonding arrangements for carbon.

Your Turn 10.4   Satisfying the Octet Rule

Examine each carbon atom in Figure 10.2. Does each atom follow the octet rule?

Chemical formulas such as C_4H_{10} indicate the kinds and numbers of atoms present in a molecule, but do not show how the atoms are arranged or connected. To get that higher level of detail, **structural formulas** are used to show the atoms and their arrangement with respect to one another in a molecule. Here is the structural formula for normal butane, or *n*-butane (C_4H_{10}), a hydrocarbon fuel used in cigarette lighters and camp stoves.

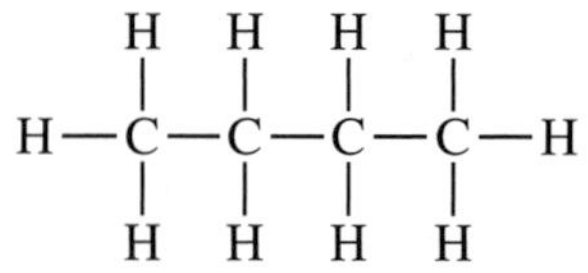

A drawback to writing structural formulas, at least in a textbook, is that they take up considerable space. To convey the same information in a format that is easily typeset into a single line, we use **condensed structural formulas** where carbon-to-hydrogen bonds are not drawn out explicitly, but simply understood to be single bonds. Here are two condensed structural formulas for *n*-butane.

$$CH_3—CH_2—CH_2—CH_3 \quad \text{or} \quad CH_3CH_2CH_2CH_3$$

Note that even though these condensed structural formulas give the opposite impression, the carbons are really bonded directly to other carbon atoms, and the hydrogen atoms do not intervene in the chain. Rather, two or three hydrogens are attached to each carbon atom, depending on its position in the molecule.

> Although a CH_3 group at the left end of a chain could be drawn as $H_3C–$, for ease we often reverse the order to $CH_3–$ with the understanding that the bond is to the C atom, not to the H atoms.

The same number and kinds of atoms can be arranged in different ways, helping to explain why there are so many different organic compounds. **Isomers** are molecules with the same chemical formula (same number and kinds of atoms), but with different structures and properties. You already encountered isomers with the chemical formula C_8H_{18} in Chapter 4.

Isomers were also described in Section 4.7.

Here we illustrate the concept of isomers with C_4H_{10}. One way to arrange the carbon atoms is in a chain to form *n*-butane. Another arrangement is possible in which the four carbon atoms are not all in a line. This isomer is known as *iso*butane. The linear *n*-butane is shown for comparison, now represented in a more realistic zigzag form.

n-butane   *iso*butane

The chemical formulas of these two isomers are the same; the way the atoms are connected is different. Note that the central carbon atom in *iso*butane has three carbon atoms connected to it; *n*-butane has no such carbon atom. You cannot rotate the bonds of one of these representations to make it look like the other; rotating the bonds won't change how the atoms are connected.

Just like its linear isomer, *iso*butane can be represented with a condensed structural formula.

$$CH_3-\underset{\underset{CH_3}{|}}{CH}-CH_3 \quad or \quad CH_3CH(CH_3)CH_3 \quad or \quad CH_3CH(CH_3)_2$$

Here, the parentheses around the $-CH_3$ groups indicate that they are attached to the C atom to their left. Note that with three $-CH_3$ groups attached to the central C atom, a "branch" has been introduced into the molecule.

Figure 10.3 shows three representations of *n*-butane and *iso*butane. The first column shows the simple structural formula, the second a ball-and-stick model. The third column shows a space-filling model that presents a more realistic view of the molecular shape.

Only two isomers of C_4H_{10} exist. As the number of atoms in a hydrocarbon increases, so does the number of possible isomers. Thus, C_8H_{18} has 18 isomers and

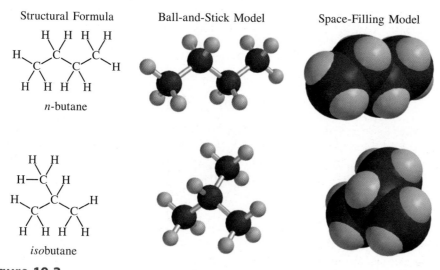

Structural Formula   Ball-and-Stick Model   Space-Filling Model

n-butane

*iso*butane

Figure 10.3

Various representations of the isomers *n*-butane and *iso*butane.

$C_{10}H_{22}$ has 75. Given a chemical formula, no simple calculation can be performed to obtain the number of isomers.

Your Turn 10.5   Drawing Structures

Draw a structural formula for each of these condensed structural formulas.

a. $CH_3CH_2CH_2CH(CH_3)_2$

b. $CH_3CH(CH_3)CH_2CH_3$

c. $CH_3CH_2C(CH_3)_3$

d. $CH_3CH(CH_2CH_3)CH_3$

Answers

a.

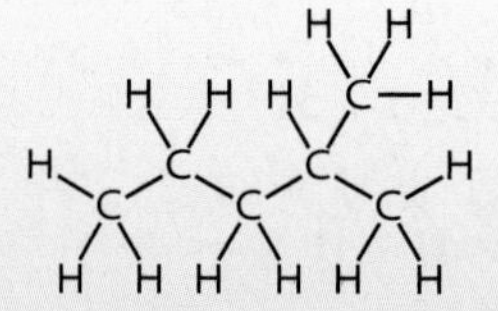

b.

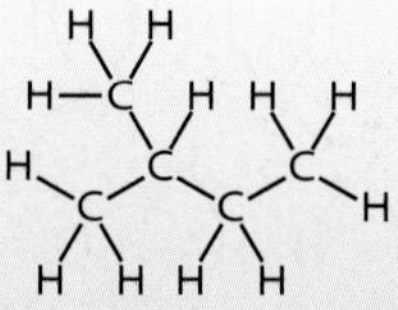

Chemists also routinely use a **line-angle drawing** to represent the structure of a molecule. This is a simplified version of a structural formula that is most useful for representing larger molecules. A line-angle drawing helps you to focus on the backbone of carbon atoms. One carbon atom is assumed to occupy each vertex position. Any line extending from the backbone signifies another carbon atom (actually a $-CH_3$ group), unless the symbol for another element is given. Hydrogen atoms are not indicated in the line-angle drawing, but are implied as required by the octet rule. Remember that each carbon atom will have four bonds to it, sharing a total of eight electrons. Line-angle drawings for *n*-butane, *iso*butane, and two other simple molecules are shown in Table 10.2.

Table 10.2   Molecular Representations

Compound	Chemical Formula	Structural Formula	Line-Angle Drawing
n-butane	C_4H_{10}	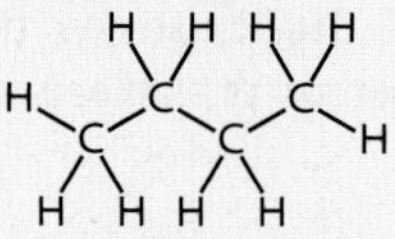	
*iso*butane	C_4H_{10}	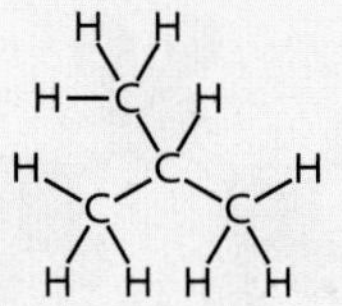	
n-hexane	C_6H_{14}	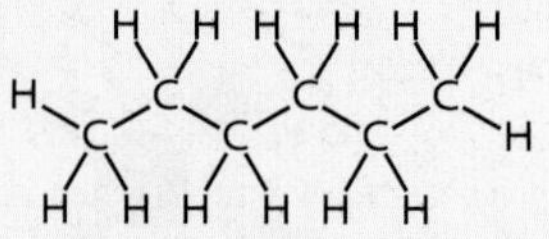	
cyclohexane	C_6H_{12}	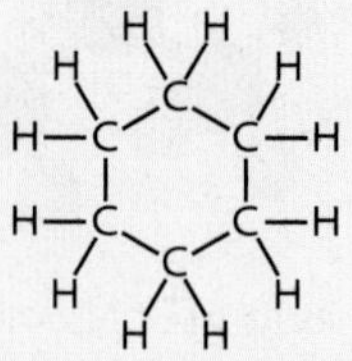	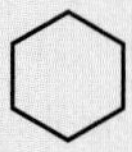

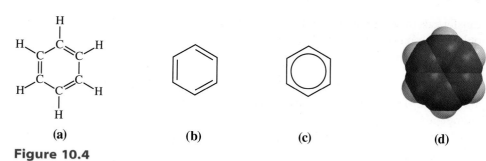

Figure 10.4

Representations of benzene, C_6H_6.

Your Turn 10.6   **Structural Isomers**

a. Are *n*-butane and *iso*butane isomers? Explain.
b. Are *n*-hexane and cyclohexane isomers? Explain.

Your Turn 10.7   **Isomers of C_5H_{12}**

Three isomers have the formula C_5H_{12}. For each, draw a structural formula, a condensed structural formula, and a line-angle drawing.

Many molecules, including aspirin, have carbon atoms arranged in a ring. For example, examine the structure of cyclohexane, C_6H_{12}, in Table 10.2. The ring in cyclohexane has six carbons, and rings most commonly contain five or six carbon atoms. In aspirin, however, the six-membered ring is based on benzene, C_6H_6, rather than on cyclohexane. The structural formula for benzene is shown in Figure 10.4a.

Structure (b) in Figure 10.4 is a line-angle drawing for benzene. Although this structure has alternating single and double bonds, experimental evidence indicates that all the C-to-C bonds in benzene actually have the same length. Since C–C single bonds are longer than C=C double bonds, benzene cannot have alternating single and double bonds. Consequently, the electrons must be uniformly distributed around the ring. The circle within the hexagon in structure (c) is an effort to convey this. Both structures (c) and (d) represent the uniformly distributed electrons around the ring, as described by resonance theory (see Section 2.3). This same hexagonal structure is found in the $-C_6H_5$ phenyl group that is part of many molecules, including styrene and polystyrene (Section 9.4).

> A C–C single bond length is 0.154 nm, and a C=C double bond length is 0.134 nm. In benzene, the C-to-C bond lengths are 0.139 nm.

10.3 | Functional Groups

Central to the study of drug discovery and interactions are functional groups. **Functional groups** are distinctive arrangements of groups of atoms that impart characteristic physical and chemical properties to the molecules that contain them. Indeed, these groups are so important that we often show them in structural formulas and represent the remainder of the molecule with an "R." The R is generally assumed to include at least one carbon or hydrogen atom. You already encountered some functional groups in Chapter 9. The generic formula for an alcohol is ROH, as in methanol, CH_3OH (an alcohol derived from degradation of wood), and ethanol, CH_3CH_2OH (alcohol derived from fermentation of grains and sugar). The presence of the –OH group attached to a carbon makes the compound an alcohol.

> Section 9.5 discussed alcohols in polymerization reactions.

> An alcohol has an –OH group *covalently* bound to the rest of the molecule. This is different from the hydroxide ion, OH^-, which is *ionically* bonded to a cation.

Similarly, a carboxylic acid group, commonly written as
$$
\begin{array}{c}
O \\
\parallel \\
C \\
\diagup \ \diagdown \\
\ \ \ \ O \ \ \ \ H
\end{array}
$$, –COOH,
or $-CO_2H$, confers acidic properties. In aqueous solution, a H^+ ion (a proton) is

transferred from the –COOH group to an H_2O molecule to form a hydronium ion, H_3O^+. We represent an organic acid with the general formula RCOOH, or RCO_2H. In acetic acid (CH_3COOH), the acid in vinegar, the R group is $-CH_3$, the methyl group.

Table 10.3 lists eight functional groups found in drugs and other organic compounds. Each functional group is characteristic of an important class of compounds.

Table 10.3	Some Important Organic Functional Groups			
			SPECIFIC EXAMPLES	
Functional Group	Generic Formula	Name*	Structural Formula	Condensed Structural Formula
hydroxyl	$\diagup^{O}\diagdown_{H}$	ethanol (ethyl alcohol)		CH_3CH_2OH
ether	$C\diagdown^{O}\diagup C$	dimethyl ether		CH_3-O-CH_3 or CH_3OCH_3
aldehyde		propanal		$CH_3CH_2-\overset{\overset{\displaystyle O}{\|}}{C}-H$ or CH_3CH_2CHO
ketone		2-propanone (dimethyl ketone, acetone)		$CH_3-\overset{\overset{\displaystyle O}{\|}}{C}-CH_3$ or CH_3COCH_3
carboxylic acid		ethanoic acid (acetic acid)		$CH_3-\overset{\overset{\displaystyle O}{\|}}{C}-OH$ or CH_3CO_2H or CH_3COOH
ester		methyl ethanoate (methyl acetate)		$CH_3-\overset{\overset{\displaystyle O}{\|}}{C}-OCH_3$ or CH_3COOCH_3
amine		ethylamine		$CH_3CH_2NH_2$
amide		propanamide		$CH_3CH_2-\overset{\overset{\displaystyle O}{\|}}{C}-NH_2$ or $CH_3CH_2CONH_2$

*IUPAC names, common names in parentheses.

Your Turn 10.8   Line-Angle Drawings

For each of these condensed structural formulas, make a line-angle drawing. Name the functional group in each one.

a. $CH_3CH_2CH_2COCH_3$ **b.** $CH_3CH_2CH(CH_3)CH_2OH$
c. $CH_3CH(NH_2)CH_2CH_3$ **d.** $CH_3COOCH_2CH_3$
e. CH_3CH_2CHO

Answers

a. 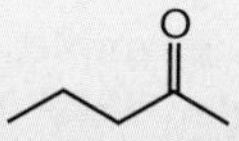 (ketone)   **b.** 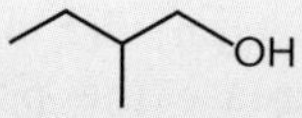 (hydroxyl)

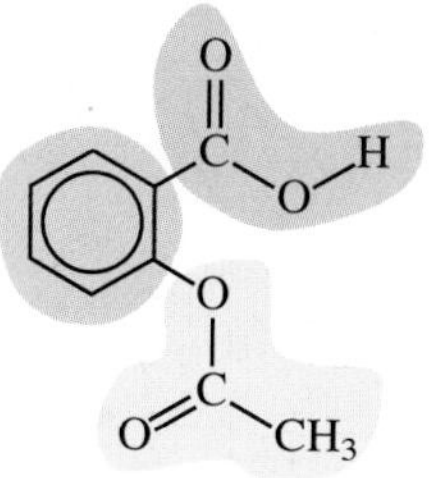

Figure 10.5
Structural formula of aspirin.

The presence and properties of functional groups are responsible for the action of all drugs. Aspirin has three functional groups, which are shown in Figure 10.5. You will recognize that the green area encloses a benzene ring. Its presence makes aspirin soluble in fatty compounds that are important cell membrane components. The other two functional groups are responsible for the drug's activity. You have just been reminded that the –COOH group indicates a carboxylic acid (blue area). The remaining functional group (yellow area) is an ester. An ester may be formed by reacting an acid and an alcohol; a water molecule is eliminated in the process. A catalytic amount of a stronger acid like H_2SO_4 speeds the reaction up; this is denoted in the reaction equation by placing an H^+ over the arrow (equation 10.1).

Felix Hoffmann prepared aspirin by modifying the structure of salicylic acid. But note that he did not modify the carboxylic acid group on the molecule. Salicylic acid also contains an –OH group that Hoffmann reacted with acetic acid as shown in equation 10.1. The product was an ester of acetic acid and salicylic acid, which accounts for one of aspirin's names: acetylsalicylic acid.

Formation of an ester was shown for condensation polymers in Section 9.5.

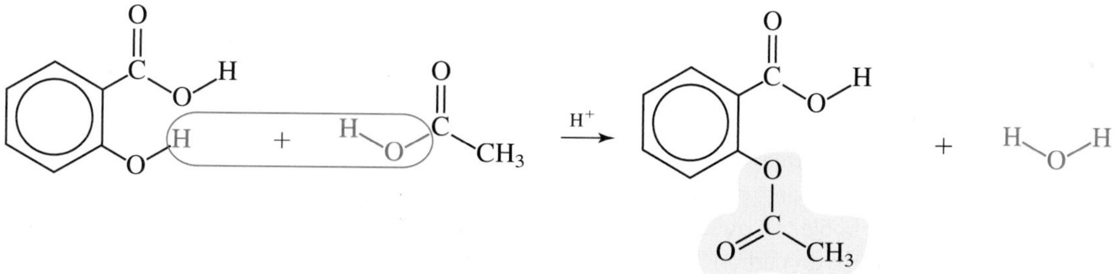

salicylic acid   acetic acid   acetylsalicylic acid   water   **[10.1]**

Because aspirin retains the –COOH group of the original salicylic acid, it still has some of the undesirable acidic properties of the parent compound. However, the ester group (yellow area in Figure 10.5) makes the compound more palatable and less irritating to the stomach lining. Once aspirin is ingested and reaches the site of its action, equation 10.1 is reversed. The ester splits into acetic acid and salicylic acid, and the latter compound exerts its antipyretic (fever-reducing) and analgesic (pain-reducing) properties.

Your Turn 10.9   Ester Formation

Draw structural formulas for the esters that form when these acid and alcohol pairs react.

a. CH_3CH_2OH + 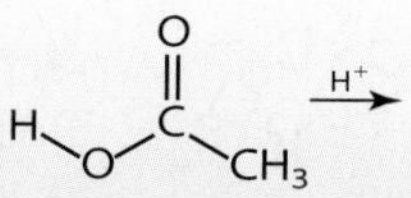   **b.**

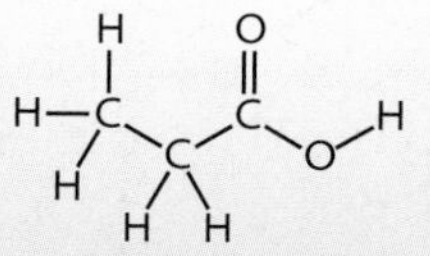

Functional groups can play a role in the solubility of a compound, an important consideration in the uptake, rate of reaction, and residence time of drugs in the body. The general solubility rule, "like dissolves like," applies in the body as well as in the test tube. A polar molecule has a nonsymmetrical distribution of electric charge. This means that a partial negative charge builds up on some part (or parts) of the molecule, and other regions of the molecule bear a partial positive charge. Water is an excellent example of a polar molecule. Relatively speaking, the oxygen atom is slightly negatively charged and the hydrogen atoms are slightly positive. Because the molecule is bent, it has a nonsymmetrical charge distribution. This is represented in Figure 10.6; the δ^+ and δ^- symbols represent partial charges. Functional groups containing oxygen and nitrogen atoms (for example, –OH, –COOH, and –NH$_2$) usually increase the polarity of a molecule. This in turn enhances its solubility in a polar substance such as water, which is advantageous for drug molecules.

The concept of *like dissolves like* was introduced in Section 5.9.

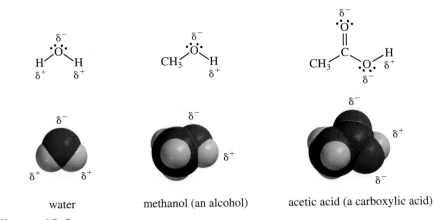

water   methanol (an alcohol)   acetic acid (a carboxylic acid)

Figure 10.6

Examples of polar molecules.

By contrast, hydrocarbons that do not contain such functional groups are typically nonpolar and will not dissolve in polar solvents. For example, *n*-octane, C_8H_{18} is nonpolar and insoluble in water. However, it does dissolve in nonpolar solvents such as hexane (C_6H_{14}) and dichloromethane (CH_2Cl_2). For the same reasons, drugs with significant nonpolar character tend to accumulate in cell membranes and fatty tissues that are largely hydrocarbon and nonpolar.

The water solubility of a drug that is either acidic or basic can be improved by neutralizing it and forming a salt. For example, many drugs contain nitrogen and are basic. When such a drug is neutralized with HCl or H_2SO_4, the nitrogen accepts an H^+ from the acid and is protonated. As a result, the nitrogen becomes positively charged and paired with the negative charge on the chloride or hydrogen sulfate (HSO_4^-) ion. Prior to being protonated, the compound is electronically neutral and is said to be in its freebase form. A **freebase** is a nitrogen-containing molecule in which the nitrogen is in possession of its lone pair of electrons.

Consider the drug pseudoephedrine, a common decongestant used in over-the-counter remedies for the common cold:

pseudoephedrine (freebase)   pseudoephedrine hydrochloride salt   **[10.2]**

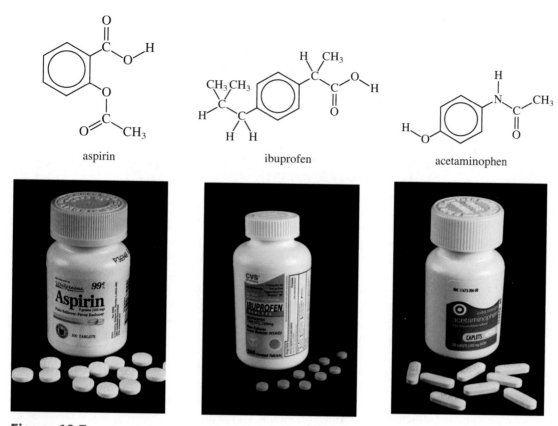

Figure 10.7

Structural formulas and samples of some common analgesics.

The nitrogen of the amino group reacts as a base when treated with hydrochloric acid. Pseudoephedrine can thus be converted to its hydrochloride salt (an ionic compound) in which the nitrogen bears a positive charge and the chloride ion a negative charge. The salt form of pseudoephedrine is preferable as a drug because it is more stable, has less of an odor, and is water-soluble. An estimated half of all drug molecules used in medicine are administered as salts that improve their water-solubility and stability, which in turn increase their shelf life. Conversion of a salt back into its freebase form may be accomplished by treating the salt with a base such as NaOH.

Drugs with similar physiological properties often have similar molecular structures and include some of the same functional groups. Of the approximately 40 alternatives to aspirin that have been produced, ibuprofen and acetaminophen are the most familiar. Figure 10.7 gives the structural formulas of the three leading analgesics. All are based on a benzene ring with two **substituents,** an atom or functional group that has been substituted for a hydrogen atom, but these substituents differ. In Your Turn 10.10, you have an opportunity to identify the structural similarities and differences of these analgesics.

Your Turn 10.10   Common Structural Features of Analgesics

Look at the structural formulas in Figure 10.7. Identify the structural features and functional groups that aspirin, ibuprofen, and acetaminophen have in common.

The current commercial method for producing ibuprofen is a stunning application of green chemistry. Previous methods of ibuprofen production required six steps, used large amounts of solvents, and generated significant quantities of waste. By using a catalyst that also serves as a solvent, BHC Company, a 1997 Presidential Green

Chemistry Challenge Award winner, makes ibuprofen in just three steps with a minimum of solvents and waste. In the BHC process, virtually all the reactants are converted to ibuprofen or another usable by-product; any unreacted starting materials are recovered and recycled. Nearly 8 million pounds of ibuprofen, enough to make 18 billion 200-mg pills, is produced annually in Bishop, Texas, at the BHC facility, built specifically for the commercial production of the drug.

10.4 | How Aspirin Works: Function Follows Form

To understand the action of aspirin, you need to know something about the body's chemical communication system. We normally think of internal communication as consisting of electrical impulses traveling along a network of nerves. This is true for the system that triggers movement, breathing, heartbeats, and reflex actions. Most of the body's messages, however, are conveyed not by electrical impulses, but through chemical processes. In fact, your very first communication with your mother was a chemical signal saying "I'm here; better get your body ready for me." It is much more efficient to release chemical messengers into the bloodstream where they can be circulated to appropriate body cells, than to "hardwire" each individual cell with nerve endings.

The chemical messengers produced by the body's endocrine glands are called **hormones.** Figure 10.8 is a representation of such chemical communication. Hormones encompass a wide range of functions and a similarly wide range of chemical composition and structure. Thyroxine, a hormone secreted by the thyroid gland, is essential for regulating metabolism. The ability of the body to use glucose (blood sugar) for energy depends on insulin. This hormone, a small protein built from only 51 polymerized amino acids, is secreted by the pancreas. Persons who suffer from diabetes are often required to take daily injections of insulin. Yet another well-known hormone is adrenaline (epinephrine), a small molecule that prepares the body to "fight or flight" in the face of danger.

Aspirin and other drugs that are physiologically active, but not anti-infectious agents, are almost always involved in altering the chemical communication system of the body. A significant problem is that this system is very complex, allowing many

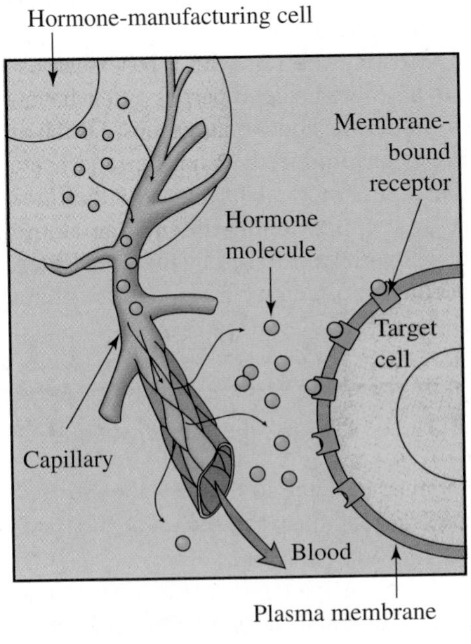

Section 12.7 describes the use of genetic engineering to obtain human insulin from bacteria.

Figure 10.8

Chemical communication in the body. Hormone molecules travel through the bloodstream from the cell where they are made to the target cell.

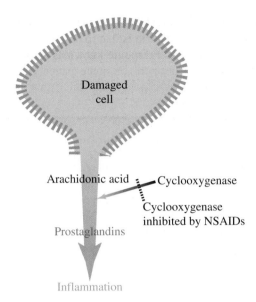

Figure 10.9
Aspirin's mode of action.

compounds to be used to send more than one message simultaneously. The wide range of aspirin's therapeutic properties, as well as its side effects, is clear evidence that the drug is involved in several chemical communication systems. It works in the brain to reduce fever and pain, it relieves inflammation in muscles and joints, and it appears to decrease the chances of stroke and heart attack. It may even lessen the likelihood of colon, stomach, and rectal cancer.

In large measure, the versatility of aspirin and similar nonsteroidal anti-inflammatory drugs (NSAIDs; steroids are covered in Section 10.7) is related to their remarkable ability to block the actions of other molecules. Research on the activity of aspirin indicates that one of its modes of action involves blocking cyclooxygenase (COX) enzymes. **Enzymes** are proteins that act as biochemical catalysts, influencing the rates of chemical reactions. Most enzymes speed up reactions and channel them so that only one product (or a set of related products) is formed. Cyclooxygenases catalyze the synthesis of a series of hormone-like compounds called prostaglandins from arachidonic acid (Figure 10.9).

Prostaglandins cause a variety of effects. They produce fever and swelling, increase the sensitivity of pain receptors, inhibit blood vessel dilation, regulate the production of acid and mucus in the stomach, and assist kidney functions. By preventing prostaglandin production, aspirin reduces fever and swelling. It also suppresses pain receptors and so functions as a painkiller. Because the benzene ring conveys high fat solubility, aspirin is also taken up into cell membranes. In certain specialized cells, the drug blocks the transmission of chemical signals that trigger inflammation. This process also appears to be related to aspirin's effectiveness as a pain reliever.

The NSAIDs exhibit these same properties in varying degrees. For example, because acetaminophen blocks COX enzymes, but does not affect the specialized cells, it reduces fever but has little anti-inflammatory action. On the other hand, ibuprofen is a better enzyme blocker and specialized cell inhibitor. Consequently, ibuprofen is both a better pain reliever and fever reducer than aspirin. Ibuprofen has fewer functional groups than aspirin, which may be the reason why ibuprofen has fewer side effects. With fewer polar functional groups, ibuprofen is more fat-soluble than aspirin. Its anti-inflammatory activity is 5 to 50 times that of aspirin.

Interestingly, aspirin is unique among these three compounds in its ability to inhibit blood clotting. This property has led to the suggestion that low regular doses of aspirin can help prevent strokes or heart attacks. Of course, these anticoagulation characteristics also mean that aspirin is not the painkiller of choice for surgical patients, or those suffering from ulcers or blood-clotting problems. That was the rationale behind the

You encountered catalysts in several other contexts, including automobile emissions control (Section 1.11), petroleum refining (Section 4.7), and addition polymerization (Section 9.3).

ad campaign that "more hospitals use Tylenol"; doctors didn't want their patients to bleed unnecessarily. Another drawback of aspirin is that in rare cases in the presence of certain viruses, it can trigger a sometimes fatal response known as Reye's syndrome, particularly in children younger than 15. Furthermore, in some patients aspirin can trigger acute episodes of asthma.

Consider This 10.11   COX-2 Inhibitors

In 1992 researchers discovered that there were two separate types of COX enzymes, COX-1 and COX-2. The COX-2 enzyme is responsible for the production of the prostaglandins that regulate inflammation, pain, and fever, thereby reducing these symptoms. The COX-1 enzyme catalyzes the production of the prostaglandins responsible for maintaining proper kidney function and for keeping the stomach lining intact. Aspirin blocks the activity of both COX enzymes so it has the unwanted side effects of producing stomach pain and bleeding.

After this discovery, new "super aspirins" that would affect only the COX-1 enzymes were developed. Touted as drugs that would cause fewer gastrointestinal problems, they entered the market with great fanfare. They became instant blockbuster drugs, selling in the billions of dollars per year. Use the web to find the names of two of these drugs. Why have they disappeared from the pharmaceutical radar screen?

A few final comments about NSAIDs seem appropriate. Because it is a specific chemical compound, aspirin is aspirin—acetylsalicylic acid, regardless of its brand. Indeed, about 70% of all acetylsalicylic acid produced in the United States is made by a single manufacturer. But, although all aspirin molecules are identical, not all aspirin tablets are the same. The products are mixtures of various components, including inert fillers and bonding agents that hold the tablet together. Buffered aspirin tablets also include weak bases that counteract the natural acidity of the aspirin. Some coated aspirins keep the tablet intact until it leaves the stomach and enters the intestine. These differences in formulation can influence the rate of uptake of the drug and hence, how fast it acts, and the extent of stomach irritation it produces. Furthermore, although standards for quality control are high, it is conceivable that individual lots of aspirin may vary slightly in purity. Aspirin also decomposes with time, and the smell of vinegar can signify that such a process has begun. Fortunately, none of this poses a significant threat to health, and the benefits of aspirin far outweigh the risks for the great majority of people.

Consider This 10.12   Supersize My Aspirin

A friend who suffers from heart disease has been told by the doctor to take one aspirin tablet a day. To save money, your friend often buys the large 300-tablet bottle of aspirin. You, on the other hand, rarely take aspirin, but cannot pass up a good bargain. You also buy the large bottle.

a. Why is the "giant economy size" bottle of aspirin not as good a deal for you as it is for your friend?
b. What chemical evidence supports your opinion?

10.5 | Modern Drug Design

The evolution of "willow bark tea" to aspirin and further modifications to this painkiller's structure to enhance its beneficial effects and decrease its side effects represent stages in historical drug design. Penicillin is another example of a miracle drug whose origin lies in "natural" sources. Molds had been used for treating infections for 2500 years, although their effects were unpredictable and sometimes toxic. The penicillin story includes an accidental discovery by the British bacteriologist Alexander Fleming in 1928. Fleming's curiosity was aroused by the chance observation that in a container of bacterial colonies, the area contaminated by the mold *Penicillium notatum* was largely free of bacteria (Figure 10.10). He correctly concluded that the mold produced a substance that inhibited bacterial growth, and he named this biologically active material penicillin.

A careful reconstruction has indicated that a series of critical, but fortuitous events had to occur for the discovery to be made. Spores from the mold, part of an experiment in a nearby lab, drifted into Fleming's laboratory and accidentally contaminated some Petri dishes containing *Staphylococcus* (bacteria) growing on a nutrient medium. Then came a series of chance incidents involving poor laboratory housekeeping, a vacation, and the effects of weather. Fleming fortunately noticed among a pile of dirty glassware the dish in which the *Staphylococcus* had been killed. His experience allowed him to interpret the phenomenon, recognizing that some unknown substance produced by the *Penicillium* was a potential antibacterial agent. "The story of penicillin," Fleming wrote, "has a certain romance in it and helps to illustrate the amount of chance, of fortune, of fate, or destiny, call it what you will, in anybody's career." But of course, the discovery would not have happened without Fleming's powers of observation and insight. The episode illustrates the often misquoted maxim of the great French scientist, Louis Pasteur: "In the fields of observation, chance favors only the prepared mind." Most versions of this famous aphorism neglect the "only." It was only because Fleming's mind was prepared that he was able to capitalize on this chain of unlikely events.

The process of taking penicillin from the Petri dish to the pharmacy was not much different from what is done today. The first step was a systematic effort to isolate the active agent produced by *Penicillium notatum*. Once identified, the substance had to be purified and concentrated by sophisticated techniques. Also, the efficacy of penicillin in treating humans had to be demonstrated. World War II gave increased impetus to this research and to the development of new methods for preparing large quantities of penicillin. Because the scientists were successful in doing so, thousands of lives were saved during the war, and millions since then (Figure 10.11).

Treatments of infections once considered incurable—pneumonia, scarlet fever, tetanus, gangrene, and syphilis—were revolutionized by Fleming's discovery. The discovery of penicillin may have been serendipitous, but the development of the next several "generations" of antibiotics in this class involved systematic and careful research. Small changes are made to a drug and the resulting substances are tested with the goal of optimizing desired activity and decreasing side effects. More than a dozen different penicillins are currently in clinical use including: penicillin G (the original discovered by Fleming and the form that causes an allergic reaction in about 20% of the population), ampicillin, oxacillin, cloxacillin, penicillin O, and amoxicillin (the pink, bubble-gum-flavored concoction you might have been given as a child). Amoxicillin is still available in capsule form; it is commonly prescribed for being effective against a broad spectrum of bacteria and is usually well tolerated.

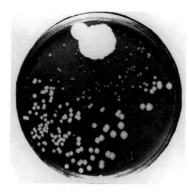

Figure 10.10

Photograph of the original culture plate of the fungus *Penicillium notatum*. This image was photographed by Sir Fleming for his 1929 paper on penicillin. The large white area at 12 o'clock is the mold *Penicillium notatum*; the smaller white spots are areas of bacterial growth.

Figure 10.11

A sign pasted at the entrance of a new facility for penicillin production during World War II.

 Consider This 10.13   **Drugs by Chance**

Modern methods of drug discovery involve structure–activity relationship studies and computer modeling, among other techniques. Sometimes side effects of a drug may open the door for its usefulness in treating other illnesses. There are many examples where a new drug was discovered by "chance." Use the resources of the web to find an example of a drug that was discovered by unusual circumstances.

The effectiveness of penicillin has unfortunately led to extreme overuse. As a result, cunning bacterial bugs have developed mechanisms for rendering penicillin (along with other antibiotics) useless. We are now witnessing strains of resistant bacteria or "superbugs," a phenomenon Fleming predicted back in 1945. Bacteria develop resistance to penicillin by secreting an enzyme that attacks the penicillin molecule before it can act. Some of the newer antibiotics differ in their effectiveness at killing certain bacteria and their susceptibility to the enzymes the organisms produce. Closely related to the penicillins are the cephalosporins (cephalexin, or Keflex) that are particularly effective against some resistant strains of bacteria. Careful research on structural modifications has led to other important medicines like cyclosporine, a drug that prevents tissue rejection. Its development made possible the revolutionary success of organ transplant surgery.

So how do chemists know which structural features are important to a drug's function? The modern approach to chemotherapy and drug design probably began early in the 20th century with Paul Ehrlich's search for an arsenic compound that would cure syphilis without doing serious damage to the patient. His quest was for a "magic bullet," a drug that would affect only the diseased site and nothing else. He systematically varied the structure of many arsenic compounds, simultaneously testing each new compound for activity and toxicity on experimental animals. He finally achieved success with arsphenamine (Salvarsan 606), so named because it was the 606th compound investigated. Since then, medicinal chemists have adopted Ehrlich's strategy of carefully relating chemical structure and drug activity. Systematic changes made to a drug molecule and assessment of the resulting changes in activity is known as a **structure–activity relationship (SAR) study.**

Drugs can be broadly classified into two groups: those that produce a physiological response in the body and those that inhibit the growth of substances that cause infections. You already learned that aspirin falls in the first group. So do synthetic hormones and psychologically active drugs. These compounds typically initiate or block a chemical action that generates a cellular response, such as a nerve impulse or the synthesis of a protein. Antibiotics exemplify drugs that prevent the reproduction of foreign invaders. They do so by inhibiting an essential chemical process in the infecting organism. Thus, they are particularly effective against bacteria.

Consider This 10.14   Friend or Foe?

Make two lists of drugs for each of the two broadly classified groups: those that bring about a desired physiological response and those that kill foreign invaders. Propose three drugs for each list, using examples not given in this section.

Although drugs vary in their versatility, many of them act only against particular diseases or infections. This specificity is consistent with the relationship that exists between the chemical structure of a drug and its therapeutic properties. Both the general shape of the molecule and the nature and location of its functional groups are important factors in determining its physiological efficacy. This correlation between form and function can be explained in terms of the interaction between biologically important molecules. Although many of these molecules are very large, consisting of hundreds of atoms, each molecule often contains a relatively small active site or receptor site that is of crucial importance in the biochemical function of the molecule. A drug is often designed to either initiate or inhibit this function by interacting with the receptor site.

An example is provided by a receptor site that controls whether a cell membrane is permeable to certain chemicals. In effect, such a site acts as a lock on a cellular door. The key to this lock may be a hormone or drug molecule. The drug or hormone bonds to the receptor site, opening or closing a channel through the cell membrane.

Whether the channel is open or closed can significantly influence the chemistry that occurs in the cell. In fact, under some circumstances, the cell may be killed, which may or may not be beneficial to the organism.

This lock-and-key analogy is often used to describe the interaction of drugs and receptor sites. Just as specific keys fit only specific locks, a molecular match between a drug and its receptor site is required for physiological function. The process is illustrated in Figure 10.12. But if a perfect lock-and-key match were required in the body, it would mean that each of the millions of physiological functions would have a unique receptor site and a specific molecular segment to fit it. Simple logic suggests that such rigid demands would not be very efficient. Consequently, the lock-and-key model, although a good starting point that works in a limited number of cases, must be modified.

Using another analogy, a receptor site is like a size 9 right footprint in the sand. Only one foot will fit it exactly, and many feet (all left feet and any right feet larger than size 9) will not fit. But many other right feet can fit into the print reasonably well. So it is with receptor sites and the molecules (or their functional groups) that bind to them. Some active sites can accommodate a variety of molecules including drugs. Indeed, the way most drugs function is by replacing a normal protein, hormone, or other substance in the invading organism. The general term **substrate** refers to the substances whose reactions are catalyzed by an enzyme. In the substrate inhibition model of enzyme activity, the presence of the drug molecule prevents the enzyme from carrying out its required chemistry. As a result, the growth of an invading bacterium is inhibited, or the synthesis of a particular molecule is turned off (Figure 10.13).

Generally speaking, the drug that best fits the receptor site has the highest therapeutic activity. In some cases, however, a drug molecule does not need to fit the receptor site particularly well. The bonding of functional groups of the drug to the receptor site may even alter the shape of the drug, the site, or both. Often what counts is for the drug to have functional groups of the proper polarity in the right places. Therefore, one important strategy in designing drugs is to determine its **pharmacophore,** the three-dimensional arrangement of atoms or groups of atoms responsible for the biological activity of a drug molecule. Medicinal chemists then synthesize a molecule having that specific active portion, but with a much simpler, nonactive remainder. These researchers custom design the molecule to meet the requirements of the receptor site. In effect, they design feet to fit footprints.

An outstanding example of this approach is provided by opiate drugs such as morphine. Morphine, a complex molecule, is difficult to synthesize. However, the pharmacophore responsible for opiate activity has been identified and is highlighted in Figure 10.14. The flat benzene ring fits into a corresponding flat area of the receptor, and the nitrogen atom binds the drug molecule to the site. Incorporating this particular portion into other less complex molecules, such as meperidine (Demerol), creates opiate activity. Demerol is much less addictive than morphine but also less potent.

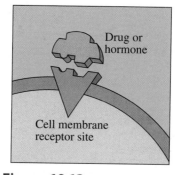

The lock-and-key analogy was first proposed in 1894 by the biochemist Emil Fischer.

Figure 10.12

Lock-and-key model of biological interactions.

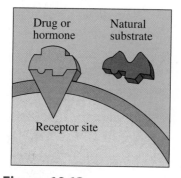

Figure 10.13

Drug molecule displacing a natural substrate on receptor site.

The term *pharmacophore* was originally described by Paul Ehrlich more than 100 years ago. Ehrlich was a German physician and biochemist who won the Nobel Prize in Physiology or Medicine in 1908 for his work on immunization.

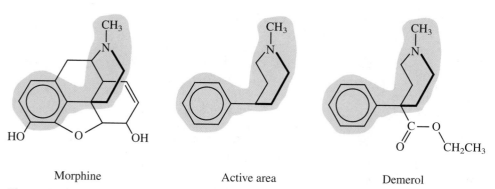

Morphine   Active area   Demerol

Figure 10.14

Molecular structures of morphine and Demerol. The highlighted "active areas," or pharmacophores, are the portions of the molecule that interact with the receptor. The darker lines indicate that these bonds are in front of the rings, or coming out in front of the plane of the page.

Consider This 10.15   3D Drugs

See for yourself how drug molecules appear in three dimensions by visiting the textbook's website, where you will find a collection of biologically active drugs.

a. Select several drugs and examine their three-dimensional structure. How do these computer representations differ from the structural formulas of drugs shown in this chapter?

b. What advantages do the computer representations have over two-dimensional drawings? What are their limitations compared with "real" molecules? Are there any disadvantages?

The discovery that only certain functional groups are responsible for the therapeutic properties of pharmaceutical molecules was an important breakthrough in drug design. Sophisticated computer graphics are now used to model potential drugs and receptor sites. Thanks to these representations with their three-dimensional character, medicinal chemists can "see" how drugs interact with a receptor site. Computers can then be used to search for compounds that have structures similar to that of an active drug. Chemists can also modify structures in computer models and visualize how the new compounds will function.

Such techniques help to minimize the cost and time it takes to prepare a so-called **lead compound,** a drug (or a modified version of that drug) that shows high promise for becoming an approved drug. An important new methodology is **combinatorial chemistry,** the systematic creation of large numbers of molecules in "libraries" that can be rapidly screened in the lab for biological activity and the potential for becoming new drugs. Drug companies have created large populations of molecules, or libraries, with a sort of a "shotgun" approach. The sheer volume of compounds in the libraries increases the likelihood that lead compounds will be discovered. Advances in automation, robotics, and computer programming have refined the technique, making it one that no drug company can afford to ignore (Figure 10.15).

Although complex protocols for combinatorial chemistry exist, let's look at the concept in its simplest form. Consider a molecule with three different functional groups as shown in Figure 10.16. When a sample of this compound is reacted with some reagent (step 1), a set of products is formed. After the addition of a second reagent (step 2) it is easy to see that many compounds can be formed rapidly. Again, different protocols may involve splitting products at various points, but simple statistics show that a great number of new compounds can be made in short order. The process can be repeated numerous

The term *lead compound,* pronounced differently, would refer to a compound containing the element Pb.

Figure 10.15

A scientist working with a combinatorial synthesis instrument.

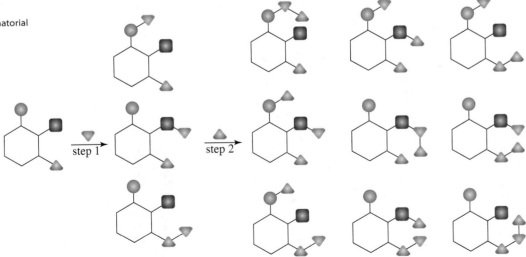

Figure 10.16

Illustration of a combinatorial synthesis process. Different colored shapes represent different functional groups.

times, with the products being screened for desired activity at each step. Unpromising reactions can be screened out quickly. Used in conjunction with computers, combinatorial chemistry can minimize the trial-and-error aspects and expense, thus speeding up drug design and development. Using traditional methods, a medicinal chemist could prepare perhaps four lead compounds per month at an estimated cost of $7000 each. With combinatorial chemical methods, the chemist can prepare nearly 3300 compounds in that same time for about $12 each.

10.6 | Give These Molecules a Hand!

Drug design is further complicated when drug–receptor interaction involves a common but subtle phenomenon called optical isomerism, or chirality. **Chiral,** or **optical, isomers** have the same chemical formula, but they differ in their three-dimensional molecular structure and their interaction with polarized light. Chirality most frequently arises when four different atoms or groups of atoms are attached to a carbon atom. A compound having such a carbon atom can exist in two different molecular forms that are nonsuperimposable mirror images of each other. One optical isomer will rotate polarized light in a clockwise manner, and this is called the dextro or (+) isomer. The other isomer is called the levo or (−) isomer, and it rotates polarized light in a counterclockwise manner.

Nonsuperimposable mirror images should be familiar to you. You carry two of them around with you all the time—your hands. If you hold them palms up, you can recognize them as mirror images. For example, the thumb is on the left side of the left hand and on the right side of the right hand. Your left hand looks like the reflection of your right hand in a mirror. But your two hands are not identical. Figure 10.17 illustrates this relationship for both hands and molecules.

Polarized light waves move in a single plane; nonpolarized light waves move in many planes.

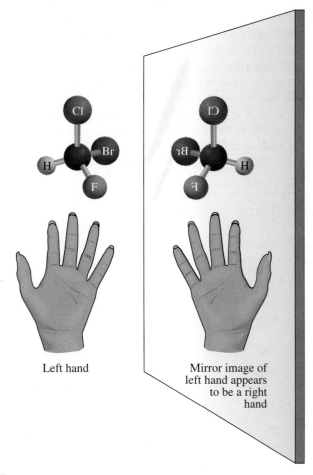

Left hand

Mirror image of left hand appears to be a right hand

Figure 10.17

Mirror image of a molecular model and a hand. The molecule CHBrClF is chiral and its shape is tetrahedral.

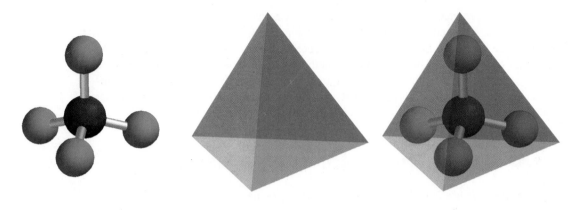

Tetrahedral molecule   Tetrahedron   Tetrahedral molecule inside tetrahedron

Figure 10.18

A tetrahedron has four equilateral triangular faces.

Cl
|
H······C—Br
F

Wedge–dash drawing of one of the molecules in Figure 10.17.

Sugars and amino acids are discussed in Sections 11.5 and 11.7.

Note that the four atoms or groups of atoms bonded to the central carbon atom are in a tetrahedral arrangement (Figure 10.18). The positions of these four atoms correspond to the corners of a three-dimensional figure with equal triangular faces. The "handedness" of these molecules gives rise to the term *chiral,* from the Greek word for hand.

Chemists often use a formalism called a wedge–dash drawing when representing a central chiral carbon atom. For example, the molecule in Figure 10.17 could be drawn as the figure on the left. Here, the Cl and the Br are in the plane of the page. The dashed line to the H indicates that the H atom is behind the page, extending away from the viewer. The solid wedge going to the F indicates that the F atom is in front of the page, oriented toward the viewer. Similar to a line-angle drawing, the central carbon atom may be implied but not drawn.

Many biologically important molecules, including sugars and amino acids, exhibit chirality. This is significant because, although most chemical and physical properties of a pair of optical isomers are very nearly identical, their biological behavior can differ markedly. Generally, the explanation for this difference is related to the necessity of a good molecular fit between a molecule and its receptor site. Maybe Lewis Carroll's Alice had some inkling of this when, in *Through the Looking Glass,* she remarked to her cat, "Perhaps looking-glass milk isn't good to drink."

Your Turn 10.16   Chiral Molecules

Use an asterisk to identify the chiral carbon in these molecules. Then draw both chiral isomers using wedge–dash drawings. Place the chiral carbon as the central atom.

a. [structure]   phenylalanine, an essential amino acid

b. $CH_3CH(OH)CH_2CH_3$ 2-butanol, an alcohol
c. $CHClFCH_3$ 1-chloro-1-fluoroethane, a hydrochlorofluorocarbon

d. [structure]   methamphetamine, a notoriously dangerous street drug

Answer

a. [structures]

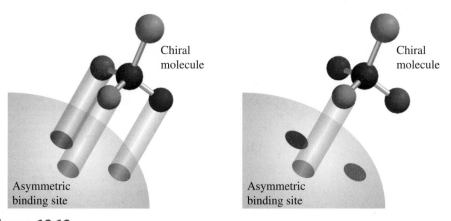

Figure 10.19

A chiral molecule binding (*left*) or not binding (*right*) to an asymmetrical site.

You can illustrate the relationship between chirality and biological activity by taking things in your own hands. Your right hand fits only a right-handed glove, not a left-handed one. Similarly, a right-handed drug molecule fits only a receptor site that complements and accommodates it. Any drug containing a carbon atom with four different atoms or groups attached to it will exist in chiral isomers, only one of which usually fits into a particular asymmetrical receptor site (Figure 10.19).

The extreme molecular specificity created by chirality complicates the medicinal chemist's task of synthesizing drugs. A drug molecule must include the appropriate functional groups, and these groups must have the three-dimensional configuration that gives the molecule its desired biological activity. In many chemical reactions, the "right" and "left" optical isomers are produced simultaneously. Such a situation results in a **racemic mixture** ($\pm$) consisting of equal amounts of each optical isomer. But frequently, only one optical isomer is pharmaceutically active. For example, many opiate drugs exist as optical isomers, only one of which may have opiate activity. In Figure 10.20, levomethorphan, the left-handed (levo, or $-$) isomer of methorphan, is an addictive opiate. On the other hand, its right-handed (dextro, or $+$) mirror image is a nonaddictive cough suppressant. This permits the use of dextromethorphan in many over-the-counter cough remedies, but the right-handed isomer must either be synthesized in pure dextro form or separated from a mixture with its levo isomer. The latter task can prove to be very difficult since the physical properties of the isomers are often identical.

Many other drugs exhibit chirality and are active only in one of the isomeric forms. This is true for some antibiotics and hormones and for certain drugs used to treat a wide range of conditions: inflammation, cardiovascular disease, central nervous system disorders, cancer, high cholesterol levels, and attention deficit disorder. Among the widely used chiral drugs are ibuprofen, cyclosporine (the drug used to prevent rejection in organ transplants), and the lipid-reducing drug atorvastatin (Lipitor). Ibuprofen is sold as a racemic

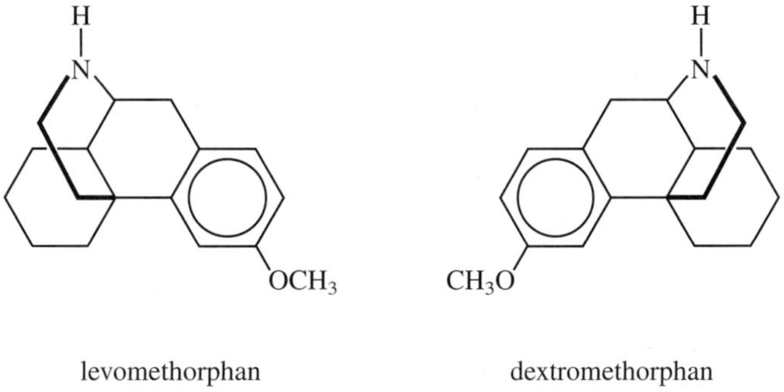

The vitamin E sold in stores is generally a racemic mixture of $+$ and $-$ isomers. The $+$ isomer is the physiologically active one that can be purchased in pure form at a significantly higher price.

levomethorphan   dextromethorphan

Figure 10.20

Levo- and dextromethorphan.

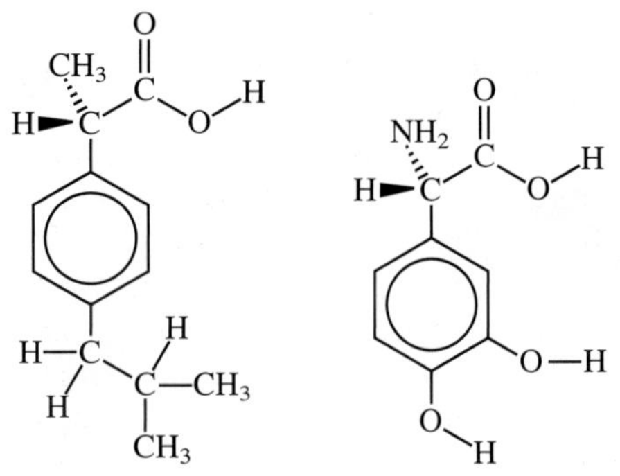

Figure 10.21
Two chiral drug molecules.

mixture of (+) and (−) isomers (Figure 10.21 shows the (+) isomer). Only (−)-ibuprofen acts as a pain reliever; (+)-ibuprofen does not. However, in the body the (+) isomer is converted to the (−) form. Therefore, it is likely that someone taking ibuprofen is just as well off taking the racemic mixture rather than the more expensive (−)-ibuprofen.

Naproxen, a common pain reliever, is one example of many in which one isomer is preferred, even required. One form of naproxen relieves pain; the other causes liver damage. One last example involves a treatment for Parkinson's disease. The initial use of racemic dopa for treatment of this disease brought on adverse effects such as anorexia, nausea, and vomiting. The use of the single isomer (−)-dopa greatly reduced these side effects and the desired effect was achieved using half the initial doses.

Your Turn 10.17   Examining (+)-Ibuprofen and (−)-Dopa

Carefully examine the structural formulas for (+)-ibuprofen and (−)-dopa given in Figure 10.21.

  a. For each drug, which is the chiral carbon atom?
  b. Identify all of the functional groups present in both drugs.
  c. Draw the structural formula of (−)-ibuprofen and (+)-dopa.

William Knowles, Barry Sharpless, and Ryoji Noyori shared the 2001 Nobel Prize in Chemistry for their research that developed new catalytic methods for synthesizing chiral drugs.

Consequently, drug companies have active research programs designed to create chirally "pure" drugs, those having only the beneficial isomer of a drug as a single chiral form. Although making the proper, single isomer might seem like an exercise of interest only to chemists, it is big business. The majority of the most successful prescription drugs sold worldwide are single-isomer drugs, with total annual sales around $50 billion and rising, a trend that will no doubt continue into the future. These include the chiral "blockbuster" drugs Lipitor, simvastatin (Zocor), esomeprazole (Nexium), and sertraline (Zoloft), with global sales over $1 billion per year.

Lipitor, a chiral drug classified as a statin, lowers cholesterol by preventing its synthesis in the liver. It is a phenomenal bestseller with annual worldwide sales topping $10 billion. Producing Lipitor requires the synthesis of hydroxynitrile (HN), another chiral molecule. Until recently, HN had been produced only as a racemic mixture, thus requiring the separation of the two isomers. Worse still, this process involved large quantities of hydrogen bromide and cyanide. These chemicals are toxic to say the least, and the process produced enormous quantities of unwanted side products.

Enter Codexis, a company that won one of the 2006 Presidential Green Chemistry Challenge Awards. Chemists at Codexis developed an elegantly green route to the HN intermediate using enzymes to carry out highly specific reactions. Their green process increases yields, reduces the formation of by-products, reduces the generation of waste and use of solvents, reduces the use of purification equipment, and increases worker safety.

The widespread use of Lipitor creates an annual demand for the HN intermediate of about 200 metric tons. This indeed is an invention worthy of a Presidential award!

New chiral drugs are big business, and drug companies have recently devised a novel strategy for increasing profits in this area. A number of proven medicines formerly sold as racemates (± mixtures) are being reevaluated and remarketed as single isomers, a strategy known as a "chiral switch." The economic reasons for this are obvious: pharmaceutical companies are able to extend patent protection on their bestseller drugs and give them a hedge against generic competition. The therapeutic rationale for the switch is that the single isomer may provide benefits such as a wider margin of safety, fewer side effects, and more simple interactions in the body. Methylphenidate, sold under the trade name Ritalin, is a drug used for treating attention-deficit hyperactivity disorder (ADHD). Formerly prescribed as a racemate, it has made the chiral switch and is now marketed as a single isomer; it is reported to be equally effective at half the dose compared with the racemate, and has an improved side effect profile.

Consider This 10.18   Chiral Switch

In principle, the strategy of chiral switching should afford a therapeutic advantage for the single-isomer drug over the racemic mixture. But is this always the case? Use the web to identify one drug (other than Ritalin) that has made the chiral switch and report on the relative merits of the single isomer versus the racemic mixture.

10.7 | Steroids

Consider contraceptives, muscle-mass enhancers, and abortive agents. What do these chemically have in common? The surprising answer is that they are all **steroids,** a class of naturally occurring or synthetic fat-soluble organic compounds that share a common carbon skeleton arranged in four rings.

As a family of compounds, steroids arguably best illustrate the relationship of form and function. Certainly no other group of chemicals is more controversial because their uses range from contraception to vanity promoters. The naturally occurring members of this ubiquitous group of substances include structural cell components, metabolic regulators, and the hormones responsible for secondary sexual characteristics and reproduction. Among the synthetic steroids are drugs for birth control, abortion, and bodybuilding. Table 10.4 lists some of the functions.

In spite of their tremendous range of physiological functions, all steroids are built on the same molecular skeleton. Thus, these compounds also provide a marvelous example of the economy with which living systems use and reuse certain fundamental structural units for many different purposes. The common characteristic of steroids is

Table 10.4	Steroid Functions
Function	**Examples**
Regulation of secondary sexual characteristics	estradiol (an estrogen), testosterone (an androgen)
Regulation of the female reproductive cycle	progesterone, RU-486 (the "abortion pill")
Regulation of metabolism	cortisol, cortisone derivatives
Digestion of fat	cholic acid
Component of cell membranes	cholesterol
Stimulation of muscle and bone growth	gestrinone, trenbolone

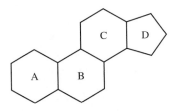

Figure 10.22

Figures Alive! Representations of cholesterol. Visit the textbook's website to learn more about the structure of cholesterol and other steroids.

a molecular framework (nucleus) consisting of 17 carbon atoms arranged in four rings. The steroid nucleus is illustrated here.

Recall that in such a representation (a line-angle drawing), carbon atoms are assumed to occupy the vertices of the rings but are not explicitly drawn. The three six-membered carbon rings of the steroid nucleus are designated A, B, and C, and the five-membered ring is designated D. Although the steroid nucleus appears flat as drawn, it actually is three-dimensional in shape. The dozens of natural and synthetic steroids are all variations on this theme. Some differ only slightly in structural detail, but have radically different physiological function. Extra carbon atoms or functional groups at critical positions on the rings are responsible for this variation.

The steroid cholesterol is a major component of cell membranes and is shown in Figure 10.22. The figure on the left includes all the atoms in the molecule; the one on the right gives the skeletal representation using a line-angle drawing.

Careful examination of Figure 10.23 illustrates how subtle molecular differences can result in profoundly altered physiological properties. The difference between a molecule of estradiol and one of testosterone lies only in one of the rings. Are the only differences between men and women due to a carbon atom and a few hydrogen atoms? You be the judge.

Some other common natural steroids are the hormones cortisone and corticosterone (Figure 10.24). You may have used some form of cortisone preparation for treating rashes and other minor skin disorders. Corticosteroids are produced in the adrenal cortex and play a wide range of roles such as regulation of inflammation, immune response, stress response, and carbohydrate metabolism. Steroidal anti-inflammatory drugs are the most often used treatment option for people with asthma; one synthetic choice is prednisone.

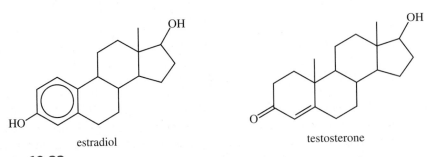

estradiol   testosterone

Figure 10.23

Check the activities at Figures Alive! Estradiol and testosterone.

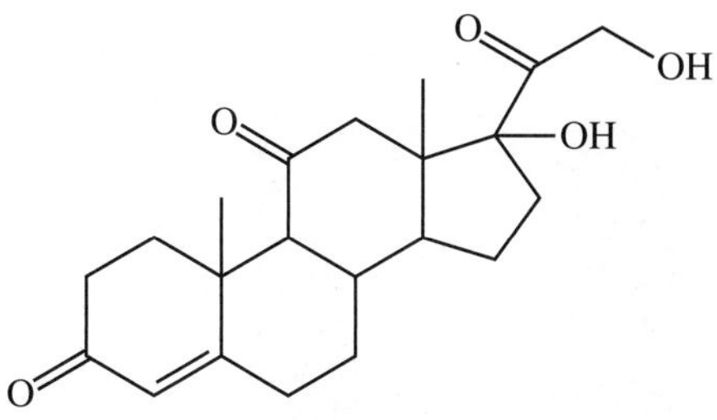

cortisone 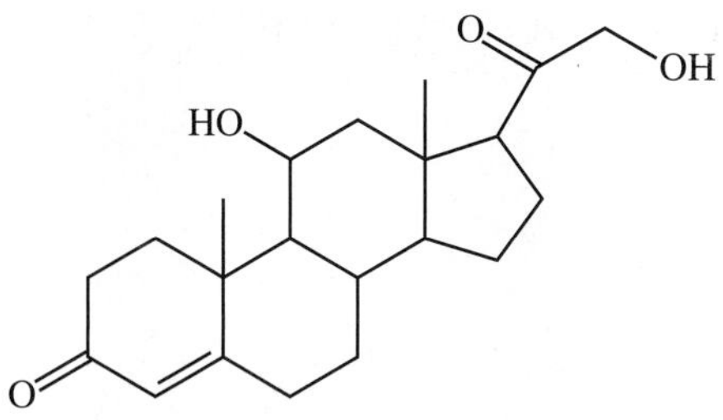   corticosterone 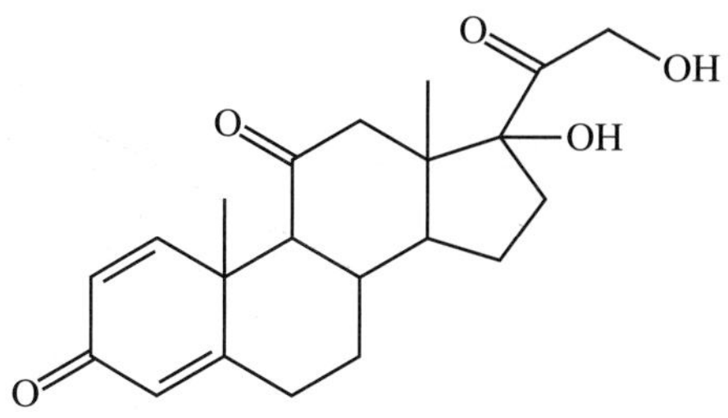   prednisone

 Figure 10.24

Check the activities at Figures Alive! Cortisone, corticosterone, and prednisone.

Your Turn 10.19   Structural Similarities of Steroids

Here are pairs of steroids. Their structures are given either here in the text or in Figures Alive! on the textbook's website.

a. Identify the structural similarities in each pair.

estradiol and progesterone   corticosterone and cortisone
cholic acid and cholesterol   prednisone and cortisone
estradiol and testosterone

b. Write the chemical formula for five of the drugs in part **a.**

10.8 | Prescription, Generic, and Over-the-Counter Medicines

Enter customer. The pharmacist asks "Brand name or generic?" This scenario is played out daily in thousands of pharmacies across the country. How is the person to decide? For millions of Americans, the cheaper generic version can mean the difference between getting the necessary medication and not being able to afford it, although not all approved drugs are available in generic form.

The two forms can be differentiated rather simply. A pioneer drug is the first version of a drug that is marketed under a brand name, such as Xanax, an antianxiety, or sedative drug. A **generic drug** is chemically equivalent to the pioneer drug, but cannot be marketed until the patent protection on the pioneer drug has run out after 20 years. The lower priced drug is commonly marketed under its generic name, in this case alprazolam instead of Xanax. The 20-year patent protection on the pioneer drug begins when it is patented, not when it's first put on the market. In cases requiring a long preapproval time, the actual marketing period can be relatively short, even less than six years. In such a situation, a drug company has very little time to recoup its research and development costs before a generic competitor can be manufactured. However, almost 80% of generic drugs are produced in brand-name equivalents (Figure 10.25). Like pioneer drugs, generic drugs must also be approved by the FDA.

In 1984, Congress passed the Drug Price Competition and Patent Restoration Act that greatly expanded the number of drugs eligible for generic status. This act eliminated the need for generics to duplicate the efficacy and safety testing done on counterpart pioneer drugs. Doing so saves drug manufacturers considerable time and money. The FDA also issued specific guidelines for a generic drug's comparability to the pioneer drug. By FDA mandate, the generic and pioneer versions must be bioequivalent in dosage form, safety, strength, route of administration, quality, performance characteristics, and intended use. In other words, they must deliver the same amount of active ingredient into a patient's bloodstream at the same rate. Once proven, companies can begin to drive the high price of a brand-name drug down through competition.

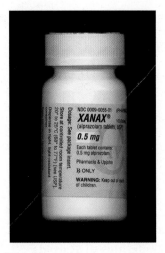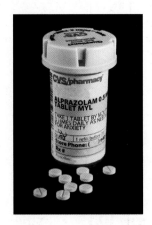

Figure 10.25

A brand-name drug (Xanax) and its generic counterpart (alprazolam).

Consider This 10.20   Your Local Pharmacist

When it comes to generic drugs, pharmacists are on the front line. With your classmates, draft two or three short questions that relate to generic drugs. Then interview a local pharmacist and share your findings with your classmates. Note: If you have a school of pharmacy nearby, you might also want to interview pharmacy students.

Over-the-counter (OTC) drugs allow people to relieve many annoying symptoms and cure some ailments without the need to see a physician. Nonprescription medications now account for about 60% of all medications used in the United States. More than 80 therapeutic categories of OTC drugs exist, ranging from acne products to weight control products. Table 10.5 contains several major categories of OTC products and their chief components. In accordance with the laws and regulations prevailing in this country, drugs including OTC drugs are subjected to an intensive, extensive, and expensive screening process before they can be approved for sale and public use. Just as in the case of prescription drugs, the ultimate question to be answered with over-the-counter drugs is "Do the benefits outweigh the risks?" The answer for an OTC drug depends on whether a consumer is using it properly. Therefore, the FDA and pharmaceutical manufacturers must try to balance OTC safety and efficacy.

Table 10.5	Examples of Over-the-Counter Drugs
Analgesics and Anti-Inflammatory Drugs	**Cough Remedies**
aspirin	guaifenesin, expectorant
ibuprofen (Advil)*	dextromethorphan, suppressant
naproxen (Aleve)	benzonatate, suppressant
acetaminophen (Tylenol)	
Antacids and Indigestion Aids	**Antihistamines**
aluminum and magnesium salts (Maalox)	brompheniramine
calcium carbonate (Tums)	chlorpheniramine
calcium and magnesium salts (Rolaids)	diphenhydramine

*Names are shown as chemical name followed by common or brand name in parentheses.

Earlier we described OTC pain relievers and NSAIDs and their mode of action. Their side effects include increased stomach bleeding (aspirin), gastrointestinal upset (aspirin, ibuprofen), aggravating asthma (aspirin), and kidney or liver damage (acetaminophen) at high dosage or with chronic use.

More than 100 viruses are responsible for the misery accompanying the common cold. A whole host of cold remedies, most with multiple components, are designed to help the sufferer. Decongestants reduce swelling when viruses invade the mucous membranes, but the adverse effects include nervousness and insomnia. Nasal sprays relieve the swollen nasal tissues, but their use beyond a three-day limit often leads to a rebound effect, or a return of the runny nose.

Antihistamines relieve the runny nose and sneezing associated with allergies, but cause drowsiness and often light-headedness. Because they induce drowsiness, it is not surprising that approved sleep aids are often antihistamines. However, children often experience insomnia and hyperactivity after taking them.

Coughing is a natural way to rid the lungs of excess secretions. Expectorants make the phlegm thinner and therefore easier to cough up, while suppressants provide relief and restful sleep. The presence of both in most cough remedy preparations seems senseless! Codeine and dextromethorphan have equally good cough-suppressing potential, but the former has a reputation for being habit-forming, a property that limits its over-the-counter availability in some states.

Heartburn, indigestion, and "acid" stomach are targets of antacids and related drugs. Antacids are basic compounds containing aluminum, magnesium, or calcium hydroxides (or combinations of them) that neutralize excess stomach acid. The popularity of the calcium-containing alternatives has grown with the promotion of the need for younger and older adults to maintain a regular supply of dietary calcium to prevent degradation of bones (osteoporosis).

Consider This 10.21 Using Common Sense

Many people are of the opinion that it is best to only take the medicines that are indicated for a particular illness. This makes plain sense. Yet some companies selling OTC preparations seem to throw in everything but "the kitchen sink," hoping to give their product some sort of "value-added" feature. Check the contents of several cold remedies found at your local drug or grocery store (or go online to find the same information). In particular, look at several cough remedy preparations. Do you find the use of both guaifenesin and dextromethorphan in one product? List the names of the products that do, and then explain why a product containing both of these medicines makes no sense.

The self-care revolution of the last several decades has encouraged the availability of safe and effective OTC drugs and has provided additional pressure for the reclassification of many prescription drugs to OTC status. According to the Consumer Healthcare Products Association, about 80 ingredients or reduced dosages of drugs have made the OTC switch since 1976. Recent prescription to OTC switches include loratidine and cetirizine (Claritin and Zyrtec, antihistamines, 2002 and 2008, respectively), and omeprazole (Prilosec, acid reducer, 2003), and lovastatin (cholesterol control, 2005). Right now, more than 700 OTC products have ingredients that were once only available by prescription. Additional pressure to have the switch occur comes from the health insurance industry. Changing widely used prescription drugs to OTC status greatly diminishes the insurance companies' share of payments. The conditions for which the drugs are prescribed must be common, non-life-threatening, and self-diagnosable by the average consumer. The FDA can change the status of an OTC drug back to prescription status if significant safety problems are uncovered.

From the drug manufacturer's standpoint, changing the status of a product from prescription to OTC often allows the manufacturer to market the product for several more years without generic competition. Sales volume also increases when a product is reclassified to OTC status. For the consumer, out-of-pocket expense in some cases may actually increase when going with OTC therapeutics because few third-party health insurance payers provide reimbursement for OTC products. But overall it is estimated that prescription to OTC switches save the American public over $20 million each year.

10.9 | Herbal Medicine

Worldwide, a growing number of people are using herbal products for preventive and therapeutic purposes. Herbal remedies and folk medicines abound in most cultures. This should not be surprising or astounding; nature is a very good chemist. Some (but not all) compounds found in plants and simple organisms are likely to have positive physiological effects in humans. U.S. sales of such popular herbal remedies as ginkgo biloba, St. John's wort, echinacea, ginseng, garlic, and kava kava have steadily risen over the past decade to $4 billion in 2003. Table 10.6 lists some herbs and plants and reasons for ingesting them.

Many people have been taking St. John's wort for its reported mood-elevating activity (Figure 10.26). The Herb Research Foundation reports data from over 2000

Table 10.6	Common Herbs and Their Possible Benefits
Herb or Plant	**Symptoms to Be Relieved**
Valerian, passion flower	anxiety
Licorice, wild cherry bark, thyme	coughs
Echinacea, garlic, goldenseal root	colds, flu
St. John's wort	depression
Chamomile, peppermint, ginger	nausea, digestive problems
Valerian, passion flower, hops, lemon balm	insomnia
Ginkgo biloba	memory loss
Valerian, passion flower, kava kava, Siberian ginseng	stress, tension

Figure 10.26

St. John's wort plant (*Hypericum perforatum*) and an extract in tablet form.

patients in 23 clinical studies that have consistently found that a preparation of St. John's wort, a plant, is just as effective against mild to moderate depression as standard antidepressant drugs. In April 2001, however, a study published in the *Journal of the American Medical Association* reported that St. John's wort is ineffective against severe depression. The herb worked no better than the placebo in over 200 adults diagnosed with severe depression. The study was funded partly by the National Institutes of Mental Health and partly by Pfizer Incorporated, which makes sertraline (Zoloft), the most commonly prescribed multibillion dollar a year antidepression drug in the United States. Dr. Richard Skelton from Vanderbilt University, a coauthor of the new study, recommended further studies on the use of the herb for mild depression, saying: "I would like to see people with mild depression studied, and see if it works in those folks. If it works, that would be great." He recommended against using the herbal medicine until further studies are done.

Consider ephedra (Figure 10.27), a naturally occurring substance derived from the Chinese herb *ma huang* as well as from other plant sources. Although ephedra has long been used to treat certain respiratory symptoms in traditional Chinese medicine, in recent years it has been heavily promoted and used for the purposes of aiding weight loss, enhancing sports performance, and increasing energy.

Ephedra contains six amphetamine-like alkaloids including ephedrine and pseudoephedrine. Ephedrine, the main constituent, is a bronchodilator (opens the airways) and stimulates the sympathetic nervous system. It has valuable antispasmodic properties, acting on the air passages by relieving swelling of the mucous membranes. Pseudoephedrine (Sudafed) is a nasal decongestant and has less stimulating effect on the heart and blood pressure.

In their synthetic form, these drugs were regulated as OTC drugs and used as a decongestant for the short-term treatment of runny nose, asthma, bronchitis, and allergic reactions by opening the air passages in the lungs. Figure 10.28 shows three related structures; methamphetamine is presented so you can see the structural similarities between ephedra drugs and this potent and dangerous stimulant. Ephedra does not contain methamphetamine.

Dietary supplements that contained ephedra made big headlines in 2003. The deaths of well-known athletes were linked to rare but serious side effects of ephedra use. Side effects reported by ephedra users included nausea and vomiting, psychiatric disturbances such as agitation and anxiety, high blood pressure, irregular heartbeat, and, more rarely, seizures, heart attack, stroke, and even death.

Figure 10.27

Ephedra's source and a common formulation.

A solid wedge indicates a group coming out toward you. A dashed wedge indicates that the group is pointing away from you. These conventions were first introduced in Section 3.3.

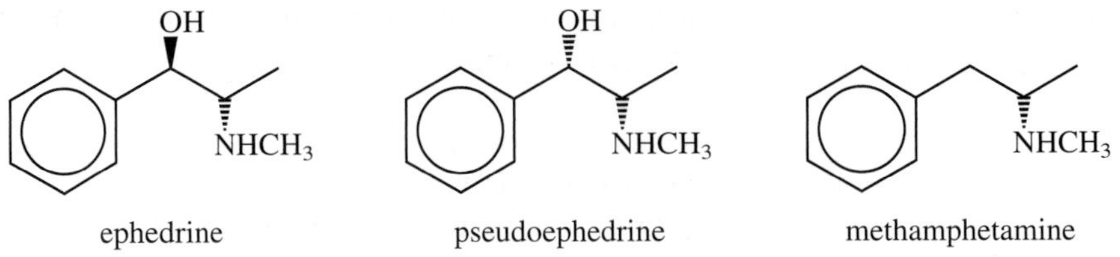

Figure 10.28

Chemical structures of ephedrine and two related drugs.

No evidence currently supports the claim that ephedra enhances athletic performance. And only preliminary evidence suggests that ephedra aids in modest, temporary weight loss. However, evidence does indicate that ephedra is associated with an increased risk of side effects, possibly even fatal ones. In 2003, the International Olympic Committee, the National Football League, the National Collegiate Athletic Association, minor league baseball, and the U.S. Armed Forces banned the use of ephedra.

In December 2003, the FDA issued a consumer alert on the safety of dietary supplements containing ephedra. Consumers were advised to immediately stop buying and using ephedra products. In February 2004, the FDA published a final rule stating that dietary supplements containing ephedra present an unreasonable risk of illness or injury. The rule effectively banned the sale of these products, which took effect 60 days after its publication. However, the FDA ban does not apply to teas that contains ephedra (these are regulated as foods) and to traditional Chinese herbal remedies when prescribed by a traditional Chinese physician.

The St. John's wort and ephedra examples illustrate important concerns about herbal medicines. Are they effective, and are they safe? Psychiatrists report that many of their patients with depression have tried St. John's wort before coming for medical help. One estimate suggests that over a million people in the United States alone have tried it or are using St. John's wort. Irrespective of the accuracy of the estimate, large numbers of Americans are apparently using herbal medicines under circumstances with little or no medical supervision.

Herbal remedies are only loosely regulated by the FDA. The Dietary Supplement Health and Education Act (DSHEA) of 1994 changed their classification from "food or drug" to "dietary supplement." **Dietary supplements** by definition include vitamins, minerals, amino acids, enzymes, and herbs and other botanicals. Many dietary supplements have shown no adverse effects and in fact have proven to be beneficial to good health. Others, like ephedra, have been shown to be problematic.

Under the DSHEA, the FDA does not review dietary supplements for safety and effectiveness before they are marketed. Rather, the law allows the FDA to prohibit sale of a dietary supplement if it "presents a significant or unreasonable risk of injury." When the FDA seeks to take regulatory action against a supplement (such as in the ephedra case), the burden of proof for establishing harm falls on the government. Unlike prescription or OTC drugs, there is no assessment of purity of preparations or concentrations of active ingredients, set amounts, or delivery protocols. Furthermore, there are no requirements for studies of interactions among herbal medicines or between them and traditional medicines. Since the manufacturers of herbal remedies are not required to submit proof of safety and efficacy to the FDA before marketing, information regarding the interactions between herbal remedies and other drugs is largely unknown.

In the spring of 2001, representatives from the American Society of Anesthesiologists reported concerns that patients undergoing surgery may risk unexpected bleeding when they take certain herbs within two weeks before surgery. To this point, no scientific studies linking the bleeding and a specific herb have been published. According to Dr. John Neeldt, president of the American Society of Anesthesiologists, the familiar question "Are you taking any medications?" should be augmented with, "Are you taking any herbal remedies?"

Consider This 10.22   **Does Natural Mean Safer?**

The legal standard of "significant or unreasonable risk" implies a risk–benefit calculation based on the best available scientific evidence. This suggests that the FDA must determine if a product's known or supposed risks outweigh any known or suspected benefits, based on the available scientific evidence. This must be done in light of the claims the manufacturer makes and with the understanding that the product is being sold directly to consumers without medical supervision.

When deciding to take any medication a consumer makes a risk–benefit analysis, sometimes unconsciously. One element of such an analysis is the *perceived* risk. Do you think that the general population perceives naturally occurring drugs as safer than synthetic ones? Explain using examples of your choice.

10.10 | Drugs of Abuse

Before concluding our foray into the world of drugs, we should take time to examine their abuse. According to the National Survey on Drug Use and Health (NSDUH), the number of people illegally using one or more drugs ranks in the millions. In 2008, 8% of the population, or an estimated 20 million Americans of age 12 or older, reported that they used an illicit drug—one not sanctioned by law or custom—during the month prior to the survey. Details from the 2008 survey included that:

- Marijuana was the most common illicit drug, with 15 million users. However, the extent of use among youths has declined slowly in the past decade.
- 2 million people (0.7% of the population) were cocaine users, 467,000 of whom used the more addictive "crack" form that can be inhaled.
- Hallucinogens were used by about 1 million people, including 525,000 users of ecstasy, a highly addictive drug that also acts as a stimulant.
- During 2008, 10 million people reported that they had operated a vehicle while under the influence of an illicit drug.

Although tobacco and alcohol are not illegal in the United States, their use was surveyed as well.

- Just over half of the U.S. population, aged 12 and older, drank alcohol, or about 130 million people. About one fifth of these were binge drinkers, meaning that they consumed five or more drinks at a time.
- About 70 million people were tobacco users, including 60 million cigarette smokers. The overall rate of tobacco use was about 28% of the population.

Consider This 10.23   **Consequences**

Legal or illegal, the drugs that humans use have consequences. People usually can point a finger at a drug that is disruptive in the home and workplace. Which drug would you pick? Make your choice and be prepared to defend it in a group discussion.

In 1970, the Comprehensive Drug Abuse Prevention and Control Act was passed into law. Title II of this law, the Controlled Substances Act, is the legal foundation of narcotics enforcement in the United States. The Controlled Substance Act regulates the manufacture and distribution of drugs and places all drugs into one of five schedules (Table 10.7).

Table 10.7	Drug Schedules		
Class	**Has Current Accepted Medical Uses**	**Potential for Abuse**	**Examples**
Schedule I	no	high	heroin LSD marijuana, hashish mescaline MDMA ("ecstasy")
Schedule II	yes	high	oxycodone (in OxyContin and Percocet) morphine, opium methadone cocaine (as a topical anesthetic) methamphetamine
Schedule III	yes	medium	hydrocodone with acetaminophen (Vicodin) codeine with acetaminophen anabolic steroids
Schedule IV	yes	low	alprazolam (Xanax) propoxyphene and acetaminophen (Darvocet) diazepam (Valium)
Schedule V	yes	lowest	cough suppressants with small amounts of codeine diphenoxylate and atropine (Lomotil) promethazine (Phenergan)

Consider This 10.24   Low Potential for Abuse?

The sedatives diazepam (Valium) and alprazolam (Xanax) are currently listed as Schedule IV drugs, indicating that they have a low potential for abuse. Search the web for information on addiction to these powerful sedative drugs. Do you agree with the current scheduling? Explain.

Cannabis sativa means useful (*sativa*) hemp (*cannabis*).

Figure 10.29

Pot plants.

Marijuana, an example of a Schedule I drug, is the most commonly used illicit drug in the United States. It is a mixture of the dried leaves, stems, seeds, and flowers of the hemp plant, *Cannabis sativa* (Figure 10.29). The drug is usually smoked and occasionally eaten. The first known record of marijuana use dates back to the time of the Chinese Emperor Shen Nung (ca. 2737 BCE), who prescribed use of the plant for the treatment of malaria, gas pains, and absentmindedness.

Hemp is the plant whose botanical name is *Cannabis sativa*. Other plants are called hemp, but *Cannabis* hemp is the most useful of these plants. Fiber is its most well-known product, and the word *hemp* can mean the rope or twine made from the hemp plant, as well as just the stalk of the plant that produced it. The major psychoactive drug in *Cannabis sativa* is concentrated in the leaves and flowers of the hemp plant, so one cannot get high from smoking hemp rope or wearing clothing woven from hemp fibers.

Extracts of marijuana were employed by physicians in the early 1800s for a tonic and euphoriant. But in 1937, the Marijuana Tax Act prohibited its use as an intoxicant and its medical use was regulated as national concern of its use emerged. The Marijuana Tax Act required anyone producing, distributing, or using marijuana for medical purposes to register and pay a tax that effectively prohibited nonmedical use of the drug. Although the act did not make medical use of marijuana illegal, it did make it expensive and inconvenient.

In 1942, marijuana was removed from the U.S. Pharmacopoeia, the official government compendium of drugs, because it was believed to be harmful, addictive, and cause psychosis, mental deterioration, and violent behavior. The current legal status of marijuana was established in 1970 with the passage of the Controlled Substances Act.

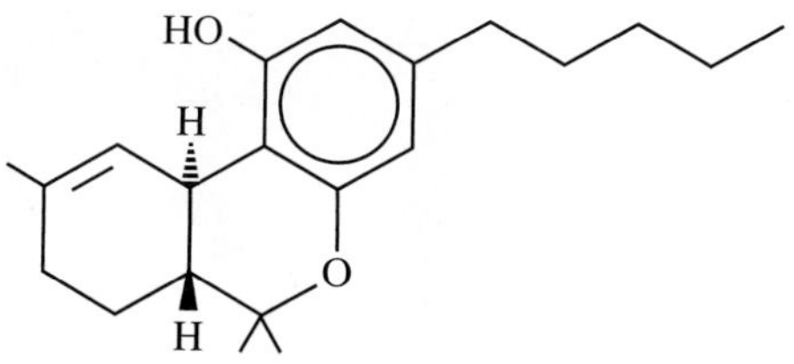

Figure 10.30

Structural formula for THC.

The major psychoactive chemical in marijuana is Δ^9-tetrahydrocannabinol, or THC, as shown in Figure 10.30. The concentration of THC varies depending on how the hemp is grown: temperature, amount of sunlight, and soil moisture and fertility. High-potency varieties of marijuana are grown across the United States, with THC levels as high as 7%. In hashish, or hash, the dried and pressed flowers and resin of the plant, THC concentrations can be as high as 12%.

When marijuana smoke is inhaled, THC rapidly passes from the lungs into the bloodstream, which carries the chemical to organs throughout the body, including the brain. In the brain, THC connects to specific sites called cannabinoid receptors on nerve cells and influences the activity of those cells. Some brain areas have many cannabinoid receptors; others have few or none. Many cannabinoid receptors are found in the parts of the brain that influence pleasure, memory, thought, concentration, sensory and time perception, and coordinated movement. THC leaves the blood rapidly through metabolism and uptake into the tissues. The chemical may remain stored in body fat for long periods; research has indicated that a single dose can take up to 30 days for complete elimination. The short-term effects of marijuana use can include problems with memory and learning, distorted perception, difficulty in thinking and problem solving, loss of coordination, decreased blood pressure, and increased heart rate.

Medicinal marijuana may be indicated for treatment of nausea, glaucoma, pain management, and appetite stimulation. Such treatment may be a last resort when all other medications have failed, such as with the unrelenting nausea and vomiting that may accompany weeks of chemotherapy in treating diseases such as leukemia and AIDS. Clinical studies on the usefulness of marijuana are difficult to conduct as many barriers discourage researchers. The scarcity of funding combined with complicated regulations enforced by both federal and state agencies make research in this area daunting.

At its heart, the debate over the medical use of marijuana pits benefits against risks. Underlying this debate, however, are the complex moral and social judgments relating to drug control policies in the United States. Supporters of medical marijuana argue that it is far less risky than some drugs currently in use and may succeed at relieving unpleasant symptoms where other drugs fail. Opponents argue that it is dangerous, unnecessary, and that its use leads to the use of other drugs.

In the United States, the Institute of Medicine of the National Academies has weighed in on the issue. In 1999, the Institute published a review that found marijuana to be "moderately well suited for particular conditions, such as chemotherapy-induced nausea and vomiting, and AIDS wasting." As of 2010, 14 states plus the District of Columbia had laws allowing physicians to prescribe marijuana for medical purposes.

Consider This 10.25   THC in a Legal Form

A legal form of marijuana exists in the form of a prescription drug, dronabinol (Marinol). It is a synthetic form of THC.

a. Marinol is swallowed rather than smoked. The 1999 report from the Institute of Medicine labels smoking "a primitive drug delivery system." What are the disadvantages of smoking (as opposed to swallowing) a drug such as marijuana?

b. Even if primitive, inhaling a drug does have key advantages in treating nausea or vomiting. Name at least one.

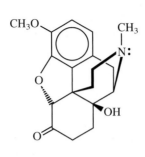
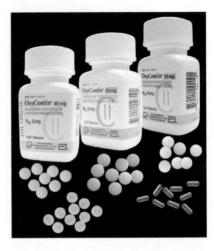

Figure 10.31

Oxycodone molecular structure and OxyContin pills.

Source: Photo © 2004, Publishers Group.

Oxycodone is a Schedule II drug under the Controlled Substances Act because of its high propensity to cause dependence and abuse.

OxyContin is an example of a Schedule II drug (Figure 10.31). Known on the street as "oxy," OxyContin is the trade name for oxycodone hydrochloride, a morphine-like narcotic. OxyContin tablets contain up to 80 mg of oxycodone with a time release mechanism. For those enduring acute and long-term pain, this drug delivers much-needed relief. Percocet and Percodan contain the same chemical, but in much smaller amounts (5–7.5 mg/tablet).

Abusers of OxyContin are able to get around the time release mechanism by crushing the tablet and either snorting it or dissolving it in water and injecting the solution. The effect is similar to that achieved from using heroin. Obviously, absorbing this much of the powerful narcotic can have dramatic consequences. Since the drug arrived on the market, emergency room overdoses linked to this narcotic increased dramatically.

The problem was first observed in rural areas of Kentucky, Virginia, West Virginia, and Maine and was given the derogatory slang names of "hillbilly heroin" and "poor man's heroin." But it has since spread to other areas in the United States, and in March 2002, an 18-year-old female became the first U.K. fatality attributed to OxyContin.

Purdue Pharma, the manufacturer of the product, has come under fire for allegedly turning a blind eye to the mounting reports of abuse of the drug. In an effort to educate health care providers about these risks, the drug manufacturer issued a warning in the form of a "Dear Health care Professional" letter. The company's other efforts have included discontinuing the most powerful pill, a 160-mg form of the drug; stamping pills from Mexico and Canada to help authorities trace illicit supplies; passing out tamper-proof prescription pads to doctors; and instituting educational programs aimed at doctors as well as potential abusers, with a particular focus in the Appalachian region.

Your Turn 10.26   Oxycodone Formulation

The oxycodone in an OxyContin pill is not a freebase, but rather is formulated as a salt. Which acid is neutralized to form this salt? Draw a structural formula for the drug in its salt form.

Morphine, oxycodone, hydrocodone (found in Vicodin, Lortab, and Lorcet), and codeine belong to a class of drugs known as opiates. Morphine and codeine are extracted from *Papaver somniferum*, the opium poppy. The other opiates listed are synthesized from morphine. They vary in their strength and thus their drug scheduling. But they all have the same mode of action. These drugs bind to a chemical receptor called *mu* that interrupts the transmission of pain in the spinal cord. Opiates also stimulate areas of the brain involved in pleasure, called the reward, or endorphin, pathways. In a normal brain, the chemical dopamine crosses between brain cells in the reward pathway, producing pleasurable feelings. Opiates stimulate the release of higher levels of dopamine, strengthening reward signals, and producing intense euphoria. Repeated use of opiates causes the brain of a user to become accustomed to an overstimulated reward pathway. This in turn brings about the phenomenon of tolerance—greater amounts of opiates are required to achieve the euphoria the user once experienced.

Consider This 10.27   House!

Gregory House, MD, the doctor in a popular TV series, in some episodes is addicted to Vicodin. An ABC News interview reports "In watching Dr. House on TV, one gets the impression that his addiction has no consequences other than affecting his medical judgment."

a. Research the effects of Vicodin abuse. Then interview one of the millions of viewers who watched this series. Do their impressions of Dr. House match what you found out about the realities of Vicodin abuse?

b. Is there a connection between Vicodin and hearing loss? Report your findings.

The drugs just described are but a few of today's commonly abused substances. There are many more, including alcohol, which is possibly the most damaging of all drugs when taken to excess. Just as for prescribed medicines, the choice to take illicit drugs involves a risk–benefit analysis. All drug users should have this basic information before making this choice.

Conclusion

Molecular modifications by chemists have created a vast new pharmacopoeia of wonder drugs that have significantly increased the number and quality of our days. Prior to penicillin, an infection could mean a death sentence. Now thanks to penicillin, sulfa drugs, and other antibiotics, the great majority of bacterial infections are easily controlled. Dreaded killers such as typhoid, cholera, and pneumonia have been largely eliminated—at least in many nations of the world. New methods of drug discovery are allowing vast libraries of novel compounds to be built. These can be stored for future testing by assays that have not yet been dreamed of.

But no drug can be completely safe, and almost any drug can be misused. These issues become the focus when the FDA is asked to change the status of a drug from prescription to over-the-counter. Taking any medication (generic or brand name) is a conscious choice between the benefits derived from the drug and the risks associated with its side effects and limits of safety. Because most drugs have very wide, carefully established margins of safety, their benefits far outweigh their risks for the general population. For some drugs, however, the trade-off between effectiveness and safety involves a different balance. A drug with severe side effects may be the only treatment available for a life-threatening disease. Someone suffering from HIV/AIDS or advanced, inoperable cancer understandably has a different perspective on drug risks and benefits than a person with a cold. And the impersonal anonymity of averages involved in clinical drug trials takes on new meaning at the bedside of a loved one.

Herbal and alternative medicines raise new questions. Who is responsible for defining their efficacy, monitoring their purity, and developing contraindications to their use with other drugs? Doctors must be advised of all medicines their patients are taking, including herbal remedies. The abuse of herbal and synthetic drugs continues to cause major problems throughout our society. Advances are being made in all of these areas. When chemistry is applied to medicine, science must be guided by morality, and reason must be tempered with compassion.

Chapter Summary

Having studied this chapter, you should be able to:

- Describe the discovery, development, and physiological properties of aspirin (10.1)
- Understand bonding in carbon-containing (organic) compounds (10.2)
- Apply the concept of isomerism to organic molecules (10.2)
- Convert chemical formulas of carbon-containing compounds to structural formulas, condensed structural formulas, and line-angle drawings (10.2)
- Recognize functional groups and the classes of organic compounds that contain them; draw structural formulas for organic molecules containing various functional groups (10.3)
- Understand that functional groups may be chemically modified to change a molecule's properties (10.3)
- Predict the products of ester formation reactions and describe how amines may be converted to their salt forms (10.3)
- Relate the molecular structure of aspirin to other analgesics (10.3)
- Understand the mode of action of aspirin and other analgesics (10.4)
- Describe the discovery of penicillin (10.5)

- Explain the lock-and-key mechanism of drug action (10.5)
- Describe how combinatorial synthesis can be employed in the creation of large collections of new drugs at lower costs than previous methods (10.5)
- Understand differences in molecular structure between a pair of chiral (optical) isomers (10.6)
- Appreciate the economic effect of chiral drugs (10.6)
- Identify the basic carbon skeleton arrangement of steroids (10.7)
- Recognize that minor changes in steroid structure may result in large changes in bioactivity (10.7)
- Compare and contrast brand-name and generic drugs (10.8)
- Identify some of the over-the-counter drug categories and their uses (10.8)
- Understand the process of a drug going from prescription to OTC (10.8)
- Describe some of the potential benefits and risks of herbal medicines (10.9)
- Explain the scheduling of prescription drugs (10.10)
- Discuss the use of marijuana and oxycodone in terms of their physiological and social effects (10.10)

Questions

Emphasizing Essentials

1. Give the intended effect of each. Can one drug exhibit all of these effects?
   a. an antipyretic drug
   b. an analgesic drug
   c. an anti-inflammatory drug

2. The field of chemistry has many subdisciplines. What do organic chemists study?

3. Write condensed structural formulas and line-angle drawings for the three isomers of C_5H_{12} assigned in Your Turn 10.7.

4. Write the structural formula and line-angle drawings for each different isomer of C_6H_{14}. *Hint:* Watch out for duplicate structures.

5. Consider the isomers of C_4H_{10}. How many different isomers could be formed by replacing a single hydrogen atom with an –OH group? How many different alcohols have the formula C_4H_9OH? Draw the structural formula for each.

6. For each compound, identify the functional group present.

7. Draw the simplest compound that can contain each of these functional groups. In some cases, only one carbon atom is required; in other cases two.
   a. an alcohol   d. an ester
   b. an aldehyde   e. an ether
   c. a carboxylic acid   f. a ketone

8. For each of these, identify the functional group. Then, draw an isomer that contains a different functional group.

9. In allergy sufferers, histamine causes runny noses, red eyes, and other symptoms. Here is its structural formula.

a. Give the chemical formula for this compound.

b. Circle the amine functional groups in histamine.

c. Which part (or parts) of the molecule make the compound water-soluble?

10. Figure 10.7 shows a somewhat condensed structural formula for acetaminophen, the active ingredient in Tylenol.

a. Draw the structural formula for acetaminophen, showing all atoms and all bonds.

b. Give the chemical formula for this compound.

c. Children's Tylenol is a flavored aqueous solution of acetaminophen. Predict what part (or parts) of the molecule make acetaminophen water-soluble.

11. Identify the functional groups in each of these.

a. Barbital (a sedative)

b. Penicillin-G (an antibiotic)

c. Amyl dimethylaminobenzoate (an ingredient in sunscreens)

12. Ibuprofen is relatively insoluble in water but readily soluble in most organic solvents. Explain this solubility behavior based on its structural formula. *Hint:* See Figures 10.6 and 10.7.

13. Here is the structural formula for diazepam, the sedative found in Valium.

Judging from its structure, do you expect it to be more soluble in fats or in aqueous solutions? Explain.

14. Draw structural formulas for the esters formed when acetic acid reacts with these alcohols.

a. *n*-propanol, $CH_3CH_2CH_2OH$

b. *iso*propanol, $(CH_3)_2CHOH$

c. *t*-butanol, $(CH_3)_3COH$

15. Interpret this sentence by giving the meaning of each acronym and explaining the effect. "NSAIDs have an effect on COX enzymes."

16. Usually carbon forms four covalent bonds, nitrogen three, oxygen two, and hydrogen only one bond. Use this information to draw structural formulas for:

a. A compound that contains one carbon atom, one nitrogen atom, and as many hydrogen atoms as needed.

b. A compound that contains one carbon atom, one oxygen atom, and as many hydrogen atoms as needed.

17. Would aspirin be more active if it were to interact with prostaglandins directly, rather than by blocking the activity of COX enzymes? Explain your reasoning.

18. Examine the combinatorial synthesis process diagrammed in Figure 10.16. If you started with a molecule that had two active functional groups, how many products could form after two synthetic steps, where each step adds a molecule with one functional group? How many products could be formed if the reagent in the first step had two reactive functional groups itself?

19. The text states that 80 billion tablets of aspirin a year are consumed in the United States. If the average tablet contains 500 mg of aspirin, how many pounds of aspirin does this consumption represent?

20. Identify the functional groups in morphine and meperidine (Demerol). Can these molecules be assigned to a particular class of compound (i.e., an alcohol, ketone, or amine)? Explain. *Hint:* See Figure 10.14 for structural formulas.

21. What is meant by the term *pharmacophore*?

22. Sulfanilamide is the simplest sulfa drug, a type of antibiotic. It appears to act against bacteria by replacing *para*-aminobenzoic acid, an essential nutrient for bacteria, with sulfanilamide. Use these structural formulas to explain why this substitution is likely to occur.

sulfanilamide

para-aminobenzoic acid

23. Which of these molecules has chiral forms?

a. $CH_3-\underset{\underset{OH}{|}}{\overset{\overset{NH_2}{|}}{C}}-CH_3$

b. $H-\underset{\underset{CH_3}{|}}{\overset{\overset{OH}{|}}{C}}-CO_2H$

c. $CH_3-\underset{\underset{\underset{N}{\overset{|||}{C}}}{|}}{\overset{\overset{NH_2}{|}}{C}}-CO_2H$

d. $CH_3-\underset{\underset{CH_3}{|}}{\overset{\overset{OH}{|}}{C}}-CO_2H$

24. Which of these molecules has chiral forms?

a. $CH_3-\underset{\underset{OH}{|}}{\overset{\overset{NH_2}{|}}{C}}-CH_2CH_3$

b. $H-\underset{\underset{H}{|}}{\overset{\overset{OH}{|}}{C}}-C_2H_5$

c. $CH_3-\underset{\underset{CH_2OH}{|}}{\overset{\overset{NH_2}{|}}{C}}-CO_2H$

d. $CH_3-\underset{\underset{CH_2SH}{|}}{\overset{\overset{OH}{|}}{C}}-CO_2H$

25. Methamphetamine hydrochloride is a powerful stimulant that is dangerous and highly addictive. This drug also goes by the street names "crystal," "crank," or "meth." This structural formula shows its salt form.

The freebase form of this drug, called "ice," is also abused. What does the term *freebase* mean, and how might the drug be converted to this form?

26. Examine the structures of levo- and dextromethorphan (see Figure 10.20). Identify any chiral carbons and list any functional groups present.

27. Molecules as diverse as cholesterol, sex hormones, and cortisone contain common structural elements. Use a line-angle drawing to show the structure they share.

Concentrating on Concepts

28. The text states that some remedies based on the medications of earlier cultures contain chemicals that are effective against disease, others are ineffective but harmless, and still others are potentially harmful. How might it be determined into which of these three categories a recently discovered substance fits?

29. Draw structural formulas for each of these molecules and determine the number and type of bonds (single, double, or triple) for each carbon atom.

   a. H_3CCN (acetonitrile, used to make a type of plastic)

   b. $H_2NC(O)NH_2$ (urea, an important fertilizer)

   c. C_6H_5COOH (benzoic acid, a food preservative)

30. Compare the physiological effects of aspirin with those of acetaminophen and ibuprofen. Relate differences to the nature of each compound at the molecular and cellular levels.

31. In Your Turn 10.7, you were asked to draw structural formulas for the three isomers of C_5H_{12}. One student submitted this set, with a note saying that six isomers had been found. (*Note:* The hydrogen atoms have been omitted for clarity.) Help this student see why some of the answers are incorrect.

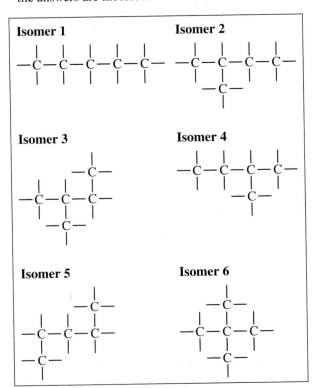

32. Styrene, $C_6H_5CH=CH_2$, the monomer for polystyrene described in Chapter 9, contains the phenyl group, $-C_6H_5$. Draw structural formulas to show that this molecule, like benzene, has resonance structures.

33. Aspirin is a specific compound, so what justifies the claims for the superiority of one brand of aspirin over another?

34. Figure 10.8 represents chemical communication within the body. Write a paragraph explaining what this figure means to you in helping to explain chemical communication.

35. Consider this statement. "Drugs can be broadly classed into two groups: those that produce a physiological response in the body and those that inhibit the growth of substances that cause infections." Into which class does each of these drugs fall?

    a. aspirin   c. (Keflex) antibiotic   e. amphetamine
    b. morphine   d. estrogen   f. penicillin

36. Consider the structure of morphine in Figure 10.14. Codeine, another strong analgesic with narcotic action, has a very similar structure in which the $-OH$ group attached to the benzene ring is replaced by an $-OCH_3$ group.

    a. Draw the structural formula for codeine and label its functional groups.

    b. The analgesic action of codeine is only about 20% as effective as morphine. However, codeine is less addictive than morphine. Is this enough evidence to conclude that replacement of $-OH$ groups with $-OCH_3$ groups in this class of drugs will always change the properties in this way? Explain.

37. Dopamine is found naturally in the brain. The drug $(-)$-dopa is found to be effective against the tremors and muscular rigidity associated with Parkinson's disease. Identify the chiral carbon in $(-)$-dopa, and comment on why $(-)$-dopa is effective, whereas $(+)$-dopa is not.

$(-)$-dopa

38. Vitamin E is often sold as a racemic mixture of $(+)$ and $(-)$ isomers. Use the web to find answers to these questions.

    a. Which is the more physiologically active isomer?

    b. How does the cost of the racemic mixture compare with the price of the pure, physiologically active isomer?

39. Consider the fact that levomethorphan is an addictive opiate, but dextromethorphan is safe enough to be sold in many over-the-counter cough remedies. From a molecular point of view, how is this possible?

40. Describe the lock-and-key analogy for the interaction between drugs and receptor sites. Use the analogy in a discussion as if you were explaining this to a friend.

41. Why are many projects to isolate or synthesize new drugs started in this country, but few drugs actually receive FDA approval for general use?

42. Until the early 19th century, it was believed that organic compounds had some sort of "life force" and could only be produced by living organisms. This was the basis of a concept called vitalism. This view was dispelled in the late 1800s. Use the web to find out what changed that opinion.

Exploring Extensions

43. One avenue for successful drug discovery is to use the initial drug as a prototype for the development of other similar compounds called analogs. The text states that cyclosporine, a major antirejection drug used in organ transplant surgery, is an example of a drug discovered in this way. Research the discovery of this drug to verify this statement. Write a brief report describing your findings, citing your sources.

44. Dorothy Crowfoot Hodgkin first determined the structure of a naturally occurring penicillin compound. What in her background prepared her to make this discovery? Write a short report on the results of your findings, citing your sources.

45. Before the cyclic structure of benzene was determined (see Figure 10.4), there was a great deal of controversy about how the atoms in this compound were arranged.

    a. Count outer electrons for C and H in C_6H_6. Then draw the structural formula for a possible linear isomer.

    b. Give the condensed structural formula for your answer in part **a**.

    c. Compare your structure with those drawn by classmates. Are they all the same? Why or why not?

46. Antihistamines are widely used drugs for treating symptoms of allergies caused by reactions to histamine compounds. This class of drug competes with histamine, occupying receptor sites on cells normally occupied by histamine. Here is the structure for a particular antihistamine.

a. Give the chemical formula for this compound.

b. What similarities do you see between this structure and that of histamine (shown in Question 9) that would allow the antihistamine to compete with histamine?

47. Over the next few years, the FDA may consider deregulating more than a dozen drugs, nearly as many as have already been approved for over-the-counter sales during the past decade. The products that have led this trend have been the widely advertised drugs for heartburn.

a. What questions need to be answered before a drug is deregulated?

b. Will these questions change if you are considering this need from the viewpoint of the FDA, a pharmaceutical company, or as a consumer?

48. Find out more about the new process for the manufacture of hydroxynitrile (HN), a precursor to Lipitor that won a 2006 Presidential Green Chemistry Challenge Award. How does this process differ from the earlier one for manufacturing HN? Write a brief report on your research, citing your sources.

49. The steroids testosterone and estrone were first isolated from animal tissue. One ton of bull testicles was needed to obtain 5 mg of testosterone and 4 tons of pig ovaries was processed to yield 12 mg of estrone.

a. Assuming complete isolation of the hormones was achieved, calculate the mass percentage of each steroid in the original tissue.

b. Explain why the calculated result very likely is incorrect.

50. Herbal remedies are prominently displayed in supermarkets, drug stores, and discount stores.

a. What influences your decision to buy one of these remedies?

b. Choose a remedy and carefully examine its label for information about the active ingredients, inert ingredients, anticipated side effects, the suggested dosage, and the cost per dose.

c. How confident are you that the safety and efficacy of these remedies are ensured? Explain your answer.

51. Habitrol was the most successful smoking-cessation prescription until the introduction of bupropion (Zyban) in 1997, which quickly claimed 50% of the market. How are these two approaches to drug therapy different?

52. Over-the-counter drugs allow consumers to treat a myriad of symptoms and ailments. An advantage is that the user can purchase and administer the treatment without the effort or the expense of consulting a physician. To provide a wider margin of safety for these circumstances, the OTC versions of drugs are often administered in lower doses. See if this is true by looking at information for prescription and OTC versions of painkillers like ibuprofen (Motrin) and heartburn treatments like nizatidine (Axid) and famotidine (Pepcid). Report on your findings.

53. Herbal or alternative medicines are not regulated in the same way as prescription or OTC medicines. In particular, the issues of concern are identification and quantification of the active ingredient, quality control in manufacture, and side effects when the herbal remedy is used in conjunction with another alternative or prescription medicine. Look for evidence from herbal supplement manufacturers that address these issues, and write a report documenting your findings, giving your references.

54. Danco Laboratories, the U.S. company that produces RU-486 (mifeprex, the "abortion pill") makes claims about the safety of this steroidal drug including a comparison to aspirin. Do some research on RU-486, and write a short report on the drug. Include its structure, mode of action, and safety record.

55. The antibiotic ciprofloxacin hydrochloride (Cipro) treats bacterial infections in many different parts of the body. This drug made headlines in 2001 for use in patients who had been exposed to the inhaled form of anthrax. Use the web or another source to obtain the structure of Cipro. Draw its structure and identify the functional groups.

56. Direct-to-consumer advertising of prescription drugs has proved to be a successful marketing tool for pharmaceutical companies. Twenty percent of consumers say that advertisements prompted them to call or visit their doctor to discuss the drug, according to PharmTrends, a patient-level syndicated tracking study of consumer behavior by market research organization Ipsos-NPD. Make a list of the pros and cons of this type of marketing from both the patient's and physician's point of view.

57. In 2003, a series of spot examinations of mail shipments of foreign drugs to U.S. consumers conducted by the FDA and U.S. Customs and Border Protection revealed that these shipments often contain unapproved or counterfeit drugs that pose serious safety problems. Although many drugs obtained from foreign sources purport to be, and may even appear to be, the same as FDA-approved medications, these examinations showed that many are of unknown quality or origin. Of the 1153 imported drug products examined, the overwhelming majority, 1019 (88%), were illegal because they contained unapproved drugs. Many of these imported drugs could pose clear safety problems. Use the FDA website to determine which drugs were most commonly counterfeited and their countries of origin.

58. Thalidomide was first marketed in Europe in the late 1950s. It was used as a sleeping pill and to treat morning sickness during pregnancy. At that time it was not known to cause any adverse effects. By the late 1960s, however, the drug was banned after it was found to be a teratogen, causing deformed limbs in the children of women who took it early in pregnancy. Use the web for information to write a short paper that describes the optical isomers of thalidomide and why the FDA did not approve thalidomide for use in the United States until recently. For what purpose has the FDA recently approved the use of thalidomide?

59. Conventional acne treatments are maintenance therapies. Antibiotics, such as tetracycline and erythromycin, are not expected to result in long-term improvement once they are stopped. Some individuals may not respond to any of the conventional medications for acne. A possible solution for those patients is the vitamin A derivative known as isotretinoin (Accutane). However, Accutane is by no means a frontline therapy. Using the web, find the serious side effects that may accompany the use of Accutane.

11 Nutrition: Food for Thought

"Even if you do not want to become a vegetarian, it's a desirable direction to move towards. Going vegetarian one day a week is a good beginning."

Dr. Andrew A. Weil, Director, Center for Integrative Medicine, University of Arizona.

Imagine never eating another hamburger. Put the thought out of your mind of picking up a piece of fried chicken. And definitely your eggs will come without bacon from now on. Would this be your worst nightmare? To some, being a vegetarian is akin to being deprived of those foods they most crave; well, with the exception of coffee and chocolate.

All too often, though, we set up our food selection in terms of all or nothing. No ice cream, because it has too many calories. No red meat, because it is unhealthy. Actually, no meat at all because feeding animals requires an inordinate amount of grain that could better be used to feed people. No soft drinks, because they are loaded with sugar. And no diet soft drinks either, because they contain artificial sweeteners. No, no, NO!

Could your choices possibly be more nuanced? Unless dictated by an allergy, a specific health concern, or a deeply held belief, what you eat need not be an all-or-nothing proposition. For example, a person who eats meat "sometimes" is called a flexatarian. If you were to become a flexatarian and eat no meat one day a week, you could improve the odds that as you age, you would have a higher quality of life. Why? As a vegetarian, you are likely to get more of the whole grains, fruits, and vegetables that your body needs. At the same time, you are likely to get less saturated fat. Do the math—one day a week does matter. One day in seven is just short of a 15% change in your food intake. Make that two days a week and you have changed your diet by over 25%.

Look for more about saturated fats in Section 11.3.

Consider also that if you switch to eating less meat, you are taking a necessary step to improve the health of the planet. Why? Food not only has a price tag in the supermarket, but also it may be costly for the environment. For example, in the context of biofuels (see Chapter 4), we mentioned the energy and environmental costs of growing and harvesting corn and other grains. In the context of water use (see Chapter 5), we made the case for the high "water footprint" costs of producing food items.

In this chapter, we will make the case that both what you dine on and what you skip not only has an effect on your health, but also affects the health of our planet. To get started, let's examine what you ate for breakfast, lunch, dinner yesterday. While you are at it, add in your snack breaks as well.

Consider This 11.1   Take a Bite

This activity gives you the opportunity to reflect on your food choices. What did you eat yesterday . . . ?

a. Make a list starting with the first cup of coffee (or however you chose to start your day).
b. From your list, select the three food items that you believe ranked *highest* in promoting your health. Name the criteria on which you based your ranking.
c. Select from your list the three food items that you believe ranked *highest* in terms of promoting the health of the land, air, and water on our planet. Again, name your criteria.

Hint: Refer to Chapter 5 about water footprints; refer to Chapter 1 about air prints.

If you are one of the people on the planet fortunate enough to have adequate food to eat, some very good reasons exist to eat simply and to vary your diet. In fact, both *what* you eat and *how much* you eat may be two of the most important decisions you make over the course of your life. At stake here is both your health and the health of the planet. In the next section, we explain why.

11.1 | Food and the Planet

Throughout history, food has played a pivotal role in human health and well being. Millions have starved from a lack of food, and others have died from eating too much of it. People have become rich by producing and selling food. Wars have been fought over food; countries have been destabilized by the lack of it. We previously have discussed the air we breathe and the water we drink, as both are essential to sustain human life. In this section, we give the same attention to the food we eat.

Our ancestors were hunter–gatherers, spending most of each day searching for their next meal. About 10,000 years ago, humans learned to grow crops and domesticate animals, thus launching the agricultural revolution. At that time, the population of the entire Earth was estimated to be 4 million people, which corresponds roughly to a city the size of Los Angeles today.

Over the next 8000 years, the global population grew to 170 million people or about half the size of the current U.S. population. By 1000 CE, the population had risen to 310 million, and 200 years ago the population finally topped 1 billion people. Today, the world population is almost 7 billion, and within the next 40 years will likely reach 9 billion. If you know someone who was born before 1960, this person has seen a doubling of the world population in his or her lifetime. Clearly, we have many mouths to feed.

Producing and transporting the massive amount of food that we need takes a toll on the planet. This toll varies by region, depending on which foods we eat. For example, in Chapter 5 we saw that it takes about 15,000 liters of water to bring a kilogram of beef protein to the table. But more than just water goes into producing the foods we eat.

Land is another factor to consider, both in terms of raising crops and affording grazing space for animals. The Food and Agriculture Organization of the United Nations estimates that 30% of the Earth's land surface is used for production of livestock. For example, it requires about 20 kg of grain and 245 m^2 of land to bring a kilogram of beef to your dinner table (Figure 11.1). Although eating pork and chicken have lower ecological costs, in most cases the lowest cost is for us to directly consume the different grains ourselves.

With rising affluence, people across the globe have had access to resources that enable them to change their diets. Indeed, we on this planet have been changing our diets! Over the past 40 years, the per capita consumption of meat has more than

According to the USDA, in 2005 people in the United States ate about 67 pounds (30 kg) of beef yearly, or about 3 ounces daily. Visit the USDA website for some amazing statistics!

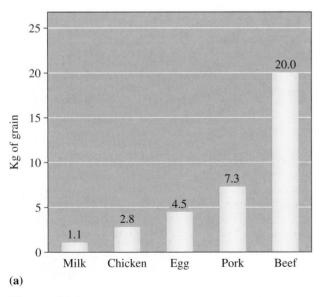

(a)

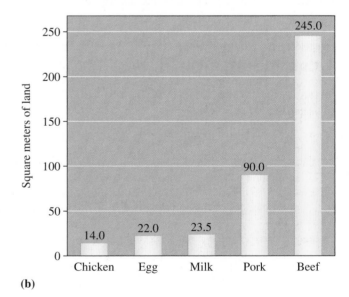

(b)

Figure 11.1

a. Kilograms of grain required to bring 1 kilogram of different foods to the table. **b.** Square meters of land required to bring 1 kilogram of different foods to the table.

Source: From Smil Vaclav. *Feeding the World: A Challenge for the Twenty-first Century.* figure: grain and land required to bring one kilogram of different food to the table, © 2000 Massachusetts Institute of Technology, by permission of The MIT Press.

doubled. In the next 40 years meat production is projected to double again. Can the Earth sustain its current population of about 7 billion if we are to continue on this path? The next activity sheds some light on possible answers to this question.

Consider This 11.2   Food for How Many People?

In 2006, the world grain production was approximately 2 billion kg (2000 million metric tons).

a. Assume that a person eats a serving or two of meat every day. This adds up to eating the equivalent of about 800 kg of grain per year. Given the current world grain production, how many people would this feed?

b. Now assume that everybody on the planet requires 800 kg of grain per year. How many kilograms of grain would be needed to support this population?

c. Answer parts **a** and **b** again on the basis of 200 kg of grain per year—a primarily vegetarian diet.

The previous activity indicates that a diet in which people eat the equivalent of 800 kg of grain per year cannot currently work for everybody on the planet, because not enough grain is currently produced on Earth. However, you might be wondering if we could increase world grain production in order to meet the desire of so many people to increase their consumption of meat. Stay tuned—we address this question in the final section of this chapter.

The previous activity also leads us to the concept of a "foodprint." This term, credited in part to researchers at Cornell University, is now turning up in news articles. A **foodprint** refers to the amount land required per year to provide the nutritional resources for one person. In 2009, these researchers analyzed about 40 different diets, all with the same number of calories. The data for people in the state of New York are assembled in Figure 11.2. As you can see, the low-fat vegetarian diet with 0 grams of meat per day required the use of about a half acre of land. In contrast, a diet high in meat and fat of 381 grams (over 13 ounces) required almost four times this, or 1.9 acres! Interestingly enough, eating 100% vegetarian did not turn out to be ideal for land use

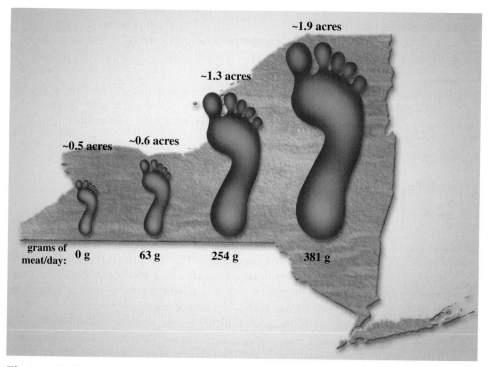

Figure 11.2

For New York State, a representation of how dietary meat is connected to the area of agricultural land required, on average, to feed a person for a year.

in this state, as high-quality fruits, vegetables, and grain require high-quality agricultural land. Rather, a diet with a small amount of meat (63 grams, or about 2.2 ounces each day) appeared to be the most efficient for feeding the most people. This is true as long as the land used for meat production is limited to land that is only suitable for pasture land. Again, a stance of all-or-nothing does not necessarily gain you anything, unless dictated by some other set of needs. Think flexatarian!

Think even more broadly than land use. Ultimately, we should take into consideration the resources required to grow, process, package, and transport food, along with the environmental consequences of these activities. For example:

- Depletion of an aquifer from pumping out water for irrigation
- Rivers drying up downstream when irrigating upstream
- Loss of forests when creating farmland
- Erosion and loss of topsoil as a result of overplanting and overgrazing
- Loss of biodiversity due to the planting of single crops
- Fossil fuel consumption due to planting, harvesting, packaging, and transporting foods
- Contamination of land and waters from using insecticides and herbicides
- Increase in BOD of waterways resulting from fertilizer runoff
- Waste accumulation resulting from crop residues and livestock

In the United States, the Rio Grande and the Colorado River no longer flow into the ocean. Water is diverted upstream, thus reducing the volume downstream.

BOD, biological oxygen demand, was described in Section 5.11.

Do you want to lower your foodprint? Changing what you eat would be a place to start. Reducing the amount of food you discard in the trash (or compost bin) might come next. And if you are so lucky, what you can grow in a garden should go on your list as well. As we mentioned in the opening to this chapter, in most cases you don't need to choose an all-or-nothing diet. Eat more grains and less meat. Minimize the food you throw away. Produce locally what you can. Choices such as these can have a positive effect not only on your health but also on the health of our planet.

Your Turn 11.3   Your Foodprint

How might each of these affect your foodprint?

a. Composting your food scraps.
b. Diluting your morning orange (or grapefruit) juice with water.
c. Eating a veggie burger rather than one of ground beef.

Answer
b. Diluting your juice won't change your foodprint, assuming that you start with the same amount of juice and dilute it. But if you pour less juice and dilute it, this would lower your foodprint. In addition, you would have saved a few calories and most likely still ingested enough vitamin C.

11.2 | You Are What You Eat

Whether you sit down to a gourmet meal or gobble junk food on the run, you eat because you need the water, energy sources, raw materials, and micronutrients that food provides. Yes, water! This compound serves both as a reactant and a product in metabolic reactions, as a coolant and thermal regulator, and as a solvent for the countless substances that are essential for life. Our human bodies are approximately 60% water.

Water, however, cannot be burned as a fuel in the body or anywhere else, for that matter. We need food as an energy source to power muscles, to send nerve impulses, and to transport molecules and ions in our bodies. In addition, food serves as the raw material for bodies, including new bone, blood cells, enzymes, and hair. Food also supplies micronutrients essential for **metabolism,** the complex set of chemical processes that are essential in maintaining life.

These and other interesting properties of water were discussed in Chapter 5.

Eating properly means more than filling your stomach. It is possible to eat, even to the point of being overweight, and still be malnourished. **Malnutrition** is caused by a diet lacking in proper nutrients, even though the energy content of the food may be adequate. Contrast malnutrition with **undernourishment,** a condition in which a person's daily caloric intake is insufficient to meet metabolic needs. Today, people worldwide are malnourished and undernourished, yet increasingly others are more overweight than ever before. In 2006, the Centers for Disease Control and Prevention reported in the *Journal of the American Medical Association* that 66% of all adults in the United States are classified as overweight with nearly half of that population classified as obese. This epidemic of obesity is caused by several factors, including eating the wrong foods, eating too much of any foods, and the lack of physical activity.

Skeptical Chemist 11.4   A Lifetime of Food

During your lifetime, it has been claimed that you will eat about 700 times your adult body weight. Is this statement in the ballpark? Do a calculation to find out. State your assumptions clearly.

Hint: You might assume a life span of 78 years and that your present weight is your adult weight. Estimate the weight of food eaten daily at present, and use these data to project your lifetime food consumption.

Think about the foods you ate yesterday. Did they come to you with minimal or no processing, such as an apple, a baked potato, or a juicy pork chop? Or did you obtain these same foods as apple sauce, a bag of frozen French fries, or sliced smoked bacon? The latter are **processed foods,** foods that have been altered from their natural state by techniques such as canning, cooking, freezing, and adding chemicals. The typical diet in many countries contains numerous processed foods.

Processed foods in the United States must list nutritional information on their labels. These labels, such as the one shown in Figure 11.3, include the amounts of fat, carbohydrate, and protein—the **macronutrients**—that provide essentially all of the energy and most of the raw material for body repair and synthesis. Sodium and potassium ions are present in much lower concentrations, but these ions are essential for the proper electrolyte balance in the body. Several other minerals and an alphabet soup of vitamins (see Section 11.8) are listed in terms of the percent of recommended daily requirements supplied by a single serving of the product. All these substances, whether naturally occurring or added during processing, are chemicals. In fact, all food is inescapably and intrinsically chemical, even food claiming to be organic or "natural."

Table 11.1 indicates the mass percentages (grams of component per 100 g of food item) of water, fats, carbohydrates, and proteins in several familiar foods. For this particular

Figure 11.3

Nutrition facts from a can of mixed nuts.

Actually, the term *fat* is too narrow a term. The next section describes both fats and oils, also termed *triglycerides.*

Table 11.1	Percent Water, Fat, Carbohydrates, and Protein in Selected Foods			
Food	Water	Fat	Carbohydrates	Protein
white bread	37	4	48	8
2% milk	89	2	5	3
chocolate chip cookies	3	23	69	4
peanut butter	1	50	19	25
sirloin steak	57	15	0	28
tuna fish	63	2	0	30
black beans (cooked)	66	<1	23	9

Source: U.S. Department of Agriculture, Agricultural Research Service, *Home and Garden Bulletin* 72, 2002.

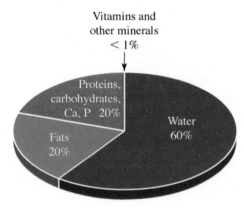

Figure 11.4

Composition of the human body.

selection of foods, the variation in composition is considerable. But in every case, these four components account for almost all of the mass present. Water ranges from a high of 89% in 2% milk to a low of 1% in peanut butter. Peanut butter is comparable to steak and fish in percent of protein. In this table, it leads in fat content. Chocolate chip cookies are the highest in carbohydrates because of their high sugar and refined flour content.

Compare this table with similar data for the human body (Figure 11.4). You are what you eat, but only to a certain extent. You are more like steak than chocolate chip cookies. You are wetter and fatter than bread, and you contain more protein than milk. From these data, you can determine that a 150-pound (68-kg) person consists of about 90 pounds (41 kg) of water and about 30 pounds (14 kg) of fat. The remaining 30 pounds is almost all protein, carbohydrates, and the calcium and phosphorus in the bones. The other minerals and the vitamins weigh less than 1 pound (0.5 kg), indicating that a little bit of each goes a long way, a point that will be discussed in Section 11.8.

In the next three sections, we take a look at each macronutrient in turn—fats, carbohydrates, and proteins. As you will see, each one is unique in several regards.

11.3 | Fats and Oils

From your experiences with ice cream, butter, and cheese, you probably know that fats can help impart a desirable flavor and texture to food. More generally, fats are greasy, slippery, and low-melting solids that are not soluble in water. When melted, they float on the top of broth or soup. Sour cream, frosting, and most pastries are loaded with fats (and calories). Most fats are of animal origin, although meats vary in their fat content.

You may also know about oils, such as those obtained from corn and soybeans. You may have seen peanut oil forming a layer on top of your peanut butter. You may enjoy eating bread dipped in olive oil. Or you may prepare a loaf of nut bread using canola oil as the shortening. Many oils are of plant origin. Oils exhibit many of the properties of animal-based fats; but unlike fats, they are liquids at room temperature.

The molecules that make up fats and oils share a common feature. They both are **triglycerides,** that is, molecules that contain three ester functional groups. They are formed from a chemical reaction between three fatty acids and the alcohol glycerol. **Fats** are triglycerides that are solids at room temperature, whereas **oils** are triglycerides that are liquids. In turn, all triglycerides are **lipids,** a class of compounds that includes not only triglycerides, but also related compounds such as cholesterol and other steroids. Figure 11.5 shows the lipid family tree.

We introduced several new terms in the previous paragraph, and we will now work through them with you one by one. First up is *fatty acids,* an interesting class of compounds. Examine Figure 11.6 to see an example of a fatty acid, stearic acid. Like all fatty acids, the stearic acid molecule has two important characteristics. One is a

This group is characteristic of an ester:

For more about esters, see Section 9.5.

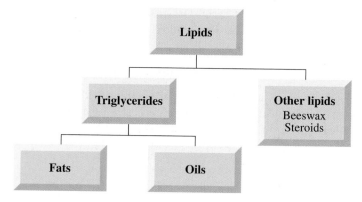

Figure 11.5

Types of lipids.

long nonpolar hydrocarbon chain of carbon atoms, typically 12 to 24 atoms in length. This hydrocarbon chain gives fats and oils their characteristic greasiness. The other is the carboxylic acid group, –COOH, at the end of the hydrocarbon chain. The carboxylic acid group accounts for the "acid" in the name fatty acid.

You can review hydrocarbons in Section 4.4 and carboxylic acids in Section 9.5.

$CH_3(CH_2)_{16}COOH$
condensed structural formula

$CH_3CH_2CH_2CH_2CH_2CH_2CH_2CH_2CH_2CH_2CH_2CH_2CH_2CH_2CH_2CH_2CH_2$—C $\begin{smallmatrix} O \\ \\ OH \end{smallmatrix}$

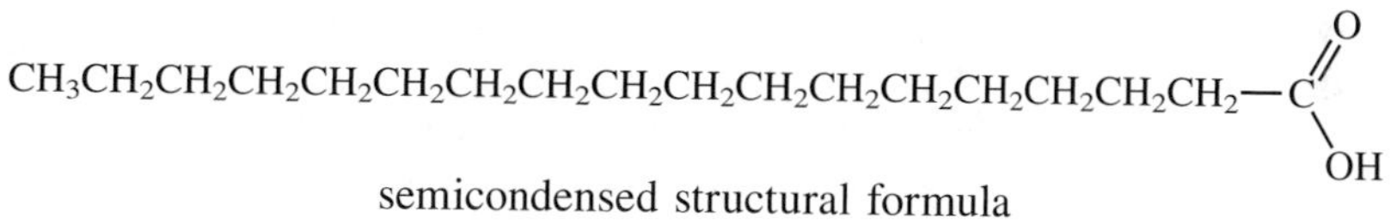

semicondensed structural formula

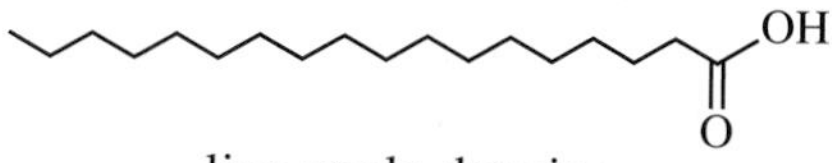

line-angle drawing

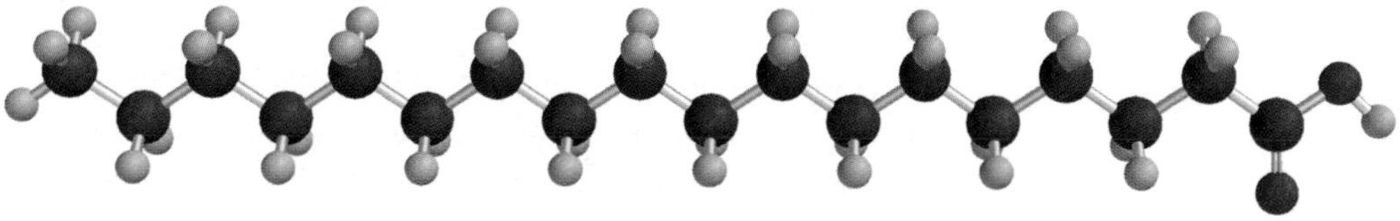

ball-and-stick model

Figure 11.6

Representations of stearic acid, $C_{17}H_{35}COOH$, an example of a fatty acid.

Line-angle drawings were introduced in Section 10.2.

Next we turn to the term *glycerol*, an alcohol that we briefly mentioned in Chapter 4 when describing biodiesel. Glycerol is a sticky, syrupy liquid that is sometimes added to soaps and hand lotions. The representations of a molecule of glycerol (Figure 11.7) show that it is an alcohol with three –OH groups.

$CH_2(OH)CH(OH)CH_2OH$

condensed structural formula

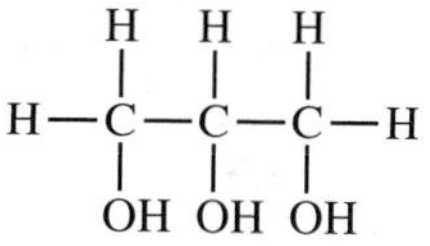

structural formula

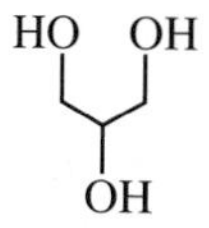

line-angle drawing

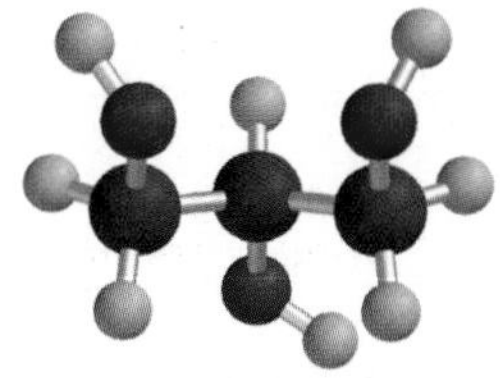

ball-and-stick model

Figure 11.7

Representations of glycerol, an alcohol.

Of interest to us here is that each –OH group of the glycerol molecule can form an ester with a fatty acid molecule. The result is a triglyceride.

3 fatty acid molecules + 1 glycerol molecule ⟶

1 triglyceride molecule + 3 water molecules **[11.1]**

For example, three stearic acid molecules can combine with a glycerol molecule to form glyceryl tristearate, a triglyceride. The three ester functional groups in glyceryl tristearate are highlighted in red in equation 11.2.

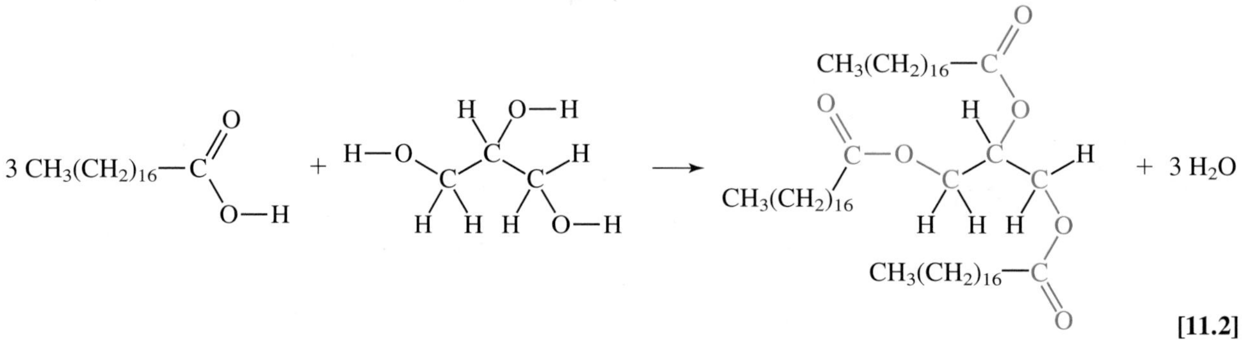

$$3\ CH_3(CH_2)_{16}-C\overset{O}{\underset{O-H}{}} + H-O\ \underset{H\ H}{\overset{H}{C}}\ \underset{H}{\overset{O-H}{C}}\ \underset{O-H}{\overset{H}{C}} \longrightarrow CH_3(CH_2)_{16}-C\overset{O}{} \quad \underset{CH_3(CH_2)_{16}}{\overset{O}{C}}-O\ \underset{H\ H}{\overset{H}{C}}\ \underset{H}{\overset{O}{C}}\ \underset{O}{\overset{H}{C}}\ \underset{CH_3(CH_2)_{16}-C}{} + 3\ H_2O$$

[11.2]

The process shown in equation 11.2 is the basis for forming most animal fats and vegetable oils. In most cases, enzymes catalyze this process. Almost all of the fatty acids in our bodies are transported and stored in the form of triglycerides.

Your Turn 11.5   Triglyceride Formation

Like stearic acid, palmitic acid is a component of animal fats.

a. Construct a line-angle drawing for palmitic acid, $CH_3(CH_2)_{14}COOH$.
b. Name the functional group responsible for the acidic properties of both fatty acids.

Answers

a.

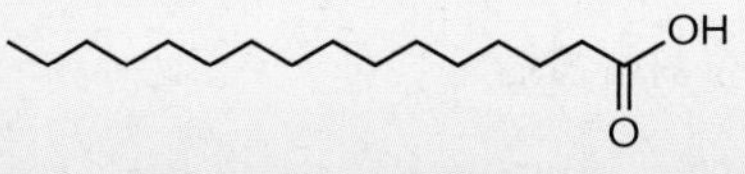

b. carboxylic acid group, –COOH

Finally, we need to more carefully distinguish the terms *fat* and *oil*. As we mentioned earlier, fats are triglycerides that are solids at room temperature, whereas oils are triglycerides that are liquids. Why the difference? The properties of a particular fat or oil depend on the nature of the fatty acids incorporated into the triglyceride. Of key importance is whether the fatty acid molecule contains one or more C=C double bonds.

A fatty acid is **saturated** if the hydrocarbon chain contains only single bonds between the carbon atoms. In a saturated hydrocarbon chain, the C atoms contain the maximum number of H atoms that can be accommodated and therefore is saturated in hydrogen. This is the case with stearic acid. In contrast, fatty acids are **unsaturated** if they contain one or more C=C double bonds.

$$CH_3(CH_2)_7CH=CH(CH_2)_7COOH$$

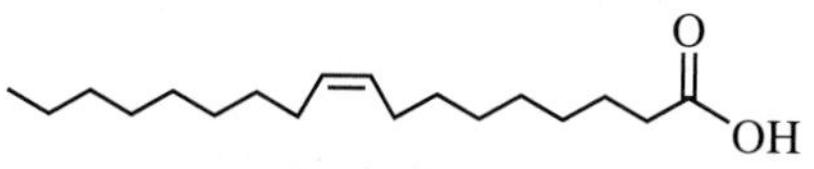

oleic acid, a **monounsaturated** fatty acid

$$CH_3(CH_2)_4CH=CHCH_2CH=CH(CH_2)_7COOH$$

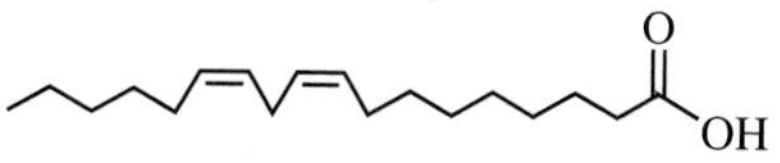

linoleic acid, a **polyunsaturated** fatty acid

$$CH_3CH_2CH=CHCH_2CH=CHCH_2CH=CH(CH_2)_7COOH$$

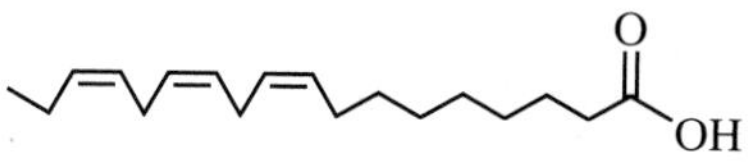

linolenic acid, a **polyunsaturated** fatty acid

Figure 11.8

 Examples of unsaturated fatty acids. See **Figures Alive!** for interactive activities relating to fatty acids.

Fatty acids are either monounsaturated or polyunsaturated. For example, oleic acid, with only one double bond between carbon atoms per molecule, is classified as **monounsaturated.** In contrast, linoleic acid (two C=C double bonds per molecule), and linolenic acid (three C=C double bonds per molecule) are both examples of polyunsaturated fatty acids. A **polyunsaturated** fatty acid contains more than one double bond between carbon atoms. In Figure 11.8, each of the unsaturated fatty acids contains 18 carbon atoms, but differs in the number and placement of the C=C double bonds. The next activity gives you a chance to work with different unsaturated fatty acids.

Your Turn 11.6   Unsaturated Fatty Acids and a Triglyceride

a. What structural feature identifies oleic, linoleic, and linolenic acids as unsaturated fatty acids?

b. Lauric acid, $CH_3(CH_2)_{10}COOH$, is a component of palm oil. Draw the line-angle representation for lauric acid and classify it as saturated or unsaturated.

Answers

a. Oleic, linoleic, and linolenic acid all have at least one C=C double bond.

b. It is a saturated fatty acid.

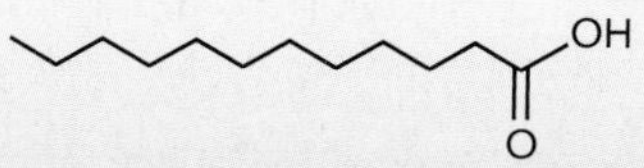

The three fatty acids that form a triglyceride molecule can be identical, two can be the same, or all three can be different. Moreover, these fatty acids can be saturated or unsaturated. They also can be sequenced differently in the molecule.

Table 11.2	Comparing Fatty Acids		
Name	Number of C Atoms per Molecule	Number of C=C Double Bonds per Molecule	Melting Point, °C
Saturated Fatty Acids			
capric acid	10	0	32
lauric acid	12	0	44
myristic acid	14	0	54
palmitic acid	16	0	63
stearic acid	18	0	70
Unsaturated Fatty Acids			
oleic acid	18	1	16
linoleic acid	18	2	−5
linolenic acid	18	3	−11

All of these factors contribute to the variety of fats and oils that we find in animals and plants. Solid or semisolid animal fats, such as lard and beef tallow, tend to be high in saturated fats. In contrast, olive, safflower, and other vegetable oils consist mostly of unsaturated fats.

Table 11.2 indicates some trends within a given family of fatty acids. For example, in saturated fatty acids, the melting points increase as the number of carbon atoms per molecule (and the molecular mass) increases. On the other hand, in a series of fatty acids with a similar number of carbon atoms, increasing the number of C=C double bonds decreases the melting point. Thus, when the melting points of the 18-carbon fatty acids are compared, saturated stearic acid (no C=C double bonds) is found to melt at 70 °C, oleic acid (one C=C double bond per molecule) melts at 16 °C, and linoleic acid (two C=C double bonds per molecule) melts at −5 °C. These trends carry over to the triglycerides containing the fatty acids and explain why fats rich in saturated fatty acids are solids at room or body temperature, whereas ones with a high degree of unsaturation are liquids.

Stearic acid is a solid at body temperature, whereas oleic and linoleic acids are liquids.

Normal body temperature is 37 °C; room temperature is approximately 20 °C.

Your Turn 11.7   Hydrocarbons, Triglycerides, and Biodiesel

a. How does a molecule of octane compare with one of a fat or oil? List two differences.
b. How does a biodiesel molecule compare with one of a fat or oil?
   Hint: See Chapter 4.

Answer
a. A molecule of octane is smaller than one of any fat or oil. It does not contain the ester functional group. Octane contains no C=C double bonds. Fats and oils may.

As you might guess, fats and oils not only differ in their physical properties, but they also differ in how they affect your health. We turn to this topic in the next section.

11.4 | Fats, Oils, and Your Diet

People tend to be preoccupied with dietary fat, as fats pack more calories than any other nutrient. But fats are far more than just a fuel. Fats enhance our enjoyment of food, improve "mouth feel," and intensify certain flavors. Almost every dessert tastes better with a bit of whipped cream! Fats also are essential for life. They provide insulation that retains body heat and that helps to cushion internal organs. Moreover, triglycerides and other lipids, including cholesterol, are the primary components of cell membranes and nerve sheaths and our brains are rich in lipids.

Fortunately, our bodies can synthesize almost all fatty acids from the foods we eat. The exceptions are linoleic and linolenic acids. These two fatty acids must be present in our diet; our bodies cannot produce them. Generally this does not pose a problem because many foods, including plant oils, fish, and leafy vegetables, contain linoleic and linolenic acid.

Figure 11.9 reveals some surprising differences in the composition of fats and oils we consume. For example, flaxseed oil is particularly rich in alpha-linolenic acid (α-linolenic acid, or ALA), a polyunsaturated acid that is being studied for its health benefits. Palm kernel and coconut oils contain much more saturated fat than corn and canola oil. Ironically, the coconut oil used in some nondairy creamers contains about 87% saturated fat, far more than the percentage found in the cream it replaces. In fact, coconut oil contains more saturated fat than pure butterfat. Concern over the high degree of saturation in coconut and palm oil accounts for the statement sometimes printed on food labels: "Contains no tropical oils."

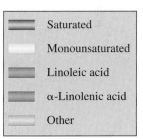

The solid form of coconut oil is called coconut butter. It melts to form an oil around room temperature.

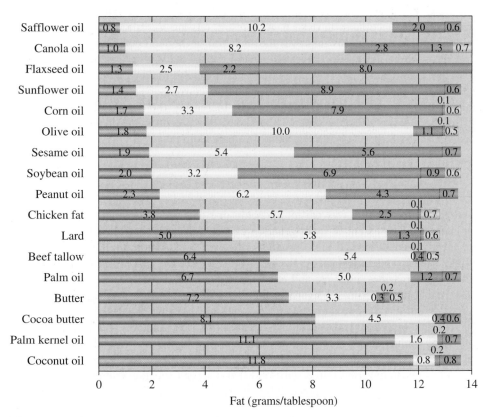

Figure 11.9

Saturated and unsaturated oils and fats. See **Figures Alive!** for more about this chart.

Source: From *Nutrition Action Health Letter* by Center for Science in the Public Interest, December 2005, Vol. 32, No. 10, p. 8. Copyright © 2005 by Center for Science in the Public Interest. Reproduced with permission of Center for Science in the Public Interest via Copyright Clearance Center.

However, the higher degree of unsaturation in oils comes with a drawback. You may have noticed the slight rancid odor that oils acquire over time. The reason for this odor is because C=C double bonds are more susceptible to reaction with the oxygen in the air than are C–C single bonds. The "off-flavor" that you may detect in an oil most likely is a result of such reactions with oxygen. As a result, oils are sometimes treated to increase their saturation. In turn, this improves the shelf life of the food containing the oil.

One way to more fully saturate an oil or a fat is by **hydrogenation,** a process in which hydrogen gas, in the presence of a metallic catalyst, adds to a C=C double bond and converts it to a single bond. Hydrogenation converts one or more of the C=C double bonds in an oil to C–C single bonds, increasing the degree of saturation and raising the melting point. Equation 11.3a shows this process with linoleic acid, one of the fatty acids in the triglycerides found in peanut oil.

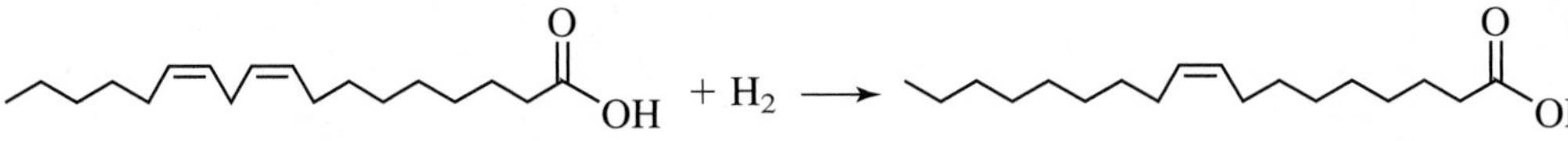

$$\text{(structure of linoleic acid)} + H_2 \longrightarrow \text{(structure of partially hydrogenated product)} \quad \text{[11.3a]}$$

Here, the hydrogenation was partial and only one of the double bonds in linoleic acid was affected. The extent of hydrogenation can be carefully controlled by temperature and pressure conditions to yield products of desired saturation and therefore melting point, softness, and spreadability. Such customized fats and oils are in many products, including margarines, cookies, and candy bars.

Peanut oil sometimes is hydrogenated by food manufacturers. For example, unless you eat "natural" peanut butter, the jar on your shelf probably is labeled something like "oil modified by partial hydrogenation." Peanut oil is rich in mono- and polyunsaturated fatty acids. You usually can see a layer of peanut oil in jars of "natural" peanut butter. But the processed "creamy-style" peanut butter does not have this layer of oil. Rather, the oil has been converted into a semisolid that does not separate from peanut butter.

Although shelf life and spreadability are important considerations, they are not the only ones. It turns out that some fats and oils are healthier for your heart than others. To understand why, we need to look more closely at the geometry of the hydrogen atoms attached to the C=C double bonds of triglycerides. It turns out that in most natural unsaturated fatty acids, the hydrogen atoms attached to the carbon atoms are on the *same* side of the C=C double bond. We call this bonding arrangement *cis*.

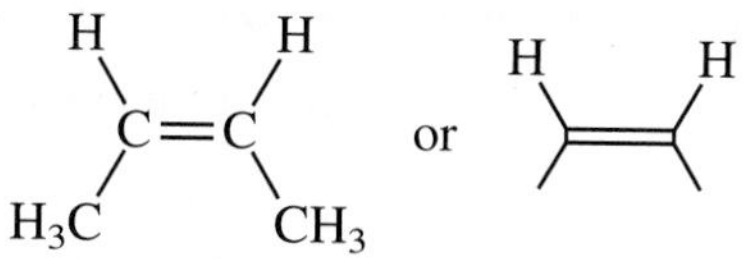

cis-2-butene

Alternatively, the hydrogen atoms can be across from each other on the C=C double bond. We call this bonding arrangement *trans*.

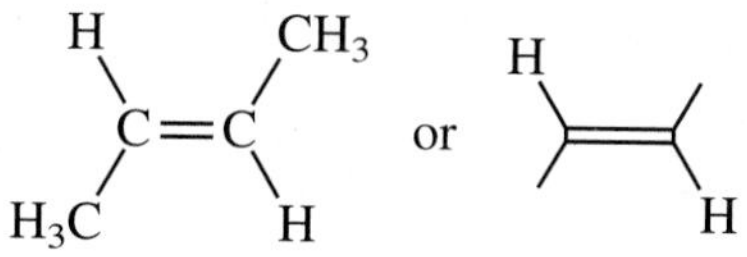

trans-2-butene

For example, oleic acid and elaidic acid are monounsaturated fatty acids that both have the same chemical formula. However, their properties, uses, and health effects are different. Oleic acid, a *cis* fatty acid, is a major component of the triglycerides in olive oil. In contrast, elaidic acid, a *trans* fatty acid, is found in some soft margarines made via hydrogenation. Compare the structures shown in Figure 11.10.

Trans fats are triglycerides that are composed of one or more *trans* fatty acids. Scientific studies show that trans fats raise the level of triglycerides and "bad" cholesterol in the blood. This finding came as somewhat of a surprise because partially hydrogenated fats still contain some C=C double bonds, and unsaturation is definitely a plus in a healthy diet. However, trans fats are similar in properties to saturated fats. With their long "straight" hydrocarbon chains, saturated fat molecules tend to pack well together, one reason why they are solids at room temperature. With their *cis* geometries, the molecules of naturally occurring unsaturated edible oils have "bends" that do not pack as well, one reason they are liquids at room temperature. Natural oils all have these bends (see Figure 11.8). When partial hydrogenation takes place, the process not only results in the addition of hydrogen but also converts some of the *cis* fatty acids to *trans* fatty acids. We can rewrite equation 11.3a to better show the *cis* and *trans* configurations.

In a *cis* isomer, the H atoms are on the *same* side of the double bond:

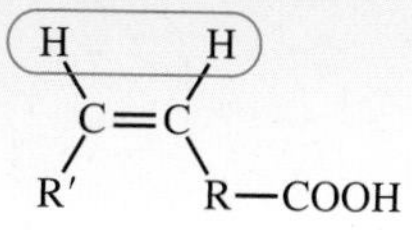

In a *trans* isomer, the H atoms are on the *opposite* side of the double bond:

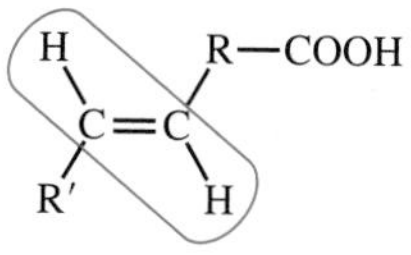

LDL (low-density lipoprotein) cholesterol is referred to as "bad" because of its tendency to build up on artery walls, causing heart attacks and strokes.

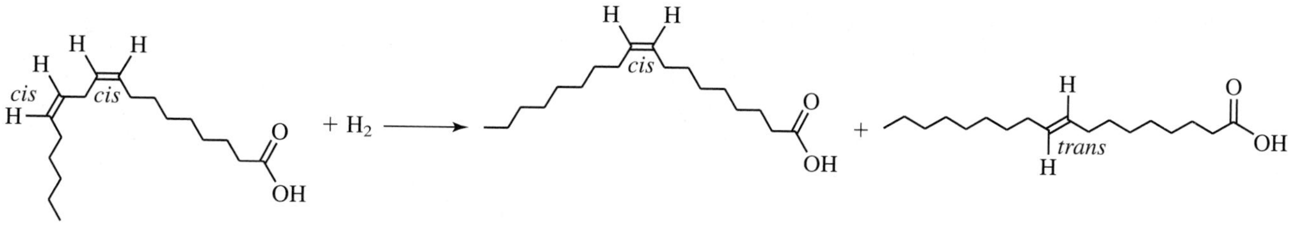

[11.3b]

The fats containing these *trans* fatty acids more closely resemble the shape of saturated fats and therefore behave in a similar manner in the body.

What types of fats and oils are in your margarine? Check out the next activity.

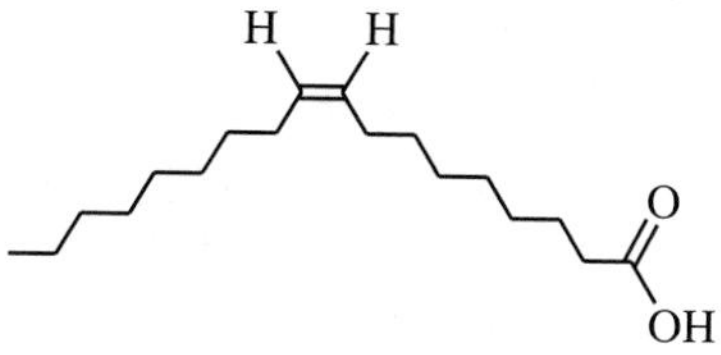

oleic acid, a *cis* fatty acid

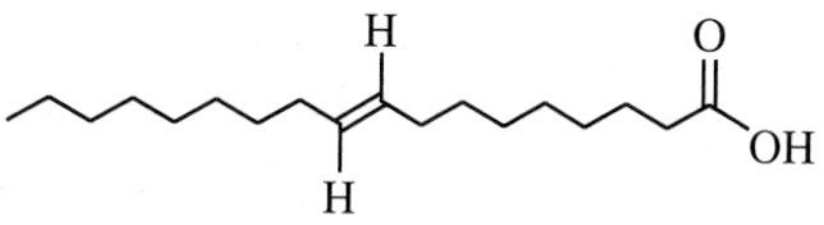

elaidic acid, a *trans* fatty acid

Figure 11.10

Line-angle drawings of a *cis* and a *trans* fatty acid. Both have the same chemical formula, $CH_3(CH_2)_7CH=CH(CH_2)_7COOH$.

Consider This 11.9   Margarines and Fat Content

This table lists the fat content for butter and three margarines. You will need to find other products to answer part **c**.

	Butter	Land O' Lakes (stick)	I Can't Believe It's Not Butter (tub)	Benecol Spread (tub)
Serving size	1 tbsp = 14 g	1 tbsp = 14 g	1 tbsp = 14 g	1 tbsp = 14 g
total fat (g)	11 g	11 g	9 g	8 g
saturated	7 g	2 g	2 g	1 g
trans	0 g	2.5 g	0 g	0 g
polyunsaturated	1 g	3.5 g	3.5 g	2 g
monounsaturated	3 g	2.5 g	2 g	4.5 g

a. Which margarine has the highest percent of saturated fat? How does it compare with butter?

b. In butter, what percent of the total fat is polyunsaturated?

c. Conduct a minisurvey. List the fat content for three different butters or margarines.

In March 2003, Denmark became the first country to strictly regulate foods containing trans fats. Canada followed suit in 2004. Starting in January 2006, the U.S. FDA required that food labels include trans fat information. Also in 2006, New York City banned restaurants from selling foods with anything but trace amounts of trans fats. Foods containing more than 0.5 g of trans fat had to be removed from the menu. In 2008, California became the first state to ban trans fats in restaurants. Consumer pressure has forced major food suppliers to reduce or eliminate trans fats from their products even in regions where there are not regulations on trans fats.

Most nutritionists now recommend that consumers go easy on any product with either trans fats or saturated fats. Manufacturers are responding by looking for substitutes for trans fats. Using tropical oils such as palm or coconut oil would not be acceptable because of their high percentage of saturated fats. Some manufacturers are adding other oils to their products that are polyunsaturated, such as sunflower oil or flaxseed oil.

Food chemists also have been busy discovering alternatives to hydrogenation that produce semisolids but do not produce trans fats. **Interesterification** is any process in which the fatty acids on two or more triglycerides are scrambled to produce a set of different triglycerides (Figure 11.11). If you perform this process with a low-melting triglyceride (an oil) and a high-melting triglyceride (a fat), the result is a mixture of triglycerides with an intermediate melting point, a semisolid fat.

One way of carrying out an interesterification reaction uses a strong base as a catalyst. The use of this base raises concerns for worker safety, results in significant loss of the oils, requires large amounts of water, and produces wastes that are high in biological oxygen demand, BOD.

Fortunately, enzymes also can catalyze this reaction. However, using enzymes came with a high price tag until Novozymes and the Archer Daniels Midland Company teamed up to refine the reaction. For their efforts, they shared a Presidential Green Chemistry Challenge Award in 2005. Not only is their method more cost-effective, but also it has environmental benefits. These include a large reduction in both water usage and in the BOD of the aqueous waste streams, less decomposition of oils during the process, and the elimination of the base catalyst.

> Catalysts were introduced in Chapter 1 in the context of catalytic converters.

Your Turn 11.10   Bring Home the Green

Revisit the principles of Green Chemistry listed on the inside front cover of this book. Which of the principles are met by the use of enzymes for interesterification developed by Novozymes and Archer Daniels Midland?

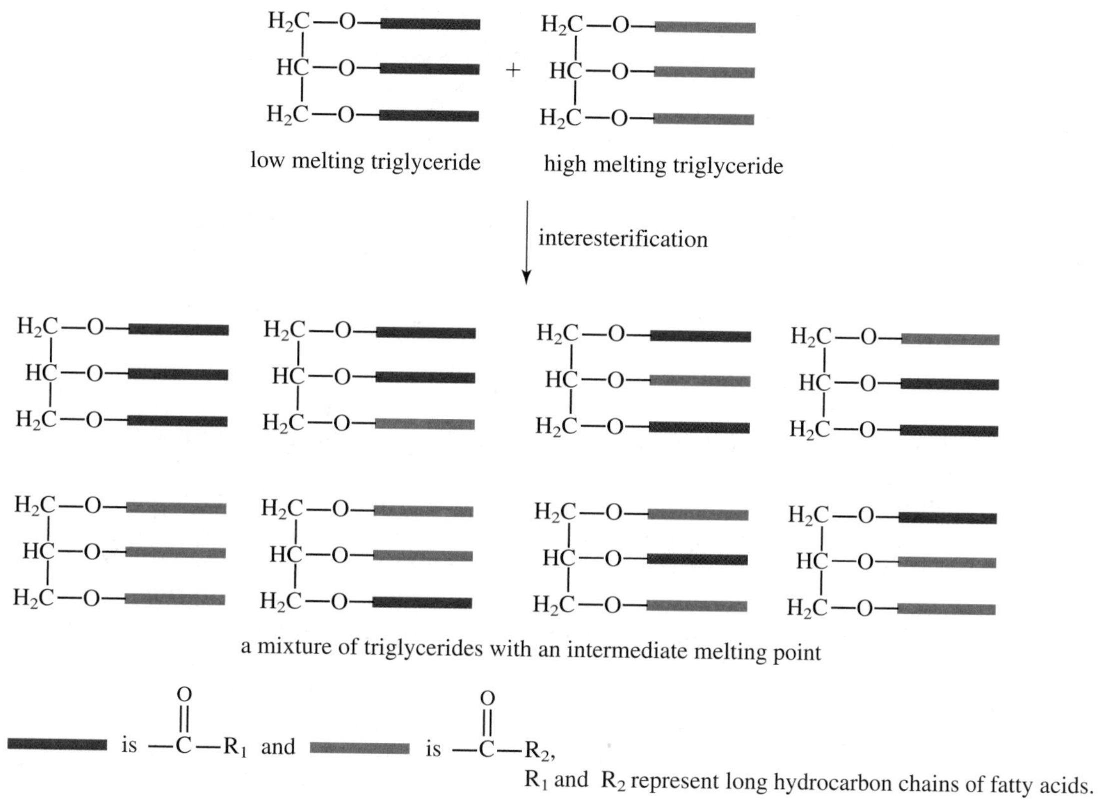

low melting triglyceride   high melting triglyceride

interesterification

a mixture of triglycerides with an intermediate melting point

━━━ is $-\overset{\overset{\displaystyle O}{\|}}{C}-R_1$ and ▬▬ is $-\overset{\overset{\displaystyle O}{\|}}{C}-R_2$,

R_1 and R_2 represent long hydrocarbon chains of fatty acids.

Figure 11.11

Interesterification: Scrambling of the fatty acids in lipids to produce a mixture of triglycerides.

Source: Adapted from *Real-World Cases in Green Chemistry,* Volume 2, Cann, M. and Umile, T., American Chemical Society, 2008.

With this, we end our discussion of fats and oils. The next section tells the sweet tale of sugars and their not-as-sweet relatives, starches.

11.5 | Carbohydrates: Sweet and Starchy

Some of the best known dietary carbohydrates are sugars such as those found in fruits and processed foods. Sugars are recognized by their sweet taste and are easily digested by the body. Another well-known carbohydrate is starch, found in nearly all plant-based foods such as grains, potatoes, and rice. Although starch is pleasant to the taste buds, it does not taste sweet and takes a bit longer to digest. Whether sweet or starchy, carbohydrates have the job of providing the cells in our bodies with the energy they need.

Chemically, **carbohydrates** are compounds containing carbon, hydrogen, and oxygen, with hydrogen and oxygen found in the same 2:1 atomic ratio as in water. This composition gives rise to the name, *carbohydrate,* which implies "carbon plus water." But the hydrogen and oxygen atoms are not bonded together to form water molecules. Rather, carbohydrate molecules are usually built of rings containing carbon atoms and an oxygen atom. The hydrogen atoms and –OH groups are attached to the carbon atoms.

These rings allow for many variations in molecular structure. For example, 32 chiral isomers are possible for the ring structure in glucose. The isomers differ in many of their properties, including intensity of sweetness, as you will see in Your Turn 11.11.

Figure 11.12 shows the structures of three simple sugars. These **monosaccharides,** or "single sugars" of fructose and glucose, both have the formula $C_6H_{12}O_6$.

From Figure 11.12, you can see that each monosaccharide contains a single ring consisting of four or five carbon atoms and one oxygen atom. The best way to visualize

Chirality was discussed in Section 10.6.

The ending -*ose* is typical for sugars.

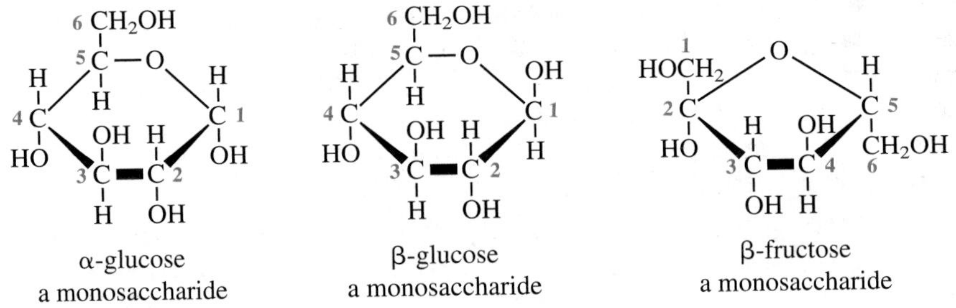

Figure 11.12

Molecular structures of three monosaccharides.

its three-dimensional structure is to imagine that the ring is perpendicular to the plane of the paper, with the bold print edges facing you. The H atoms and –OH groups fall either above or below the plane of the ring. The result is two forms of glucose: alpha (α) and beta (β). In α-glucose, the –OH group on carbon 1 is on the opposite side of the ring from the –CH₂OH group attached to carbon 5.

Ordinary table sugar, sucrose, is an example of a **disaccharide, a "double sugar," formed by joining two monosaccharide units.** In forming a sucrose molecule, an α-glucose and a β-fructose unit are connected by a C–O–C linkage created when an H and an OH are split from the monosaccharides to form a water molecule. This is a condensation reaction. Water produced in this way is used for other reactions in the body. This reaction and the structure of the sucrose molecule are shown in Figure 11.13.

Condensation reactions were described in Sections 9.5 and 10.3.

Your Turn 11.11   The Sweet Story

Here is a comparison of different sugars. See Table 11.3 for actual sweetness values.

Sugar	Also Known as	Sweetness	Calories/gram	Chemical Formula
glucose	"blood sugar" "grape sugar" "corn sugar"	least sweet	3.87 Cal/g	$C_6H_{12}O_6$
fructose	"fruit sugar"	sweetest	3.87 Cal/g	$C_6H_{12}O_6$
sucrose	"table sugar" "granulated sugar"	intermediate	4.01 Cal/g	$C_{12}H_{22}O_{11}$

a. Propose a reason why these three sugars differ in their degree of sweetness.
b. Explain why these sugars are almost identical in their calories per gram.

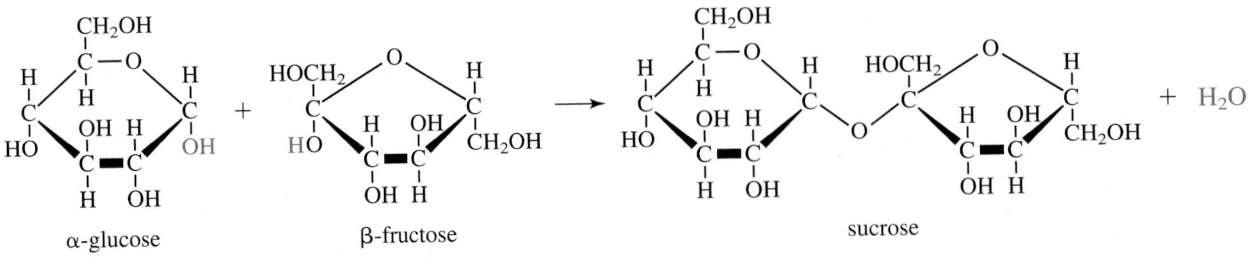

Figure 11.13

Formation of sucrose, a disaccharide.

As you can see from the previous activity, the sugars we consume are remarkably similar in their chemical composition, in their calories, and even in their sweet taste. Although sweetness varies, the difference between glucose and fructose is only about a factor of 2. So does it matter which sugars you eat? Actually there is a better question to ask. How *much* sugar are you eating? The quick answer may be "too much." We will explore this question in the next section.

Not only can monosaccharide molecules link to form disaccharides but also much bigger molecules as well. Some of the most common and abundant carbohydrates are **polysaccharides,** condensation polymers made up of thousands of monosaccharide units. As the name implies, these macromolecules consist of "many sugar units." When monosaccharide monomers combine, a water molecule is released each time a monomer is incorporated into the chain. Starch, cellulose, and glycogen are three familiar examples of polysaccharides.

Our bodies can digest starch by breaking it down into individual glucose units; in contrast, we cannot digest cellulose. Consequently, we depend on starchy foods such as potatoes or pasta as carbohydrate sources rather than devouring toothpicks or textbooks. The reason is a subtle difference in how the glucose units are joined. In the alpha (α) linkage in starch, the bonds connecting the glucose units have a particular orientation, whereas the beta (β) linkage between glucose units in cellulose is in a different orientation (Figure 11.14). Our enzymes and the enzymes of many mammals are unable to catalyze the breaking of beta linkages in cellulose. Consequently, we can't dine on grass or trees.

In contrast, cows, goats, sheep, and other ruminants manage to break down cellulose with a little help. Their digestive tracts contain bacteria that decompose cellulose into glucose monomers. The animals' own metabolic systems then take over. Termites also contain cellulose-hungry bacteria, which means that wooden structures are sometimes at risk.

When we have excess glucose in our bodies, it is polymerized to glycogen with the help of insulin and stored in our muscles and liver. When our glucose levels slip below normal the glycogen is converted back into glucose. Glycogen has a molecular structure similar to that of starch. The chains of glucose units in glycogen, however, are longer and more branched than those in starch. Glycogen is vitally important

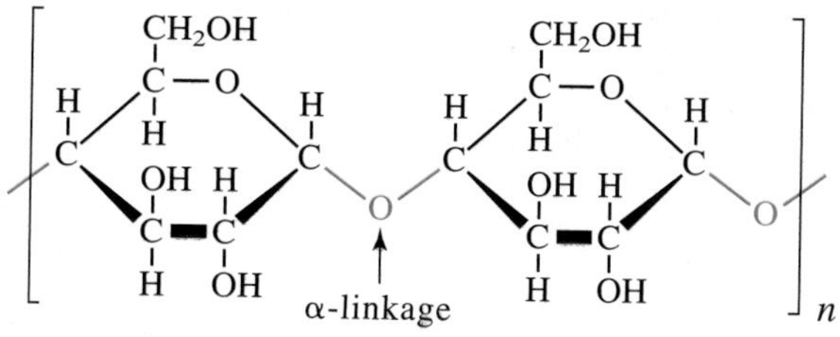

A 2009 survey of the American Dietary Association pointed out the need to reduce "added sugars" from our diets; that is, those sugars added to processed foods.

(a) starch

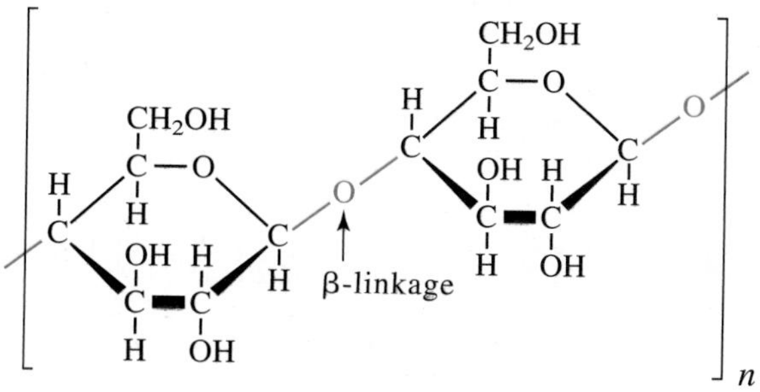

(b) cellulose

Figure 11.14

The bonding between glucose units in **(a)** starch and **(b)** cellulose.

because it stores energy for use in our bodies. It accumulates in muscles and especially in the liver, where it is available as a quick source of internal energy.

As you may know, a healthful diet derives more of its carbohydrates from polysaccharides than from simple sugars, such as mono- and disaccharides. We now delve into topics related to sweetness.

11.6 | How Sweet It Is: Sugars and Sugar Substitutes

Got a sweet tooth? We seem to be born with a preference for sweets, and for most us, this lasts throughout our lives. Sweeteners come as syrups, small crystals in packages, and as cubes or tablets that you can drop into a cup of coffee. Some of these sweeteners are natural; others are synthetic ("artificial"). Some are very sweet; others less so. Some have been available since antiquity; others are relatively new on the market. Like all of the foods we eat, each one can affect both your health and the health of the planet.

How sweet is sweet? Table 11.3 indicates the relative sweetness of some common natural sweeteners, all compared with sucrose, which is assigned a value of 100. From this table, one would have to use less fructose to equal the sweetness of a teaspoon of sucrose (table sugar). In contrast, lactose (milk sugar) would require more than 6 teaspoons.

Does it matter which sugar you consume? Yes and no. As we pointed out earlier, for most people the issue is not which sugar. Rather, it is too much of all of them. Again, a healthful diet derives more of its carbohydrates from polysaccharides than from simple sugars. If you overindulge in sugar, you increase your risk both for becoming obese and for the diseases that accompany obesity, such as diabetes and high blood pressure.

Let's begin by getting a handle on how much sugar you consume. The next activity allows you to explore one possible source of sugar intake.

Consider This 11.12   Your Favorite Cola or UnCola

Soda has been referred to as liquid candy. Is this a fair characterization? Make an argument pro or con. In either case, cite the grams of sugar involved.

One reason that you consume sugars is because foods naturally contain them. All the sugars listed in Table 11.3 occur naturally. For example, fructose occurs in many fruits and lactose occurs in milk. Another reason that you consume sugar is because you add it during cooking or at the table or because food companies add it. For example, you may add sugar to your coffee or sprinkle it on your cereal. Manufacturers of processed foods add sugar to peanut butter, spaghetti sauce, and bread. These and many other products have small amounts of sugar added to improve the taste, the texture, or the shelf-life. According to data released in the 2001–2004 National Health and Nutrition Examination Survey, people in the United States consume about

Honey is primarily composed of fructose and glucose.

Table 11.3	Approximate Sweetness Values				
NATURAL SWEETENERS					
lactose	maltose	glucose	honey	sucrose	fructose
16	32.5	74.3	97	100	173

Source: International Food Information Council.

22 teaspoons of added sugar daily. At about 4 grams of sugar per teaspoon and 4 Calories a gram, this translates to about 350 Calories daily from added sugar!

High-fructose corn syrup is an example of a sugar that is added to beverages and foods. Depending on where you live, high-fructose corn syrup goes by different names. In Europe, it is called isoglucose, and in Canada it goes by glucose–fructose. In this textbook we refer to it as HFCS.

Corn syrup primarily contains glucose. But if you treat corn syrup with enzymes, you can convert the glucose to fructose, which is sweeter. Several different "blends" of HFCS exist, depending on the end use. For example, a typical blend used in soft drinks is about 55% fructose, with the remainder being glucose.

Why is HFCS added to foods? Actually, many reasons exist. In 2009, Audrae Erickson, President of the Corn Refiners Association, pointed out that:

> *"High fructose corn syrup is used in the food supply because of its many functional benefits. For example, it retains moisture in bran cereals, helps keep breakfast and energy bars moist, maintains consistent flavors in beverages and keeps ingredients evenly dispersed in condiments. High fructose corn syrup enhances spice and fruit flavors in yogurts and marinades. In addition to its excellent browning characteristics for breads and baked goods, it is a highly fermentable nutritive sweetener and prolongs product freshness."*

Opponents of HFCS have suggested that this corn sweetener is metabolized differently from sucrose. However, this does not appear to be true. The American Medical Association (AMA) has concluded that HFCS "does *not* appear to contribute more to obesity than other caloric sweeteners." Metabolically, HFCS is similar to sucrose.

So wherein lies the argument? Again, it lies in the amount of sugar—any type of added sugar—that we consume daily. The AMA recommendation has been that people limit the amount of added sugars to 32 grams or less, based on a 2000-Calorie diet. This translates to 8 teaspoons of sugar or about 128 Calories daily. In August 2009, the American Heart Association released a journal article that fine-tuned these values by age and gender:

> *"Most women should consume no more than 100 Calories (about 25 grams) of added sugars per day. Most men should consume no more than 150 Calories (about 37.5 grams) each day. That's about six teaspoons of added sugar a day for women and nine for men."*

In contrast, the Corn Refiners Association quotes a recommendation of no more than 25% added sugar calories a day, based on a 2002 report from the Institute of Medicine at the U.S. National Academies. This would translate to a maximum of 500 Calories of added sugar.

Bottom line? For personal health, we suggest moderation. No absolute number is likely to ever be set. Clearly, though, lower is better in terms of weight gain and diabetes.

In terms of the health of the planet, the issues are more complex. More so than any other major crop, corn requires the use of fertilizer, herbicides, and pesticides, all of which require the burning of fossil fuels to produce. The runoff from cornfields has the potential to pollute rivers and streams, including those far downstream.

For example, the fertilizer runoff is largely responsible for a huge dead zone in the Gulf of Mexico at the mouth of the Mississippi River (Figure 11.15). When the nitrogen- and phosphorus-rich waters reach the ocean, they cause algae blooms. In turn, the algae deplete the oxygen in the waters, killing aquatic life. So although HFCS may appear to be a bargain, its price does not include the high environmental costs related to growing corn.

We end this section by turning to artificial (synthetic) sweeteners, also called sugar substitutes. As you can see from Table 11.4, these are far sweeter than natural sugars. The Calories per gram of aspartame is about the same as sucrose, but since it is about 200 times sweeter than sucrose, 1/200th of a teaspoon of aspartame is used as the equivalent of a teaspoon of sucrose. Saccharin, aspartame, sucralose, neotame,

Both cane sugar and beet sugar are primarily sucrose.

A blend of fructose (55%) and glucose (45%) has a sweetness comparable to that of sucrose, table sugar.

In the United States, HFCS is less expensive than sucrose. Reasons include government subsidies for growing corn and tariffs for imported sugar.

Figure 11.15

Brown, nutrient-rich water from the Mississippi River meets the Gulf of Mexico, creating a "dead zone." The zone is approximately 6000–7000 square miles, varying in size seasonally.

Source: Nancy Rabalais, Louisiana Universities Marine Consortium.

and acesufame potassium are the five artificial sweeteners approved in the United States. Other countries have a slightly different list.

So are artificial sweeteners the way to go? These compounds certainly do offer the advantage of fewer Calories. For example, the sugar in a 12-ounce can of a soda pop accounts for about 140 Calories. In terms of a 2000-Calorie diet, this translates to about 7%. By contrast, the same beverage sweetened with a synthetic sweetener would have 0 Calories. Although people have been concerned about the health effects of using artificial sweeteners, studies indicate that the sweeteners currently on the market are safe for most people. There is one notable exception, aspartame, which we will describe in the next section.

Consider This 11.13   Chemicals!

A well-known U.S. medical clinic posted this statement online: "Artificial sweeteners are chemicals or natural compounds that offer the sweetness of sugar without as many calories."

a. What is the problem with the categories "chemicals or natural compounds"?
b. How might you rewrite this language?

Table 11.4	Approximate Sweetness Values			
SYNTHETIC SWEETENERS				
acesulfame potassium	aspartame	neotame	saccharin	sucralose
200	200	7,000–13,000	300	600

Source: International Food Information Council.

11.7 | Proteins: First Among Equals

The word *protein* derives from *protos,* Greek for "first." The name is misleading. Life depends on the interaction of thousands of chemicals, and to assign primary importance to any single compound or class of compounds is simplistic. Nevertheless, proteins are an essential part of every living cell. They are major components in hair, skin, and muscle. They also transport oxygen, nutrients, and minerals through the bloodstream. Many of the hormones that act as chemical messengers are proteins, as are most of the enzymes that catalyze the chemistry of life.

A **protein** is a polyamide or polypeptide; that is, a polymer built from amino acid monomers. The great majority of proteins are made from various combinations of the 20 different naturally occurring amino acids. Molecules of amino acids share a common structure. Four chemical species are attached to a carbon atom: a carboxylic acid group, an amine group, a hydrogen atom, and a side chain designated R, all shown in Figure 11.16.

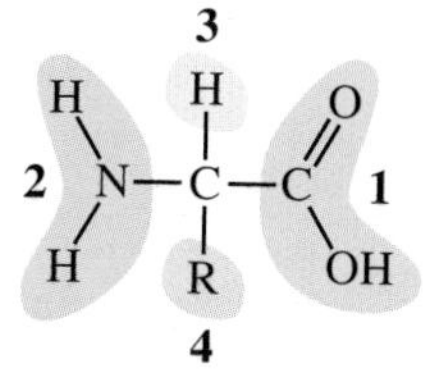

Figure 11.16

General structure for an amino acid. **1** is a carboxylic acid group; **2** an amine group; **3** a hydrogen atom; and **4** an R group that stands for any of the possible side chains.

Variations in the R side chain differentiate individual amino acids, as shown in Figure 11.17. For example, in the simplest amino acid (glycine), R is a hydrogen atom. In alanine, R is a $-CH_3$ group; in aspartic acid (found in asparagus), it is $-CH_2COOH$; and in phenylalanine, $-CH_2(C_6H_5)$. Two of the 20 naturally occurring amino acids have R groups that bear a second –COOH functional group, three have R groups containing amine groups, and two others contain sulfur atoms.

$-C_6H_5$ designates the phenyl group, first introduced in Table 9.2.

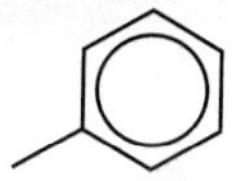

Combining amino acids to form proteins takes place through a reaction between an amine group and a carboxylic acid group. Equation 11.4 represents the reaction of glycine with alanine to form a **dipeptide,** a compound formed from two amino acids. Here, the acidic –COOH group of a glycine molecule reacts with the $-NH_2$ group of alanine, and an H_2O molecule is produced. In the process, the two amino acids link through the C–N peptide bond shown in the blue shaded area. Once incorporated into the peptide chain, the amino acids are called **amino acid residues.**

The peptide bond that forms when two amino acids react was defined in Section 9.6 and will be further explored in the context of protein synthesis in Chapter 12.

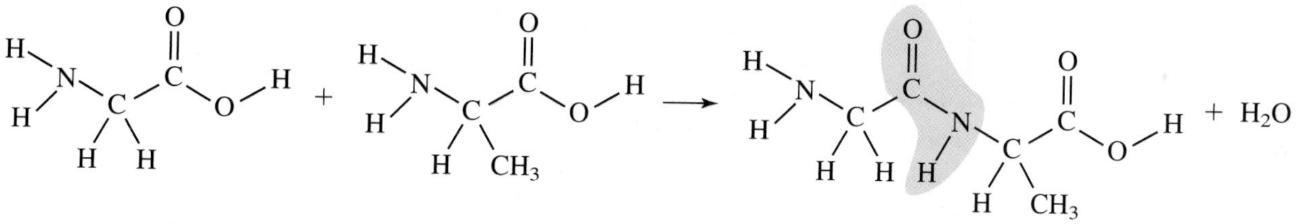

glycine alanine dipeptide water

[11.4]

Glycine and alanine can form more than one dipeptide. We illustrate the options with simple block diagrams for each amino acid. The first case is shown in equation 11.4 in which glycine provides the carboxylic acid and alanine provides the amine. This is an example of condensation polymerization, the same process that you encountered in the previous section with polysaccharides.

glycine alanine aspartic acid phenylalanine

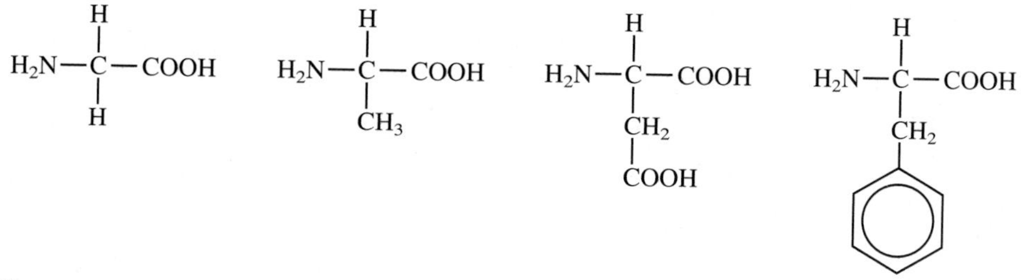

Figure 11.17

Examples of amino acids with different side chains.

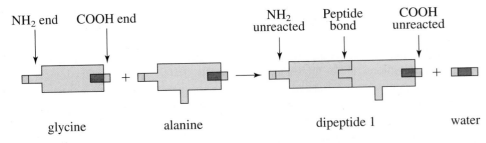

In the second case, the amino acids reverse roles; alanine provides the –COOH and glycine the –NH₂.

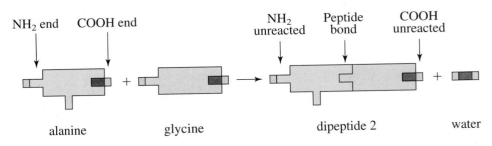

Look closely to see that the two dipeptides are different. In the first dipeptide, the unreacted amine group is on the glycine residue and the unreacted acid group is on the alanine residue; in the second dipeptide, the –NH₂ is on the alanine residue and the –COOH is on the glycine residue.

    The point of all this is that the order of amino acid residues in a peptide makes a difference. The particular protein formed depends not only on which amino acids are present, but also on their sequence in the protein chain. Assembling the correct amino acid sequence to make a particular protein is like putting letters in a word; if they are in a different order, a completely new meaning results. Thus, a tripeptide consisting of three different amino acids is like a three-letter word containing the letters *a, e,* and *t.* There are six possible combinations of these letters. Three of them—*ate, eat,* and *tea*—form recognizable English words; the other three—*aet, eta,* and *tae*—do not. Similarly, some sequences of amino acids may be biological nonsense.

    Still restricting ourselves to three-letter words and only the letters *a, e,* and *t,* but allowing the duplication of letters, we can make perfectly good words such as *tee* and *tat,* and lots of meaningless combinations such as *aaa* and *tte.* There are, in fact, a total of 27 possibilities, including the 6 identified earlier. Just as words can use letters more than once, most proteins contain specific amino acids more than once.

Putting the amino acids of a protein into their proper order is like assembling a train correctly by placing each car in the right sequence.

See Section 12.4 for more information about the structure and synthesis of proteins.

Your Turn 11.14   Making Tripeptides

The block diagrams in this section show that glycine (Gly) and alanine (Ala) can combine to form two dipeptides: GlyAla and AlaGly. If these two amino acids can be used more than once, two other dipeptides are possible: GlyGly and AlaAla. Thus, four different dipeptides can be made from two different amino acids. Eight different tripeptides can be made from supplies of two different amino acids, assuming that each amino acid can be used once, twice, three times, or not at all. Use the symbols Gly and Ala to write down representations of the amino acid sequence in all eight of these tripeptides.
Hint: Start with GlyGlyGly.

    Normally, the body does not store a reserve supply of protein, so foods containing protein must be eaten regularly. As the principal source of nitrogen for the body, proteins are constantly being broken down and reconstructed. A healthy adult on a balanced diet is in nitrogen balance, excreting as much nitrogen (primarily as urea in

Table 11.5	The Essential Amino Acids	
histidine	lysine	threonine
isoleucine	methionine	tryptophan
leucine	phenylalanine	valine

The *-ine* ending is used for naming most amino acids.

the urine) as she or he ingests. Growing children, pregnant women, and persons recovering from long-term debilitating illness or burns have a positive nitrogen balance. This means that they consume more nitrogen than they excrete because they are using the element to synthesize additional protein. A negative nitrogen balance exists when more protein is being decomposed than is being made. This occurs in starvation, when the energy needs of the body are unmet from the diet, and muscle is metabolized to maintain physiological functions. In effect, the body feeds on itself.

Another cause of a negative nitrogen balance may be a diet that does not include enough of the **essential amino acids,** those required for protein synthesis but that must be obtained from the diet because the body cannot synthesize them. Of the 20 natural amino acids that make up our proteins, we can synthesize 11 in our bodies from simpler molecules. We must obtain the other 9 from the foods we eat. If your diet is missing any of the nine essential amino acids identified in Table 11.5, the result can be severe malnutrition.

Good nutrition thus requires protein in sufficient quantity and suitable quality. Beef, fish, and poultry contain all the essential amino acids in approximately the same proportions found in the human body. Therefore, all of these are "complete" proteins. However, most people of the world depend on grains and other vegetable crops rather than on meat or fish. If such a diet is not sufficiently diversified, some essential amino acids may be lacking. For example, Mexican and Latin American diets tend to be rich in corn and corn products, a protein source that is *incomplete* because corn is low in tryptophan, an essential amino acid. A person may eat enough corn to meet the total protein requirement, but still be malnourished because of insufficient tryptophan.

Fortunately for millions of vegetarians, a reliance on vegetable protein does not doom them to malnutrition. The trick is to apply a principle nutritionists call **protein complementarity,** combining foods that complement essential amino acid content so that the total diet provides a complete supply of amino acids for protein synthesis. Although some may worry that vegetarians need to strictly adhere to this principle, most are likely to do it automatically. Say, for example, you eat a peanut butter sandwich. Bread is deficient in lysine and isoleucine, but peanut butter supplies these amino acids. On the other hand, peanut butter is low in methionine, but it is provided by the bread. The traditional diets in many countries also tend to meet protein requirements. For example, in Latin America, beans are used to complement corn tortillas; soy foods are eaten with rice in parts of Southeast Asia and Japan. People in the Middle East combine bulgur wheat with chickpeas or eat hummus, a paste made from sesame seeds and chickpeas, with pita bread. In India, lentils and yogurt are eaten with unleavened bread. Thus it is likely that if you follow a balanced vegetarian diet, you will ingest sufficient quantities of essential amino acids, assuming that you are eating an adequate number of calories.

We end this section by revisiting sweetness, the topic of the previous section. It may surprise you to learn that aspartame, a sugar substitute, is a dipeptide! Aspartame is composed primarily of the amino acids aspartic acid and phenyl alanine (Figure 11.18).

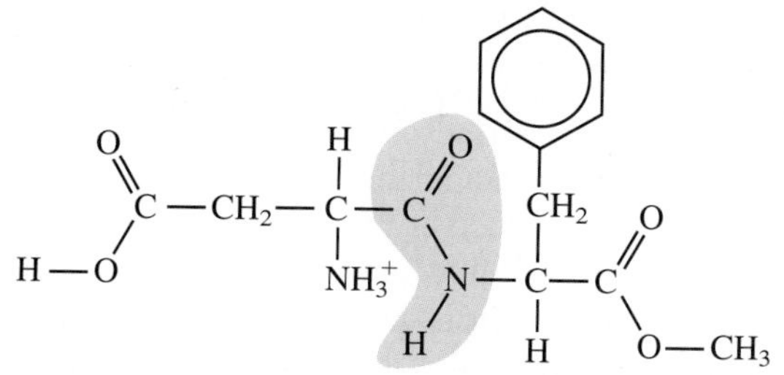

Figure 11.18
The structural formula of aspartame.

It is one of the most highly studied food additives and for the vast majority of consumers, aspartame is a safe alternative to sugar. One group of people, however, definitely should not use aspartame. The warning on packets of artificial sweeteners and products containing aspartame is explicit: "Phenylketonurics: Contains Phenylalanine."

This is a case where one person's treat is another person's poison. Phenylalanine is an essential amino acid converted in the body to tyrosine, a different amino acid. Individuals with phenylketonuria, a genetically transmitted disease, lack the enzyme that catalyzes this transformation. Consequently, the conversion of dietary phenylalanine to tyrosine is blocked and the phenylalanine concentration rises. To compensate for the elevated phenylalanine, the body converts it to phenylpyruvic acid, excreting large quantities of this acid in the urine. Phenylpyruvic acid is termed a *keto* acid because of its molecular structure; hence, the disease is known as phenyl*keto*nuria or PKU. People with the disease are called phenylketonurics.

Excess phenylpyruvic acid causes severe mental retardation. Therefore, the urine of newborn babies is tested for this compound using special test paper placed in the diaper. Infants diagnosed with PKU must be put on a diet severely limited in phenylalanine. This means avoiding excess phenylalanine from milk, meats, and other sources rich in protein. Commercial food products are available for such diets, their composition adjusted to the age of the user. Because phenylalanine is an essential amino acid, a minimum amount of it must still be available, even in phenylketonurics. Supplemental tyrosine may also be needed to compensate for the absence of the normal conversion of phenylalanine to tyrosine. A phenylalanine-restricted diet is recommended for phenylketonurics at least through adolescence. Adult phenylketonurics also must limit their phenylalanine intake and hence curtail their use of aspartame.

This last discussion has demonstrated that even small quantities of a substance can make a difference in your diet. The next section looks at other substances that are found in your diet in small quantity.

11.8 | Vitamins and Minerals: The Other Essentials

Yes, essential! Vitamins and minerals are **micronutrients,** substances that are needed only in miniscule amounts but still are essential to life. Nearly everyone in the United States knows that vitamins and minerals are important, but a thriving multimillion-dollar supplement industry reminds any who forget. Unfortunately, many processed foods that are high in sugars and fats lack essential micronutrients.

Only relatively recently have we come to understand the role of vitamins and minerals in our diet. Over the ages, humans learned that they became ill if certain foods were lacking. Systematic studies began early in the 20th century with the discovery of "Vitamine B_1" (thiamine). The particular designation, B_1, was the label on the test tube in which the sample was collected. The general term *vitamin* was chosen because the compound, which is vital for life, contains the amine group. The final "e" disappeared with the discovery that not all vitamins are amines.

Vitamins are organic compounds with a wide range of physiological functions. Although only small amounts are needed in the diet, vitamins are essential for good health, proper metabolic functioning, and preventing disease. In general, vitamins are not sources of energy for the body although some help break down macronutrients. They can be classified on the basis of being water- or fat-soluble. For example, examine the structural formula of vitamin A shown in Figure 11.19 to see that it contains almost exclusively C and H atoms. As a result, vitamin A is a nonpolar compound, lipid-soluble, and similar to the hydrocarbons derived from petroleum. Water-soluble vitamins often contain several —OH groups that can hydrogen bond with water molecules. Vitamin C is a case in point, shown in the same figure.

The ketone functional group was shown in Table 10.3.

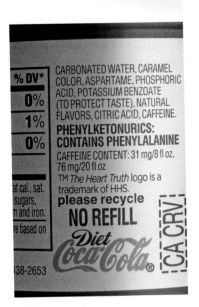

Hydrogen bonds were discussed in Section 5.2. The relationship between molecular structure and solubility was explored in Section 5.5.

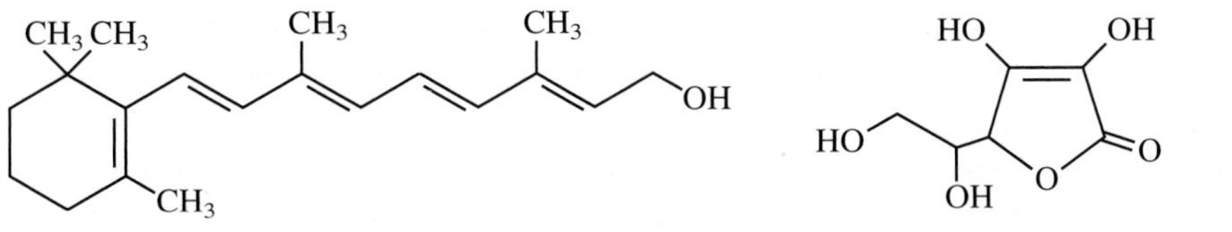

vitamin A, a lipid-soluble vitamin   vitamin C, a water-soluble vitamin

Figure 11.19

An example of a lipid-soluble and water-soluble vitamin.

Consider This 11.15 **Classifying Vitamins**

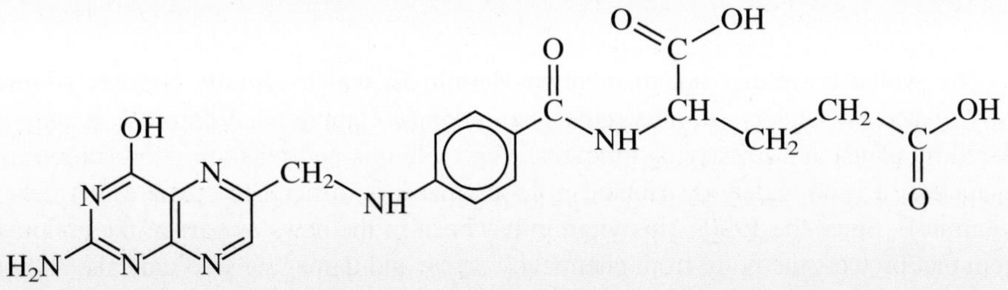

Folic acid helps prevent certain types of anemia and aids in nucleic acid synthesis. This vitamin is particularly important for pregnant women. Do you expect that it would be soluble in fat tissue (lipids) or in the bloodstream and cell tissue (water)? Explain your reasoning.

The solubility of vitamins has significant implications for health. Because of their fat-solubility, vitamins A, D, E, and K are stored in cells rich in lipids, where they are available on biological demand. If swallowed in excess, fat-soluble vitamins can build up to a toxic level. For example, high doses of vitamin A can result in both troublesome symptoms such as fatigue and headache and in more serious ones such as blurred vision and liver damage. Although the toxic level of vitamin D is not known, similarly illness can result if too much vitamin D is ingested, including heart and kidney damage. High levels of these vitamins are not reached via diet; rather, they are a result of excessive use of vitamin supplements.

In contrast, water-soluble vitamins are excreted in the urine rather than stored in the body. As a result, you need to eat foods containing these vitamins frequently. Unfortunately, even water-soluble vitamins can accumulate at toxic levels when taken in large doses, although such cases are rare. For most people, a balanced diet provides the necessary vitamins and minerals, making vitamin supplements unnecessary. The one exception seems to be vitamin D, which is synthesized in the skin by using the energy of sunlight, rather than ingested. Recent research on vitamin D has led more physicians to check vitamin D blood levels as part of an annual physical exam and to use the results to determine whether taking a supplement is necessary.

Many of the water-soluble vitamins serve as **coenzymes,** molecules that work in conjunction with enzymes to enhance their activity. Members of the vitamin B family are particularly adept in acting as coenzymes. Niacin plays an essential role in energy transfer during glucose and fat metabolism. The synthesis of niacin in the body requires the essential amino acid tryptophan. Thus, a diet deficient in tryptophan may lead to niacin deficiency. Such a deficiency causes pellagra, a serious condition that is characterized by "the 4Ds" of diarrhea, dermatitis, dementia, and death. This disease is still common today in parts of the world, including several African nations.

Some vitamins were discovered when observers correlated diseases with the lack of specific foods. For example, vitamin C (ascorbic acid) must be supplied in the diet, typically via citrus fruits and green vegetables. An insufficient supply of the vitamin leads to scurvy, a disease in which collagen, an important structural protein, is broken down. The link between citrus fruits and scurvy was discovered more than 200 years

This practice also led to British sailors being called "limeys."

ago when it was found that feeding British sailors limes or lime juice on long sea voyages prevented the disease. Thanks to Nobel laureate Linus Pauling who in 1970 authored *Vitamin C and the Common Cold*, vitamin C continues to be in the public eye.

 Consider This 11.16   **Megadoses of Vitamin C**

Decades ago, Linus Pauling claimed that large doses of vitamin C were therapeutic in preventing the common cold.

  a. What range of vitamin C daily constitutes a megadose?
  b. Find evidence to either support or refute the claim of preventing the common cold. Cite your sources.
  c. Interview three people of different age groups, including a nurse or physician if you are able. Ask if they take vitamin C and if so why.

We would be remiss not to mention vitamin E, which actually consists of several closely related fat-soluble vitamins rather than a single compound. Vitamin E is only synthesized by plants and in varying amounts. Vegetable oils and nuts are good sources of it. Nonetheless, it is so widely distributed in foods that it is difficult to create a diet deficient in vitamin E. Since the 1990s, this vitamin has been in the news as part of the antioxidant system that protects the body from chemically active and damaging free radicals. Although at one time taking vitamin E supplements was recommended, this no longer is the case. Skin preparations are another matter, though. Many products contain vitamin E and claim that it prevents or helps heal skin damage. Investigate for yourself in the next activity.

 Consider This 11.17   **Vitamin E and Your Skin**

Check the advertisements and you will see that many hand lotions and beauty creams contain vitamin E.

  a. Identify three skin products that contain vitamin E.
  b. How is vitamin E thought to work in helping protect your skin?
  c. Although it might seem logical that vitamin E would be good for the skin, it is difficult to find the evidence. Investigate this topic to see for yourself. The textbook's website provides some links to get you started.

Minerals are ions or ionic compounds that, like vitamins, have a wide range of physiological functions. You may be familiar with minerals such as sodium and calcium, but actually the list is much longer. Depending on how much of them you need, minerals are classified as either macro, micro, or trace.

Macrominerals are still micronutrients.

- **Macrominerals** Ca, P, Cl, K, S, Na, and Mg
  These elements are necessary for life but not nearly as abundant in our bodies as O, C, H, and N. You need to ingest macrominerals daily, typically in the range of 1 to 2 g.
- **Microminerals** Fe, Cu, and Zn
  The body requires lesser amounts of these. You may recognize iron as a component of hemoglobin, a protein in the blood that carries oxygen.
- **Trace minerals** I, F, Se, V, Cr, Mn, Co, Ni, Mo, B, Si, and Sn
  These usually are measured in microgram quantities. Although the total amount of trace elements in the body is only about 25–30 g, their small quantity belies the importance they have in good health.

Revisit Section 5.6 for more about ions.

The periodic table in Figure 11.20 displays the essential dietary minerals. The metals exist in the body as cations, for example, Ca^{2+} (calcium ion), Mg^{2+} (magnesium ion), K^+ (potassium ion), and Na^+ (sodium ion). The nonmetals typically are present as anions. For example, chlorine is found as Cl^- (chloride ion) and phosphorus appears in PO_4^{3-} (phosphate ion).

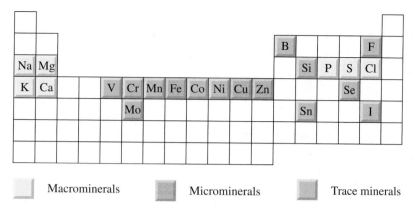

Macrominerals   Microminerals   Trace minerals

Figure 11.20

Periodic table indicating dietary minerals necessary for human life.

The physiological functions of minerals are widely diverse. Calcium is the most abundant mineral in the body. Along with phosphorus and smaller amounts of fluorine, it is a major constituent of bones and teeth. Blood clotting, muscle contraction, and transmission of nerve impulses all require the calcium ion, Ca^{2+}.

Sodium is also essential for life but not in the excessive amounts supplied by the diets of most people today. The salt we eat does not necessarily come from the salt-shaker. Rather, you perhaps unknowingly add salt to your diet from sauces, snack foods, fast foods, and even canned soup! Labels now are required to list the "sodium" content, meaning the number of milligrams of Na^+ per serving. For example, different brands of tomato soup may have between 700 and 1260 mg of Na^+ per serving. Compare this with the recommended daily value of no more than 2400 mg (2.4 grams) of Na^+ per day. The major concern with excess dietary sodium is its correlation with high blood pressure for some individuals. They may be advised by their doctors to limit their sodium intake.

> The Latin word for salt is *sal*. Salt was so highly valued in Roman times that soldiers were paid in *sal*, thereby forming the root for the modern word *salary*.

Your Turn 11.18   Sodium in Your Diet

Compare sodium content for foods in the same category, such as different brands of pretzels, bread, frozen pizza, salad dressing, or even tomato soup. Have your findings surprised you or influenced your future choices? Recall that 1 g = 1000 mg.

Oranges, bananas, tomatoes, and potatoes all help supply the recommended daily requirement of 2 grams of potassium (in the form of K^+), another essential mineral. You may have heard both K^+ and Na^+ referred to as the "electrolytes" of sports drinks. Sodium and potassium ions are close chemical cousins, as they are both in Group 1A of the periodic table. They have similar chemical properties and physiological functions. Within cells, the concentration of K^+ is considerably greater than that of Na^+. The reverse situation holds true in the lymph and blood serum outside the cells in which the concentration of K^+ is low and that of Na^+ is high. The relative concentrations of K^+ and Na^+ are especially important for the rhythmic beating of the heart. Individuals who take diuretics to control high blood pressure may also take potassium supplements to replace potassium excreted in the urine. However, such supplements should be taken only under a physician's directions because they also can dramatically alter the potassium–sodium balance in the body and lead to cardiac complications.

> Electrolytes were described earlier in Section 5.6.

In most instances, microminerals and trace elements have very specific biological functions and are incorporated in relatively few biomolecules. Iodine is an example. Most of the body's iodine is found in the thyroid gland incorporated into thyroxine, a hormone that regulates metabolism. Excess thyroxine production is associated with hyperthyroidism (Graves' disease) in which basal metabolism is accelerated to an

> Seafood is one rich source of iodine. Another is iodized salt, sodium chloride (NaCl) to which 0.02% of potassium iodide (KI) has been added.

The explosion at Chernobyl released I-131 into the surrounding countryside. See Section 7.5 for more about the uptake of I-131 by children and the resulting thyroid cancer.

unhealthy level, rather like a racing engine. In contrast, a thyroxine deficiency, sometimes caused by a lack of dietary iodine, slows metabolism and results in tiredness and listlessness. The tendency of the thyroid gland to concentrate iodine makes possible the use of radioactive I-131 in treating thyroid disorders and in imaging the thyroid gland for diagnostic purposes. The next activity gives you the opportunity to combine what you learned about radioactive I-131 and iodine as a trace mineral.

🕸 Consider This 11.19   Radioactive Iodine

Some individuals have an overactive thyroid gland (hyperthyroidism).

a. I-131 is used to treat hyperthyroidism. Explain how ingestion of this radioisotope can lead to a reduction in the function of the thyroid gland.

b. I-131 treatment has both risks and benefits for a patient. List two of each.

c. Patients treated with I-131 temporarily carry a source of radioactivity in their bodies. After 10 half-lives, a radioisotope can be said to be "gone." How much time is this for a patient treated with I-131? *Hint:* See Table 7.4.

11.9 | Energy from Food

The energy needed to keep our bodies warm and to run our complex chemical, mechanical, and electrical systems comes from the food we eat: fats, carbohydrates, and proteins. As noted in several earlier chapters, this energy initially arrives on Earth in the form of sunlight and then is absorbed by green plants. In the process of photosynthesis, CO_2 and H_2O are combined to form $C_6H_{12}O_6$. Hence the Sun's energy is stored in chemical bonds of the monosaccharide we know as glucose.

$$\text{energy (from sunshine)} + 6\,CO_2 + 6\,H_2O \xrightarrow{\text{chlorophyll}} C_6H_{12}O_6 + 6\,O_2 \quad \textbf{[11.5]}$$

During respiration, the outcome from photosynthesis is reversed. Glucose is converted into simpler substances (ultimately in most cases to CO_2 and H_2O), and the energy is released.

$$C_6H_{12}O_6 + 6\,O_2 \longrightarrow 6\,CO_2 + 6\,H_2O + \text{energy (from respiration)} \quad \textbf{[11.6]}$$

If you need a refresher on energy changes at the molecular level, see Section 4.6.

The energy balance between equations 11.5 and 11.6 can be schematically represented (Figure 11.21).

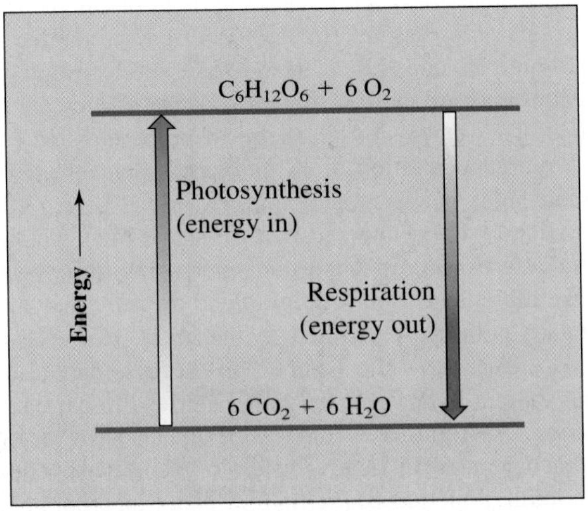

Figure 11.21

Glucose energy balance (photosynthesis and respiration).

Table 11.6	Average Energy Content of Macronutrients
fats	9 Cal/g
carbohydrates	4 Cal/g
proteins	4 Cal/g

In addition to having a supply of sufficient energy, our bodies must have some way of regulating the rate at which the energy is released. Without such control, your body temperature would wildly fluctuate. The automobile provides an analogy. Dropping a lighted match into the fuel tank would burn all the gasoline at once and possibly the car as well. Under normal operating conditions, just enough fuel is delivered to the ignition system to supply the automobile with the energy it needs without raising the temperature of the car and its occupants beyond reason. By releasing a little energy at a time, the efficiency of the process is enhanced. So it is with the body. The conversion of foods ultimately into carbon dioxide and water occurs over many small steps, each one involving enzymes, enzyme regulators, and hormones. As a result, energy is released gradually, as needed, and body temperature is maintained within normal limits. The energy in calories associated with the metabolism of a gram of fats, carbohydrates, and proteins is given in Table 11.6.

1 dietary calorie = 1 Cal = 1 kcal = 1000 calories

On a Calorie-per-gram basis, fats provide about 2.5 times as much energy as do proteins and carbohydrates. This observation makes it easy to understand the popularity of low-fat diets for losing weight. Although proteins, like carbohydrates, yield about 4 Cal/g if metabolized, proteins are not used in the body primarily as an energy source. Rather proteins are used for building skin, muscles, tendons, ligaments, blood, and enzymes.

The reason for the difference in energy content between fats and carbohydrates is evident from their chemical composition. Compare the formula of a fatty acid, lauric acid, $C_{12}H_{24}O_2$, with that of sucrose (table sugar), $C_{12}H_{22}O_{11}$. Both compounds have the same number of carbon atoms per molecule and very nearly the same number of hydrogen atoms. When molecules such as these "burn" as fuel in your body, the C and H atoms that they contain combine with oxygen to form CO_2 and H_2O, respectively. But more oxygen is required to burn a gram of lauric acid, $C_{12}H_{24}O_2$, than a gram of sucrose, $C_{12}H_{22}O_{11}$. Examine these two reactions.

More information about lauric acid is revealed by writing its condensed structural formula of $CH_3(CH_2)_{10}COOH$.

$$C_{12}H_{24}O_2 + 17\,O_2 \longrightarrow 12\,CO_2 + 12\,H_2O + 8.8\ \text{Cal/g} \qquad \textbf{[11.7]}$$
lauric acid

$$C_{12}H_{22}O_{11} + 12\,O_2 \longrightarrow 12\,CO_2 + 11\,H_2O + 3.8\ \text{Cal/g} \qquad \textbf{[11.8]}$$
sucrose

In the language of chemistry, the sugar is already more "oxygenated" or more "oxidized" than the fatty acid. Weaker C–H bonds (416 kJ/mol) have already been replaced by stronger O–H bonds (467 kJ/mol) in the sucrose. The result is that even though fewer O=O double bonds (498 kJ/mol) must be broken for sucrose to combine with O_2, less energy overall is released than is the case for the combustion of lauric acid.

Oxygenated fuels were discussed in Section 4.7.

Given how many tasty foods contain fat, it is easy to get an unhealthy percentage of our daily calories from fats. The problem is illustrated by considering Skeptical Chemist 11.20. In accordance with the *Dietary Guidelines for Americans* released by the U.S. Department of Agriculture (USDA) and the U.S. Department of Health and Human Services (HHS) in January 2005, 20–35% of calories should come from fats, for adults. Furthermore, the guidelines recommend that less than 10% of calories come from saturated fatty acids and that trans fat consumption be kept as low as possible.

Table 11.7	Estimated Calorie Requirements (United States)		
	ACTIVITY LEVEL		
Age (yr)	Sedentary*	Moderately Active†	Active‡
Females			
14–18	1800	2000	2400
19–30	2000	2000–2200	2400
31–50	1800	2000	2200
51+	1600	1800	2000–2200
Males			
14–18	2200	2400–2800	2800–3200
19–30	2400	2600–2800	3000
31–50	2200	2400–2600	2800–3000
51+	2000	2200–2400	2400–2800

Sedentary means a lifestyle that includes only the light physical activity associated with typical day-to-day life.

† *Moderately active* means a lifestyle that includes physical activity equivalent to walking about 1–3 miles per day at 3–4 miles per hour, in addition to the light physical activity associated with typical day-to-day life.

‡ *Active* means a lifestyle that includes physical activity equivalent to walking more than 3 miles per day at 3–4 miles per hour, in addition to the light physical activity associated with typical day-to-day life.

Source: Dietary Guidelines for Americans, 2005, USDA.

Skeptical Chemist 11.20   Low-Fat Cheese

A popular brand of low-fat shredded cheddar cheese advertises that it provides 1.5 g of fat per serving. Of this 1.5 g of fat, 1.0 g is saturated fat. In addition, a serving of this cheese is 28 grams (or 1/4 cup) and has 50 Calories, with 15 of these coming from fat. Is this a "low-fat" cheese? Support your answer with some numbers. Remember that the dietary recommendation is that 20–35% of calories should come from fat.

So how many calories does a person need? The answer is "It depends." The number of calories your diet should supply each day is a function of your level of activity, the state of your health, your gender, age, body size, and a few other factors. Table 11.7 summarizes the daily food energy intakes that have been recommended for people in the United States. The estimated calorie requirements are presented by gender and age groups at three different activity levels. Growing children (not included in the table) need a larger energy intake, both to fuel their high level of activity and to provide raw material for building muscle and bone. Children are particularly susceptible to undernourishment and malnutrition. Indeed, mortality rates among infants and young children are disproportionately high in famine-stricken countries.

Your Turn 11.21   Calories by Gender and Age

Consider the information in Table 11.7 and the *Dietary Guidelines for Americans, 2005.* A link is provided on the textbook's website.

a. Do males and females of the same age require the same number of calories for the same level of activity? Explain.

b. As an active male or female grows older, how does the estimated calorie requirement change?

Table 11.8	Energy Expenditure for Common Physical Activities*		
Moderate Physical Activity	**Cal/hr**	**Vigorous Physical Activity**	**Cal/hr**
hiking	370	jogging (5 mph)	590
light gardening/yard work	330	heavy yard work (chopping wood)	440
dancing	330	swimming (freestyle laps)	510
golf (walking, carrying clubs)	330	aerobics	480
bicycling (<10 mph)	290	bicycling (>10 mph)	590
walking (3.5 mph)	280	walking (4.5 mph)	460
weight lifting (light workout)	220	weightlifting (vigorous workout)	440
stretching	180	basketball (vigorous)	440

* Values include both resting metabolic rate and activity expenditure for a 70-kg (154-pound) person. Calories burned per hour are higher for persons heavier than 154 pounds and lower for persons who weigh less.

Where does all this food energy go? The first call on the calories you consume is to keep your heart beating, your lungs pumping air, your brain active, all major organs working, and your body temperature at about 37 °C. These requirements define the **basal metabolism rate (BMR),** the minimum amount of energy required daily to support basic body functions. This corresponds to approximately 1 Calorie per kilogram (2.2 pounds) of body mass per hour, although it varies with size and age.

To put this on a personal basis, consider a 20-year-old female weighing 55 kg (121 pounds). If her body has a minimum requirement of 1 Cal/(kg·h), her daily basal metabolism rate will be 1 Cal/(kg·h) × 55 kg × 24 h/day, or about 1300 Cal/day. According to Table 11.7, the recommended daily energy intake for a woman of this age and weight is a maximum of 2200 Cal if she is moderately active. This means that 59% of the energy derived from this food goes just to keep her body systems going.

Where do the rest of her calories go? The law of conservation of energy decrees that the energy must go somewhere. If she "burns off" the extra calories through exercise, none will be stored as added fat and glycogen. But, if the excess energy is not expended, it will accumulate in chemical form. Putting it more bluntly, "those who indulge, bulge."

How hard and how long we have to work (or play) to burn dietary calories is reported in Table 11.8. In Table 11.9, exercise is related in readily recognizable units such as hamburgers, potato chips, and beer. Of course by combining the information in this section with the information in earlier parts of this chapter about the types of nutrients in food, it should be clear that a healthful diet cannot be achieved simply by consuming the correct number of calories. A 2000-Calorie diet of only potato chips and beer would leave a person malnourished. Proper nutrition is not simply a matter of how much, but also of what kind of food a person consumes.

Each heartbeat uses about one joule (1 J) of energy.

Your basal metabolism rate is approximately 1 Cal/kg body mass per hour.

$$\frac{1300 \text{ Cal}}{2200 \text{ Cal}} \times 100 = 59\%$$

Your Turn 11.22 Basketball and Calories

A 70-kg person consumes a meal consisting of two hamburgers, 3 oz of potato chips, 8 oz of ice cream, and a 12-oz beer. Calculate the number of calories in the meal and the number of minutes the person would have to vigorously play basketball in order to "work off" the meal.

Answer
about 1500 Calories, about 200 min

Table 11.9	How Much Must I Exercise if I Eat This Cookie?*		
Food	Calories	Walking at 3.5 mph Time (min)	Jogging at 5 mph Time (min)
apple	125	27	13
beer (regular) 8 ounces	100	21	10
chocolate chip cookie	50	11	5
hamburger	350	75	35
ice cream, 4 ounces	175	38	18
pizza, cheese, 1 slice	180	39	18
potato chips, 1 ounce	108	23	11

* Values include both resting metabolic rate and activity expenditure for a 70-kg (154-pound) person.

11.10 | Quality Versus Quantity: Dietary Advice

Despite great advances in healthcare in the last century, obesity is now on the rise in the United States. In the last 30 years scientific research on nutrition has made tremendous progress. However, with advertising and the mass media, people seem more frustrated than ever in trying to understand dietary advice. One day the "experts" say one thing; the next they seem to say the opposite.

Indeed, dietary advice has been changing. If you are a young adult, your parents or grandparents may remember "The Basic Four" and "The Food Pyramid." Figure 11.22 gives a glimpse of dietary advice from the past, courtesy of the U.S. Department of Agriculture.

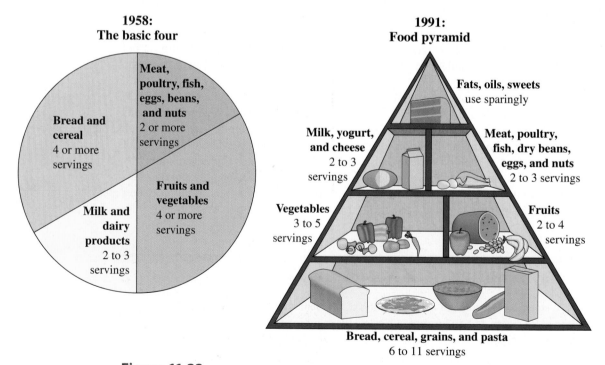

Figure 11.22

The USDA basic four food groups and the food pyramid.

Source: United States Department of Agriculture.

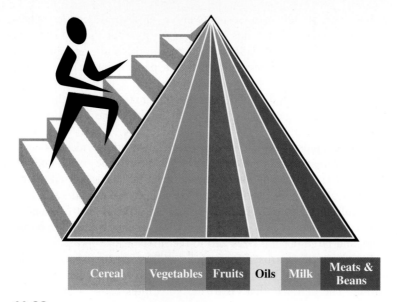

Figure 11.23

MyPyramid logo, "Steps to a Healthier You."

Source: United States Department of Agriculture, MyPyramid.gov, 2009.

An even greater change took place in April 2005 when the U.S. Department of Agriculture introduced a new type of pyramid (Figure 11.23). The 1991 pyramid was turned sideways and stairs were added together with a person climbing these stairs. In line with the current fashion, it is called MyPyramid. The accompanying website ("Steps to a Healthier You") allows individuals to complete an in-depth assessment of their diet quality and physical activity level. It also offers personalized advice for different types of diets, with tips ranging from cutting down on saturated fats to increasing exercise or eating out without busting a diet. You can explore this resource in the next activity.

Consider This 11.23   **Step up the Pyramid**

According to the USDA website, "MyPyramid offers personalized eating plans and interactive tools to help you plan and assess your food choices." Explore MyPyramid using the links provided on the textbook's website.

a. Enter your age, sex, and physical activity level into "My Pyramid Plan." What did you learn about what you need to eat?

b. For a more detailed assessment of your food intake and physical activity level, click on "MyPyramid Tracker." Again, what did you learn?

History can help us understand the proliferation of diets. In the 1960s, experiments based on controlled feeding of particular food items to participants for several weeks showed that saturated fat increased cholesterol levels. But these studies also showed that polyunsaturated fats—found in vegetables and fish—reduced cholesterol. Advice in the following decades was to replace saturated fats with unsaturated ones rather than reducing total fat. The "sat fat is bad" movement led to the greatly expanded use of vegetable oils. Unfortunately, as was noted in Section 11.4, the partial hydrogenation of these oils to create products such as margarine gave rise to trans fats that pose more health risks.

Two of the largest and longest running studies of diet are the Harvard-based Nurses' Health Study and Health Professionals Follow Up Study that followed over

90,000 women and 50,000 men, respectively, over several decades. These studies provided information that often has led to suggestions for dietary changes (including development of the 1991 USDA food pyramid), and sometimes fueled diet trends. The Harvard studies have shown that a participant's risk of heart disease was strongly influenced by the type of dietary fat consumed. Eating trans fat increases the risk substantially, and saturated fat increases it slightly. Unsaturated fats decrease the risk. Therefore, the total fat intake alone is not associated with heart disease risk.

Furthermore, epidemiological studies have shown little evidence that total fat or specific fats affect the risks of developing several cancers. The rise in obesity has been blamed on fat, but U.S. consumption of calories from fat has decreased since the 1980s while the rate of obesity still continues to grow. All of this would seem to imply that fat is not the culprit it was once believed to be. Or, at least, that the solution to U.S. dietary woes is not quite so clear-cut. Certainly the sedentary lifestyles of many in the United States and large portion sizes add to the mix.

So what about the "carbs"? Diet books addressing the issue of carbohydrates have spanned the spectrum from banning all carbs to a "good carb versus bad carb" differentiation. The claim is that "bad" carbohydrates cause a quick rise in blood sugar, followed by a spike in blood insulin level and a weight gain caused by making the body store fat or making the person feel hungry again from the low blood sugar. Insulin is a hormone secreted by the pancreas that allows the cells in your body to absorb and store sugar that is in the blood. The sugar that is not immediately burned for energy is converted to and stored as fat in your cells. Furthermore, glucagon is another hormone secreted by the pancreas that, essentially, has the opposite effect of insulin: it promotes the use of stored glucose in cells. The release of glucagon decreases after a "glucose spike," such as results from eating "bad carbs," but it is known to increase after consumption of proteins. Therefore, many of the "low-carb" diets promote the consumption of increased amounts of protein as a way to use up stored calories. The long-term health effects of this approach have yet to be seen.

Clearly, nutrition and dieting involve a complex bit of chemistry. However, research in this area increasingly is contributing to our understanding of these issues. In the United States, a concern is how to keep people from dying from complications relating to obesity. But in the United States and many parts of the world as well, another concern has been how to keep people from dying of starvation.

Speaking of books, read Barbara Kingsolver's *Animal, Vegetable, Miracle* for an informative personal journal about eating locally.

11.11 | Eat Local . . . ? Eat Veggies . . . ?

In this chapter, we have been making the case that what you dine on (and what you skip) affects not only your health but also that of the planet. We defined the concept of "foodprint" as the amount of land required per year to provide the nutritional resources for a person. We also described a case study in which those who ate less meat, but were not 100% vegetarian, had the most favorable foodprint. We now comment more broadly on the connections between food production, energy use, and global climate change, fully realizing that this section would more appropriately be the length of a book.

"Eat Local" . . . ? Yes, at least to some extent. Reasons for this include supporting local growers and getting to eat fresh produce. Table 11.10 offers a longer list. Of particular concern to us here are reasons that connect to the energy cost including transportation. Moving food any distance—but especially such a long one—clearly requires an energy input (accompanied by a production of waste), most likely from fossil fuels. As you have learned, burning fuels releases not only air pollutants such as NO, SO_2, and particulate matter, but also huge quantities of CO_2, a greenhouse gas.

Let's do some of the math connected with eating locally. The next activity gives you the opportunity to delve into some actual numbers.

Table 11.10	Reasons to "Eat Local"

1. **Eating local means more for the local economy.** A dollar spent locally generates twice as much income for the local economy. When businesses are not owned locally, money leaves the community at every transaction.

2. **Locally grown produce is fresher and tastes better.** Produce at your local farmer's market often has been picked within 24 hours of your purchase. This freshness not only affects the taste of your food, but also the nutritional value. Ever tried a tomato that was picked within 24 hours? 'Nuff said.

3. **Locally grown fruits and vegetables have longer to ripen.** Because the produce is handled less, locally grown fruit does not have to stand up to the rigors of shipping. You will get peaches so ripe that they fall apart as you eat them, and melons that were allowed to ripen until the last possible minute on the vine.

4. **Eating local is better for air quality and pollution than eating organic.** In a March 2005 study published in *Food Policy*, researchers found that the miles that organic food may travel to our plate can outweigh the benefit of buying organic. Consider buying local organic food.

5. **Buying local food keeps us in touch with the seasons.** By eating with the seasons, our foods are at their peak taste, are the most abundant, and are the least expensive.

6. **Supporting local providers supports responsible land development.** When you buy locally, you give those with local open space—farms and pastures—an economic reason to stay open and undeveloped.

Source: Adapted from "10 Reasons to Eat Local Food," Jennifer Maizer, EatLocalChallenge.com, 2009.

Consider This 11.24   Food Miles Detective

The USDA has recently required a country of origin label for certain foods. In addition, many foods produced in the United States indicate the state of origin. Head to your local grocery store and see if you can determine the origin of these foods. Then estimate the miles each traveled to reach your grocery store. Calculate the average.

a. apples, bananas, and a fruit of your choice
b. lettuce and other salad greens of your choice
c. shrimp and a fish of your choice
d. coffee
e. a steak or other cut of beef

What did you learn? By some estimates, the average bite of food in the United States has traveled 1500 miles before it reaches you. But depending on where you live and which foods you select, the average distance will differ. For example, those who live in Melbourne, Australia, are likely to have a longer average distance than those in Paris, France.

Given these large distances, should all food be produced and consumed locally? Not necessarily. Foods have energy costs other than transportation. For example, it may be more energy-efficient to import tomatoes from a warmer climate than to grow them in a greenhouse in a chilly one. Similarly, beef that is grass-fed may be more energy-efficient to import than to raise the cattle and feed them grain in the region where you live.

This sort of argument, however, is a bit reminiscent of the paper-versus-plastic debate in the earlier section on recycling (see Chapter 9). Given the choice of having your groceries bagged in paper or plastic, the better answer is not to use either. Rather, you should bring and reuse your own bag. A similar line of reasoning holds here as well. Should you eat local or global tomatoes, melons, oranges, or beef? The better answer may be not necessarily either. Rather, it might make more sense not to eat very much corn-fed beef at all. And it might make more sense to eat tomatoes and melons primarily when in season. Admittedly, this logic has its limits. You need certain foods to maintain your health. For example, those people who live where citrus fruits do not grow still need a source of vitamin C.

If your primary reason for eating local is to reduce fossil fuel consumption, you may be interested in a 2008 study done at Carnegie Mellon University. It showed that the transportation of U.S. foods from producer to retailer accounted for only 4% of food-related greenhouse gases. An all-local diet saved the equivalent in greenhouse gases of driving 1000 fewer miles each year. In contrast, eating a vegetarian diet just one day a week translated to driving about 1200 fewer miles per year!

Why such a big difference? A larger proportion is related to the production of food and this varies significantly with the type of food. The reason is because the production of meat results in larger greenhouse gas emissions. This is primarily due to the inefficient conversion of the energy from plants to edible meat. Remember it takes 20 kilograms of grain to produce a kilogram of beef. Large amounts of greenhouse gas emitting fossil fuels are required to run farm machinery, to produce fertilizers and pesticides, and for transportation before the food goes to your local store. Ruminant animals also produce significant amounts of the greenhouse gas methane.

> Ruminant animals "chew their cud" after it has been partially digested in their first stomach. Many animals do this, including cows, sheep, and goats. But so do giraffes, camels, yaks, and llamas!

Your Turn 11.25   Local Produce and CO₂ Emissions

In the Skeptical Chemist 3.18 activity, we noted that a clean-burning automobile engine emits about 5 pounds of C (or ~2000 g) in the form of CO_2 for every gallon of gasoline it consumes.

a. How many pounds of CO_2 is this?
b. If you saved the equivalent of 1000 fewer miles each year by eating locally, approximately how many pounds of CO_2 emissions would this translate to? List any assumptions you make.
c. If you ate vegetarian one day a week, what would be the change in your personal CO_2 emissions?

"Eat Veggies". . . ? If this means eating leaves and plants in the broadest sense of the word, again yes. At the start of this chapter, we suggested that being a vegetarian was not an all-or-nothing proposition. Rather, you could cut back on the meat you eat without eliminating it entirely. One reason to do this is for your own health. In earlier sections, we described the benefits of eating plenty of fruits, grains, and vegetables. Now we offer another rationale for a plant-based diet; namely, to reduce greenhouse gas emissions. Some foods, most notably animals that are grain-fed, are costly in terms of the energy and water required to bring them to your table. So by reducing the meat you eat, you reduce greenhouse gas emissions.

Not only does food production affect the climate, but climate also affects food production. The two are intimately connected. For example, carbon dioxide is the raw material that plants use in the process of photosynthesis. Evidence exists that plants grow faster as the concentration of CO_2 in air increases. This may sound like a boon, but it has farmers and scientists alike worried. Although crops may grow faster, so do the weeds that compete with them for nutrients and moisture.

Another connection of climate change and food production relates to the melting of glaciers. People worldwide depend on melting snowpack for their water supply. For example, glaciers in the Himalayas provide the water that in turn ensures the food supply for over a third of the world's population. The regions most negatively affected by climate change are likely to be India and many areas of Africa and China. This adds insult to injury, as these areas already face significant challenges in feeding existing populations.

Other effects of climate change that could negatively affect food production include more drastic weather patterns such as storms and droughts, as well as expanding insect ranges and desertification. These factors present a daunting task to those who would try to predict the effects of climate change on food production. With the changes we are causing in the Earth's atmosphere, we are conducting a giant experiment on our planet. The effects of climate change on food production will only be known in the decades and centuries to come.

11.12 | Feeding a Hungry World

In the first section of this chapter, we asked if it were feasible to raise world grain production to meet the demands of a growing world population. As promised, we now return to this question. To get started, we note that Thomas Malthus and more recently entomologist Paul Ehrlich predicted that the human population of the Earth would out-strip food production. During the first 100 years after Malthus published his essay, the population of the Earth grew by 60% to 1.6 billion. During the next 100 years (the 20th century), a massive population explosion took place with the addition of 4.4 billion people to the planet. In 2005, the Food and Agriculture Organization of the United Nations (FAO), estimated that more than 852 million people worldwide are undernourished.

In 1798, Thomas Malthus wrote his *Essay on Population*. In 1978, Paul Ehrlich wrote the book *The Population Bomb*.

Remarkably, we have managed to increase the world food supply to meet the consumption needs of over 85% of our rapidly increasing population. How did we do this? In part, we simply planted more land. We also increased crop yields. Can this continue? No, as most of the world's biologically productive land is in use. Furthermore, in some places cropland is actually shrinking due to such factors as desertification and development. Other pressures on food production include diversion of foods to fuels (e.g., corn to ethanol) and diminished water for irrigation due to overpumping of aquifers and loss of water stored and released by glaciers.

What are our options? The 1940s saw the beginning of the Green Revolution. From 1950–1985, total worldwide grain production increased from about 600 million metric tons in 1950 to more than 1600 million metric tons by 1985. Agricultural productivity per acre of corn, rice, and wheat more than doubled. An amalgamation of factors was responsible for the revolution, including the use of fertilizers and pesticides, irrigation, mechanization, double cropping, and most importantly development of high-yielding strains of grain. Billions of people across the world have benefited. Since 1985 worldwide grain production has continued to increase but at a slower pace.

The Green Revolution is neither a panacea nor the ultimate answer. In spite of its successes, the Green Revolution is not universally applicable and has not been without costs. It works best in areas where water for irrigation is abundant, where money is available for supplemental fertilizers such as ammonia, urea, or nitrates, and where technological understanding and application exist. It also requires the use of large amounts of fossil fuels. The Green Revolution comes with significant environmental costs, as we brought to light earlier with the concept of a "foodprint."

Like water resources, food resources are not equally distributed around the globe. Many local and regional food shortages exist. Of the undernourished people worldwide, almost all live in developing countries. Figure 11.24 maps the location of the countries facing serious food shortages.

Food shortages have multiple and interrelated causes. Some relate to geography and climate, as certain areas of the world simply do not have enough arable land and adequate soil to produce sufficient food for their people. Prolonged droughts can reduce crop yields; floods can wash away soil and crops. Other causes are economic. For example, fertilizers may not be available because of their high cost. Furthermore, animals for working the fields, to say nothing of tractors, may be too expensive for farmers in some regions. Still other causes are political or military. Civil strife in some countries can block the flow of food and other agricultural products. Finally, food shortages are compounded by disease. For example, the HIV-AIDS pandemic has affected food production, marketing, and transportation.

The situation is changing in many parts of the world, as seen in Figure 11.25. To meet growing populations, global grain and cereal production has doubled over the past quarter century, and supplies of vegetables, fruits, milk, meat, and fish have also increased. In many developing countries, food production has generally not kept pace with growing populations. The exception has been Asia, where the rate of increase in per capita food production has been greater than that even in the developed world. In stark contrast, sub-Saharan Africa has experienced a long-term, continuing decline in per capita food production. One major reason for the increase in Asian crop yields has been greater use of fertilizers and pesticides. The application of both has often been criticized as being harmful to the environment, but the fact remains that millions have been saved from starvation, with fertilizers and pesticides playing a role.

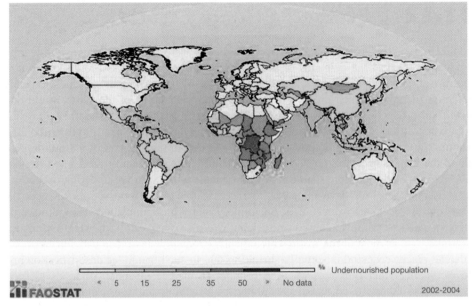

Figure 11.24

Countries facing food shortages and an undernourished population.

Source: The Food and Agriculture Organization of the United Nations, from *The State of Food and Agriculture,* 2002–2004.

Consider This 11.26   **In Defense of Food**

In his book, *In Defense of Food,* Michael Pollan writes: "Eat food. Not too much. Mostly plants."

a. Give three reasons why his words connect to your own health.

b. Explain why his words connect to the health of the planet.

Researchers estimate that, within 20 years, global demand for the world's three most important crops—rice, maize (a type of corn), and wheat—will have increased by 40% simply to keep pace with global food requirements. However agricultural productivity increases have slowed to only 1–2% per year, a rate that is insufficient to meet the

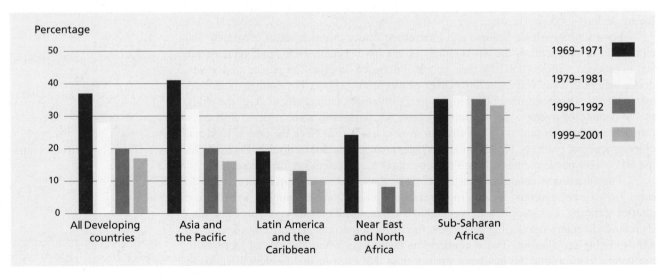

Figure 11.25

Percentage of population that is undernourished in developing countries, by region.

Source: The Food and Agriculture Organization of the United Nations, from *The State of Food and Agriculture,* 2002–2004.

growing population and the desire of so many to move up the food chain. Some believe that a second green revolution will be fostered by genetically engineered crops, but in many areas of the world there is significant opposition to transgenic foods. In the next and final chapter of this book, we tackle the chemistry of genes and genetic engineering.

> Look for more about genetic engineering for food production in Section 12.8.

Conclusion

Even though our individual tastes vary, our biological needs are much the same. We need carbohydrates and fats as our energy sources; fats for cell membranes, synthesis, and lubrication; proteins to build muscle and create the enzymes that catalyze the intricate chemistry of life; and vitamins and minerals to help make that chemistry happen. Many people with an abundant supply of food eat too much and don't pay enough attention to what they eat. Of course, the millions of hungry and starving people think of little else beyond how to find their next meal.

What and how much we eat surely affects our health! But it also affects the health of the planet. In this chapter we saw some of the environmental and human health consequences of our food choices. Some foods, such as most meats, disproportionately require water, grain, fuel, and land to produce. Other foods, such as corn, not only stress the farmer's field, but also stress ecosystems downstream.

How to sustainably meet all the dietary needs of the people on our planet, regardless of region or wealth, now and in the future, is one of the great challenges of our time. A knowledge of chemistry provides part of the solution, but individual and community choices, which are determined by economic, social, religious, and political factors, are key elements in this endeavor.

Chapter Summary

Having studied this chapter, you should be able to:

- Discuss the concept of "foodprint" and its implications on the health of the planet (11.1)
- Differentiate between malnutrition and undernourishment (11.2)
- Describe what makes a food "processed" (11.2)
- Describe the distribution of water, fats, carbohydrates, and proteins in the human body and some typical foods (11.2)
- Recognize lipids, carbohydrates, and proteins by their structural formulas (11.3, 11.5, 11.6)
- Identify sources of saturated and unsaturated fats and their significance in the diet (11.3)
- Show how fatty acids and glycerol can combine to form a triglyceride (11.3)
- Understand how hydrogenation leads to the formation of trans fats (11.4)
- Discuss sources of cholesterol and its significance in the diet (11.4)
- Describe the green chemistry associated with interesterification (11.4)
- Explain the differences between sugars, starch, and cellulose (11.5)
- Draw the general structural formula for an amino acid and explain how amino acids combine to form proteins (11.6)

- Discuss the importance of essential amino acids and their dietary significance (11.6)
- Explain the principle of protein complementarity (11.6)
- Describe the symptoms and cause of phenylketonuria (11.7)
- Discuss the effects of selected vitamins on human health, differentiating between fat-soluble and water-soluble vitamins (11.8)
- Discuss the connections between minerals (macrominerals, microminerals, and trace minerals) and human health (11.8)
- Explain why carbohydrates, fats, and proteins differ as energy sources (11.9)
- Identify and use basal metabolism rate (BMR) (11.9)
- Know appropriate resources for obtaining up-to-date dietary advice (11.10)
- Use resources to determine personal diet plans (11.10)
- Discuss how climate change influences food production and vice versa (11.11)
- Discuss the problems of undernourishment, identifying contributing factors to observed trends (11.12)
- Describe various strategies for sustainably feeding the world's growing population (11.12)
- Discuss the effects of the Green Revolution on food production and the health of the planet (11.12)

Questions

Emphasizing Essentials

1. The theme of this chapter is that what you eat affects both your health and the health of the planet. Give an example of each.

2. Select a profession of your choice, possibly the one you intend to pursue. Name at least one way in which a person in this profession could have a positive influence on food sustainability.

3. How, if at all, might each of these affect your "foodprint"?
   a. eating vegetarian two nights a week instead of eating meat
   b. walking to the supermarket and carrying your food home in your own bag

4. List a reason why it makes sense not to think of your diet in "all-or-nothing" terms, unless for some reason you have to.

5. List two reasons why eating less beef is perceived as a choice that is more sustainable, long-term, for the planet.

6. It takes about 10 pounds of grain to produce a pound of beef on the hoof. When cattle are slaughtered, only about 40% of each pound of the animal is eaten. How much grain is needed to produce a pound of edible meat?

7. What's the difference between malnutrition and undernourishment?

8. What is a processed food? Make a list of 10 foods, perhaps selecting from those in your kitchen, and indicate which ones are processed.

9. Inspect the ingredients label on a bag of potato chips. Would you classify this as a processed food? Explain.

10. a. What is a macronutrient?
    b. What role do macronutrients play in keeping us healthy?
    c. Name the three classes of macronutrients.

11. Water is not considered a macronutrient, but it clearly is essential to maintaining health. Name three roles that water plays in our bodies. *Hint:* You may want to refer to Chapter 5.

12. Consider this pie chart.

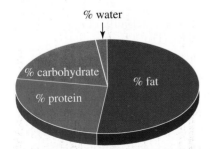

Based on the relative percentages of protein, carbohydrate, water, and fat given, is it more likely to represent steak, peanut butter, or a chocolate chip cookie? Justify your choice.

13. Of the foods listed in Table 11.1,
    a. identify the best sources of carbohydrates and arrange them in decreasing order.
    b. identify the best sources of protein and arrange them in decreasing order.
    c. identify two that should be avoided if you are controlling dietary intake of fat.

14. At a local restaurant, an 18-ounce steak is the manager's special. Use Table 11.1 to calculate the ounces of protein, fat, and water that would be consumed.

15. List the similarities and differences between edible fats and oils.

16. Lactic acid has the condensed structural formula of $CH_3CH(OH)COOH$.
    a. Draw the structural formula for lactic acid that shows all bonds and atoms.
    b. Is lactic acid a saturated or unsaturated compound? Explain.
    c. Is lactic acid a fatty acid? Explain.

17. Identify the fat or oil in Figure 11.9 that contains the highest number of grams per tablespoon of:
    a. polyunsaturated fat.
    b. monounsaturated fat.
    c. total unsaturated fat.
    d. saturated fat.

18. The label of a popular brand of soft margarine lists "partially hydrogenated soybean oil" as an ingredient. Explain the term *partially hydrogenated*. Why must the label report partially hydrogenated soybean oil rather than simply soybean oil?

19. Your friend wants to cut food costs and has learned that peanut butter is a good protein source. What additional information should your friend consider before eating peanut butter as a major dietary protein source? *Hint:* See Table 11.1.

20. Explain each term and give an example.
    a. monosaccharide
    b. disaccharide
    c. polysaccharide

21. What health problems can arise from regularly consuming excess dietary servings of carbohydrates?

22. Fructose, $C_6H_{12}O_6$, is a carbohydrate.
    a. Rewrite this chemical formula to demonstrate how a *carbohydrate* can be thought of as "carbon plus water."
    b. Draw the structural formula for any isomer of fructose.
    c. Do you expect different isomers of fructose to have the same sweetness? Explain.

23. Fructose and glucose both have the chemical formula $C_6H_{12}O_6$. How do their molecular structures differ?

24. Chemical names, especially for organic compounds, can give information about the structure of the molecules that these compounds contain. What does the term *amino acid* suggest about its molecular structure?

25. Certain amino acids are called "essential amino acids." Explain why.

26. Why should people who are phenylketonurics not drink beverages containing aspartame but can drink those sweetened with sucralose?

27. Explain the nutritional significance of the elements shaded on this periodic table.

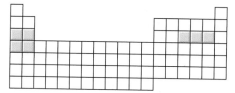

28. Why is it safer to take large doses of vitamin C than vitamin D?

29. Here is a schematic diagram of the 1991 food pyramid.

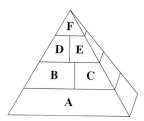

Correlate each lettered region on the pyramid with the colors of the wedges on the 2005 "Steps to a Healthier You" logo.

Concentrating on Concepts

30. A piece of sausage pizza contains items from several food groups. Identify each group and name the part of the pizza that is responsible for representing that particular food group.

31. Explain to a friend why it is impossible to go on a highly advertised "all organic, chemical-free" diet.

32. The data in this table are excerpted from a 2006 report by the Centers for Disease Control and Prevention.

Percentage of People Who Are Obese and Overweight in the United States

Years	1999–2000		2003–2004	
	Obese	Overweight	Obese	Overweight
all adults	30.5	64.5	32	66
men	27.5	67	31	71
women	33	62	33	62
children	14	28	17	34

a. When comparing the 1999–2000 with the 2003–2004 data, which group (men, women, or children) has shown the greatest percentage increase in being overweight? Explain.

b. One news source reported these data with the headline "U.S. obesity epidemic may be leveling off." Do you think these data support such a headline? Explain.

c. These data were gathered through in-person examinations that include actual weight measurements. Offer some possible reasons why this approach is considered more reliable than telephone surveys.

33. For each statement, indicate whether it is always true, may be true, or cannot be true. Explain your reasoning.

a. Plant oils are lower in saturated fat than are animal fats.

b. Lard is more healthful than butterfat.

c. There is no need to include fats in our diets because our bodies can manufacture fats from other substances we eat.

34. Experimental evidence suggests that some physiological effects of saturated fats, compared with unsaturated fats, may be caused by differences in packing the molecules. The hydrocarbon chains in saturated fatty acids can pack more tightly than those of unsaturated or polyunsaturated fatty acids.

a. Explain why saturated fatty acid molecules are able to pack more tightly than molecules of unsaturated or polyunsaturated fatty acids. *Hint:* You could use molecular models to help you see the effect single or double bonds can have on the ease of packing.

b. Explain why the extent of molecular packing influences the melting points of stearic, oleic, linoleic, and linolenic acids. See Table 11.2 for melting point values.

c. How is structural packing related to the undesirable health effects of trans fats?

35. Some people prefer to use nondairy creamer rather than real cream or milk. Some, but not all nondairy creamers, use coconut oil derivatives to replace the butterfat in cream. Is a person trying to reduce dietary saturated fats wise to use nondairy creamer such as these? Explain.

36. An avocado is a tropical fruit, but is unlike the tropical coconut or palm in its type of fat/oil composition. What is the primary type of fat found in avocados?

37. A mom, wanting to serve her family healthy nutritious foods, made these two comments about milk. Evaluate the logic of each.

a. "Milk contains a lot of sugar. Because of this, I don't serve it all that often."

b. Different types of milk—whole, 2%, and skim—contain different amounts of sugar. You need to check the labels carefully to be sure what you are getting."

38. A yellow packet of Splenda sugar substitute contains the compound sucralose. Use the resources of the web to answer these questions.

   a. How many calories does a packet of Splenda contain?

   b. Splenda's slogan is "Made from Sugar, So It Tastes Like Sugar." Comment on this. For example is it a correct statement or false advertising? Explain.

39. Substitutes have been developed for fat ("fake fats" such as Olean) and sugar (sucralose and aspartame, for example). Why have there not been attempts to develop a comparable substitute for protein?

40. Here is information about the sugar content of different foods.

Food Product	Sugar	Calories	Serving Size
Altoids, peppermint	2 g	10	3 pieces (2 g)
Ginger snaps	9 g	120	4 cookies (28 g)
Critic's Choice Tomato Ketchup	3 g	15	1 tbsp (13 g)
Del Monte Pineapple Cup	13 g	50	Individual cup (113 g)
Dr Pepper soft drink	40 g	150	1.5 cups
French Vanilla Coffee Mate	5 g	40	1 tbsp (15 mL)
Hostess Twinkies	14 g	150	1.5 ounces
LifeSavers, Wint O Green	15 g	60	4 mints (16 g)
Tropicana Homestyle Orange Juice	22 g	110	8 ounces (1 cup)
Snickers bar	29 g	200	2.1 ounces
Sunkist orange soda	52 g	190	1.5 cups
Wheatables crackers	4 g	130	13 crackers (29 g)

   a. Examine this list. Which item has the highest ratio of grams of sugar to the number of calories (g sugar/Cal) in one serving?

   b. The sugar content of some of these foods may surprise you. If so, which ones?

   c. Do you predict that the type(s) of sugar found in Dr Pepper would be the same as those found in Sunkist orange soda? In the cranberry juice or in the pineapple cup? Explain.

   d. The complete label for Wint O Green Lifesavers shows 16 g of total carbohydrates per serving, 15 g

of which is sugars. What might account for the other 1 g of carbohydrates?

41. Here is the label information from a popular brand of canned chicken noodle soup.

   Serving size: 1/2 cup (4 ounces; 120 g)

   Servings per container: about 2.5

   Amount per serving

   Calories 75

   Calories from fat 25

	Amount	% Daily Value
total fat	2.5 g	4
saturated fat	1.5 g	8
cholesterol	20 mg	7
sodium	970 mg	40
total carbohydrates	9 g	3
dietary fiber	1 g	4
sugars	1 g	
protein	4 g	
vitamin A		15
vitamin C		2
calcium		2
iron		4

   a. Analyze this information to see if the soup conforms to 2005 *Dietary Guidelines for Americans.*

   b. Is the serving size recommended on the label a reasonable size for you? Explain.

   c. What effect would changing the serving size have on your answer to part **a**?

42. When the USDA makes a decision to change dietary recommendations, it also changes the way the information is visually displayed to consumers. The earlier pie chart was replaced by a food pyramid, and now by a more symbolic pyramid and an interactive website. What are the advantages and disadvantages of each approach?

43. According to one USDA study, nearly 40% of the food that the average American eats each day consists of milk or dairy products. Would such a diet be possible and still meet the 2005 *Dietary Guidelines for Americans*?

44. Consider this structure for one form of vitamin K. Do you expect it to be water-soluble or lipid-soluble? Explain.

45. Consider the structure for riboflavin, one of the B vitamins found in leafy green vegetables, milk, and eggs. Why is it somewhat safer to take large doses of vitamin B than vitamin D?

46. American diets depend heavily on bread and other wheat products. A slice of whole wheat bread (36 g) contains approximately 1.5 g of fat (with 0 g saturated fat), 17 g of carbohydrate (with about 1 g of sugar), and 3 g of protein.

    a. Calculate the total calorie content in a slice of this bread.

    b. Calculate the percent calories from fat.

    c. Do you consider bread a highly nutritious food? Explain your reasoning.

47. Revisit the definition of sustainability provided in Chapter 0. With this in mind, write a one-page essay discussing how feeding the world now and in the future can be done sustainably.

Exploring Extensions

48. Use the lock-and-key model discussed in Section 10.5 to offer a possible explanation as to why individuals who suffer from lactose intolerance can digest sugars such as sucrose and maltose, but not lactose. Use the web to find the structure of lactose.

49. The composition of a fast-food meal is given here. Do calculations to determine whether the meal eaten meets the guideline that only 8–10% of total calories should come from saturated fats.

	Cheeseburger	French Fries	Shake
calories	330	540	360
calories from fat	130	230	80
total fat (g)	14	26	9
saturated fat (g)	6	4.5	6
cholesterol (mg)	45	0	40
sodium (mg)	830	350	250
carbohydrates (g)	38	68	60
sugars (g)	7	0	54
proteins (g)	15	8	11

50. In September of 2004, the U.S. FDA gave "qualified health claim" status to two omega-3 fatty acids, stating that "supportive but not conclusive research shows that consumption of these omega-3 fatty acids may reduce the risk of coronary heart disease." Find the structure and make a line-angle drawing of any omega-3 fatty acid. What are the food sources for this fatty acid? What is the current thinking about the health benefits of omega-3 fatty acids?

51. Compare these two pie charts for the percentage of macronutrients in soybeans and wheat.

    a. Explain why the World Health Organization has helped develop soy rather than wheat-based food products for distribution in parts of the world where protein deficiency is a problem.

    b. Suggest cultural reasons why soy might be preferable to wheat for some areas of the world.

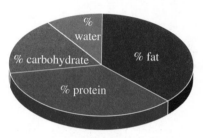

Soybeans

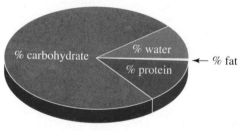

Wheat

52. To use the *Dietary Guidelines for Americans, 2005,* the consumer must know what constitutes a reasonable serving size. Investigate what constitutes reasonable serving sizes for one of the food groups, and then prepare a poster with your results to share with others who are investigating the reasonable serving sizes for other food groups. Which of the serving sizes surprised you?

53. Not everyone considers milk nature's "perfect food." Compare and contrast the viewpoints of the dairy industry with groups that work against the dairy industry. What are some of the specific benefits attributed to milk, and what are some of the reasons that milk has been called "nature's not so perfect food"?

12 Genetic Engineering and the Molecules of Life

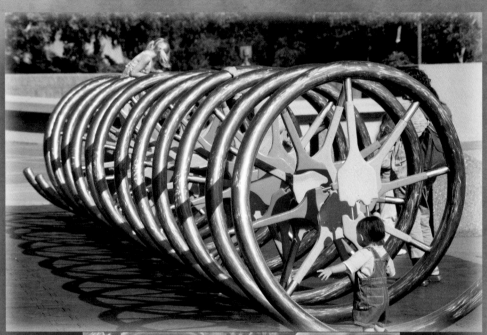

A DNA sculpture becomes an intersection of art, science, and play. Lawrence Hall of Science, Berkeley, CA.

"No branch of science has created more acute or more subtle and interesting ethical dilemmas than genetics. . . . it is genetics that makes us recall, not simply our responsibilities to the world and to one another, but our responsibilities for how people will be in the future. For the first time we can begin to determine not simply who will live and who will die, but what all those in the future will be like."

Justine Burley and John Harris, Eds.
Companion to Genethics, 2002

Have you thought of a future in which your food was tastier and more nutritious? Of a future in which growing crops was easier and yields were higher? Imagine a world in which our fields produce better biofuels, our farms create our drugs and vaccines, and our bacteria clean our wastewater, all for us.

Geneticists may have the power to create this utopia. Advocates of genetic engineering argue that using our knowledge of genetics, we can solve world hunger, produce less expensive drugs, and have a cleaner environment. But is tampering with genes worth the risks? Opponents point to superbugs, super-resistant weeds, and a sharp decrease in the biodiversity of the planet. Perhaps, this price is too high, even for utopia.

In part, the debate over genetic engineering stems from the complexities of the science involved. If a single cell is complex, whole organisms are *really* complex, and ecosystems are complex beyond description. If we make a microscopic genetic change in one seed, we may solve a problem, perhaps even a global problem. But that small change may create additional unforeseen changes in entire ecosystems. The bottom line is that it is difficult to predict all of the consequences of our actions, especially in the genetic arena.

But the debate also stems from social and ethical concerns. We humans have a responsibility to act wisely, both for current and future generations. A reasonable premise on which to decide whether to act or not to act is to "do no harm." But our needs are great and our societal problems are pressing. Therefore we must also consider whether our *inaction* might result in more harm than good. What are your thoughts about genetic engineering and, more specifically, about genetically modified or "GM" food? The next activity allows you to explore and record your point of view.

Consider This 12.1   Your Opinion of GM Foods

One way or another, you probably have heard of GM (genetically modified) food. Even if you have very little knowledge, you still may be able to answer these questions.

a. Given a choice between a GM food and one that was not, do you have a preference as to which one you would eat? Explain.

b. What circumstances, if any, would cause you to rethink your point of view? List at least two. Save your answers so that you can revisit them at the end of the chapter.

This chapter is about the chemistry of life. Can we and should we control this chemistry? If so, to what extent and with which safeguards in place? Once again, we can offer you good questions, but not always simple or immediately satisfying answers.

12.1 | Stronger and Better Corn Plants?

Ours is a wet planet. Oceans, rivers, lakes, and ice cover over 70% of our planet's surface. This leaves less than a third of the surface for the forests, grasslands, deserts, mountains, and fields that we know as land. Recent satellite data tell us that about 40% of this land is farmed. Why have we converted forests and grasslands to farm fields? In large part, we do this to feed a hungry planet. Wheat is the number one crop worldwide, followed closely by corn.

In an attempt to find better and easier-to-grow corn, scientists have genetically modified it. The same is true for soybeans, tomatoes, potatoes, cotton, and papayas, as all of these exist in genetically modified forms for one reason or another. In this section, we'll focus on corn, as it is the food that many of us consume in large amounts directly through high-fructose corn syrup, corn meal, or cornstarch, as well as indirectly by eating animals that were cornfed.

See Chapter 5 for a thorough discussion of our wet planet.

In 1700, less than 10% of the land was used for agriculture.

Corn was first cultivated in the Americas, perhaps as long as 7000 years ago.

Figure 12.1

(a) A cornfield, not quite as uniform as it may appear. (b) A European corn borer busily at work.

To get started, examine Figure 12.1a. A cornfield is never simply a field of corn plants, although it may appear so. Rather, cornfields are small unbalanced ecosystems. From the perspective of the corn plant, the cornfield is home, a place with nutrients in the soil, the sunshine above, and lots of water. From the perspective of the weeds, the cornfield is a clear stretch of fertile land ready to be taken over. The weeds grow quickly, take up the nutrients, and some may block the sunlight. From the perspective of the insects, a cornfield is a nice home, an incubator for the young, and a very tasty meal (Figure 12.1b). Many insects have evolved to take advantage of the cornfields we plant.

If the corn is to grow and thrive, farmers must expend time, effort, money, and fuel to actively nurture the plants and protect them against insects and weeds. In the process, they may inadvertently damage ecosystems, both locally and further down the watershed. For example, consider what happens when farmers spray insecticides and herbicides to control insects and weeds. Using pesticides not only is expensive, but also, without enough care in the choice or application, these chemicals may damage the plants and the surrounding ecosystem. The combination of expense and trouble for the farmer with environmental risk seems difficult to justify, but growing corn is extremely profitable. It is sold for use as food, animal fodder, or biofuel. With such a wide range of applications, we are unlikely to see an end to corn farms. Clearly, everybody would benefit if growing corn were easier and less harmful to the environment.

Are you curious about what it takes to grow corn? Two recent college graduates were. In 2004, they joined with a producer to create the film, *King Corn*.

Recall the green chemistry principle: It is better to use materials that degrade into innocuous products at the end of their useful life. Many pesticides fail to meet this criterion.

Consider This 12.2   *King Corn*

a. Check the *King Corn* website and you will find what two recent college graduates did: ". . . with the help of friendly neighbors, *genetically modified seeds*, *nitrogen fertilizers*, and *powerful herbicides*, they planted and grew a bumper crop of America's most productive, most subsidized grain on one acre of Iowa soil." Explain how each of the italicized terms was connected to growing the corn. The textbook's website has a link to *King Corn*.
Hint: Revisit Section 6.9 on nitrogen.

b. This film also connects with the content of the previous chapter on nutrition. Give one example of a connection, either from viewing the film or from reading the content of the website.

What if you could offer farmers a bag of seed to grow corn plants that are resistant to insects and weeds? Well, it turns out that you can. Two very common genetic modifications in corn can provide the farmer with a crop that is resistant to both insects, such as the European corn borer or Western corn rootworm, and an herbicide, such as Roundup. The corn plant produces its own insecticide, allowing the farmer to spray less pesticide. And the corn plant is resistant to one general herbicide, meaning the farmer can spray that herbicide rather than others that deposit more toxins in the watershed.

How can we get the corn plant to resist the herbicide or produce its own insecticide? We do this by "teaching" it to make some new chemicals. Inside each cell of the corn plant is the complete set of instructions, a guidebook if you like, on how to grow and reproduce. The guidebook passes from one generation to the next, often completely unchanged. This guidebook, termed the **genome,** is the complete set of inheritable traits of an organism.

These inheritable characteristics are divided into short sections of instructions to produce specific reactions, chemicals, or events in the cell. These specific pieces are the basic units of heredity, **genes,** short pieces of code for the production of proteins. A change within a gene changes an inheritable trait. For a corn plant, a change in the gene for color may switch the corn kernels from light yellow to white. But small changes within a gene are not enough to make the plant produce peas instead of corn or even to produce a new chemical, such as an insecticide. We need a more dramatic change.

What we really need to do is to insert a whole new set of instructions (that is, a gene) in the genome of the corn plant. Rather than create these instructions ourselves, instead we search for another organism that already has the instructions we want. For the European corn borer (Figure 12.1b) and Western corn rootworm, our search brings us to a protein that is toxic to these insects but is considered safe for humans. The protein occurs naturally and can even be used in organic agriculture. A group of small organisms, a soil bacterium called *Bacillus thuringiensis,* already has the instructions to make this protein. By taking a gene out of the bacterium and inserting it into the corn plant, we create corn plants that can produce an insecticidal protein.

The most common strain of corn, B73, contains over 32,000 genes in its genome, more genes than are in human DNA. Researchers spent four years cataloging the massive amount of information so that we may understand the genes behind higher yields, disease-resistance, and drought-resistance in various crop strains.

Often called *Bt*, these bacteria produce *Bt* toxins, a wide variety of proteins that perform the same way against different insects.

Your Turn 12.3   Proteins and Carbohydrates

a. What is a protein? We just used the term *insecticidal protein*. Describe the features of a protein molecule.
   Hint: You can review proteins in Sections 9.7 and 11.7.
b. What is a carbohydrate? Corn primarily is composed of carbohydrates. Describe carbohydrates at a molecular level.

The corn plant, the bacterium, and the cornfield are more complex than you might think. What exactly are we modifying when we genetically modify something? We turn to this topic in the next section.

12.2 | A Chemical That Codes Life

With each passing second, the corn plant is host to millions of chemical reactions. Some of these reactions decompose compounds; others synthesize them. Some reactions transfer chemical signals; others process them. Some reactions release energy; others utilize it. One very special chemical lies at the heart of this dazzling chemical complexity.

We require a lot from this very special chemical. As cells grow and multiply, this chemical must replicate itself without error. It must remain largely unharmed and unchanged by its environment. This one chemical must organize and securely store a lot of information. This information is context-sensitive, as some reactions are always going; others start and stop depending on specific signals. In short, we need a highly advanced database in chemical form.

The chemical we have just described is **deoxyribonucleic acid,** or **DNA,** the biological polymer that carries genetic information in all species. DNA is the template of life, containing all of the biochemical information to make a full corn plant. DNA can replicate easily, transfer information, and respond to feedback within the cell.

Like the corn plant, you have a special template of life written on a tightly coiled thread of DNA. Unraveled, the DNA in *each* of your cells is about 2 meters (roughly 2 yards) long. If all of the DNA in all 100 trillion of your cells were placed end to end, the resulting ribbon would stretch from here to the Sun and back, more than 600 times! But as you will soon discover, this astronomical figure is far from the most astounding feature of this amazing molecule.

Any strand of DNA—long or short—consists of three fundamental chemical units: nitrogen-containing bases, deoxyribose sugars, and phosphate groups. All are illustrated in Figure 12.2.

DNA contains not one nitrogen-containing base, but four. Each one differs slightly from the others. The larger bases, adenine (A) and guanine (G), have a six-membered ring and a five-membered ring of atoms fused together. The smaller bases, cytosine (C) and thymine (T), have only a six-membered ring. Notice that all of these compounds have nitrogen atoms embedded in their rings, leading to the name "nitrogen-containing bases." These bases also contain oxygen atoms that can participate in hydrogen bonding, as we will see in the next activity.

The estimates for the number of cells in your body range from 50 to 100 trillion. The bacterial cells outnumber the human cells by an estimated 10 times.

Revisit Section 6.2 to review our first nitrogen-containing base, ammonia.

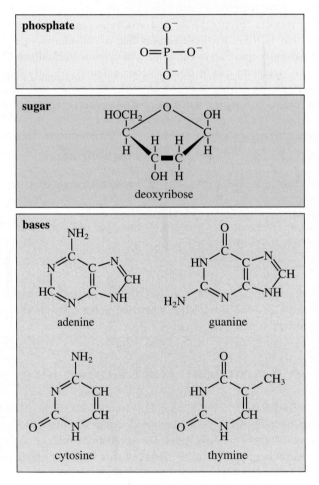

Figure 12.2

The components of deoxyribonucleic acid, DNA.

 Figures Alive! Visit the textbook's website to learn more about the structure of DNA.

Your Turn 12.6   **Another Nucleotide**

Analogous to Figure 12.3, draw the structural formula for the nucleotide containing cytosine.

12.3 | The Double Helix of DNA

DNA is a gorgeous molecule. If you look at the opening photos of this chapter, you find a sculptor's rendition of DNA with two silvery strands curving in a gentle spiral, both elegant and simple. Hidden within the structural simplicity is a powerful chemical code for information. The structure, both how the nucleotides are covalently bonded and how the strands pack together, contributes to the function of DNA. Understanding how DNA performs its many functions required solving the puzzle of the DNA structure.

To see the structure of such a small object, scientists turned to the technique of X-ray diffraction. This technique has revolutionized our understanding of molecular structures and chemistry by helping us visualize chemical shapes. **X-ray diffraction** is an analytical technique in which a crystal is hit by a beam of X-rays to generate a pattern that reveals the positions of the atoms in the crystal. The X-ray photons interact with the electrons of the atoms in the crystal and are diffracted, or scattered. The crucial point is that the X-rays are only scattered at certain angles that are related to the distance between atoms, and that information can be used to determine the structures of a wide variety of crystalline materials. The X-ray diffraction pattern of a DNA fiber was obtained in late 1952 by the British crystallographer Rosalind Franklin (Figure 12.5).

James Watson and Francis Crick (see Figure 12.5) combined Franklin's X-ray diffraction data with earlier chemical and biological analyses to finally create a model of the structure of DNA. The pattern in Franklin's diffraction photograph was consistent with a repeating helical arrangement of atoms, similar to a loosely coiled spring. Moreover, the X-ray photographs contained evidence of a repeated pattern separated by 0.34 nm within a DNA molecule. The Watson–Crick model explained this repetition by twisting the strands of DNA into a **double helix,** a spiral consisting of two strands that coil around a central axis as shown in Figure 12.6. The base pairs

Return to Section 2.4 to find X-rays on the electromagnetic spectrum.

(a)   (b)   (c)   (d)

Figure 12.5

James Watson **(a)**, Francis Crick **(b)**, and Maurice Wilkins **(c)** shared the 1962 Nobel Prize in Physiology or Medicine for their contributions to the structural understanding of DNA. Though crystallography data from Rosalind Franklin **(d)** was vital, she died in 1958 and was not eligible for the Nobel Prize in 1962.

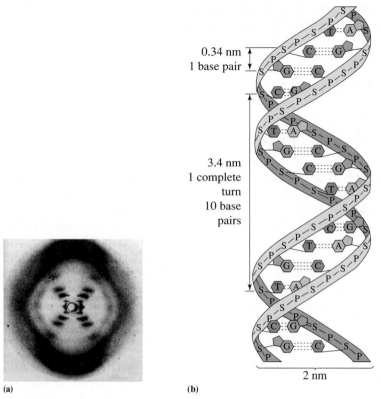

0.34 nm
1 base pair

3.4 nm
1 complete
turn
10 base
pairs

2 nm

(a)   (b)

Figure 12.6

(a) Rosalind Franklin's X-ray diffraction photo of a hydrated DNA fiber. The cross in the center is indicative of a helical structure, and the darkened arcs at the top and bottom are due to the stack of base pairs. **(b)** A model of DNA with P = phosphate group; S = the sugar, deoxyribose; and the bases A = adenine; T = thymine; C = cytosine; G = guanine. The sugar and phosphate groups alternate on the backbone, and the four bases attach to this backbone.

are parallel to each other, perpendicular to the axis of the DNA molecule, and separated by 0.34 nm, the same distance calculated from the diffraction pattern. In addition, Franklin's results also suggested a second repetition pattern separated by 3.4 nm. Watson and Crick took this to be the length of a complete helical turn consisting of 10 base pairs.

The letters in our English words have directionality. For example, the words *ward* and *draw* have the same letters in the same order, but the meaning is different because the direction is different. The same is true of the DNA polymer, with the bases being analogous to letters. For example, the base string TAC does not have the same meaning as CAT. The structure of the DNA backbone defines the directionality. Look carefully at the alternating phosphate and deoxyribose groups in the DNA backbone (see Figure 12.4) to see how the deoxyribose ring connects directly to the phosphate below it, and the one above it links through another carbon. The different types of chemical bonds make one direction different from the other. When the two strands of the DNA double helix come together, one strand must run in the opposite direction from the other.

Early chemical analyses showed that the nitrogen-containing bases in DNA come in pairs. No matter the species, the percent of A almost exactly equals the percent of T (Table 12.1). Similarly, the percent of G is essentially identical to the percent of C. The structural model of DNA validated these rules. Adenine and thymine bases fit almost perfectly together, like pieces in a jigsaw puzzle. A closer look shows these two bases linked by two hydrogen bonds (Figure 12.7). Similarly, cytosine and guanine are linked by three hydrogen bonds. This base pairing is the molecular basis underlying both the structure and much of the function of DNA. To repeat: A pairs with T, and G pairs with C.

Palindromes are the exception and can be read in either direction. For example, the word RACECAR reads the same in either direction.

The base pairing rules are termed *Chargaff's rules* after the discoverer, Austrian chemist Erwin Chargaff.

Table 12.1	The Percent Base Compositions of DNA for Various Species				
Scientific Name	Common Name	Adenine	Thymine	Guanine	Cytosine
Homo sapiens	human	31.0	31.5	19.1	18.4
Drosophila melanogaster	fruit fly	27.3	27.6	22.5	22.5
Zea mays	corn	25.6	25.3	24.5	24.6
Neurospora crassa	mold	23.0	23.3	27.1	26.6
Escherichia coli	bacterium	24.6	24.3	25.5	25.6
Bacillus subtilis	bacterium	28.4	29.0	21.0	21.6

Source: From I. Edward Alcamo, *DNA Technology: The Awesome Skill,* 2E © 2000 The McGraw-Hill Companies, Inc. All rights reserved. Reprinted with permission.

Your Turn 12.7   Complementary Base Sequences

Adenine and thymine are said to be complementary bases. So are cytosine and guanine. In both cases, the bases form hydrogen bonds when they pair. Using one-letter codes, write out the base sequences that are complementary to each of these.

a. ATACCTGC

b. GATCCTA

Answers

a. TATGGACG

b. CTAGGAT

The structure of DNA and the puzzle-piece pairing of its nucleotides inspired another vital discovery. One side of the DNA strand contains all the information required to generate its partner strand! Thus a single strand of DNA can guide the generation of its complement as well as correct it. **Replication** is the process of cell reproduction in which the cell must copy and transmit its genetic information to its progeny. The process is well understood and is diagrammed in Figure 12.8.

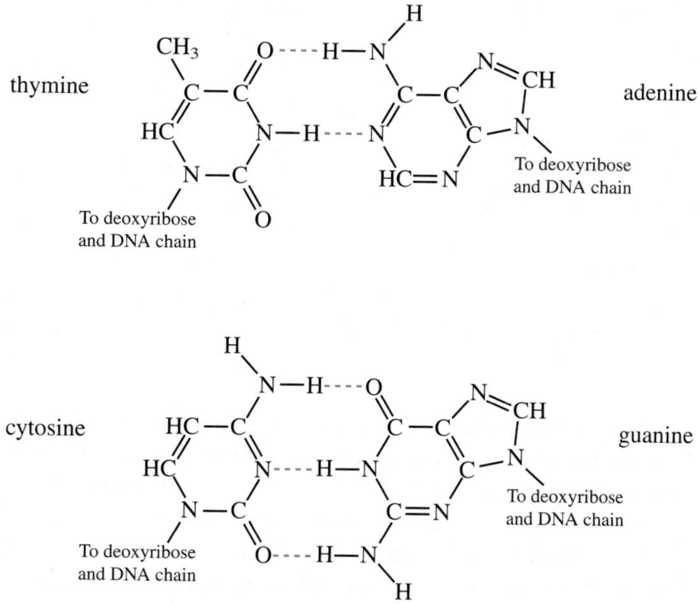

Figure 12.7

Base pairing of adenine with thymine and cytosine with guanine in DNA. Chemical bonds are solid black lines, and the hydrogen bonds are dashed red lines.

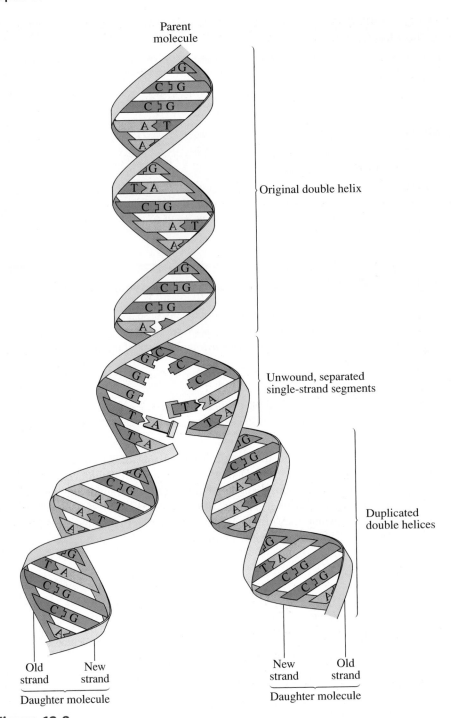

Parent molecule

Original double helix

Unwound, separated single-strand segments

Duplicated double helices

Old strand	New strand

Daughter molecule

New strand	Old strand

Daughter molecule

Figure 12.8

 Diagram of DNA replication. The original DNA double helix (*top portion of figure*) partially unwinds, and the two complementary portions separate (*middle*). Each of the strands serves as a template for the synthesis of a complementary strand (*bottom*). The result is two complete and identical DNA molecules.

Before a cell divides, the double helix rapidly, but only partially, unwinds. This results in a region of separated strands of DNA, as pictured in the middle portion of Figure 12.8. Free nucleotides in the cell are selectively hydrogen-bonded to these two single strands that serve as templates for a new DNA molecule: A to T, T to A, C to G, and G to C. Held in these positions, the nucleotides bond together by the action of an enzyme, a biological catalyst. By this mechanism, each strand of the original DNA generates a complementary copy of itself. The original strand and its newly synthesized complement coil to form a new daughter molecule identical to the first. Similarly, the other separated strand of the original molecule twines around its new partner, forming

Return to Section 10.4 to review enzymes.

another new daughter molecule. Thus, where there was originally one double helix, now there are two identical copies.

 Consider This 12.8   **A Gorgeous Molecule**

Revisit the photograph that opens the chapter. It shows a sculpture of DNA that people can climb on. As a piece of art, it represents some parts of the DNA molecule better than others.

   a. List three disadvantages of this DNA representation. What chemical details are omitted? What information is lost?
   b. Now list three advantages. What information is highlighted? What is gained?
   c. Find another artistic rendition of DNA, possibly on your campus or on the web, and repeat parts **a** and **b**. Cite your sources.

 Consider This 12.9   **DNA Replication**

To see Figure 12.8 in motion, visit Figures Alive! at the textbook's website. You will find several interactive activities relating to DNA, and the final one shows the process of replication. The replication animation was difficult to script and we invite you to critique it.

   a. Similar to Figure 12.8, the Figures Alive! DNA shows no atoms. List its strengths and weaknesses.
   b. Play and replay the DNA replication animation so that you can observe both what it shows and what it fails to show. Make a list of both points.

In most organisms, the DNA is left not extended as a double helix, but must be very tightly coiled. This not only saves space, but also it further organizes and protects the genetic information. The coiling is carefully regulated so that small portions of DNA can be accessed when specific stored information is needed. This complete set of genetic information is packaged into **chromosomes,** rod-shaped, compact coils of DNA and specialized proteins packed in the nucleus of cells.

Your Turn 12.10   **Is Your DNA Doin' the Twist?**

The distance between base pairs in a molecule of DNA is 0.34 nm.

   a. Calculate the length (in centimeters, cm) of the shortest human chromosome, one that consists of 50,000,000 base pairs.
   b. If this is the length of the uncoiled DNA molecule, what does this imply about the organization of DNA in the chromosome?

Answers

   a. $\dfrac{0.34 \text{ nm}}{1 \text{ base pair}} \times \dfrac{1 \text{ m}}{1 \times 10^9 \text{ nm}} \times \dfrac{1 \times 10^2 \text{ cm}}{1 \text{ m}} \times \dfrac{5 \times 10^7 \text{ base pairs}}{1 \text{ chromosome}} = \dfrac{1.7 \text{ cm}}{1 \text{ chromosome}}$

   b. This is a small distance (about two thirds of an inch), but a typical human cell is only 10 μm in diameter. The best way to fit 50 million base pairs into such a small space is to somehow pack them tightly.

As the nucleus splits and the cell divides into two daughter cells, each chromosome is rapidly duplicated. A complete set of chromosomes is incorporated into each daughter cell. This process is repeated again and again, so that each cell in an organism contains an identical set of chromosomes. In rapidly dividing cells, such as skin and cancer, the process occurs frequently. These cells are more susceptible to collecting and passing on DNA damaged by ionizing radiation, free radicals, and many chemical agents.

Return to Chapter 7 to review ionizing radiation and free radicals.

12.4 | Cracking the Chemical Code

Remember all of the complex chemistry happening in your cells every minute—the molecule of DNA organizes a lot of information. The billions of base pairs repeated in every corn cell provide the blueprint for producing one corn plant. The base pairs are ordered into specific sequences and grouped, sometimes into genes, to code for the production of proteins. Other information is present in the DNA, but our understanding of how it is used is in its infancy.

Although the information is carried in DNA, it is expressed in other, smaller, molecules. The best understood are proteins. Proteins are found throughout our bodies, in skin, muscle, hair, blood, and the thousands of enzymes that regulate the chemistry of life. By directing the synthesis of proteins, DNA can dictate many of the characteristics of the organism.

> Section 11.7 defined proteins in the context of the foods we eat. Section 9.6 described proteins as polymers (polypeptides or polyamides).

Proteins are large molecules formed by the linking of amino acids. Recall that the 20 amino acids that commonly occur in proteins can be represented by this general structural formula that we reproduce from Chapter 11.

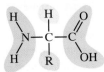

> Refer back to Figure 11.16 in Chapter 11 for more about the general structure of an amino acid.

The amine or amino group is $-NH_2$, the acid group is $-COOH$, and R represents a side chain that is different for each of the 20 amino acids. In a condensation reaction, the $-COOH$ group of one amino acid reacts with the $-NH_2$ group of another. In this process a peptide bond is formed and a molecule of H_2O is formed. When many amino acids are connected, the result is a protein; that is, a polymer built from amino acid monomers. We also can describe a protein as a long chain of amino acid residues. **Amino acid residue** is the term used for amino acids that have lost a $-H$ atom and an $-OH$ group in the process of bonding together.

> The discovery of the molecular code for genetic information arguably is history's most amazing example of cryptography, the science of writing in secret code.

The information in a sequence of DNA nucleotides translates via a code into a sequence of specific amino acids in a protein. The code cannot be a simple one-to-one correlation between bases and amino acids. There are only four bases in DNA. If each base corresponded to an individual amino acid, DNA could encode for only four amino acids. But 20 amino acids appear in our proteins. Therefore, the DNA code must consist of at least 20 distinct code "words," each word representing a different amino acid. And the words must be made up of only four letters—A, T, C, and G—or, more accurately, the bases corresponding to those letters.

Some simple statistics can help us determine the minimum length of these code words. To find out how many words of a given length can be made from an alphabet of known size, raise the number of letters available to a power n, corresponding to the number of letters per word.

$$words = (letters)^n$$

For example, using 4 letters to make two-letter words generates 4^2 or 16 different words. Thus, DNA bases taken in pairs (akin to 2 letters per word) could only code for 16 amino acids, insufficient to provide a unique representation for each of the 20 amino acids. So we repeat the calculation and assume that the code is based on 3 sequential bases or, if you prefer, three-letter words. Now the number of different triplet-base combinations is 4^3, or $4 \times 4 \times 4 = 64$. This system provides more than enough capacity to do the job.

Your Turn 12.11   Quadruplet-Base Code

Suppose that the DNA code used four sequential base pairs instead of a triplet-base code. How many different four-base sequences would result?

Answer

$4 \times 4 \times 4 \times 4 = 4^4 = 256$ different four-base sequences

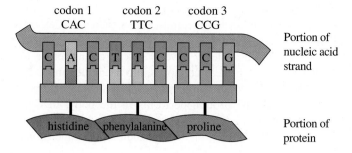

Figure 12.9

A nine-base nucleic acid sequence showing three codons.

The three letter groupings of nucleotides are the basis of the information transfer from DNA to proteins. Each grouping, or **codon,** is a sequence of three adjacent nucleotides that either guides the insertion of a specific amino acid or signals the start or end of protein synthesis. If you were to use the letters A, T, C, and G in a game of Scrabble, you could generate 64 different three-letter combinations. A few, CAT, TAG, and ACT, for example, make sense. Most are like AGC, TCT, and GGG and are meaningless—at least in English. Nature does far better than that; 61 of the 64 possible triplet codons specify amino acids. Thus, the codon sequence CAC in a DNA molecule signals that a molecule of the amino acid histidine should be incorporated into the protein, TTC codes for phenylalanine, and CCG stands for proline. The three-base sequences that do not correspond to amino acids are signals to start or stop the synthesis of the protein chain. An example of a nine-base nucleic acid segment and how it codes for three amino acids is shown in Figure 12.9.

Our calculations showed that three-letter groupings are the minimum necessary to cover the 20 amino acids, but did not show us how all 64 codons are used. The code has redundancy. Many amino acids have more than one codon. For example, leucine, serine, and arginine have six codons each. Also, three different codons tell protein synthesis to "stop." On the other hand, two amino acids (tryptophan and methionine) and the signal to start protein synthesis are represented by only a single codon.

Your Turn 12.12   Duplicate Codons

Suggest some advantages of a genetic code in which several codons represent the same amino acid.

The genetic code is identical in all living things. With only a handful of exceptions, the instructions to make people, bacteria, and trees are written in the same molecular language of those 64 codons. In the genetic code, we have a Rosetta Stone to translate any genetic sequence from any organism. The significance of this statement may not be immediately apparent. Looking back to our original example of genetically engineered corn, this Rosetta Stone means that the gene sequence for the *Bt* toxin makes the same bug-destroying protein in the original bacteria *and* in the corn plant.

The markings on the Rosetta Stone were in several languages and helped to decipher Egyptian hieroglyphs.

12.5 | Proteins: Form to Function

Proteins are polymers. Admittedly they don't much resemble the clear polymer PET that may hold your favorite soft drink. Similarly, they don't seem to have much in common with the tough polypropylene that is used to make carpets. But nonetheless, proteins are big molecules built from little ones.

More specifically, proteins are polyamides. Like nylon, they are built by the chemical reaction of carboxylic acids and amines. Unlike nylon, though, a protein is built from 20 different amino acid monomers. Comparing proteins to nylon may leave you with the image of proteins as long strands, rather than complex three-dimensional molecules. This image is far too simple. Proteins exist in a complicated environment—the cell. Just like necklaces in a messy jewelry box, proteins refuse to stay extended and neat. Unlike jewelry, each protein collapses into a unique, and often very specific, three-dimensional form. The final shape may look a mess, but that exact shape is necessary for the chemistry the protein performs in the body. In Chapter 9, we discussed proteins as polymers. In Chapter 11, we discussed them as nutrients. In this chapter, we discuss the function of proteins and their three-dimensional shapes.

Your Turn 12.13   How is Hamburger Like Nylon?

Take a moment to refresh your knowledge about two polyamides: the nylon from a sports jersey and the protein found in hamburger.

a. What functional group do nylon and meat proteins have in common?
b. Nylon is usually synthesized from two types of monomers; in contrast, proteins are synthesized from one type of monomer. What functional group(s) do the monomers for each contain?
c. Are nylons and proteins addition polymers or condensation polymers?

Answer
b. Typically, nylon is synthesized by the reaction of a monomer containing two carboxylic acids with a monomer containing two amine groups. For example, refer back to equation 9.7. In contrast, proteins are synthesized from a monomer (an amino acid) that contains one carboxylic acid and one amine group.

We begin our discussion of protein shapes with the **primary structure** of a protein; that is, the unique identity and sequence of the amino acids that make up each protein (Figure 12.10). Primary structure is the first and most basic identifier of a protein, the list of amino acids read over the length of the polymer. Knowing that a short protein contains 3 valines (val), 2 glutamic acids (glu), and 1 histidine (his) may tell you the molecular weight and a few other details, but is not sufficient to specify a protein and explore its shape and function. The order and sequence of the amino acids matters. For example, val-glu-val-his-glu-val is a different protein from val-val-val-his-glu-glu. These short peptides behave differently as well!

Recall that each amino acid has a side chain group. This side chain interacts with other side chains or the molecules around and inside of the protein. Side chains can attract and "lock" together, and in so doing, they hold a protein in a particular overall shape. The order and identity of the amino acids defines how and where those side chain links can form. Each amino acid plays a role; changing just one can change the shape and, as a result, the function of a protein.

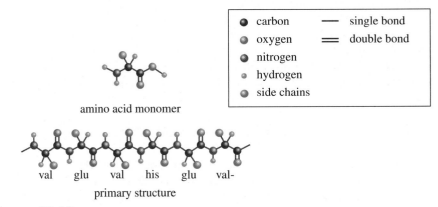

● carbon	— single bond
○ oxygen	═ double bond
● nitrogen	
◦ hydrogen	
○ side chains	

amino acid monomer

val   glu   val   his   glu   val-

primary structure

Figure 12.10

Representation of the primary structure of a protein.

Luckily, you do not need to memorize all 20 amino acids. We will group the side chains into two categories: polar (either charged or neutral) or nonpolar. Like oil and water, nonpolar and polar side chains tend to separate. In water (the typical environment for a protein), the polar side chains can stay in the water and nonpolar side chains tend to group inside the protein, away from the water.

Nonpolar and polar side chains play different roles in protein structure. Nonpolar side chains group tightly together to avoid unfavorable interactions with water. Polar side chains can make more types of interactions: ionic or hydrogen bonds. Side chains that contain acidic or basic groups (such as carboxylic acids or amines) often become charged ions and attract their opposites in salt bridges. Noncharged yet polar side chains often contain hydroxyl groups or amides. Many of these side chains can form hydrogen bonds.

One very special amino acid contains a thiol group (–SH) in its side chain. In proteins, thiol groups perform an important and highly specialized function; namely, a thiol group can react with a second thiol group to form disulfide (S–S) bonds between the sulfur atoms. These strong bonds covalently link two different regions of a protein together. All of these tendencies, the nonpolar side chains grouping together, the polar groups making salt bridges or hydrogen bonds, and the thiol groups making disulfide bonds, serve to give each sequence of amino acids a unique signature.

We continue our discussion by examining the **secondary structure** of a protein; that is, the folding pattern within a segment of the protein chain. Many, but not all, protein chains form regular, repeating structures from the particular bond angles and attractions between neighboring amino acids. The two most common are the α-helix, a spiraling strand, and the β-pleated sheet, extended strands stretching alongside each other with a slight zigzag.

Both forms of secondary structure depend on the tendency of the protein backbone to form intramolecular hydrogen bonds. Figure 12.11 shows the hydrogen bonds between the backbone O and the N–H of the amide group as dotted lines. The number and regular spacing of these hydrogen bonds can zip up a protein strand, stabilizing the secondary structure. The choice of secondary structure, or even the complete lack of it, can be loosely predicted from the primary structure. Some side chains tend to pack well into β-pleated sheets, others tend toward the helix, and others even predispose the chain to disorder.

Proteins are large, three-dimensional molecules, but both primary and secondary structures are relatively flat. We need a more "global" description of their shape, the **tertiary structure,** or the overall molecular shape of the protein defined by the interactions

Revisit Section 5.1 for more about polar and nonpolar molecules.

The importance of hydrogen bond formation in proteins was first mentioned in Section 5.2

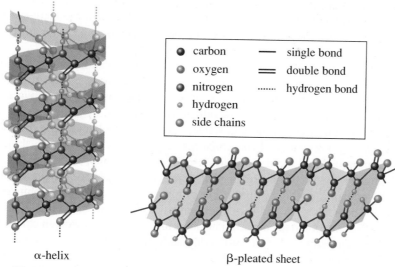

● carbon	— single bond
● oxygen	= double bond
● nitrogen	⋯⋯ hydrogen bond
● hydrogen	
● side chains	

α-helix   β-pleated sheet

Figure 12.11

Representations of the secondary structures of a protein. The two major types of secondary structures are the alpha (α)-helix and the beta (β)-pleated sheet.

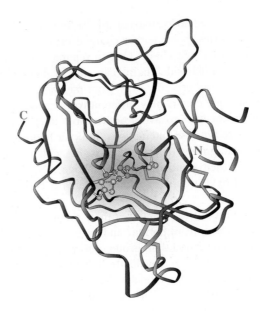

Figure 12.12

Tertiary structure of the enzyme chymotrypsin showing its active site. The "ribbon" portion represents the amino acid chain; the central colored portion is the active site at which the enzymatic chemistry takes place.

between amino acids far apart in sequence, but close in space. If you imagine a protein as a jumbled charm bracelet, the secondary structure might be the kinks in the chain itself, but the tertiary structure forms from the contact between the Eiffel Tower charm near the clasp and the Statue of Liberty charm near the middle. The overall fold results in a net increase in stability, maintained through hydrogen bonds, salt bridges, disulfide bonds, and interactions or avoidance of interactions between the side chains and water.

From only 20 amino acids, proteins make a wide variety of shapes and serve a large array of functions. The three-dimensional shapes are well-suited to the functions. Enzymes catalyze chemical reactions. The shape of enzymes must create an **active site,** the catalytic region, often a crevice, in an enzyme that binds only specific reactants and accelerates the desired reaction (Figure 12.12). Enzymes are the most commonly discussed type of protein, but a number of other examples exist. Some proteins bind DNA, to protect it or send a signal. Again, form suits function. When these proteins fold, they display positively charged side chains to attract and bind the negatively charged DNA. Another type of protein funnels material through the membrane of the cell, forming channels that shuttle a specific chemical across the outer layers of the cell while keeping the cell impermeable to undesirable chemicals.

A subtle change in the primary structure of a protein can have a profound effect on its properties. Notice the word *can* in the previous statement. Sometimes, a change in an amino acid leaves both the protein's shape and function unchanged. For instance, a nonpolar leucine can be switched to a valine, also nonpolar, by just removing a $-CH_3$ group. The protein may be a little less stable, but on the whole the same. But, instead, change the wrong glutamic acid (in which the side chain is often negatively charged) to a nonpolar valine, and the disease sickle-cell anemia occurs.

Hemoglobin is the blood protein that transports oxygen. The single alteration of a particular glutamic acid to valine in the primary structure of hemoglobin creates a variant called hemoglobin S and the condition called sickle-cell anemia. This substitution causes the hemoglobin to convert to an abnormal shape at low oxygen concentration, forcing red blood cells to distort into rigid sickle or crescent shapes (Figure 12.13). Because these cells lose their normal flexibility, they cannot pass through the tiny openings of the capillaries in the spleen and other organs. Some of the sickle cells are destroyed and anemia results. Others clog organs so badly that the blood supply to these organs is reduced.

Sickle-cell disease affects a sizable population, over 70,000 individuals in the United States alone, and lowers life expectancy to an average of 40 years.

Capillaries are the smallest vessels in the circulatory system of the blood at only about one red blood cell wide.

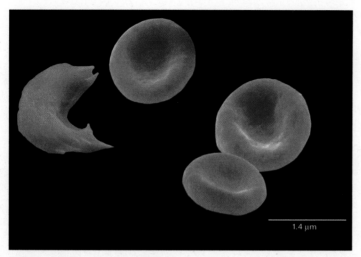

1.4 μm

Figure 12.13

A scanning electron micrograph of normal shaped red blood cells compared to a collapsed red blood cell showing the effect of sickle-cell disease.

Consider This 12.14   Function Follows Form

As we just mentioned, in sickle-cell anemia, a glutamic acid residue in the sequence of hemoglobin is replaced with a valine residue.

valine   glutamic acid

a. Describe the structural difference between these two amino acids: valine and glutamic acid.
b. Predict the solubility for each of these amino acids in water.
c. Explain how these differences could give rise to the deformed cells typified by sickle-cell anemia.

12.6 | The Process of Genetic Engineering

We started this chapter with thoughts of utopia; specifically, we dreamt of a perfect corn seed that could resist both herbicides and pests. We noted that this dream may be attainable with genetic engineering. We posed the question "What are we modifying when we genetically engineer something?"

By now, we hope you know the answer. We modify the DNA in the cell. If we change the genes, we change the proteins that are synthesized by these genes. Ultimately, we change the chemistry of the cell. As the cell grows and develops, we generate a plant with new characteristics bestowed by different DNA.

Throughout history, humans have manipulated genes. This may surprise you, as you might think that our ability to modify genes has come about only recently. But consider, for example, how we have cultivated plants. We tended to grow ones that carried specific traits, such as a better taste or appearance. The others we rejected. To produce these strains with new and unique traits, we crossbred different strains. The process took many years, but eventually we "domesticated" plants and created the crops

Figure 12.14
Corn's early ancestor, teosinte, below modern ears of corn.

that feed us today. Our crops are so far removed from their wild forbearers that we hardly would recognize them.

Corn is an excellent example. As we noted earlier, corn was native to the Americas. The people native to the region manipulated the genes of the teosinte plant, one that bore seeds on the end of its stalks rather than on the body of the plant, a growth pattern seen in today's corncob (Figure 12.14). Domesticating the teosinte plant led to a food that was both more nutritious and more abundant.

Domesticating a plant is a process of genetic modification. Even without understanding the chemistry, we selected plants with certain sequences of DNA and rejected ones with other sequences. Over time—slowly over time—we encouraged changes in the DNA, allowing one sample of DNA to carry on, spread, and survive. Traits that conferred fitness that we could not see, taste, smell, or feel were often lost. Fast-growing yet scrawny plants were discarded. So were deep, persistent roots that were impossible to till. Similarly, disease resistance not immediately required was lost. Modern-day crops, like those found in the cornfield at the opening of the chapter, are now unable to survive without humans to tend to them.

Nature carries out genetic modification as well. For example, all fields contain bacteria, and plants are susceptible to the different strains of bacteria to greater or lesser degrees. Let's suppose that a new, highly virulent bacterium appears in the field. Perhaps this bacterium arose through a mutation, a small random change in its genome. Over time, most of the wild grain plants in the field fall to the new pathogenic bacterium strain. But one particular plant contains the cell chemistry to resist the new onslaught. Suddenly, a gene that may not have mattered before is the key to survival. That plant is able to resist the bacterium while others cannot. The plant spreads and the gene survives.

The stress on the plant may take other forms as well, such as a three-year drought or a more aggressive weed. But the process is the same. Nature generally selects for plants that are more self-sufficient; humans tend to select for plants that look and taste better. In either case, the result is changes in the genome of the plant. The process of selective breeding, either in the wild or in human agriculture, is a long, slow, and somewhat random route toward modifying the gene pool. Neither natural nor artificial selection is genetic engineering. **Genetic engineering,** as we know it, is the direct manipulation of the DNA in an organism.

The easiest organisms to manipulate are small single-celled bacteria. These bacteria contain **plasmids,** or rings of DNA, in addition to their chromosome. Scientists

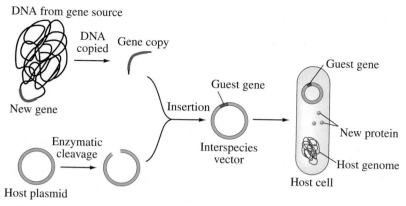

DNA from gene source

DNA copied

Gene copy

New gene

Enzymatic cleavage

Host plasmid

Insertion

Guest gene

Interspecies vector

Guest gene

New protein

Host genome

Host cell

Figure 12.15

A representation of the process of genetic engineering.

can easily remove, change, and replace the plasmids to create new chemistry in the bacteria. They use special enzymes to cut open the plasmid at specific sites. Scientists then copy the DNA containing a promising gene from another organism, and the DNA from this organism is inserted into the plasmid ring from the bacterium. The result is a new interspecies DNA plasmid, or **vector,** a modified plasmid used to carry DNA back into the bacterial host (Figure 12.15). Once inside the cell, the biochemistry of the bacterium takes over. The bacterium grows rapidly and produces new cells. Soon, the scientist has millions of copies of the "guest" gene and its protein product.

Inserting foreign DNA becomes a bit trickier as one climbs the evolutionary ladder from bacteria to plants. Higher organisms are better at protecting themselves against foreign DNA. For example, plants have thick cell walls. For another, many organisms have chemical mechanisms designed to detect and destroy foreign DNA. Even so, scientists have found ways to get around such defenses. One is to hijack a special soil bacterium that has the ability to infect plants. This bacterium creates a bridge into the plant cell and, in the process, transfers its own DNA into the genome of the plant. The bacterial genes induce points of large abnormal growth (Figure 12.16). You may have seen these growths (tumors) on a wide variety of plants and trees, including apple trees, rose bushes, and some vegetable plants.

Agrobacterium tumefaciens is commonly used, particularly in the commercial seed industry.

Scientists are able to disable the bacterial gene and use the bacterium to insert a different gene of interest. If the gene of interest is the specific sequence from the soil bacterium, *Bacillus thuringiensis* (or *Bt*), that produces a *Bt* toxin, then the plant acquires the instructions to make the toxin itself. Earlier in the chapter, we described

Figure 12.16

A crown gall tumor (*Agrobacterium radiobacter*) on a chrysanthemum plant.

how this process of genetic transfer worked to create corn plants that were resistant to insects. Not only is there *Bt* corn, but also *Bt* cotton, *Bt* potatoes, and *Bt* rice, all producing toxins against crop-destroying pests!

The *Bt* toxin is but one example of genetically engineered possibilities. Not all examples of genetically engineered crops are **transgenic,** or an organism resulting from the transfer of genes across species. Genetic engineering is used to achieve the same aims as crossbreeding, but with far more speed and control. Imagine you have a rice crop that grows well, tastes great, and cooks with a perfect level of stickiness, but it is being destroyed by a local plant virus. At the same time, a wild relative of your crop grows just beside it, resistant to the virus. Instead of securing the resistance trait by crossbreeding, find and copy just the resistance gene from the wild plant, then splice it into your weak but otherwise perfect crop. Technically, the genetic engineering process is identical, but the result is a plant with only rice genes.

Transgenic rice plants have been developed for use in Africa where the yellow mottle virus destroys much of the rice crop each year (Figure 12.17). At first, the gene that offered resistance to the rice plants was taken from a surprising source: the virus itself! The result was a transgenic plant with a small amount of viral code. The gene source is perhaps a concern, particularly to those that worry that consuming the viral proteins may result in allergic reactions. A preferable gene source is in development. Scientists are finding new potential genes by exploring the reason for viral resistance found in some rice varieties.

Perhaps you have a trait in mind, but have no source for the gene. In such a case, both traditional crossbreeding and whole-gene transfer won't work. Neither of these can function if a trait does not already exist. Although in theory you could wait for a trait to evolve through natural, random mutations, most likely you would run low on patience. To speed up the process of evolution, scientists use either chemicals or ionizing radiation to produce random mutations in a batch of seeds. Upon planting, some seeds do not grow at all, others grow but with no apparent changes, and still others show unique traits. Once a useful trait appears, scientists can use plant breeding to isolate and refine a new crop or they can copy the gene that gives rise to the trait and engineer it into a different plant.

Your Turn 12.15   Ionizing Radiation

Ionizing radiation is a term commonly used by both scientists and health professionals. We first introduced it in Chapter 7 and would like you to review it here.

a. What is an ion? Give two examples.
b. *Radiation* can have more than one meaning. Explain.
   Hint: See Your Turn 7.12.
c. How does radiation produce ions?
   Hint: See Section 7.6 on "Radioactivity and You."
d. How do these ions lead to random mutations in DNA?

Figure 12.17
Virus-resistant transgenic rice.

Early in this chapter, we pointed out that as humans, we have a responsibility to act wisely, both for current and future generations. We also mentioned that our needs are great and our societal problems are pressing, asking whether our *inaction* might result in more harm than good. The sheer power of genetic engineering can frighten us. You may be automatically concerned at even the idea of transgenic organisms. The next section describes the benefits of genetic modification to chemical industry, both current and projected. In the one after, we delve into potential risks.

12.7 | Making Chemical Synthesis Green from Genetic Modification

Pest and herbicide resistance are not the only reasons for genetic modification. Biochemists also have modified the genes of corn, soybeans, and wheat with the aims of making them more resistant to disease; more tolerant of stresses such as salt, heat, or drought; and to improve the nutritional quality of the food itself. Although farmers have benefitted, many others have as well. Biochemists have designed plants to absorb toxic metals from contaminated soil. Some are developing crops such as soybeans that produce high yields of biofuel per acre. Others are engineering bacteria to detect and remediate radioactive contamination. Of interest to us in this section is that genetic technology can incorporate the principles of green chemistry for large-scale chemical production.

 A synthetic route to a desired chemical may require toxic chemicals, large amounts of solvents, and high temperatures. Although such processes yield many useful chemicals, they also produce a staggering amount of waste—up to 100 times the weight of the compound! One way to reduce the ecological footprint is to use enzymes, the biological catalysts described earlier. These "biological machines" perform reactions just as you would conventionally within flasks or beakers, but faster and safer with fewer toxic reagents, at lower temperatures, and with less waste. Another plus is that enzymes can be used over and over again. Minimal waste, reusable reagents, and low toxicity are green chemistry standards.

Via genetic engineering, biochemists use enzymes to create new drugs or more efficiently build existing drugs. The gene coding for the drug is introduced into a host organism that, in turn, synthesizes the desired product. This application of recombinant DNA technology is not only one of the most rapidly growing fields but also the oldest.

Consider the production of human insulin. Insulin is quite a small protein built from 51 amino acids, but its chemical synthesis is far from trivial. An insufficient supply of insulin to regulate blood sugar levels results in diabetes. Untreated, the disease may lead to kidney failure, cardiovascular problems, blindness, and even death. Luckily, those with diabetes can control the disease through proper diet, exercise, and insulin injections—a genetically engineered product.

Before 1982, all insulin used was isolated from the pancreatic glands of cows and pigs. But the insulin gathered from animals is not identical to that of humans. The insulin of cows ("bovine insulin") differs from the human hormone in 3 out of 51 amino acids; the insulin of pigs ("porcine insulin") and human insulin differ in only one amino acid. The slight differences are nonetheless significant. Some patients developed antibodies against the foreign insulin protein and rejected it.

In short, humans need human insulin. Although chemists were able to synthesize the insulin protein in the laboratory, the process is far too complex and costly for large-scale production. Luckily, in 1982, the common bacterium, *E. coli,* was coaxed into making human insulin. The gene for human insulin was placed in a plasmid which in turn was placed in the bacterium. The result was what we can depend on today—a steady and less expensive supply of human insulin.

Growing bacteria with natural or artificial genes is now a step in manufacturing small-molecule drugs. The process is particularly easy when the gene, or something similar, already exists. Sometimes, however, special tricks are required, as was the case in the synthesis of the drug Atorvastatin (Figure 12.18). Enzymes for specific reaction steps could not be found, so they had to be evolved. Scientists mimic natural selection

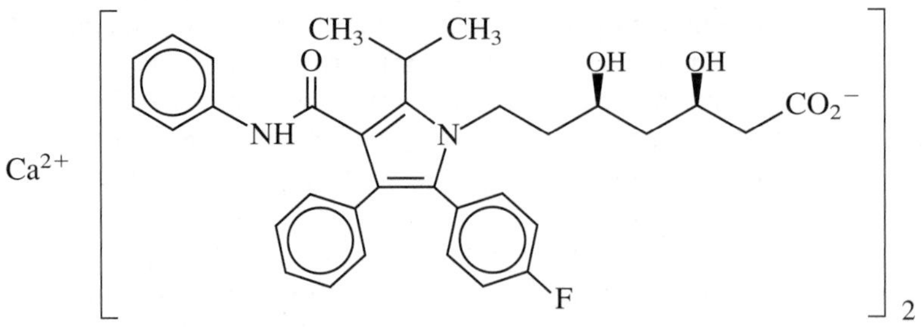

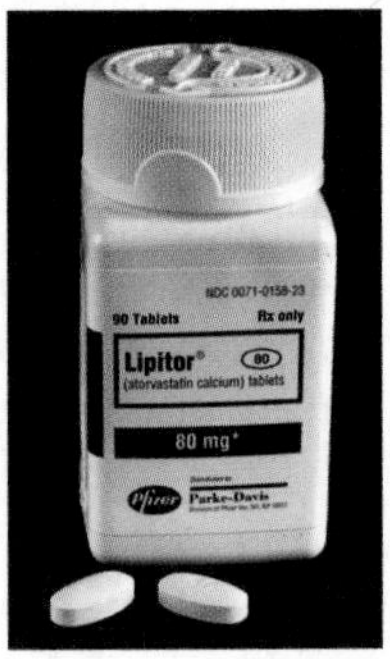

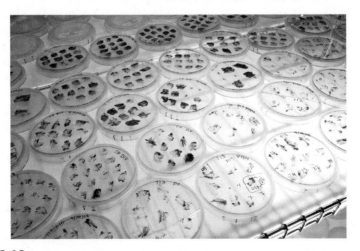

Figure 12.18

Atorvastatin, the active ingredient in Lipitor, requires enzyme-generated building blocks. The cholesterol-lowering drug produced by Pfizer has annual sales exceeding $10 billion.

"Directed evolution" can also be performed without using bacteria by selecting, mutating, and multiplying DNA with enzymes.

Return to Chapter 9 for a review on plastic production, uses, and consequences.

Scientists and engineers at Metabolix earned a Presidential Green Chemistry Challenge Award in 2005 for their innovative and sustainable bioplastic technology.

by creating an environment where the bacteria must evolve a new trait in order to survive; the process is termed "directed evolution." Typically, the scientists start with a random variety of DNA sequences transferred into a population of bacteria. Over multiple generations grown under certain conditions, such as providing only certain chemicals as food, a new enzyme emerges.

Bioengineered organisms can even produce plastics. Most synthetic polymers are synthesized in large chemical plants using a process that consumes large amounts of chemical reagents and energy. Furthermore, these chemical reagents often are derived from petroleum. Scientists at Metabolix bioengineered organisms to ameliorate the problems. These organisms produce monomers from renewable materials such as corn, sugarcane, and vegetable oil, and then catalyze the polymerization reaction. One example of a resulting polymer is polyhydroxybutyrate (PHB). This bioplastic is used to produce items such as plastic utensils and coatings for cups, much like polypropylene (PP). However, unlike PP, PHB is biodegradable. The process uses materials of low toxicity, is extremely efficient, and lowers greenhouse gas emissions.

Consider This 12.16   GM Algae for New Biofuels

Algae can be an excellent source of biodiesel (Figure 12.19). Using the web, explore the production of biodiesel and other biofuels from algae.

a. Describe the biofuel(s) that can be produced from algae with *no* genetic engineering.
b. List three advantages of algae over other fuel sources, such as corn and switchgrass. *Hint:* Revisit Sections 4.9 and 4.10.
c. How might genetically modifying algae lead to the better production of biofuels?

Figure 12.19

Cultivating strains of algae for new biofuels.

To finish our discussion, we ask you to revisit one of the lines from the opening paragraph of this chapter: "*our farms create our drugs and vaccines.*" This is not a utopian dream, as our transgenic plants now can produce both. For example, vaccines against infectious diseases of the intestinal tract have been produced in potatoes and bananas. Anticancer antibodies have been expressed and introduced into wheat. Peptide drugs against HIV/AIDS have been produced from tobacco fields. Also, most vaccines require refrigeration or other special handling together with trained professionals to administer them. In some countries, health professionals cannot afford even the needles to inoculate people, leading to possible infections from reused needles. Vaccines produced within edible products may be difficult to correctly dose but would be easy to administer and transfer. Rather than yielding crops for food, these fields yield the hope of low-cost, readily available vaccines. Thus, these fields of transgenic plants can go hand-in-hand with good public health policy.

The technology is still developing, but the bonuses are already both apparent and immense. Compared with traditional methods, a genetically engineered enzyme, microorganism, or whole field can yield highly pure products while reducing waste and by-products. The genetically engineered routes increase the yield with fewer steps and often eliminate labor, energy, and resource-intensive purification processes. They avoid the use of toxic and corrosive chemicals, which is better for the environment and increases overall worker safety.

This all being said, we end this section in a manner similar to the way in which we ended the previous one. We need to act wisely, both for current and future generations. The promise of developing technology and society's desperate needs can lead to rash choices. The risks of genetically engineered cures may be large, perhaps unacceptably so. In the next section, we delve into the potential risks.

12.8 | The New Frankenstein

In Mary Shelley's work of science fiction, Dr. Frankenstein created a man from the best "parts" of other men. However, he then lost control of his creation. Today, are we the new Dr. Frankenstein? In this era of genetic engineering, many have asked this very question.

Over recent years, people have rallied against genetically modified (GM) foods (Figure 12.20). These protests have occurred worldwide, including in Germany, Australia, Spain, the United Kingdom, and the United States. Have you run across the term *Frankenfood*? The next activity offers you a chance to explore it as you learn more about the opposition to GM food.

Figure 12.20
Greenpeace activists dumping papayas during a protest in Bangkok.

Consider This 12.17   Frankenfood

a. Explain the term *Frankenfood*.
b. Search the web to find one context in which GM food is protested. Give the country, the type of food, and the reasons for opposition.
c. Outline three arguments of those in opposition to GM food.

As this activity most likely revealed, opposition takes many forms. We explore some of them here.

GM plants will spread into the wild. For example, take the new transgenic corn we described at the start of the chapter. The corn may spread outside its field by the accidental release of seeds or pollen, taking its new genetic trait along with it. The corn's distant, weedy relatives are still found in wild fields. And the wild and domesticated corn, though far removed, may still breed. Their hybrid offspring may carry the new gene; perhaps the trait could confer an advantage and lead to a new population of superweeds, resistant to the bug or pesticide of our design.

Response: The new trait will only spread in the wild if the GM plants and their offspring are competitive with wild plants. Remember though, offspring do not take just one or two traits from each parent; rather, they acquire many. A hybrid of domesticated corn with a weedy relative most likely will show the GM resistance to the corn borer together with many of the other traits of the domestic corn, including its weaknesses. In experiments, hybrids have not been strong enough to survive in the wild because of these weak domesticated genes. To further prevent the spread, many countries regulate the planting location of genetic crops. No GM crop may be grown in the vicinity of weedy relatives. Of course, the regulations do not apply to illegal or accidental growth. For instance, GM corn has been found in Mexico, an area with both a diverse population of specialized domestic corn and wild relatives. Chances remain slim, although not impossible, that a gene will transfer and survive in the wild.

GM crops threaten the natural balance, even from within the field. In laboratory experiments, pollen from GM plants has harmed caterpillars and other insects distinct from the target pest. A farm field will have less pollen than in the experiments, but the exposure time will be significantly longer, even over generations. The true effects are still unknown. The concern is not limited to the bugs; the damage may have consequences higher up the food chain, by poisoning or contaminating the food supply for birds or destroying populations important for pollination or soil rejuvenation.

Response: Ideally, genetically engineered pest resistance reduces general pesticide use. A healthy variety of bugs is allowed to flourish by selectively killing only a few. If the new genetically engineered trait is chosen well, the risk to other insects can be minimized. In addition, advanced genetic engineering may control where and when the plant expresses an artificial gene. If the pesticide is only produced in the plant leaves, the risk to insects interacting with the pollen decreases. Methods such as these can alleviate the risks. Again though, most experiments have shown little or no ill effects on the natural insect life.

Use of genetically engineered crops will force weeds, bugs, and bacteria to evolve. Natural evolution did not create the diversity of the world and stop. It continues. The evidence is easily seen in the increasing numbers of antibiotic-resistant bacterial strains. In our genetically engineered corn crop, the weeds, bugs, or bacteria must move, die, or evolve. The genetically engineered corn has been engineered for one narrow purpose. Once a new bug or disease evolves and attacks, the GM corn crop will pose little resistance. Historically, when a new disease arose, a natural trait was found within the varieties of corn. Breeding created a new stronger hybrid. If we come to depend on a narrow set of GM corn crops, we will have no genetic variety to utilize.

Response: The easiest proposed solution is to continue to plant nonengineered crops alongside our genetically engineered varieties. The mix should prevent the resistance trait from dominating as it will not be a requirement for the pests to survive. Regulators in both the United States and in the European Union require this very condition, but still debate how much of the nonengineered crop is enough. Although new resistance traits evolve in insects or pests in laboratories, the same has not been commonly found in genetically engineered crops, particularly if nonengineered crops are planted nearby. Likely, the resistance is not yet a survival requirement. But, as the engineered crops become widespread, the risk of new resistance in our target increases.

We have many regulations to keep control. Currently, the regulations appear to work, as the described risks are not yet an observed reality. But GM crop use is fast expanding. The possibilities we just listed focused on the consequences to the environment. But with each new genetically engineered crop, we need to list other possible effects as well. For example, what about people who might experience an allergic reaction after accidentally eating genetically engineered material?

Relevant to this discussion is the proposition that *none* of these risks are worth taking. At the beginning of this chapter, we held up the image of a genetically engineered utopia. This has alternatively been discussed as a second "Green Revolution." Genetic engineering has been developing and spreading for decades, and yet, the utopia has not arrived. Why? In a time of rising food prices and continued starvation, genetic engineering has not provided the much hoped for second "Green Revolution." Perhaps it cannot or will not. Some argue that genetic engineering has provided nothing that traditional breeding methods could not. They claim that the research dollars are wasted and the environment unduly risked. Others would point out the already promising successes, but claim that current civic policies block true progress. In an era in which a genetic sequence is patent-worthy, what hope do academic researchers or small beneficent companies have for ground-breaking progress?

Look to Section 11.12 for a thorough discussion on the agriculture developments leading to the first "Green Revolution."

Consider This 12.18   **Your Opinion of GM Foods (revisited)**

As promised, we now return to the question posed at the opening of this chapter.

a. Given a choice between a GM food and one that was not, do you have a preference as to which one you would eat? Explain.
b. Whichever preference you expressed, now make an argument for the other side.

Concern or cynicism about the rapid application of a new science is justifiable. Any decision that attempts to fix a problem in the short term without regard for the future is fraught with danger. As with many controversies, it is easy to spot the positions at the poles. However, it is much harder to find the middle ground. We hope this discussion has both given you some answers and, more importantly, new questions.

Conclusion

This final chapter of the book, like almost all that preceded it, ends with a dilemma: How can we balance the great benefits of modern chemical sciences and technology and the risks that seem inevitably to accompany them? The consequences of losing control are perhaps more complicated and wide-ranging than the science of genetic engineering itself.

Throughout this text, the authors have occasionally looked, with myopic professorial vision, into the cloudy crystal ball of the future. It is in the nature of science that we cannot confidently predict what new discoveries will be made by tomorrow's scientists. Nor can we know the applications of those discoveries, good or bad. Such uncertainty is one of the delights of our discipline. A chemist must learn to live with ambiguity, indeed, to thrive on it, in the search to better understand the nature of atoms and their intricate combinations, in all their various guises.

But all citizens of this planet must at least develop a tolerance for ambiguity and a willingness to take reasonable risks, especially considering that life itself is a biological, intellectual, and emotional risk. Of course, we all seek to maximize benefits, but we must recognize that individual gain must sometimes be sacrificed to benefit society. We live in multiple contexts—the context of our families and friends, our towns and cities, our states, our countries, our special planet. We have responsibilities to all. *You*, the readers of this book, will help create the context of the future. We wish you well.

Chapter Summary

Having studied this chapter, you should be able to:

- Discuss the complications of corn farming as an example of inspiration for genetic engineering (12.1)
- Understand that cells function by a complex series of chemical reactions (12.2)
- Discuss DNA as a storage device of information to run chemical reactions (12.2)
- Understand the chemical composition of deoxyribonucleic acid (DNA), a polymer of nitrogen-containing bases, deoxyribose, and phosphate groups (12.2)
- Interpret evidence for the double-helical structure of DNA and its base pairing (12.3)
- Understand the structural basis of DNA replication (12.3)
- Explain how the genetic code is written in groupings of three DNA bases called codons (12.4)
- Understand how the codons relate to amino acids across organisms (12.4)
- Discuss the primary, secondary, and tertiary structure of proteins (12.5)

- Recognize the general properties of amino acid side chains (12.5)
- Relate the properties of amino acids to the interactions formed in protein structure (12.5)
- Discuss, with examples, how small changes in protein sequence may result in disease (12.5)
- Understand the essential steps in carrying out recombinant DNA techniques (12.6)
- Describe what is meant by transgenic organisms and give examples (12.6)
- Give brief examples of natural selection, selective breeding, and genetic engineering (12.6)
- Discuss, with examples, how genetic modification has changed the chemical industry (12.7)
- Discuss controversial issues associated with transgenic organisms and the fears about Frankenfood (12.8)
- Debate issues associated with the prudent and ethical applications of genetic engineering (12.8)

Questions

Emphasizing Essentials

1. The theme of this chapter is that DNA guides the chemistry of every living organism on the planet. Name three traits that you possess that were determined by your DNA.

2. The first section in this chapter is called "Stronger and Better Corn Plants?"

   a. Name three ways in which a corn plant could be stronger or better.

   b. Why is there a question mark in the title of the section? In your explanation, use the word *genome*.

3. What is the difference between a genome and a gene?

4. Consider the structural formulas in Figure 12.2.

   a. What functional group(s) are found in the adenine molecule?

   b. What functional group(s) are found in the deoxyribose molecule?

   c. From what you learned in Section 11.5, why is deoxyribose a sugar and adenine is not?

5. a. What three units must be present in a nucleotide?

   b. What type of bonding holds these three units together?

6. All species use the four bases: adenine, cytosine, guanine, and thymine. Name two similarities among the four bases. Highlight one feature unique to each.

7. Circle and name the functional groups in this nucleotide. Also label the sugar, the base, and the phosphate group.

8. Compare the DNA segment in Figure 12.4 with the nucleotide shown in the previous question. Two functional groups from the nucleotide react to form a polymer similar to DNA. Circle these functional groups in the nucleotide.

9. What does each letter in DNA stand for? Examine Figure 12.4. Which aspects of the DNA molecule does the name DNA highlight? Which aspects of the DNA molecule are not part of the name?

10. Here is the structural formula for the base thymine.

a. Label the H atoms that can hydrogen bond with a water molecule.

b. Use electronegativity differences to explain why only these H atoms can form hydrogen bonds.

11. Explain why the base sequence ATG is different from the base sequence GTA.

12. Given a short sequence of DNA: TATCTAG

a. Identify the base sequence that is complementary to the sequence.

b. Draw lines between the sequences to show each hydrogen bond between the base pairs.

13. A typical chromosome, stretched completely, is about 2.0 cm in length. Each base pair in a chromosome is 0.34 nm from its neighboring base pair.

a. Calculate the number of base pairs found in this chromosome. *Hint:* 1 m = 10^2 cm; 1 m = 10^9 nm.

b. How much smaller does the chromosome need to be to fit in a cell? Calculate the ratio of the diameter of a cell to the length of a stretched out chromosome. *Hint:* A typical human cell is 10,000 nm in diameter.

14. What is a codon and what is its role in the genetic code?

15. Only 61 of the possible 64 codons code for an amino acid. What is the function of the other three codons?

16. In relation to DNA, what is a gene?

17. Amino acids are the monomers used to build proteins.

a. Draw the *general* structural formula for an amino acid.

b. Name the functional groups in the structural formula you just drew.

18. Amino acids can be classified as nonpolar or polar. The polar amino acids can be further grouped as acidic, basic, or neutral.

a. Draw an example of an amino acid for each category.

b. Describe each category in more detail. What functional groups are found in the side chains?

19. Give two examples of amino acids that have side chains that are nonpolar. Explain what characteristics make them nonpolar.

20. Describe what is meant by the primary, secondary, and tertiary structure of a protein.

21. Explain how an error in the primary structure of a protein in hemoglobin causes sickle-cell anemia.

22. List two advantages and two disadvantages of transgenic crops.

Concentrating on Concepts

23. This chapter opens stating "Geneticists may have the power to create a utopia." Explain.

24. Figure 12.4 represents a segment of DNA.

a. What part of each nucleotide becomes the backbone of the polymer?

b. What part hangs off the backbone?

c. How is this represented in sculpture at the beginning of the chapter and Figure 12.7?

25. Compare the two representations of a segment of DNA in Figure 12.4. Discuss one strength and one weakness for each.

26. Use Figure 12.7 to help explain why stable base pairing does *not* occur between adenine and cytosine. Name another base pair for which this is also true. Explain why.

27. Use Figure 12.7 to explain why adenosine–thymine pairs are less stable than cytosine–guanine base pairs.

28. Nuclear radiation can damage molecules by breaking the bonds they contain. In DNA, the radiation may break one or both strands by destroying the nucleotide or the backbone of the DNA. A single-strand break is easier for a cell to repair than a double-strand break. Explain.

29. UV-C light is germicidal. In strands of DNA, two thymine residues absorb UV-C light and react together to form a cross-link.

a. List a potential consequence of cross-linking between two thymine molecules on a single strand of DNA.

b. List a potential consequence of cross-linking between two thymine molecules from two complementary strands.

30. Many compounds damage DNA and very effectively kill bacterial cells in culture dishes. Why are these compounds not typically used as antibacterial drugs?

31. Errors sometimes occur in the base sequence of a strand of DNA. But not all of these errors result in the incorporation of an incorrect amino acid in a protein for which the DNA codes. Explain why a single base change may not change the amino acid. Why is this advantageous?

32. Several diseases are caused by a single amino acid change in a specific enzyme, for example familial amyotrophic lateral sclerosis (FALS), an inheritable form of Lou Gehrig's disease, can be caused by the alteration of a single amino acid in the enzyme superoxide dismutase.

   a. For any protein, how many base pairs are responsible for specifying an amino acid in the gene that codes for the protein? What is the minimum number of base pairs that would have to be changed to switch one amino acid for another?

   b. One hereditary mutant superoxide dismutase replaces a glycine with alanine. Find a table that specifies the codons for each amino acid, then list the codon(s) that code for glycine and the alanine. Identify the minimum number of base pairs changed.

33. Almost all species use the same four bases and the same codons. Why is this necessary for gene transfer to work?

34. A clone is an identical copy of a cell or organism, often created by transferring the nucleus from an adult cell into an egg cell lacking a nucleus.

   a. Explain why the nucleus from an adult cell can be used to create an identical organism.

   b. Explain why the egg cell must lack a nucleus.

   c. Explain two ways in which this process differs from genetic engineering.

35. Human insulin and human growth hormone have both been made through the use of recombinant DNA technology through the genetic engineering of bacteria. List two possible reasons this has not received the same amount of public concern as the use of genetic engineering in plants.

36. Create a graph showing the relative amounts of GM or transgenic crops grown by six different countries. Include with your graph the source of your data.

37. Consider the idea of mixing genes as an improvement on nature.

   a. Describe what is meant by the term *transgenic organisms*.

   b. Why is the alteration of the genetic makeup of plants by genetic engineering preferred to traditional crossbreeding methods?

Exploring Extensions

38. In the opening section of the chapter, the genetic engineering of corn was discussed. In light of the projected environmental and population changes discussed in other chapters, name three traits not mentioned in this chapter you would propose to add to corn or other common crops. Discuss the advantages and disadvantages of each of your traits.

39. The human genome contains about 3 billion base pairs, but only about 2% of this DNA consists of unique genes. The number of genes is estimated at 30,000. Use this information to calculate the average number of base pairs per gene. What does this new information suggest about the amount of information in our genome?

40. Consider the structural formula of deoxyribose shown in Figure 12.2. Draw another isomer of deoxyribose, one that would still form a nucleoside with a base and with phosphate. Compare your isomer side-by-side with the structural formula of deoxyribose.

41. Perhaps you have learned some memory aids (mnemonics) when taking music lessons (Every Good Boy Does Fine), memorizing the names of the Great Lakes (HOMES), or learning about oxidation and reduction (OIL RIG). One of the authors learned "All-Together, Go-California" as the mnemonic to remember the correct base pairings in DNA.

   a. What is the relationship in this mnemonic to DNA base pairing?

   b. Design a different mnemonic to help you remember such base pairings.

42. Of the major players in the discovery of the structure of DNA, only Rosalind Franklin had a degree in chemistry. What was her background and experience that enabled her to make significant contributions? Did her contributions receive adequate credit and recognition? Write a short report, citing your sources.

43. a. What are X-rays? *Hint:* Revisit Chapters 2 and 7.

   b. How are X-rays used in X-ray diffraction to determine the structure of a molecule?

   c. The first X-ray diffraction patterns were of simple salts, such as sodium chloride. The X-ray diffraction studies of nucleic acids and proteins did not come until much later. Suggest two reasons why.

44. Act as a Skeptical Chemist in checking the correctness of the claim that the DNA in an adult human would stretch from Earth to the Moon and back more than a million times.

   Distance from Earth to the Moon = 3.8×10^5 km
   Number of DNA-nucleated cells in an adult human = 1×10^{13}
   Length of stretched human DNA strand = 2 m

45. The genetic traits leading to sickle-cell disease are more common in people of African, African-American, or Mediterranean heritage. Using the web, explain a proposed reason that the sickle-cell trait has persisted rather than being discarded through evolution.

46. In order to fold, proteins may search through a number of possible configurations before finding the right one. Explore folding@HOME (http://folding.stanford.edu). How does the computer simulation repeat the natural process? Describe why your computer is necessary. Finally, explore one of the results from the program and write a single-paragraph report.

47. GM foods are debated particularly due to the patenting of genes or whole plants. The patenting of GM plants

will change the availability of new technology. List two advantages and two disadvantages of this approach. Do you favor the patenting of genes?

48. To clone or to breed? Grieving pet owners may find the opportunity to clone a beloved dog too tantalizing to resist. Even so, many professional dog breeders and their organizations advise against cloning. Phil Buckley, a spokesperson for the Kennel Club in England, argues, "Canine cloning runs contrary to the Kennel Club's objective to promote in every way the general improvement of dogs." Explain why cloning cannot improve dog breeds.

49. Transgenic plants have not been widely accepted in all countries.

    a. For the European Union, find two examples of transgenic plants that have been banned. Then compare them to two that have been allowed. Discuss the differences between those allowed and those rejected.

    b. Create a timeline of five events key for the rapid increase or subsequent leveling (or both) of the adoption of transgenic crops in the United States. Briefly discuss each of your choices.

50. One reason why science fiction is successful is that it starts with a known scientific principle and extends, elaborates, and sometimes embroiders it. *Jurassic Park* began with the scientific principle of copying and manipulating DNA and was extended. Now it is your turn. Take any scientific principle from this text. Then write a one- or two-page outline for a story based on that principle. Be sure to identify the chemical concepts and any pseudoscience that you might employ.

51. Gene therapy involves the use of recombinant DNA techniques. Use the web to gather information on this medical tool that has met with mixed results. Write a one- to two-page report on gene therapy, including specific examples of diseases that are being treated and how the patients are faring.

52. A recent focus of participants in the Human Genome Project has been to determine the base sequence for many different microbes (that is, bacteria and viruses). Use the web to document recent efforts and progress. Suggest three reasons why these sequences would generate so much interest.

53. Find a transgenic organism not discussed in the text. Describe the motivation, gene source, and the process for the genetic modification.

54. You are the head of a government facing another year of a long drought and a serious risk of famine. Another nation has offered you a supply of genetically modified corn to feed your people. List two pros and two cons of accepting the aid. Decide whether you would accept it.

Appendix 1

Measure for Measure

Conversion Factors and Constants

Metric Prefixes

deci (d)	$1/10 = 10^{-1}$	deka (da) $\quad 10 = 10^{1}$
centi (c)	$1/100 = 10^{-2}$	hecto (h) $\quad 100 = 10^{2}$
milli (m)	$1/1000 = 10^{-3}$	kilo (k) $\quad 1000 = 10^{3}$
micro (μ)	$1/10^{6} = 10^{-6}$	mega (M) $\quad = 10^{6}$
nano (n)	$1/10^{9} = 10^{-9}$	giga (G) $\quad = 10^{9}$

Length

1 centimeter (cm) = 0.394 inch (in.)

1 meter (m) = 39.4 in. = 3.28 feet (ft) = 1.08 yard (yd)

1 kilometer (km) = 0.621 miles (mi)

1 in. = 2.54 cm = 0.0833 ft

1 ft = 30.5 cm = 0.305 m = 12 in.

1 yd = 91.44 cm = 0.9144 m = 3 ft = 36 in.

1 mi = 1.61 km

Volume

1 cubic centimeter (cm^3) = 1 milliliter (mL)

1 liter (L) = 1000 mL = 1000 cm^3 = 1.057 quarts (qt)

1 qt = 0.946 L

1 gallon (gal) = 4 qt = 3.78 L

Mass

1 gram (g) = 0.0352 ounce (oz) = 0.00220 pound (lb)

1 kilogram (kg) = 1000 g = 2.20 lb

1 lb = 454 g = 0.454 kg

1 metric ton (t) = 1 tonne = 1 long ton = 1000 kg = 2200 lb

$\qquad\qquad$ = 1.10 ton

1 ton = 1 short ton = 2000 lb = 909 kg = 0.909 t

Time

1 year (yr) = 365.24 days (d)

1 day = 24 hours (hr or h)

1 hr = 60 minutes (min)

1 min = 60 seconds (s)

Energy

1 joule (J) = 0.239 calorie (cal)

1 cal = 4.184 joule (J)

1 exajoule (EJ) = 10^{18} J

1 kilocalorie (kcal) = 1 dietary Calorie (Cal)

$\qquad\qquad\qquad$ = 4184 J = 4.184 kilojoule (kJ)

1 kilowatt-hour (kWh) = 3,600,000 J = 3.60×10^{6} J

Constants

Speed of light (*c*) = 3.00×10^{8} m/s

Planck's constant (*h*) = 6.63×10^{-34} J · s

Avogadro's number (N_A) = 6.02×10^{23} objects per mole

Atomic mass unit (*m*) = 1.66×10^{-24} g

The Power of Exponents

Scientific (or exponential) notation provides a compact and convenient way of writing very large and very small numbers. The idea is to use positive and negative powers of 10. Positive exponents are used to represent large numbers. The exponent, written as a superscript, indicates how many times 10 is multiplied by itself. For example,

$$10^1 = 10$$

$$10^2 = 10 \times 10 = 100$$

$$10^3 = 10 \times 10 \times 10 = 1000$$

Note that the positive exponent is equal to the number of zeros between the 1 and the decimal point. Thus, 10^6 corresponds to 1 followed by six zeros, or 1,000,000. This same rule applies to 10^0, which equals 1. One billion, 1,000,000,000, can be written as 10^9.

When 10 is raised to a negative exponent, the number being represented is always less than 1. This is because a negative exponent implies a reciprocal, that is, 1 over 10 raised to the corresponding positive exponent. For example,

$$10^{-1} = 1/10^1 = 1/10 = 0.1$$

$$10^{-2} = 1/10^2 = 1/100 = 0.01$$

$$10^{-3} = 1/10^3 = 1/1000 = 0.001$$

It follows that the larger the negative exponent, the smaller the number. The negative exponent is always one more than the number of zeros between the decimal point and the 1. Thus, 1×10^{-4} is equal to 0.0001. Conversely, 0.000001 in scientific notation is 1×10^{-6}.

Of course, most of the quantities and constants used in chemistry are not simple whole-number powers of 10. For example, Avogadro's number is 6.02×10^{23}, or 6.02 multiplied by a number equal to 1 followed by 23 zeros. Written out, this corresponds to $6.02 \times 100,000,000,000,000,000,000,000$, or 602,000,000,000,000,000,000,000. Switching to very small numbers, a wavelength at which carbon dioxide absorbs infrared radiation is 4.257×10^{-6} m. This number is the same as 4.257×0.000001, or 0.000004257 m.

Your Turn   **Appendix 2.1**

Express these numbers in scientific notation.

   a. 10,000   **b.** 430   **c.** 9876.54

   d. 0.000001   **e.** 0.007   **f.** 0.05339

Answers

   a. 1×10^4 **b.** 4.3×10^2 **c.** 9.87654×10^3

   d. 1×10^{-6} **e.** 7×10^{-3} **f.** 5.339×10^{-2}

Your Turn   **Appendix 2.2**

Express these numbers in conventional decimal notation.

   a. 1×10^6 **b.** 3.123×10^6 **c.** 25×10^5

   d. 1×10^{-5} **e.** 6.023×10^{-7} **f.** 1.723×10^{-16}

Answers

   a. 1,000,000   **b.** 3,123,000

   c. 2,500,000   **d.** 0.00001

   e. 0.0000006023   **f.** 0.0000000000000001723

Appendix 3

Clearing the Logjam

You may have encountered logarithms in mathematics courses but wondered if you would ever use them. In fact, logarithms (or "logs" for short) are extremely useful in many areas of science. The essential idea is that they make it much easier to deal with very large *ranges* of numbers, for example, moving by powers of 10 from 0.0001 to 1,000,000.

It is likely that you have met logarithmic scales without necessarily knowing it. The Richter scale for expressing magnitudes of earthquakes is one example. On this scale, an earthquake of magnitude 6 is 10 times more powerful than one of magnitude 5. An earthquake of magnitude 8 would be 100 times more powerful than one of magnitude 6. Another example is the decibel (dB) scale. Each increase of 10 units represents a 10-fold increase in sound level. Therefore, a normal conversation between two people 1 m apart (60 dB) is 10 times louder than quiet music (50 dB) at the same distance. Loud music (70 dB) and extremely loud music (80 dB) are 10 times and 100 times as loud, respectively, as a normal conversation.

A simple exercise using a pocket calculator can be a good way to learn about logs. You will need a calculator that "does" logs and preferably has a "scientific notation" option. Start by finding the logarithm of 10. Simply enter 10 and press the "log" button. The answer should be 1. Next find the log of 100 and then the log of 1000. Write down the answers. What pattern do you see? (The pattern may be more obvious if you recall that 100 can be written as 10^2 and 1000 is the same as 10^3.) Predict the log of 10,000 and then check it out. Then try the log of 0.1, or 10^{-1}, and the log of 0.01 (10^{-2}). Predict the log of 0.0001 and check it out.

So far so good, but we have only been considering whole-number powers of 10. It would be helpful to be able to obtain the logarithm of any number. Once again, your handy little calculator comes to the rescue. Try calculating the logs of 20 and 200, then 50 (5×10^1) and 500 (5×10^2). Predict the log of 5×10^3, or 5000. Now for something slightly trickier: the log of 0.05. Finally, try the log of 2473 and the log of 0.000404. In each of the three cases, does the answer seem to be in the right ballpark? Remember that your calculator will happily provide you with many more digits than have any meaning, so you will need to do some reasonable rounding.

In Chapter 6, the concept of pH is introduced as a quantitative way to describe the acidity of a substance. A pH value is simply a special case of a logarithmic relationship. It is defined as the negative of the logarithm of the H^+ concentration, expressed in units of molarity (M). Square brackets are used to indicate molar concentrations. The mathematical relationship is given by the equation $pH = -\log [H^+]$. The negative sign indicates an inverse relationship; as the H^+ concentration diminishes, the pH increases. Let us apply the equation by using it to calculate the pH of a beverage with a hydrogen ion concentration of 0.000546 M. We first set up the mathematical equation and substitute the hydrogen ion concentration into it.

$$pH = -\log [H^+] = -\log (5.46 \times 10^{-4}M)$$

Next, we take the negative logarithm of the H^+ concentration by entering it into a calculator and pressing the log button, then the "plus/minus" key to change the sign. This gives 3.26 as the pH of the beverage. (It may display 3.262807357 if you have not preset the number of digits, but common sense prompts you to round the displayed value.) Apply the same procedure to calculate the pH of milk with a hydrogen ion concentration of 2.20×10^{-7} M.

If we can convert hydrogen ion concentration into pH, how do we go in the reverse direction, that is, how to convert pH into a hydrogen ion concentration? Your calculator can do this for you if it has a button labeled "10^x." (Alternatively, it may use two buttons: first "Inv" and then "log." To demonstrate the procedure, suppose you wish to find the hydrogen ion concentration of human blood with a pH of 7.40. Proceed as follows: Enter 7.40, use the "plus/minus" key to change the sign to negative, and then hit 10^x (or follow whatever steps are appropriate for your calculator). The display should give the hydrogen ion concentration as 3.98×10^{-8} M. Now apply the same procedure to calculate the H^+ concentration of an acid rain sample with a pH of 3.6.

Your Turn — Appendix 3.1

Find the pH concentration in each sample.

a. tap water, $[H^+] = 1.0 \times 10^{-6}$ M
b. milk of magnesia, $[H^+] = 3.2 \times 10^{-11}$ M
c. lemon juice, $[H^+] = 5.0 \times 10^{-3}$ M
d. saliva, $[H^+] = 2.0 \times 10^{-7}$ M

Answers

a. 6.0   b. 10.5
c. 2.3   d. 6.7

Your Turn — Appendix 3.2

Find the H^+ concentration in each sample.

a. tomato juice, pH = 4.5
b. acid fog, pH = 3.3
c. vinegar, pH = 2.5
d. blood, pH = 7.6

Answers

a. 3.2×10^{-5} M   b. 5.0×10^{-4} M
c. 3.2×10^{-3} M   d. 2.5×10^{-8} M

Answers to Your Turn Questions Not Answered in the Text

Chapter 0

0.3 c. Given that so many drinks are now sold in plastic bottles (see Chapter 9), the total yearly mass may not be that high. If you drink two canned beverages per week, this translates to about 100 cans per year, or 1500 grams of aluminum. However, when you consider how many people are drinking canned beverages, the mass of aluminum is indeed quite large.

d. Search the web for "aluminum can recycling" to find some amazing statistics. Sites that you might wish to explore include Earth 911 and the Aluminum Association.

0.5 a. In 2010, the population of the United States was about 310 million.

c. Doing the math, 3 billion hectares divided by 11 billion hectares is about 0.27, or 27% of the biologically productive land available. For purposes of reference, the United States has about 4.5% of the world's population.

Chapter 1

1.7 b. Yes. Judging by Table 1.2, $PM_{2.5}$ must have more serious health consequences than PM_{10} because the concentrations are set at lower limits.

1.8 a. $\dfrac{1050\ \mu g\ SO_2}{15\ m^3} = \dfrac{70\ \mu g\ SO_2}{1\ m^3}$

b. The woman's exposure does not exceed either the 24-hr standard of 365 $\mu g/m^3$ or the annual standard of 80 $\mu g/m^3$.

1.9 Examples of preventing air pollution include (1) not burning leaves (produces smoke and particulate matter), but rather letting them compost or otherwise decompose, (2) burning low-sulfur coal rather than high-sulfur coal, using a scrubber to remove SO_2 if burning high-sulfur coal, or conserving so as to burn less coal of any type, (3) choosing methods of transportation, such as bicycling or walking, that do not release air pollutants.

1.11 a. Hydrogen (H_2) and helium (He) are elements.

b. Other substances found in the air include nitrogen, N_2 (element), oxygen, O_2 (element), argon, Ar (element), carbon dioxide, CO_2 (compound), and water vapor, H_2O (compound).

c. Hydrogen (0.54 ppm, or 0.000054%), helium (5 ppm, or 0.0005%), and methane (17 ppm, or 0.00017%).

1.13 b. SO_2 is sulfur dioxide; SO_3 is sulfur trioxide.

1.14 a. The *eth-* in ethanol indicates 2 C atoms in the chemical formula.

c. The *prop-* in propanol indicates 3 C atoms in the chemical formula.

1.15 b.

Balanced equation: $N_2 + 2\,O_2 \longrightarrow 2\,NO_2$

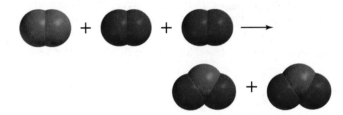

1.17 Equation 1.7 contains 16 C, 36 H, and 34 O on each side.

1.19 Your list should include: O_2, N_2, CO_2, CO, H_2O, NO, soot (particulate matter), and VOCs. The exhaust also contains tiny amounts of Ar and even tinier amounts of He, but we usually omit these gases as they are inert and at low concentration.

1.20 b. $CuS + O_2 \longrightarrow Cu + SO_2$

1.21 a. Other gasoline-powered machines or vehicles include some lawn mowers, leaf blowers, forklifts, chain saws, snow blowers, and electrical generators.

b. One example is lawn and garden equipment, such as mowers and blowers. The U.S. EPA has set emissions reductions for 2012. The regulations will reduce both fuel evaporation and exhaust gas pollutants. For the latter, emissions control technology will be used, just as currently done for larger engines.

1.27 a. A concentration of 1193 μg of particulate matter per cubic meter of air exceeds the national ambient air quality standards for both PM_{10} and for $PM_{2.5}$.

b. Breathing fine particles at this level is hazardous for everybody. The primary danger is to the cardiovascular system, as the particles, when inhaled, pass into the bloodstream and cause or further aggravate heart disease.

1.28 Indoor activities that generate pollutants include burning incense, cigarette or cigar smoking, frying foods (especially when something burns), operating a faulty furnace or space heater, painting or varnishing (except low-VOC paints), using some cleaning products such as ammonia or spray oven cleaner, using aerosol hair sprays and some hair-coloring products, using some furniture polishes, using spray insecticides, and many others.

1.30 The Green Chemistry principles met by the new coalescents include that it is better to prevent waste than to treat or clean up waste after it is formed (i.e., not to put VOCs into the air), to use renewable materials (i.e., produced from vegetable oils), to use materials that degrade into innocuous products at the end of their useful life. Those met by the new undercoat include that it is better to prevent waste than to treat or clean up waste after it is formed (i.e., not to put VOCs into the air) and that it is better to use less energy (i.e., to cure the undercoat without using heat).

1.32 b. No, it wouldn't be more valid. The additional digits would not provide any real information, as the meter is not able to determine values with this level of precision. Compare this to dividing $20 evenly among seven people. Even if your calculator displayed the result of $2.857142857, each person would not receive this amount because we have no coins in the United States smaller than pennies.

Chapter 2

2.2 b. The maximum is 12,000 ozone molecules per billion molecules and atoms of all gases that make up the stratosphere (see paragraph that precedes this activity).

   c. The EPA limit is 0.075 ppm for an 8-hr average, equivalent to 75 ppb, or 75 ozone molecules per billion molecules and atoms of all gases that make up the troposphere (see Table 1.2).

2.4 c. 17 protons, 17 electrons

   d. 24 protons, 24 electrons

2.5 c. Group 5A; 5 outer electrons

   d. Group 8A; 8 outer electrons

2.6 b. Beryllium (Be), magnesium (Mg), calcium (Ca), strontium (Sr), barium (Ba), and radium (Ra) all have two outer electrons and are members of Group 2A.

2.7 c. 53 protons, 53 electrons, 78 neutrons

2.8 b. 2 Br atoms (:B̈r·) × 7 outer electrons per atom
                                        = 14 outer electrons
Here is the Lewis structure for Br_2.

:B̈r:B̈r: or   :B̈r—B̈r:

2.9 b. 1 C atom (·C̈·) × 4 outer electrons per atom
                                        = 4 outer electrons

   2 Cl atoms (:C̈l·) × 7 outer electrons per atom
                                        = 14 outer electrons

   2 F atoms (:F̈·) × 7 outer electrons per atom
                                        = 14 outer electrons

                     Total = 32 outer electrons

Here is the Lewis structure for CCl_2F_2.

2.10 b. 1 S atom (·S̈·) × 7 outer electrons per atom
                                        = 14 outer electrons

   2 O atoms (·Ö·) × 6 outer electrons per atom
                                        = 12 outer electrons

                     Total = 26 outer electrons

Here are two resonance forms for the Lewis structure of SO_2.

2.12 b. Radio waves have an average wavelength of about 10^1 m. In contrast, the average wavelength for an X-ray is about 10^{-9} m, thus making a radio wave approximately 10^{10} times longer.

2.14 a. UV-A < UV-B < UV-C

   b. No, the order is not the same. Rather, it is reversed because wavelength and energy are inversely proportional.

2.16 a. O_3 forms from O and O_2.

$$O + O_2 \longrightarrow O_3$$

   b. The source of the O atoms is the breakdown of the oxygen molecule (naturally occurring in the atmosphere) in the presence of UV light.

$$O_2 \longrightarrow 2\,O$$

   c. O_3 molecules in the stratosphere break down by several mechanisms, both part of the Chapman cycle. One is that they decompose into O_2 and O.

$$O_3 \longrightarrow O_2 + O$$

   The other is that they combine with atomic oxygen to form 2 molecules of O_2.

$$O_3 + O \longrightarrow 2\,O_2$$

2.23 a. 1 H atom (H·) × 1 outer electron per atom
                                        = 1 outer electron

   1 O atom (·Ö·) × 6 outer electrons per atom
                                        = 6 outer electrons

                     Total = 7 outer electrons

Here is the Lewis structure for the ·OH radical.

·Ö:H   or   ·Ö—H

b. 1 N atom ($\cdot \overset{\cdot\cdot}{\underset{\cdot}{N}}\cdot$) × 5 outer electrons per atom
= 5 outer electrons

1 O atom ($\cdot \overset{\cdot\cdot}{\underset{\cdot\cdot}{O}}\cdot$) × 6 outer electrons per atom
= 6 outer electrons

Total = 11 outer electrons

Here is the Lewis structure for the ·NO radical. Note the unpaired electron on the N atom.

$$\overset{\cdot}{N}=\overset{\cdot\cdot}{\underset{\cdot\cdot}{O}}$$

c. 2 N atoms ($\cdot \overset{\cdot\cdot}{\underset{}{N}}\cdot$) × 5 outer electrons per atom
= 10 outer electrons

1 O atom ($\cdot \overset{\cdot\cdot}{\underset{\cdot\cdot}{O}}\cdot$) × 6 outer electrons per atom
= 6 outer electrons

Total = 16 outer electrons

Here is the Lewis structure for N_2O.

$$:N\equiv N-\overset{\cdot\cdot}{\underset{\cdot\cdot}{O}}:$$

2.24 The chemical equations with bromine, analogous to equations 2.10 and 2.15 are

$$2\,Br\cdot + 2\,O_3 \longrightarrow 2\,ClO\cdot + 2\,O_2$$
$$Br\cdot + O_3 \longrightarrow ClO\cdot + O_2$$

2.29 **a.** Halon-1301 and HFC-23 have chemical formulas of $CBrF_3$ and CHF_3, respectively. Halon-1301 would be expected to deplete ozone because it contains bromine. Since HFC-23 contains neither bromine nor chlorine, it does not deplete ozone.

b. Figure 2.21 shows that HFC-23 has a high global warming potential. For this reason, it is unlikely to be used in future years.

Chapter 3

3.2 **b.** infrared, visible, ultraviolet

3.3 **a.** 25% of energy from the Sun is reflected from the atmosphere, 6% is reflected from the surface, 9% is emitted from the surface, and 60% is emitted from the atmosphere. These percents add up to 100%.

b. 23% of the incoming radiation is directly absorbed in the atmosphere, and an additional 37% is absorbed in the atmosphere after the Earth radiates longer wavelength heat energy. These values add up to 60%, the total energy emitted from the atmosphere.

c. The colors differentiate the wavelengths of light emitted from the Sun and Earth. Yellow represents a mixture of all incoming wavelengths, blue represents the shorter wavelength (higher energy) UV radiation, and red represents the longer wavelength (lower energy) IR radiation.

3.5 **a.** The atmospheric CO_2 concentration in 1960 was about 315 ppm; in 2010 it was about 390 ppm. The percent increase is

$$\frac{390\ ppm - 315\ ppm}{315\ ppm} \times 100 = 23.8\%$$

b. Within a given year, the atmospheric CO_2 concentration varies by 6–7 ppm.

3.8 **b.** Total outer electrons: $4 + 2(7) + 2(7) = 32$. Eight go around the central C atom to form four single bonds, one to each Cl atom and one to each F atom. The other 24 outer electrons are nonbonding pairs on the Cl or F atoms. The CCl_2F_2 molecule is tetrahedral, the same shape as CH_4 and CCl_4.

c. Total outer electrons: $2(1) + 6 = 8$. These eight all go around the central S atom to form two single bonds, one to each H atom, and two nonbonded pairs. The H_2S molecule is bent, like H_2O.

3.9 **a.** Total outer electrons: $6 + 2(6) = 18$. Eight go around the central S atom to form one single bond to an O atom, one double bond to an O atom, and one nonbonded pair. The other 10 outer electrons are nonbonding pairs on the O atoms. The SO_2 molecule is bent, just as the O_3 molecule is bent. Only one resonance form is shown.

b. Total outer electrons: $6 + 3(6) = 24$. Eight go around the central S atom to form one double bond to an O atom and two single bonds to two additional O atoms. The other 18 outer electrons are nonbonding pairs on O atoms. The SO_3 molecule is triangular planar. Only one of three resonance forms is shown.

3.11 **a.** The processes that remove carbon from the atmosphere include emissions from oceans, deforestation, and the burning of fossil fuels.

b. Photosynthesis, absorption in oceans, and reforestation are processes that remove carbon from the atmosphere.

c. The two largest reservoirs of carbon are fossil fuels and carbonate minerals.

d. The parts of the carbon cycle most influenced by human activities are burning fossil fuels and deforestation. The extra anthropogenic atmospheric CO_2 also increases the rate of CO_2 absorption into the oceans.

3.12 **a.** The atomic number of N is 7 and the atomic mass is 14.01.

b. A neutral N-14 atom has 7 protons, 7 neutrons, and 7 electrons.

c. A neutral N-15 atom has 7 protons, 8 neutrons, and 7 electrons. Only the number of neutrons differs.

d. N-14 is the most abundant natural isotope, given that the atomic mass is 14.01.

3.14 b. 5×10^9 N atoms $\times \dfrac{2.34 \times 10^{-23} \text{ g N}}{1 \text{ N atom}}$

$= 1.17 \times 10^{-13}$ g N

c. 6×10^{15} N atoms $\times \dfrac{2.34 \times 10^{-23} \text{ g N}}{1 \text{ N atom}}$

$= 1.40 \times 10^{-7}$ g N

3.15 b. 1 mol N_2O = 44.0 g N_2O

c. 1 mol CCl_3F = 137.5 g CCl_3F

3.16 c. The mass ratio compares the molar mass of N to the molar mass of N_2O.

$$\frac{1.0 \text{ mol } N_2O}{44.0 \text{ g } N_2O} \times \frac{2 \text{ mol N}}{1 \text{ mol } N_2O} \times \frac{14 \text{ g N}}{1 \text{ mol N}} = \frac{0.636 \text{ g N}}{1 \text{ g } N_2O}$$

To find the mass percent of N in N_2O, multiply the mass ratio by 100.

$$\frac{0.636 \text{ g N}}{1.00 \text{ g } N_2O} \times 100 = 63.6\% \text{ N in } N_2O$$

3.17 b. The mass ratio of S to SO_2 is known from Your Turn 3.16.

$$142 \times 10^6 \text{ t } SO_2 \times \frac{32.1 \times 10^6 \text{ t S}}{64.1 \times 10^6 \text{ t } SO_2} = 7.11 \times 10^7 \text{ t S}$$

3.19 For CO_2: $\dfrac{390 \text{ ppm} - 270 \text{ ppm}}{270 \text{ ppm}} \times 100 = 44\%$

For CH_4: $\dfrac{1.76 \text{ ppm} - 0.70 \text{ ppm}}{0.70 \text{ ppm}} \times 100 = 151\%$

For N_2O: $\dfrac{0.322 \text{ ppm} - 0.275 \text{ ppm}}{0.270 \text{ ppm}} \times 100 = 17\%$

Order: $CH_4 > CO_2 > N_2O$

3.21 a. The global warming potential of CFC-12 is 8100 times greater than CO_2. However, CFC-12 has a much lower abundance in the troposphere, 0.56 parts per billion (ppb) compared with the 2010 value for CO_2 of 390,000 ppb (or 390 ppm). Therefore CO_2 is a more effective greenhouse gas.

b. The global warming potential of HFC-134a is 1300 times greater than CO_2. However, HFC-134a has a far lower abundance in the troposphere, 7.5 parts per trillion (ppt) compared with 390,000,000 ppt (390 ppm) for CO_2. The atmospheric lifetime of HFC-134a also is shorter than that of CO_2. Therefore CO_2 is a more effective greenhouse gas.

c. Global atmospheric lifetime is important because this is a measure of how long a greenhouse gas persists in the atmosphere. Freon-12 has a higher GWP, a higher abundance, and a longer atmospheric lifetime than HFC-134a. Therefore, it is able to have a larger effect on global warming over a longer period of time, a major reason it has been banned.

3.22 Infrared, visible, and ultraviolet radiation are the main types of solar radiation striking the Earth. Visible light accounts for the largest percentage.

3.24 b. Increases in greenhouse gas concentration and changes in the Earth's albedo are additional forcings required to accurately recreate the temperature data for the 20th century.

3.29 (5×10^6 metric tons/6×10^9 metric tons) $\times$ 100% = 0.083%.

3.30 a. 19 tons $CO_2 \times \dfrac{2000 \text{ lb } CO_2}{1 \text{ ton } CO_2} \times \dfrac{1 \text{ tree}}{25 \text{ lb } CO_2}$

$= 1520$ trees

If 50 lb per tree is used, the answer is 760 trees.

b. Using 50 lb CO_2 per tree:

$$7 \times 10^9 \text{ trees} \times \frac{50 \text{ lb } CO_2}{1 \text{ tree}} \times \frac{1 \text{ ton } CO_2}{2000 \text{ lb } CO_2}$$

$= 1.7 \times 10^8$ tons CO_2 absorbed

$$\frac{1.7 \times 10^8 \text{ tons } CO_2 \text{ absorbed}}{6.0 \times 10^9 \text{ tons } CO_2 \text{ emitted}} \times 100\% = 2.8\%$$

Chapter 4

4.3 During the composting process, one of the compounds produced is water. The process also gives off heat that may convert the water from a liquid to a gas. On a cool day, the water vapor may come in contact with the chilly air and condense to form "steam." *Note:* The "steam" that you can see is actually condensed water vapor, such as in fog or a cloud. Water vapor itself is invisible.

4.4 b. Plant A burns 4.4×10^8 g of coal per day, creating 1.6×10^9 g of CO_2. Plant B burns 3.6×10^8 g of coal per day, creating 1.3×10^9 g of CO_2. Therefore, Plant B emits 3.0×10^8 fewer grams of CO_2 each day than Plant A.

4.8 d. 1.0 g $C_{135}H_{96}O_9NS \times \dfrac{1 \text{ mol } C_{135}H_{96}O_9NS}{1906 \text{ g } C_{135}H_{96}O_9NS} \times$

$\dfrac{135 \text{ mol } CO_2}{1 \text{ mol } C_{135}H_{96}O_9NS} \times \dfrac{6.02 \times 10^{23} CO_2 \text{ molecules}}{1 \text{ mol}}$

$= 4.2 \times 10^{22} CO_2$ molecules

4.9 a. Coal contains small amounts of sulfur that combine with oxygen during combustion to produce SO_2. Volcanoes are a natural source of SO_2 in the atmosphere.

b. Although coal does contain small amounts of nitrogen that combine with oxygen during combustion, the major portion of NO is formed from the reaction of N_2 and O_2 from the air in the high temperatures produced in the combustion process. Other sources of NO include engine exhaust, lightning, and grain silos.

4.12 a. 1500 kJ $\times \dfrac{1 \text{ mol } CH_4}{50.7 \text{ kJ}} \times \dfrac{1 \text{ mol } CO_2}{1 \text{ mol } CH_4} \times$

$\dfrac{44 \text{ g } CO_2}{1 \text{ mol } CO_2} = 1300$ g CO_2

b. For bituminous coal from Maryland, which releases 30.7 kJ/g, 2150 g of CO_2 is released when 1500 kJ of heat is produced.

4.17 From Section 2.3, here are the resonance forms for ozone:

$$\ddot{O}{=}\ddot{O}{-}\ddot{\underset{\cdot\cdot}{O}}{:} \longleftrightarrow {:}\ddot{\underset{\cdot\cdot}{O}}{-}\ddot{O}{=}\ddot{O}$$

And here is the Lewis structure for oxygen:

$$\ddot{\underset{\cdot\cdot}{O}}{::}\ddot{\underset{\cdot\cdot}{O}} \quad \text{or} \quad \ddot{O}{=}\ddot{O}$$

The bond energy for O_3 is *intermediate* between the O–O single bond (147 kJ/mol) and the O=O double bond (498 kJ/mol) values; that is, it is less than the bond energy for the O=O double bond in O_2 (498 kJ/mol). Energy is inversely proportional to wavelength. Therefore, the *higher* bond energy of O_2 requires radiation of *shorter* wavelength to break its bonds.

4.18 a. One possible pair of products is C_8H_{18} and C_8H_{16}. Their structural formulas are:

```
    H   H   H   H   H   H   H   H
    |   |   |   |   |   |   |   |
H — C — C — C — C — C — C — C — C — H
    |   |   |   |   |   |   |   |
    H   H   H   H   H   H   H   H


    H   H   H   H   H   H   H   H
    |   |   |   |   |   |   |   |
H — C — C — C — C — C — C — C = C
    |   |   |   |   |   |   |   |
    H   H   H   H   H   H   H   H
```

c. The generic chemical formula is C_nH_{2n}, where n is an integer.

4.20 Every 100 g of E85 contains 85 g of ethanol and 15 g of gasoline. The energy content for 100 g of E85 is 3200 kJ, whereas the energy content for 100 g of pure gasoline (assume pure octane) is 4890 kJ. Therefore, you would expect roughly a 35% decrease in gas mileage.

Chapter 5

5.3 b. The O–H bond is more polar. The electron pair is more strongly attracted to the O atom.

c. The S–O bond is more polar. The electron pair is more strongly attracted to the O atom.

5.5 a. The dashed lines represent hydrogen bonds; that is, a weak attraction between a H atom bonded to an electronegative atom (F, N, or O) and a neighboring electronegative atom on another molecule.

b. Any of the H atoms on the water molecules could be labeled δ^+. Similarly, any of the O atoms could be labeled δ^-. The molecules are oriented so that the opposite charges are in close proximity.

c. Hydrogen bonds are intermolecular forces because they occur *between* two molecules.

5.12 a. 1.6×10^{-2} ppm; 16 ppb

b. No, the mercury concentration is eight times higher than the standard of 2 ppb.

5.13 a. 8.0×10^{-8} M

b. A 500 mL sample of 1.5 M NaCl(*aq*) contains 0.75 mol of solute. A 500 mL sample of 0.15 M NaCl(*aq*) contains 0.075 mol of solute.

c. The solution made from 0.60 mol NaCl is more concentrated. Converting to molarity, it has a concentration of 3.0 M compared with 2.0 M for the other solution.

d. No, the resulting solution is not 2.0 M. When 40.0 g of $CuSO_4$ is dissolved in a 1.0 L of water, the solution contains only 0.25 mol of solute.

5.14 a. Li and K form cations; S and N form anions.

b. Mg^{2+}, Lewis structures: $\cdot Mg \cdot$ and Mg^{2+}

O^{2-}, Lewis structures: $\cdot \ddot{\underset{\cdot\cdot}{O}} \cdot$ and $\left[{:}\ddot{\underset{\cdot\cdot}{O}}{:} \right]^{2-}$

Al^{3+}, Lewis structures: $\cdot \dot{Al} \cdot$ and Al^{3+}

5.15 a. CaS, calcium sulfide

b. KF, potassium fluoride

c. From Figure 5.19, Mn can form two ions: Mn^{2+} and Mn^{4+}. O forms the oxide ion, O^{2-}. The possible chemical formulas are MnO, manganese(II) oxide and MnO_2, manganese(IV) oxide.

d. $AlCl_3$, aluminum chloride

5.16 c. $Al(C_2H_3O_2)_3$ **d.** K_2CO_3

5.17 c. Sodium hydrogen carbonate or sodium bicarbonate

d. Calcium carbonate

e. Magnesium phosphate

5.18 a. NaClO   **b.** $MgCO_3$ **c.** NH_4NO_3

5.20 b. Soluble. All sodium salts are soluble.

c. Insoluble. Most sulfides are insoluble (except those with Group 1A or NH_4^+ cations).

d. Insoluble. Most hydroxides are insoluble (except those with Group 1A or NH_4^+ cations).

5.21 All of the oxygen atoms should be labeled δ^-. Only the hydrogen atoms attached to the oxygen atoms should be labeled as δ^+. The hydrogen atoms bonded to carbon atoms do not carry a partial positive charge, as the C–H bond is nonpolar.

5.22 C–C bonds are nonpolar. C–H bonds are also nonpolar, as the electronegativity difference between C and H is small.

5.26 a. The sample with a concentration of 20 ppb lead ion (20 mg/L) is higher than that of 0.003 mg/L lead (3 mg/L or 3 ppb).

b. A concentration of 20 ppb Pb^{2+} is greater than the 15 ppb standard, and 3 ppb Pb^{2+} is less than the 15 ppb standard.

5.28 a. Sulfate ion is SO_4^{2-}, hydroxide ion is OH^-, calcium ion is Ca^{2+}, and aluminum ion is Al^{3+}.

b. Calcium sulfate, $CaSO_4$; calcium hydroxide, $Ca(OH)_2$, aluminum sulfate, $Al_2(SO_4)_3$; aluminum hydroxide, $Al(OH)_3$

c. Sodium hypochlorite $NaClO$, calcium hypochlorite $Ca(ClO)_2$

5.29 **a.** Possibilities include $CHCl_3$, $CHBr_3$, $CHBrCl_2$, and $CHBr_2Cl$. The Lewis structures for these are analogous to the one shown here for CHF_2Cl.

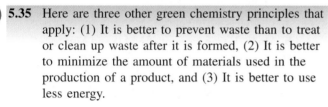

b. THMs do not contain fluorine atoms.

5.35 Here are three other green chemistry principles that apply: (1) It is better to prevent waste than to treat or clean up waste after it is formed, (2) It is better to minimize the amount of materials used in the production of a product, and (3) It is better to use less energy.

Chapter 6

6.1 **a.** Possibilities include hydrochloric acid, sulfuric acid, acetic acid, nitric acid, phosphoric acid, lactic acid, and citric acid.

b. Possibilities include having used any of these acids in chemistry experiments, having heard about lactic acid buildup in muscles, and having seen phosphoric acid and citric acid listed as an ingredient in soft drinks.

6.2 **a.** $HI(aq) \longrightarrow H^+(aq) + I^-(aq)$

b. $HNO_3(aq) \longrightarrow H^+(aq) + NO_3^-(aq)$

6.4 **a.** $KOH(s) \xrightarrow{H_2O} K^+(aq) + OH^-(aq)$

b. $LiOH(s) \xrightarrow{H_2O} Li^+(aq) + OH^-(aq)$

c. $Ca(OH)_2(s) \xrightarrow{H_2O} Ca^{2+}(aq) + 2\,OH^-(aq)$

6.5 **a.** $HNO_3(aq) + KOH(aq) \longrightarrow KNO_3(aq) + H_2O(l)$

$H^+(aq) + NO_3^-(aq) + K^+(aq) + OH^-(aq) \longrightarrow$
$K^+(aq) + NO_3^-(aq) + H_2O(l)$

$H^+(aq) + OH^-(aq) \longrightarrow H_2O(l)$

b.
$H_2SO_4(aq) + 2\,NH_4OH(aq) \longrightarrow (NH_4)_2SO_4(aq) + 2\,H_2O(l)$

$2\,H^+(aq) + SO_4^{2-}(aq) + 2\,NH_4^+(aq) + 2\,OH^-(aq) \longrightarrow$
$2\,NH_4^+(aq) + SO_4^{2-}(aq) + 2\,H_2O(l)$

$2\,H^+(aq) + 2\,OH^-(aq) \longrightarrow 2\,H_2O(l)$

$H^+(aq) + OH^-(aq) \longrightarrow H_2O(l)$

6.6 **b.** The solution is basic because $[OH^-] > [H^+]$.
$[H^+][OH^-] = 1 \times 10^{-14}$.
Solving, $[H^+] = 1 \times 10^{-8}$ M.

c. The solution is basic because $[OH^-] > [H^+]$.
$[H^+][OH^-] = 1 \times 10^{-14}$.
Solving, $[OH^-] = 1 \times 10^{-4}$ M.

6.7 **a.** When potassium hydroxide dissociates, one hydroxide ion is released for each potassium ion. The basic solution contains much more OH^- than H^+. $OH^-(aq) = K^+(aq) > H^+(aq)$

b. When nitric acid dissociates, one hydrogen ion is released for each nitrate ion. The acidic solution contains much more H^+ than OH^-. $H^+(aq) = NO_3^-(aq) > OH^-(aq)$

c. When sulfuric acid dissociates, two hydrogen ions are released for each sulfate ion. The acidic solution contains much more H^+ than OH^-. $H^+(aq) > SO_4^{2-}(aq) > OH^-(aq)$

6.9 **a.** Although the pH values differ only by 1, the lake water sample is 10 times more acidic and (for an equal volume) would have 10 times more H^+ than the rain sample.

b. Although the pH values differ only by 3, the tap water sample is 1000 times more acidic and (for an equal volume) would have 1000 times more H^+ than the ocean water sample.

6.13 $H_2SO_3(aq) \longrightarrow H^+(aq) + HSO_3^-(aq)$
$HSO_3^-(aq) \longrightarrow H^+(aq) + SO_3^{2-}(aq)$
$H_2SO_3(aq) \longrightarrow 2\,H^+(aq) + SO_3^{2-}(aq)$

6.14 **c.** $1.68 \times 10^4 \text{ tons SO}_2 \times \dfrac{64 \text{ tons SO}_2}{32 \text{ tons S}}$
$= 3.36 \times 10^4 \text{ tons SO}_2$

d. The SO_3 can react with water to form sulfuric acid: $SO_3(g) + H_2O(l) \longrightarrow H_2SO_4(aq)$

6.15 It requires 946 kJ of energy to break a mole of triple bonds between nitrogen atoms. This is one of the highest bond energies in Table 4.4.

6.16 Nitrogen monoxide (NO) is a gas that forms in automobile combustion engines. As you saw in Chapter 1, once released in the exhaust, it reacts in the atmosphere to produce NO_2.

$$N_2 + O_2 \xrightarrow{\text{high temperature}} 2\,NO$$
$$2\,NO + O_2 \longrightarrow 2\,NO_2$$

NO also reacts with ozone to form nitrogen dioxide (NO_2).

$$NO(g) + O_3(g) \longrightarrow NO_2(g) + O_2(g)$$

Nitrogen dioxide (NO_2) is a toxic brown gas. One indication of its reactivity is the damage it does to your lungs if you breathe it. If NO_2 were inert, it would not be such a health hazard. Ammonia (NH_3) is a gas that you may have encountered in the context of "household ammonia," an aqueous solution of ammonia. One indication of the reactivity of NH_3 is that it is a good cleaning agent. In addition, ammonia is a base that reacts quickly with acids.

6.17 Industrial processing and miscellaneous sources of NO_x did not change appreciably between 1940 and 1995. In contrast, fuel combustion (the burning of coal) and transportation (automobiles) did. The large increase in emissions of NO_x in the 1970s had two sources. The first was electric power plants (large stationary sources) and the second was automobiles (small mobile sources). In both cases, nitrogen monoxide was produced in the presence of heat from N_2 and O_2 in the air.

6.18 **a.** Iron (Fe) is a metal.

b. Aluminum (Al) is a metal.

c. Fluorine (F) is a nonmetal.

d. Calcium (Ca) is a metal.

e. Zinc (Zn) is a metal.

f. Oxygen (O) is a nonmetal.

6.19 $4\,Fe(s) + 2\,O_2(g) + \cancel{8\,H^+(aq)} \longrightarrow$

$$\cancel{4\,Fe^{2+}(aq)} + \cancel{4\,H_2O(l)}$$

$\cancel{4\,Fe^{2+}(aq)} + O_2(g) + \cancel{4\,H_2O(l)} \longrightarrow$

$$2\,Fe_2O_3(s) + \cancel{8\,H^+(aq)}$$

$$\overline{4\,Fe(s) + 3\,O_2(g) \longrightarrow 2\,Fe_2O_3(s)}$$

6.20 **a.** Fe is the metal form of iron.

c. Fe^{2+} and Fe^{3+} have lost two and three valence electrons, respectively.

6.21 **a.** $MgCO_3(s) + 2\,H^+(aq) \longrightarrow$

$$Mg^{2+}(aq) + CO_2(g) + H_2O(l)$$

b. Sodium bicarbonate is soluble in water. Section 5.8 does not contain this information explicitly, but any experience using baking soda (i.e., sodium bicarbonate) in cooking will show that it is indeed soluble in water.

6.22 $CaCO_3(s) + H_2SO_4(aq) \longrightarrow$

$$CaSO_4(s) + CO_2(g) + H_2O(l)$$

6.25 $S(s) + O_2(g) \longrightarrow SO_2(g)$

$SO_2(g) + \frac{1}{2}\,O_2(g) \longrightarrow SO_3(g)$

$SO_3(g) + H_2O(l) \longrightarrow H_2SO_4(aq)$

6.26 **a.** $H_2SO_4(aq) + NH_4OH(aq) \longrightarrow$

$$NH_4HSO_4(aq) + H_2O(l)$$

b. $H_2SO_4(aq) + 2NH_4OH(aq) \longrightarrow$

$$(NH_4)_2SO_4(aq) + 2\,H_2O(l)$$

6.28 **b.** The hydrogen carbonate ion is acting as a base because it is reacting with an acid, that is, a hydrogen ion.

Chapter 7

7.5 U-234 has 92 protons and 142 neutrons.

7.6 **b.** $_0^1n + _{92}^{235}U \longrightarrow _{52}^{137}Te + _{40}^{97}Zr + 2\,_0^1n$

7.7 For anthracite coal:

$$9.0 \times 10^{10}\,kJ \times \frac{1.0\,g\ anthracite\ coal}{30.5\,kJ} \times \frac{1\,kg}{1000\,g}$$

$$= 3.0 \times 10^6\,kg\ anthracite\ coal$$

The equivalent masses for the other grades are calculated in the same way: bituminous coal, 2.9×10^6 kg; subbituminous coal, 3.8×10^6 kg; lignite (brown coal), 5.6×10^6 kg.

7.8 $_{95}^{241}Am \longrightarrow _{93}^{237}Np + _2^4He$

$_4^9Be + _2^4He \longrightarrow _6^{12}C + _0^1n + _0^0\gamma$

7.9 In an emergency such as an earthquake, the fuel rods should be inserted in preparation for shutdown in order to slow the nuclear fission reaction in the reactor core.

7.10 The cloud is tiny droplets of condensed water vapor. People sometimes call this "steam," but technically this is not correct. Steam (water vapor) is invisible until it condenses. The cloud does not contain any nuclear fission products.

7.12 **c.** "Radiation" refers to UV radiation, a region of the electromagnetic spectrum.

d. "Radiation" refers to the nuclear radiation (alpha particles) emitted by uranium nuclei.

7.13 **b.** $_{94}^{239}Pu \longrightarrow _{92}^{235}U + _2^4He$

7.14 **a.** $\ddot{\underset{..}{I}}\cdot \qquad :\ddot{\underset{..}{I}}:\ddot{\underset{..}{I}}: \qquad \left[:\ddot{\underset{..}{I}}:\right]^-$
    atom   molecule   ion

b. The iodine atom is the most reactive here because it has one unpaired electron, that is, it is a free radical.

7.20 **a.** Answers vary because of cosmic ray exposure (elevation) and medical exposures. As of 2010, the EPA link was http://www.epa.gov/rpdweb00/understand/calculate.html and the Los Alamos National Laboratory (LANL) link was http://newnet.lanl.gov/info/dosecalc.asp.

b. The results are not identical because the questions are different. For example, the LANL calculator asks more detailed questions about lifestyle.

7.21 $\dfrac{3.0 \times 10^{14}\ C\text{-}14\ atoms}{3.5 \times 10^{26}\ all\ C\ atoms} \times 100 = 8.57$

$$\times 10^{-11}\%\ of\ carbon\text{-}14$$

7.23 **a.** The other isotope is U-238. A trace amount of U-234 is present as well.

b. U-238 is not fissionable under the conditions present in a nuclear reactor.

c. These other radioisotopes are fission products. U-235 splits in many different ways, so many radioisotopes are formed.

7.25 After 10 half-lives, 0.0975% of the original sample remains.

# of half-lives	% decayed	% remaining
0	0	100
1	50	50
2	75	25
3	87.5	12.5
4	93.75	6.25
5	97.88	3.12
6	98.44	1.56
7	99.22	0.78
8	99.61	0.39
9	99.805	0.195
10	99.9025	0.0975

7.26 After 36.9 years (3 half-lives), 12.5% of the original tritium will remain.

7.27 a. Radon-222 is a product in the natural decay series of uranium-238, the isotope of uranium of highest natural abundance. Therefore if uranium is present in the rocks and soils, radon will be present as well.

    b. Four half-lives (4 × 3.8 days = 15.2 days) are required for the level of radioactivity to drop from 16 pCi to 1 pCi.

    c. Radon will continue to enter your basement because the uranium in the soils and rocks underneath your home continuously produces it. See part **a**.

7.28 $^{1}_{0}n + {}^{235}_{92}U \longrightarrow {}^{143}_{54}Xe + {}^{90}_{38}Sr + 3\,{}^{1}_{0}n$

7.34 a. The heat is released from the fission of the nuclear fuel (U-235).

    b. The reactor vessel and its components (control rods, fuel rods) are exposed to the highest temperatures.

Chapter 8

8.2 Equations **c** and **e** are reduction half-reactions because electrons are gained, appearing on the reactant side of the equations. Equations **b** and **d** are oxidation half-reactions because electrons are lost, appearing on the product side.

8.3 a. Each equation is balanced from the standpoint of number and type of atoms.

    b. Each equation is balanced from the standpoint of charge, but the total charge does not have to be zero on each side, just the same.

    c. Electrons do not appear in the overall cell reaction, but are shown in half-reactions.

8.4 a. $Zn \longrightarrow Zn^{2+} + 2\,e^-$

    b. $Cu^{2+} + 2\,e^- \longrightarrow Cu$

    c. To be rechargeable, the products need to remain near the reactants rather than dispersing. In the case of this cell, the ions formed are mobile and disperse throughout the solution rather than staying near the electrodes.

8.6 b. Metallic lead, Pb(s) loses two electrons to form aqueous lead ion, $Pb^{2+}(aq)$. Metallic lead is oxidized.

    c. The battery discharge is the forward reaction of equation 8.12. As in part **b**, when metallic lead, Pb(s), is oxidized to form lead ion, $Pb^{2+}(aq)$.

8.7 a. Metals are elements that are good conductors of electricity and heat. They have a shiny appearance. In contrast, nonmetals are poor conductors of electricity and heat and are not shiny. See Section 1.6. Metals have low electronegativity values in comparison to nonmetals. This explains why metals tend to lose electrons to form cations (while nonmetals tend to gain electrons to form anions). See Chapter 5 for more information about electronegativity.

    c. $CdS(s) + O_2(g) \longrightarrow Cd(s) + SO_2(g)$

    In CdS(s), the cadmium ion (Cd^{2+}) gains 2 electrons to form Cd(s) and is reduced.

    d. Sulfur dioxide released from smelting containing ores contributes air pollution. The EPA monitors the level of SO_2 as part of ambient air quality standards due to the serious nature of this air pollutant as a respiratory irritant.

8.15 b. The reactions are endothermic. More energy is required to break bonds than is released in bond formation.

    c. Although there is general agreement (+182 kJ vs. +165 kJ), remember that Table 4.4 gives *average* bond energies, not specific energies associated with the bonds in these compounds (with the exception of CO_2).

8.18 a. Doping with phosphorus forms an *n*-type semiconductor. Phosphorus is in Group VA and has an additional electron per atom than a silicon atom.

    b. Doping with boron forms a *p*-type semiconductor. Boron is in Group IIIA and has one fewer electron per atom than a silicon atom.

8.19 b. Uses include providing electricity to power lighting, heating, and other electrical needs in a building or in a process used by the business. For example, PVs have been used to generate enough electricity to power the electrical needs of a small microbrewery.

    c. Uses include supplying electricity to part or all of a home. At present, some homes that have a PV system are even able to sell electricity back to electric power companies. A PV system also can be used for electrical back-up in locations in which storms occur that disrupt the power.

8.22 a. A solar collector with a heat exchange fluid system would replace the burner/boiler systems.

b. Advantages of a distributed generation for collecting solar energy include (1) a distributed generation system does not require a condenser and does not generate thermal pollution, and (2) it is suitable for locations not reached by power lines. Advantages of a centralized location for conventional energy generation include (1) the efficiency of scale—a concentrated energy source servicing many users, and (2) the user is not required to install the energy generation equipment.

Chapter 9

9.4 a.

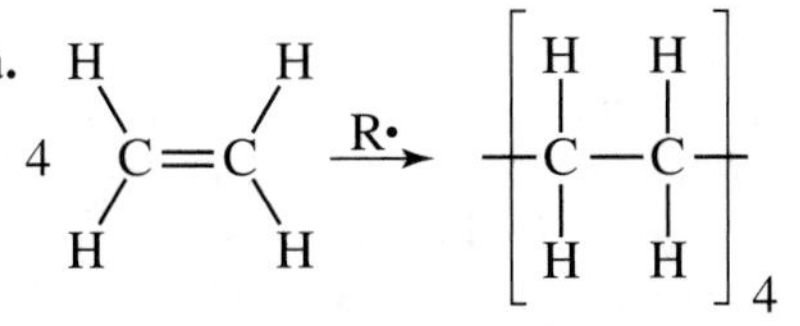

b.

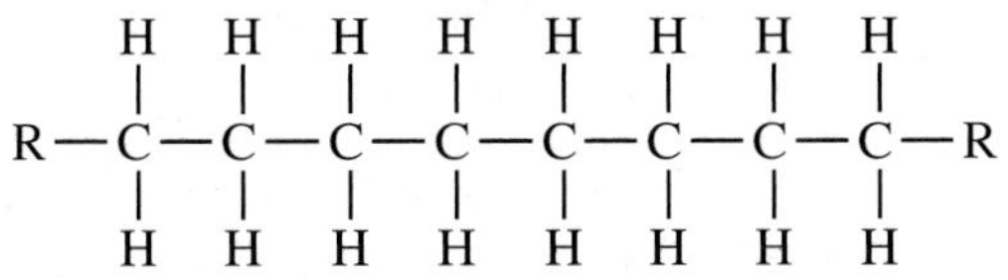

9.5 Pearly shampoo bottles, brightly colored detergent bottles, translucent milk jugs, squeeze glue bottles—these and many other plastic containers that hold soaps and nonoily foods typically are made of HDPE. This plastic is usually opaque and sometimes brightly colored (because it is mixed with a pigment). In contrast, you are more likely to find LDPE as a plastic packing sheet material or plastic baggie. It is usually soft and flexible.

9.11 a.

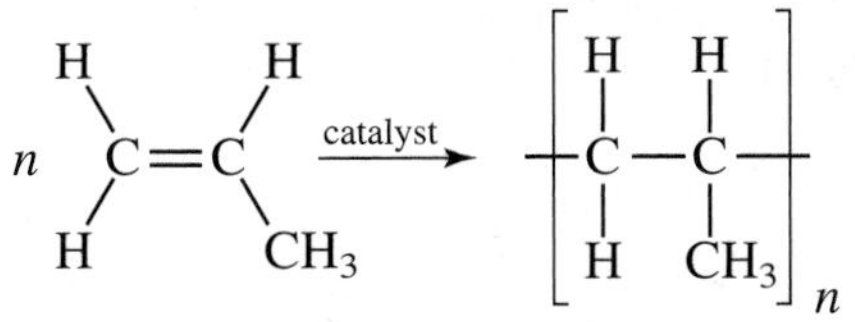

b.

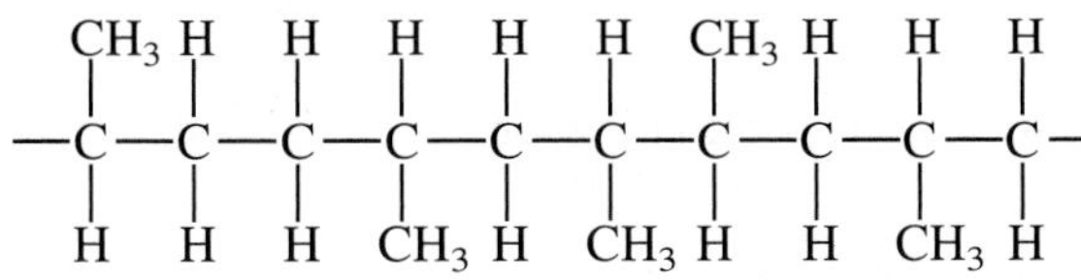

9.13

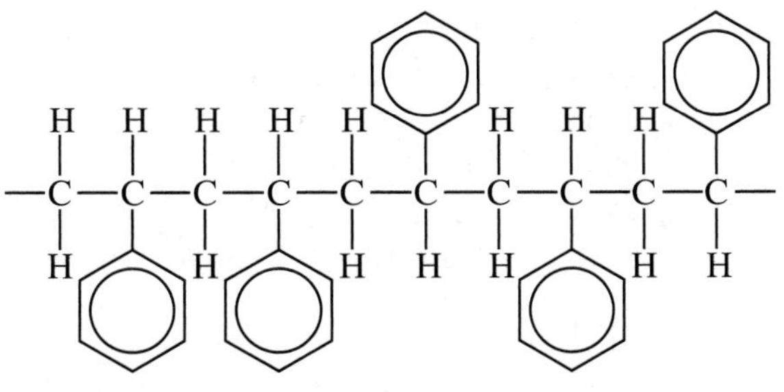

The head-to-tail arrangement is favored because this spaces the benzene rings as far apart as possible, minimizing repulsions between them.

In the head-to-head, tail-to-tail arrangement, the rings would fall on carbons that are right next to each other.

9.14 a.

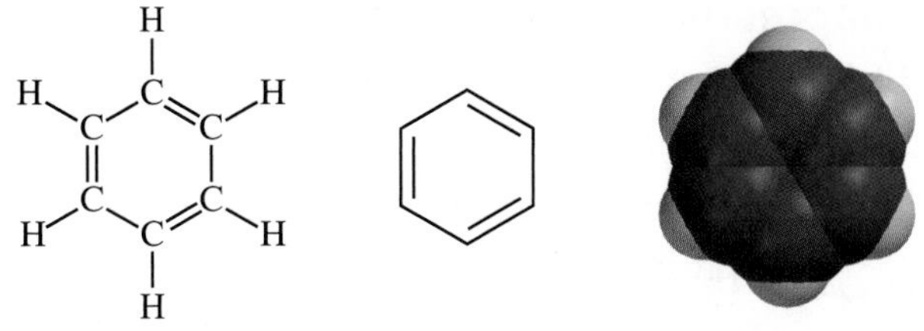

b.

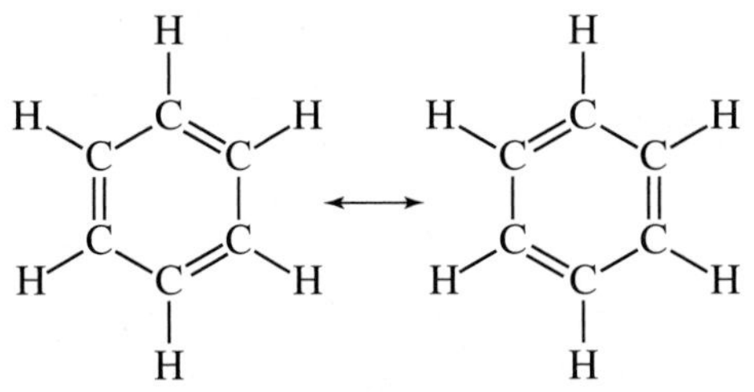

benzene

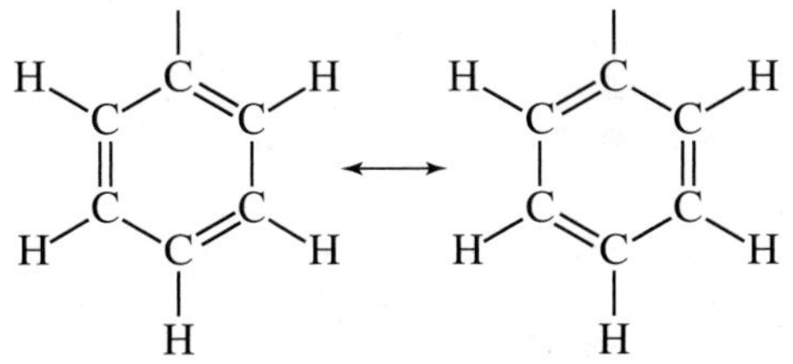

phenyl group

9.17

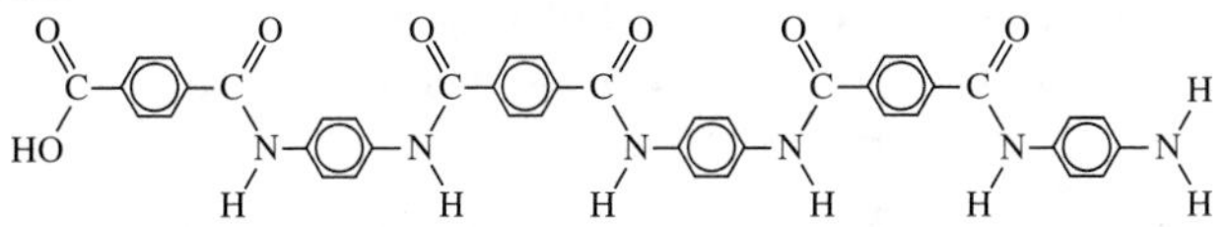

9.22 Propylene is $H_2C=CHCH_3$ or C_3H_6.

$$2500\ C_3H_6 + 11250\ O_2 \longrightarrow 7500\ CO_2 + 75000\ H_2O$$

9.25 When returned for a deposit, most glass bottles are cleaned, refilled, and reused. This is the most preferred scenario (at the top of the solid waste management hierarchy). Deposits on plastic containers are uncommon. However, some food stores, particularly food cooperatives, sell items in bulk. These stores may charge for plastic containers, thus encouraging customers to bring them back and refill them, time after time. This, too, is a preferred scenario.

9.27 Recyclable items that you might purchase include many plastic containers and jugs, beverages in aluminum cans, and newspapers. Examples of recycled-content items include paper products (some paper napkins, towels, toilet paper), and plastic items (some outdoor picnic tables, park benches, lumber, and even railroad ties). In theory, anything that contains postconsumer waste can be recycled again. In practice, however, this may not occur because municipalities do not offer recycling services for all potentially recyclable products.

Chapter 10

10.4 Each carbon atom in Figure 10.2 is surrounded by 8 electrons (4 bonds), so the octet rule is followed.

10.5 c.

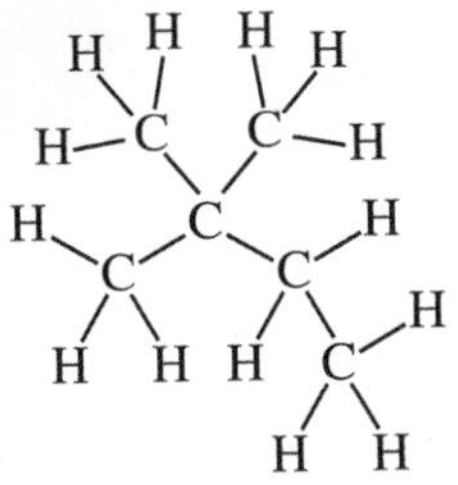

d.

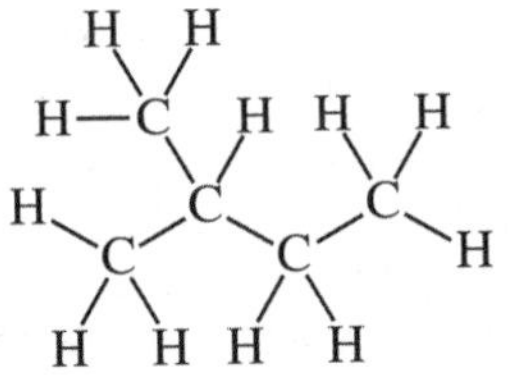

10.6 a. Yes, *n*-butane and *iso*butane are isomers. They have the same formula, C_4H_{10}, but different structures.

b. No, *n*-hexane (C_6H_{14}) and cyclohexane (C_6H_{12}) are not isomers. They have different formulas as well as different structures.

10.7

Structural Formula	Condensed Structural Formula and Line-Angle Drawing
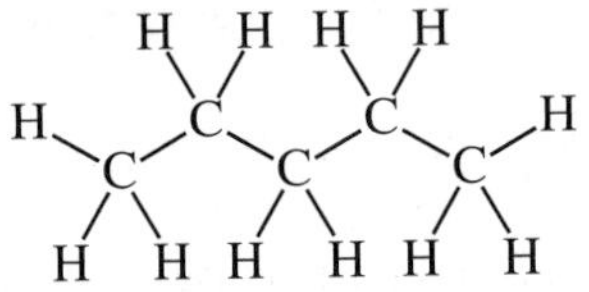	$CH_3CH_2CH_2CH_2CH_3$ 
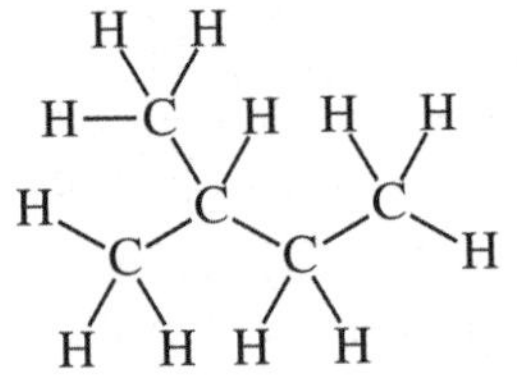	$CH_3CH(CH_3)CH_2CH_3$ 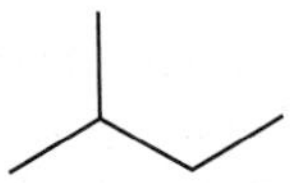
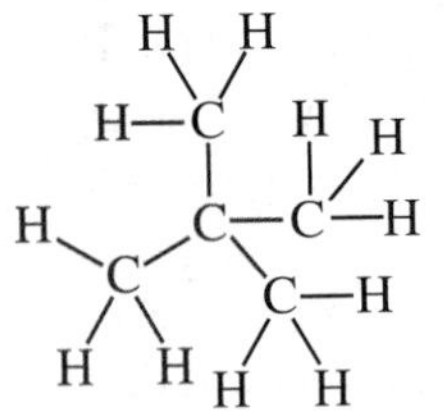	$C(CH_3)_4$ 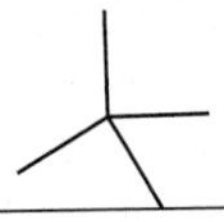

10.8 c. 　NH$_2$

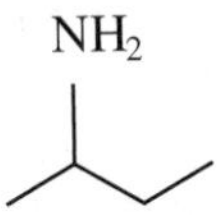

amine

d. 　　ester

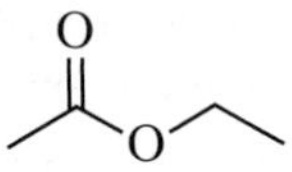

e. 　　aldehyde

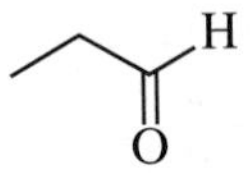

10.9 a.

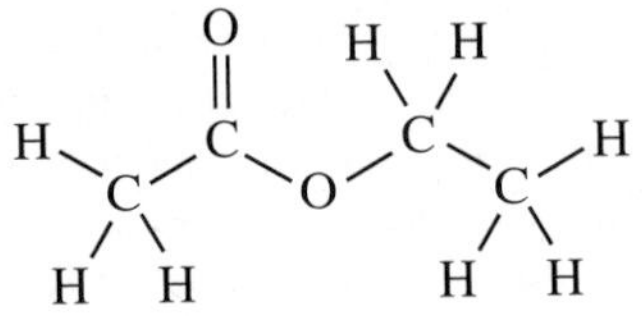

b.

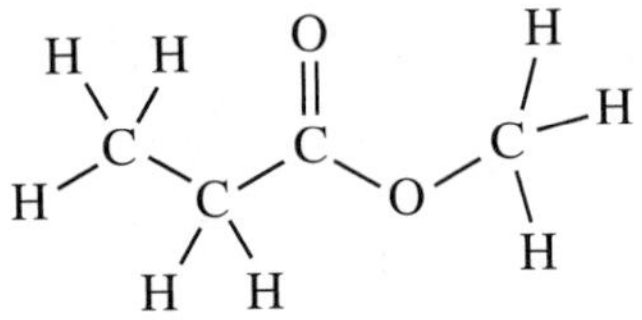

10.10 All three molecules have a benzene ring with attached functional groups that can form hydrogen bonds with water. Both aspirin and ibuprofen have a carboxylic acid group attached.

10.16 b. 　OH　　　　H　OH　　　HO　H

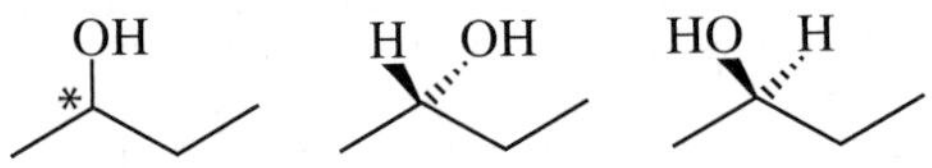

c. 　*CHClFCH$_3$　　　Cl F　　　　F　Cl
　　　　　　　　　　H　CH$_3$　　　H　CH$_3$

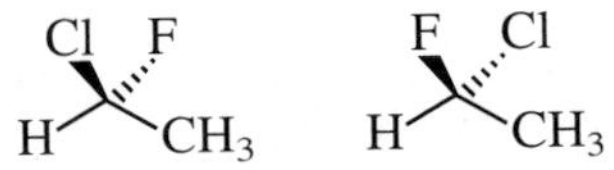

d. 　　　　　HNCH$_3$

　　　　　CH$_3$NH H

　　　　　H HNCH$_3$

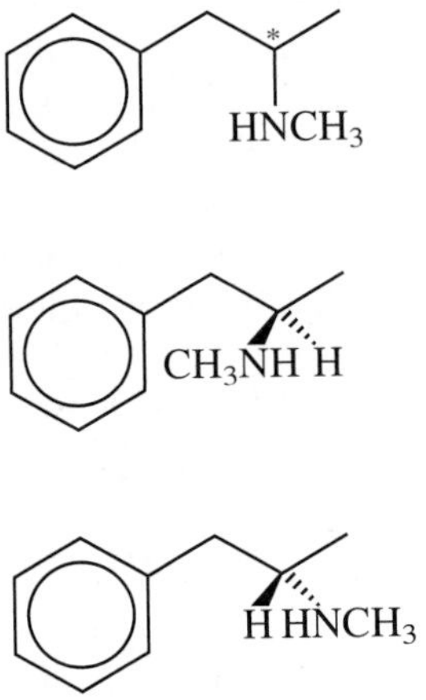

10.17 a. The chiral carbons are indicated with an asterisk in these structures.

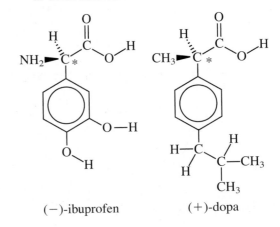

(−)-ibuprofen　　　(+)-dopa

b. Ibuprofen has a phenyl ring and a carboxylic acid group. The dopa molecule has a phenyl ring, an amine, and a carboxylic acid group.

c. Drawings are given in part **a**.

10.19 a.

estradiol	progesterone
–OH group on the D ring	C=O group off the D ring
–CH$_3$ group on CD ring intersection	C=O and C=C on the A ring
	–CH$_3$ groups on CD and AB ring intersections

estradiol	testosterone
–OH group on the D ring	–OH group on the D ring
–CH$_3$ group on CD ring intersection	–CH$_3$ groups on CD and AD ring intersections
	C=O and C=C on the A ring

cholic acid	cholesterol
–OH groups on the A, B, and C rings	–OH group on the A ring
carboxylic acid group on D ring side chain	double bond in A ring
–CH$_3$ groups on CD and AD ring intersections	–CH$_3$ groups on CD and AD ring intersections

corticosterone	cortisone
C=O and C=C on the A ring	C=O and C=C on the A ring
–OH group on C ring	C=O group on C ring, and also on D ring side chain
–OH group on D ring side chain	–OH groups on D ring and D ring side chain
–CH$_3$ groups on CD and AD ring intersections	–CH$_3$ groups on CD and AD ring intersections

prednisone	cortisone
C=O and 2 C=C bonds on the A ring	C=O and C=C on the A ring
–OH group on C ring	C=O group on C ring, and also on D ring side chain
–OH group on D ring side chain	–OH groups on D ring and D ring side chain
–CH$_3$ groups on CD and AD ring intersections	–CH$_3$ groups on CD and AD ring intersections

b. estradiol $C_{18}H_{24}O_2$ testosterone $C_{19}H_{28}O_2$
progesterone $C_{21}H_{30}O_2$ cholic acid $C_{24}H_{40}O_5$
prednisone $C_{21}H_{26}O_5$ corticosterone $C_{21}H_{30}O_4$
cortisone $C_{21}H_{28}O_5$ cholesterol $C_{27}H_{46}O$

10.26 Hydrochloric acid, HCl(*aq*), is used to form the salt of oxycodone. This is the structure for the salt. The

N atom no longer has a lone pair of electrons, which is what characterizes the freebase form.

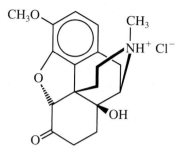

Chapter 11

11.3 a. Composting your food scraps lowers the amount of waste that you send to the landfill. If you use the resulting soil to grow your own crops, your foodprint is lowered even more.

c. This will significantly lower your foodprint because the grain required to produce beef requires a large amount of land (and water). Eating the grain directly lowers your foodprint.

11.7 b. Biodiesel is the methyl ester of a fatty acid. Fats and oils are triglycerides, that is, triple esters of glycerol.

11.10 The process of interesterification allows food producers to use fats and oils for food without trans fats. This follows green chemistry principle #3: It is better to use and generate substances that are not toxic.

11.11 a. Sweetness is related to the molecular structure (and shape). Since the sugars have different structures, they will interact with the receptors in your tongue in a different manner, resulting in a different degree of sweetness.

b. Although these sugars differ in their chemical structures, their chemical composition is essentially the same. Fructose and glucose have the same chemical formula, and sucrose is nearly the same. With the same chemical formula, the same amount of heat is liberated when these sugars are metabolized (or burned).

11.14 GlyGlyGly, GlyGlyAla, GlyAlaAla, GlyAlaGly, AlaAlaAla, AlaAlaGly, AlaGlyGly, AlaGlyAla

11.18 Processed food tends to be "heavy" in salt, that is, have a sodium content per serving that is relatively high compared with the daily requirement. For some people, this tastes quite normal. But for others, this is noticeably salty. For example, a can of soup can be 800 mg or more of sodium per serving. The low salt choices have less sodium but still include some salt.

11.21 a. No. Males require more calories than females.

b. As males and females grow through their teens and into their 20s, their estimated calorie requirement increases. After age 30 and beyond, the calorie requirement declines.

11.25 a. $2000 \text{ g C} \times \dfrac{1 \text{ mol CO}_2}{12 \text{ g C}} \times \dfrac{44 \text{ g CO}_2}{1 \text{ mol CO}_2} \times \dfrac{1 \text{ lb CO}_2}{453 \text{ g CO}_2}$

= about 16 pounds of CO_2 emitted per gallon of gas.

b. Let's assume that your car gets 30 miles to the gallon. 1000 miles would then translate to 33 gallons of gas burned.

33 gallons $\times$ 16 pounds CO_2 per gallon = 530 pounds of CO_2

c. Multiply your answer from part **a** by 1200 miles, equivalent to eating vegetarian one day a week. About 630 pounds of CO_2 would not be emitted.

Chapter 12

12.3 a. A protein is a large molecule (a polymer) built from amino acids (smaller molecules, monomers). Protein molecules are characterized by features such as their composition, their distinctive shape, and their biological function.

b. A carbohydrate is a molecule composed of only C, H, and O in the ratio of CH_2O. Many carbohydrates, such as starch and cellulose, are large molecules (polymers) built from smaller molecules (sugar monomers).

12.6

12.12 If several codons represent the same amino acid, then changes in single base pairs are less likely to result in a change to the amino acid sequence the gene encodes. The redundancy in the genetic code makes it more robust.

12.13 a. Nylon and proteins (such as those in meat) both contain the amide functional group.

c. Both nylons and proteins are condensation polymers. When they form, a small molecule (water for proteins, water for some nylons) is released.

12.15 a. An ion is an atom or group of atoms with a positive or negative charge. Examples include the hydroxide ion (OH^-) and the sodium ion (Na^+).

b. Radiation refers both to electromagnetic radiation and to nuclear radiation. The former refers to all types of "light"—infrared radiation, microwave radiation, and ultraviolet radiation, for example. The latter refers to alpha, beta, and gamma radiation.

c. When radiation of high enough energy collides with a molecule, it knocks out an electron, producing an ion.

d. In some cases, the ions produced by ionizing radiation are highly reactive because they have an unpaired electron. This is the case with an ionized water molecule. See Chapter 7.

Answers to Selected End-of-Chapter Questions Indicated in Color in the Text

Chapter 0

4. These representations are similar in that they both contain the same elements: the economy, the society, and the environment. Both also depict these elements as overlapping. They differ in how these three elements overlap. As drawn for this question, the environment encompasses both the economy and society. Some would argue that this is essential, as without the environment, the other two cannot exist. Others, however, represent the environment, the economy, and the society as equally important (with the same area), as is the case with Figure 0.1.

6. When you are riding the bus, your footprints are reduced in the sense that your feet don't hit the pavement. In addition, your ecological footprint is lower, because you are using public transportation rather than your own vehicle. Of course, if you normally would walk rather than take the bus, you are "reducing your footprints" only in the first sense.

9. For example, as a teacher you could work with students and administrators in your school to conserve energy (power down computers when not in use) or produce less waste (compost coffee grounds and food scraps). As a gardener, you might want to consult with those who practice sustainable gardening in your local community. These folks have given much thought to alternative practices, such as ways to reduce runoff with rain barrels and rain gardens.

Chapter 1

1. a. $\dfrac{0.5 \text{ L}}{1 \text{ breath}} \times \dfrac{15 \text{ breaths}}{1 \text{ min}} \times \dfrac{60 \text{ min}}{1 \text{ hr}} \times 8 \text{ hr}$

$= 3600 \text{ L}$

b. Possibilities include burning less (wood, vegetation, cooking fuels, gasoline, incense), using products that pollute less (low-emission paints), and using motor-less appliances and tools (hand-operated lawnmower, egg beater, broom, rake).

3. a. $Rn < CO < CO_2 < Ar < O_2 < N_2$

b. CO and CO_2

5. Like the other noble gases, radon is colorless, tasteless, odorless, and relatively unreactive. However, radon is radioactive and the others are not.

6. a. 0.9 parts per hundred

$\times \dfrac{1{,}000{,}000 \text{ parts per million}}{100 \text{ parts per hundred}}$

$= 9000 \text{ parts per million}$

(Move the decimal 4 places to the right.)

7. a. Compound (2 molecules of one compound made up of two different elements)

b. Mixture (2 atoms of one element plus 2 atoms of another)

c. Mixture (three different substances, two elements and one compound)

d. Element (4 atoms of the same element)

10. a. 85,000 g   **b.** 10,000,000 gallons

11. a. $2.2 \times 10^{-4} \text{ g/m}^3$

b. No, because CO is odorless.

13. a. Group 1A and Group 7A

b. 1A: hydrogen, lithium, sodium, potassium, rubidium, cesium, francium 7A: fluorine, chlorine, bromine, iodine, astatine

16. a. The chemical formula CH_4 indicates two types of atoms (C and H). A molecule of methane consists of 1 carbon atom and 4 hydrogen atoms. The chemical formula SO_2 indicates two types of atoms (S and O). A molecule of sulfur dioxide consists of 1 sulfur atom and 2 oxygen atoms. The chemical formula O_3 indicates only one type of atom. A molecule of ozone consists of 3 oxygen atoms.

b. CH_4 (methane), SO_2 (sulfur dioxide), O_3 (ozone).

18. a. $N_2(g) + O_2(g) \longrightarrow 2 \text{ NO}(g)$
(occurs at high temperature)

b. $O_3(g) \longrightarrow O_2(g) + O(g)$
(occurs in the presence of UV light)

c. $2 \text{ S}(s) + 3 O_2(g) \longrightarrow 2 \text{ SO}_3(g)$

24. a. platinum (Pt), palladium (Pd), rhodium (Rh)

b. All three metals are in Group 8B on the periodic table. Platinum is directly under palladium, and rhodium is just to the left of palladium.

c. These metals are solids at the temperature of the exhaust gases, so they must have relatively high melting points. Also, they do not undergo permanent chemical change when catalyzing the reaction of CO to CO_2 in the exhaust stream.

26. The Hilo photograph shows a wall, presumably built by humans (requiring fuel to be burned in the process of moving the stones). If the stones are cemented, the cement also emits gases as it dries. The cut lawn implies that a lawn mower was used, probably a power one, given the size of the lawn. The vegetation may have herbicides or pesticides applied, all of which can be carried by the air. The airport photograph is easier. The jet engines and the vehicles on the tarmac both leave air prints as they burn fuel. Any spilled fuel may evaporate into the air. The haze is most likely the result of emissions from vehicles and industry.

31. a. Normally, the exhaust gases are released to the atmosphere through the tailpipe and don't find their way back into the interior of the car. The tailpipe does not have a connection to the interior. However, if the gases are released into an enclosed space like a snow bank, they may seep back into the car as there is no easy escape path into the wider environment.

    b. CO is an odorless, colorless, and tasteless gas.

34. CO is termed the "silent killer" because your senses cannot detect this colorless, tasteless, and odorless gas. The same term cannot be applied to pollutants such as O_3 or SO_2 because each has a distinctive odor that can be detected at concentrations below the level of toxicity.

37. Ozone up high is "good" because it protects us by absorbing incoming ultraviolet radiation from the Sun. Ozone nearby is "bad" because it can be dangerous to breathe and damages vegetation and materials such as rubber.

38. a. The elderly, the young, and people with respiratory diseases such as asthma and emphysema are most affected by ozone.

    b. 7 days

    c. Ozone is highly reactive, thus does not persist long in the atmosphere. Since no ozone is produced at night when the Sun isn't shining, its concentration falls.

    d. Answers will vary. Possibilities include overcast skies, rain, or high winds. It could be a day when fewer people are driving or that industries are shut down.

    e. Ozone levels in London, Ontario, are lower in December because there is less daylight in the winter months.

40. a. 15 ppm is 0.0015% and 2% is 20,000 ppm. 15 ppm is roughly 1300 times smaller than 20,000.

47. CO is a hazard when present at the part per million level, and instruments can easily detect CO at this concentration. In contrast, radon levels are much lower, on the order of parts per 10^{20}. Most radon detection kits sample the air over a period of time in order to get a high enough reading.

Chapter 2

3. a. Yes, this is over the detection minimum of 10 ppb.

$$\frac{0.118 \text{ parts } O_3}{1,000,000 \text{ parts air}} = \frac{118 \text{ parts } O_3}{1,000,000,000 \text{ parts air}} \text{ or } 118 \text{ ppb}$$

    b. Yes, this is well over the detection minimum of 10 ppb.

$$\frac{25 \text{ parts } O_3}{1,000,000 \text{ parts air}} = \frac{25,000 \text{ parts } O_3}{1,000,000,000 \text{ parts air}} \text{ or } 25,000 \text{ ppb}$$

8. a. A neutral atom of oxygen has 8 protons and 8 electrons.

    b. A neutral atom of nitrogen has 7 protons and 7 electrons.

10. a. helium, He

    b. potassium, K

    c. copper, Cu

11. c. 92 protons, 146 neutrons, and 92 electrons

    f. 88 protons, 138 neutrons, and 88 electrons

12. c. $^{222}_{86}\text{Rn}$

13. a. ·Ca·

    b. ·N̈·

    c. :C̈l·

    d. He:

14. b. There are $2(1) + 2(6) = 14$ outer electrons. The Lewis structure is

$$\text{H:}\ddot{\underset{\cdot\cdot}{\text{O}}}\text{:}\ddot{\underset{\cdot\cdot}{\text{O}}}\text{:H} \quad \text{and} \quad \text{H}-\ddot{\underset{\cdot\cdot}{\text{O}}}-\ddot{\underset{\cdot\cdot}{\text{O}}}-\text{H}$$

    c. There are $2(1) + 6 = 8$ outer electrons. The Lewis structure is

$$\text{H:}\ddot{\underset{\cdot\cdot}{\text{S}}}\text{:H} \quad \text{and} \quad \text{H}-\ddot{\underset{\cdot\cdot}{\text{S}}}-\text{H}$$

16. a. Wave 1 has longer wavelength than wave 2.

    b. Wave 1 has lower frequency than wave 2.

    c. Both waves travel at the same speed.

19. In order of increasing energy per photon: radiowaves < infrared < visible < gamma rays

21. a. In order of increasing wavelength: UV-C < UV-B < UV-A

    b. In order of increasing energy: UV-A < UV-B < UV-C

    c. In order of increasing potential for biological damage: UV-A < UV-B < UV-C

25. a. Methane, CH_4, has $4 + 4(1) = 8$ outer electrons.

$$\text{H}-\underset{\underset{\displaystyle\text{H}}{|}}{\overset{\overset{\displaystyle\text{H}}{|}}{\text{C}}}-\text{H}$$

Ethane, C_2H_6, has $2(4) + 6(1) = 14$ outer electrons.

$$H-\overset{\overset{\displaystyle H}{|}}{\underset{\underset{\displaystyle H}{|}}{C}}-\overset{\overset{\displaystyle H}{|}}{\underset{\underset{\displaystyle H}{|}}{C}}-H$$

b. Three different CFCs are based on methane. They are: CF_3Cl, CCl_3F, and CCl_2F_2.

26. $Cl\cdot$ has 7 outer electrons. Its Lewis structure is: $:\ddot{\underset{..}{C}l}\cdot$

$NO_2\cdot$ has $5 + 2(6) = 17$ outer electrons. Its Lewis structure is:

$$\ddot{\underset{..}{O}}::\dot{N}:\ddot{\underset{..}{O}}: \text{ or } \ddot{\underset{..}{O}}=\dot{N}-\ddot{\underset{..}{O}}:$$

$ClO\cdot$ has $7 + 6 = 13$ outer electrons. Its Lewis structure is:

$$\cdot\ddot{\underset{..}{C}l}:\ddot{\underset{..}{O}}\cdot \text{ or } \cdot\ddot{\underset{..}{C}l}-\ddot{\underset{..}{O}}\cdot$$

$\cdot OH$ has $6 + 1 = 7$ outer electrons. Its Lewis structure is: $\cdot\ddot{\underset{..}{O}}:H$ or $\cdot\ddot{\underset{..}{O}}-H$

b. They all contain an unpaired electron.

29. The message is that ground-level ozone is a harmful air pollutant. Ozone in the stratosphere, on the other hand, is beneficial because it can absorb harmful UV-B before it reaches the surface of the Earth.

31. a. The most energetic fraction is the UV light.

b. Up in the stratosphere where the air is very thin, UV-C splits oxygen molecules, O_2, into two oxygen atoms, O. These in turn react with other oxygen molecules to produce ozone, O_3. See the reactions of the Chapman cycle. Without the UV-C light (which does not reach the surface of our planet), the ozone layer would not form.

41. Although UV-C radiation causes damage to both animals and plants, it is completely absorbed by the O_2 in our atmosphere before it can reach the surface of the Earth.

44. $ClO\cdot$ acts as a catalyst in the series of reactions in which stratospheric O_3 molecules are changed to O_2 molecules. As it is not consumed in the reaction, $ClO\cdot$ can continue to catalyze the breakdown of O_3.

52. O_2, O_3, and N_2 all have an even number of valence electrons. In contrast, N_3 has 15 valence electrons. Molecules with odd numbers of electrons cannot follow the octet rule and are more reactive.

53. a. $90 + 12 = 102$. The compound contains one carbon atom, no hydrogen atoms, and two fluorine atoms. The chemical formula for CFC-12 is CCl_2F_2.

b. CCl_4 contains 1 carbon atom, 0 hydrogen atoms, and 0 fluorine atoms. Therefore, the code number for CCl_4 is 100 or $90 + 10$. The name is CFC-10.

c. Yes the "90" method will work for HCFCs. $90 + 22 = 112$, so HCFC-22 would be composed of 1 carbon atom, 1 hydrogen atom, and 2 fluorine atoms and its chemical formula would be CHF_2Cl.

d. No, this method will not work for halons as there is no guideline for handling bromine.

Chapter 3

4. a. $6\,CO_2 + 6\,H_2O \longrightarrow C_6H_{12}O_6 + 6\,O_2$

b. The number of atoms of each element on either side is the same. $C = 6$, $O = 18$, $H = 12$.

c. No, the number of molecules is not the same. There are 12 on the left, but only 7 on the right. The glucose molecule on the product side contains 24 atoms!

6. a. The rest of the Sun's energy is absorbed or reflected by the atmosphere.

7. a. As of 2010, the atmospheric concentration of CO_2 was about 390 ppm; however, 20,000 years ago, the concentration was only about 190 ppm. Looking back to 120,000 years ago, the concentration was about 270 ppm, still 40% below current levels.

b. The mean atmospheric temperature at present is somewhat above the 1950–1980 mean atmospheric temperature. The mean atmospheric temperature 20,000 years ago was lower by about 9 °C. However, 120,000 years ago the mean atmospheric temperature was lower than the present temperature by only about 1 °C.

c. Although there appears to be a *correlation* between mean atmospheric temperature and CO_2 concentration, this graph does not prove *causation* of either factor by the other.

11. a. $H-\ddot{S}-H$ bent

b. $:\ddot{\underset{..}{C}l}-\ddot{\underset{..}{O}}-\ddot{\underset{..}{C}l}:$ bent

c. $:\ddot{N}=N=\ddot{\underset{..}{O}}:$ or $:N\equiv N-\ddot{\underset{..}{O}}:$ or $:\ddot{N}-N\equiv O:$ linear

13. a. $3(1) + 4 + 6 + 1 = 14$ outer electrons. This is the Lewis structure.

$$H-\overset{\overset{\displaystyle H}{|}}{\underset{\underset{\displaystyle H}{|}}{C}}-\ddot{\underset{..}{O}}-H$$

b. The geometry around the C atom is tetrahedral, and there are no lone pairs. A H–C–H bond angle of about 109.5° is predicted.

c. There are four pairs of electrons around the O atom, two of which are bonding pairs and two are nonbonded pairs. Repulsion between the two nonbonded electron pairs and their repulsion of the bonding pairs is predicted to cause the H–O–C bond angle to be slightly less than 109.5°.

15. All can contribute to the greenhouse effect. In each case, the atoms move as the bond stretches or bends, and therefore the charge distribution changes. Unlike the linear CO_2 molecule, the water molecule is bent and so its polarity changes with each of these modes of vibration.

use $E = \dfrac{hc}{\lambda}$ to calculate the energies.

$$E = \dfrac{(6.63 \times 10^{-34}\,\text{J} \cdot \text{s}) \times (3.00 \times 10^8\,\text{m} \cdot \text{s}^{-1})}{4.26\,\mu\text{m} \times \dfrac{1\,\text{m}}{10^6\,\mu\text{m}}}$$

$$= 4.67 \times 10^{-20}\,\text{J}$$

$$E = \dfrac{(6.63 \times 10^{-34}\,\text{J} \cdot \text{s}) \times (3.00 \times 10^8\,\text{m} \cdot \text{s}^{-1})}{15.00\,\mu\text{m} \times \dfrac{1\,\text{m}}{10^6\,\mu\text{m}}}$$

$$= 1.33 \times 10^{-20}\,\text{J}$$

19. a. $C_6H_{12}O_6 \longrightarrow 3\,H_2O + 3\,CO_2$

b. in one day:

$$1.0\,\text{mg glucose} \times \dfrac{1\,\text{g}}{1000\,\text{mg}} \times \dfrac{1\,\text{mol glucose}}{180\,\text{g glucose}} \times$$

$$\dfrac{3\,\text{mol CO}_2}{1\,\text{mol glucose}} \times \dfrac{44\,\text{g CO}_2}{1\,\text{mol CO}_2} = 7.33 \times 10^{-4}\,\text{g CO}_2$$

in one year: 7.33×10^{-4} g/day $\times$ 365 days/year
$= 0.27$ g/year

21. a. A neutral atom of Ag-107 has 47 protons, 60 neutrons, and 47 electrons.

b. A neutral atom of Ag-109 has 47 protons, 62 neutrons, and 47 electrons. Only the number of neutrons has changed.

24. a. $2(1.0) + 16.0 = 18.0$ g/mol

b. $12.0 + 2(19.0) + 2(35.5) = 121.0$ g/mol

28. a. CO_2

b. Although CO_2 accounts for the largest percentage of greenhouse gas contributions, the case could be made that N_2O has the greatest effect. Its global warming potential is almost 300 times greater than that of CO_2, but its concentration is far lower than that of either CO_2 or CH_4. However, the net effectiveness of N_2O is the largest of these three gases.

Gas	% Contribution (graph)	GWP (Table 3.5)	Net Effectiveness (product)
CO_2	55	1	0.6
CH_4	15	23	3.5
N_2O	5	296	14.8

33. Scientists can analyze the deuterium-to-hydrogen ratio in ice cores and recreate the temperatures in the distant past. Drilled ocean cores can be analyzed for the number or type of microorganisms present. Another correlating piece of evidence is the changing alignment of the magnetic field in particles in the sediment over time.

39. a. $C_2H_5OH + 3\,O_2 \longrightarrow 3\,H_2O + 2\,CO_2$

b. 2 mol CO_2

c. 30 mol O_2

44. 73×10^6 metric tons $CH_4 \times \dfrac{12\,\text{metric tons C}}{16\,\text{metric tons CH}_4}$

$$= 5.5 \times 10^7\,\text{metric tons C}$$

Chapter 4

1. a. Coal, oil, and natural gas

b. Fossil fuels originated hundreds of millions of years ago from plant and animal matter. Under the action of heat and pressure, these materials were transformed into the compounds that make up fossil fuels.

c. No. The length of time required to form these fuels means they are not renewable.

5. a. $\dfrac{5.00 \times 10^8\,\text{J}}{\text{s}} \times \dfrac{3600\,\text{s}}{\text{hr}} \times \dfrac{24\,\text{hr}}{\text{day}} \times \dfrac{365\,\text{days}}{\text{year}}$

$$= 1.58 \times 10^{16}\,\text{J generated per year}$$

1.58×10^{16} J $= 4.2 \times 10^{16}$ J of heat for electricity generation

b. 4.2×10^{16} J $\times \dfrac{1\,\text{kJ}}{1000\,\text{J}} \times \dfrac{1\,\text{g}}{30\,\text{kJ}} = 1.4 \times 10^{12}$ g

1.4×10^{12} g $\times \dfrac{1000\,\text{kg}}{10^6\,\text{g}} = 1.4 \times 10^9$ metric tons

6. a. $2\,C_2H_6 + 7\,O_2 \longrightarrow 4\,CO_2 + 6\,H_2O$

b.

c. $\dfrac{52.0\,\text{kJ}}{1\text{g C}_2\text{H}_6} \times \dfrac{30.1\,\text{g C}_2\text{H}_6}{1\,\text{mol C}_2\text{H}_6} = \dfrac{1570\,\text{kJ}}{1\,\text{mol C}_2\text{H}_6}$

8. A typical power plant burns 1.5 million tons of coal each year. The first calculation is for coal with 50 ppb mercury; the second is for 200 ppb.

$x = 0.075$ ton Hg

$x = 0.3$ ton Hg

The plant releases between 0.075 and 0.3 ton Hg a year.

9. b. $\dfrac{2.4 \times 10^8\,\text{kcal}}{1\,\text{yr}} \times \dfrac{1\,\text{yr}}{65\,\text{barrel oil}} \times \dfrac{1\,\text{barrel oil}}{42\,\text{gal}}$

$$= \dfrac{8.8 \times 10^4\,\text{kcal}}{1\,\text{gal}}$$

c. $\dfrac{2.4 \times 10^8\,\text{kcal}}{1\,\text{yr}} \times \dfrac{1\,\text{yr}}{16\,\text{ton coal}} = \dfrac{1.5 \times 10^7\,\text{kcal}}{1\,\text{ton coal}}$

11. Assuming that room temperature is 20–25 °C, pentane should be a liquid because room temperature is below its boiling point (36 °C) but above its melting point (–130 °C). Triacontane should be solid at room temperature because room temperature is below its melting point (66 °C).

Octane should be a liquid at room temperature for the same reason as pentane.

14. $70 \text{ Cal} \times \dfrac{4.184 \text{ kJ}}{1 \text{ Cal}} \times \dfrac{1000 \text{ J}}{1 \text{ kJ}} \times \dfrac{1 \text{ beat}}{1 \text{ J}} \times \dfrac{1 \text{ min}}{80 \text{ beats}}$

$$= 3700 \text{ min}$$

17. a. Exothermic. A charcoal briquette releases heat as it burns.

b. Endothermic. Water absorbs the heat necessary for evaporation from your skin, and your skin feels cooler.

19. a. Bonds broken in the reactants
1 mol N≡N triple bonds = 1(946 kJ) = 946 kJ
3 mol H–H single bonds = 3(436 kJ) = 1308 kJ
Total energy *absorbed* in breaking bonds = 2254 kJ

Bonds formed in the products
6 mol N–H single bonds = 6(391 kJ) = 2346 kJ
Total energy *released* in forming bonds = 2346 kJ
Net energy change is (+2254 kJ)
$$+ (-2346 \text{ kJ}) = -92 \text{ kJ}$$
The overall energy change is negative, characteristic of an exothermic reaction.

b. Bonds broken in the reactants
12 mol C–H single bonds = 12(416 kJ)
$$= 4,992 \text{ kJ}$$
4 mol C–C single bonds = 4(356 kJ) = 1424 kJ
11 mol O=O double bonds = 11(498 kJ)
$$= 5478 \text{ kJ}$$
Total energy *absorbed* in breaking bonds
$$= 11,894 \text{ kJ}$$

Bonds formed in the products
10 mol C≡O triple bonds = 10(1073 kJ)
$$= 10,730 \text{ kJ}$$
24 mol O–H single bonds = 24(467 kJ)
$$= 11,208 \text{ kJ}$$
Total energy *released* in forming bonds
$$= 21,938 \text{ kJ}$$

Net energy change is (+11,894 kJ) +
$$(-21,938 \text{ kJ}) = -10,044 \text{ kJ}$$
The overall energy change is negative, characteristic of an exothermic reaction.

c. Bonds broken in the reactants
1 mol H–H single bonds = 1(436 kJ) = 436 kJ
1 mol Cl–Cl single bonds = 1(242 kJ) = 242 kJ
Total energy *absorbed* in breaking bonds = 678 kJ
Bonds formed in the products
2 mol H–Cl single bonds = 2(431 kJ) = 862 kJ
Total energy *released* in forming bonds = 862 kJ

Net energy change is (+678 kJ) + (−862 kJ)
$$= -184 \text{ kJ}$$
The overall energy change is negative, characteristic of an exothermic reaction.

24. a.

b.

29. Considering only the molar heats of combustion, octane, with more atoms and more chemical bonds, has a greater heat of combustion than butane. However, comparisons should be based on the same amount of each substance, such as the heat released per gram of each fuel.

5070 kJ/mol octane × (1 mol octane/114 g octane)
= 44.5 kJ/g octane

(heat released per gram octane burned)

2658 kJ/mol butane × (1 mol butane/58 g butane)
= 45.8 kJ/g butane

(heat released per gram butane burned)

Here the values are much closer. Because heat comparisons should be made based on the same mass of fuel, you will have to educate your friend on this point.

30. a. The C–F single bond requires 485 kJ/mol, the C–Cl single bond requires 327 kJ/mol, and the C–Br single bond requires 285 kJ/mol to break the bond. The C–Br bond is the weakest. Thus, when halon-1211 absorbs UV radiation, bromine atoms are likely to form and react with ozone.

b.

bond most easily broken

In this molecule, the C–Cl bond has the lowest bond energy and thus is broken most easily.

31. a. When $n = 1$, the balanced equation is
$$\text{CO} + 3\,\text{H}_2 \longrightarrow \text{CH}_4 + \text{H}_2\text{O}$$
To calculate the heat evolved we use the same method as in Problem 4.19.

Bonds broken in the reactants:
1 mol C≡O triple bonds = 1(1073 kJ) = 1073 kJ
3 mol H–H single bonds = 3(436 kJ) = 1308 kJ
Total energy *absorbed* in breaking bonds
$$= 2381 \text{ kJ}$$

Bonds formed in the products
4 mol C–H single bonds = 4(416 kJ) = 1664 kJ
2 mol O–H single bonds = 2(467 kJ) = 934 kJ
Total energy *released* in forming bonds = 2598 kJ
Net energy change is (+2381 kJ) + (−2598 kJ)
$$= -217 \text{ kJ}$$

b. Reactions with *n* greater than 1 release more energy as *n* becomes larger, assuming that we are viewing the energy per mole of the hydrocarbon formed (not per gram). There will always be *n* C≡O triple bonds to break and $(2n + 1)$ H–H single bonds to break. The number of C–H bonds forming is $(2n + 2)$, and the number of O–H bonds forming is $2n$. As *n* becomes larger, more and more energy is released.

35. Section 1.11 described the catalytic converters in automobiles. Section 2.9 described the catalytic destruction of ozone by chlorine free radicals.

43. Figure 4.16 gives the energy released per gram for the combustion of several fuels. Assuming the densities of octane and ethanol are similar (a good assumption) one gallon of gasoline releases more energy (47.8 kJ/mol) than one gallon of ethanol (29.7 kJ/mol). This makes sense, because ethanol is an oxygenated fuel, that is, it contains oxygen and thus is already "partially burned."

45. A royal flush is an ace, king, queen, jack, and 10 of the same suit. It is a highly improbable hand in poker (1 in about 650,000 five-card hands). It exhibits a higher degree of order (low entropy) and is more highly valued than a simple high-card hand (a higher degree of entropy). The hand with the least entropy wins!

Chapter 5

5. a. N and C, $3.0 - 2.5 = 0.5$

O and S, $3.5 - 2.5 = 1.0$

N and H, $3.0 - 2.1 = 0.9$

S and F, $4.0 - 2.5 = 1.5$

b. N more strongly than C

O more strongly than S

N more strongly than H

F more strongly than S

6. a. The Lewis structure for ammonia is

$$H—\overset{\cdot\cdot}{N}—H$$
$$|$$
$$H$$

b. Each N–H bond is polar. The difference in electronegativity between N and H is $3.0 - 2.1 = 0.9$.

c. The ammonia molecule is triangular pyramidal. This shape, together with the polarity of the N–H bonds, causes the molecule to be polar.

8. a. Examples of nonmetals include C, H, O, S, Cl, and N. Nonmetals generally have higher electronegativity values than metals.

13. a. Partially soluble. Orange juice concentrate contains some solids (pulp) that do not dissolve in water.

b. Very soluble. Note that ammonia is a gas. If you ever have seen a lecture demonstration involving ammonia and water (e.g., "the ammonia fountain"), you know that ammonia dissolves almost instantly in water. Ammonia dissolves in water in any proportion.

d. Very soluble. When you add laundry detergent to your load of wash, it dissolves in the water (or at least it should dissolve—sometimes solid laundry detergent cakes together).

16. a. A chlorine atom gains one electron to form a chloride ion with a charge of 1−. The octet rule is satisfied.

$$:\overset{\cdot\cdot}{\underset{\cdot\cdot}{Cl}}\cdot \text{ and } \left[:\overset{\cdot\cdot}{\underset{\cdot\cdot}{Cl}}:\right]^{-}$$

d. A barium atom loses two electrons to form a barium ion, with a charge of 2+. The octet rule is satisfied.

$$\cdot Ba\cdot \text{ and } \left[Ba\right]^{2+}$$

17. a. NaBr sodium bromide

d. Al_2O_3 aluminum oxide

18. a. $Ca(HCO_3)_2$

b. $CaCO_3$

19. a. Potassium acetate

e. Calcium hypochlorite

24. a. The solution will conduct electricity and the bulb will light. Based on Table 5.9, $CaCl_2$ is a soluble salt and therefore releases ions (Ca^{2+} and Cl^- when it dissolves). These ions carry the current.

b. The solution will not conduct electricity. Although ethanol (C_2H_5OH) is soluble in water, it is a covalent compound and does not form ions.

27. The concentration of Mg^{2+} is 2.5 M, and the concentration of NO_3^- is 5.0 M.

28. a. To prepare 2 liters of 1.50 M KOH, weigh out 168 g of KOH and place it into a 2-L volumetric flask. Add distilled (or deionized) water to fill the flask to the mark. *Note:* If you don't have a 2-L volumetric flask, you will need to repeat the procedure twice with a 1-L flask.

b. To prepare 1 liter of 0.050 M NaBr, weigh out 5.2 g of NaBr and place it into a 1-L volumetric flask. Add water as in part **a.**

30. Growing the potato (25 L) < growing the orange (50 L) < producing the beer (168 L) < producing the hamburger (2400 L). Potato crops require less irrigation than orange groves. The beer requires grain, which in turn requires water. But the cow that was used to produce the beef requires far more grain. *Note:* The water footprint values in liters are approximate.

35. a. The electronegativities of the elements generally increase from left to right across a period (until Group 8A is reached) and from bottom to top within any group. Thus, the element in position 2 is predicted to have the highest electronegativity.

b. Ranking the other elements is not straightforward. Element 1 is expected to be more electronegative than element 3, based on their relative positions in the same group. Element 4 will likely be more electronegative than elements 1 and 3 and less electronegative than 2. However, because element 4 is not in the same period with any other element, this prediction cannot be made with certainty. Here are the values found in references (they do not appear in Table 5.1): 0.8 for element 1, 2.4 for element 2, 0.7 for element 3, and 1.9 for element 4. These values confirm the relative ranking in order of decreasing electronegativity: 2, 4, 1, 3.

38. Like water, NH_3 is a polar molecule. It has polar N–H bonds and a triangular pyramidal geometry. Therefore, despite its low molar mass, considerable energy must be added to liquid NH_3 to overcome the intermolecular forces (hydrogen bonding) among NH_3 molecules.

39. a. A single covalent bond holds together the two H atoms in H_2.

b. Hydrogen bonding is a type of *inter*molecular force, a force of attraction between a H atom on one molecule and an electronegative atom on another molecule (or in some cases, between a H atom and an electronegative atom on a different part of the same molecule).

42. For a given contaminant, the MCLG (a goal) and the MCL (a legal limit) are usually the same. However, the levels may differ when it is not practical or possible to achieve the health goal as set by the MCLG. This sometimes is the case for carcinogens, for which the MCLG is set at zero (under the assumption that any exposure presents a cancer risk).

44. a. Nitrate ion (NO_3^-) and nitrite ion (NO_2^-)

b. In the body, oxygen is needed to metabolize ("burn") glucose to produce energy.

c. The nitrate ion is not volatile. It is a solute that does not evaporate or decompose with heat.

Chapter 6

3. a. As of 2011, the approximate atmospheric concentration of CO_2 is 390 ppm.

b. The concentration of carbon dioxide in the atmosphere is increasing because humans are burning fossil fuels and cutting down forests which absorb CO_2.

c. Here is the Lewis structure. $\ddot{O}=C=\ddot{O}$

d. No, you would not. Carbon dioxide is a nonpolar compound, and seawater is a polar solution of primarily water and sodium chloride. "Like dissolves like." When CO_2 dissolves in seawater, it forms carbonic acid, H_2CO_3.

4. Anthropogenic emissions, such as those of carbon dioxide, nitrogen oxides, and sulfur dioxide, are generated by human activities.

6. a. Possibilities include nitric acid (HNO_3), hydrochloric acid (HCl), sulfuric acid (H_2SO_4), sulfurous acid (H_2SO_3), phosphoric acid (H_3PO_4), carbonic acid (H_2CO_3), and hydrobromic acid (HBr).

b. In general, acids taste sour, turn litmus paper red (and have characteristic color changes with other indicators), are corrosive to metals such as iron and aluminum, and release carbon dioxide ("fizz") from a carbonate. These properties may not be observed if the acid is not sufficiently concentrated.

7. a. $HBr(aq) \longrightarrow H^+(aq) + Br^-(aq)$

b. $H_2SO_3(aq) \longrightarrow H^+(aq) + HSO_3^-(aq)$

8. a. Possibilities include sodium hydroxide (NaOH), potassium hydroxide (KOH), ammonium hydroxide (NH_4OH), magnesium hydroxide ($Mg(OH)_2$), and calcium hydroxide ($Ca(OH)_2$).

b. In general, bases taste bitter, turn litmus paper blue (and have characteristic color changes with other indicators), have a slippery feel in water, and are caustic to your skin.

9. a. $KOH(s) \longrightarrow K^+(aq) + OH^-(aq)$

12. a.
$$KOH(aq) + HNO_3(aq) \longrightarrow KNO_3(aq) + H_2O(l)$$

13. a. The solution of pH = 6 has 100 times more $[H^+]$ than the solution of pH = 8.

d. The solution with $[OH^-] = 1 \times 10^{-2}$ M has 10 times more $[OH^-]$ than the solution with $[OH^-] = 1 \times 10^{-3}$ M.

17. $S(s) + O_2(g) \longrightarrow SO_2(g)$

21. Sulfur dioxide reacts with calcium carbonate as follows:

$$CaCO_3(s) + SO_2(g) + H_2O(l) \longrightarrow$$
$$Ca^{2+}(aq) + HCO_3^-(aq) + HSO_3^-(aq)$$

The molar masses of SO_2 and $CaCO_3$ are needed to solve the problem. Note that it is not necessary to change tons to grams. The ratio of the number of grams per mole is the same as the ratio of the number of kilograms per kilomole or the ratio of the number of tons per ton-mole.

$$1.00 \text{ ton } SO_2 \times \frac{1 \text{ ton-mole } SO_2}{64.1 \text{ tons } SO_2} \times \frac{1 \text{ ton-mole } CaCO_3}{1 \text{ ton-mole } SO_2}$$
$$\times \frac{100 \text{ tons } CaCO_3}{1 \text{ ton-mole } CaCO_3} = 1.56 \text{ tons } CaCO_3$$

26. a. If we compare Coca-Cola (pH = 2.7) with rainwater (average pH = 5.5), we find that the beverage is about 1000 times more acidic.

29. c. The chemical formula $HC_2H_3O_2$ has the advantage of being written in the same format we have used for other acids; that is, with the acidic H atom written first. The advantage of using CH_3COOH is that it gives a better indication of how the atoms in the molecule are bonded.

31. a. $[H_2O] > [Na^+] = [OH^-] > [H^+]$

32. a. Combustion engines, such as those associated with jet aircraft, directly emit CO, CO_2, and NO. If small amounts of sulfur are present in the fuel, SO_2 and SO_3 are emitted as well.

33. Fuel combustion contributes greatly to the emissions of SO_2, with the major source being the burning of coal. Although fuel combustion contributes less to the emissions of NO_3, the percentage still is substantial. In contrast, transportation makes a small contribution to the emissions of SO_2 and a much larger one to the emissions of NO_3. Nitrogen monoxide is produced from N_2 and O_2 in the air whenever there is a high temperature. Thus, both automobile engines and power plants are big contributors of NO_3.

39. a. The slightly higher pH values in the lab indicate that the acidity decreased slightly between collection and measurement.

b. One possibility is that the field samples contained naturally occurring acids that were not stable over time. They decomposed, thus making the solution less acidic. Another possibility is that some of the acids present in the sample reacted with other molecules in the sample or with the sample container itself between the time the sample was collected and analyzed.

Chapter 7

1. One carbon atom can differ from another in the number of neutrons (such as C-12 and C-13) and in the number of electrons (carbon ions do exist, but we do not discuss them in this text). All carbon atoms differ from all uranium atoms in the number of protons, neutrons, and electrons. Carbon atoms also differ from uranium atoms in their chemical properties.

3. a. 94 protons

b. Np (neptunium), Pu (plutonium)

4. a. C-14 has 6 protons and 8 neutrons.

10. Neutrons are needed to initiate the process of nuclear fission of U-235.

$$^1_0n + {}^{235}_{92}U \longrightarrow [{}^{236}_{92}U] \longrightarrow {}^{141}_{56}Ba + {}^{92}_{36}Kr + 3\,{}^1_0n$$

The fission products include two or three neutrons that can initiate more fission reactions. In this manner, a self-sustaining chain reaction can be established in which the products of one reaction can initiate another.

12. **A** is the control rod assembly, **B** is the cooling water out of the core, **C** is the control rods, **D** is the cooling water into the core, and **E** is the fuel rods.

15. a. $^1_0n + {}^{10}_5B \longrightarrow [{}^{11}_5B] \longrightarrow {}^4_2He + {}^7_3Li$

b. Boron can be used in control rods because it is a good neutron absorber.

17. a. $^{239}_{94}Pu \longrightarrow {}^{235}_{92}U + {}^4_2He$

$^{131}_{53}I \longrightarrow {}^{131}_{54}Xe + {}^{\ 0}_{-1}e + {}^0_0\gamma$

b. In a particulate form such as a powder or a dust, plutonium can be inhaled. If plutonium particles become lodged in the lungs, the ionizing radiation they emit (alpha particles) can damage lung cells. The decay products also are radioactive and can damage tissue.

c. Iodine accumulates in the thyroid gland.

d. After about 10 half-lives, samples have decayed to very low levels. The half-life of Pu-239 is about 24,000 years, so the timescale for a decrease to background level is on the order of hundreds of thousands of years. The half-life of I-131 is 8.5 days, so 10 half-lives is 85 days or about 3 months. A sample of I-131 decays to low levels on a timescale of months.

21. For this type of question, it is helpful to construct a chart.

# of half-lives	% remaining	% decayed
0	100	0
1	50	50
2	25	75
3	12.5	87.5
4	6.25	93.75
5	3.12	97.88
6	1.56	98.44

26. The natural abundances of U-238 and U-235 are 99.3% and 0.7%, respectively. U-235 can be induced to undergo nuclear fission and thus is suitable as a fuel for nuclear power plants and nuclear weapons. Because U-235 has such a low natural abundance, it is more difficult to procure in large quantities. Also, it is *extremely* difficult to separate U-235 from U-238. Had U-235 been readily available, far more countries would have had access to nuclear weapons.

34. See question 21. After 7 half-lives, 99% of a sample has decayed which is a reasonable approximation of being "gone." However, actually the radioactivity is *not* gone, as 0.78% of the radioactive sample still remains. Thus, if you start with a large amount of a radioactive substance (for example, 2000 pounds), after 7 half-lives you still have about 10 pounds left!

49. a. The sum of the masses of the reactants is 5.02838 g, and the sum for the products is 5.00878 g. The difference is 0.0196 g.

b. To use Einstein's equation, $E = mc^2$, you need to pay close attention to the units. The speed of light is 3.00×10^8 meters/second. In order to have joules (J) as the unit of energy, the mass must be in kilograms (kg), so you need to convert g to kg. In addition, you need the conversion factor that $1\ J \times 1$ kg-meter^2/second^2. Whew! Here is the calculation.

$$E = 0.0196\ g \times \frac{1\ kg}{10^3 g} \times \left[\frac{3.00 \times 10^8\ m}{s}\right]^2 \times \frac{1\ J}{kg\text{-}m^2/s^2}$$

$1\ J = kg \times m^2/s^2.$
$E = 1.76 \times 10^{12}\ J$

Chapter 8

1. a. Oxidation is a process in which an atom, ion, or molecule *loses* one or more electrons. Reduction is a process in which an atom, ion, or molecule *gains* one or more electrons.

  b. Electrons are transferred from the species losing electrons to the species gaining electrons.

3. $Zn(s)$ is oxidized to Zn^{2+} in zinc oxide. $O_2(g)$ is reduced to O^{2-} in zinc oxide.

4. A galvanic cell is a type of electrochemical cell that converts the energy released in a chemical reaction into electrical energy. An example of a galvanic cell, an alkaline cell, is shown in Figure 8.5. A battery is a series of galvanic cells connected together. An example is the lead–acid storage battery made of several galvanic cells. *Note:* the term *battery* commonly is used interchangeably with *cell* (galvanic cell). For example, sometimes the D cell (a type of galvanic cell) is referred to as a D cell battery.

6. a. The anode is $Zn(s)$. The oxidation half-reaction is:

$$Zn(s) \longrightarrow Zn^{2+}(aq) + 2\,e^-$$

  b. The cathode is $Ag(s)$. The reduction half-reaction is:

$$2\,Ag^+(aq) + 2\,e^- \longrightarrow 2\,Ag(s)$$

11. a. The electrolyte completes the electrical circuit. It provides a medium for transport of ions, thus allowing charge to be transferred.

  b. $KOH(aq)$ (in a paste-like form)

  c. $H_2SO_4(aq)$ (concentrated sulfuric acid)

13. It represents the reduction half-reaction. The conversion of O_2 to H_2O requires a supply of electrons.

17. Oxidation half-reaction:

$$CH_4 + 8\,OH^- \longrightarrow CO_2 + 6\,H_2O + 8\,e^-$$

Reduction half-reaction:

$$2\,O_2 + 4\,H_2O + 8\,e^- \longrightarrow 8\,OH^-$$

Overall reaction:

$$CH_4 + 2\,O_2 \longrightarrow CO_2 + 2\,H_2O$$

23. Each Si atom is surrounded by 8 electrons, but the atom in the center (the one that is doping the semiconductor) is surrounded by 9 electrons. Silicon is in Group 4A and has 4 outer electrons. Thus central atom in the figure must have 5 outer electrons. This is consistent with an element in Group 5A such as arsenic, so this is an *n*-type silicon semiconductor.

25. In every electrochemical process described in this chapter, energy is produced through electron transfer. Chemical reactions (such as those that take place in galvanic cells, batteries, and fuel cells) produce electrons that can do work because the anode and cathode are physically separated in space. The transfer of electrons also may be initiated when light strikes a photovoltaic cell.

29. The primary difference is that these produce electricity using different chemical reactions. In addition, a lead–acid storage battery converts chemical energy into electrical energy by means of a reversible reaction. No reactants or products leave the "storage" battery, and the reactants can be reformed during the recharging cycle. A fuel cell also converts chemical energy into electrical energy, but the reaction is not reversible. A fuel cell continues to operate only if fuel and oxidant are continuously added, which is why it is classed as a "flow" battery.

40. a. Conversion of fuel:

$$C_8H_{18}(l) + 4\,O_2(g) \longrightarrow 9\,H_2(g) + 8\,CO(g)$$

  Fuel cell reaction:

$$CO(g) + H_2O(g) \xrightarrow{\text{catalyst}} CO_2(g) + H_2(g)$$

  b. This type of fuel cell is convenient because it runs on a liquid fuel, gasoline, rather than using gaseous hydrogen. Currently, most nations have an infrastructure for gasoline refueling. However, the liquid fuel is still petroleum-based and therefore nonrenewable. It also burns to produce CO_2, a greenhouse gas. Therefore, although such fuel cells may find specialty applications in the near future, their long-term prospects are not promising.

Chapter 9

1. Cotton, silk, rubber, wool, and DNA are examples of natural polymers. Synthetic polymers include Kevlar, polyvinyl chloride (PVC), Dacron, polyethylene, polypropylene, and polyethylene terephthalate.

3. The *n* on the left side of the equation gives the number of monomers that react to form the polymer. This *n* is a coefficient. In contrast, the *n* on the right side is a subscript and represents the number of repeating units in the polymer.

6. The bottle on the left most likely is made of low-density polyethylene; the one on the right high-density polyethylene. The molecular structures of LDPE and HDPE help explain at a molecular level the difference in properties. LDPE is a more highly branched polymer, lessening molecular attractions between the chains and causing the plastic to be softer and more easily deformed. HDPE molecules, with fewer branches, can more closely approach each other, increasing molecular attractions.

9. Each ethylene monomer has a molar mass of 28.052 grams. To determine the number of monomers in the polymer, divide 40,000 (the molar mass of the polymer) by 28.052 (the molar mass of the monomer). The result is 1426 monomers. To determine the number of carbon atoms present in the polymer, note that each monomer contains two carbon atoms ($H_2C{=}CH_2$). Accordingly, the polymer contains 2×1426 carbon atoms, or 2852 carbon atoms.

11. This is the head-to-head, tail-to-tail arrangement of PVC formed from three monomer units.

22. a. A blowing agent is a gas (or a substance capable of producing a gas) used to manufacture a foamed plastic. For example, a blowing agent produces Styrofoam from PVC.

   b. Carbon dioxide replaces the CFCs or the HCFCs that once were used as blowing agents. Although CO_2 is a greenhouse gas, it still is preferable because CFCs and HCFCs both deplete the ozone layer and are greenhouse gases.

23. a. In 1997, 8.9×10^{10} pounds of plastic was produced in the United States. In contrast, 1.07×10^{11} pounds of plastic was produced in 2003.

   b. $\dfrac{1.07 \times 10^{11} \text{ lb plastic}}{2.90 \times 10^{8} \text{ people}}$ = 370 lb/person in 2003

   $\dfrac{8.9 \times 10^{10} \text{ lb plastic}}{2.69 \times 10^{8} \text{ people}}$ = 330 lb/person in 1997

   c. $\dfrac{370 \text{ lb/person} - 330 \text{ lb/person}}{330 \text{ lb/person}}$

   $\times 100 = 12\%$ change

28. Factors other than the chemical composition of the monomer(s) influence the properties of the polymer. These include length of the chain (the number of monomer units), the three-dimensional arrangement of the chains, the degree of branching in the chain, and orientation of monomer units within the chain.

30. For addition polymerization, the monomer must have a C=C double bond. Although some monomers have benzene rings as part of their structures (styrene, for example), the double bond involved in addition polymerization must not be in the ring. An example is the formation of PP from propylene.

For condensation polymerization, each monomer must have two functional groups that can react and eliminate a small molecule such as water. For example, an alcohol and a carboxylic acid can react to eliminate water. An example is the formation of PET from ethylene glycol and terephthalic acid.

31. In vinyl chloride, there are three bonds (two single and a double) around each carbon. Thus the geometry is trigonal planar and the Cl–C–H bond angle is 120°. In the polymer, each carbon atom is connected to other atoms by four single bonds, and the geometry is tetrahedral, with a bond angle of 109.5°.

32. a. This is the Lewis structure.

b. When Acrilan fibers burn, one of the combustion products is the poisonous gas hydrogen cyanide, HCN.

34. It requires 598 kJ/mol to break C=C bonds. The formation of C–C single bonds releases 356 kJ/mol. If we consider the reaction of two ethylene monomers, two double bonds are broken and replaced with four single bonds (two bonds between the C atoms of the monomers, one between the first monomer and the second, and a bond extending to what would be the third ethylene monomer). The calculation is $(2 \times 598 \text{ kJ/mol}) - (4 \times 356 \text{ kJ/mol}) = -228 \text{ kJ/mol}$. Thus, the reaction is exothermic.

38. a. The approximate increase in plastic production for any five-year period is between 8 and 12 billion pounds.

   b. Plastics production in 1977 was 34 billion pounds. A doubling of this production requires an output of 68 billion pounds, a level that, according to the graph, occurred around 1992. Thus it took 15 years to double the plastics output of 1977.

43. a. Branched LDPE cannot be used in this application because it would not be strong enough. It would not protect against accidental cuts or punctures.

   b. The linear HDPE polymer Spectra is very resistant to being cut or punctured. With an extremely thin liner, the surgeon still is able to have protection from cuts or punctures while retaining the flexibility needed.

45. The "Big Six" polymers are generally nonpolar molecules and therefore do not dissolve in polar solvents such as water. The generalization, developed in Chapter 5, is that "like dissolves like." Some of the "Big Six" dissolve or soften in hydrocarbons or chlorinated hydrocarbons because these nonpolar solvents interact with the nonpolar polymeric chains.

Chapter 10

1. a. An antipyretic drug is intended to reduce fever.

   b. An analgesic drug is intended to reduce pain.

   c. An anti-inflammatory drug is intended to reduce inflammation; that is, swelling and pain caused by irritation, injury, or infection.

2. Organic chemists study the chemistry of carbon compounds.

3. The condensed formulas are $CH_3CH_2CH_2CH_2CH_3$ [or $CH_3(CH_2)_3CH_3$], $CH_3CH_2CH(CH_3)CH_3$, and $CH_3C(CH_3)_2CH_3$. Here are the line-angle drawings.

5. For the chemical formula C_4H_9OH, four isomers exist. Here are the structural formulas, with the H atoms omitted for clarity.

6. a. contains the C–O–C functional group and is an ether.

b. contains the functional group and is a carboxylic acid.

c. contains the functional group and is a ketone.

d. contains the functional group and is an amide.

e. contains the functional group and is an ester.

7. One-carbon examples exist for alcohol, aldehydes, and acids. The other functional groups require more than one C atom.

a. Alcohol. The simplest compound is methanol, CH_3OH.

b. Aldehyde. The simplest example is methanal (commonly called formaldehyde), CH_2O.

d. Ester. The simplest example has two carbons: $HCOOCH_3$. It is called methyl methanoate, or methyl formate (but these names may be beyond the scope of your study).

8. a. The compound is an alcohol (ethanol). An isomer with a different functional group is an ether:

b. The compound is an aldehyde (propanal). An isomer with a different functional group is a ketone:

c. The compound is an ester (propyl formate). An isomer with a different functional group is a carboxylic acid:

10. a. Here is the structural formula for acetaminophen.

b. The chemical formula is $C_8H_9NO_2$.

11. a. There are two amide groups,

14. a. n-propanol, $CH_3CH_2CH_2OH$

b. isopropanol, $(CH_3)_2CHOH$

c. *t*-butanol, $(CH_3)_3COH$

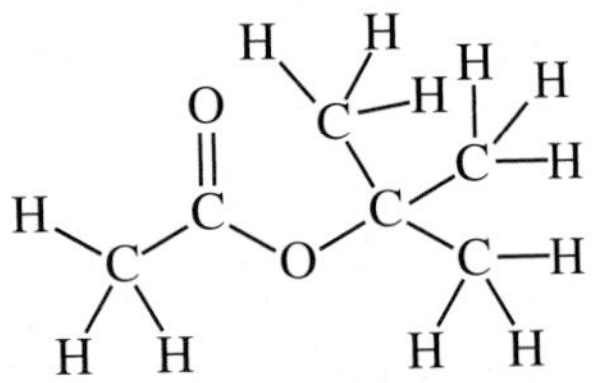

16. a. $H—C≡N:$

17. No, aspirin would not be more active if it were to interact with prostaglandins directly. Increasing its effectiveness would require a direct correspondence between the number of molecules of aspirin and prostaglandin. When aspirin blocks a COX enzyme, it prevents the synthesis of many prostaglandin molecules, since one enzyme is responsible for increasing the rate of synthesis of the prostaglandins.

18. If you started with 2 active functional groups, you would have 4 different products after 2 synthetic steps. If the reagent used in the first step had 2 reactive groups itself, you would produce 8 different products after the 2 synthetic steps (assuming the second step had a reagent with only 1 reactive group).

21. A pharmacophore is the three-dimensional arrangement of atoms, or groups of atoms, responsible for the biological activity of a drug molecule.

22. Sulfanilamide has the same basic shape and contains similar functional groups in the same regions as *para*-aminobenzoic acid, so it replaces the nutrient in some biologically important process. Without the key nutrient, the bacteria die.

23. a. This compound cannot exist in chiral forms. The central carbon atom is bonded to two equivalent –CH₃ groups.

b. This compound can exist in chiral forms. The four groups attached to the central carbon atom are all different.

c. This compound can exist in chiral forms. The four groups attached to the central carbon atom are all different.

d. This compound cannot exist in chiral forms. The central carbon atom is bonded to two equivalent –CH₃ groups.

25. A freebase is a nitrogen-containing molecule in which the nitrogen is in possession of its lone pair of electrons. Treating methamphetamine hydrochloride with a base (e.g., hydroxide ions, OH⁻) strips off one of the hydrogen atoms attached to the nitrogen atom, freeing up its lone pair.

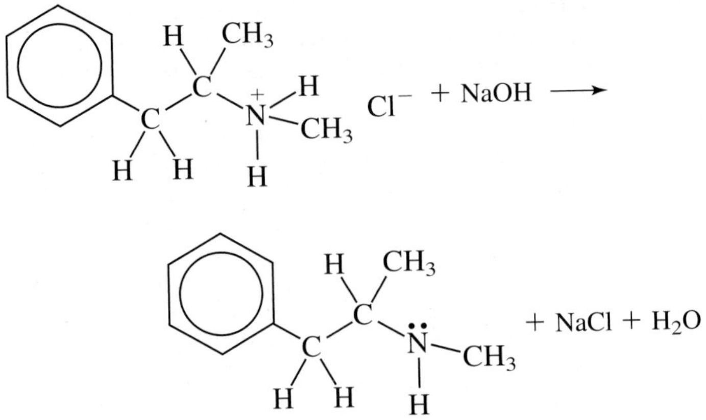

27.

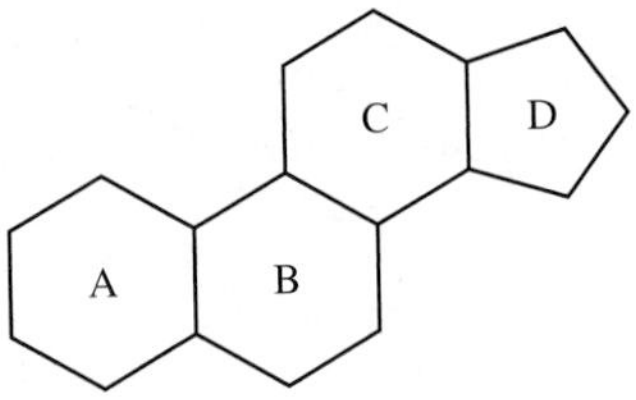

29. a. Four single bonds, one triple bond

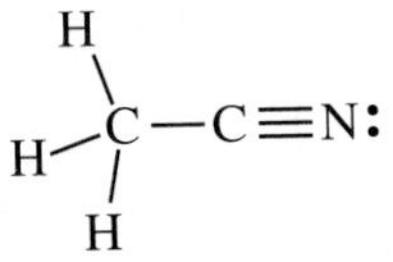

b. Six single bonds, one double bond

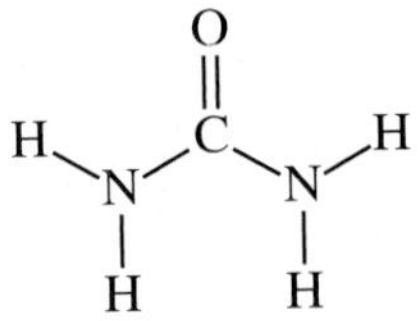

31. Only three isomers shown here, because some structures are duplicates. Numbers 1 and 5 are different paper-and-pencil representations of the *same* isomer. Numbers 2, 3, and 4 are all different paper-and-pencil representations of the *same* isomer. Number 6 is an isomer *different* from numbers 1 and 5, and from numbers 2–4.

32.

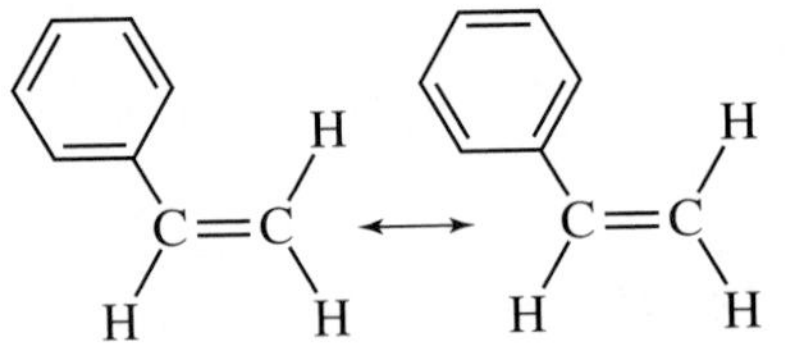

35. a. Aspirin produces a physiological response in the body.

b. Morphine produces a physiological response in the body.

c. Antibiotics kill or inhibit the growth of bacteria that cause infections.

e. Amphetamine produces a physiological response in the body.

37. The drug (−)-dopa is effective because the molecule fits in the receptor site, but the nonsuperimposable mirror image form, (+)-dopa, does not. This is the structure of (−)-dopa, with the chiral carbon atom marked in red. Note that there are four different groups attached to the starred carbon atom.

$$
\begin{array}{c}
\text{HO} \\
\text{HO}
\end{array}
\bigcirc
\begin{array}{c}
\text{H} \\
\text{C} \\
\text{H}
\end{array}
\begin{array}{c}
\text{NH}_2 \\
\text{H—C}^*\text{—C} \\
\end{array}
\begin{array}{c}
\text{O} \\
\text{OH}
\end{array}
$$

39. There is a very specific fit between a chiral molecule and its asymmetrical binding site. It must be that the (−)-methorphan has a much better fit and is able to act as a narcotic, but (+)-methorphan does not fit as well, and is therefore not as potent a drug. Because (+)-methorphan has less activity, it can be added to over-the-counter medicines with some safety, assuming that consumers follow the label directions on the over-the-counter drugs containing this compound.

45. a. There are 6(4) + 6(1) or 30 electrons available. Here is a possible linear isomer for benzene.

Structural formula:

$$
\begin{array}{c}
\text{H} \\
\text{C} \\
\text{H}
\end{array}
= \text{C} =
\begin{array}{c}
\text{H} \\
\text{C} \\
\end{array}
\quad
\begin{array}{c}
\text{H} \\
\text{C} \\
\text{H}
\end{array}
= \text{C} =
\begin{array}{c}
\text{H} \\
\text{C} \\
\text{H}
\end{array}
$$

b. This is the condensed formula:
$CH_2{=}C{=}CH{-}CH{=}C{=}CH_2$

c. First check to see if all of the structures correctly represent C_6H_6 and that each carbon has four bonds. If these conditions are met, the structures with double bonds should differ only in the placement of the C–C bond. However, structures including carbon–carbon triple bonds can also be drawn; these would be distinctly different.

Chapter 11

1. Both what you eat and how much you eat affects your health. Some of the factors just involve common sense. For example, excessive food intake correlates with obesity; a diet rich in sugars and fats is high in calories and thus is not a wise choice for either children or adults. Less obvious is the effect of food choices on ecosystems and water use. As this chapter points out, eating beef that is corn-fed and then sent to slaughter connects to both local degradation of the land (corn usually is grown with herbicides and fertilizers) and regional deterioration of the water ways (in the United States, check the effluent of the Mississippi River into the "Dead Zone" in the Gulf of Mexico). Other ways in which your food choices affect wider regions include (1) the effects of irrigation (depletion of aquifers, rivers drying up downstream), (2) the loss of forests when creating farmland, and (3) the energy used to plant, grow, and harvest a particular crop.

4. An all-or-nothing diet, such as eating no meat ever, leaves no middle ground. If you are not careful in your food choices, the absence of all meat can lead to an unbalanced diet depriving you of the proper nutrients that you need to stay healthy. Additionally, it doesn't leave you the option of significantly reducing your consumption of certain meats while not giving it up entirely.

5. In most cases, raising beef requires large amounts of grain to feed the animals and land to raise the grain and also pasture the animals. In turn, the grain requires both water and energy to produce. Eating beef that comes from animals that are grass-fed and free ranging is a more sustainable possibility, but for many people this is either not possible or not practical.

9. Yes. The potatoes have been sliced and cooked in partially hydrogenated vegetable oil, and ingredients such as salt and preservatives have been added.

10. a. Macronutrients are materials that are consumed in relatively large amounts from the foods we eat.

b. They provide essentially all of the energy and most of the raw material for repair and synthesis.

c. The three major classes of macronutrients are carbohydrates, fats, and proteins.

12. The chart indicates more carbohydrate present than would be found in steak, and more protein than would be found in chocolate chip cookies. The chart is likely to be a representation of peanut butter. (See Table 11.1 for confirmation.)

15. Similarities: Both oils and fats feel greasy and are insoluble in water. Both can go rancid. On a molecular level, fats and oils are both triglycerides, that is, triple esters. These molecules are characterized by the presence of long, nonpolar hydrocarbon chains. Edible fats and oils both contain some oxygen.

Differences: Fats are solid and oils are liquid at room temperature. Oils tend to contain more polyunsaturated fatty acids and smaller fatty acids than fats.

16. a. Here is the structural formula for lactic acid.

$$
\begin{array}{c}
\text{H} \quad \text{OH} \\
| \quad\quad | \\
\text{H—C—C—C} \\
| \quad\quad | \\
\text{H} \quad \text{H}
\end{array}
\begin{array}{c}
\text{O} \\
\\
\text{OH}
\end{array}
$$

b. Lactic acid is saturated because the hydrocarbon chain contains only single bonds between the carbon atoms.

c. No, lactic acid is not a fatty acid. Although it has the carboxylic acid group of an acid, it lacks the long hydrocarbon chain.

19. According to Table 11.1, peanut butter is a good source of protein. However, it is also 50% fat, which is quite high if one needs to limit fat in the diet. On the positive side, much of the fat in peanut butter is unsaturated, unless the peanut butter has been hydrogenated.

23. See Figure 11.13 for structures of fructose and glucose. Observe that the structure of fructose is based on a five-membered ring composed of 4 C atoms and 1 O atom. The structure of glucose is based on a six-membered ring composed of 5 C atoms and 1 O atom. Glucose has one $-CH_2OH$ side chain, and fructose has two.

24. The "amino" in *amino acid* indicates that an amine functional group is present. The "acid" indicates there is an acidic functional group, in this case, a carboxylic acid.

30. Meat group—sausage

    Bread and grains group—crust

    Dairy group—cheese

    Vegetable group—tomato sauce

    Fats, oils, and sweets group—sausage, cheese

34. **a.** The molecules of saturated fats, with their long hydrocarbon chains, can pack more closely to each other. There is free rotation of the atoms around C–C bonds that allows for twisting and turning until the molecules approach each other as closely as possible. In contrast, the C=C double bonds in unsaturated fats introduce "bends" in the hydrocarbon chain and prevent these molecules from such close packing. The C=C double bonds do not allow for free rotation. This is easier to understand with the help of molecular models.

    b. The melting points decrease as the number of C=C bonds increases. With no C=C bonds, stearic acid molecules can pack well, increasing the extent of intermolecular attractions between molecules and therefore increasing the melting point. Molecules with C=C double bonds don't pack as well, resulting in reduced intermolecular interactions and lower melting points.

    c. Trans fats behave more like saturated fats than oils, and thus share some of the health effects of saturated fats. The reason stems from the differences in molecular structure. In unsaturated fats, the two H atoms are on the same side of the bond (*cis*), the bond serves as a "bend" in the long hydrocarbon chain. In trans fats, the two H atoms are on the opposite side of the double bond (*trans*), such as in elaidic acid shown in Figure 11.10, the hydrocarbon chain lacks this bend and is structurally more like that of a saturated hydrocarbon chain.

37. **a.** Milk does contain the disaccharide lactose, which contributes about 40% of the calories to whole cow's milk. However, milk also contains valuable nutrients including proteins, calcium, and vitamin C.

    b. You certainly should read food labels carefully. In this case mom is wrong in that the whole milk, 2% and skim all contain the same amount of carbohydrates (sugar). Mom has confused sugar with fat content. Whole cow's milk contains about 3.3% fat, "2% milk" contains 2% fat, and skim milk has 0% fat.

43. The Dietary Guidelines of 2005 include a recommendation that 30% or less of total calories should come from fat. For a typical 2000 Cal/day diet, that would mean only 600 Cal from fat. Meeting this recommendation while also maintaining a diet low in harmful saturated fats could be possible if all of the milk or dairy products consumed were low-fat or no-fat. For example, 1 oz of regular cheddar cheese has 6.0 g of saturated fat and 114 Cal. However, 1 oz of low-fat cheddar cheese has only 1.2 g of saturated fat and 49 Cal. One cup of whole milk has 4.6 g of saturated fat and 146 Cal, but switching to 1% milk means consuming 1.5 g saturated fat and 102 Cal per cup.

Chapter 12

1. Traits determined by the DNA of an individual include hair and eye color, fingerprints, the shape of earlobes (attached or hanging), and the shape of the nose. Some traits, such as body type, result from your diet and environment as well as from your genes.

3. A genome contains all of the genetic information in a cell. In contrast, a gene is a small subsection of the genome that codes for a single protein.

5. **a.** A nucleotide must contain a base, a deoxyribose molecule, and a phosphate group linked together.

   b. Covalent bonds hold the units together.

7.

9. DNA stands for **d**eoxyribo**n**ucleic **a**cid. The name highlights the deoxyribose sugar, the acidic nature of the phosphate groups, and the fact that it contains nucleotides. The name does not highlight the amines important in the bases. Nor does the name suggest the polymeric nature of DNA, in contrast to **poly**amides or **poly**esters.

12. a. The complementary base sequence is ATAGATC.

b. Your answer should have two lines between each A and T, and three lines between each C and G in the sequence.

14. Codons are the basis of the genetic code. Each codon consists of a three-nucleotide sequence that is specific for an amino acid or the start/stop of protein synthesis. All of the codons together make up the code for translating a sequence of DNA into the amino acid sequence of a protein.

17. a. This is the general formula for an amino acid, where R represents a side chain that is different in each of the 20 amino acids.

$$\begin{array}{ccc} H & H & O \\ | & | & \parallel \\ N & - C - & C \\ | & | & \diagdown \\ H & R & OH \end{array}$$

b. The functional groups are –COOH, the carboxylic acid group, and –NH$_2$, the amine group.

21. The primary structure of a protein is its sequence of amino acids. In the hemoglobin S chain, a glutamic acid is replaced with valine. This seemingly innocuous change in the protein's primary structure dramatically affects the shape of the protein because glutamic acid has a polar R group. In contrast, the valine has a nonpolar one. The change in protein shape causes blood cells to become "sickled" under certain conditions.

24. a. The phosphates and sugars combine to form the backbone of the DNA molecule. In Figure 12.4, these can be found along the left.

b. The nitrogen bases hang off the backbone, never connecting to each other. In the figure, these are found along the right.

c. In Figure 12.7 the backbone is represented by ribbons with the letters "s" and "p" for the alternating sugar and phosphate groups within. The nitrogen bases are shown as colored hexagons hanging off these two ribbons. The sculpture at the beginning of the chapter represents the backbones with the silver coils without distinguishing sugars from phosphate. Here the bases between the ribbons with colored plastic pieces for each type of base.

26. Figure 12.7 gives clear evidence of the importance of "fit" for the bases as they interact through hydrogen bonding. Adenosine and thymine match up with each other to form two hydrogen bonds. The first bond pairs an O atom from thymine with an N–H on adenine; the second pairs an N–H on thymine with an N atom on adenine. Cytosine and guanine match up with each other to form three hydrogen bonds. Here, the pairings are different.

Adenosine and cytosine would not fit match efficiently because the bases are not in the right positions to permit the hydrogen bonds to form. This is also true of thymine and guanine.

28. The two DNA strands are complimentary, meaning that the sequence on one strand can be reconstructed from the other. If a single strand breaks, the other strand can be used to correctly replace any lost nucleotides. If both strands break, the information on how to repair the break is lost. Furthermore, the strands may not reattach to each other in the same way, or these strands may attach to other strands of DNA. No matter which scenario ensues, the situation is more complicated.

30. The DNA-damaging agents that kill bacteria can also damage the DNA in other species, including humans. The risks involved with such drugs outweigh any potential benefits.

34. a. The nucleus of all cells in an adult contains all of the genetic information necessary to create the organism. A skin cell has all of the same genetic information as a nerve cell, even though much of the material is not in use.

b. The egg cell must have its own nuclear material removed, otherwise the genes from the egg's source may be used instead of the new material from the adult cell.

c. This differs from genetic engineering as (1) the host's genetic material is removed in cloning but is only altered in genetic engineering, and (2) cloning transfers all genes while genetic engineering only transfers a few.

37. a. Transgenic organisms are plants and animals whose genome contains genes from another species.

b. Traditional crossbreeding mechanisms take more time and the results are less predictable than using genetic engineering to alter the genetic makeup of plants.

43. a. In the electromagnetic spectrum, X-rays have high energies and short wavelengths. They are similar to gamma rays.

b. A beam of X-rays is directed at an unknown substance. The nuclei in the substance scatter the X-rays. A detector measures the intensity and pattern of scattered X-rays. If the atoms in the substance are arranged in a regular pattern, the diffracted X-rays can be used to calculate the distance between atoms.

c. One reason is that salts, such as sodium chloride, easily form crystals. In contrast, nucleic acids and proteins are much larger and do not easily form crystals. Another reason is that when nucleic acids and proteins do crystallize, the interpretation of their X-ray diffraction pattern is more difficult.

44. The statement is false. Calculations show that the DNA in an adult would stretch just about 26,000 times to the Moon and back.

$$\frac{2 \text{ m}}{1 \text{ DNA thread}} \times \frac{1 \times 10^{13} \text{ cells}}{1 \text{ adult}}$$
$$= 2 \times 10^{13} \text{ m of DNA in an adult}$$

$$3.8 \times 10^5 \text{ km} \times \frac{1 \times 10^3 \text{ m}}{1 \text{ km}} \times 2 = 7.6 \times 10^8 \text{ m},$$

the distance to the Moon and back.

The ratio of these two distances is

$$\frac{2 \times 10^{13} \text{ m}}{7.6 \times 10^8 \text{ m}} = \frac{26,000}{1}$$

48. Cloning produces an identical dog containing all of the same genetic traits. It does not allow for improving a breed, as is possible by breeding dogs. Crossbreeding can be more of a gamble, but you will create unique dogs with some traits from each parent. A dog with more strengths and fewer problems can result from breeding, but cloning can only produce the same dog repeatedly.

Glossary

The numbers at the end of each entry indicate the section in the text where the term first appears.

A

acid a compound that releases hydrogen ions, H^+, in aqueous solution 6.1

acid deposition a more inclusive term than acid rain that includes wet forms such as rain, snow, fog, and cloud-like suspensions of microscopic water droplets as well as the "dry" forms of acids. 6.6

acid rain rain with a pH below 5 that is more acidic than "normal" rain. The causes are both natural and anthropogenic. 6.6

acid-neutralizing capacity the capacity of a lake or other body of water to resist a decrease in pH 6.13

activation energy the energy necessary to initiate a chemical reaction 4.8

aerosols liquid or solid particles that remain suspended in the air rather than settling out 1.11

albedo a measure of the reflectivity of a surface, the ratio of electromagnetic radiation reflected from a surface relative to the amount of radiation incident on it 3.9

alkanes hydrocarbons with only single bonds between carbon atoms 4.4

alpha particle (α) a type of nuclear radiation. An alpha particle is a positively charged particle emitted from the nucleus. It consists of 2p and 2n (the nucleus of a He atom) and has a 2+ charge since no electrons accompany the helium nucleus. 7.4

ambient air the air surrounding us, usually meaning the outside air 1.3

amino acid residues amino acids which have been incorporated into the peptide chain 11.7

amino acids monomer from which our body builds proteins. Each amino acid molecule contains two functional groups: an amine group ($-NH_2$) and a carboxylic acid group ($-COOH$). 9.6

amorphous regions in a polymer, a region in which the long polymer molecules are in a random, disordered arrangement 9.4

anaerobic bacteria bacteria that can function without the use of molecular oxygen 3.8

anion a negatively charged ion 5.6

anode the electrode at which oxidation takes place 8.1

anthropogenic human activities, such as industry, transportation, mining, and agriculture 3.1

aqueous solution a solution in which water is the solvent 5.5

atom the smallest unit of an element that can exist as a stable, independent entity 1.7

atomic mass the mass (in grams) of the same number of atoms that are found in exactly 12 g of carbon-12 3.6

atomic number the number of protons in the nucleus of an atom 2.2

Avogadro's number the number of atoms in exactly 12 grams of C-12 3.6

B

background radiation the level of radiation, on average, that is present at a particular location. It can arise from both natural and human-made sources. 7.6

basal metabolism rate (BMR) the minimum amount of energy required daily to support basic body functions 11.9

base a compound that releases hydroxide ions, OH^-, in aqueous solution 6.2

beta particle (β) a type of nuclear radiation. A beta particle is a high-speed electron emitted from the nucleus. 7.4

biofuels the generic term for renewable fuels derived from plant matter such as trees, grasses, agricultural crops, or other biological material 4.9

biological oxygen demand (BOD) a measure of the amount of dissolved O_2 that microorganisms use up as they decompose the organic wastes found in water. A low BOD is one indicator of good water quality. 5.11

biomagnification the increase in concentration of certain persistent chemicals in successively higher levels of a food chain 5.7

blowing agent either a gas or a substance capable of producing a gas used to manufacture a foamed plastic 9.4

bond energy the amount of energy that must be absorbed to break a specific chemical bond 4.6

breeder reactor a nuclear reactor that can produce more fissionable fuel (usually Pu-239) than it consumes (usually U-235). 7.9

C

calorie (cal) the amount of heat necessary to raise the temperature of one gram of water by one degree Celsius 4.2

calorimeter a device for experimentally measure the quantity of heat energy released in a combustion reaction 4.5

carbohydrate a compound containing carbon, hydrogen, and oxygen, with hydrogen and oxygen found in the same 2:1 atomic ratio as in water 11.5

carbon capture and storage (CCS) the process of separating CO_2 from other combustion products and storing (sequestering) it in a variety of geologic locations 3.11

carbon footprint an estimate of the amount of CO_2 and other greenhouse gas emissions in a given time frame, usually a year 3.9

carbon sinks natural processes that remove CO_2 from the atmosphere 3.5

carcinogenic capable of causing cancer 1.13

catalyst a chemical substance that participates in a chemical reaction and influences its rate without undergoing permanent change 1.11

catalytic cracking a process in which catalysts are used to crack larger hydrocarbon molecules into smaller ones at relatively low temperatures 4.7

catalytic reforming a process in which the atoms within a molecule are rearranged, usually starting with linear molecules and producing ones with more branches 4.7

cathode the electrode at which reduction takes place. The cathode receives the electrons produced at the anode. 8.1

cation a positively charged ion 5.6

cellulosic ethanol ethanol produced from cornstalks, switchgrass, wood chips, and other materials that are non-edible by humans 4.9

chain reaction a term that generally refers to any reaction in which one of the products becomes a reactant and thus makes it possible for the reaction to become self-sustaining. 7.1

Chapman cycle the first set of natural steady-state reactions proposed for stratospheric ozone 2.6

chemical equation a representation of a chemical reaction using chemical formulas. To students, chemical equations are probably better known as "the thing with the arrow in it" 1.9

chemical formula a symbolic way to represent the elementary composition of a substance 1.7

chemical reaction a process whereby substances described as reactants are transformed into different substances called products 1.9

chemical symbol a one- or two-letter abbreviation for an element. Also sometimes called an atomic symbol 1.6

chiral (optical) isomers compounds with the same chemical formula but different three-dimensional molecular structures and different interaction with plane polarized light 10.6

chlorofluorocarbons (CFCs) compounds composed of the elements chlorine, fluorine, and carbon (but do not contain the element hydrogen) 2.9

climate a term that describes regional temperatures, humidity, winds, rain, and snowfall over decades, not days. Contrast with weather 3.10

climate adaptation the ability of a system to adjust to climate change (including climate variability and extremes) to moderate potential damage, to take advantage of opportunities, or to cope with the consequences 3.11

climate mitigation any action taken to permanently eliminate or reduce the long-term risk and hazards of climate change to human life, property, or the environment 3.11

coalescents chemicals added to soften the latex particles in paint so that the paint spreads to form a continuous film of uniform thickness. 1.13

coenzymes molecules that work in conjunction with enzymes to enhance the enzyme's activity 11.8

combinatorial chemistry systematic creation of large numbers of molecules in "libraries" that can be rapidly screened in the lab for biological activity and the potential for becoming new drugs 10.5

combustion the chemical process of burning; that is, the rapid combination of fuel with oxygen to release energy in the form of heat and light 1.9

compound a pure substance made up of two or more different elements in a fixed, characteristic chemical combination. Compounds contain two or more different types of atoms. 1.6

concentration the ratio of the amount of solute to the amount of solution 5.5

condensation polymerization a type of polymerization in which a small molecule such as water is split out (eliminated) when the monomers join to form a polymer 9.5

condensed structural formula a chemical structure in which all bonds are not drawn out explicitly; rather, the structure is understood to contain an appropriate number of bonds 10.2

conductivity meter an apparatus that produces a signal to indicate that electricity is being conducted 5.6

control rods in a nuclear reactor, rods that are composed primarily of an excellent neutron absorber, such as cadmium or boron, that can be positioned to absorb fewer or more neutrons. 7.3

copolymer a polymer built from two or more different monomers 9.5

covalent bond a bond formed when electrons are shared between two atoms 2.3

cradle-to-cradle a term coined in the 1970s that refers to a regenerative approach to the use of things in which the end of the life cycle of one item dovetails with the beginning of the life cycle of another, so that everything is reused rather than disposed of as waste 0.4

cradle-to-grave an approach to analyzing the life cycle of an item, starting with the raw materials from which it came and ending with its ultimate disposal someplace, presumably on Earth 0.4

critical mass the amount of fissionable fuel required to sustain a chain reaction 7.1

crystalline regions in a polymer, a region in which the long polymer molecules are arranged neatly and tightly in a regular pattern 9.4

curie (Ci) a unit, named in honor of Marie Curie, that is a measure of the radioactivity of a sample (3.7×10^{10} nuclear decays per second) 7.6

current (electrical) the rate of electron flow through a circuit 8.2

D

denitrification the process of converting nitrates back to nitrogen gas 6.9

density the mass per unit volume 5.2

depleted uranium uranium composed almost entirely of U-238 (~99.8%) because much of the U-235 that it once naturally contained has been removed. Nicknamed DU. 7.7

desalination any process that removes ions from salt water 5.12

diatomic molecule a molecule consisting of two atoms 1.7

dietary supplements vitamins, minerals, amino acids, enzymes, herbs, and other botanicals 10.8

dipeptide a compound formed from two amino acids 11.7

disaccharide a "double sugar" formed by joining two monosaccharide units, such as sucrose (table sugar) 11.5

dispersion forces attractions between molecules that result from a distortion of the electron cloud that causes an uneven distribution of the negative charge 9.3

distillation a separation process in which a solution is heated to its boiling point and the vapors are condensed and collected 4.4

distributed generation generating electricity on-site where it is used (i.e., with a fuel cell), thus avoiding the losses of energy that occur over long electric transmission lines 8.6

doping the process of intentionally adding small amounts of other elements to pure silicon to modify its semiconductor properties 8.7

double bond a covalent bond consisting of two pairs of shared electrons 2.3

E

ecological footprint a means of estimating the amount of biologically productive space (land and water) necessary to support a particular standard of living or lifestyle. 0.5

effective stratospheric chlorine a term reflecting both chlorine and bromine-containing gases in the stratosphere 2.11

electricity the flow of electrons from one region to another that is driven by a difference in potential energy 8.1

electrodes in an electrochemical cell, the electrical conductors (anode and cathode) that serve as sites for chemical reactions 8.1

electrolysis the process of passing a direct current of electricity of sufficient voltage to cause a chemical reaction to occur. For example, the electrolysis of water decomposes it into H_2 and O_2. 8.6

electrolyte a solute that conducts electricity in an aqueous solution 5.6

electrolytic cell a type or electrochemical cell in which electrical energy is converted to chemical energy 8.6

electromagnetic spectrum continuum of waves that ranges from short, high-energy X-rays and gamma rays to long, low-energy radio waves 2.4

electron a subatomic particle with a mass much smaller than that of a proton or neutron and with a negative charge equal in magnitude to that of a proton, but opposite in sign 2.2

electronegativity a measure of the attraction of an atom for an electron in a chemical bond 5.1

element one of the 100 or so pure substances in our world from which compounds are formed. Elements contain only one type of atom 1.6

endothermic a term to describe any chemical or physical change that absorbs energy 4.5

enhanced greenhouse effect the process in which atmospheric gases trap and return *more than* 80% of the heat energy radiated by the Earth 3.1

enriched uranium uranium, typically to fuel nuclear reactors or nuclear weapons, that has a higher percent of U-235 than its natural abundance of about 0.7%. 7.7

entropy a measure of how much energy gets dispersed in a given process 4.2

enzymes an amino acid that is required for protein synthesis but that must be obtained from the diet because the body cannot synthesize it 10.4

essential amino acids those amino acids required for protein synthesis but that must be obtained from the diet because the body cannot synthesize them 11.7

exothermic a term to describe any chemical or physical change accompanied by the release of heat 4.5

exposure the amount of the substance encountered 1.3

F

fats triglycerides that are solids at room temperature 11.2

first law of thermodynamics also called the law of conservation of energy, states that energy is neither created nor destroyed 4.1

foodprint the amount of land required per year to provide the nutritional resources for one person 11.1

fossil fuel combustible substances derived from the remnants of prehistoric origins. The most common examples are coal, petroleum, and natural gas. 3.2

free radical a highly reactive chemical species with one or more unpaired electrons 2.8

freebase nitrogen-containing molecule in which the nitrogen is in possession of its lone pair of electrons 10.3

frequency the number of waves passing a fixed point in 1 second 2.4

fuel cell a galvanic cell that produces electricity by converting the chemical energy of a fuel directly into electricity without burning the fuel 8.5

functional groups distinctive arrangement of groups of atoms that impart characteristic physical and chemical properties to the molecules that contain them 9.5

G

galvanic cell a type of electrochemical cell that converts the energy released in a spontaneous chemical reaction into electrical energy 8.1

gamma ray (γ) a type of nuclear radiation. A gamma ray is emitted from a radioactive nucleus and has no charge or mass. It is a short-wavelength high energy photon. 7.4

gaseous diffusion a process in which gases with different molecular weights are forced through a series of permeable membranes. One use is to enrich uranium. 7.7

generic drug a drug that is the chemical equivalent of a pioneer drug but that cannot be marketed until the patent protection on the pioneer drug has run out after 20 years 10.8

genetic engineering the direct manipulation of DNA in an organism 12.6

genome the complete set of inheritable traits of an organism 12.1

global atmospheric lifetime characterizes the time required for a gas added to the atmosphere to be removed. It is also referred to as the "turnover time." 3.8

global climate change sometimes used interchangeably with global warming, this term refers to changes in the weather over time 3.1

global warming a popular term used to describe the increase in average global temperatures as a result of an enhanced greenhouse effect 3.1

global warming potential (GWP) a number that represents the relative contribution of a molecule of the atmospheric gas to global warming 3.8

green chemistry the design of chemical products and processes to use less energy, fewer hazardous materials, and renewable resources whenever possible. The desired outcome is to produce less waste, especially toxic waste, and to use fewer resources. 0.6

greenhouse effect the natural process by which atmospheric gases trap a major portion (about 80%) of the infrared radiation radiated by the Earth 3.1

greenhouse gases gases capable of absorbing and trapping infrared radiation, thereby warming the atmosphere. Examples include water vapor, carbon dioxide, methane, nitrous oxide, ozone, and chlorofluorocarbons. 3.1

groundwater fresh water found in underground reservoirs, also known as aquifers 5.4

group a column in the periodic table that organizes elements according to the important properties they have in common. Groups are numbered left to right. 1.6

H

half-life ($t_{1/2}$) the time required for the level of radioactivity to fall to one half of its initial value 7.8

half-reaction a type of chemical equation that shows the electrons either lost or gained by the reactants 8.1

halogen one of the reactive nonmetals in Group 7A, such as fluorine (F), chlorine (Cl), bromine (Br), or iodine (I) 1.6

halons inert, nontoxic compounds similar to CFCs, but in addition they contain bromine. Depending on the definition, other compounds may be included as well. 2.9

heat the kinetic energy that flows from a hotter object to a colder one 4.5

heat of combustion the quantity of heat energy given off when a specified amount of a substance burns in oxygen 4.5

high-level radioactive waste (HLW) nuclear waste that has high levels of radioactivity and, because of the long half-lives of the radioisotopes involved, requires essentially permanent isolation from the biosphere 7.9

hormesis the concept that low doses of a harmful substance (such as radiation) may actually be beneficial 7.6

hormones chemical messengers produced by the body's endocrine glands 10.4

hybrid electric vehicle (HEV) a vehicle propelled by a combination of a conventional gasoline engine and an electric motor run by batteries 8.4

hydrocarbon a compound made up only of the elements of hydrogen and carbon 1.8

hydrochlorofluorocarbons (HCFCs) compounds of hydrogen, chlorine, fluorine, and carbon used as replacements for CFCs 2.12

hydrofluorocarbons (HFCs) compounds of hydrogen, fluorine, and carbon used as replacements for CFCs and HCFCs 2.12

hydrogen bond an electrostatic attraction between a H atom bonded to a highly electronegative atom (O, N, or F) and a neighboring O, N, or F atom, either in another molecule or in a different part of the same molecule 5.2

hydrogenation a process in which hydrogen gas, in the presence of a metallic catalyst, adds to a C=C double bond and converts it to a single bond 11.4

hygroscopic describes a substance that readily absorbs water from the atmosphere and retains it 6.12

I

infrared (IR) a region of the electromagnetic spectrum that lies adjacent to red light, but at longer wavelength 2.4

interesterification any process in which the fatty acids on two or more triglycerides are scrambled to produce a set of different triglycerides 11.4

intermolecular force a force that occurs between molecules 5.2

intramolecular force a force that exists within a molecule 5.1

ion an atom or group of atoms that has acquired a net electric charge as a result of gaining or losing one or more electrons 5.6

ionic bond a chemical bond formed when oppositely charged ions attract 5.6

ionic compound a compound composed of ions that are present in fixed proportions and arranged in a regular, geometric structure 5.6

isomers molecules with the same chemical formula (same number and kinds of atoms), but with different structures and properties 4.7

isotopes two or more forms of the same element (same number of protons) whose nuclei differ in number of neutrons and hence in mass 2.2

J

joule (J) a unit of energy equal to 0.239 cal 4.2

K

kinetic energy the energy of motion 4.1

L

law of conservation of matter and mass a law stating that in a chemical reaction, matter and mass are conserved, that is, neither created nor destroyed 1.9

lead compound drug (or a modified version of that drug) that shows high promise for becoming an approved drug 10.5

Lewis structure a representation of an atom or molecule that shows its outer electrons 2.3

line-angle drawing simplified version of a structural formula that is most useful for representing larger molecules 10.2

linear, nonthreshold model a model that assumes a linear relationship between the adverse effects and the radiation dose, with radiation being harmful at all doses, even low ones. 7.6

low-level radioactive waste (LLW) nuclear waste that contains smaller quantities of radioactive materials than HLW and specifically excludes spent nuclear fuel 7.9

M

macrominerals elements that are necessary for life (Ca, P, Cl, K, S, Na, and Mg) but not nearly as abundant in the body as O, C, H, and N 11.8

macromolecules molecules of high molecular mass that have characteristic properties because of their large size 9.1

mass number the sum of the number of protons and neutrons in the nucleus of an atom 2.2

maximum contaminant level (MCL) the legal limit for the concentration of a contaminant expressed in parts per million or parts per billion 5.11

maximum contaminant level goal (MCLG) the maximum level of a contaminant in drinking water at which no known or anticipated adverse effect on human health would occur 5.7

megacity an urban area with over 10 million people, such as Tokyo, New York City, Mexico City, or Mumbai 1.5

metabolism the complex set of chemical processes that are essential in maintaining life 11.2

metal an element that is shiny and conducts electricity and heat well, such as copper, iron, and magnesium 1.6

metalloid an element that lies between metals and nonmetals on the periodic table and does not fall cleanly into either group. Sometimes called a semimetal 1.6

microgram (μg) a millionth of a gram (g) or 10^{-6} g 1.3

micrometer (μm) a millionth (10^{-6}) of a meter (m), sometimes referred to simply as a micron 1.3

microminerals nutrients that the body requires lesser amounts of, such as Fe, Cu, and Zn 11.8

micronutrients substances that are needed only in miniscule amounts, but remain essential for the body to produce enzymes, hormones, and other substances needed for proper growth and development 11.8

minerals ions or ionic compounds that, like vitamins, have a wide range of physiological functions 11.8

mixture a physical combination of two or more pure substances present in variable amounts 1.1

moderator in a nuclear reactor, slows the neutrons, thus making them more effective in producing fission 7.3

molar mass is the mass of one Avogadro's number, or mole, of whatever particles are specified 3.7

molarity (M) a unit of concentration represented by the number of moles of solute present in one liter of solution 5.5

mole (mol) an Avogadro's number of objects 3.7

molecule two or more atoms held together by chemical bonds in a certain spatial arrangement 1.7

monomer (from *mono* meaning "one" and *meros* meaning "unit"). A small molecule used to synthesize a larger polymer 9.1

monosaccharide a single sugar, such as fructose or glucose 11.5

monounsaturated fatty acid a fatty acid containing only one double bond between carbon atoms in the hydrocarbon chain 11.3

municipal solid waste (MSW) garbage, that is, everything you discard or throw into your trash, including food scraps, grass clippings, and old appliances. MSW does not include industrial waste or waste from construction sites. 9.7

N

nanometer (nm) one billionth of a meter (m) 2.4

nanotechnology a subdiscipline that relates to the creation of materials at the atomic and molecular (nanometer) scale: 1 nanometer (nm) = 1×10^{-9} m 1.7

neutral solution a solution that is neither acidic nor basic, that is, it has equal concentrations of H^+ and OH^- ions 6.3

neutralization (acid-base) a chemical reaction in which the hydrogen ions from an acid combine with the hydroxide ions from a base to form water molecules 6.3

neutron a subatomic electrically neutral particle having almost exactly the same mass as a proton 2.2

nitrification the process of converting ammonia in the soil to the nitrate ion 6.9

nitrogen cycle a set of chemical pathways whereby nitrogen moves through the biosphere. This cycle could be described in biological and geological terms as well. 6.9

nitrogen-fixing bacteria bacteria that remove nitrogen from the air and convert it to ammonia 6.9

noble gas one of the inert elements in Group 8A that undergoes few, if any, chemical reactions 1.6

nonelectrolyte a solute that is nonconducting in aqueous solutions 5.6

nonmetal an element that does not conduct heat or electricity well, such as sulfur, chlorine, and oxygen. Nonmetals have no one characteristic appearance. 1.6

n-type semiconductor a semiconductor in which there are freely moving negative charges (electrons) 8.7

nuclear fission the splitting of a large nucleus into smaller ones with the release of energy 7.1

nuclear fuel cycle a way of conceptualizing all the different processes that can happen when uranium ore is mined, processed, used to fuel a reactor, and then dealt with as waste 7.7

nucleus the minuscule and very dense center of an atom composed of protons and neutrons 2.2

O

ocean acidification the lowering of the ocean pH due to increased atmospheric carbon dioxide 6.5

octet rule a generalization that electrons are arranged around atoms so that these atoms have a share in eight electrons. Hydrogen is an exception. 2.3

oils triglycerides that are liquids at room temperature 11.3

organic chemistry the branch of chemistry devoted to the study of carbon compounds 10.2

organic compound a compound that contains primarily carbon and hydrogen, but may contain other elements as well, such as oxygen and nitrogen 1.11

osmosis the passage of water through a semipermeable membrane from a solution that is less concentrated to a solution that is more concentrated 5.12

outer (valence) electrons the electrons found in the highest energy level that help to account for many of the observed trends in chemical properties. 2.2

oxidation a process in which a chemical species loses electrons 8.1

oxygenated gasoline a blend of petroleum-derived hydrocarbons with added oxygen-containing compounds such as MTBE, ethanol, or methanol (CH_3OH) 4.7

ozone layer a designated region in the stratosphere of maximum ozone concentration 2.1

P

parts per billion (ppb) one part out of one billion, or 1000 times less concentrated than 1 part per million 1.3

parts per million (ppm) a concentration of one part out of a million. One ppm is a unit of concentration 10,000 times smaller than 1% (one part per hundred). 1.2

peptide bond the covalent bond that forms when the –COOH group of one amino acid reacts with the $-NH_2$ group of another, thus joining the two amino acids 9.6

percent (%) parts per hundred. For example, 15% is 15 parts out of 100. 1.1

periodic table an orderly arrangement of all the elements based on similarities in their properties 1.6

pH a number, usually between 0 and 14, that indicates the acidity (or basicity) of a solution. A low pH is acidic; a pH of 7 is neutral; a higher pH is basic. 6.4

pharmacophore three-dimensional arrangement of atoms or groups of atoms responsible for the biological activity of a drug molecule 10.5

photon a way of conceptualizing light as a particle that has energy but no mass 2.5

photovoltaic cell (PV) a device that converts light energy directly to electrical energy, sometimes called a solar cell 8.7

plasticizer a compound added in small amounts to a polymer to make the polymer softer and more pliable 9.4

plug-in hybrid electric vehicle (PHEV) a vehicle for short daily commutes that uses rechargeable batteries to run an electric motor and can switch to a combustion engine to travel longer distances 8.4

PM_{10} (particulate matter) particles with an average diameter of 10 μm or less, a length on the order of 0.0004 inches 1.2

$PM_{2.5}$ (particulate matter) a subset of PM_{10} and includes particles with an average diameter of less than 2.5 μm. Sometimes called "fine particles" 1.2

polar covalent bond a covalent bond in which the electrons are not equally shared but rather are closer to the more electronegative atom 5.1

polar stratospheric clouds (PSCs) thin clouds are composed of tiny ice crystals formed from the small amount of water vapor present in the stratosphere 2.10

polyamide a condensation polymer that contains the amide functional group 9.6

polyatomic ion two or more atoms covalently bound together that have an overall positive or negative charge 5.7

polymer a large molecule built from smaller ones (monomers) that consists of a long chain or chains of atoms covalently bonded together 9.1

polysaccharide a condensation polymer made up of thousands of monosaccharide units. Examples include starch and cellulose. 11.5

polyunsaturated fatty acid a fatty acid containing more than one double bond between carbon atoms in the hydrocarbon chain 11.3

postconsumer content used material that would otherwise have been discarded as waste 9.8

potable water water that is safe to drink and to cook with 5.3

potential energy stored energy or the energy of position 4.1

precautionary principle stresses the wisdom of acting, even in the absence of full scientific data, before the adverse affects to human health or the environment become significant or irrevocable 2

preconsumer content waste left over from the manufacturing process itself, such as scraps and clippings 9.8

primary coolant in a nuclear reactor, a liquid that comes in direct contact with the fuel bundles and control rods and carries away heat. See also secondary coolant. 7.3

processed foods foods that have been altered from their natural state by techniques such as canning, cooking, freezing, or adding chemicals such as thickeners or preservatives. 11.2

protein a polyamide or polypeptide; that is, a polymer built from amino acid monomers 11.7

protein complementarity combining foods that complement essential amino acid content so that the total diet provides a complete supply of amino acids for protein synthesis 11.7

proton a subatomic positively-charged particle with approximately the same mass as a neutron 2.2

***p*-type semiconductor** a semiconductor in which there are freely moving positive charges, or holes 8.7

Q

quantized an energy distribution that is not continuous, but rather consists of many individual steps. 2.5

R

racemic mixture (±) mixture consisting of equal amounts of each optical isomer of a compound 10.6

rad a unit, short for "radiation absorbed dose"; a measure of the energy deposited in tissue defined as the absorption of 0.01 joule of radiant energy per kilogram of tissue 7.6

radiant energy the entire collection of different wavelengths, each with its own energy 2.4

radiation sickness the illness characterized by early symptoms of anemia, nausea, malaise, and susceptibility to infection that are the result of a large exposure to radiation 7.6

radiative forcings factors (both natural and anthropogenic) that influence the balance of Earth's incoming and outgoing radiation 3.9

radioactive decay series a characteristic pathway of radioactive decay that begins with a radioisotope and progresses through a series of steps to eventually produce a stable isotope 7.4

radioactivity the spontaneous emission of radiation (alpha particles, beta particles, or gamma rays) by certain elements 7.4

recyclable products products that can be recycled. They do not necessarily contain any recycled materials. 9.8

recycled-content products products made from material that otherwise would have been in the waste stream 9.8

reduction a process in which a chemical species gains electrons 8.1

reformulated gasoline (RFG) an oxygenated gasoline that also contains a lower percentage of certain more volatile hydrocarbons found in nonoxygenated conventional gasoline 4.7

rem a unit, short for "roentgen equivalent man"; a measure of the dose of radiation that takes into account the damage caused to human tissue. It is calculated by multiplying Q by the number of rads. 7.6

residual chlorine the name given to chlorine-containing chemicals that remain in the water after the chlorination step. These include hypochlorous acid (HClO), the hypochlorite ion (ClO^-), and dissolved elemental chlorine (Cl_2). 5.11

resonance forms Lewis structures that represent hypothetical extremes of electron arrangements in a molecule 2.3

respiration the process of metabolizing the foods we eat to produce carbon dioxide and water and to release the energy that powers other chemical reactions in our bodies 1.1

reverse osmosis a process that uses pressure to force the movement of water through a semipermeable membrane from a solution that is more concentrated to a solution that is less concentrated 5.12

risk assessment the process of evaluating scientific data and making predictions in an organized manner about the probabilities of an outcome 1.3

S

scientific notation is a system for writing numbers as the product of a number and 10 raised to the appropriate power 1.3

second law of thermodynamics the entropy of the universe is constantly increasing. 4.2

secondary coolant in a nuclear reactor, the water in the steam generators that does not come in contact with the reactor. See also primary coolant. 7.3

secondary pollutant a pollutant produced from chemical reactions involving one or more other pollutants (e.g., ozone). In contrast, primary pollutants are produced directly in one step (e.g., CO and NO). 1.12

semiconductor a material that does not normally conduct electricity well, but can do so under certain conditions, such as exposure to sunlight 8.7

shifting baseline the idea that what people expect as "normal" on our planet has changed over time, especially with regard to ecosystems 0.1

sievert (Sv) a unit used internationally equal to 100 rem 7.6

significant figure a number that correctly represents the accuracy with which an experimental quantity is known 1.14

single covalent bond a bond formed when two electrons (one pair) are shared between two atoms 2.3

solute the solid, liquid, or gas that dissolves in a solvent 5.5

solution a homogeneous (of uniform composition) mixture of a solvent and one or more solutes 5.5

solvent a substance, often a liquid, capable of dissolving one or more pure substances 5.5

specific heat the quantity of heat energy that must be absorbed to increase the temperature of one gram of a substance by one degree Celsius 5.2

spent nuclear fuel (SNF) the radioactive material remaining in fuel rods after they have been used to generate power in a nuclear reactor. 7.9

steady state a condition in which a dynamic system is in balance so that there is no net change in concentration of the major species involved 2.6

steroid a class of naturally occurring or synthetic fat-soluble organic compounds that share a common carbon skeleton arranged in four rings 10.7

strong acid an acid that dissociates completely in aqueous solution 6.1

strong base a base that dissociates completely in aqueous solution 6.2

structural formula a representation of how the atoms in a molecule are connected. It is a Lewis structure from which the nonbonding electrons have been removed. 2.3

structure-activity relationship (SAR) study a study in which systematic changes are made to a drug molecule followed by an assessment of the resulting changes in activity 10.5

substituent an atom or functional group substituted for a hydrogen atom in a molecule 10.3

substrate a molecule (or molecules) being acted on, often catalytically by an enzyme 10.5

surface water fresh water found in lakes, rivers, and streams 5.4

surfactant a molecule that has both polar and nonpolar regions that allows it to help solubilize different classes of molecules 5.7

sustainability "meeting the needs of the present without compromising the ability of future generations to meet their needs." (from *Our Common Future*, a 1987 report of the United Nations) 0.2

T

temperature a measure of the average kinetic energy of the atoms or molecules present in a substance 4.5

tetrahedron a four-cornered geometric shape with four equal triangular sides, also sometimes called a triangular pyramid 3.3

thermal cracking a process that breaks large hydrocarbon molecules into smaller ones by heating them to a high temperature 4.7

thermoplastic polymer a plastic that can be melted and reshaped over and over again 9.4

toxicity the intrinsic health hazard of a substance 1.3

trace mineral an element present in the body, usually at microgram levels, such as I, F, Se, V, Cr, Mn, Co, Ni, Mo, B, Si, and Sn 11.8

tragedy of the commons the situation in which a resource is common to all and used by many, but has no one in particular responsible for it. As a result, the resource may be destroyed by overuse to the detriment of all that use it. 1.12

trans fat a triglyceride that is composed of one or more *trans* fatty acids 11.4

transgenic the result of a transfer of genes across species 12.6

triglycerides molecules that contain three ester functional groups. They are formed from a chemical reaction between three fatty acids and the alcohol glycerol. 11.3

trihalomethanes (THMs) compounds such as $CHCl_3$ (chloroform), $CHBr_3$ (bromoform), $CHBrCl_2$ (bromodichloromethane), and $CHBr_2Cl$ (dibromochloromethane) that are formed from the reaction of chlorine or bromine with organic matter in drinking water 5.11

triple bond a covalent bond made up of three pairs of shared electrons 2.3

Triple Bottom Line a three-way measure of the success of a business based on its benefits to the economy, to society, and to the environment 0.3

troposphere the lower region of the atmosphere in which we live that lies directly above the surface of the Earth 1.5

U

ultraviolet (UV) region a region of the electromagnetic spectrum that lies adjacent to violet light, but at shorter wavelength 2.4

unsaturated fatty acid a fatty acid in which the hydrocarbon chain contains one or more double bonds between carbon atoms 11.3

V

vitamin a compound with a wide range of physiological functions and essential for good health, proper metabolic functioning, and disease prevention 11.8

vitrification a process in which the spent fuel elements or other mixed waste are encased in ceramic or glass. 7.9

volatile a substance that readily passes into the vapor phase; that is, it evaporates easily 1.11

volatile organic compounds (VOCs) compounds that pass easily into the vapor phase. All contain carbon and hydrogen and originate from a variety of sources. 1.11

voltage the difference in electrochemical potential between the two electrodes 8.1

volumetric flask a type of flask that contains a precise amount of solution when filled to the mark on its neck 5.5

W

water footprint an estimate (for an individual or a nation) of the amount of water required to sustain the consumption of goods and services 5.3

wavelength the distance between successive peaks in a sinusoidal wave 2.4

weak acid an acid that dissociates only to a small extent in aqueous solution 6.1

weak base a base that dissociates only to a small extent in aqueous solution 6.2

weather includes the daily highs and lows, the drizzles and downpours, the blizzards and heat waves, and the fall breezes and hot summer winds, all of which have relatively short durations. Contrast with climate. 3.10

CHAPTER 0

Opener: Image by Reto Stockli, NASA, Goddard Space Flight Center. Enhancements by Robert Simmon; Page 3, Page 4, Page 5(both): Cathy Middlecamp; Page 7 & 8: Published 2008 American Chemical Society; Page 9: © 2002, William McDonough and Michael Braungart. Published by North Point Press; Page 13: © Fahd Shadeed/AFP/Getty Images; PAGE 14: Image by Reto Stockli, NASA, Goddard Space Flight Center. Enhancements by Robert Simmon.

CHAPTER 1

Opener: Page 17, Page 18, 1.3, 1.4: Cathy Middlecamp; 1.5: © Jill Braaten; 1.6: Courtesy Missouri Botanical Garden Plant Finder 1.7: Cathy Middlecamp; Page 29: NASA; 1.10b: © Tom Smart/Deseret News; 1.13: Image originally created by IBM Corporation 1.14: © The McGraw-Hill Companies, Inc./Bob Coyle photographer; 1.15: Cathy Middlecamp; 1.17b: © Courtesy, Corning Incorporated; 1.19: Environment Canada, Meteorological Service of Canada. Reproduced with the permission of the Minister of Public Works and Government Services Canada, 2010; Page 50 (Bark): Cathy Middlecamp; Page 50 (Cigar): © Sascha Burkhard/Shutterstock.com; 1.20a: © Image Source/Corbis RF; 1.20b: © Digital Vision Vol. DV384/ Getty Images RF; 1.21: © YOLO Colorhouse®; 1.22: © The McGraw-Hill Companies, Inc./Ken Karp photographer; 1.23: © David M. Grossman/Photo Researchers, Inc.; 1.24: © Sheila Terry/Photo Researchers, Inc.; Page 59, Page 60 top & bottom: Cathy Middlecamp; Page 63 left & right: National Science Foundation.

CHAPTER 2

Opener: NASA/Goddard Space Flight Center; 2.1: © Galen Rowell/Corbis; Page 70: Earth Sciences and Image Analysis Laboratory/Johnson Space Center/NASA; 2.3(all): © The McGraw-Hill Companies, Inc./Stephen Frisch photographer; 2.6: © Philip McAulay/Shutterstock.com; 2.13: Courtesy Blue Lizard Products; Page 84: © Image Source/Alamy; 2.15(all): © The McGraw-Hill Companies, Inc./Stephen Frisch photographer Page 92: Courtesy Carlye Calvin; 2.18: © Ross J. Salawitch, University of Maryland; Page 96: © Jill Braaten; 2.22: © Courtesy Pyrocool Technologies, Inc.

CHAPTER 3

Opener: © Duncan Walker/iStockphoto.com; 3.1: NASA; 3.4a: © Lonnie G. Thompson, Ohio State University; 3.4b: © Vin Morgan/AFP/Getty Images; 3.4c: © W. Berner, 1978, PhD Thesis University of Bern, Switzerland. (D. Lüthi, M. Le Floch, B. Bereiter, T. Blunier, J.-M. Barnola, U. Siegenthaler, D. Raynaud, J. Jouzel, H. Fischer, K. Kawamura, and T.F. Stocker, High-resolution carbon dioxide concentration record 650,000-800,000 years before present, *Nature*, 453, 379–382, 2008.); 3.8: NASA image by Robert Simmon, based on GISS surface temperature analysis data; 3.10: © Michael Newman/PhotoEdit, Inc.; 3.22: Courtesy Conrad Stanitski; 3.23a-b: Ocean Drilling Program; 3.24: Courtesy of the Oak Ridge National Laboratory, managed by the U.S. Department of Energy by UT-Battelle, LLC; Page 136: © Photodisc/Getty Images RF; 3.28a-b: NASA GSFC Scientific Visualization Studio; Page 141 (Coral): © Ove Hoegh-Guidberg/AFP/Getty Images; Page 141 (Butterfly): www.wisconsinbutterflies.org; 3.29: Cathy Middlecamp.

CHAPTER 4

Opener: © Péter Gudella/Shutterstock.com; 4.1a-d: Cathy Middlecamp; 4.5: © Charles D. Winters/Photo Researchers, Inc.; 4.7 left: © Martin Shields/Photo Researchers, Inc.; 4.7 right: © Mark A. Schneider/Photo Researchers, Inc.; 4.8: © J. Miles Carey/Knoxville News Sentinel/Polaris Images; 4.10: © Claudius/zefa/Corbis; Page 172: © Digital Vision, Vol. DV418/ Getty Images RF; 4.19: © Justin Sullivan/Getty Images; Page 183: © Sustainability Institute; 4.21: © Khuong Huang/ iStockphoto.com; 4.22a: © & Courtesy of David & Associates, Hastings, NE; 4.22b: © Photo by J. Emilio Flores for General Motors; 4.23a-b: © AP/Wide World Photos; 4.24: Courtesy A. Truman Schwartz.

CHAPTER 5

Opener: © Rich Armstrong; 5.6, 5.9: Cathy Middlecamp; 5.10a: © Jaimie Duplass/Shutterstock.com; 5.10b: © Adek Berry/AFP/ Getty Images; 5.11a: © Noah Seelam/AFP/Getty Images; 5.11b: Andrea Booher/FEMA; 5.14a-c: EROS Data Center, U.S.G.S.; 5.14d: NASA image created by Jesse Allen; 5.16: © Laurence Gough/iStockphoto.com; 5.17a-c: © Tom Pantages; 5.25a-b: © Charles D. Winters/Photo Researchers, Inc.; 5.29: airviewonline.com.au; 5.30b: © & Courtesy of SolAqua; 5.32: Courtesy of Katadyn; 5.33a-b: © Vestergaard Frandsen; 5.34: USDA/National Resources Conservation Service; 5.35: Artwork © Robert Schiller. Photograph by Sally Mitchell.

CHAPTER 6

Opener: © Manamana/Shutterstock.com; 6.1: © PhotoDisc Vol. 77/Getty Images RF; 6.2: © The McGraw-Hill Companies, Inc./Photo by Eric Misko, Elite Images Photography; 6.3: © The McGraw-Hill Companies, Inc./C. P. Hammond photographer; 6.7: © Eric Maston, Australian Institute of Marine Science; 6.8: © Charles D. Winters/Photo Researchers, Inc.; 6.9a, 6.10a-c: Cathy Middlecamp; 6.13: © E. R. Degginger/Color Pic; 6.15: © Dan Chenier; 6.17: © M. Kaleb/Custom Medical Stock Photo; 6.20: Cathy Middlecamp; 6.22a: © NYC Parks Photo Archive/

Fundamental Photographs NYC; 6.22b: © Kristen Brochmann/Fundamental Photographs NYC; 6.23: © A. J. Copley/Visuals Unlimited; 6.24a: © iPhotos/Shutterstock.com; 6.24b: © Anton Hazewinkel/Flickr/Getty Images; 6.25b: © Pittsburgh Post Gazette Archives, 2004. All rights reserved. Reprinted with permission; Page 278: Cathy Middlecamp; Page 279 (Jet): © Stock Portfolio/Stock Connection/Picture Quest RF; Page 279 (Hail): Cathy Middlecamp.

CHAPTER 7

Opener (Power Plant): © Brand X Pictures/PunchStock RF; Opener (In Tray): © Image Club RF; Page 283: © The McGraw-Hill Companies, Inc./Photo by Eric Misko. Elite Images Photography; Page 285: © Rob Crandall/The Image Works; 7.4: © Bettmann/Corbis; 7.6: U.S. Dept. of Energy; 7.8: © The McGraw-Hill Companies, Inc./C.P. Hammond photographer; 7.9, 7.10: © AP/Wide World Photos; Page 297: © Corbis RF; 7.11: © Hulton-Deutsch Collection/Corbis; 7.14: © AP/Wide World Photos; 7.15: © Chuck Nacke/Time Life Pictures/Getty Images; 7.16: © James Hill/Contact Press Images; 7.17: © Southern Illinois University/Photo Researchers, Inc.; 7.20: © & Courtesy of URENCO Ltd.; 7.22: © AP/Wide World Photos; 7.24: U.S. Dept. of Energy; 7.25: U.S. Nuclear Regulatory Commission; 7.27: © Science Source/Photo Researchers, Inc.; 7.28b: U.S. Dept. of Energy; Page 320: © J. Miles Carey/Knoxville News Sentinel/Polaris Images; 7.30: Cathy Middlecamp; Page 322: © Earth Policy Institute; 7.31: Cover logo and mechanicals reproduced with permission from Chem. Eng. News, August 24, 2009, 87(34). Copyright 2009 American Chemical Society. Cover photo © & Courtesy of Warren Wright/Areva; Page 327: Cathy Middlecamp.

CHAPTER 8

Opener: © Tomonari Tsuji/Amana Images/Getty Images; 8.1: © 2009 Thaves. Reprinted with permission. Newspaper dist. By NEA, Inc.; 8.2: © The McGraw-Hill Companies, Inc./Jill Braaten, photographer; 8.4: © The McGraw-Hill Companies, Inc./Photo by Eric Misko, Elite Images Photography; Page 337: © Steven Good/Shutterstock.com; 8.7(all): © The McGraw-Hill Companies, Inc./Stephen Frisch photographer; 8.8: © mazoncini/iStockphoto.com; 8.9a-c: © Jill Braaten; 8.13: © National Hydrogen Association (www.HydrogenAssociation.org); 8.18: © Michael Barnes, University of California; 8.20: Warren Gretz/NREL/U.S. Dept. of Energy; 8.21b: NREL/U.S. Dept. of Energy; 8.21c: © Daniel Karmann/DPA/Corbis; 8.26, 8.28: NREL/U.S. Dept. of Energy; 8.29a-b: Solar Millennium AG/Paul Langrock; 8.30a: © AP Photo/Tim Wright; Page 367: © Arctic Images/Corbis.

CHAPTER 9

Opener: © Mark Richards/PhotoEdit, Inc.; Page 369: © Dynamic Graphics/JupiterImages RF; 9.2a-b, 9.3a: Cathy Middlecamp; 9.5a-d: © The McGraw-Hill Companies, Inc./Jill Braaten, photographer; 9.6a: © Bill Aron/PhotoEdit, Inc.; 9.9b, 9.11: © The McGraw-Hill Companies, Inc./Jill Braaten, photographer; 9.12, 9.13: Courtesy DuPont; 9.16: © The Garbage Project, University of Arizona; 9.23: Courtesy A. N. Wyeth; 9.24: © Gayna Hoffman/Stock Boston.

CHAPTER 10

Opener (top): © Eduardo Rivero/Shutterstock.com; Opener (bottom): © Steve Gorton/Dorling Kindersley/Getty Images; 10.1: © Terry Wild Studio; 10.7a-c: © The McGraw-Hill Companies, Inc./Jill Braaten, photographer; 10.10: Courtesy of the Alexander Fleming Laboratory Museum, St. Mary's Hospital, Paddington, London; 10.11: Library of Congress; 10.15: Photo courtesy John M. Rimoldi, University of Mississippi; 10.25 left: © Bill Aron/PhotoEdit, Inc.; 10.25 right: © The McGraw-Hill Companies, Inc./Jill Braaten, photographer; 10.26a-b: © Michael P. Gadomski/Photo Researchers, Inc.; 10.27 left: © Gerald & Buff Corsi/Visuals Unlimited; 10.27 right: © James Leynse/Corbis; 10.29: © Chris Knapton/Photo Researchers, Inc.; 10.31: © 2004, Publishers Group www.streetdrugs.org.

CHAPTER 11

Opener: © Jonelle Weaver/Brand X Pictures/Getty Images RF; Page 447: Cathy Middlecamp; 11.15: © Nancy Rabalais, Louisiana Universities Marine Consortium; Page 470: © Jill Braaten; Page 472: © SunnyS/Shutterstock.com; Page 480: Mike McCann; Page 481: © Steve Bower/Shutterstock.com.

CHAPTER 12

Opener top: © Michael Halberstadt/Siliconvalleystock.com; Opener bottom: © Eric Huang; 12.1a: © Brand X Pictures/PunchStock RF; 12.1b: © Scott Camazine/Photo Researchers, Inc.; 12.5a-b: © Bettmann/Corbis; 12.5c: History of Medicine/National Library of Medicine; 12.5d: Courtesy of National Library of Medicine; 12.6a: King's College London Archives; 12.13: © Dr. Stanley Flegler/Visuals Unlimited; 12.14: © & Courtesy of Hugh Iltis/The Doebley Lab; 12.16: © Nigel Cattlin/Photo Researchers, Inc.; 12.17: © Dung Vo Trung/Corbis; 12.18: © The McGraw-Hill Companies, Inc./Jill Braaten, photographer; 12.19: Courtesy of Solazyme; 12.20: © AFP/Greenpeace/Getty Images; Page 513: © Bettmann/Corbis.

Index

References followed by t and f refer to tables and figures respectively.